THEMATIC CARTOGRAPHY
AND GEOGRAPHIC VISUALIZATION

**Prentice-Hall Series
in Geographic
Information Science**

KEITH C. CLARKE,
Series Advisor

Avery/Berlin, *Fundamentals of Remote Sensing and Air Photo Interpretation*

Clarke, *Analytical and Computer Cartography*

Clarke, *Getting Started with Geographic Information Systems*

Clarke/Parks/Crane, *Geographic Information Systems and Environmental Modeling*

Heywood/Cornelius/Carver, *Introduction to Geographical Information Systems*

Jensen, *Introductory Digital Image Processing: A Remote Sensing Perspective*

Jensen, *Remote Sensing of the Environment: An Earth Resource Perspective*

Slocum/McMaster/Kessler/Howard, *Thematic Cartography and Geographic Visualization*

Lo/Yeung, *Concepts and Techniques in Geographic Information Systems*

THEMATIC CARTOGRAPHY AND GEOGRAPHIC VISUALIZATION

Second Edition

TERRY A. SLOCUM
ROBERT B. McMASTER
FRITZ C. KESSLER
HUGH H. HOWARD

PEARSON

Prentice
Hall

Upper Saddle River, NJ 07458

Library of Congress Cataloging-in-Publication Data

Thematic cartography and geographic visualization/Terry A. Slocum. . . [et al.].—2nd ed.
    p. cm.—(Prentice Hall series in geographic information science)
 Includes bibliographical references and index.
  ISBN 0-13-035123-7
  1. Cartography   2. Visualization.   I. Slocum, Terry A.   II. Series

GA 108.7.T48 2005
526—dc22

2004002166

Editor-in-Chief, Science: *John Challice*
Executive Editor: *Daniel Kaveney*
Associate Editor: *Amanda Griffith*
Editorial Assistant: *Margaret Ziegler*
Vice President and Director of Production and Manufacturing, ESM: *David W. Riccardi*
Production Editor: *Beth Lew*
Art Editor: *Abby Bass*
Art Production Coordinator: *Dan Missildine*
Marketing Manager: *Robin Farrar*
Manufacturing Manager: *Trudy Pisciotti*
Manufacturing Buyer: *Lynda Castillo*
Art Director: *Jayne Conte*
Cover Designer: *Kiwi Design*

About the Cover

Three-dimensional visualization of Santa Cruz, California, incorporating thematic mapping techniques and a virtual environment. Vegetation is derived from the National Landcover Dataset, while thematic data are derived from the 2000 U.S. Census. Darker census block-group polygons represent greater population densities; larger circles represent greater median age per block group. Circles have been vertically offset from the landscape in order to prevent larger circles from obscuring the block groups they are associated with.

Caution must be exercised when comparing circle sizes in this static three-dimensional view, as the apparent size of all circles decreases naturally with increased distance from the viewer. Ideally, this visualization would be viewed in an interactive computing environment, where the user would have the ability to navigate around the geographic region in real time, revealing the landscape and thematic information from an infinite number of perspectives.

This image was produced by Matthew D. Dunbar and Hugh H. Howard, May 2003, using 3D Nature *Visual Nature Studio*, ESRI *ArcGIS*, and Adobe *Photoshop*.

Pearson Education Ltd., *London*
Pearson Education Australia Pty. Ltd., *Sydney*
Pearson Education Singapore, Pte. Ltd.
Pearson Education North Asia Ltd., *Hong Kong*

Pearson Education Canada, Inc., *Toronto*
Pearson Educación de Mexico, S. A. de C. V.
Pearson Education—Japan, *Tokyo*
Pearson Education Malaysia, Pte. Ltd.

Contents

Preface

Like the first edition, this one is a blend of the old and the new. By old, we refer to traditional approaches for creating thematic maps. For this edition, we have divided traditional approaches into two major sections: Part I ("Principles of Cartography") and Part II ("Mapping Techniques"). Part I provides a foundation of the history of academic cartography in the United States, statistics and graphics, basic symbolization approaches (choropleth vs. isopleth, proportional symbol, and dot), data classification, generalization, projections, color, map design, and production and dissemination. Techniques covered in Part II include choropleth, isarithmic, proportional symbol, dot, and dasymetric mapping. Additionally, we cover bivariate and multivariate mapping, techniques for symbolizing topography, and we touch briefly on cartograms, flow maps, and approaches for mapping true 3-D phenomena.

By new, we refer to approaches that have been developed in the area of geographic visualization (or geovisualization). Given the growing importance of geographic visualization, we have included separate chapters in Part III on animation, data exploration, electronic atlases and multimedia, visualizing data quality, and virtual and mixed environments. Also, given the rapid changes that are taking place, we have included a chapter on ongoing developments, briefly touching on such topics as collaborative visualization, spatial data mining, information visualization, multimodal interfaces, and the use of sound to depict spatial data.

Some other key distinguishing features of this book are as follows:

- The content clearly contrasts different approaches for symbolizing spatial data. Many texts present individual mapping techniques, but they generally fail to contrast the different approaches (see section 4.4 of Chapter 4).
- Our chapter on data classification (Chapter 5) clearly explains (and illustrates) the differences among various data classification techniques; the chapter also includes a section on multivariate cluster analysis.

- A separate chapter on generalization (Chapter 6) considers a variety of basic generalization operations.
- We provide an extensive introduction to map projections (Chapter 8), and a complete chapter on selecting an appropriate projection for a particular mapping situation (Chapter 9).
- Our chapter on map design (Chapter 11) provides the student with clear descriptions of various aspects of effective, efficient map design, with an emphasis on the practical application of design theories. This chapter includes the most comprehensive description of map elements (and their appropriate use) of any cartography textbook.
- The text discusses approaches for selecting appropriate color schemes for choropleth maps. Other books cover this material, but they do not consider the broad range of factors, nor do they include as many sample maps (see section 13.3 of Chapter 13).
- We discuss various algorithms for interpolating spatial data, including kriging (see Chapter 14). Generally, cartographic or GIS texts do not cover this material in the depth that we do. An exception would be Burrough and McDonnell's GIS text *Principles of Geographical Information Systems*.
- Chapter 15 discusses differences among various approaches for symbolizing the Earth's topography, including recently developed techniques such as those of Tom Patterson and the physical models developed by Solid Terrain Modeling.
- We include an extensive discussion of bivariate and multivariate mapping (Chapter 18), which is not covered in other cartographic texts.

In comparing this edition with the previous one, you will note numerous new chapters, and some restructuring of earlier material. The chapters on history of academic cartography in the United States (Chapter 2), scale and cartographic generalization (Chapter 6), projections (Chapters 7–9), map production and dissemination (Chapter 12), and virtual and mixed environments (Chapter 24) are entirely new. The chapter on map design (Chapter 11) contains a bit of material from Chapter 2 of the earlier edition, but the bulk of the material is new.

Material related to choropleth maps has been combined into a single chapter called "Choropleth Mapping." We have chosen, however, to keep the data classification material as a separate chapter to emphasize that classification can be applied to data associated with other sorts of maps. The data classification chapter now includes a section on cluster analysis. The "intepolation methods" and "symbolization" chapters associated with "smooth continuous phenomena" in the earlier edition (Chapters 8 and 9) have been restructured and enhanced to create chapters on isarithmic mapping (Chapter 14) and symbolizing topography (Chapter 15), hopefully making this material more accessible for both the instructor and the student. Material related to visualizing data quality, formerly included in the recent developments chapter (Chapter 16 in the earlier edition) now appears as a separate chapter (Chapter 23), reflecting the importance of this concept.

We have revised the chapter on electronic atlases (Chapter 22) and provided a brief section on multimedia.

We have chosen not to include a section on tools for developing your own software (see section 15.3 of the earlier edition), as we perceive this to be of less interest for the introductory student, and it is difficult to ensure that the material is up-to-date.

There is ample material here for either a one- or two-semester course in thematic cartography. For a one-semester course, we suggest Chapters 1, 4, 5 (excluding cluster analysis), 7, 8, 11, 13, 14 (excluding kriging), 16, and 20 or 21. For a two-semester course, the bulk of Part I and a portion of Part II (possibly through choropleth and isarithmic mapping) could be covered in the first semester, with the remainder of the book covered in the second semester. We have tried to write each chapter so that it can stand independently; thus, skipping a chapter generally should not prevent subsequent chapters from being understood. Where preceding material is essential, we have referred the reader back to the appropriate sections.

Numerous Web sites are mentioned throughout the text. The home page for the book *www.prenhall.com/slocum* summarizes these sites on a chapter-by-chapter basis and provides links to additional topics and references that space did not permit including in the book, or materials that have become available since the book was written.

ACKNOWLEDGMENTS

This book would not have been possible without the assistance of many people and organizations. First, we would like to thank the University of Kansas (KU) for providing sabbatical support for Professor Slocum—projects of this magnitude definitely require a release from the usual teaching and service commitments. Darin Grauberger and his Cartographic Services staff at KU were extremely helpful in ensuring that the graphics were of high quality; staff that assisted included Justin Busboom, Angela Gray, Matthew Harman, John Kostelnick, and Matthew Stratton. Others at KU who provided assistance included Matt Dunbar (for his effort in designing the cover), and Barbara and Pete Shortridge (for editorial suggestions).

Numerous people at Prentice Hall were of considerable assistance. We thank Dan Kaveney, Executive Editor, for his willingness to undertake this project. Margaret Ziegler graciously provided editorial assistance whenever we needed it. Beth Lew and her production staff ensured that the book went to press in a timely fashion. Beth, we appreciate your attention to detail, your good humor, and of course your endless patience.

Those who edited or provided feedback on individual chapters included Sven Fuhrmann, David Hermann, Jonathan Lawton, Hans-J. Meihoefer, James Miller, and daan Strebe. Those who kindly provided assistance in editing sections pertaining to their work included Keith Clarke, Jason Dykes, Nick Hedley, Patrick Kennelly, David Martin, and Elisabeth Nelson. Numerous people graciously provided illustrations—we would especially like to thank Natalia and Gennady Andrienko, Cynthia Brewer, Daniel Carr, Keith Clarke, Jerome Dobson, Jason Dykes, Robert Edsall, Sara Fabrikant, Sven Fuhrmann, Mark Harrower, Christopher Healey, Nick Hedley, Patrick Kennelly, Ryan Koehnen, Victoria Interrante, John Lomax, Alan MacEachren, David Martin, Susanna McMaster, James Miller, Kate Moore, Alan Murray, Elisabeth Nelson, Alex Pang, Tom Patterson, Rajeev Sharma, André Skupin, Philippe Thibault, and Craig Wittenbrink.

Terry Slocum thanks Michael Dobson and George Jenks for nurturing his interest in cartography and his family (Arlene, Diane, Kevin, and Danny) for their patience during those long evening and weekend writing sessions. Terry also thanks the Department of Geography at the University of South Carolina for providing a home while the first edition of this text was written. Finally, Terry thanks Master Ki-June Park for the punishing workouts that permitted him to return to the book with a fresh mind.

Robert McMaster thanks the continued interest in cartography from many of his University of Minnesota colleagues, in particular Philip Porter, Mark Lindberg, John Adams, and Dwight Brown, who actually still feel maps are a critical part of the geographers' life, and to Mark Monmonier, who helped to launch his career in mapping. He also thanks the three women at home, Susanna, Keiko, and Katherine, who run his life, which is a good thing.

Fritz Kessler thanks Terry Slocum, Alan MacEachren, and Hugh Blömer for their irreplaceable stewardship in cultivating his excitement for and awareness about cartography. Gratitude is also extended to the late John Snyder, who made the realm of map projections accessible and fascinating. Ultimately, without the unending support of Loretta, my accomplishments would be considerably less than what they are.

Hugh Howard thanks Hans-J. Meihoefer for acting as mentor and role model in cartography and geographic education. Hugh also thanks his present and former students for constantly re-igniting his love of maps.

In memory of
George F. Jenks

1

Introduction

OVERVIEW

This book covers thematic mapping and the associated expanding area of geographic visualization (or geovisualization). A **thematic map** (or **statistical map**) is used to display the spatial pattern of a theme or **attribute**. A familiar example is the temperature map shown in daily newspapers; the theme (or attribute) in this case is the predicted high temperature for the day. The notion of a thematic map is described in section 1.1 and contrasted with the **general-reference map**, which focuses on geographic location as opposed to spatial pattern (e.g., a topographic map might show the location of rivers). In section 1.2 the different uses for thematic maps are described: to provide **specific information** about particular locations, to provide **general information** about spatial patterns, and to **compare patterns** on two or more maps.

An important function of this book is to assist you in selecting appropriate techniques for representing spatial data. For example, imagine that you wish to depict the forest cleared for agriculture in each country during the preceding year, and you have been told that the number of acres of forest cleared is available by country on the World Wide Web (WWW). You wonder whether additional data should be collected (e.g., the total acres of land in each country) and how the resulting data should be symbolized. Section 1.3 considers steps that assist you in tackling such problems, and ultimately enable you to communicate the desired information to map readers. These steps are as follows: (1) Consider the real-world distribution of the phenomenon, (2) determine the purpose for making the map, (3) collect data appropriate for the map purpose, (4) design and construct the map, and (5) determine whether users find the map useful and informative. Although some have criticized the appropriateness of such steps, they are helpful in avoiding design blunders that can result from using the most readily available data and software.

Like many disciplines, the field of cartography has undergone major technological changes. As recently as the 1970s, most maps were still produced by manual and photomechanical methods, whereas today nearly all maps are produced using computer technology. Section 1.4 considers some of the consequences of this technological change, including (1) the ability of virtually anyone to create maps using personal computers; (2) new mapping methods, such as **animation**; (3) the ability to explore geographic data in an interactive graphics environment; (4) the ability to link maps, text, pictures, video, and sound in **multimedia** presentations; (5) the ability to create realistic representations of the environment (**virtual environments** or **virtual reality**) and the related notion of **augmented reality** (enhancing our view of the real world through computer-based information); and (6) the ability to access maps and related information via the Web.

In section 1.5 we consider the origin and definition of geographic visualization. The term visualization has its roots in **scientific visualization**, which was developed outside geography to explore large multivariate data sets, such as those associated with medical imaging, molecular structure, and fluid flows. Borrowing from these ideas, geographers have created the notion of **geographic visualization**, which can be defined as a private activity in which unknowns are revealed in a highly interactive environment. **Communication** on traditional printed maps involves the opposite: It is a public activity in which knowns are presented in a noninteractive environment.

Although our emphasis in this book is on cartography, we should be aware of developments in the related techniques of geographic information systems (GIS), remote sensing, and quantitative methods. In section 1.6, we consider the increased capability provided by GIS and remote sensing, which allow us to create detailed maps *more easily than was possible with manual techniques. GIS accomplishes this through its extensive spatial analysis capabilities, and remote sensing allows us to "sense" the environment, particularly outside our normal visual capabilities (e.g., detecting previsual levels of vegetation stress). The major development in quantitative methods relevant to cartography is that of exploratory spatial data analysis (ESDA), which has close ties with the notion of data exploration that cartographers utilize. We also need to keep in mind that those working in these related areas can produce more effective displays through sound training in cartography.*

*Just as technological advances have had a major impact on cartography, the discipline has also experienced changes in its philosophical outlook. Section 1.7 deals with the increasing role that **cognition** now plays in cartography. Traditionally, cartographers approached mapping with a behaviorist view, in which the human mind was treated like a black box. The trend today is toward a cognitive view, in which cartographers hope to find why symbols work effectively. Section 1.8 deals with social and ethical issues in cartography—here we see that maps often have hidden agendas and meanings, and that our increasing technological capability provides tremendous opportunity, but also is fraught with potential problems (e.g., the notion of **geoslavery**).*

1.1 WHAT IS A THEMATIC MAP?

Cartographers commonly distinguish between two types of maps: general-reference and thematic. **General-reference maps** are used to emphasize the location of spatial phenomena. For instance, topographic maps, such as those produced by the United States Geological Survey (USGS), are general-reference maps. On topographic maps readers can determine the location of streams, roads, houses, and many other natural and cultural features. **Thematic maps** (or **statistical maps**) are used to emphasize the spatial pattern of one or more geographic **attributes** (or *variables*), such as population density, family income, and daily temperature maximums. A common thematic map is the **choropleth map**, in which **enumeration units** (or data collection units such as states) are shaded to represent different magnitudes of an attribute (Color Plate 1.1). A variety of thematic maps are possible, including proportional symbol, isarithmic, dot, and flow maps (Figure 1.1). A major purpose of this book is to introduce these and other types of thematic maps, as well as methods used in constructing them.

Although cartographers commonly distinguish between general-reference and thematic maps, they do so largely for the convenience of categorizing maps. The

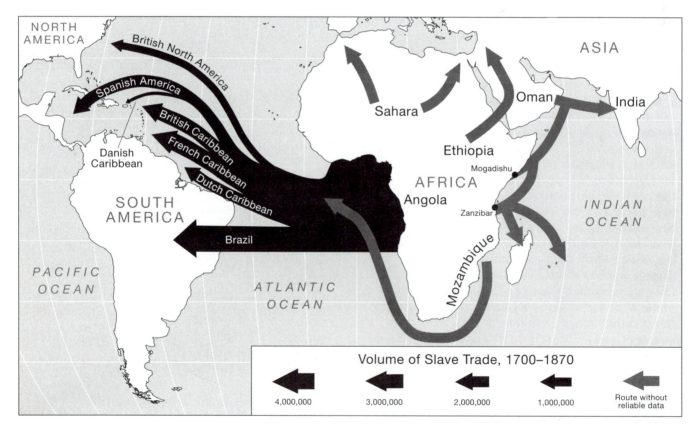

FIGURE 1.1   A flow map: an example of a thematic map. (From *Human Geography: Culture, Connections and Landscapes* by Bergman, Edward, © 1995. Adapted by permission of Prentice-Hall, Inc., Upper Saddle River, NJ.)

general-reference map also can be viewed as a thematic map in which multiple attributes are displayed simultaneously; thus, the general-reference map can be termed a multivariate thematic map. Furthermore, although the major emphasis of general-reference maps is on *location* of spatial phenomena, they can also portray the *spatial pattern* of a particular attribute (e.g., the pattern of drainage on a USGS topographic sheet).

1.2 HOW ARE THEMATIC MAPS USED?

Thematic maps can be used in three basic ways: to provide specific information about particular locations, to provide general information about spatial patterns, and to compare patterns on two or more maps. As an example of specific information, map A of Color Plate 1.1 indicates that between 8.8 and 12.0 percent of the people in Louisiana voted for Ross Perot in the 1992 U.S. presidential election. As another example, Figure 1.1 indicates that approximately 2 million slaves were transported from Africa to Spanish America between 1700 and 1870. Obtaining general information requires an overall analysis of the map. For example, map B of Color Plate 1.1 illustrates that a low percentage of people voted for Perot in the southeastern part of the United States, whereas a higher percentage voted for him in the central and northwestern states; and Figure 1.1 indicates that the bulk of the slave trade between 1700 and 1870 occurred outside North America.

A pitfall for naive mapmakers is that they often place inordinate emphasis on specific information. Map A of Color Plate 1.1 is illustrative of this problem. Here one can discriminate the data classes based on strikingly different colors and thus determine which class each state belongs in (as we did for Louisiana), but it is difficult to acquire general information without carefully examining the legend. In map B, the reverse is the case: Determining class membership is more difficult because the map is all blue, but the spatial pattern of voting is readily apparent because there is a logical progression of legend colors.

As an illustration of pattern comparison, consider the **dot maps** of corn and wheat shown in Figure 1.2. Note that the patterns on these two maps are quite different. Corn is concentrated in the traditional Corn Belt region of the Midwest, whereas wheat is concentrated in the Great Plains, with a less notable focus in the Palouse region of eastern Washington. Conventionally, a comparison of patterns such as these was limited by their fixed placement on pages of paper atlases, but interactive graphics now allow us to readily compare arbitrarily selected distributions.

Two further issues are important when considering the ways in which thematic maps are used. First, one should distinguish between **information acquisition** and **memory for mapped information**.[*] Thus far, we have focused on information acquisition, or acquiring information while the map is being used. We can also consider memory for map information and how that memory is integrated with other spatial information (obtained through either maps or field work). For example, a cultural geographer might note that houses in a particular area are built predominantly of limestone. Recalling a geologic map of bedrock, the geographer might mentally correlate the spatial pattern of limestone in the bedrock with the pattern of limestone houses.

A second issue is that terms other than *specific* and *general* can be found in the cartographic literature. We have used these terms (developed by Alan MacEachren 1982b) because they appear frequently in the literature. Others have developed a more complex set of terms. For example, Philip Robertson (1991, 61) distinguished among *three* kinds of information: values at a point, local distributions characterized by "gradients and features," and the global distribution characterized by "trends and structure"; Robertson argued that these levels corresponded closely with Jacques Bertin's (1981) elementary, intermediate, and superior levels.[†]

1.3 BASIC STEPS FOR COMMUNICATING MAP INFORMATION

In this section, we consider basic steps involved in communicating map information to others. For instance, imagine that you wish to create a map for a term paper or that you are working for a local newspaper and need to create a map illustrating the spatial pattern of crime rates within a city. Traditionally, basic steps for communicating map information were taught within the framework of map communication models (e.g., Dent 1996, 12–14; Robinson et al. 1984, 15–16). Although such models have received criticism (e.g., MacEachren 1995, 3–11), their use can often lead to better designed maps. The map communication model that we use is shown as a set of five idealized steps in Figure 1.3. Let's examine these steps by assuming that you wish to map the distribution of total population in the United States from the 1990 decennial census.

Step 1. Consider what the real-world distribution of the phenomenon might look like. One way to implement this step is to ask yourself "What would the distribution of the phenomenon look like if I were to view it while traveling across the landscape?" In the case of

---

[*] Technically, memory for mapped information would be equivalent to what psychologists term long-term memory, but for simplicity the word *memory* is normally used.

[†] Additional terminology can be found in Olson (1976a) and Board (1984).

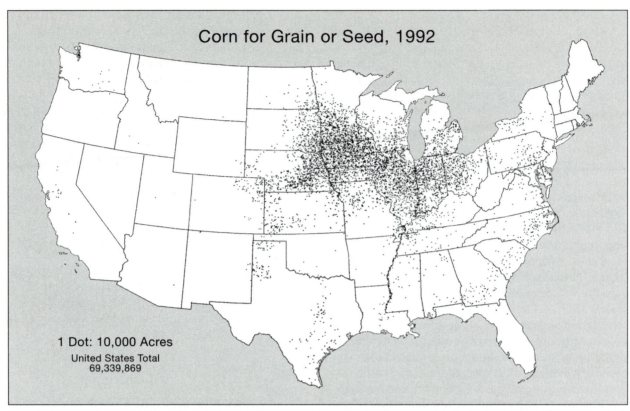

Corn for Grain or Seed, 1992

1 Dot: 10,000 Acres

United States Total
69,339,869

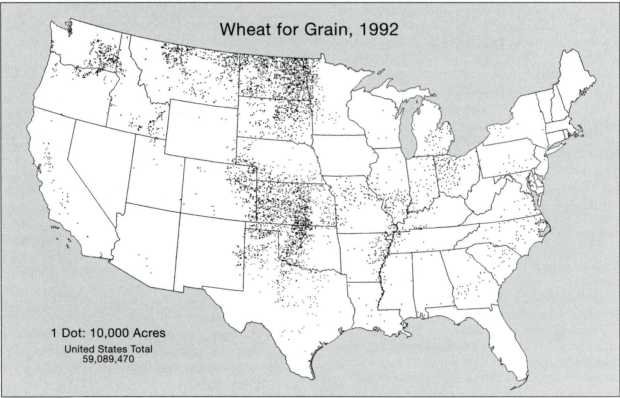

Wheat for Grain, 1992

1 Dot: 10,000 Acres

United States Total
59,089,470

FIGURE 1.2   An illustration of pattern comparison, one of the fundamental ways in which thematic maps are used. (From U. S. Bureau of the Census 1995.)

4

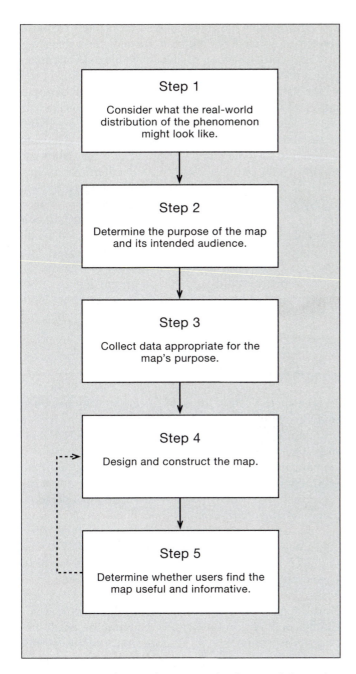

FIGURE 1.3 Basic steps for communicating map information to others. See text for discussion.

officials dispute the census figures for their city. Also, census figures do not necessarily count the homeless or illegal aliens (the latter would account for a significant percentage of the population in some areas of the United States, such as California). Another problem is that population obviously varies locationally during the day and throughout the year—people commute to work, travel to the beach on weekends, and take vacations far away from home.* In spite of these problems, it is useful to think of a "real-world" distribution. Such an approach forces you to think about the distribution at its most detailed level, and then decide what degree of generalization meets the purpose of the map.

Step 2. Determine the purpose for making the map. One purpose would be to attempt to match the real world as closely as possible (within the constraints of the map scale used). In the case of population, you might want to distinguish clearly between urban and rural population. Another purpose might be to map the distribution at a particular geographic level (say, the county level); such views are often sought by government officials for political reasons. From the viewpoint of the mapmaker, it is important to realize that mapping at a particular geographic level can introduce error into the resulting map because each enumeration unit is represented by a single value, and thus the variation within units cannot be portrayed. This error might be unimportant if the focus is on how one enumeration unit compares to another, but it can be a serious problem if readers infer more from the map than was intended; for example, readers might erroneously assume that the population density is uniform across a county on a choropleth map. A key point is that mapmakers often display data at the level of a convenient political unit (county, state, or nation) because data are available for that level, rather than considering the purpose of the map.

Step 3. Collect data appropriate for the map's purpose. In general, spatial data can be collected from primary sources (e.g., field studies) or secondary sources (e.g., Census data). For something close to the real-world view of population, you would likely consult the U.S. Census of Population for information on urban and rural population; additionally, you would collect ancillary data that could assist in locating the population data within rural areas. For a county-level view of population, the Census figures for individual counties would suffice.

Step 4. Design and construct the map. This step is a complex one that involves assessing the following questions:

1. How will the map be used? Will it be used to portray general or specific information?

our population example, you might know (based on your travels or knowledge as a geographer) that a large percentage of people were concentrated in major cities and that such cities were much more densely populated than rural areas.

Often, however, it is unrealistic to presume a single "objective" real world. In the case of population, "correct" population values are unknown for several reasons. The U.S. Bureau of the Census is never able to make an exact count of population; after every census, some city

---

* For a statistical approach for handling mobile populations, see Ii (1998).

2. What is the spatial dimension of the data? For instance, are the data available at *points*, do they extend along *lines*, or are they *areal* in nature?
3. At what level are the data measured—nominal, ordinal, interval, or ratio?
4. Is data standardization necessary? If the data are raw totals, do they need to be adjusted?
5. How many attributes are to be mapped?
6. Is there a temporal component to the data?
7. Are there any technical limitations? For example, a journal might not be willing to reproduce maps in color.
8. What are the characteristics of the intended audience? Is the map intended for the general public or professional geographers?
9. What are the time and monetary constraints? For example, creating a high-quality dot map will cost more than a choropleth map, regardless of the technical capabilities available.

A full consideration of these questions will occupy the rest of this book. The following, however, are two maps that could result from efforts to construct a population map of the United States: a combined proportional symbol–dot map (Color Plate 1.2) for the real-world view, and a choropleth map for the county-level view (map A in Figure 1.4). The proportional symbol–dot map is particularly illustrative of how one can attempt to match the mapped spatial distribution to the real world. Note that the overall population is split into urban and rural categories, and that urban population is further subdivided into "urbanized areas" and "places outside urban areas."

Step 5. Determine whether users find the map useful and informative. Possibly the most important point to keep in mind is that you are designing the map for others, not for yourself. For example, you might find a particular color scheme pleasing, but you should ask yourself whether others also will find it attractive. Ideally, you should answer such questions by getting feedback about the map from potential users. Admittedly, time and monetary constraints might make this task difficult, but it is desirable because you could discover not only whether a particular mapping technique works, but also the nature of information that users acquire from the map. Moreover, if you plan on employing the map to illustrate a *particular* concept (as for a class lecture), then you would want to know whether users acquired this concept when using the map.

If your analysis reveals that the map is not useful and informative, then the map will have to be redesigned. This possibility is shown as a dashed line in Figure 1.3. It is conceivable that you might also have to return to an earlier step, but it is more likely that you will have to modify some design aspect, such as choosing a different color scheme.

Unfortunately, naive mapmakers are unlikely to follow the five steps we have outlined. Instead, their decisions are frequently based on readily available data and mapping software. As an example, imagine that for a term paper a student wished to map the distribution of population we have been discussing. Rather than considering steps 1 and 2 of the model, the student might simply use state totals, either because fewer numbers would have to be entered into the computer or because the data were readily available (as at a Web site). Furthermore, in step 4 the student might choose a choropleth map (map B of Figure 1.4) because software for creating such a map is readily available. Presuming that the student had collected data in raw total form, the choropleth map would be a poor choice because it requires standardized data (as we will see later in this chapter and in Chapter 4). Cartographers would argue that a proportional symbol map (map C of Figure 1.4) would be a better choice if raw totals were mapped.

1.4 CONSEQUENCES OF TECHNOLOGICAL CHANGE IN CARTOGRAPHY

Ever since the 1960s, the field of cartography has been undergoing major technological change, evolving from a discipline based on pen and ink to one based on computer technology. In this section we consider several consequences of this technological change. One is that map production is no longer the sole province of trained cartographers, as virtually anyone with access to a personal computer can create maps. Although this provides more people with the opportunity to make maps, there is no guarantee that the resulting maps will be well designed and accurate. The maps shown in Color Plate 1.1 illustrate a good example of design problems that can arise. Map A uses an illogical set of unordered hues, and map B uses logically ordered shades from the same hue. Although map A allows users to discriminate easily between individual states, it does not readily permit perception of the overall spatial pattern, which is one of the major reasons for creating a map.

Another error commonly committed by naive mapmakers is illustrated in Figure 1.5. Map A suggests that forested land is more likely to be found in the western part of the United States, whereas map B suggests another pattern with an eastern dominance. Map B portrays the more accurate distribution because it is based on **standardized data** (the number of acres of forested land relative to the area of each state); in contrast, map A is based on **raw totals** (the number of acres of forested land). Choropleth maps of raw totals tend to portray large areas as having high values of a mapped attribute; in the case of map A, readers might incorrectly interpret the dark shades as indicating a high proportion of land in forests.

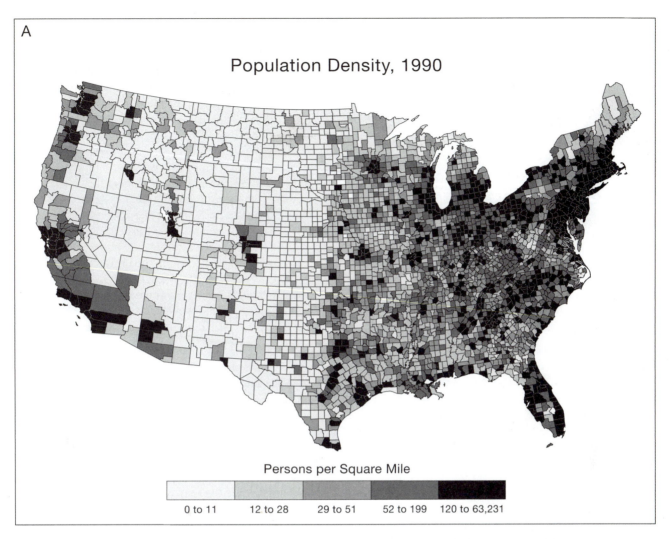

Population Density, 1990

Persons per Square Mile

| 0 to 11 | 12 to 28 | 29 to 51 | 52 to 199 | 120 to 63,231 |

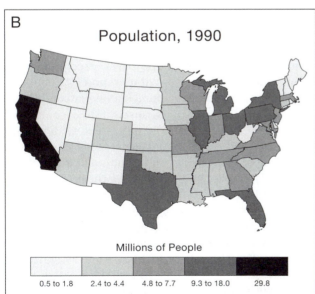

Population, 1990

Millions of People

| 0.5 to 1.8 | 2.4 to 4.4 | 4.8 to 7.7 | 9.3 to 18.0 | 29.8 |

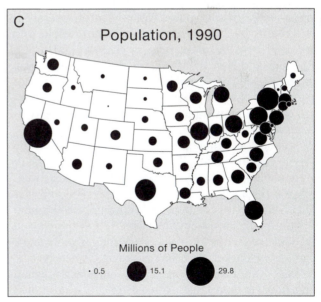

Population, 1990

Millions of People

· 0.5   15.1   29.8

FIGURE 1.4 Potential maps of the population distribution in the United States in 1990: (A) a standardized choropleth map at the county level; (B) an unstandardized choropleth map at the state level; and (C) a state-level proportional symbol map. (Data source: U. S. Bureau of the Census 1994.)

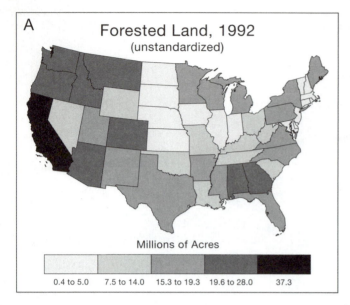

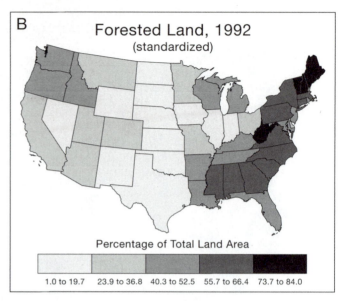

FIGURE 1.5   A comparison of the effect of data standardization. Map A is based on raw totals (the number of acres of forested land), whereas map B is based on standardized data (the number of acres of forested land relative to the area of each state). Map A is misleading because states with large areas tend to have more forest. (Data source: Powell et al. 1992.)

One purpose of this book is to explain how to avoid these and other map symbolization problems. For example, Chapter 13 discusses how to select appropriate color schemes for choropleth maps, and Chapter 4 discusses the need for data standardization. An alternative to cartographic instruction is the development of **expert systems**, in which a computer automatically makes decisions on symbolization by using a knowledge base provided by experienced cartographers. Although prototypical expert systems have been developed (Buttenfield and Mark, 1991, provide an overview; see Zhan and Buttenfield, 1995, and Forrest, 1999a, for more recent work), commonly used cartographic and GIS software has not implemented expert systems to date. In part, this is because cartographers do not agree on the rules for symbolization (Wang and Ormeling 1996).

A second consequence of technological change is the ability to produce maps that would have been difficult or impossible to create by manual methods. An early example was the unclassed map, introduced by Waldo Tobler (1973).* Figure 1.6 compares a traditional classed map (A) with its unclassed counterpart (B). On the **classed map**, data are grouped into classes of similar value and a progressively darker gray tone is assigned to each class, whereas on the **unclassed map**, gray tones are assigned proportional to data values. Some people have promoted unclassed maps, arguing that they more accurately reflect the real-world distribution, whereas others have pro-

moted classed maps on the grounds that that they are easier to interpret. We consider this issue in more detail in Chapter 13.

Animated maps (or maps characterized by continuous change while the map is viewed) are particularly representative of the capability of modern computer technology. Although the notion of animated mapping has been around since at least the 1930s (Thrower 1959; 1961), only recently have cartographers begun to recognize its full potential (Campbell and Egbert 1990). Probably the most common forms of animation are those representing changing cloud cover, precipitation, and fronts on daily television weather reports. Animations of spatial data are now also found in popular computerized encyclopedias, such as Grolier's (DiBiase 1994). In Chapter 20, we consider a variety of animations that geographers and other spatial scientists have developed; for example, we consider an animation that Lloyd Treinish (1992) developed for visualizing the formation of the ozone hole over Antarctica (see Color Plate 20.1).

A third consequence of technological change is that it alters our fundamental way of using maps. With the communication model approach, cartographers generally created one "best" map for users. In contrast, interactive graphics now permit users to examine spatial data dynamically and thus develop several different representations of the data—a process termed **data exploration**. The software package MapTime (described in more detail in Chapter 21) exemplifies the nature of data exploration. MapTime permits users to explore the data using three approaches: animation, **small multiples** (individual maps

---

* The earliest choropleth maps were actually unclassed, but they rapidly gave way to classed maps (see Robinson 1982, 199).

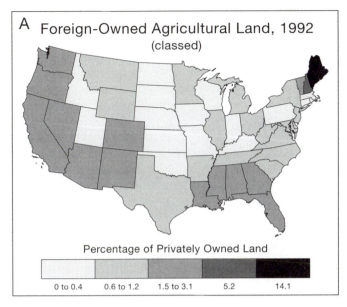
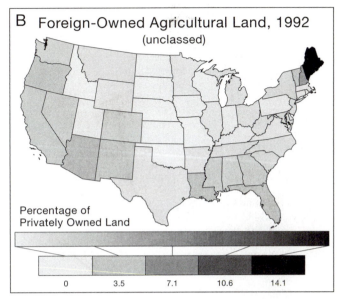

FIGURE 1.6 A comparison of (A) classed and (B) unclassed maps. On the classed map, states are grouped into classes of similar value and a gray tone is assigned to each class, whereas on the unclassed map, gray tones are selected proportional to the data value associated with each state. (Data source: DeBraal 1992.)

are shown for each time period), and **change maps** (an individual map depicts the difference between two points in time). Examination of population values for 196 U.S. cities from 1790 to 1990 (a data set distributed with MapTime) illustrates the different perspectives provided by these approaches. For example, an animation reveals major growth in northeastern cities over most of the period, with an apparent drop for only some of the largest cities from about 1950 to 1990. In contrast, a change map, showing the percent population gains or losses between 1950 and 1990, reveals a distinctive pattern of population decrease throughout the Northeast (Color Plate 1.3). One of the keys to data exploration is that displays such as these can be created in a matter of seconds.

A fourth consequence of technological change is that it enables mapmakers to link maps, text, pictures, video, and sound in **multimedia** presentations. For example, in the Grolier encyclopedia already mentioned, animations include sound clips to assist in understanding. These animations are integrated, of course, with the rest of the encyclopedia, which includes text, pictures, and videos. In a more sophisticated form of multimedia known as **hypermedia**, the various forms of media can be linked transparently in ways not anticipated by system designers (Buttenfield and Weber 1994).

Closely associated with animation, data exploration, and multimedia is the **electronic atlas**. Initially, electronic atlases emulated the appearance of paper atlases (e.g., the *Electronic Atlas of Arkansas* developed by Richard Smith in 1987). More recently, however, electronic atlases have begun to incorporate animation, data exploration,

and multimedia capability. We discuss electronic atlases in Chapter 22.

A fifth consequence of technological change is the capability to create realistic representations of the Earth's natural and built environments. To illustrate, Color Plate 1.4 is a frame taken from a fly-through animation of a portion of Lawrence, Kansas, that was created by geography students at the University of Kansas. Such realistic fly-throughs provide users a sense of being immersed in the 3-D landscape. When users are able to navigate through and interact with the realistic 3-D environment, we term this a **virtual environment** or **virtual reality**. For instance, in Chapter 24 we describe Virtual Puget Sound, a system in which users don a **head-mounted display (HMD)** and examine water movement, the behavior of tides, and salinity levels in Puget Sound. The notions of visual realism and virtual environments certainly challenge our traditional thoughts about cartography. Normally, we have thought of maps as consisting of abstract symbols—now we need to extend our notion of maps to include realism.

Closely allied with virtual environments is the notion of **augmented reality**, in which computer-based information is used to enhance our view of the real world. For instance, imagine that you are a physical geographer studying vegetation changes in a particular region, and you wish to examine the current vegetation in the field and compare it with past vegetation patterns. Traditionally, you would accomplish this by taking maps into the field and comparing them with current vegetation patterns. With augmented reality, you can don a *wearable computer* (and associated specialized viewing devices)

TABLE 1.1   Internet uses for cartography

Function	Example	Address (URL)
Locational data	Longitude and latitude of major U.S. cities	*http://www.census.gov/cgi-bin/gazetteer*
Attribute data	Population estimates for counties	*http://eire.census.gov/popest/data/national.php*
Maps		
Static	3-D representation of pollution in New York Harbor	*http://www.rpi.edu/locker/69/000469/dx/harbor.www/maxus.dieldrin.cr.gif*
Animated	Animation of urban sprawl in the San Francisco Bay area	*http://www.ncgia.ucsb.edu/projects/gig/v2/About/abApps.htm*
Software for creating static maps	Software for creating choropleth maps at the state level	*http://maps.esri.com/esri/mapobjects/tmap/tmap.htm*
Software for exploring data	CommonGIS	*http://www.ais.fhg.de/SPADE/products/index_eng.html*
Electronic atlases	The Atlas of Canada	*http://atlas.gc.ca/site/index.html*
Tools for developing software	GeoVISTA Studio	*http://geovistastudio.sourceforge.net/*
Teaching materials for students	The Geographer's Craft	*http://www.colorado.edu/geography/gcraft/contents.html*
Teaching materials for instructors	Cartography: The Virtual Geography Department	*http://www.csun.edu/~hfgeg005/cwg/*

and actually *see* historic vegetation patterns overlaid on the present-day landscape. We consider this evolving technology in Chapter 24.

A sixth consequence of technological change is the ability to access maps and related information via the World Wide Web (WWW, or simply the Web), which is part of the larger Internet.* The Web has, of course, changed our daily lives as evidenced by the common listing of Web addresses in newspaper and television advertisements. From the standpoint of cartography, the Web can serve as a source of data, maps, software, electronic atlases, tools for developing software, and teaching materials (Table 1.1). In the course of reading this text, you will find reference to numerous Web sites. To access these sites, please utilize the home page for this book, which provides links for individual pages in the book, along with broad categories of sites relevant to cartography.

In addition to providing a wealth of maps and related information, the Web also changes the way in which maps are produced and used. Prior to the existence of the Web, cartography could be characterized as *supply-driven*—cartographers created maps and mapping software that they and other experts thought would be useful. With the advent of the Web, cartography can be considered *demand-driven*, as the nature of maps and mapping software is driven by what users seem to want (e.g., a Web site containing maps that receives few "hits" probably is not particularly useful).[†] As part of this demand-driven character, users might want to design their own maps. User-designed maps were possible with mapping software prior to the Web, but now there are millions of Web users and their cartographic abilities obviously vary. As we have already suggested, this raises the possibility of millions of poorly designed maps. Poorly designed maps are problematic even if only the individual making the map will use that map, as the individual might derive incorrect information as a function of the poor design.

A distinct advantage of the Web is that it provides the general public with access to spatial information that they normally might not have access to. For example, imagine that officials are planning to build a bypass around the city that you live in, and that the city has hired a consulting firm to model the anticipated impact on the city. If the consulting firm makes its predictions available via the Web, you should be able to examine them from the comfort of your home; you might even be able to run their models under different scenarios—for example, you might wonder what will

---

* For a technical discussion of the World Wide Web and Internet, see Dodge and Kitchin (2001b, 1–13).

[†] The terms *supply-driven* and *demand-driven* were taken from the work of Kraak and Brown (2001).

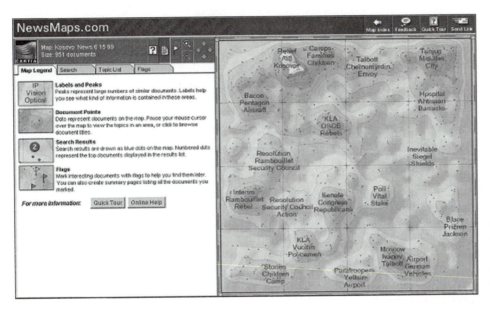

FIGURE 1.7 An example of mapping the information space of the Internet. The relationship and frequency of topics associated with more than 900 online news reports of the Kosovo crisis are depicted using the isarithmic method. (Originally produced by Newsmaps, Cartia, Inc.)

happen if the bypass is located north as opposed to south of the city. The ability of the general public to work with such spatial information is commonly termed *public participation GIS* (PPGIS).*

One aspect of the Web (and associated Internet technology) that has received considerable interest is mapping the technology and its associated information spaces. For instance, Color Plate 1.5 is an example of mapping the technology itself—in this case, early Internet traffic. The red balls and white lines connecting them represent the backbone of NSFNET in 1991 (NSFNET formed the basis for the present-day Internet). The colored vertical lines depict inbound data to each of the nodes of the backbone, with low and high amounts of data depicted by purple and white, respectively. Figure 1.7 is an example of mapping the information space of the Internet. Here more than 900 online news reports of the Kosovo crisis have been "mapped" using the isarithmic method; dark peaks represent similar commonly occurring news topics. Note that this is not a map in the conventional sense because we are mapping abstract space. Numerous examples of mapping the Internet and its associated information spaces can be found in Martin Dodge and Rob Kitchin's *Atlas of Cyberspace* (2001a) and on Martin Dodge's Web site (*http://www.cybergeography.org/atlas/*).

1.5 GEOGRAPHIC VISUALIZATION

Outside geography, the term *visualization* has its origins in a special issue of *Computer Graphics* authored by Bruce McCormick and his colleagues (1987). To McCormick and

his colleagues, the objective of **scientific visualization** was "to leverage existing scientific methods by providing … insight through visual methods" (p. 3). Today, scientific visualization generally involves using sophisticated workstations to explore large multivariate data sets (Color Plate 1.6). Work in scientific visualization extends far beyond the realm of spatial data, which geographers deal with, to include topics such as medical imaging and visualization of molecular structure and fluid flows. Peter and Mary Keller (1993) provided numerous examples of the use of scientific visualization. The most recent developments in scientific visualization can be found in the proceedings of the Institute of Electrical and Electronics Engineers (IEEE) Visualization conference, which has been held every year since 1990.

The notion of **information visualization** is closely related to scientific visualization. Information visualization focuses on the visual representation and analysis of nonnumerical abstract information. The "map" of online news reports of the Kosovo crisis shown in Figure 1.7 is an example of information visualization. In Chapter 1 of their book *Readings in Information Visualization: Using Vision to Think*, Stuart Card and his colleagues (1999) provided numerous additional examples of information visualization. In Chapter 25, we look at some ongoing developments in information visualization in geography.

Although those outside geography were the first to popularize visualization, the idea has existed in cartography since at least the 1950s (MacEachren and Taylor 1994, 2). As a result, cartographers have struggled to define the term. Thus far, two basic definitions have emerged. The first is a broad one that encompasses both paper and computer-displayed maps. According to Alan MacEachren and his colleagues (1992, 101):

---

* For an extensive discussion of PPGIS, see the November 2001 issue of *Environment and Planning B: Planning and Design.*

Geographic visualization [can be defined] as the use of concrete visual representations—whether on paper or through computer displays or other media—to make spatial contexts and problems visible, so as to engage the most powerful human information-processing abilities, those associated with vision.

Using this definition, geographic visualization could be applied to the visual analysis of a paper map created by pen-and-ink methods or to the visual analysis of a map created on an interactive graphic display.

The second, and narrower definition, is based on MacEachren's (1994b, 6) cartography-cubed representation of how maps are used, which is shown in Figure 1.8.* In this graphic, visualization is contrasted with communication along three dimensions: private versus public, revealing unknowns versus presenting knowns, and the degree of human–map interaction. Based on this diagram, MacEachren argued that **geographic visualization** is a private activity in which unknowns are revealed in a highly interactive environment, whereas **communication** involves the opposite: a public activity in which knowns are presented in a noninteractive environment.

As an example of geographic visualization, MacEachren employed Joseph Ferreira and Lyna Wiggin's (1990) "density dial," in which class break points on choropleth

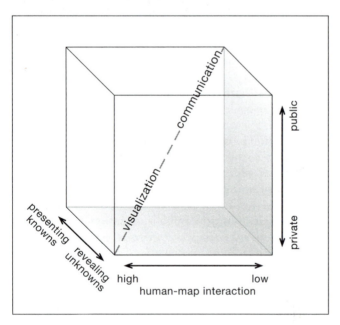

FIGURE 1.8 A graphical representation of how maps are used. Note that visualization is contrasted with communication along three dimensions: private versus public, revealing unknowns versus presenting knowns, and the degree of human–map interaction. (After MacEachren 1994b, 7.)

_____

* MacEachren used the term (Cartography)[3]; we use the word *cubed* to avoid confusion with the superscript 3.

maps were manipulated to identify and enhance spatial patterns. The idea is that users of this density dial could discover previously unknown spatial patterns. In contrast, MacEachren argued communication is exemplified by "you are here" maps used to locate oneself in a shopping mall. MacEachren stressed that certain map uses do not fit neatly into either category (thus the need for the cartography-cubed representation). For example, Gail Thelin and Richard Pike's (1991) dramatic shaded relief map shown in Figure 1.9 (to be discussed in Chapter 15) fits into the communication realm because it is available to a wide readership (making it "public") and the user cannot interact with the paper version of it. The Thelin and Pike map, however, fits visualization in the sense that it can reveal unknowns; for instance, many readers would be unfamiliar with the flatiron-shaped plateau in the north central part of the map—the Coteau des Prairies. The net result is that the Thelin and Pike map falls in the upper right corner of MacEachren's diagram.

The notions of map communication models and data exploration introduced previously can be associated with communication and visualization, respectively, in MacEachren's cartography-cubed representation. Thus, when using the five steps of the communication model presented in section 1.3, the intention generally is that the map is being made for the general public, there will be low human–map interaction, and the focus is on presenting knowns. In data exploration, the emphasis is on revealing unknowns via high human–map interaction in a private setting; in this sense the word "exploration" could easily be substituted for "visualization" in MacEachren's graphic.

On an informal basis, visualization has been used to describe any recently developed novel method for displaying data. Thus, cartographers have placed animation and virtual environments under the rubric of "visualization." Additionally, novel methods that might result in static maps (e.g., Daniel Dorling's cartograms; see Chapter 19) are also placed under the heading of visualization. Recently, MacEachren has simplified the term geographic visualization to **geovisualization** (e.g., MacEachren et al. 1999c). We generally have chosen to use the term geographic visualization in this book, but you should be aware that the term geovisualization frequently is used in the refereed literature.

1.6 RELATED TECHNIQUES

The picture of cartography today would not be complete without considering the related techniques of geographic information systems (GIS), remote sensing, and quantitative methods. In this section, we consider not only how cartographers can use these techniques, but also how

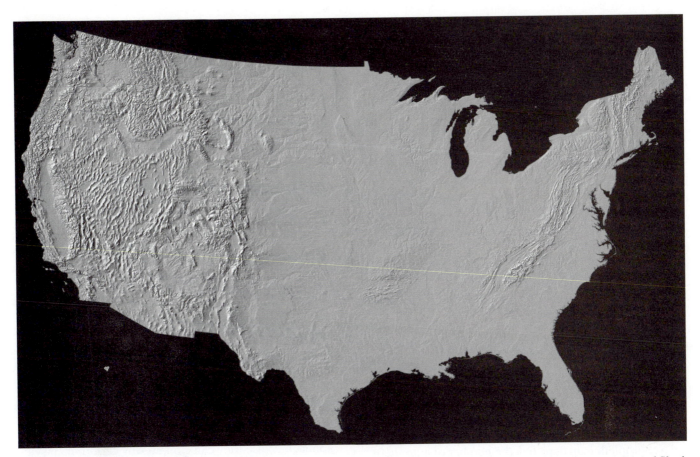

FIGURE 1.9 A small-scale version (1:21,400,000) of the USGS map *Landforms of the Conterminous United States: A Digital Shaded-Relief Portrayal*, developed by Thelin and Pike (1991). (Courtesy of Gail P. Thelin.)

knowledge of cartography can provide those working with these techniques the ability to create more effective cartographic products.

GISs are computer-based systems that are used to analyze spatial problems. For example, we might use a GIS to determine optimal bus routes in a school district, or to predict the likely location of a criminal's residence based on a series of apparently related crimes. Traditionally, most digital thematic maps were created using so-called **thematic mapping software** (e.g., MapViewer).* Today, GIS software typically has considerable thematic mapping capability, in addition to its inherent spatial analysis capabilities (e.g., solving the bus routing problem mentioned earlier). Its spatial analysis capabilities also enable GIS to handle more sophisticated mapping problems than traditional thematic mapping software. Dot maps (such as those shown in Figure 1.2) provide a good illustration of the capability of GIS software. Ideally, dots on a dot map should be located as accurately as

possible to reflect the underlying phenomenon. Thus, in the case of the wheat map, we would want to place dots where wheat is most likely to be grown (say, on level terrain with fertile soil) and not where it cannot be grown (in water bodies). Traditional thematic mapping software did not accomplish this, as dots would normally be based solely on the basis of enumeration units (e.g., counties). In contrast, GIS software enables a large number of factors (or layers, in GIS terminology) to be accounted for. We consider this process further in Chapter 17.

The basic purpose of remote sensing is to record information about the Earth's surface from a distance (e.g., via satellites and aircraft). For instance, we might use remote sensing to determine temporal changes resulting from forest fires (acreage burned, effects of erosion, and regrowth) or the health of crops. The importance of remote sensing to cartography can be illustrated by again presuming that we wish to map the distribution of wheat across the United States. Rather than use a GIS approach in which layers of related information are considered, we could use remote sensing to directly determine the precise location of wheat fields. For instance, Stephen Egbert

---

* MapViewer is still marketed as a thematic mapping tool, but now has some GIS analytic capability.

and his colleagues (1995) found that wheat fields could be identified with accuracy as high as 99 percent by using remotely sensed imagery for three time periods.

Quantitative methods are used in the statistical analysis of spatial data. For instance, we might develop an equation that relates deaths due to drunk driving to various attributes that we think might explain the spatial variation in the death rate, such as the severity of laws that penalize drunk driving, the extent to which the laws are enforced, the percentage of the population that are members of religious groups, the traffic density, and many others. The major development in quantitative methods relevant to cartography is that of exploratory spatial data analysis (ESDA), which refers to data exploration techniques that accompany a statistical analysis of the data. As an example, we might explore a map of predicted deaths due to drunk driving to see how the pattern is affected by various attributes that we include in the model. We only touch on ESDA in this book (in Chapter 21) because a complete understanding requires a more complete background in statistics than is typical for the introductory cartography student.

Thus far, we have considered how cartographers can benefit from knowledge of other geographic techniques. Those working with these other techniques can also benefit from a knowledge of cartography. For example, imagine that you are working in the GIS department of a city—you wish to examine the distribution of auto thefts, and so create a choropleth map depicting the number of auto thefts for each census tract in the city. Based on the high incidence of auto thefts in a contiguous set of census tracts, you recommend to the police department that they focus their patrols in those areas. Unfortunately, your solution might be inappropriate because you failed to consider the population (and possibly the number of cars owned) in each census tract. Instead of mapping the raw number of auto thefts, you probably would want to adjust for the population (or number of cars owned) in each tract. This is another example of the data standardization problem that we mentioned earlier. Effective use of GIS requires an understanding of such basic cartographic principles.

As another example of how those working in other areas can benefit from cartography, imagine that you are a remote sensor working on the GreenReport (*http://www.kars. ukans.edu/products/greenreport.shtml*), which uses remotely sensed images to depict the health of crops and natural vegetation throughout the United States. Color Plate 1.7A illustrates a basic greenness map that might be used to represent current vegetation conditions. One can argue that this is a logical color scheme as dark green represents "High Biomass" whereas dark brown indicates "Low Biomass" Color Plate 1.7B depicts a change map that might be used to compare current conditions to two weeks earlier. Note that in this case the "Little or no Change" category is similar to some of the categories on the basic greenness map, which

could confuse a map reader. An improved symbology would be to use a gray for the "Little or no Change" category, as shown in Color Plate 1.7C. An alternative symbology would be to use completely different colors to represent the changes in greenness; for example, Color Plate 1.7D uses shades of blue to represent increased greenness and orange and red shades to represent decreased greenness. The latter symbology would be desirable because users could associate certain colors with the raw image and a different set of colors with the changes.

1.7 COGNITIVE ISSUES IN CARTOGRAPHY

Understanding the role that cognition plays in cartography requires contrasting it with perception. *Perception* deals with our initial reaction to map symbols (that a symbol is there, that one symbol is larger or smaller than another, or that symbols have different colors). In contrast, *cognition* deals not only with perception but also with our thought processes, prior experience, and memory.* For example, contour lines on a topographic map can be interpreted without looking at a legend because of one's past experience with such maps. Alternatively, one might correlate the pattern of soils on a particular map with the distribution of vegetation seen on a previous map.

The principles of cognition are important to cartographers because they can explain why certain map symbols work (i.e., communicate information effectively). Traditionally, cartographers were not so concerned with why symbols worked but rather with determining which worked best. This was known as a *behaviorist view*, in which the human mind was treated like a black box. The trend today is toward a *cognitive view*, in which cartographers hope to find *why* symbols work effectively. A cognitive view should enable a theoretical basis for map symbol processing that will not only assist in telling us why particular symbols work, but provide a basis for evaluating other map symbols, even those that have not yet been developed.

To illustrate the difference between the behaviorist and cognitive views, consider an experiment in which a cartographer wishes to compare the effectiveness of two color schemes for numerical data: five hues from a yellow-to-red progression and five hues from the electromagnetic spectrum (red, orange, yellow, green, blue). Let us presume that the results of such an experiment reveal that the yellow–red progression works best. The traditional behaviorist would report these results but probably not provide any indication as to why one sequence worked better than another. In contrast, the cognitivist would consider how color is processed by the eye–brain system, possibly theorizing that

---

* For a more in-depth discussion of perception and cognition, see Goldstein (2002) and Matlin (2002), respectively.

the yellow–red progression works best because of opponent process theory (see Chapter 10). Effectiveness of the spectral hues might be argued for on the grounds that different hues will appear to be at slightly different distances from the eye and thus form a logical progression (e.g., red will appear nearer than, say, blue, as discussed in Chapter 14). Spectral hues might also be considered effective because of their common use on maps, and the likelihood that readers have experience in using them.

An important concept of cognition is the three types of memory: iconic memory, short-term visual store, and long-term visual store (Peterson 1995). **Iconic memory** deals with the initial perception of an object (in our case, a map or portion thereof) by the retina of the eye (see Chapter 10 for a detailed discussion of the retina). Calling this "memory" is somewhat of a misnomer, because it exists for only about one-half second and we have no control over it. Visual information initially recorded in iconic memory is passed on to the **short-term visual store**. Only selected information is passed on at this stage; for example, the boundary of Texas shown in Figure 1.10 will likely be simplified to some extent in moving from iconic to short-term visual store. Keeping information in short-term visual store requires constant attention (or activation). This is accomplished by rehearsal of the items being memorized (e.g., staring at the map of Texas and telling yourself to remember its shape).

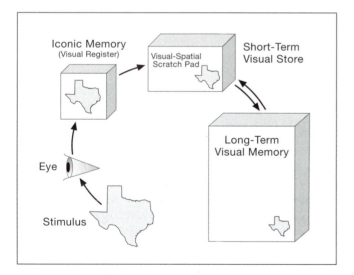

FIGURE 1.10 Three forms of memory used in cartography. A perceived map is initially stored in iconic memory (the retina of the eye). The image is then passed to the short-term visual store in the brain (where it is rehearsed). Finally, information is stored for later use in the long-term visual memory in the brain. (From *Interactive & Animated Cartography* by Peterson, M. P., © 1995. Adapted by permission of Prentice-Hall, Inc., Upper Saddle River, NJ.)

After the information has been rehearsed in the short-term visual store, it is ultimately passed on to **long-term visual memory** (Figure 1.10). Note that arrows are shown in both directions between the short-term visual store and long-term visual memory. When something is initially memorized, information must be moved from the short-term store to long-term memory; when something is retrieved from memory, the opposite is the case. As an example of the latter, imagine that you are shown a map of Texas and asked to indicate which state it is. To make your decision, you must retrieve the image of Texas from long-term memory and compare it with the image in the short-term store to make your decision. We have covered here only some of the basic concepts of cognitive psychology that are necessary for understanding this text. For a broader overview, see Peterson (1995, Chapter 2) and MacEachren (1995).

1.8 SOCIAL AND ETHICAL ISSUES IN CARTOGRAPHY

Although maps, and especially interactive digital maps, have tremendous potential for visualizing spatial data, we should also consider various social and ethical issues associated with their use. The notion of social and ethical issues in cartography was first developed in the context of *postmodernism*. Those who subscribe to a postmodernist view believe that problems can best be approached from multiple perspectives or viewpoints; for example, in a study of an urban neighborhood, a postmodernist would want to acquire not only the perspective of those in positions of political and economic power, but as wide a sample of inhabitants as possible: men and women, children, the elderly, each of the social classes, ethnic and racial groups, or any others who might shed light on the dynamics of the neighborhood (Cloke et al. 1991, 170–201).

An important notion of postmodernism is that a text (or map) is not an objective form of knowledge, but rather has numerous hidden agendas or meanings. **Deconstruction** enables us to uncover these hidden agendas and meanings. Brian Harley (1989, 3) stated:

> Deconstruction urges us to read between the lines of the map … to discover the silences and contradictions that challenge the apparent honesty of the image. We begin to learn that cartographic facts are only facts within a specific cultural perspective.

To illustrate deconstruction, consider Mark Bockenhauer's (1994) examination of the official state highway maps of Wisconsin over a seven-decade period. Although the major purpose of highway maps is presumably to assist a motorist in getting from one place to another, Bockenhauer (17) argued that "three dominant 'cultures' [could] be identified as factors influencing the map product: a transportation/modernization culture, a culture of promotion, and a subtle, beneath-the-surface culture of dominion." As an example of the transportation/ modernization culture, Bockenhauer recounted the removal of surfaced county roads from recent editions of the map, and thus the greater emphasis on "getting us from here to there by encouraging travelers onto the freeways" (p. 21). An illustration of highway maps as promotional devices was their use by government officials; for example, in the 1989–1990 edition, Governor Tommy Thompson showcased himself next to a replica of a Duesenberg automobile. Finally, an example of the culture of dominion was the portrayal of women on the maps: "Among the most common and prominent images appearing on the … maps … are those of women in swim suits and fishermen. Nearly all of the photos of people enjoying Wisconsin fishing … are of white men. … [The] women seem to be part of the package of 'pleasure' offered to white men in Wisconsin" (p. 24). Although some would disagree with Bockenhauer's interpretations of these maps, it is clear that maps can convey information other than their supposed primary purpose.

Although we do not focus on postmodernism and map deconstruction in this book, both mapmakers and map users need to recognize their importance. Mapmakers must realize that maps can communicate unintended messages, and that the data they have chosen to map or the method of symbolizing that data might be a function of the culture of which they are a part. Conversely, map users must recognize that a single map might depict only one representation of a spatial phenomenon (e.g., a map of percent forest cover is one representation of vegetation).

Note that there is some overlap between the notions of data exploration and postmodernism, because both promote the notion of multiple representations of data. In the context of data exploration, "multiple representations" refers to the various methods of symbolizing the data (e.g., using MapTime to display population data as both an animation and a change map). The postmodernist would likely support this approach because it concurs with the notion that there is no single "correct" way of visualizing data. Additionally, however, the postmodernist would be interested in the multiple meanings and potentially hidden agendas found in a particular thematic map.

More recently, a consideration of social and ethical issues has considered the role that sophisticated digital technologies might play in violating our personal privacy. For instance, the Digital Angel Corporation (*http://www.digitalangel.net/*) markets a wristband that can be locked to an individual, enabling another party to track the movement of that individual in real time via the global positioning system (GPS) and GIS. Although such technology is arguably useful for tracking children and the elderly, its availability raises a number of interesting ethical questions. For instance, what if a child does not wish to be tracked (not surprising for a teenager)? Should spouses be able to track one another? Jerry Dobson and Peter Fisher (2003) were particularly wary of such technology, noting that if a transponder is added to the wristband, it would be possible to administer a form of punishment to the individual. They term the net result **geoslavery**, suggesting that the results would be far worse than George Orwell's *1984*, as one "master" could potentially monitor and enslave thousands of people. Although geoslavery is not to our knowledge a problem at the present time, hopefully this example will encourage you to consider social and ethical issues associated with the use of modern digital technologies. In *Spying with Maps: Surveillance Technologies and the Future of Privacy*, Mark Monmonier (2002) provided an overview of numerous other ways in which GPS and GIS technologies can be used that raise issues of privacy.

FURTHER READING

Board, C. (1984) "Higher-order map-using tasks: Geographical lessons in danger of being forgotten." *Cartographica* 21, no. 1:85–97.

   Discusses issues relevant to the kinds of information that can be acquired from maps.

Buttenfield, B. P., and Mark, D. M. (1991) "Expert systems in cartographic design." In *Geographic Information Systems: The Microcomputer and Modern Cartography*, ed. by D. R. F. Taylor, pp. 129–150. Oxford: Pergamon.

   Although a bit dated, this work provides a good summary of the potential for using expert systems in map design; for more recent work, see Zhan and Buttenfield (1995) and Forrest (1999a).

Campbell, C. S., and Egbert, S. L. (1990) "Animated cartography: Thirty years of scratching the surface." *Cartographica* 27, no. 2:24–46.

   An overview of early work in animated cartography, along with some suggestions for the potential of animation.

DiBiase, D. (1990) "Visualization in the earth sciences." *Earth and Mineral Sciences* 59, no. 2:13–18.

   Discusses visualization and its role in geographic research.

Dodge, M., and Kitchin, R. (2001a) *Atlas of Cyberspace*. Harlow, England: Addison-Wesley.

   Provides numerous examples of mapping the Internet and its associated information spaces.

Environment and Planning B: Planning and Design, 28, no. 6, 2001 (entire issue).

This special issue deals with public participation in GIS. Also see vol. 25, no. 2, 1998 of *Cartography and Geographic Information Science*.

Keller, P. R., and Keller, M. M. (1993) *Visual Cues: Practical Data Visualization*. Los Alamitos, CA: IEEE Computer Society Press.

Provides examples of visualization from a wide variety of disciplines.

Koláčný, A. (1969) "Cartographic information: A fundamental concept and term in modern cartography." *The Cartographic Journal* 6, no. 1:47–49.

Presents a cartographic communication model that MacEachren (1995) claimed had "the greatest initial impact on cartography" (p. 4).

Kraak, M.-J., and Brown, A. (eds.) (2001) *Web Cartography: Developments and Prospects*. London: Taylor & Francis.

Provides a broad overview of issues related to the Web.

MacEachren, A. M. (1994b) "Visualization in modern cartography: Setting the agenda." In *Visualization in Modern Cartography*, ed. by A. M. MacEachren and D. R. F. Taylor, pp. 1–12. Oxford: Pergamon.

Discusses definitions for visualization.

MacEachren, A. M. (1995) *How Maps Work: Representation, Visualization, and Design*. New York: Guilford.

An advanced treatment of cognitive issues in cartography.

MacEachren, A. M. (1999) "Cartography, GIS, and the World Wide Web." *Progress in Human Geography* 22, no. 4:575–585.

Covers potential research issues associated with maps and mapping software available via the Web.

MacEachren, A. M., and Kraak, M.-J. (1997) "Exploratory cartographic visualization: Advancing the agenda." *Computers & Geosciences* 23, no. 4:335–343.

Provides an alternative view of MacEachren's cartography-cubed model.

McCormick, B. H., DeFanti, T. A., and Brown, M. D. (1987) "Visualization in scientific computing." *Computer Graphics* 21, no. 6 (entire issue).

A classic early work on visualization.

Monmonier, M. S. (1985) *Technological Transition in Cartography*. Madison, WI: University of Wisconsin Press.

A text dealing with technological change in cartography.

Monmonier, M. (2002) *Spying with Maps: Surveillance Technologies and the Future of Privacy*. Chicago: University of Chicago Press.

Presents myriad ways in which modern GPS and GIS technologies can be used for surveillance and discusses related privacy issues.

Peterson, M. P. (1997) "Cartography and the Internet: Introduction and research agenda." *Cartographic Perspectives* no. 26:3–12.

An overview of the Internet and its role in cartography. Other articles in this issue also deal with the Internet. The complete issue can be found at *http://maps.unomaha.edu/NACIS/cp26*. For more recent work, see Peterson (2003).

Rundstrom, R. A. (Ed.) (1993) "Introducing cultural and social cartography." *Cartographica* 30, no 1 (entire issue).

A set of articles dealing with postmodern issues in cartography.

Sheppard, E., and Poiker, T. (eds.) (1995) "GIS and society." *Cartography and Geographic Information Systems* 22, no. 1 (entire issue).

A set of articles covering postmodern issues in GIS.

Wood, D., and Fels, J. (1992) *The Power of Maps*. New York: Guilford.

An extensive essay on postmodern issues in cartography. Also see Wood and Fels (1986).

2

History of U.S. Academic Cartography*

OVERVIEW

This chapter details the history and development of U.S. academic cartography in the 20th century. Developments in the history of U.S. academic cartography are first reviewed by identifying and discussing four major periods (Table 2.1). The incipient period, spanning from the early part of the century to the 1940s, represents what might be called nodal activity, where academic cartography was centered at only two to three institutions with individuals who were not necessarily educated in cartography. An example was J. Paul Goode at the University of Chicago. A second period, from the 1940s to the 1960s, saw the building of core programs with multiple faculty, strong graduate programs, and PhD students who ventured off to create their own programs. Three core programs stand out—the University of Wisconsin, University of Kansas, and University of Washington. Other universities developed cartographic programs in the third period, including those at UCLA, Michigan, and Syracuse. This third period, from the 1960s to the 1980s, also witnessed rapid growth in academic cartography in terms of faculty hired, students trained, journals started, and development within professional societies. It is in this period where cartography emerged as a true academic subdiscipline, nurtured within academic geography departments with strong research programs and well-established graduate education. The pinnacle of academic cartography in the United States occurred in the mid-1980s when cartography had reached its maximum growth, but the effect of the emerging discipline of geographic information science (GIS) had not yet been felt. Finally, a fourth period is one of transition, where cartography has become

increasingly integrated within GIS curricula. The result has been fewer academic positions in cartography, fewer students educated as thoroughly in thematic cartography, and growth in geographic visualization. From the perspective of academic geography as we start the 21st century, cartography witnessed remarkable growth from the 1940s to the 1980s, but has, in the past decade, seen a decline as a direct result of the rapid rise of the new related discipline, GIS. However, as we approach the next millennium, it appears that a synthesis of the two is slowly emerging with the development of an integrated cartography–GIS curricula.

We recognize many important developments in other countries such as the United Kingdom, France, the Netherlands, Germany, Austria, Switzerland, China, Japan, and Russia; however, these are not addressed in this chapter to allow for a more detailed discussion of the U.S. case. Sara Fabrikant (2003) provided a brief overview of important developments that have taken place in some other countries.

*In the latter part of the chapter, we discuss the nature of **analytical cartography** developed by Waldo Tobler. Analytical cartography includes the topics of cartographic data models, digital cartographic data collection methods and standards, coordinate transformations and map projections, geographic data interpolation, analytical generalization, and numerical map analysis and interpretation.*

2.1 FOUR MAJOR PERIODS OF U.S. ACADEMIC CARTOGRAPHY

2.1.1 Period 1: The Incipient Period

John Paul Goode

According to McMaster and Thrower (1987), although basic training in cartography started as early as 1900 in

---

* Portions of this chapter were taken from McMaster, R. and McMaster, S. (2002) "A history of twentieth-century American academic cartography." *Cartography and Geographic Information Science* 29, no. 3: 305–321. Reprinted with permission from the American Congress on Surveying and Mapping and Mark Monmonier.

TABLE 2.1 The four major periods of U.S. academic cartography

Incipient period (early 1900s to mid-1940s)
  J. Paul Goode
  Erwin Raisz
  Guy-Harold Smith
  Richard Edes Harrison
Post-war era of building core programs (mid-1940s to mid-1960s)
  University of Wisconsin
  University of Kansas
  University of Washington
Growth of secondary programs (late 1960s to late 1980s)
  UCLA
  University of Michigan
  University of South Carolina
  Syracuse University
Integrated curricula with geographic information systems/science (1990s)
  University of California, Santa Barbara
  SUNY Buffalo
  University of South Carolina

FIGURE 2.1 J. Paul Goode—arguably the first genuine American academic cartographer. (After McMaster and Thrower 1991. First published in *Cartography and Geographic Information Systems* 18(3), p. 151. Reprinted with permission from the American Congress on Surveying and Mapping.)

the United States, "It could be argued that the first genuine American academic cartographer was John Paul Goode at the University of Chicago" (p. 346). Goode (Figure 2.1), who graduated from the University of

Minnesota in 1889, taught at Minnesota State Normal School at Moorhead until 1898, and received a PhD in economic geography from the University of Pennsylvania in 1903, was one of the most professionally active geographers of his time. During the early part of the century, Goode became a charter member of the Association of American Geographers (AAG), served as co-editor of the *Journal of Geography* from 1901 through 1904, helped organize the Geographic Society of Chicago, and was appointed by President Taft to assist in leading a U.S. tour for a distinguished group of Japanese financiers. However, Goode is best known for the development of Goode's Homolosine Projection, first presented at the AAG's meetings in 1923, and the development of *Goode's World Atlas*, also published in 1923. Goode's AAG presidential address, "The Map as a Record of Progress in Geography," given at the 1926 meetings, illustrated the importance of maps to Goode's philosophy of geography.

In 1924, Goode taught a course at the University of Chicago entitled "A Course in Graphics," which included four major themes: The Graph, The Picture, Preparation of Illustrative Material for Reproduction, and Processes of Reproduction (Table 2.2). It is clear that Goode used this course as a prerequisite to the more traditional cartography class that he taught. Although he provided information on both statistical presentation and the creation of base maps (through the projection process), this was certainly not what we would consider a comprehensive course in cartography. One can see, however, his approach to statistical mapping in Section I. D of the course (Table 2.2), where he focused on "The areal distribution of magnitudes; application of the graph to the map." One can also see from the nature of Goode's exercises that he did, in fact, teach basic principles of statistical mapping. His exercise Number 1 in the 1924 course asked students to map the total mineral output of the United States for the latest year on record, with the requirement that the circles be proportional to production in the area of each state. A similar exercise required students to map the great seaports of the world by net register tonnage entered and cleared, with circles centered on each port, and the area of the circle proportional to the traffic. He had similar types of exercises with isometers (isolines) of intensity applied to the map and the areal distribution of intensity shown by small uniform areas of dots.

Goode's influence was extended through his students at Chicago. Most of these students did not devote themselves to cartography specifically, but some were able to influence the course of the field through positions in the private sector, government, and academia. Two of Goode's students at the University of Chicago, Henry Leppard (University of Chicago, Washington, and UCLA) and Edward Espenshade (Northwestern University) devoted their careers to cartographic education,

TABLE 2.2 Outline for J. Paul Goode's 1924 course in Graphics

I.   The Graph
     A. Showing comparative magnitudes in one, two, or three dimensions, the elements of time omitted
         1. Magnitudes compared in one direction
         2. Magnitudes compared in two directions
         3. Magnitudes compared in three dimensions, as in cubes, or spheres, or bales, or barrels, or ships, or men, etc.
     B. The composition of a variable with a constant
     C. Comparison of variables through time, as in relation of two or more changing series
     D. The areal distribution of magnitudes; application of the graph to the map
         1. Comparison of regional areas
         2. Distribution of a series of magnitudes in countries or states are areal units
         3. Distribution of two or more series of magnitudes by political unit areas
         4. Isometers of intensity applied to the map
         5. Areal distribution of intensity shown by small uniform unit areas, e.g. dots
     E. The problem of color in the graph or map
         1. Pastel pencils
         2. Liquid colors
         3. Color equipment
         4. Studies in the use of color
     F. The problem of legibility
     G. The art of lettering

II. The Picture
     A. The photograph
     B. The lantern slide and the lantern
     C. The stereograph
     D. The cinema

III. Preparation of Illustrative Material for Reproduction
         1. The line drawing
         2. The photograph

IV. Processes of Reproduction: Engraving and Printing

and continued Goode's work with both base map development and the many generations of the *Goode's World Atlas*, published by Rand McNally. Goode's successor in cartography at the University of Chicago was Leppard, who stayed there until after World War II, when he went to the University of Washington (where he worked with John Sherman); later Leppard finished his career at UCLA. Espenshade spent his entire career at Northwestern University where he continued to edit *Goode's World Atlas*.

Erwin Raisz

Erwin Raisz (Figure 2.2) was the leading American academic cartographer between the time of Goode's death and the emergence of major graduate programs at Wisconsin, Kansas, and Washington. Raisz, born at Locse, Hungary on March 1, 1893, received a degree in civil engineering

FIGURE 2.2 Irwin Raisz—a leading American academic cartographer between the time of Goode's death and the emergence of major graduate programs in cartography. (After McMaster and Thrower 1991. First published in *Cartography and Geographic Information Systems* 18(3), p. 153. Reprinted with permission from the American Congress on Surveying and Mapping.)

before serving as a lieutenant in the Sappeaurs of the Imperial Austro-Hungarian Army during World War I. Following World War I, he immigrated to the United States, where he simultaneously was employed by a map company in New York City and worked on his master's degree at Columbia, which he received in 1923. While working on his PhD in geology at Columbia, Raisz studied with Douglas Johnson, a William Morris Davis–trained geomorphologist at Harvard who had strong interests in the construction of block diagrams and the representation of landscapes.* As an instructor at Columbia, Raisz offered the first cartography course there. Thus, the seeds for Raisz's approach to landscape representation had been acquired from multiple disciplines, including civil engineering and geology.

After receiving his PhD from Columbia in 1929, Raisz published his dissertation, "The Scenery of Mount Desert Island: Its Origin and Development," in the *Annals of the New York Academy of Science*. In 1931, he published

---

* One of Davis's students, A.K. Lobeck, carried on the tradition of block diagrams for representing the landscape. For an example, see Figure 15.17.

a paper, "The Physiographic Method of Representing Scenery on Maps," which expressed his own individualized approach to landscape representation, in the *Geographical Review*. Based on this work and the recommendation of his mentor Johnson, he was offered the position of lecturer in cartography in the Institute of Geographical Exploration at Harvard by Davis himself.

During the 1930s, Raisz continued to publish and work on his techniques. In 1938, he published the first edition of the influential book *General Cartography*, which was to remain the only general English textbook on cartography for 15 years. This singular event indicates that, at that time, there was a large enough interest in academic cartography to warrant the publication of a specific book on the subject. The book was published as part of the McGraw-Hill series in geography, which at the same time was publishing such classics as Finch and Trewartha's *Physical Elements of Geography*, Platt's *Latin America*, and Whitbeck and Finch's *Economic Geography*. Clearly, this major publisher was willing to "invest" in a substantial cartographic project. Besides the obvious emphasis on "representation," the contents of *General Cartography* were what one would expect to find at this time period: projections, lettering, composition, and drafting. A significant part of Raisz's book was also spent on the history of cartography, with sections on manuscript maps, the renaissance of maps, the reformation of cartography, and American cartography. What is most interesting is the relatively small amount of text that was devoted to "statistical mapping" as we know it today.

Although the geography department at Harvard collapsed in 1947 (Smith 1987a), Raisz remained in the Boston area, teaching at Clark University until 1961. Raisz finished his career at the University of Florida, where he published his second textbook, *Principles of Cartography*, in 1962, and the *Atlas of Florida*. He was also influential in establishing the AAG Committee on Cartography in the early 1950s, from which he received the AAG Meritorious Service Award in 1955.

What is curious about Raisz is that he never held a regular academic appointment. Thus, he was unable to produce a generation of students that would perpetuate his brand of cartography. It was through his textbooks, his role with the AAG, and mostly through his maps that Raisz's influence has been felt. Raisz is best known for the production of his "landforms" maps of various parts of the world. His "Landform Outline Map of the United States" (1954), perhaps one of the best examples of cartography from the 20th century, has become a standard reference in United States geography classes (see section 15.2.3 for a discussion of this map). Other Raisz landform maps include England (1948), Central America (1953), and the Greater Antilles (1953). Raisz, of course, was continuing a tradition of landform mapping in the United States. Robinson and Sale (1969) asserted that the landform map or physiographic diagram, such as those created by William Morris Davis, A. K. Lobeck, Guy-Harold Smith, and Raisz, was "one of the more distinctive contributions of American cartography" (p. 187).

Guy-Harold Smith

Cartography at The Ohio State University was taught as early as 1925—a class in Map Construction and Interpretation offered by Fred Carlson—making this one of the oldest cartography courses in the country. In 1927, Guy-Harold Smith (Figure 2.3), a recent PhD from the University of Wisconsin (A. K. Lobeck was his advisor), took over the cartography program, where he taught for nearly 40 years until his retirement in 1965. Although chair of the department for 29 years, Smith was a prolific thematic cartographer, producing his famous "Relative Relief Map of Ohio" and "Population Map of Ohio" using graduated spheres (Figure 16.4). A talented teacher, his best-known student was Arthur Robinson, who created the influential program

FIGURE 2.3 Guy-Harold Smith—an early thematic cartographer who produced a famous map of graduated spheres. Arthur Robinson was one of Smith's students. (After McMaster and Thrower 1991. First published in *Cartography and Geographic Information Systems* 18(3), p. 153. Reprinted with permission from the American Congress on Surveying and Mapping.)

in cartography at the University of Wisconsin, which we describe later.

Richard Edes Harrison

Richard Edes Harrison (Figure 2.4), born in Baltimore in 1901, was the son of Ross Granville Harrison, one of the most distinguished biologists of his time. Although Harrison graduated with a degree in architecture from Yale University in 1930, his interests soon turned to scientific illustration where, in the years after completing his degree, he eventually was attracted to cartography. He drew his first map for *Time* magazine in 1932. This initial exposure to mapping piqued his curiosity, and he soon became a freelance cartographer for *Time* and *Fortune* magazines. In the late 1940s, Harrison would fly to Syracuse University once a week to teach the course in cartography (George Jenks was one of his students), and he also lectured at Clark, Trinity, and Columbia University. Although not formally an educator, Harrison nonetheless influenced the discipline of cartography through his specific technique and intrinsic cartographic abilities. He can also be considered one of the first "popular" cartographers for his work in media mapping.

There are few accounts of what exactly was taught by Harrison at Syracuse University. Monmonier (1991a, 205), however, wrote that Harrison remembered Jenks well, and that Jenks seemed more interested in the "nuts

FIGURE 2.4   Richard Edes Harrison—one of the first "popular" cartographers for his work in media mapping; George Jenks was one of his students.

and bolts" of cartography than the artistic component that was a major focus of Harrison's work. A sense of art must have been a critical part of Harrison's lectures. In a 1991 paper, Jenks described Harrison's classes, which he attended during the 1946–47 year, as a mixture of lectures, demonstrations, drafting, and hand lettering. It was Harrison's demonstrations that Jenks found most useful. Jenks (1991, 161) wrote:

> In one demonstration he discussed editorial sessions with the editors of Fortune magazine. The editors were interested in military movements in Africa, Europe, Japan, and East Asia, and as they talked Harrison would sketch maps of these areas from memory. His memory of geographic features was phenomenal. His lectures on map projections led to various exercises we conducted between his visits to the campus. One exercise was the construction of an azimuthal equidistant projection centered on Syracuse, which took us a full semester to complete.

In the early 1950s Harrison completed a survey of 24 cartographers in the New York City area to determine what their training had been. He found that virtually none had been trained in cartography but had drifted into the discipline. He also observed that none of those interviewed had a background in geography. Harrison's conclusions were quite emphatic about the state of American academic cartography: "There remains only the necessity of stating the dismal fact that cartography, as a well-rounded profession, does not exist in this country" (Harrison 1950, 15).

It is clear that Harrison was extremely well connected in American geography. In his seminal work *Look at the World: The Fortune Atlas for World Strategy* (1944), he acknowledged the influence of Wallace W. Atwood, S. Whittemore Boggs, William Briesemeister, George B. Cressey, Richard Hartshorne, Colonial Lawrence Martin, O. M. Miller, Erwin Raisz, Arthur Robinson, John K. Wright, and others. Harrison's impact was mostly through his design and technique, not necessarily his students, but he was active in the professional cartographic community, helping out with the incipient Committee on Cartography of the AAG.

2.1.2 Period 2: Post-World War II Graduate Education Centers of Excellence

The period following World War II was associated with a great expansion of geography departments in many U.S. universities and colleges, especially Wisconsin, Kansas, and Washington, as well as a decline at others such as Harvard, which dissolved its geography program in 1947 (Smith 1987a). It was after the war that Erwin Raisz and other members of the AAG sought to establish a more permanent base for cartography within that organization. A seminal event—the first meeting of the

Committee on Cartography—in the evolution of American academic cartography was organized by Erwin Raisz on April 6, 1950, at Clark University during the national meetings of the AAG. Five speakers gave presentations at this meeting, including Erwin Raisz (an "Introduction"), Richard Edes Harrison ("Cartography in Art and Advertising"), Carl Mapes ("Cartography in Map Companies"), Clarence Odell ("Cartography and Cartographers in Commercial Map Companies"), and George Kish ("Teaching of Cartography in the United States and Canada"). The papers and presentations were published in a special 1950 issue of *The Professional Geographer* (November), as were the notes from a Discussion and Question Period that followed. Raisz wrote:

> This is for us a historic event—our first official meeting as a distinct Committee in Cartography, and may the sapling become a strong tree with many branches and with rich and abundant fruit. It all started before the war, when we first discussed the necessity of a national organization. During the war the profession grew by leaps and bounds, but academic cartography was not quite prepared to lead the way. More and more geographers became interested in cartography and time was ripe for some consolidation of the profession. (Raisz 1950, 9)

Perhaps one can point to this as one of the first philosophical discussions as to what cartography really was. Raisz felt that cartographers fell into two categories: "geographer cartographers," who wish to express their ideas with graphs, charts, maps, globes, models, and bird's-eye views; and "cartotechnicians," who "help produce maps, models and globes by doing the color-separation or cardboard cutting" (Raisz 1950, 10). He proposed the idea of different types of cartographers, including the cartologist, cartosophist, toponymists, map compilers, map designers, draftsmen, letterists, engravers, map printers, and cartothecarian.

What we see at this juncture is the attempt by cartography to position itself in relation to geography and other disciplines. Raisz attempted to "bound" the geographic cartographer, differentiate him or her from surveyors, and provide the essence of the modern mapmaker. Unquestionably at this point in the history of academic cartography, Raisz was a national and even international leader. His two editions of *General Cartography* had been published, and he was now in a position of organizing others. Yet cartography at this moment could be seen as atheoretical and mostly descriptive. The significant problems were associated with drafting media and production techniques. Most of the methods for symbolization, including the dot, graduated symbol, isarithmic, choropleth, and even dasymetric map, were developed in the 19th century or before. Fortunately, a series of academic cartographers with strong interests in more conceptual

and theoretical issues emerged during the 1950s and led the development of basic research programs.

In the early 1950s, George Kish reported on the first "scientific" survey designed to determine the status of academic cartography in America. The survey, distributed in 1950 by Erwin Raisz, requested information on the following (Kish 1950, 20):

1. Is cartography taught as an independent course?
2. Is the institution giving advanced degrees (MA or PhD) in cartography as a major or minor subject?
3. Names of cartography concentrations.
4. Is cartography taught as a part of other courses?
5. Is the institution interested in establishing a course in cartography?

Amazingly, 94 institutions in the United States and Canada returned questionnaires (along with Oxford University and the University of Sydney).

The 1950 special issue of *The Professional Geographer* also provides us with a sense of current cartographic activity in the United States at that time. Books reviewed included: *An Introduction to Map Projections, Maps and Map-Makers, Modern Cartography*, and *Base Maps for World Needs*. News from Cartographic Centers detailed activity at Michigan State College, Yale University, Denoyer-Gepper Company, and Rand McNally. Perhaps most interesting, the current activities of several American cartographers were provided, from which we learn of the work of Edward Espenshade, Jr., Richard Edes Harrison, Donald Hudson, A. K. Lobeck, Allen Philbrick, Erwin Raisz, Arthur Robinson, and John K. Wright.

The status of cartographic education at the beginning of the 1950s was promising. From a sporadic set of institutions offering courses in cartography before World War II, the demands of the war accelerated the development of cartographic curricula. Over the following decade (1950–1960) three major programs would emerge—Wisconsin, Kansas, and Washington.

University of Wisconsin

Although Arthur Robinson (Figure 2.5) traces the teaching of cartography at the University of Wisconsin, Madison, back to 1937 (Robinson 1991, 156), the key point was when Robinson was hired in 1945 after completing a stint in the Geography Division of the Office of Strategic Services (OSS) in World War II. Robinson (1991) noted that an independent map division was created within the OSS that worked closely with geographers, historians, economists, and regional specialists, and grew to a staff of 100, with at least 50 professional cartographers.

Robinson's faculty position at Wisconsin included the responsibility of establishing a cartography and map use instructional program that at the outset included two basic

FIGURE 2.5   Arthur Robinson—responsible for developing the cartography program at the University of Wisconsin.

cartography courses (i.e., introductory and intermediate cartography) as well as an aerial photo interpretation course. Later, other courses were added, including Seminar in Cartography, Cartographic Production, and Use and Evaluation of Maps. These were followed by another series of courses in Map Projections and Coordinate Systems, Problems in Cartography, Computer Cartography, History of Map Making, and Cartographic Design. In the late 1960s, the staff in cartography was increased when Randall Sale became associate director of the University of Wisconsin Cartographic Laboratory.

In a 1979 paper on the influence of World War II on cartography, Robinson wrote about these early years:

> In the development of cartography in American academic geography, probably the most notable event prior to World War II was the publication of Erwin Raisz's *General Cartography*, in 1938. By the mid-1930s the majority of graduate students in geography (probably few if any undergraduates) took one course in cartography. Mine, at Wisconsin, came before Raisz's book appeared and our "textbooks" were Deetz and Adams' *Elements of Map Projections* and Loebeck's *Block Diagrams*. Besides constructing map projections and making crude dot map, isopleth, and pie chart maps, we were taught how to tint glass lantern slides with Japanese water colors. When I transferred to Ohio State University for doctoral work, I was not allowed

to take the cartography course because I already had one! Raisz's pioneering book provided a small beginning for an academic program in cartography, but nothing really got started before World War II began. (Robinson 1979, 97)

Arthur Robinson established himself as the unofficial "Dean" of American academic cartographers, and built the program in cartography at the University of Wisconsin into the very best in the United States during the 1970s and early 1980s. His seminal volume, *The Look of Maps*, based on his doctoral dissertation at The Ohio State University, was arguably the seed for three decades of cartographic research. Robinson established the first American journal in cartography, *The American Cartographer*, in 1974, wrote six volumes of *Elements of Cartography*, and was president of the International Cartographic Association. Robinson also influenced several generations of students who themselves ventured off and established graduate programs in cartography. Robinson and Sale guided the cartography program at Madison until 1968, when Joel Morrison, a Robinson PhD, and later Philip Muehrcke joined them. Both brought strong mathematical expertise to the program. Thus in the mid-1970s, when many geography departments struggled to maintain a cartography program with a single faculty member, Wisconsin had four individuals, separate BS and MS degrees in cartography, and the very best cartography laboratory (within a geography department). It was a cartographic tour de force.

The cartography program at Wisconsin has produced more than 100 students with master's degrees and 20 students with doctoral degrees. The first master's degree with a cartography specialization was awarded in 1949 and the first doctoral degree in 1956 (James Flannery's graduate circle dissertation; see Chapter 16 for a discussion of his work). Another important factor in the development of cartographic instruction at Wisconsin was the awarding of several National Defense Education Act fellowships in the 1960s to support graduate work in cartography. Each fellowship, which included a generous three-year stipend and a grant to support the development of the cartography instructional program, attracted some of the very best graduate students who, on completing their PhDs, created their own undergraduate and graduate programs in cartography. PhDs from Wisconsin included Norman J. W. Thrower (UCLA), Richard Dahlberg (Syracuse University and Northern Illinois University), Henry Castner (Queens University), Mei-Ling Hsu (University of Minnesota), George McCleary (Clark University and the University of Kansas), David Woodward (Newberry Library and Wisconsin), Barbara Bartz Petchenik (R. R. Donnelly), Joel Morrison (Wisconsin, USGS, the Bureau of the Census, and the Center for Mapping at The Ohio State University), Judy Olson (Boston University and

Michigan State University), and A. Jon Kimerling (Oregon State University).

With the retirement of Robinson in 1979, David Woodward, a former Robinson student who specialized in the history of the discipline, was hired. Sale retired in 1981 and Morrison left in 1983 for the USGS. In the early 1980s, James Burt—a climatologist out of UCLA who specialized in computer graphics—and Barbara Buttenfield—a John Sherman PhD out of Washington—were hired. Buttenfield left in 1987 for SUNY Buffalo, and Lynn Usery replaced her in 1988. Phillip Muehrcke, the last of the four core cartographers at Wisconsin (Robinson, Sale, Morrison, and Muehrcke) retired in 1998. Recently, Mark Harrower, a cartographer with a strong interest in animation (see Chapter 20), was hired.

University of Kansas

The cartography program at the University of Kansas was started and nurtured for more than 35 years by George Jenks (Figure 2.6). Jenks received his PhD in agricultural geography at Syracuse University and also studied with Richard Edes Harrison, the cartographer for *Time* and *Fortune* magazines, at Syracuse. As Jenks discussed in a 1991 paper, "I attended Harrison's courses in cartography during 1946 and 1947. They were a mixture of lectures,

FIGURE 2.6   George Jenks—responsible for developing the cartography program at the University of Kansas.

demonstrations, drafting, and hand lettering. In the spring of 1946 there were five of us in his class, but attendance grew rapidly the following years. While his courses were interesting, I recall his demonstrations with fondness" (Jenks 1991, 162). After a single year at the University of Arkansas, in 1949 Jenks arrived at a small but talent-laden department at Kansas and started building the cartography program. A significant event in Jenks's career, and for the program itself, was a grant from the Fund for the Advancement of Science that allowed him to visit all major mapmaking establishments of the federal government as well as a number of quasi-public laboratories in 1951–52. The information collected during this grant year was incorporated into an *Annals* of the Association of American Geographers paper entitled "An Improved Curriculum for Cartographic Training at the College and University Level," and adopted in the cartography program at Kansas.

Through the grant, Jenks (1953a) identified a series of key problems for cartographers, including: (1) mass production techniques had to be improved; (2) new inks, papers, and other materials were needed; and (3) additional personnel had to be trained. Jenks (1953a, 317) wrote:

> Increased demands for map-makers have induced many American colleges and universities to add cartography courses to their curricula. Prior to World War II very few courses in cartography were offered in the United States, but now well over one hundred institutions of higher learning offer training in the subject. Unfortunately, increasing the number of courses does not solve the problem of poorly trained map-makers. That cartographic instructors are cognizant of the need for improved cartographic training in map-making is evidenced by numerous articles in recent issues of professional journals and by the repeated attention this problem receives at national meetings.

Jenks pointed out that, at the time, several factors served to impede cartographic training: the use of inexperienced instructors, poorly equipped cartographic facilities and map libraries, both limited research and limited access to research, and too much of an emphasis on theory. It is interesting that Jenks, who spent much of his research career building cartographic theory in design, symbolization, and classification, would make such an argument. But he wrote, "too little time and effort has been spent on the practical application of theory. Theorizing on art does not make an artist, knowledge of medical theory does not make a qualified doctor, and talking about maps (and listening to lectures on cartography) does not mean that the student can execute a map" (Jenks 1953a, 319).

Jenks's grant led him to identify four key objectives of cartographic training:

1. Cartographic training should stress the fundamental principles of the field as a whole.

2. Cartographic training must include numerous opportunities for applying theory to actual map problems.

3. Cartographic training should encompass a wide range of general and technical courses in allied fields.

4. Cartographic training should be available to students in many disciplines and varying degrees of intensity.

During this time period, Jenks discovered that some were arguing for separate departments of cartography. For example, George Harding (1951) felt that cartography would need to leave civil engineering and establish its own home, and Wilbur Zelinski argued for a school of cartography (Cartographic Panel 1950). However, most realized the impracticality—both politically and financially—of creating a separate department of cartography. Because the objectives of Jenks's study were to determine what subject matter should be included in a cartographic curriculum, a critical aspect involved the interviewing of 88 individuals. The results of the question "What subject matter should be included in a college cartographic training program?" are tabulated in his *Professional Geographer* paper. Finally, Jenks proposed a five-course core sequence in cartography, including:

Course 1. Elementary training in projections, grids, scales, lettering, symbolization, and simple map drafting.

Course 2. The use, availability, and evaluation of maps.

Course 3. Planning, compiling, and constructing small-scale maps, primarily subject matter maps.

Course 4. Planning, compiling, and constructing large-scale maps, primarily topographic maps.

Course 5. Nontechnical training in the preparation of simple manuscript maps for persons wishing the minimum in the manipulative aspects of cartography.

The importance of Jenks's landmark study cannot be overemphasized. Cartography had emerged from World War II as a true discipline, in part due to the great demand for war-effort maps and mapping. Both those who had been practicing before the war, such as Arthur Robinson and Erwin Raisz, and those who emerged after, such as George Jenks and John Sherman, realized that comprehensive cartographic curricula could be maintained within geography departments. Jenks's study, in parallel with the previously described efforts by Erwin Raisz and the AAG, provided the intellectual infrastructure for those attempting to build cartography in universities.

Another significant influence on Jenks's early career was his relationship with John Sherman of the University of Washington. In the summer of 1956 Sherman came to Kansas to teach, and later Jenks was in residence at Seattle. An important event during the 1960s

was the establishment of the National Science Foundation–funded Summer Institutes for College Teachers. These summer institutes, organized under the direction of Sherman and Jenks, were first offered in Seattle in 1963. The institutes, nine weeks each in length, were designed to educate college professors in the modern techniques of cartography. Jenks wrote, "We were surprised to find that a number of professors had been assigned arbitrarily by their deans or chairman to teach mapmaking that fall. Moreover, several were going to have to teach without a laboratory, equipment, or supplies. These activities greatly enhanced my teaching and were the basis for numerous changes in our curriculum" (Jenks 1963, 163).

Despite faculty at Kansas with interests closely related to cartography—in particular statistics and remote sensing—at the end of the 1960s Jenks was still the only cartographer on the staff. Robert Aangeenbrug, with strong interests in computer cartography and urban cartography and the director of two of the International Symposium on Computer-Assisted Cartography (Auto-Carto) conferences, joined the Kansas faculty in the 1960s. Thomas Smith, who had arrived in the department as its second hire in 1947, established coursework in the history of cartography during the 1970s and 1980s.

The Kansas program experienced rapid growth in the 1970s. As explained by Jenks, "George McCleary joined the staff, and with his help we renovated and broadened the offerings in cartography. More emphasis was placed on map design and map production, and new courses at the freshman and sophomore level were added. Greater numbers of students with undergraduate training in other departments enrolled in our M.A. and Ph.D. programs in cartography" (Jenks 1991, 164). During this period, Jenks initiated research projects on 3-D maps, eye-movement research, thematic map communication, and geostatistics. By the end of the 1970s, Jenks had turned his attention to cartographic line generalization. Also during this period, he supervised 10 PhDs, 15 MA candidates, and four postdoctoral cartographers. Many of these individuals accepted academic appointments and continued the "Jenks school," including Richard Wright (San Diego State University), Paul Crawford (Bowling Green), Michael Dobson (SUNY Albany and Rand McNally), Ted Steinke and Patricia Gilmartin (University of South Carolina), Carl Youngmann (University of Washington), Jean-Claude Muller (University of Alberta, the International Training Center (ITC), and the University of Bochum), Barbara Shortridge (University of Kansas), Terry Slocum (University of Kansas), Joseph Poracsky (Portland State University), and Robert McMaster (UCLA, Syracuse, and the University of Minnesota). Jenks continued to teach and be engaged in research until his retirement at Kansas in 1986. Terry Slocum, one of Jenks's PhDs who joined the faculty in 1982, remains on the faculty, along with George McCleary.

Jenks continued to teach and complete research even after his formal retirement in 1986. In fact, in a 1987 Festschrift for Jenks—published as a special issue of the journal *Cartographica*—he revisited his 1953 curriculum. Based on three tenets on which a revised curriculum should stand (cartographers should be trained in geography departments, freshman–sophomore coursework in geography should be broad-based, and that technical training not be allowed to dominate), Jenks (1991) provided detailed outlines for four modern cartography courses, including:

Course Number 1: Map Use and Appreciation.
Course Number 2: Visualization and Planning of Thematic Maps.
Course Number 3: Map Symbolization and Compilation.
Course Number 4: Map Composition.

Thus, there is one major thread throughout Jenks's career: Cartographers should have a broad base of geographic education and a clear understanding of cartographic communication. Most of his major research projects, including the data model concept, eye-movement studies, research into statistical mapping, and automated generalization, were based on this principle. Although Jenks never synthesized his work into a text, his influence was felt both through his careful seminal research papers, and through his graduate students.

University of Washington

Although the first formally identified "cartography" course at the University of Washington was offered by William Pierson in the Geography Department during the 1937–38 academic year, John Sherman (Figure 2.7) is primarily associated with developing the cartography program there. Sherman received his BA degree from the University of Michigan in 1937, his MA from Clark University in 1944, and the PhD at Washington in 1947. Interestingly, unlike both Jenks and Robinson, who had both received formal training in cartography, Sherman never had coursework in cartography.

When Donald Hudson came to the University of Washington in 1951, he implemented a new program for the Geography Department with concentrations in Anglo-America, the Far East, economic geography, and cartography. He asked Sherman, who was appointed to the faculty at Washington in 1950, to lead the new cartography concentration and also invited Henry Leppard, recently retired from Chicago, to join the department (Velikonja 1997). Leppard, who had studied under J. Paul Goode at the University of Chicago, had remained at Chicago after Goode's death. As explained by Sherman:

> By 1953 six cartography courses were in place, including Maps and Map Reading, Introductory Cartography, Intermediate Cartography, Techniques in the Social Sciences, Map Reproduction, and Map Intelligence. In 1954, Leppard left for UCLA and in 1958 Willis Heath, having completed his Ph.D. in the department, joined Sherman in carrying on the cartography program. (Sherman 1991, 169)

One seminal event in the early history of the program was Heath and Sherman's participation in the Rand McNally-sponsored Second International Cartographic Conference at Northwestern University, held in June 1958. According to Sherman (1991, 169), a group of some 50 international cartographers were able to discuss "the graphic philosophy, functional analysis, and technological developments that were then influencing the field." Based on discussions at the conference, changes and additions were made to the cartography program at Washington. Another event influencing Sherman, and the program in cartography at Washington, was the Summer Institute for College Teachers in Cartography that Sherman led with Jenks (see our earlier discussion). Later, in 1968, Sherman developed a proposal to establish a National Institute of Cartography, which had been requested by the National Academy of Sciences/National Research Council (NAS/NRC) Committee on Geography. A panel of prestigious cartographers, including Arch Gerlach, Norman Thrower, Richard Dahlberg, Waldo Tobler, George McCleary, George Jenks, and Arthur Robinson, assisted Sherman. Unfortunately, the proposed institute was never created.

A review of the University of Washington Bulletins shows the development of the cartography program during the building years of the 1950s. During the period between 1953 and 1955 seven cartography—and cartography-related—courses were offered (Table 2.3). In

FIGURE 2.7 John Sherman discussing lunar modeling with a group of students. Sherman was responsible for developing the cartography program at the University of Washington.

TABLE 2.3 Cartography courses offered at the
University of Washington in 1953

	Geographic Techniques	
358	Maps and Map Reading	Leppard, Sherman
360	Introductory Cartography	Leppard, Sherman
363	Aerial Photographic Interpretation	Marts
425J	Graphic Techniques in the Social Sciences	Schmid
461	Intermediate Cartography	Leppard, Sherman
462	Advanced Cartography	Leppard, Sherman
464	Map Reproduction	Sherman

the next issue of the Bulletin (1957–1959), several new courses had been added to the geographic techniques section, one had changed its number (358 to 258), one had changed its name (Advanced Cartography to Map Compilation and Design), and Heath had replaced Leppard. Further adjustments to the course offerings can be found in the 1959–1961 Bulletin. Oddly, no reference is made to the Maps and Map Reading course, which had seemed to be a foundation course in the department. Introductory Cartography had changed to Principles of Cartography, a new course entitled Experimental Cartography was added, the name "Problems in" was added to Map Compilation and Map Reproduction, and a Research Seminar was added.

Thus, at a time when cartography was still emerging as a true academic field, one of America's premier cartography programs was "self-adjusting" to maintain currency. The program put together by Sherman and Heath had a strong emphasis on both design and production. Personal correspondence with Carlos Hagen (a graduate student at Washington during this period) supports this. Hagen wrote, "One thing that particularly impressed me at that time was the importance that John Sherman and Bill Heath gave to a sort of sacred trilogy, 'Drafting-Printing-Reproduction.' In the Latin American and European traditions, these production techniques are certainly not considered part of the academic environment. They are very respected and much appreciated, but generally you will find them not in academia, but in the realm of a very professional and dedicated tradition of craftsmanship" (Carlos Hagen, personal communication 1987). He continued, "When I look back, I feel the program of cartography at UW, with that heavy emphasis on production techniques, could stand rather unchallenged because, after all, that was the trend of the times."

Whereas Sherman's main research interests were in communication and map design and tactile mapping, many of his doctoral students pursued dissertation topics related to analytical and computer cartography. Sherman (1987) wrote:

I cannot isolate and identify any one question that was the trigger for my concern for design. If we step back for a

moment, our course on methods of map production was developed not in any sense to train technicians, but to familiarize students with the principles of reproduction techniques, from black and white to multicolor, sufficiently to enable them to translate this knowledge into greater freedom in map design.

Sherman's doctoral students included Waldo Tobler (University of Michigan and University of California at Santa Barbara), Richard Taketa (University of Michigan), Everett Wingert (University of Hawaii), Jois Child (SUNY Buffalo and Eastern Washington), and Barbara Buttenfield (University of California at Santa Barbara, Wisconsin, SUNY Buffalo, and Colorado). Many others went into government and industry.

In more modern times, Washington has seen a series of cartographers come to and leave the department. Phillip Muehrcke, a student of Waldo Tobler's at the University of Michigan, joined the faculty in 1969, but soon left the department for the University of Wisconsin in 1972. While at Washington, however, "he offered our first course in computer cartography, expanded the seminar offerings, and amplified our interdisciplinary activities with computer scientists on campus and cartography-oriented computer users in state government agencies in Olympia" (Sherman 1991, 169). Heath became ill in the early 1970s and was replaced by Carl Youngmann, a Jenks-educated University of Kansas PhD. Youngmann stayed at Washington for 10 years and, after his resignation in 1983, was replaced in 1985 by Timothy Nyerges, an Ohio State PhD. In 1987 Nyerges was joined by Nicholas Chrisman, who had spent many years at the Harvard Laboratory for Spatial Analysis and Computer Graphics.

Wisconsin–Kansas–Washington Graduates

For a period of nearly 30 years—from the mid-1950s to the mid-1980s—these three programs dominated the academic landscape of cartography in the United States. The PhDs produced by Robinson, Jenks, and Sherman up to 1986 and their colleagues at other institutions were placed in many of America's very best universities, and created a second generation of graduate programs, as depicted in Figure 2.8.

2.1.3 Period 3: Diffusion of Cartographic Programs in Geography Departments

During the 1970s and 1980s, a series of what might be called secondary programs, many established by PhDs from Wisconsin, Kansas, and Washington, were created in the United States. Although not exhaustive, one can point to programs at UCLA with Norman J. W. Thrower (a Wisconsin PhD), Michigan with Waldo Tobler (a Washington PhD), South Carolina with Ted Steinke and Patricia Gilmartin (Kansas PhDs), SUNY Buffalo with Kurt Brassel (a Zurich PhD) and Duane Marble (a Washington

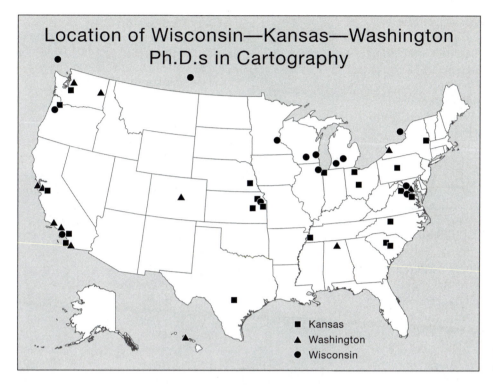

FIGURE 2.8 The placement of Wisconsin, Kansas, and Washington PhDs.

PhD), Michigan State with Richard Groop (a Kansas PhD) and Judy Olson (a Wisconsin PhD), Northern Illinois University with Richard Dahlberg (a Wisconsin PhD), Oregon State University with A. Jon Kimerling (a Wisconsin PhD), Syracuse with Mark Monmonier (a Pennsylvania State PhD), Penn State University with Alan MacEachren (a Kansas PhD), and Ohio State with Harold Moellering (a Michigan PhD). There were several key activities in these departments, including Tobler's development of analytical cartography, Thrower's expertise in the history of cartography and remote sensing, Moellering's animated cartography and emphasis on a numerical cartography, Monmonier's statistical mapping, and Olson's work in cognitive research. Each of the institutions developed its own area of expertise where, unlike the earlier days when students would pursue a general graduate program in cartography, individual graduate programs were identified for their particular research specialty, such as cognitive or analytical cartography.

2.1.4 Period 4: The Transition Period

The intellectual landscape of cartography has changed significantly over the past 10 years, in large part due to the rapid growth of geographic information science and systems. Fifteen years ago, the prognosis for a PhD in cartography acquiring an academic position was excellent, whereas today's job market seeks out the geographic information scientist. One can certainly still study cartography at most major institutions, but the number of

courses has decreased as the number of GIS-related courses has increased. Additionally, the term geographic visualization, increasingly used by many departments instead of cartography, has caused a further erosion of the professional base of cartography. However, one hope for the discipline is that as GISs become almost ubiquitous in our society, there seems to be the realization that a deeper knowledge of maps, cartography, and map symbolization and design is still a crucial and necessary skill. Kraak and Ormeling (1996), in their textbook, *Cartography: Visualization of Spatial Data*, made the following point with respect to the relationship between GIS and cartography:

> Many of the concepts and functions of GIS were first conceived by cartographers. This is not only valid for the GIS output module, but for many of the processing actions (e.g., transformations, analyses) and input functions (e.g., digitizing, scanning) of a GIS as well. There are conflicting views regarding the relations between cartography and GIS, viz. whether GIS is a technical-analytical subset of cartography, or whether cartography is just a data visualization subset of GIS. For the purpose of this book, also written for GIS analysts who have to learn to use the cartographic method, cartography will be regarded as an essential support for nearly all aspects of handling geographical information. (p. 16)

There have been major changes in the way cartography is taught in American universities during the 1990s. A survey of six different universities with a focus on cartography and

GIS education confirms the nature of the changes that have occurred in U.S. cartographic education in general. Some of the most significant changes include: (1) a closer integration with education in GIS; (2) the nearly complete transition to digital methods; (3) a lesser emphasis on procedural programming (e.g., Fortran and Pascal), and greater emphasis on object-oriented, user interface, and Windows programming; and (4) a greater emphasis on the dynamic aspects of cartography, including animation and multimedia.

2.2 THE PARADIGMS OF AMERICAN CARTOGRAPHY

In the post-World War II period, as academically oriented graduate programs emerged, basic research in cartography accelerated. Although many research paradigms could be documented, some of the more substantial efforts were in communication models, a theory of symbolization and design, experimental cartography, analytical cartography, and the recent series of debates on social and ethical issues in cartography. Because we touched on communication models and social and ethical issues in Chapter 1, and much of this book deals with symbolization and design (and related experimental studies), we discuss only the analytical cartography paradigm here.

2.2.1 Analytical Cartography

If any one paradigm within cartography has an "intellectual leader" it is analytical cartography. Waldo Tobler (Figure 2.9) originated—in the 1960s—and nurtured—in the 1970s and 1980s—the idea of a mathematical, transformational, or analytical approach to the subject. Tobler laid out the agenda for analytical cartography in his seminal 1976 paper, "Analytical Cartography," published in *American Cartographer*. This paper, and Tobler's ideas, had a profound effect on American academic cartography. What exactly is "analytical cartography"? Kimerling, in his 1989 *Geography in America* review of cartography, described it as "the mathematical concepts and methods underlying cartography, and their application in map production and the solution of geographic problems" (p. 697), which includes the topics of cartographic data models, digital cartographic data collection methods and standards, coordinate transformations and map projections, geographic data interpolation, analytical generalization, and numerical map analysis and interpretation. Tobler's original syllabus described a series of topics steeped in theory and mathematics. His goal for the course was futuristic:

> What is easy, convenient, or difficult depends on the technology, circumstances, and problem. The teaching of cartography

FIGURE 2.9   Waldo Tobler—developer of the notion of analytical cartography.

must reflect this dynamism, and the student can only remain flexible if he has command of a theoretical structure as well as specific implementation. The spirit of Analytical Cartography is to try to capture this theory, in anticipation of the many technological innovations which can be expected in the future; wrist watch latitude/longitude indicators, for example, and pocket calculators with maps displayed by colored light emitting diodes, do not seem impossible. In a university environment one should not spend too much time in describing how things are done today. (Tobler 1976, 29)

Tobler had finished his PhD in 1961 at the University of Washington under John Sherman, completing a doctoral dissertation entitled, "Map Transformations of Geographic Space." While at Washington, Tobler was not only influenced by the strong emphasis among the faculty (William Garrison, for instance) on quantification, but also the large number of graduate students interested in mathematical geography, including Duane Marble, Arthur Getis, Brian Berry, and John Nystuen, among others. In the early 1960s, the Department of Geography at Washington was ground zero for the quantitative revolution in geography. Many of these students had enrolled in J. Ross MacKay's Statistical Cartography course, taught by MacKay in the late 1950s. In a personal interview, Tobler (2001) also discussed the influence of Carlos Hagen, a graduate student at Washington who arrived from Chile in the late 1950s hoping to pursue graduate work in mathematical cartography. Tobler himself actually had little training in formal mathematics, but was self-taught and was intrigued by Hagen's work in projections. In addition to the strong influence of the faculty and graduate students at Washington, Tobler spent one year working at the Rand Company in Santa Monica, California.

After completing his dissertation at Washington, Tobler joined the faculty at the University of Michigan, where his graduate student colleague from Washington, John Nystuen, had also moved. It is at Michigan that Tobler honed his ideas on analytical cartography, in part assisted by a relatively obscure event in American geography: the meetings of the Michigan IntraUniversity Community of Mathematical Geographers (MICMOG). Many of the topics presented at these Detroit-based meetings were strongly cartographic in nature, including Gould's "Mental Maps," Perkal's "Epsilon Filtering," and Tobler's own work on generalization. His work, which had a significant influence on both the disciplines of cartography and geography, led to his election into the prestigious National Academy of Sciences, the only geographic cartographer to hold that honor.

What emerged from the concept of "analytical cartography" was a cadre of individuals working on problems that can be identified as analytical, computational-digital, and mathematical in nature. Some were Tobler's own PhD students, such as Stephen Guptill (USGS), Harold Moellering (Ohio State University), and Phil Muehrcke (University of Washington and Wisconsin). Others were immersed in the paradigm without necessarily having formal education in it, such as Mark Monmonier, the author of the first textbook on computer cartography; Carl Youngmann (a Jenks-educated cartographer at Kansas who joined Sherman at Washington); and Jean-Claude Muller (another Jenks student who worked at the University of Georgia, University of Alberta, the International Training Center in the Netherlands, and the University of Bochum in Germany). Additionally, a large group of individuals educated in the late 1970s through the early 1980s considered themselves computer or analytical cartographers, including Terry Slocum (PhD, University of Kansas), Keith Clarke (PhD, University of Michigan), Nicholas Chrisman (PhD, Bristol), Timothy Nyerges (PhD, Ohio State University), Marc Armstrong (PhD, University of Illinois), Barbara Buttenfield (PhD, University of Washington), and Robert McMaster (PhD, University of Kansas).

A strong argument can be made that the principles of numerical, analytical, and digital cartography became the "core" of modern GISs. For instance, many of the basic ideas in analytical and computer cartography developed at the Harvard Laboratory for Spatial Analysis and Computer Graphics, including the concept of topological data structures, were directly translated into modern GISs such as Environmental System Research Institute's ARC/Info product.

SUMMARY

Dividing a history into categories is problematic and subjective. However, several logical "dividing" lines can be identified in the history of U.S. academic cartography. One clear line is pre- and post-World War II. From a rather sporadic set of institutions offering one or at most two courses in cartography, the post-World War II era witnessed the creation of well-established centers of excellence. Cartography before World War II was considered a relatively minor part of geography, with few individuals focusing on the topic. The idea of teaching "topographic" mapping was rare, although certain institutions did develop some expertise in surveying, and The Ohio State University established a program in geodetic science. It is true serendipity that certain programs were established, and only possible because of the interest of key individuals such as Arthur Robinson, George Jenks, and John Sherman. A second line can be found when, on maturation of these graduate programs, the centers began sending out PhDs educated in cartography to establish other programs—a second generation of centers with intellectual children from the initial set. Finally, the discipline witnessed significant changes in the late 1980s and 1990s, as cartography has increasingly become a component—often a smaller component—of expanding programs in geographic information science.

FURTHER READING

Cartography and Geographic Information Systems 18, no. 3, 1991 (entire issue).

> Papers discuss the history and development of academic cartography in the United States.

Cartography and Geographic Information Science 29, no. 3, 2002 (entire issue).

> This special issue is dedicated to the history of cartography in the 20th century.

Fabrikant, S. I. (2003) "Commentary on 'A history of twentieth-century American academic cartography'." *Cartography and Geographic Information Science* 30, no. 1: 81–84.

> Provides a perspective on developments in thematic cartography in Germany, Austria, and Switzerland. For related discussion, see McMaster and McMaster (2003) and Thrower (2003).

Montello, D. R. (2002) "Cognitive map-design research in the twentieth century: Theoretical and empirical approaches."

Cartography and Geographic Information Science 29, no. 3:283–304.

Reviews the development of research related to cognition and map design.

Robinson, A. H. (1982) *Early Thematic Mapping in the History of Cartography*. Chicago: The University of Chicago Press.

Reviews the development of individual thematic mapping techniques.

Tobler, W. R. (1976) "Analytical cartography." *The American Cartographer* 3, no. 1:21–31.

Tobler's classical piece on analytical cartography.

3

Statistical and Graphical Foundation

OVERVIEW

One purpose of this chapter is to provide a statistical and graphical foundation for the remainder of the text. For example, in this chapter we will define basic statistical measures such as the mean and standard deviation. In Chapter 5, we will consider a method for classifying data that is based on these measures. Obviously, the logic of the classification method cannot be comprehended unless the concepts of mean and standard deviation are first understood.

A second purpose of this chapter is to introduce a range of techniques (tables, graphs, and numerical summaries) that can be used along with maps to analyze spatial data. To illustrate the need for such techniques, consider Figure 1.6B, which used an unclassed choropleth map to show the distribution of foreign-owned agricultural land in the United States. Why is this map composed entirely of light tones of gray, with the exception of a solid black tone for the state of Maine? A graphical plot of the data along a number line would reveal that 47 of the 48 states were in the range between 0 and 5.2, with a distinct **outlier** at 14.1. Only by seeing such a graph would the reader develop a full appreciation of the spatial distribution shown on the map.

Sections 3.1 and 3.2 briefly deal with the principles of population versus sample and descriptive versus inferential statistics. A **population** is the total set of elements or things that could potentially be studied, whereas a **sample** is the portion of the population that is actually examined. **Descriptive statistics** are used to summarize the character of a sample or population, and **inferential statistics** are used to make a guess (or inference) about a population based on a sample. The focus of this chapter is on descriptive statistics because these are most useful in mapping.

Section 3.3 covers a broad range of methods for analyzing data using tables, graphs, and numerical summaries.

The section is split into three major parts: analyzing the distribution of individual attributes (3.3.1), analyzing the relationship between two (or more) attributes (3.3.2), and exploratory data analysis (3.3.3). Readers with coursework in statistics will find some of this material a review (e.g., **histograms**, **measures of central tendency**, **correlation**, and **stem-and-leaf plots**), but other material will likely be new (e.g., **hexagon bin plots**, the **scatterplot matrix**, the **reduced major axis**, **scatterplot brushing**, and **parallel coordinate plots**).

One limitation of the numerical summaries covered in section 3.3 is that spatial location is not an integral part of the summaries. In contrast, section 3.4 deals with numerical summaries in which spatial location is an integral part—these are often referred to as **geostatistical methods**. Some geostatistical methods covered in section 3.4 analyze just spatial location; for example, the formula for computing the **centroid**, or balancing point for a geographic region, uses just the x and y coordinates bounding a region. Other methods consider both spatial location and the values of an attribute; for example, **spatial autocorrelation** measures the likelihood that similar attribute values occur near one another.

To illustrate many concepts in this chapter, we will analyze the relationship between murder rate (number of murders per 100,000 people) and the following attributes for 50 U.S. cities with a population of 100,000 or more in 1990: (1) percentage of families whose income was below the poverty level; (2) percentage of those 25 years and older who were at least high school graduates; (3) the drug arrest rate (number of arrests per 100,000 people); (4) population density (number of people per square mile); and (5) total population (in *thousands).* * *The raw data are shown in Table 3.1 (ordered on the basis of murder rate).*

One problem with these data is that an analysis at the city level might be inappropriate because it fails to account for the variation within a city; it might instead be desirable to look at finer geographic units, such as census tracts, or at individual murder cases. We chose the city level for analysis, however, because it is easier to relate to individual cities than to individual census tracts. Later in the chapter we will consider the effect of aggregation of enumeration units on measures of numerical correlation.

3.1 POPULATION AND SAMPLE

In statistics, a **population** is defined as the total set of elements or things that one could study, and a **sample** is the portion of the population that is actually examined (in this book, the number of elements in each will be represented by N and n, respectively). Generally, scientists collect samples because they don't have the time or money to examine the entire population. For example, a geomorphologist studying the effect of wave behavior on beach development would collect data at a series of points along a shoreline, rather than examining the entire shoreline.

The data in Table 3.1 have characteristics of either a population or sample, depending on one's perspective of the data. Consider first a perspective from the standpoint of attributes. Murder rate, drug arrest rate, population, and population density are all based on the entire population of each city (as defined by the census); for example, in the case of murder rate, all murders occurring within a city are considered relative to the entire population of that city. The other attributes, percentage of families below the poverty level and percentage of high school graduates, are based on sampling approximately one of every six housing units (U.S. Bureau of the Census 1994, A-2).

From the perspective of observations, the 50 cities shown in Table 3.1 were sampled from 200 cities. Sampling was done in two stages. The first stage involved eliminating cities with political boundaries that extended beyond the limits of where most people live within those cities (using the "extended city" definition provided by the U.S. Census Bureau). This was done because one of the attributes being analyzed, population density, was a function of city area. In the second stage, the remaining cities were ordered on the basis of total population and split into 10 classes using Jenks's (1977) optimal method

(see Chapter 5). A proportional number of cities were sampled from each of the 10 classes to obtain a broad range of city sizes that would be representative of those found in the United States.

3.2 DESCRIPTIVE VERSUS INFERENTIAL STATISTICS

Statistical methods can be split into two types: descriptive and inferential. **Descriptive statistics** describe the character of a sample or population. For example, to assess the current president's job performance, you might ask a sample of 500 people, "Is the President doing an acceptable job?" The percentage responding yes, say 52 percent, would be an example of a descriptive statistic. **Inferential statistics** are used to make an inference about a population from a sample. For example, based on the 52 percent figure just given, you might infer that 52 percent of the entire population thinks the President is doing an acceptable job. You would be surprised, of course, if the 52 percent figure truly applied to the population because the figure is based on a sample. To correct for this problem, in inferential statistics it is necessary to compute a *margin of error* (e.g., plus or minus 3 percent) around the sampled value; we often find such errors reported in media coverage of polling.

3.3 METHODS FOR ANALYZING SPATIAL DATA, IGNORING LOCATION

This section considers methods for analyzing data, ignoring the spatial location of the data. The section is split into three parts: (1) analyzing the distribution of individual attributes, (2) analyzing the relationship between two or more attributes, and (3) exploratory data analysis.

3.3.1 Analyzing the Distribution of Individual Attributes

Tables

Raw Table. The simplest form of tabular display is the **raw table** in which the data for an attribute of interest are listed from lowest to highest value, as for the murder rate data in Table 3.1. Tabular displays are useful for providing specific information about observations (e.g., Buffalo, New York, had a murder rate of 11.3 per 100,000 people in 1990), but they can provide additional information if they are examined carefully. Note, for example, that the sorted values provide the minimum and maximum of the data (0 for Irvine, California, and 77.8 for Washington, DC). With some mental arithmetic, the *range* can be calculated by simply subtracting the minimum from the maximum ($77.8 - 0 = 77.8$ in this case).

---

* The murder and drug arrest data were obtained from the *Sourcebook of Criminal Justice Statistics 1991* (Flanagan and Maguire 1992). The remaining data were taken from the *1994 City and County Data Book* (U.S. Bureau of the Census 1994). All data were for either 1989 or 1990.

TABLE 3.1   Sample data for 50 U.S. cities (sorted on murder rate)

City	Murder Rate*	Families below Poverty Level (%)	High School Graduates (%)	Drug Arrest Rate†	Population Density‡	Total Population (in Thousands)
Irvine, CA	0.0	2.6	95.1	780	2607	110
Cedar Rapids, IA	0.9	6.6	84.5	110	2034	109
Overland Park, KS	0.9	1.9	94.1	255	2007	112
Livonia, MI	1.0	1.7	84.7	665	2823	101
Lincoln, NE	1.6	6.5	88.3	294	3033	192
Madison, WI	1.6	6.6	90.6	57	3311	191
Glendale, CA	1.7	12.3	77.2	452	5882	180
Allentown, PA	1.9	9.3	69.4	1078	5934	105
Tempe, AZ	2.1	7.0	89.9	295	3590	142
Boise City, ID	2.4	6.3	88.6	512	2726	126
Lakewood, CO	2.4	5.2	88.2	216	3100	126
Mesa, AZ	3.1	6.9	84.8	223	2653	288
Pasadena, TX	3.4	11.1	69.8	370	2727	119
San Jose, CA	4.5	6.5	77.2	1289	4568	782
Waterbury, CT	4.6	9.9	66.8	1326	3815	109
Springfield, MO	5.0	11.6	77.0	446	2068	140
Chula Vista, CA	5.2	8.6	75.7	808	4661	135
St. Paul, MN	6.6	12.4	81.1	260	5157	272
Arlington, VA	7.0	4.3	87.5	758	6605	171
Alexandria, VA	7.2	4.7	86.9	834	7281	111
Portland, OR	7.6	9.7	82.9	1001	3508	437
Des Moines, IA	8.3	9.5	81.0	118	2567	193
Lansing, MI	8.7	16.5	78.3	780	3755	127
Pittsburgh, PA	9.5	16.6	72.4	723	6649	370
Yonkers, NY	9.6	9.0	73.6	917	10403	188
Riverside, CA	9.7	8.4	77.8	1703	2916	227
Elizabeth, NJ	10.0	13.7	58.5	929	8929	110
Berkeley, CA	10.7	9.4	90.3	1569	9783	103
Buffalo, NY	11.3	21.7	67.3	580	8080	328
Raleigh, NC	11.5	7.7	86.6	634	2360	208
Sacramento, CA	11.7	13.8	76.9	1555	3836	369
Tacoma, WA	14.1	12.5	79.3	673	3677	177
Knoxville, TN	15.2	15.3	70.8	328	2137	165
Beaumont, TX	16.7	16.6	75.2	693	1427	114
Winston-Salem, NC	16.8	11.6	77.0	1343	2018	143
Montgomery, AL	18.2	14.4	75.7	131	1386	187
Waco, TX	21.2	19.7	68.4	400	1367	104
Jackson, MS	22.3	18.0	75.0	693	1804	197
Savannah, GA	23.9	18.5	70.1	707	2198	138
Norfolk, VA	24.1	15.1	72.7	624	4859	261
Los Angeles, CA	28.2	14.9	67.0	1391	7426	3485
Chicago, IL	30.6	18.3	66.0	1157	12251	2784
New York, NY	30.7	16.3	68.3	1255	23701	7323
Houston, TX	34.8	17.2	70.5	555	3020	1631
Newark, NJ	40.7	22.8	51.2	1751	11554	275
Baltimore, MD	41.4	17.8	60.7	2063	9108	736
Gary, IN	55.6	26.4	64.8	261	2322	117
Detroit, MI	56.6	29.0	62.1	1052	7410	1028
Atlanta, GA	58.6	24.6	69.9	2330	2990	394
Washington, DC	77.8	13.3	73.1	1738	9883	607

*Murders per 100,000 people.

†Arrests per 100,000 people.

‡Number of people per square mile.

Raw tables can also reveal any duplicate values and outliers in the data that might have special significance. For murder rate, duplicate values (e.g., 0.9 for Cedar Rapids, Iowa, and Overland Park, Kansas) are unim-portant, as they are a function of the number of signifi-cant digits reported. For some attributes, however, du-plicates can be quite meaningful. For example, an examination of the distribution of Major League Baseball player salaries would reveal numerous duplicates because several players on each team typically earn exactly the same amount, the minimum required salary for all of Major League Baseball. **Outliers**—values that are quite unusual or atypical—can also be seen (Barnett and Lewis 1994). For murder rate, no value is considerably different from the rest (although Washington, DC, is clearly larger), but for the total population attribute, note that several cities are quite large, with New York more than twice the value of any other city in the sample.

Although raw tables are useful for providing specific information, they are not very good at providing an overview of how data are distributed along the number line. For example, in studying Table 3.1, note that roughly half the cities have murder rates below 10.0, but it is difficult to develop a feel for what the overall murder rate distribution really looks like. Grouped-frequency tables are more useful for this purpose.

Grouped-Frequency Table.   To construct **grouped-frequency tables**, we divide the data range into equal intervals and then tally the number of observations that fall in each interval. Although grouped-frequency tables generally are constructed using software (e.g., SPSS and SAS), it is useful to consider the actual steps involved so that you have a clear understanding of what the resulting table reveals. (We will use a similar approach in Chapter 5 to create equal interval classes for choropleth maps.)

Step 1.   Decide how many groups (or classes) you want to use. When grouped-frequency tables were created manually, this was an important step because of the time and thought needed to construct the table. With the ready availability of software, it is reasonable to construct tables for various numbers of classes. For this discussion, presume that 15 classes will be chosen.

Step 2.   Determine the width of each class (the class interval). The class interval is computed by dividing the range of the data by the number of classes. The following would be the computation for the murder rate data using 15 classes:

$$\frac{\text{Range}}{\text{Number of classes}} = \frac{\text{High} - \text{Low}}{\text{Number of classes}}$$
$$= \frac{77.8 - 0.0}{15} = 5.187$$

Note that we use more decimal places for the class interval than appear in the actual data to avoid rounding errors that could cause the last class not to match the highest value in the data.

Step 3.   Determine the upper limit of each class. The upper limit for each class is computed by repeatedly adding the class interval to the lowest value in the data.

The result is the right-hand side of the Limits column in Table 3.2. Note that the highest calculated limit does not match the highest data value (77.8) exactly because the class interval is not a simple fraction; a more precise class interval would be 5.186666667, but even with this interval the match would not be exact (the highest value would be 77.80000001).

Step 4.   Determine the lower limit of each class. Lower limits for each class are specified so that they are just above the highest value in a lower valued class. This is done so that any observation can fall in only one class. For example, the lowest value in the second class is 5.188, which is 0.001 larger than 5.187, the highest value in class 1.

Step 5.   Tally the number of observations falling in each class. These numbers are shown as Number of Observations in Class in Table 3.2. Also shown in the table are the percent and cumulative percent in each class.

In comparison to the raw table, the grouped-frequency table provides a somewhat better overview of the data. For example, note that 80 percent of the cities fall in the five lowest murder rate classes. Both raw and grouped-frequency tables, however, do not take full advantage of our visual processing powers; for this we turn to graphs.

Graphs

Point and Dispersion Graphs.   In a **point graph** or **one-dimensional scatterplot** each data value is represented by a small point symbol plotted along the number line (Cleveland 1994, 133; in Figure 3.1A an open circle is used). For the murder rate attribute, the point graph shows that the data are concentrated at the lower end of

TABLE 3.2   Grouped-frequency table for the murder rate data

Class	Limits	Number of Observations in Class	Percent in Class	Cumulative Percent
1	0.000–5.1870	16	32	32
2	5.188–10.374	11	22	54
3	10.375–15.561	6	12	66
4	15.562–20.748	3	6	72
5	20.749–25.935	4	8	80
6	25.936–31.122	3	6	86
7	31.123–36.309	1	2	88
8	36.310–41.496	2	4	92
9	41.497–46.683	0	0	92
10	46.684–51.870	0	0	92
11	51.871–57.057	2	4	96
12	57.058–62.244	1	2	98
13	62.245–67.431	0	0	98
14	67.432–72.618	0	0	98
15	72.619–77.805	1	2	100

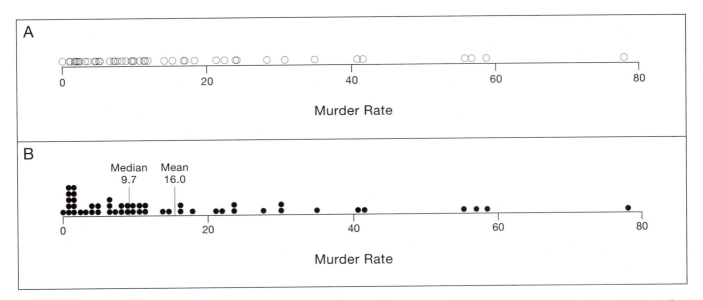

FIGURE 3.1   Point (A) and dispersion (B) graphs for the murder rate data presented in Table 3.1. Note the locations of the median and mean in (B).

the distribution (based on the graph, under a rate of about 12). One obvious problem with the point graph is that individual symbols might overlap, thus making the distribution difficult to interpret (note the "smearing" in the left-hand portion of Figure 3.1A); duplicate values, in particular, cannot be detected at all by this method.

An alternative to the point graph is the **dispersion graph** (Hammond and McCullagh 1978), in which data are grouped into classes, the number of values falling in each class are tallied, and dots are stacked at each class position (Figure 3.1B). Intervals for classes are defined in a fashion identical to the grouped-frequency table, except that a large number of classes are used. For example, 99 classes were used for Figure 3.1B. The dispersion graph for the murder rate data portrays a distribution similar to the point graph, except that potential confusion in the overlapping areas is eliminated.

Histogram.   A **histogram** is constructed in a manner analogous to the dispersion graph, except that fewer classes are generally used and bars of varying height are used to represent the number of values in each class. Because the histogram is more commonly used than either the point or dispersion graph, all of the attributes shown in Table 3.1 are graphed using the histogram in Figure 3.2. Looking first at the murder rate histogram, note that the up-and-down nature of the dispersion graph has been smoothed out; there is clearly a peak in the graph on the left, with a decreasing height in bars as one moves to the right.

Histograms are often compared with a hypothetical **normal** (or bell-shaped) **distribution**. For a normal distribution, most of the observations fall near the mean (in

the middle of the distribution), with fewer observations in the tails of the distribution (Figure 3.3). Curves representing a normal distribution have been overlaid on the histograms shown in Figure 3.2. Distributions lacking the symmetry of the normal distribution are termed **skewed**. *Positively skewed* distributions have the tallest bars concentrated on the left-hand side (as for total population, murder rate, and population density), whereas *negatively skewed* distributions have the tallest bars on the right-hand side. (There is no distinctive example of a negatively skewed distribution in Figure 3.2, although percent with high school education has a slight negative skew).

Because many inferential tests require a normal distribution, raw data are often transformed to make them approximately normal. Transformation involves applying the same mathematical operation to each data value of an attribute; for example, we might compute the $\log_{10}$ of each murder rate value, or alternatively we could compute the square root of each murder rate value.* $\log_{10}$ and square root transformations are commonly used to convert a positively skewed distribution to a normal one; to illustrate, Figure 3.4 portrays such transformations for the murder and drug arrest data, respectively.

Numerical Summaries

Although tables and graphs are useful for analyzing data, they are prone to differing subjective interpretations, and the limited space of formal printed publications often limits their use. As an alternative, statisticians frequently use numerical summaries, which typically are split into two

---

* The $\log_{10}$ for a number is the power that we would raise 10 to get that number. For example, $\log_{10} 100 = 2$ because $10^2 = 100$.

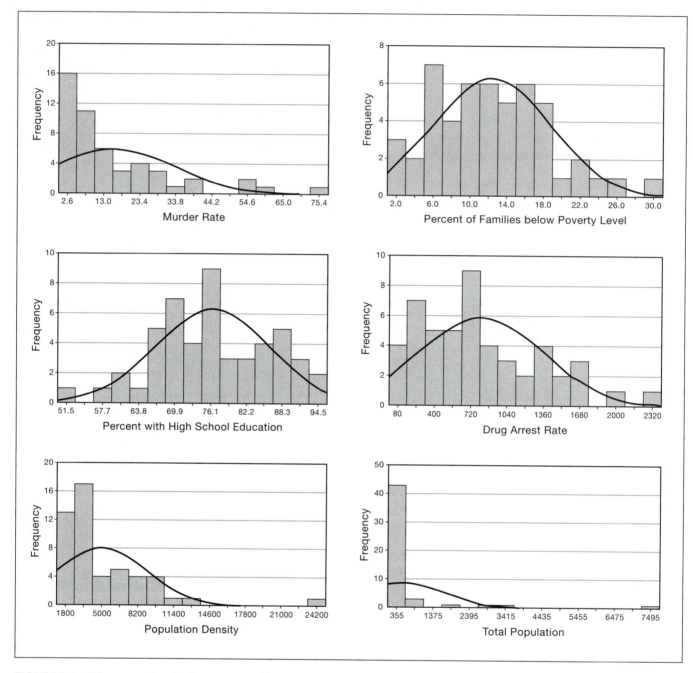

FIGURE 3.2   Histograms for the data presented in Table 3.1.

broad categories: measures of central tendency and measures of dispersion.

Measures of Central Tendency. **Measures of central tendency** are used to indicate a value around which the data are most likely concentrated. Three measures of central tendency are commonly recognized: mode, median, and mean. The **mode** is the most frequently occurring value, and is thus generally useful for only nominal data, such as on a land use/land cover map. The **median** is the middle value in an ordered set of data or, alternatively, the 50th percentile, because 50 percent of the data are below it. For the murder rate data, the median is 9.7. Note its location in Figure 3.1B. The **mean** is often referred to as the "average" of the data and is calculated by summing all values and dividing by the number of values.

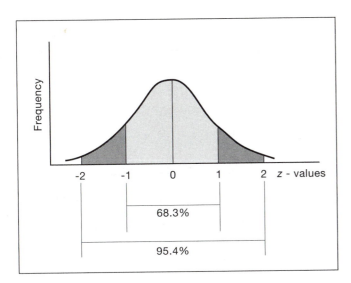

FIGURE 3.3 An example of a normal curve. Histograms will approximate this shape if the data are normal. For a perfectly normal data set, approximately 68 percent and 95 percent of the observations will fall within 1 and 2 standard deviations, respectively, of the mean.

Separate formulas are used to distinguish mean values for the sample and population, as follows:*

$$\text{Sample:} \quad \overline{X} = \frac{\sum\limits_{i=1}^{n} X_i}{n}$$

$$\text{Population:} \quad \mu = \frac{\sum\limits_{i=1}^{N} X_i}{N}$$

---

* Σ is the symbol for summation, indicating that all data values should be summed. Readers unfamiliar with summation notation should consult an introductory statistics book such as Burt and Barber 1996, 68–70.

where the X_i are individual data values. Because the data in Table 3.1 are a sample from 200 cities, the sample formula is appropriate in this case. The mean for the murder rate data is 16.0. Note its location in Figure 3.1B. One problem with the mean is that either a skew or outliers in the data affect it, whereas the median is resistant to these characteristics. We can see this in Figure 3.1B, where the median falls where most of the data are concentrated, whereas the mean is pulled to the right by the positive skew.

Measures of Dispersion. Measures of dispersion provide an indication of how data are spread along the number line. The simplest measure is the **range**, which was defined earlier as the maximum minus the minimum. Obviously, the range is of limited usefulness because it is based on only two values, the maximum and minimum of the data.

More useful measures of dispersion are the interquartile range and standard deviation, which should be used with the median and mean, respectively. The **interquartile range** is the absolute difference between the 75th and 25th percentiles, or where the middle 50 percent of the data lie. For the murder rate data, the result is $|22.700 - 3.325| = 19.4$. An important characteristic of the interquartile range is that it, like the median, is unaffected by outliers in the data. For example, if the highest murder rate were replaced by a value of 150, the interquartile range would still be 19.4.

In a fashion similar to the mean, separate formulas normally are provided for the sample and population **standard deviation**:

$$\text{Sample:} \quad s = \sqrt{\frac{\sum\limits_{i=1}^{n} (X_i - \overline{X})^2}{n - 1}}$$

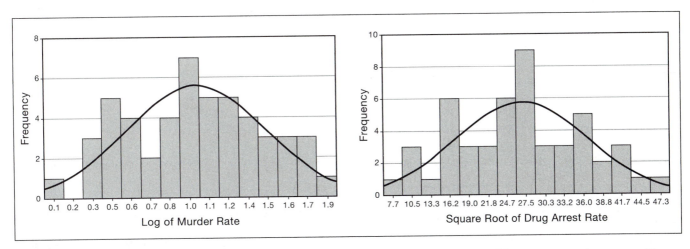

FIGURE 3.4 Histograms of two transformed attributes: log of murder rate and square root of drug arrest rate (compare with corresponding histograms in Figure 3.2).

TABLE 3.3   Tabular relationship between murder rate and percent of families below poverty level for cities in Table 3.1

Murder Rate	Percent Families below Poverty Level				
	1.70–7.16	7.17–12.62	12.63–18.08	18.09–23.54	23.55–29.00
62.25–77.80			1		
46.69–62.24					3
31.13–46.68			2	1	
15.57–31.12		1	6	3	
0.00–15.56	13	14	5	1	

Population: $\sigma = \sqrt{\dfrac{\sum\limits_{i=1}^{N}(X_i - \mu)^2}{N}}$

In comparing these formulas, note that they differ principally in the denominator: a value of 1 is subtracted in the sample case, but not in the population case. Subtracting a value of 1 is necessary because a sample estimate using just n tends to underestimate the population value (Burt and Barber 1996, 64). Using the sample formula, the standard deviation for the murder rate data is 17.5. In contrast to the interquartile range, outliers in the data do affect the standard deviation. To illustrate, replacing the highest murder rate by a value of 150 results in a standard deviation of 24.4, an increase of nearly 40 percent.

3.3.2 Analyzing the Relationship between Two or More Attributes

Tables

In the previous section, we saw that considerable information could be derived from raw tables when examining an individual attribute. When trying to relate two attributes, however, this task becomes difficult. To convince yourself of this, try using Table 3.1 to relate murder rate with percent families below the poverty level. In general, the attributes appear related (low poverty values are associated with low murder rates and high poverty values are associated with high murder rates), but it is difficult to summarize the relationship. To simplify the process, we can class both attributes using the grouped-frequency method and create a matrix of the result (Table 3.3). The same general relation between the attributes is still apparent, but it is more easily seen; also, the matrix reveals that Washington, DC (in the highest murder rate class) does not fit the general trend of the data, which extends from the lower left to the upper right of the table.

Graphs

A **scatterplot** is used to examine the relationship of attributes against one another in two-dimensional space. To illustrate, Figure 3.5 portrays scatterplots of murder rate and log of murder rate against percent of families below poverty level (also shown are best-fit regression lines, which we will consider shortly). On scatterplots, *dependent* and *independent* attributes normally are plotted on the y and x axes, respectively. Because it seems reasonable that the murder rate might depend, in part, on poverty, murder rate has been plotted on the y axis.

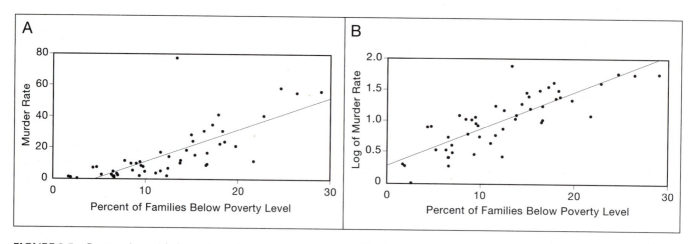

FIGURE 3.5   Scatterplots of (A) murder rate against percent of families below poverty level and (B) log of murder rate against the same attribute. Also shown are best-fit regression lines.

In examining Figure 3.5A, note the similarity of the distribution of points to the pattern of cells in Table 3.3. Also note that after transformation (Figure 3.5B), Washington, DC, is not quite so different from the rest of the data; thus, data transformations affect not only the values for individual attributes but also the relationship between attributes.

When the number of observations is large, scatterplots can become difficult to interpret, just as smearing of dots made the point graph of murder rate hard to interpret. One solution to this problem is the **hexagon bin plot**, which is shown in the right-hand portion of Figure 3.6. Such a plot is created by laying a grid of hexagons on a scatterplot (see the left-hand portion of Figure 3.6), and then filling each hexagon grid cell with a solid hexagon with size that is proportional to the number of dots falling in each cell.*

To examine the relationship among three attributes simultaneously, the basic scatterplot can be extended to a **three-dimensional scatterplot** by specifying *x*, *y*, and *z* axes. Many statistical packages have options for creating such plots, and even permit rotating them interactively. Our experience is that these plots are difficult to interpret; moreover, they cannot be extended to handle more than three attributes.

One method that can handle three or more attributes is the **scatterplot matrix** (Figure 3.7). This graph might at first appear rather complex, but the principles underlying it are actually quite simple. Note first that the relation

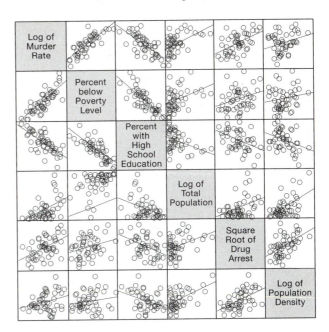

FIGURE 3.7 A scatterplot matrix for the data presented in Table 3.1. Also shown are best-fit regression lines.

between any two attributes is displayed twice, once above the diagonal of attribute names and once below it. This approach enables each attribute to be shown as either an independent or dependent attribute in relation to other attributes. By scanning a row, you can see what happens when an attribute is considered dependent, and, by scanning a column, you can see what happens when it is considered independent. If our focus were on log of murder rate as a potential dependent attribute, we would want to examine the first row, where we see that the strongest relationship appears to be between log of murder rate and percent below poverty level.

A second approach for displaying three or more attributes simultaneously is the **parallel coordinate plot**. In this graph, attributes are depicted by a set of parallel axes, and observations are depicted as a series of connected line segments passing through the axes. To illustrate, Figure 3.8 portrays a parallel coordinate plot for a subset of 10 of the 50 cities shown in Table 3.1. The six attributes in Table 3.1 are represented by six parallel vertical lines, whereas the 10 observations are represented by horizontal line segments. Figure 3.8A shows all 10 observations plotted in black, a conventional approach for black-and-white presentations; Figure 3.8B shows Washington, DC, highlighted. At the top of the figures are correlation coefficients, or *r* values, which are described in detail in the "Numerical Summaries" section later.

In interpreting the parallel coordinate plot, it is important to recognize the characteristic shape of plots associated with particular values of *r*. For instance, Figure 3.9 illustrates plots for *r* = 1 (parallel line

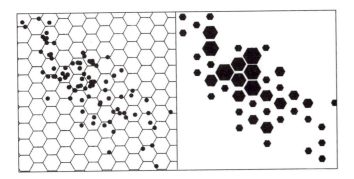

FIGURE 3.6 Hexagon bin plot: an alternative to the scatterplot. The size of hexagons shown on the right is proportional to the number of dots falling within the hexagons on the left. The *x* and *y* axes represent sulfate and nitrate deposition values for sites in the eastern United States (After Carr et al. 1992. First published in *Cartography and Geographic Information Systems* 19(4), p. 229. Reprinted with permission from the American Congress on Surveying and Mapping.)

---

* For other approaches to handling a large number of observations, see Carr 1991 and Cleveland 1994.

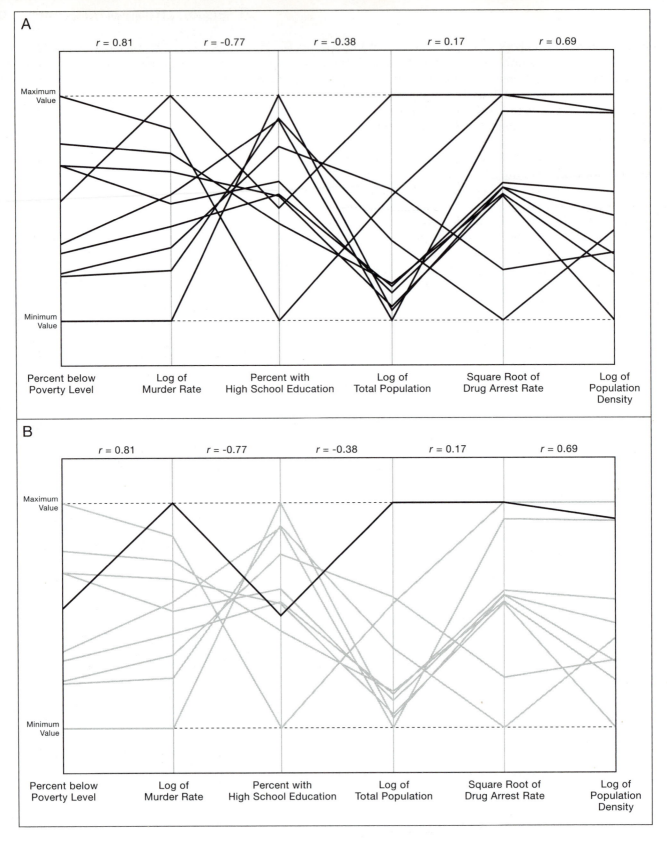

FIGURE 3.8 (A) A parallel coordinate plot for 10 of the 50 cities shown in Table 3.1; (B) the same plot, with Washington, DC, highlighted.

segments), $r = -1$ (line segments intersect one another in the middle of the graph), and $r = 0$ (line segments cross one another at different angles). In Figure 3.8, the highly positively correlated percent below poverty level and log of murder rate ($r = .81$ for the 10 cities) have relatively parallel line segments (Washington, DC, is a notable exception). In contrast, line segments for the highly negatively correlated log of murder rate and percent with high school education ($r = -.77$ for the 10 cities) tend to cross one another near the middle of the graph. Obviously, the appearance of the parallel coordinate plot will change, depending on the order in which the axes are plotted. Ideally, a successful interpretation of the plot requires an interactive program that can easily change the order of the attributes and highlight selected observations. In Chapter 21, we will consider such a program that Robert Edsall (2003a) developed for exploring parallel coordinate plots in the context of spatiotemporal health statistics data.

Numerical Summaries

Bivariate correlation-regression is the most widely used approach for summarizing the relationship between two numeric attributes. **Bivariate correlation** is used to summarize the nature and strength of the relationship, and **bivariate regression** provides an equation for a best-fit line passing through the data when shown in a scatterplot.

Bivariate Correlation. One value commonly computed in bivariate correlation is the **correlation coefficient**, or r:

$$r = \frac{\sum_{i=1}^{n}(X_i - \overline{X})(Y_i - \overline{Y})}{\sqrt{\sum_{i=1}^{n}(X_i - \overline{X})^2}\sqrt{\sum_{i=1}^{n}(Y_i - \overline{Y})^2}}$$

Table 3.4 shows r values for the transformed sample data. (Note that the table is symmetric about a diagonal extending from upper left to lower right; this occurs because the correlation between attributes A and B is identical to that between attributes B and A.)

Extreme values for r range from -1 to $+1$. A positive r value indicates a *positive relationship* in which increasing

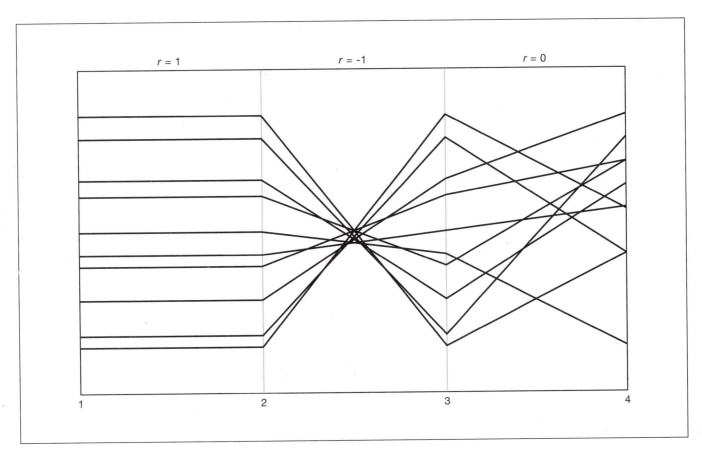

FIGURE 3.9 Characteristic parallel coordinate plots for particular values of r.

TABLE 3.4   Matrix of correlation coefficients for the transformed city data

	Log of Murder Rate	Percent below Poverty Level	Percent with High School Education	Log of Total Population	Square Root of Drug Arrest Rate	Log of Population Density
Log of murder rate	1.00	.82	−.70	.52	.49	.27
Percent below poverty level	.82	1.00	−.79	.38	.29	.17
Percent with high school education	−.70	−.79	1.00	−.39	−.44	−.36
Log of total population	.52	.38	−.39	1.00	.39	.54
Square root of drug arrest rate	.49	.29	−.44	.39	1.00	.52
Log of population density	.27	.17	−.36	.54	.52	1.00

values on one attribute are associated with increasing values on another attribute; for example, the r value for log of murder rate and percent below poverty level is .82. Conversely, a negative r value indicates a *negative relationship* in which increasing values on one attribute are associated with decreasing values on another attribute; for example, log of murder rate and percent with high school education have a correlation of −.70. Note that as r values approach −1 or +1 the points cluster more tightly about the best-fit line. (Compare Table 3.4 and Figure 3.7, and note that the graph of log of murder rate and percent below poverty level is more tightly clustered than the one for log of murder rate and square root of drug arrest.)

The value of r is primarily used to indicate the *direction* of a relationship (whether it is positive or negative). To properly compute the *strength* of relationship, one must compute the coefficient of determination, or r^2, which measures the proportion of variation in one attribute explained, or accounted for, by another attribute. For example, r^2 between log of murder rate and percent below poverty level is .67, indicating that 67 percent of the variation in murder rate can be accounted for by a linear function of the poverty attribute.

Bivariate Regression.   In general, the equation for any straight line is $Y_i = a + bX_i$, where X_i, Y_i is a point on the line, a is the y intercept, and b is the slope. In regression analysis, X_i and Y_i are raw values for the independent and dependent attributes respectively, and the line of best fit is defined as $\hat{Y}_i = a + bX_i$, where $\hat{Y}_i$ is a predicted value for the dependent attribute. The best-fit line is found by minimizing the differences between the actual and predicted values $(Y_i - \hat{Y}_i)$, the vertical distances

shown in Figure 3.10. The values of a and b that minimize the differences are:

$$a = \overline{Y} - b\overline{X}$$

$$b = \frac{n\sum_{i=1}^{n} X_i Y_i - \sum_{i=1}^{n} X_i \sum_{i=1}^{n} Y_i}{n\sum_{i=1}^{n} X_i^2 - \left(\sum_{i=1}^{n} X_i\right)^2}$$

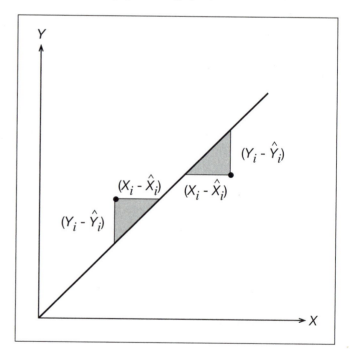

FIGURE 3.10   Possible approaches for determining best-fit lines for a set of data include (1) minimizing vertical distances, (2) minimizing horizontal distances, and (3) minimizing the area of the resulting triangles.

The best-fit lines shown in Figures 3.5 and 3.7 were derived using this approach.

Major-Axis Approach. When one does not wish to specify a dependent attribute (as when relating murder rate and race), it makes sense to minimize the vertical and horizontal distances shown in Figure 3.10 simultaneously, effectively minimizing the area of the triangles; this is known as the **reduced major-axis approach** (Davis 2002, 214–218). The equations for this approach turn out to be simpler than those for standard regression. The slope is just the ratio of the standard deviations of the attributes, or s_Y/s_X, and the intercept is still $\overline{Y} - b\overline{X}$. In Chapter 18 we will consider how Eyton (1984a) used the reduced major-axis approach to create a logical bivariate map.

Multiple Regression and Other Multivariate Techniques. When the concept of bivariate correlation-regression is extended into the multivariate realm, it is termed **multiple regression**. In multiple regression, there is still a single dependent attribute, but multiple independent attributes are possible. In a fashion similar to the preceding, it is possible to perform correlation and regression analyses that summarize the relationship between the dependent and independent attributes. For example, we might attempt to explain the murder rate as a function of percent below poverty level, percent with high school education, and the drug arrest rate. A discussion of multiple regression and other multivariate techniques, such as **principal components analysis**, is beyond the scope of this book. You should consult statistical textbooks such as Clark and Hosking (1986), Davis (2002), and Rogerson (2001) for related information.

Considerations in Using Correlation-Regression. There are several things that you should consider carefully when using bivariate correlation-regression and multiple regression. One is that high correlations do not necessarily imply a causal relationship. The high correlation between murder rate and poverty that we found could be a result of chance (although it is unlikely two random data sets could result in a correlation of this magnitude), or some other attribute or set of attributes could be influencing both of these attributes.

A second consideration is that the magnitude of r can be affected by the level at which data have been aggregated, which is known as the **modifiable areal unit problem** (Clark and Hosking 1986; Barrett 1994). Generally, coarser levels of aggregation (e.g., analyzing at the city level rather than at the census tract level) will lead to higher r values because "aggregation reduces the between unit variation in an attribute, making the attribute seem more homogeneous" (Clark and Hosking, 1986,

405). A related issue is that the magnitude of r might be a function of the size and arrangement of enumeration units when the number of enumeration units remains constant (Fotheringham 1998). A solution to the aggregation problem is to examine data at the individual level (looking at individual murders and collecting data regarding the people involved). An example is The Project on Human Development in Chicago Neighborhoods, which is analyzing data from a variety of Chicago neighborhoods; here the emphasis is on individual interviews and extensive fieldwork (see *http://phdcn.harvard.edu/about/index.html*).

A third consideration is that specialized regression techniques have been developed to handle the fact that geographical data tend to be spatially autocorrelated, meaning that like values tend to be located near one another (e.g., a high-income census tract will tend to occur near another high-income census tract). For a discussion of these specialized techniques, see Griffith (1993) and Rogerson (2001). SpaceStat (*http://www.terraseer.com/Spacestat.html*) is a software package that has been developed explicitly for handling such techniques.

Finally, we need to recognize that regression techniques can be applied both globally and locally. By global, we mean that a regression equation is applied uniformly throughout a geographic region. In contrast, local regression techniques consider the notion that a single model might not be appropriate for an entire region, as local variation might necessitate different models within subregions. As with modifications due to spatial autocorrelation, a full discussion of this issue is beyond the scope of this book. For an overview of the local–global issue applied to regression and other statistical methods, see Fotheringham (1997).

3.3.3 Exploratory Data Analysis

One of the most important advances in statistical analysis in the last 25 years was John Tukey's (1977) development of **exploratory data analysis (EDA)**. In Chapter 1, we suggested that rather than trying to make one "best" map, interactive graphics systems should provide multiple representations of a spatial data set. In much the same way, Tukey proposed that rather than trying to fit statistical data to standard forms (normal, Poisson, binomial), data should be explored, much as a detective investigates a crime. In the process of exploring data, the purpose should not be to confirm what one already suspects, but rather to develop new questions or hypotheses.

One technique representative of Tukey's approach is the **stem-and-leaf plot**, which is depicted in Figure 3.11 using the murder rate data. To construct a stem-and-leaf

```
0  *  |  0111222222233
0  ·  |  5555777889
1  *  |  000011224
1  ·  |  5778
2  *  |  1244
2  ·  |  8
3  *  |  11
3  ·  |  5
4  *  |  11
4  ·  |
5  *  |
5  ·  |  679
6  *  |
6  ·  |
7  *  |
7  ·  |  8
```

FIGURE 3.11 Stem-and-leaf plot of the murder rate data presented in Table 3.1.

plot, one first separates the digits of the data values into three classes: sorting digits, display digits, and digits not displayed because of rounding.* For the murder rate data, we chose the 10s place as a sorting digit, the 1s place as the display digit, and did not display the 10ths place because we rounded to the nearest whole percent. Sorting digits are placed to the left of the vertical line shown in Figure 3.11 and are known as *stems*, whereas display digits are placed to the right and are known as *leaves*. For Figure 3.11, Tukey's conventional system of asterisks and dots was used to split the 10s place into two parts (leaf values of 0 to 4 are plotted on one row and 5 to 9 on the next row). For example, Norfolk's murder rate of 24.1 appears as the fourth leaf ("4") in the fifth row ("2 *"), and Los Angeles's rate of 28.2 appears as the only leaf ("8") in the sixth row ("2 ·").

If you mentally rotate the stem-and-leaf plot so that the stems are on the bottom, and then compare it to those graphical methods discussed previously for individual attributes (Figures 3.1 and 3.2), you will note a great deal of similarity to the histogram. Both methods portray a peak on the left side of the graph with a distinct positive skew. Although the graphs are similar, the major advantage of the stem-and-leaf plot is that it is possible to determine approximate values for each

observation. For the murder rate data, this might not be particularly useful, but for some data sets it can be. For example, Burt and Barber (1996, 542–544) describe a stem-and-leaf plot that portrays when houses were constructed in a neighborhood. The stem-and-leaf plot reveals that seven houses were built during the post–World War II period (1945–1947), but no houses were built from 1948 to 1950.

Another technique representative of Tukey's work is the **box plot** (Figure 3.12). Here, a rectangular box represents the interquartile range (the 75th minus the 25th percentile), and the middle line within the box represents the median, or 50th percentile. The position of the median, relative to the 75th and 25th percentiles, is an indicator of whether the distribution is symmetric or skewed; for the murder rate data, the position indicates a positive skew (a tail toward higher values on the number line). The horizontal lines outside the rectangular box represent the maximum and minimum values in the data. The relative positions of these lines also permit an examination of the symmetry of the distribution, and their position relative to the box indicates how extreme the maximum and minimum values are. In this case, their relative position again suggests a positive skew.[†]

Box plots are most frequently used to compare two or more distributions having the same units of measurement.

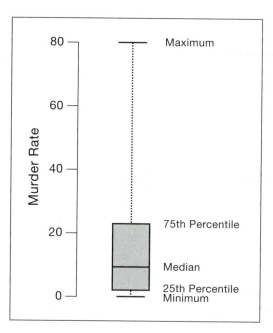

FIGURE 3.12 Box plot of the murder rate data presented in Table 3.1.

---

* Tukey (1977) indicated that digits beyond the display digit could either be used for rounding or simply ignored.

[†] Numerous variations of the box plot have been developed; for an overview and suggestions for an alternative using color, see Carr (1994).

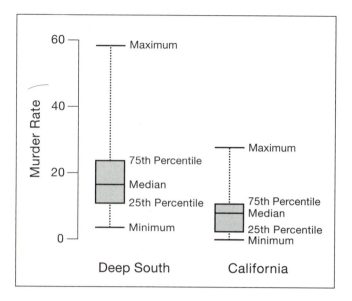

FIGURE 3.13 Box plots comparing murder rate data for cities in the Deep South and California. California cities appear to have a distinctly smaller range, a smaller maximum, and an interquartile range that is smaller and associated with lower murder rates.

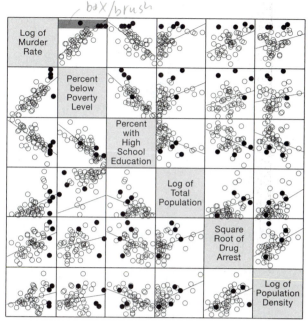

FIGURE 3.14 Scatterplot brushing. As the rectangular box (or brush) is moved, dots (cities) within it are highlighted in all scatterplots.

For example, Figure 3.13 compares murder rate data for cities falling in two distinct regions (the "Deep South" and "California," as defined by Birdsall and Florin 1992). Note that California cities have a distinctly smaller range, a smaller maximum, and an interquartile range that is smaller and shifted toward lower murder rates.

It is important to realize that many of the methods covered thus far can also be used in an exploratory manner. For example, in a fashion similar to the stem-and-leaf plot, a dispersion graph can uncover nuances in the data not revealed by a histogram. The scatterplot matrix can also be used in an exploratory fashion, especially if the number of attributes is large. Remember that the key to exploratory analysis is to reveal hidden characteristics; the broadest possible range of approaches should be considered for achieving this.

A technique that fits especially well under the heading of EDA is **scatterplot brushing**, which is illustrated in Figure 3.14 for the data for the 50 U.S. cities. Note that the scatterplot for log of murder rate–percent below poverty level contains a gray rectangular box or "brush," and that all dots (cities) within this box are highlighted by a solid fill; these same cities are also highlighted within other scatterplots in the matrix. Brushing involves using an interactive graphics display to move the box; as the box moves, observations falling within it are highlighted within all scatterplots.

Mark Monmonier (1989) developed the notion of integrating scatterplot brushing with maps. Monmonier indicated that as the brush is moved within the scatterplot, mapped areas corresponding to dots within the brush should also be highlighted (Figure 3.15). He also suggested the notion of a **geographic brush**: as areas of a map are brushed, corresponding dots in the scatterplot matrix should be highlighted. We will explore the notion of geographic brushing further in Chapter 21.

In this section we have focused on the notion of EDA developed by statisticians, and suggested how you, as a cartographer, might make use of these ideas. You should realize that statisticians within geography have also borrowed ideas from EDA, developing an area they call *exploratory spatial data analysis* (ESDA). Anselin (1998) provides a good overview of developments in ESDA.

3.4 NUMERICAL SUMMARIES IN WHICH LOCATION IS AN INTEGRAL COMPONENT

This section considers numerical summaries in which spatial location is an integral component. Some of these methods analyze just spatial location (formulas for the centroid and various shape indices), whereas others consider both spatial location and the values of an attribute (e.g., spatial autocorrelation).

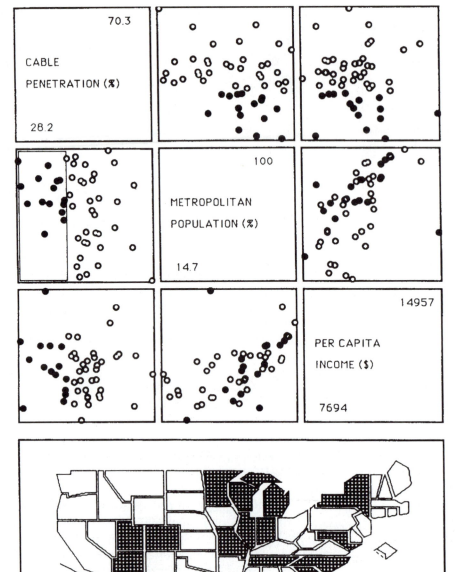

FIGURE 3.15 How scatterplot brushing can be integrated with a map. As the brush (the rectangular box) is moved in the scatterplot, dots within it are highlighted within the scatterplot and on the map (Monmonier 1989. Reprinted by permission from *Geographical Analysis*, Vol. 21, No. 1 (Jan. 1989). Copyright 1989 by Ohio State University Press. All rights reserved.).

3.4.1 Analysis of Spatial Location

Centroid

The **centroid** is defined as

$$\overline{X}_r = \frac{\sum_{i=1}^{n}(X_iY_{i+1} - X_{i+1}Y_i)(X_i + X_{i+1})}{3\sum_{i=1}^{n}(X_iY_{i+1} - X_{i+1}Y_i)}$$

$$\overline{Y}_r = \frac{\sum_{i=1}^{n}(X_{i+1}Y_i - X_iY_{i+1})(Y_i + Y_{i+1})}{3\sum_{i=1}^{n}(X_{i+1}Y_i - X_iY_{i+1})}$$

where the X_iY_i are a sequence of points defining the boundary of a region, assuming that the n + first point is identical to the first point (Bachi 1999, 106). These formulas appear rather complicated, but the concept is really quite simple. Imagine cutting the outline of the 48 contiguous states out of a thin metal sheet and then attempting to balance the sheet on the eraser end of a pencil. The point at which the sheet balances (which happens to be located near Lebanon, Kansas) would be its centroid (or center of gravity). It should be recognized, however, that the centroid derived by the preceding formula might not necessarily fall within a region if the region is highly convoluted. As a result, it might be more appropriate to consider other measures for computing the center of region, such as the center of a rectangle surrounding a region—the so-called **bounding rectangle** (Carstensen 1987).

Indexes for Measuring Shape

Geographers have developed a wide variety of indexes for measuring the shape of geographic regions. One of the simplest is the **compaction index (CI)** (Hammond and McCullagh 1978, 69–70), which is defined as the ratio of the area of a shape to the area of a circumscribing circle (a circle that just touches the bounds of the shape). Values for CI range from 0 to 1, with 0 representing the least compact shape (e.g., a narrow rectangular box) and 1 representing the most compact shape (a circle). As an example, consider the shapes of Tennessee and Arkansas relative to a circle (Figure 3.16); based on the CI, Arkansas is clearly more compact than Tennessee. One limitation of the CI is that it does not differentiate some shapes well. For example, Figure 3.17 shows two regions with shapes that differ, but which have similar CI values. This sort of problem led geographers to develop other methods for analyzing shape, many of which are more complex than the CI; examples include the Boyce-Clark index (Unwin 1981) and Moellering and Rayner's (1982) harmonic-analysis method.

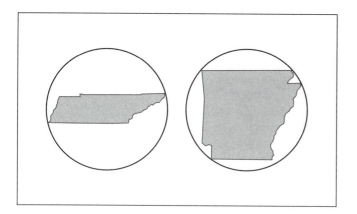

FIGURE 3.16 Computing the compaction index (CI) for two states. CI is computed as a ratio of the area of the region to a circle enclosing that region; the CI value for Arkansas clearly would be greater than that for Tennessee.

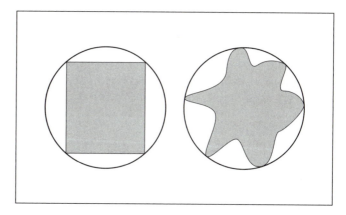

FIGURE 3.17 Two regions having different shapes but similar CI values.

One application of shape indices is **redistricting**, which involves the combination of voting precincts to form legislative or congressional districts. This process must be done at each decennial census to account for migration and natural increases and decreases in population. You might be familiar with the related term **gerrymandering**, in which districts are purposely structured for partisan benefit. Shape indexes can be used as objective measures of the degree of gerrymandering (Morrill 1981; Monmonier 2001).

3.4.2 Analyzing an Attribute in Association with Spatial Location

This section considers various methods for analyzing spatial data in which an attribute is linked with its spatial location. Topics covered include (1) weighting attribute data to account for sizes of enumeration units, (2) central tendency and dispersion for point data, and (3) spatial autocorrelation and map complexity measures.

Weighting Attribute Data to Account for Sizes of Enumeration Units

Arthur Robinson and his colleagues (1995) argued that when data are associated with enumeration units, basic summary measures such as the mean and standard deviation should be weighted to account for the differing sizes of enumeration units. The formulas for a population are as follows:

$$\mu_w = \frac{\sum_{i=1}^{n} a_i X_i}{\sum_{i=1}^{n} a_i} \qquad \sigma_w = \sqrt{\frac{\sum_{i=1}^{n} a_i X_i^2}{\sum_{i=1}^{n} a_i} - \mu_w^2}$$

where X_i is a value for an attribute for the ith enumeration unit, and a_i is the area of the ith enumeration unit.

These weighted measures are appropriate if the intention is to impart to the reader the visual impact that large enumeration units have on a distribution. You should realize, however, that different means and standard deviations will arise, depending on how enumeration units are defined. To illustrate, consider the single enumeration unit and its potential four subregions shown in Figure 3.18. Presume that within the single enumeration unit 39 out of 1,300 people, or 3 percent, are college graduates, but that in three of the four subregions 13 out of 100 people, or 13 percent, are college graduates, and in the fourth subregion there are no college graduates. The average of the four subregions is 9.75 percent, a value that is quite different than the 3 percent for the aggregated data.

Another weighting procedure relevant to enumeration units would be to modify the formula for r to account for the sizes of the enumeration units—the logic being that larger enumeration units should impact r more. The formula is as follows (Robinson 1956):

$$
r_w = \cfrac{\begin{aligned}&\sum_{i=1}^{n} a_i \sum_{i=1}^{n} a_i X_i Y_i \\ &- \sum_{i=1}^{n} a_i X_i \sum_{i=1}^{n} a_i Y_i\end{aligned}}{\begin{aligned}&\sqrt{\sum_{i=1}^{n} a_i \sum_{i=1}^{n} a_i X_i^2 - \left(\sum_{i=1}^{n} a_i X_i\right)^2} \\ &\times \sqrt{\sum_{i=1}^{n} a_i \sum_{i=1}^{n} a_i Y_i^2 - \left(\sum_{i=1}^{n} a_i Y_i\right)^2}\end{aligned}}
$$

where a_i is the area of the ith enumeration unit. Edwin Thomas and David Anderson (1965) showed that such a weighting procedure is inappropriate from a statistical perspective because unweighted correlation coefficients resulting from different arrangements of enumeration unit boundaries were not significantly different from one another. Although inappropriate from a statistical perspective, such a measure might be used as an indicator of the visual correlation between two maps. (We will consider the problem of visual correlation in greater depth in Chapter 18.)

Central Tendency and Dispersion for Point Data

Earlier in this chapter we considered measures of central tendency and dispersion that ignored spatial location. For example, in computing the mean for murder rate, we ignored the location of individual cities. We now consider analogous measures of central tendency and dispersion for explicitly spatial data. Imagine a set of x and y coordinates defining the location of small towns in a portion of Minnesota. The central tendency for such a set of points is termed the **mean center**, and is defined as

$$
\overline{X}_c = \frac{\sum_{i=1}^{n} X_i}{n} \qquad \overline{Y}_c = \frac{\sum_{i=1}^{n} Y_i}{n}
$$

Note that these formulas simply apply the nonspatial mean formula twice, once for the x axis and once for the y axis. As with the nonspatial mean, there is some danger that the mean center will not fall where the data are concentrated. This problem can be handled using a spatial version of the median (Hammond and McCullagh 1978, 48–53).

A measure of dispersion associated with the mean center is **standard distance**, **SD**, which is computed as*

$$
\mathrm{SD} = \sqrt{\frac{\sum_{i=1}^{n} d_{ic}^2}{N}} = \sqrt{\sigma_X^2 + \sigma_Y^2}
$$

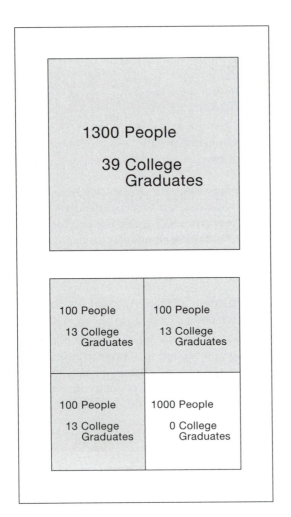

FIGURE 3.18   As explained in the text, the areally weighted mean is not necessarily a logical measure to use.

---

* The formulas given are for a population. Griffith and Amrhein (1991, 123) indicate that for standard distance there is little difference between sample and population formulas.

where d_{ic} is the distance from the ith point to the mean center. The formula on the left indicates that SD is a measure of spread about the mean center, and the one on the right is valuable when using a calculator, which generally does not include functions for spatial statistics.

For some applications, it can be argued that measures of central tendency and dispersion for points should be weighted to reflect the magnitude of an attribute at the points. For example, imagine that you have population values for point locations (e.g., the centers of counties) and wish to find the mean center and standard distance for the population. The formulas for the *weighted mean center* and *weighted standard distance* are:

$$\overline{X}_{cw} = \frac{\sum_{i=1}^{n} w_i X_i}{\sum_{i=1}^{n} w_i} \qquad \overline{Y}_{cw} = \frac{\sum_{i=1}^{n} w_i Y_i}{\sum_{i=1}^{n} w_i}$$

$$\mathrm{SD}_w = \sqrt{\frac{\sum_{i=1}^{n} w_i d_{ic}^2}{\sum_{i=1}^{n} w_i}}$$

Using formulas similar to these, the U.S. Bureau of the Census has found that the population mean center has moved from near Baltimore, Maryland, in 1790 to about 3 miles east of Edgar Springs, Missouri, in 2000 (U.S. Department of Commerce 2001).

Spatial Autocorrelation and Measuring Spatial Pattern

Although maps allow us to visually assess spatial pattern, they have two important limitations: their interpretation varies from person to person, and there is the possibility that a perceived pattern is actually the result of chance factors, and thus not meaningful. For these reasons, it makes sense to compute a numerical measure of spatial pattern, which can be accomplished using spatial autocorrelation. In this section we will consider how spatial autocorrelation can be applied to choropleth maps. For a discussion of measures appropriate for examining patterns on other types of maps, see Unwin (1981) and Davis (2002).

Spatial autocorrelation is the tendency for like things to occur near one another in geographic space. For example, expensive homes likely will be located near other expensive homes, and soil cores with high clay content likely will be found near other soil cores with high clay content. Two measures used by statisticians to express the degree of spatial autocorrelation are the Moran coefficient (MC) and the Geary ratio (GR). We will consider only the Moran coefficient because it is statistically more powerful (Griffith 1993, 21), and more commonly used.

The formula for MC is

$$\mathrm{MC} = \frac{\sum_{i=1}^{n,} \sum_{j=1}^{n} w_{ij}(X_i - \overline{X})(X_j - \overline{X}) \Big/ \sum_{i=1}^{n} \sum_{j=1}^{n} w_{ij}}{\sum_{i=1}^{n} (X_i - \overline{X})^2/n}$$

where $w_{ij} = 1$ if enumeration units i and j are adjacent (or contiguous) and 0 otherwise.* Computations for MC for a hypothetical region consisting of nine enumeration units are shown in Figure 3.19. Within each enumeration unit, the upper left number is an identifier for that unit, and the lower right number is the value for a hypothetical attribute. Note that the formula involves multiplying a weight for two enumeration units (w_{ij}) times the product of the difference between the attribute values for the enumeration units and the mean of the data $(X_i - \overline{X})(X_j - \overline{X})$. Computations are shown only for adjacent enumeration units, because w_{ij} will be 0 for nonadjacent units and thus the product $w_{ij}(X_i - \overline{X})(X_j - \overline{X})$ also will be 0.

The formula for MC resembles the formula for the correlation coefficient, r, discussed previously; for both equations, the denominator contains a measure of the variation in the attribute about the mean, and the numerator contains a measure of how adjacent enumeration units covary (compare the equation shown here with that for r in section 3.3.2). Thus, it is not surprising that MC also ranges from -1 to $+1$. A value close to $+1$ indicates that similar values are likely to occur near one another, whereas a value close to -1 indicates that unlike values are apt to occur near one another. Finally, a value near 0 is indicative of no autocorrelation, or a situation in which values of the attribute are randomly distributed. "Moderate" values of positive spatial autocorrelation are most frequently observed in the real world (Griffith 1993, 2).

To illustrate how the MC might be used, consider mapping the rates for respiratory cancer for white males in Louisiana counties (Figure 3.20). A visual assessment of this map suggests that there is a strong positive spatial autocorrelation, with high cancer rates occurring both in the southern part of the state and along an east–west strip in the northern part of the state. The MC for this pattern is .14, with a probability of less than 3 percent that this value would occur by chance (Odland 1988). The value of .14 indicates that the pattern is not quite as strongly positively autocorrelated as a visual examination of the map suggests, but the associated probability indicates the pattern is significant and worthy of further exploration.

---

* The weights can be modified to account for differing sizes of enumeration units (Odland 1988, 29–31). We have used weights of 0 and 1 for computational simplicity.

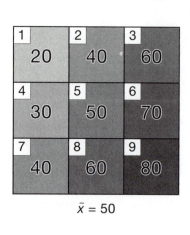

$\bar{x} = 50$

Enumeration Units

i	j	w_{ij}	$(x_i - \bar{x})(x_j - \bar{x})$	=	
1	2	1	(20–50)(40–50)	=	300
1	4	1	(20–50)(30–50)	=	600
2	1	1	(40–50)(20–50)	=	300
2	3	1	(40–50)(60–50)	=	–100
2	5	1	(40–50)(50–50)	=	0
3	2	1	(60–50)(40–50)	=	–100
3	6	1	(60–50)(70–50)	=	200
4	1	1	(30–50)(20–50)	=	600
4	5	1	(30–50)(50–50)	=	0
4	7	1	(30–50)(40–50)	=	200
5	2	1	(50–50)(40–50)	=	0
5	4	1	(50–50)(30–50)	=	0
5	6	1	(50–50)(70–50)	=	0
5	8	1	(50–50)(60–50)	=	0
6	3	1	(70–50)(60–50)	=	200
6	5	1	(70–50)(50–50)	=	0
6	9	1	(70–50)(80–50)	=	600
7	4	1	(40–50)(30–50)	=	200
7	8	1	(40–50)(60–50)	=	–100
8	5	1	(60–50)(50–50)	=	0
8	7	1	(60–50)(40–50)	=	–100
8	9	1	(60–50)(80–50)	=	300
9	6	1	(80–50)(70–50)	=	600
9	8	1	(80–50)(60–50)	=	300
					4000

$$\sum_{i=1}^{n}\sum_{j=1}^{n} w_{ij}(x_i - \bar{x})(x_j - \bar{x}) = 4000$$

$$\sum_{i=1}^{n}\sum_{j=1}^{n} w_{ij} = 24$$

$$\sum_{i=1}^{n}(x_i - \bar{x})^2/n = \sigma^2 = 333.333$$

$$MC = (4000/24)/333.333 = 0.50$$

FIGURE 3.19   Computation of the Moran coefficient (MC) for spatial autocorrelation.

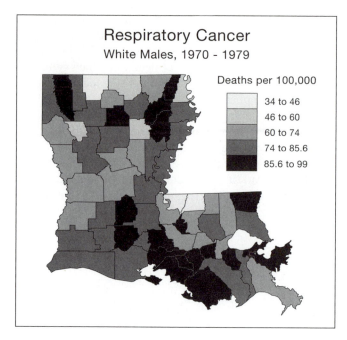

FIGURE 3.20   A map of rates for respiratory cancer for white males in Louisiana counties. The Moran coefficient (a measure of spatial autocorrelation) for this map is .14, with a probability of less than 3 percent that this value occurs by chance. Thus, the pattern is significant and worthy of further exploration. (Courtesy of John Odland.)

An important recent development in statistical geography is a consideration of spatial autocorrelation in the context of inferential statistics. We touched on this idea earlier in the chapter with respect to regression, but it should be recognized that virtually all traditional inferential statistics can be modified to account for spatial autocorrelation. For further discussion of this issue, see Griffith (1993) and Rogerson (2001).

MacEachren's Measures of Map Complexity

The notion of spatial autocorrelation and associated measures such as the MC were developed by statistical geographers. Cartographers have also been interested in measures for describing spatial pattern. As an example, we'll consider some map complexity measures developed by Alan MacEachren (1982a). MacEachren defined **map complexity** as "the degree to which the combination of map elements results in a pattern that appears to be intricate or involved" (32). He argued that when a map is used for presentation, very complex maps might hinder the communication of information.

MacEachren's measures are based on Muller's (1974) application of graph theory to choropleth maps. To compute measures of complexity, a base map is treated as a set of faces (enumeration units), edges (the boundaries of the units), and vertices (where the edges intersect; see Figure 3.21A). When mapping a distribution, edges

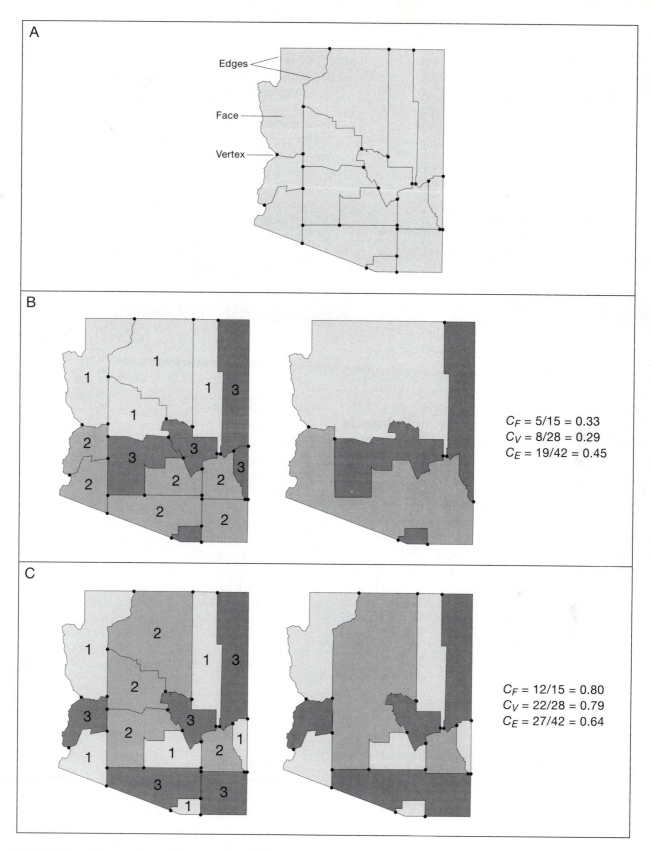

FIGURE 3.21 Computation of MacEachren's (1982a) face, vertex, and edge complexity measures. (A) Faces, edges, and vertices for a county-level map of Arizona; (B) three-class map for which low complexity values result; (C) three-class map for which high complexity values result.

and vertices between like values are omitted (Figures 3.21B and 3.21C). Complexity measures are computed by dividing the number of faces, edges, and vertices on the mapped distribution by the corresponding number on the base map:*

$$C_F = \frac{\text{Observed number of faces}}{\text{Number of original faces}}$$

$$C_V = \frac{\text{Observed number of vertices}}{\text{Number of original vertices}}$$

$$C_E = \frac{\text{Number of original edges between categories}}{\text{Number of original edges}}$$

In the example illustrated in Figure 3.21, map C has the higher complexity measures, and so would presumably be more difficult for a user to interpret.

SUMMARY

In this chapter, we have examined a variety of techniques (tables, graphs, and numerical summaries) that can be used along with maps to analyze spatial data. These techniques each have their advantages and disadvantages. In the univariate realm, **raw tables** are useful for providing specific information (e.g., the murder rate for Atlanta, Georgia, in 1990 was 58.6 per 100,000 people), but they fail to provide an overview of a data set (e.g., that the murder rate for U.S. cities in 1990 had a distinct positive skew). **Grouped-frequency tables** do provide an overview, but are not as effective as graphical methods (e.g., the **dispersion graph** and **histogram**). Potential weaknesses of graphical methods are the subjectivity of interpretation and the physical space they require. Numerical summaries (e.g., the **mean** and **standard deviation**) are a solution to these problems. A weakness of numerical summaries,

---
* MacEachren used the "number of original edges between categories" for the numerator of the edge measure to attempt to account for the size of faces.

however, is that they hide the detailed character of the data; for example, an **outlier** might be missed if only a numerical summary is used. Because of these advantages and disadvantages, data exploration software (discussed in Chapter 21) should include a broad range of methods for analyzing spatial data.

When examining two or more attributes at once, tabular displays are of limited use. Much more suitable are graphical displays, such as the **scatterplot** and **scatterplot matrix**. The numerical method of **bivariate correlation-regression** can also be useful if one wishes to summarize the relationship between two attributes by a single number or equation. As with univariate data, care should be taken in using bivariate numerical summaries, as they hide the detailed character of the relationship between attributes. An important problem with bivariate correlation is that it is a function of the level of aggregation (a different **correlation coefficient** might arise at the census tract level as opposed to the block-group level), an issue known as the **modifiable areal unit problem**.

Given the emphasis on data exploration in this book, special emphasis in this chapter was placed on **exploratory data analysis**—that data should be explored, much as a detective investigates a crime. We looked at three common methods for exploratory data analysis: the **stem-and-leaf plot**, the **box plot**, and **scatterplot brushing**. We saw that the notion of scatterplot brushing could be extended to a **geographic brush**—as areas on a map are highlighted, corresponding dots in a scatterplot matrix are highlighted. This is the sort of capability we would expect to find in data exploration software.

The latter section of the chapter considered several numerical summaries for which spatial location is an integral part; here we examined the **centroid, compaction index, mean center** (and associated **standard distance**), **spatial autocorrelation**, and **map complexity** measures. In Chapter 5, we will make use of map complexity measures when we consider a method for simplifying choropleth maps.

FURTHER READING

Anselin, L. (1998) "Exploratory spatial data analysis in a geocomputational environment." In *Geocomputation: A Primer*, ed. by P. A. Longley, S. M. Brooks, R. McDonnell, and B. MacMillan, pp. 77–94. Chichester, England: Wiley.

Summarizes developments in exploratory spatial data analysis (ESDA).

Bachi, R. (1999) *New Methods of Geostatistical Analysis and Graphical Representation: Distributions of Populations over Territories.* New York: Kluwer Academic/Plenum.

Presents geostatistical methods for summarizing the distribution of population over geographic regions.

Barrett, R. E. (1994) *Using the 1990 U.S. Census for Research.* Thousand Oaks, CA: Sage.

Summarizes data available from the U.S. Bureau of the Census and discusses problems associated with using such data.

Burt, J. E., and Barber, G. M. (1996) *Elementary Statistics for Geographers.* 2nd ed. New York: Guilford.

An introductory statistics book for geographers.

Cleveland, W. S. (1994) *The Elements of Graphing Data.* Rev. ed. Summit, NJ: Hobart Press.

Covers a broad range of graphical methods for summarizing data. The text has more of a statistical emphasis than the Kosslyn text listed below. Also see Cleveland (1993).

Cleveland, W. S., and McGill, M. E. (eds.) (1988) *Dynamic Graphics for Statistics*. Belmont, CA: Wadsworth.

Contains 16 chapters describing various researchers' efforts to explore statistical data using interactive graphics.

Davis, J. C. (2002) *Statistics and Data Analysis in Geology*. 3rd ed. New York: Wiley.

Although intended primarily for geologists, this text covers a broad range of statistical methods, many of which are of interest to cartographers.

Dykes, J. A. (1994) "Visualizing spatial association in area-value data." In *Innovations in GIS*, ed. by M. F. Worboys, pp. 149–159. Bristol, PA: Taylor & Francis.

Considers problems of using a *mathematical measure* of spatial autocorrelation to represent the *perceived* degree of autocorrelation.

Fotheringham, A. S. (1999) "Trends in quantitative methods III: Stressing the visual." *Progress in Human Geography* 23, no. 4:597–606.

Presents some directions that researchers might take in developing methods for displaying multivariate data that would be useful to geographers.

Jacoby, W. G. (1997) *Statistical Graphics for Univariate and Bivariate Data*. Thousand Oaks, CA: Sage.

An overview of graphical methods for summarizing both univariate and bivariate data. This short work (98 pages) is a useful complement to Cleveland's book mentioned above. For graphical methods for summarizing multivariate data, see Jacoby (1998).

Kosslyn, S. M. (1994) *Elements of Graph Design*. New York: W. H. Freeman.

A how-to text on constructing graphs; the text is liberally illustrated with *do* and *don't* examples.

Monmonier, M. (2001) *Bushmanders & Bullwinkles*. Chicago: The University of Chicago Press.

A cartographer's view on redistricting; Chapter 5 considers two numerical measures that can be utilized to gauge the compactness of congressional districts.

Odland, J. (1988) *Spatial Autocorrelation*. Newbury Park, CA: Sage.

A primer on spatial autocorrelation.

Rogerson, P. A. (2001) *Statistical Methods for Geography*. London: Sage.

A statistics book for geographers that covers both introductory and advanced concepts. Also see Clark and Hosking (1986).

Tufte, E. R. (1983) *The Visual Display of Quantitative Information*. Cheshire, CT: Graphics Press.

Covers a broad range of techniques (both graphs and maps) for representing numerical data.

Tukey, J. W. (1977) *Exploratory Data Analysis*. Reading, MA: Addison-Wesley.

The classic reference on exploratory data analysis.

Unwin, D. (1981) *Introductory Spatial Analysis*. London: Methuen.

Covers a broad range of methods for analyzing spatial data (i.e., the data have a distinct spatial component and thus can be mapped).

Wegman, E. J. (1990) "Hyperdimensional data analysis using parallel coordinates." *Journal of the American Statistical Association* 85, no. 411:664–675.

Explains the mathematics underlying parallel coordinate plots and how to interpret them. For a more detailed discussion of the mathematics, see Inselberg (1985).

4

Principles of Symbolization

OVERVIEW

The purpose of this chapter is to cover basic principles of symbolizing geographic phenomena. An overarching goal is to assist you in selecting among four common thematic mapping techniques: **choropleth**, **proportional symbol**, **isopleth**, and **dot**. For example, imagine that you wish to map the spatial pattern of income in Washington, DC, and that you have collected data on the annual income of all families in each census tract of the city. You might wonder which of these four techniques would be appropriate. Determining the appropriate technique will require that we first consider (1) the spatial arrangement of geographic phenomena, (2) the various levels at which we can measure geographic phenomena, and (3) the types of symbols that can be used to represent spatial data.

Section 4.1 discusses the spatial arrangement of geographic phenomena. One way to think about spatial arrangement is to consider a phenomenon's extent or **spatial dimension**—whether a phenomenon can be conceived of as **points**, **lines**, **areas**, or **volumes**. For example, water well sites in a rural area constitute a point phenomenon, whereas a city boundary is representative of a linear phenomenon. Another way of thinking about spatial arrangement is to contrast discrete and continuous phenomena. **Discrete phenomena** occur at isolated point locations, whereas **continuous phenomena** occur everywhere. For example, water towers in a city would be discrete, but the distribution of solar insolation during the month of January is continuous. Discrete and continuous phenomena can also be classified as **smooth** or **abrupt**. For instance, rainfall and sales tax rates for states are both continuous in nature, but the former is smooth, whereas the latter is abrupt (varying at state boundaries). In thinking about the arrangement of geographic phenomena, we need to distinguish between a phenomenon that exists in the real world and the data that we use to represent that phenomenon.

Section 4.2 considers **levels of measurement**, which refers to the various ways of measuring a phenomenon when a data set is created. For instance, we might specify the soil type of a region as an entisol, as opposed to a mollisol; such a categorization of soils would be termed a nominal level of measurement. We consider four basic levels of measurement: **nominal**, **ordinal**, **interval**, and **ratio**. The latter two levels are commonly combined into numerical data, which is the focus of this book.

The term **visual variables** is commonly used to describe the various perceived differences in map symbols that are used to represent spatial data. For example, the visual variable spacing involves varying the distance between evenly spaced marks (e.g., horizontal lines). Section 4.3 covers a host of visual variables, including **spacing**, **size**, **perspective height**, **orientation**, **shape**, **arrangement**, **hue**, **lightness**, **saturation**, and **location**.

Section 4.4 introduces four common thematic mapping techniques (choropleth, proportional symbol, isopleth, and dot) and considers how a mapmaker selects among them. We will see that the selection is a function of both the nature of the underlying phenomenon and the purpose for making the map. Section 4.4 also introduces the notion of **data standardization** to account for the area over which data are collected; here we will consider the most direct form of standardization, which involves dividing **raw totals** by the areas of enumeration units (e.g., dividing acres of wheat for each county by the area of each county).

Section 4.5 considers the issue of selecting an appropriate visual variable for choropleth mapping, which has traditionally been the most common thematic mapping method. Selecting an appropriate visual variable requires *creating a logical match between the level of measurement of the data and the visual variable (e.g., if data are numerical, the visual variable should appear to reflect the numerical character of the data).*

4.1 SPATIAL ARRANGEMENT OF GEOGRAPHIC PHENOMENA

4.1.1 Spatial Dimension

One way to think about the spatial arrangement of geographic phenomena is to consider their extent or **spatial dimension**. For our purposes, we consider five types of phenomena with respect to spatial dimension: point, linear, areal, $2\frac{1}{2}$-D and true 3-D.

Point phenomena are assumed to have no spatial extent and are thus termed "zero-dimensional." Examples include weather station recording devices, oil wells, and locations of nesting sites for eagles. Locations for point phenomena can be specified in either two- or three-dimensional space; for example, places of religious worship are defined by x and y coordinate pairs (longitude and latitude), whereas nesting sites for eagles are defined by x, y, and z coordinates (the z coordinate would be the height above the earth's surface).

Linear phenomena are one-dimensional in spatial extent, having length, but essentially no width. Examples include a boundary between countries and the path of a stunt plane during an air show. Locations of linear phenomena are defined as an unclosed series of x and y coordinates (in two-dimensional space), or an unclosed series of x, y, and z coordinates (in three-dimensional space).

Areal phenomena are two-dimensional in spatial extent, having both length and width. An example would be a lake (assuming that we focus on its two-dimensional surface extent). Data associated with political units (e.g., counties) can also fit into this framework, because the location of each county can be specified as an enclosed region. In two-dimensional space, areal phenomena are defined by a series of x and y coordinates that completely enclose a region (computer systems generally require that the first coordinate pair equal the last).

When we move into the realm of volumetric phenomena, it is convenient to consider two types: $2\frac{1}{2}$-D and true 3-D. The first of these, **$2\frac{1}{2}$-D phenomena**, can be thought of as a surface, in which geographic location is defined by x and y coordinate pairs and the value of the phenomenon is the height above a zero point (or depth below a zero point). Probably the easiest example to understand is elevation above sea level, because we can actually see the surface in the real world; here height above a zero point is the elevation of the land surface above sea level. A more abstract example would be precipitation

falling over a region over the course of a year; in this case the height of the surface would be the total amount of precipitation for the year.

Another way of thinking about $2\frac{1}{2}$-D surfaces is that they are single-valued in the sense that each x and y coordinate location has a single value associated with it. In contrast, **true 3-D phenomena** are multivalued because each x and y location can have multiple values associated with it. With true 3-D phenomena, any point on the surface is specified by four values: an x coordinate, a y coordinate, a z coordinate (which is the height above, or depth below, sea level), and the value of the phenomenon. Consider mapping the concentration of carbon dioxide (CO_2) in the atmosphere. At any point in the atmosphere, it is possible to define longitude, latitude, height above sea level, and an associated level of CO_2. Color Plate 4.1 illustrates a true 3-D phenomenon: geologic material underneath the earth's surface. Although we should ideally distinguish true 3-D phenomena from $2\frac{1}{2}$-D phenomena, in this book we sometimes refer to a map of either as a *3-D map*.

It is important to realize that map scale plays a major role in determining how we handle the spatial dimension of a phenomenon. For example, on a **small-scale map** (e.g., a page-size map of France) places of religious worship occur at points, but on a **large-scale map** (e.g., a map of a local neighborhood) individual buildings would be apparent, and thus the focus might be on the area covered by the place of worship. Similarly, a river could be considered a linear phenomenon on a small-scale map, but on a large-scale map, the emphasis could be on the area covered by the river.

4.1.2 Models of Geographic Phenomena

The notion of spatial dimension is just one way of thinking about how geographic phenomena are arranged in the real world. Another approach is to consider the arrangement of geographic phenomena along discrete–continuous and abrupt–smooth continua. In this section we define the terms associated with these continua and show how they provide a useful set of *models of geographic phenomena*, a notion developed by Alan MacEachren and David DiBiase (1991).

The terms "discrete" and "continuous" are often used in statistics courses to describe different types of data along a number line; here we consider their use by cartographers in a spatial context. **Discrete phenomena** are presumed to occur at distinct locations (with space in between). Individual people living in a city would be an example of a discrete phenomenon; for an instant in time, a location can be specified for each person, with space between individuals. **Continuous phenomena** occur throughout a geographic region of interest. The examples presented previously for 2½-D phenomena would also be considered continuous phenomena. For instance, when considering elevation, every longitude and latitude position has a value above or below sea level.

Discrete and continuous phenomena can also be described as either abrupt or smooth. **Abrupt phenomena** change suddenly, whereas **smooth phenomena** change in a gradual fashion. This concept is most easily understood for continuous phenomena. The number of electoral votes for each state in the United States would be considered an abrupt continuous phenomenon because although each enumeration unit (a state) has a value, there are abrupt changes at the boundaries between states. In contrast, the distribution of total precipitation over the course of a year for a humid region would be a smooth continuous phenomenon because we would not expect such a distribution to exhibit abrupt discontinuities.

Figure 4.1 provides a graphic portrayal of a variety of models of geographic phenomena that result when we combine the discrete–continuous and abrupt–smooth continua. We'll discuss these models in detail because considering the nature of geographic phenomena is extremely important in selecting an appropriate method of symbolization. First, consider the continuous phenomena shown in the bottom row of Figure 4.1. Percent sales tax is an obvious *abrupt* continuous phenomenon, as it changes suddenly at the boundary between enumeration units (e.g., one state's sales tax is different from another).

In contrast, average farm size is an example of a *smooth* continuous phenomenon because we would expect it to vary in a relatively gradual fashion (as the climate becomes drier, we would expect the average farm size to increase). Average income falls somewhere between percent sales tax and average farm size on the abruptness–smoothness continuum. In some cases, average income would exhibit the abrupt changes of percent sales tax (as at the boundary between urban neighborhoods), while in others it would exhibit a more gradual change (as one moves up a hill toward a region of more attractive views, average income should increase).

In contrast to the bottom row, the top row of Figure 4.1 represents a range of discrete phenomena. The number of employees located at county courthouses is clearly an *abrupt* discrete phenomenon, as there can be only one value for a county and it occurs at an isolated location. In contrast, the number of employed people (based on where they live, as opposed to where they work) is a *smooth* discrete phenomenon, because it gradually changes over geographic space. The number of government employees (again, based on where they live) falls somewhere between these; it might exhibit an abrupt character in the sense that government employees might live near government offices, but it will not exhibit the extreme abruptness of the courthouse example.

The middle row of Figure 4.1 represents phenomena that can be classified as not clearly continuous or discrete, and that also span the abruptness–smoothness continuum. This row is probably most easily understood by considering the influenza case first. Because influenza is an infectious disease, it should exhibit a smooth character. Although individual influenza cases could be represented at discrete locations, it makes sense to suggest some degree of continuity if we wish to stress the potential of infection. At the other end of the row is the number of subscribers to a particular telephone company. Competition between telephone companies could lead to a distribution that changes abruptly but that exhibits continuity between the lines of abrupt change. Finally, the number of people with AIDS is in the middle of the diagram. AIDS occupies a more abrupt position than influenza because of its mode of transmission (sexual intercourse, sharing of needles, and blood transfusions).

4.1.3 Phenomena versus Data

When mapping geographic phenomena, it is important to distinguish between the actual *phenomenon* and the *data* collected to represent that phenomenon. For example, imagine that we wish to map the percentage of forest cover in South Carolina. If we try to visualize the phenomenon, we can conceive of it as smooth and continuous in some portions of the state where the percentage

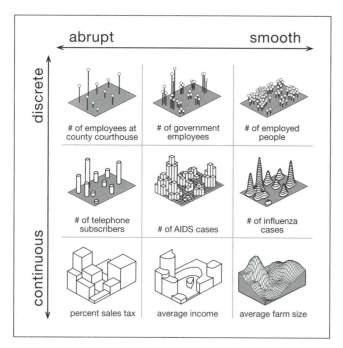

FIGURE 4.1   Models of geographic phenomena arranged along discrete–continuous and abrupt–smooth continua. (After MacEachren 1992, 16; courtesy of North American Cartographic Information Society and Alan MacEachren.)

gradually increases or decreases. In other areas, we can conceive of relatively abrupt changes where the percentage shifts very rapidly (when, say, an urban area is bounded by a hilly forested region).

One form of data that we might use to represent percentage of forest cover would be individual values for counties, which can be found in the state statistical abstract for South Carolina (South Carolina State Budget and Control Board 1994, 45). We might consider mapping these data directly by creating the **prism map** shown in Figure 4.2B. Note that in this case there are abrupt changes at the boundaries of each county. Such a map might be appropriate if we wished to provide a typical

value for each county, but it obviously hides the variation within counties and misleads the reader into thinking that changes take place only at county boundaries.

Potentially, a better approach would be the smooth, continuous map **(fishnet map)** shown in Figure 4.2A; this map indicates that the percentage of forest cover does not coincide with county boundaries, but rather changes in a gradual fashion. A still better map would be one that shows some of the abrupt changes that are likely to occur. Creating such a map would require detailed information about the location of forest within the state, as might be available from a remotely sensed image. Our purpose at this point in the text is not to create the most

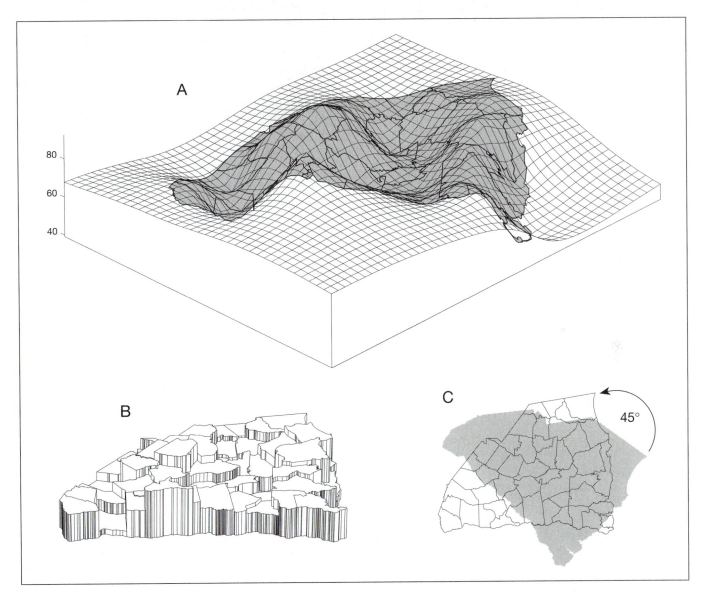

FIGURE 4.2   Approaches for mapping a data set of percentage of forest cover by county for the state of South Carolina: (A) The data are treated as coming from a smooth continuous phenomenon; (B) the data are treated as an areal phenomenon. Map C illustrates that maps A and B have been rotated 45° from a traditional north-oriented map. For A, values outside the state are extrapolated, and thus must be treated with caution. (Data Source: South Carolina State Budget and Control Board 1994.)

representative map of the phenomenon, but to stress that the mapmaker must carefully distinguish between data that have been collected and the phenomenon that is being mapped. Which type of map is used will be a function of both the nature of the underlying phenomenon and the purpose of the map. We consider this issue in greater depth in section 4.4.

4.2 LEVELS OF MEASUREMENT

When a geographic phenomenon is measured to create a data set, we commonly speak of the **level of measurement** associated with the resulting data. Conventionally, four levels of measurement are recognized—nominal, ordinal, interval, and ratio—with each subsequent level including all characteristics of the preceding levels. The **nominal** level of measurement involves grouping (or categorization), but no ordering. The classic example is religion, in which individuals might be identified as Catholic, Protestant, Jewish, or other; here each religious group is different, but one is not more or less in value than another. Another example would be classes on a land use/land cover map; for example, grassland, forest, urban, water, and cropland differ from one another, but one is not more or less in value.

The second level of measurement, **ordinal**, involves categorization plus an ordering (or ranking) of the data. For example, a geologist asked to specify the likelihood of finding oil at each of 50 well sites might be unwilling to provide numerical data, but would feel comfortable specifying a low, moderate, or high potential at each site. Here three categories (low, moderate, and high) are provided, with a distinct ordering among them. Another example of ordinal data would be rankings resulting from a map comparison experiment. Imagine that you constructed dot maps for 10 different phenomena and asked people to compare these maps with another dot map (say, of population) and to rank the maps from "most like" to "least like" the population map. The 10 maps ranked by each person would constitute a distinct ordering, and thus represent ordinal data.

An **interval** level of measurement involves an ordering of the data plus an explicit indication of the numerical difference between two categories. Classic examples are the Fahrenheit and Celsius temperature scales. Consider temperatures of 20°F and 40°F recorded in Fairbanks, Alaska, and Chattanooga, Tennessee, respectively. These two values are ordered, and they reveal the precise numerical difference between the two cities. One characteristic of interval scales is the arbitrary nature of the zero point. In the case of the Celsius scale, 0 is the freezing point for pure water, whereas on the Fahrenheit scale, 0 is the lowest temperature obtained by mixing salt and ice. A result of an *arbitrary zero point* is that ratios of two

interval values cannot be interpreted correctly; for example, 40°F is numerically twice the value of 20°F, but it is not twice as warm (in terms of the kinetic energy of the molecules). An example of an interval scale familiar to academics is SAT scores, which range from a minimum of 200 to a maximum of 800. Note that it is not possible to say that an individual scoring 800 on an SAT exam did four times better than an individual scoring 200; all that can be said is that the individual scored 600 points better. A geographical example of interval-level data is elevation, where the establishment of mean sea level represents an arbitrary zero point.

A **ratio** level of measurement has all the characteristics of the interval level, plus a *nonarbitrary zero point*. Continuing with the temperature example, the Kelvin scale is ratio in nature because at 0°K all molecular motion ceases; thus, a temperature of 40°K is twice as warm as 20°K (in terms of the kinetic energy of the molecules). Ratio data sets are more common than interval ones. For example, a perusal of maps shown in this text will reveal that most are based on ratio-level data. Because many symbol forms can be used with both interval and ratio scales, these two levels of measurement are often grouped together and referred to as **numerical data**. The basic scales that we have discussed can also be divided into *qualitative* (nominal data) and *quantitative* (ordinal, interval, and ratio data) scales.

The four levels of measurement we have considered are the only ones normally covered in geographic textbooks. Nicholas Chrisman (1998; 2002, 25–33) argues that these are insufficient for working with geographic data, and so has proposed several extensions. One is to create a separate level of measurement for data sets that are constrained to a fixed set of numbers, such as probability (the range is 0 to 1) or percentages (the range is 0 to 100). Chrisman terms this an *absolute* level of measurement, as no transformations are possible that could retain the meaning of measurement. We instead use the term **constrained ratio** recommended by Forrest (1999b).

A second extension proposed by Chrisman is the **cyclical** level of measurement, which is appropriate for phenomena that have a cyclical character. For instance, angular measurements have a cycle of 360°, with angle *x* as far from 0° as 360°–*x*. This notion is not dealt with in the linear unbounded number line associated with ratio measurements. As another example, the seasons are cyclical and can be specified with different starting points—we could specify spring–summer–fall–winter–spring or fall–winter–spring–summer–fall. A third extension is the notion of **counts**, in which individual objects, such as people, are counted. Although counts have a nonarbitrary zero point, Chrisman argues that they cannot be rescaled as readily as ratio scales because we cannot conceive of a fraction of an object (e.g., half of a person).

A final extension is the notion of **fuzzy categories**.* Normally, we think of individual items as falling wholly within a particular nominal category; for instance, an individual is either a Protestant or not (in terms of church membership). In practice, category memberships are often fuzzy, as it might not be entirely clear whether an item is within a particular category. A good illustration of fuzzy categories is some remote sensing classification procedures, which produce a probability that each pixel falls in a particular land use (e.g., a pixel might have an 85 percent probability of being wheat and a 15 percent probability of being corn). Increasingly race and ethnicity can be considered "fuzzy" in that individuals identify themselves with more than one race. In fact, the 2000 Census, for the first time, allowed for multiple ethnicity.

Another extension to the basic levels of measurement is the three kinds of numerical data proposed by J. Ronald Eastman (1986): bipolar, balanced, and unipolar. **Bipolar data** are characterized by either natural or meaningful dividing points. A *natural* dividing point is inherent to the data and can be used intuitively to divide the data into two parts. An example would be a value of 0 for percentage of population change, which would divide the data into positive and negative percent changes. A *meaningful* dividing point does not occur inherently in the data, but can logically divide the data into two parts. An example would be the mean of the data, which enables differentiating values above and below the mean. **Balanced data** are characterized by two phenomena that coexist in a complementary fashion. An example is the percentage of English and French spoken in Canadian provinces—a high percentage of English-speaking people implies a low percentage of French-speaking people (the two are in "balance" with one another). **Unipolar data** have no natural dividing points and do not involve two complementary phenomena. Per capita income associated with countries of Africa or states of the United States would be an example of unipolar data.

4.3 VISUAL VARIABLES

The term **visual variables** is commonly used to describe the various perceived differences in map symbols that are used to represent geographic phenomena. The notion of visual variables was developed by the French cartographer Jacques Bertin (1983) and subsequently modified by others, including McCleary (1983), Morrison (1984), DiBiase et al. (1991), and MacEachren (1994a). Our approach is similar to that of MacEachren, but differs primarily in the inclusion of $2\frac{1}{2}$-D and true 3-D phenomena and the use of the perspective-height visual variable.

---
* Chrisman terms this extension "graded membership in categories."

In this chapter, we consider only visual variables for static maps. Additional visual variables for animated maps and for depicting uncertainty are covered in Chapters 20 and 23, respectively. In Chapter 25, we consider **abstract sound variables**, which utilize sound to communicate spatial information (e.g., a louder tone at a particular map location could represent a greater magnitude of the phenomenon at that location). We will find that abstract sound variables can be especially useful for visually impaired map users.

The visual variables that we discuss are illustrated in Figure 4.3 and Color Plate 4.2. Note that the visual variables appear in the rows, and the columns represent the dimensions of spatial phenomena discussed in the preceding section. In discussing the visual variables, we sometimes need to distinguish between the overall *symbol* and the *marks* making up the symbol. For example, note that the spacing visual variable shown for point phenomena consists of circular symbols, and that each circle is composed of parallel horizontal marks.

4.3.1 Spacing

The **spacing** visual variable involves changes in the distance between the marks making up the symbol (Figure 4.3). Cartographers traditionally have used the term *texture* to describe these changes (e.g., Castner and Robinson 1969), but we use the term spacing because texture has varied usages in the literature.

4.3.2 Size

Cartographers have used size as a visual variable in two different ways. One has been to change the size of the entire symbol, as is shown for the point and linear phenomena (Figure 4.3). Another is to change the size of individual marks making up the symbol, as for the areal, $2\frac{1}{2}$-D, and true 3-D phenomena. This inconsistency might be a bit confusing, but the term *size* seems to reflect the visual differences that arise in each case. Note that for areal phenomena, the size of the entire areal unit could also be changed, as is done on cartograms, to be discussed in Chapter 19.

4.3.3 Perspective Height

Perspective height refers to a perspective 3-D view of a phenomenon (Figure 4.3). It is interesting to consider some of the potential applications of this visual variable. In the case of point phenomena, oil production at well locations might be represented by raised sticks (or lollipops) above each well, with the stick height proportional to well production. For linear phenomena, total traffic flow between two cities over some time period

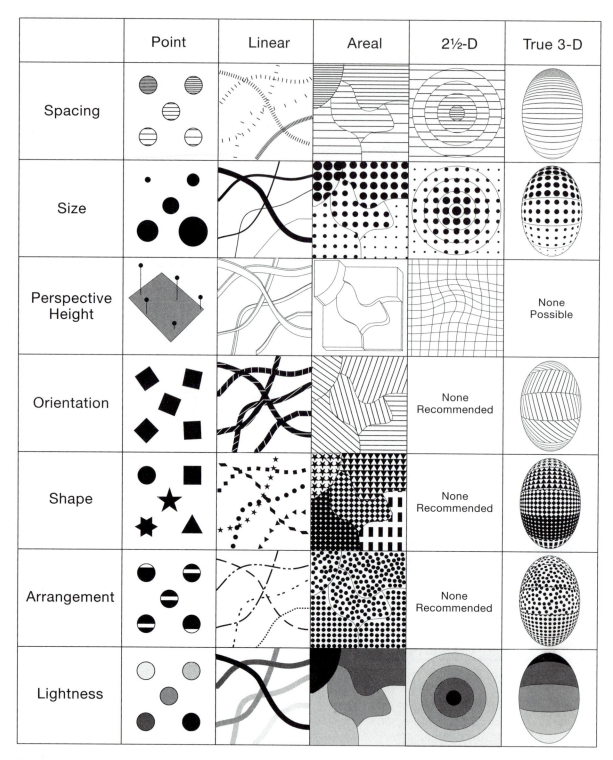

FIGURE 4.3   Visual variables for black-and-white maps. For visual variables for color maps, see Color Plate 4.2.

could be represented by a fencelike structure above each roadway, with the height of the "fence" proportional to traffic flow. In the case of areal and 2½-D phenomena, we have already discussed examples for the forest cover data in South Carolina (Figure 4.2). Perspective height cannot be used for true 3-D phenomena because three dimensions are needed to locate the phenomenon being mapped.

4.3.4 Orientation and Shape

As with the size visual variable, the character of the **orientation** visual variable is a function of the kind of spatial phenomena. For linear, areal, and true 3-D phenomena, orientation refers to the direction of individual marks making up the symbol. In contrast, for point phenomena, orientation refers to the direction of the entire point symbol (Figure 4.3). (Marks of differing direction could be applied to point symbols, but the small size of point symbols often makes it difficult to see the marks.) Because orientation is most appropriate for representing nominal data, we do not recommend using it for $2^{1}/_{2}$-D phenomena, which are inherently numerical. Note that the **shape** visual variable is handled in a fashion similar to orientation.

4.3.5 Arrangement

Understanding the **arrangement** visual variable requires a careful examination of Figure 4.3. For areal and true 3-D phenomena, note that arrangement refers to how marks making up the symbol are distributed; marks for some areas are part of a square arrangement, whereas marks for other areas appear to be randomly placed. For linear phenomena, arrangement refers to splitting lines into a series of dots and dashes, as might be found on a map of political boundaries. Finally, for point phenomena, arrangement refers to changing the position of the white marker within the black symbol.

4.3.6 Hue, Lightness, and Saturation

The visual variables hue, lightness, and saturation are commonly recognized as basic components of color.* **Hue** is the dominant wavelength of light making up a color (the notion of wavelengths of light and the associated electromagnetic spectrum will be considered in detail in Chapter 10). In everyday life, hue is the parameter of color most often used; for example, you might note that one person has on a red shirt and another a blue shirt. Color Plate 4.2 illustrates how various hues can be used to depict spatial phenomena.

Lightness (or **value**) refers to how dark or light a particular color is, while holding hue constant; for example, in Color Plate 4.2, different lightnesses of a green hue are shown. Lightness also can be shown as shades of gray (in the absence of what we commonly would call color), as in Figure 4.3.

Saturation (or **chroma**) can be thought of as a mixture of gray and a pure hue. It is the intensity of a color; for instance, we might speak of different intensities of colorful

---

* See Brewer (1994a) for a discussion of terminology associated with color.

shirts. This concept is illustrated in Color Plate 4.3, where the areal symbols shown for saturation in Color Plate 4.2 are arranged along a continuum from a desaturated red (grayish red) to a fully saturated red (while holding lightness constant).

4.3.7 Some Considerations in Working with Visual Variables

You should bear in mind that Figure 4.3 and Color Plate 4.2 depict only a fraction of the many symbols that could be used to depict the visual variables; for example, either circles or squares might be used to depict point phenomena for the size visual variable. A major group of symbols not shown in the figures are **pictographic symbols**, which are intended to look like the phenomenon being mapped (as opposed to **geometric symbols** such as circles). For instance, Figure 4.4 illustrates the use of different-sized beer mugs to represent the number of microbreweries and brewpubs in each U.S. state. Pictographic symbols are often used in children's atlases.

Also keep in mind that the visual variables can serve as basic building blocks of more complex representations. For example, Figure 4.5 illustrates how the visual

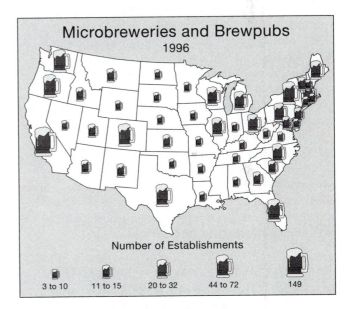

FIGURE 4.4 Using a pictographic visual variable (beer mugs) to represent the number of microbreweries and brewpubs in each U.S. state. (For similar data, see *http://brewpubzone.com.*)

FIGURE 4.5 Combining the visual variables spacing and size. (After MacEachren 1994a, p. 26.)

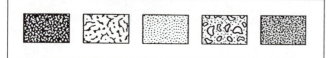

FIGURE 4.6 Different "patterns" or "textures" that can be created to portray nominal information. Note that these are not readily described in terms of the visual variables shown in Figure 4.3.

variables spacing and size might be combined. MacEachren (1994a, 27) called the resulting symbol a form of texture. In our opinion, such an approach produces a rather coarse-looking map, but it can highlight various aspects of a distribution (for an example, see Figure 2.19 of MacEachren's work).

In Figure 4.3 and Color Plate 4.2, the term for each visual variable used appears to be a clear expression of the visual differences that we see; for example, in the case of the orientation visual variable for point phenomena, we see that one square is at a different orientation than another. Moreover, if we wanted, we could compute a mathematical expression of this difference (that one square is rotated 40° from a vertical, whereas another is rotated 50°). Sometimes, describing the visual difference between symbols is not so easy. For example, try describing the differences between the symbols shown in Figure 4.6. Such symbols are often referred to as differing in "pattern" or "texture" and are frequently used to symbolize nominal data.

It also should be noted that the visual variable location was not explicitly depicted in the illustrations. **Location** refers to the position of individual symbols. We chose not to illustrate this visual variable because it is an inherent part of mapping (e.g., each symbol shown for point phenomena can be defined by the *x* and *y* coordinate values of its center). If location were illustrated, it would be represented by constant symbols (identical dots for point phenomena) that varied only in position.

4.4 COMPARISON OF CHOROPLETH, PROPORTIONAL SYMBOL, ISOPLETH, AND DOT MAPPING

In this section, we define and contrast four common thematic mapping techniques: choropleth, proportional symbol, isopleth, and dot. For illustrative purposes, we examine these techniques by mapping data for acres of wheat harvested in counties of Kansas (Table 4.1). Selecting an appropriate technique is a function of both the nature of the underlying phenomenon and the purpose for making the map. Here we consider a basic introduction to these mapping techniques; more advanced concepts are covered in

subsequent chapters (13 for choropleth, 16 for proportional symbol, 14 for isopleth, and 17 for dot).

4.4.1 Choropleth Mapping

A **choropleth map** is commonly used to portray data collected for enumeration units, such as counties or states. To construct a choropleth map, data for enumeration units are typically grouped into classes and a gray tone or color is assigned to each class. The choropleth map is clearly appropriate when values of a phenomenon change abruptly at enumeration unit boundaries, such as for state sales tax rates. Choropleth maps might also be appropriate when you want the map reader to focus on "typical" values for individual enumeration units, even though the underlying phenomenon does not change abruptly at enumeration unit boundaries. For example, politicians and government officials might use this approach when stressing how one county or state compares with another. Although choropleth maps are commonly used in this fashion, it is important to recognize two major limitations: (1) Such maps do not portray the variation that might actually occur within enumeration units, and (2) the boundaries of enumeration units are arbitrary, and thus unlikely to be associated with major discontinuities in the actual phenomenon.*

An important consideration in constructing choropleth maps is the need for **data standardization**, in which **raw totals** are adjusted for differing sizes of enumeration units. To understand the need to standardize, consider map A of Figure 4.7, which portrays a hypothetical distribution consisting of three distinct regions: S, T, and U. Note that regions S and T have equal-sized enumeration units, each 16 acres in size. In contrast, region U has enumeration units four times the size of those in S and T, or 64 acres in size.

Let's presume that the number of acres of wheat harvested from enumeration units in each region is as follows: 0 in S, 16 in T, and 64 in U (these numbers are shown within each enumeration unit in Figure 4.7A). The acres of wheat harvested from each enumeration unit represent raw totals. Mapping these raw totals with the choropleth method produces the result shown in Figure 4.7B (note that higher data values are depicted by a darker gray tone). A user examining this map would likely conclude that because region U is the darkest, it must have more wheat grown in it. Unfortunately, this conclusion would be inappropriate because we have not accounted for the size of enumeration units. One approach to adjust (or standardize) for the size of enumeration units is to divide each raw total by the area of the associated enumeration unit; the resulting values are 0/16, or 0 for region S; 16/16, or 1 for

---

* See Langford and Unwin 1994 for a more detailed discussion of these limitations.

TABLE 4.1 Wheat harvested in Kansas counties 1993

County	Acres Harvested	Area of County (in acres)	% of Land in Wheat
Allen	35200	324735	10.8
Anderson	36600	373075	9.8
Atchison	12300	278937	4.4
Barber	150200	727904	20.6
Barton	199400	576390	34.6
Bourbon	14800	409210	3.6
Brown	26600	365894	7.3
Butler	80700	926707	8.7
Chase	15000	497369	3.0
Chautauqua	15500	412610	3.8
Cherokee	78000	377881	20.6
Cheyenne	147800	651910	22.7
Clark	93100	626576	14.9
Clay	99400	420435	23.6
Cloud	107700	461204	23.4
Coffey	31400	419001	7.5
Comanche	92000	506214	18.2
Cowley	125600	725741	17.3
Crawford	41100	380157	10.8
Decatur	133200	572289	23.3
Dickinson	153600	546094	28.1
Doniphan	9000	254242	3.5
Douglas	15700	304118	5.2
Edwards	108400	397693	27.3
Elk	11600	416442	2.8
Ellis	132800	575971	23.1
Ellsworth	97700	463260	21.1
Finney	224800	833658	27.0
Ford	226100	701190	32.2
Franklin	14500	369251	3.9
Geary	17700	259175	6.8
Gove	162400	685862	23.7
Graham	96400	575048	16.8
Grant	123300	368538	33.5
Gray	190700	555865	34.3
Greeley	175900	498307	35.3
Greenwood	11100	736333	1.5
Hamilton	160000	637046	25.1
Harper	281100	512292	54.9
Harvey	125000	345324	36.2
Haskell	126800	369004	34.4
Hodgeman	136900	551464	24.8
Jackson	17500	421717	4.1
Jefferson	8900	354557	2.5
Jewell	126300	585604	21.6
Johnson	11300	306957	3.7
Kearny	144100	557439	25.9
Kingman	216800	554529	39.1
Kiowa	116300	464185	25.1
Labette	66200	421057	15.7
Lane	155200	459018	33.8
Leavenworth	11400	303572	3.8
Lincoln	106000	460879	23.0
Linn	16000	387873	4.1
Logan	179800	688192	26.1
Lyon	16000	549098	2.9
Marion	142200	610675	23.3
Marshall	92500	578662	16.0
Mcpherson	226300	576316	39.3
Meade	121000	627167	19.3
Miami	10800	378094	2.9

Continued

TABLE 4.1 Continued

County	Acres Harvested	Area of County (in acres)	% of Land in Wheat
Mitchell	150000	459022	32.7
Montgomery	54000	413768	13.1
Morris	46000	449145	10.2
Morton	78100	466625	16.7
Nemaha	31000	459537	6.7
Neosho	45000	369610	12.2
Ness	171600	687891	24.9
Norton	139400	563491	24.7
Osage	19500	459872	4.2
Osborne	107400	572513	18.8
Ottawa	103400	460748	22.4
Pawnee	153400	482433	31.8
Phillips	103700	572563	18.1
Pottawatomie	17300	551123	3.1
Pratt	163100	469889	34.7
Rawlins	160100	686637	23.3
Reno	292500	813939	35.9
Republic	96600	459717	21.0
Rice	160400	466673	34.4
Riley	17800	397742	4.5
Rooks	71500	573454	12.5
Rush	154000	459719	33.5
Russell	98100	576143	17.0
Saline	121500	459756	26.4
Scott	170500	459183	37.1
Sedgwick	262300	646618	40.6
Seward	98400	410803	24.0
Shawnee	14500	354420	4.1
Sheridan	120300	573394	21.0
Sherman	185400	675725	27.4
Smith	126000	573619	22.0
Stafford	160200	508439	31.5
Stanton	134100	436392	30.7
Stevens	101600	465830	21.8
Sumner	443700	758976	58.5
Thomas	235400	687149	34.3
Trego	119500	575635	20.8
Wabaunsee	15000	511601	2.9
Wallace	123800	584778	21.2
Washington	88000	575534	15.3
Wichita	152300	460170	33.1
Wilson	52400	367770	14.2
Woodson	18800	322672	5.8
Wyandotte	700	99196	0.7

region T; and 64/64, or 1 for region U. Mapping these values with the choropleth method results in Figure 4.7C; note that regions T and U are now identically shaded and thus have the same density of wheat harvested.

In a similar fashion, the Kansas wheat data can be standardized by dividing the acres harvested within each county by the area of the corresponding county. Because the wheat harvested data and the area of the county are both in acres, dividing the two produces a proportion (or percentage when we multiply the result by 100; see Table 4.1). Figure 4.8 portrays maps of both the unstandardized and standardized Kansas wheat data. The difference between the maps is not as dramatic as the hypothetical example because the areas of counties in Kansas are similar in size, but we can still note areas where the patterns differ.

Standardization not only adjusts for the differing sizes of counties but also provides a very useful attribute, namely, the proportion (or percentage) of each county from which wheat was harvested. Such an attribute provides an indication of the probability that one might see wheat being cut at harvest time while driving through the county. The standardized map is shown in Figure 4.9A for comparison with the other methods of mapping discussed in this section. (Note that the *lightness* visual variable has been used for the choropleth map.)

4.4.2 Proportional Symbol Maps

A **proportional symbol map** is constructed by scaling symbols in proportion to the magnitude of data occurring at

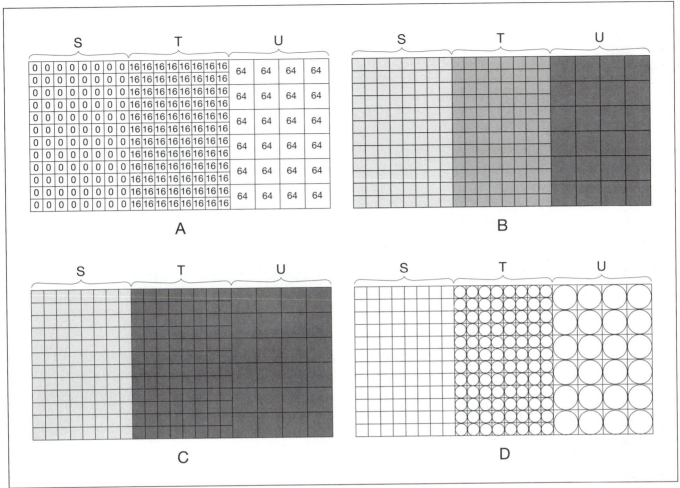

FIGURE 4.7 A hypothetical illustration of the effect of data standardization: (A) raw totals—number of acres of wheat harvested in each enumeration unit; (B) a choropleth map of the raw totals; (C) a choropleth map of standardized data achieved by dividing the raw totals by the area of the corresponding enumeration unit; and (D) a proportional circle map of the raw totals.

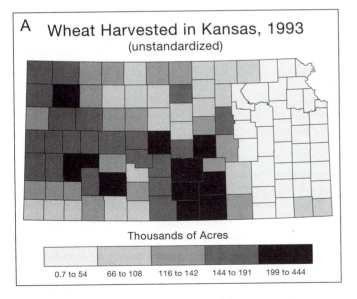

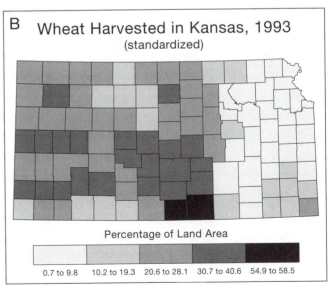

FIGURE 4.8 Standardizing wheat harvested in Kansas counties in 1993: (A) a map of the number of acres harvested; and (B) a standardized map resulting from dividing number of acres harvested by the area of each county. (Data Source: Kansas Agricultural Statistics 1994.)

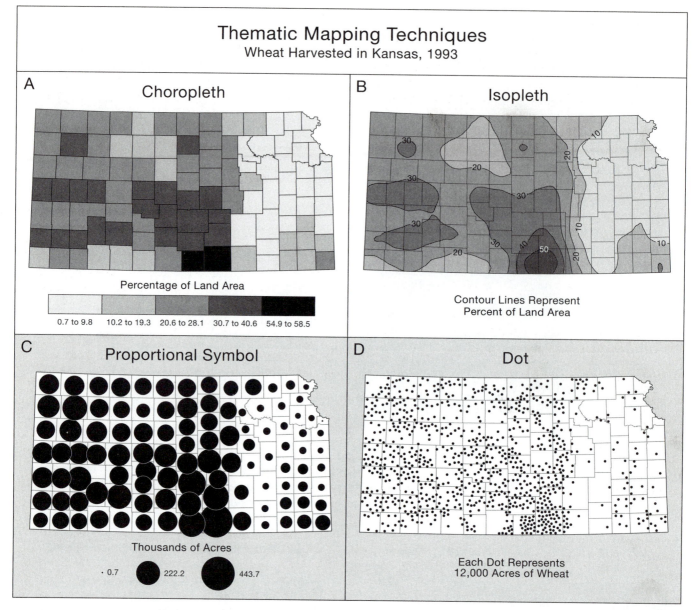

FIGURE 4.9   A comparison of basic thematic mapping techniques: (A) choropleth, (B) isopleth, (C) proportional symbol, and (D) dot maps. The choropleth and isopleth maps are based on the percentage of land area from which wheat was harvested, whereas the proportional symbol and dot maps are based on the total acres of wheat harvested.

point locations. These locations can be *true points,* such as an oil well, or *conceptual points,* such as the center of an enumeration unit for which data have been collected; the latter is the case with the wheat harvested data. In contrast to the standardized data depicted on choropleth maps, proportional symbol maps are normally used to display raw totals. Thus, the magnitudes for acres of wheat harvested are depicted as proportional circles in Figure 4.9C. (Note that the visual variable used here is *size.*)

The raw totals depicted on proportional symbol maps provide a useful complement to the standardized data shown on choropleth maps. Raw totals are important

because a high proportion or rate might not be meaningful if there is not also a high raw total. As an example, consider counties of the same size having populations of 100 and 100,000, in which 1 and 1,000 people, respectively, have some rare form of cancer. Dividing the number of cancer cases by the population yields the same proportion of people suffering from cancer (0.01), but the rate for the less populous county would be of lesser interest to the epidemiologist.

Although the proportional symbol map is a better choice than the choropleth map for depicting raw totals, care should be taken in using it. To illustrate, consider map D of Figure 4.7, which displays the hypothetical wheat data using proportional circles. Note that all circles in region U are larger than those in region T. This could lead to the mistaken impression that counties in region U are more important in terms of wheat production than those in region T. Counties in region U might be more important to a politician in assigning tax dollars (more wheat harvested indicates a greater tax is appropriate), but in terms of the density of wheat harvested, regions T and U are identical.

4.4.3 Isopleth Map

An **isarithmic map** (or **contour map**) is created by interpolating a set of isolines between sample points of known values; for example, we might draw isolines between temperatures recorded for individual weather stations. The **isopleth map** is a specialized type of isarithmic map in which the sample points are associated with enumeration units. It is an appropriate alternative to the choropleth map when one can assume that the data collected for enumeration units are part of a smooth continuous ($2\frac{1}{2}$-D) phenomenon. For example, in the case of the wheat data, it might be argued that the proportion of land in wheat changes in a relatively gradual (smooth) fashion, as opposed to changing just at county boundaries (as on the choropleth map).

In a fashion similar to a choropleth map, an isopleth map also requires standardized data. Referring again to the hypothetical raw totals shown in Figure 4.7A, imagine drawing contours through such data. High-valued contour lines would tend to occur in region U, where there are high values in the data; but as has already been shown for the choropleth case, region U is really no different from region T. Dividing the raw totals by the area of each enumeration unit would result in standardized data that could be appropriately contoured.

The isopleth map resulting from contouring the standardized Kansas wheat data is shown in Figure 4.9B. (Again, note that the visual variable lightness has been used.) Although this map might be more representative of the general distribution of wheat harvested than the choropleth map, the assumption of continuity and the use of county-level data produce some questionable results. For example, note the island of higher value near the center of the extreme southeastern county (Cherokee). In reality, it seems unlikely that you would find a higher value here; the high value is more likely a function of the fact that the centers of counties were used as a basis for contouring and Cherokee's value was higher than any of the surrounding counties. Note that a similar problem occurs within two northern counties (Figure 4.9B). The dot map could be a solution to this type of problem.

4.4.4 Dot Mapping

To create a **dot map**, one dot is set equal to a certain amount of a phenomenon, and dots are placed where that phenomenon is most likely to occur. The phenomenon might actually cover an area or areas (e.g, a field or fields of wheat), but for the sake of mapping, the phenomenon is represented as located at points. Constructing an accurate dot map requires collecting ancillary information that indicates where the phenomenon of interest (wheat, in our case) is likely found. For the wheat data, this was accomplished using the cropland category of a land use/land cover map (the detailed procedures are described in Chapter 17). The resulting dot map is shown in Figure 4.9D. (In this case, the visual variable *location* is used.) Clearly, the dot map is able to represent the underlying phenomenon with much more accuracy than any of the other methods we have discussed. Also note that parts of the distribution exhibit sharp discontinuities that would be difficult to show with the isopleth method (which presumes smooth changes).

4.4.5 Discussion

An examination of Figure 4.9 reveals that each of the four maps provides a quite different picture of wheat harvested in the state of Kansas. Which method is used should depend on the purpose of the map. If the purpose is to focus on "typical" county-level information, then the choropleth and proportional symbol maps are appropriate. The choropleth map provides standardized information, whereas the proportional symbol map provides raw total information. It must be emphasized that neither map depicts the detail of the underlying phenomenon, which is unlikely to follow enumeration unit boundaries.

When data are collected in the form of enumeration units, the dot and isopleth methods should be considered as two possible solutions for representing an underlying phenomenon that is not coincident with enumeration unit boundaries. In the case of the wheat data, the dot method is probably the more appropriate approach because it can capture some of the discontinuities in the phenomenon. The isopleth method, however, could probably be improved on with a finer grid of enumeration units (e.g., townships);* of course, this would also be true of the choropleth and proportional symbol maps.

It must be noted that we have only considered four of the more common methods of thematic mapping. One alternative would be a **dasymetric map**, which, like the dot map, can show very detailed information, but uses standardized data. We will cover the dasymetric map in

---

* Data at the township level are not released to the general public to protect the confidentiality of individual farm production.

Chapter 17. Another alternative would be to modify the proportional symbol map by making the area of the circle that is filled in proportional to the percent of land area from which wheat is harvested—this creates what is called a **pie chart**. Finally, we should keep in mind that if maps are to be viewed in an interactive graphics environment, the mapmaker will have the option of showing several of them, thus providing the user with various perspectives on the distribution of wheat harvested in Kansas.

4.5 SELECTING VISUAL VARIABLES FOR CHOROPLETH MAPS

In the preceding section, the visual variable lightness was utilized for the choropleth map. An examination of Figure 4.3 and Color Plate 4.2 reveals that there are a number of other visual variables that might be used to represent a phenomenon that is treated as areal in nature. This section considers how we might select among these visual variables. The basic solution is to select a visual variable that appears to "match" the level of measurement of the data. For illustrative purposes, we again use the Kansas wheat data.

The specific visual variables we discuss are illustrated in Figure 4.10 and Color Plate 4.4. In examining these figures, note that they depict classed maps using *maximum-contrast symbolization,* which means that symbols for classes have been selected so that they are maximally differentiated from one another. An alternative approach would be to create an unclassed map in which symbols are directly proportional to the value for each enumeration unit (as in Figure 1.6B). The maximum-contrast approach is used here because it is common and more easily constructed (particularly in the case of the size visual variable).

In addition to discussing Figure 4.10 and Color Plate 4.4, we also consider Figure 4.11, which summarizes the use of visual variables for various levels of measurement. Note that the body of this figure is shaded and labeled to indicate various levels of acceptability: Poor (P), Marginally effective (M), and Good (G). MacEachren (1994a, 33) developed a similar figure, which he appeared to apply to all kinds of spatial phenomena. We use Figure 4.11 only for areal phenomena; as an exercise, you might consider developing such a figure for other kinds of phenomena.

We'll consider the perspective height and size visual variables first because they have the greatest potential for logically representing the numerical data depicted on choropleth maps. Use of perspective height produces what is commonly termed a *prism map* (Figure 4.10A). In Figure 4.11, note that perspective height is the only visual variable receiving a "good" rating for numerical data. The justification is that an unclassed map based on perspective height can portray ratios correctly (a data value twice as large as another will be represented by a prism

twice as high), and that readers perceive the height of resulting prisms as ratios (Cuff and Bieri 1979).

There are two problems, however, that complicate the extraction of numerical information from prism maps. One is that tall prisms sometimes block smaller prisms. A solution to this problem is to rotate the map so that blockage is minimized; for example, the map in Figure 4.10A has been rotated so that the view is from the lower valued northeast. A second solution to the blockage problem is to manipulate the map in an interactive graphics environment. If a flexible program is available, it might even be possible to suppress selected portions of the distribution so that other portions can be seen. A third solution is to use the perspective height variable but also symbolize the distribution with another visual variable; for example, Figure 4.10D might be displayed in addition to Figure 4.10A.

Another problem with prism maps is that rotation might produce a view that is unfamiliar to readers who normally see maps with north at the top. This problem can be handled by showing a second map (as suggested earlier) or by using an overlay of the base to show the amount of rotation (as in Figure 4.2C).

The size visual variable is illustrated in Figure 4.10B; note that here the size of individual marks making up the areal symbol has been varied. Size can be considered appropriate for representing numerical relations because circles can be constructed in direct proportion to the data (a data value twice another can be represented by a circle twice as large in area). Furthermore, readers should see the circles in approximately the correct relations. (However, we will see in Chapter 16 that a correction factor might have to be implemented to account for underestimation of larger circles.)

Although some cartographers (most notably Bertin) have used this sort of argument to promote the use of the visual variable size on choropleth maps, two problems are apparent. First, it is questionable whether map users actually consider the sizes of circles when used as part of an areal symbol. Users might analyze circle size when trying to acquire specific information, but it seems unlikely that they would do so when analyzing the overall map pattern. Rather, it is more likely that they would perceive areas of light and dark, in a fashion similar to the lightness visual variable. Second, many cartographers (and presumably map users) find the coarseness of the resulting symbols unacceptable—they would prefer the fine tones shown in Figure 4.10D. The latter problem in particular caused us to give the size variable only a moderate rating for portraying numerical data (Figure 4.11).

Note also that we have given both perspective height and size only moderate ratings for portraying ordinal data. The logic is that if such variables are used to illustrate numerical relations, users might perceive such relations when only ordinal relations are intended.

Visual Variables
Wheat Harvested in Kansas, 1993

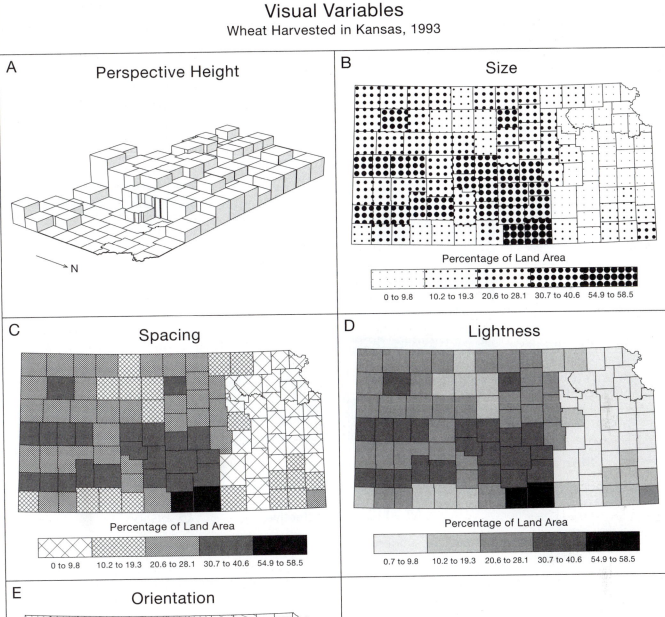

A Perspective Height

→ N

B Size

Percentage of Land Area

0 to 9.8   10.2 to 19.3   20.6 to 28.1   30.7 to 40.6   54.9 to 58.5

C Spacing

Percentage of Land Area

0 to 9.8   10.2 to 19.3   20.6 to 28.1   30.7 to 40.6   54.9 to 58.5

D Lightness

Percentage of Land Area

0.7 to 9.8   10.2 to 19.3   20.6 to 28.1   30.7 to 40.6   54.9 to 58.5

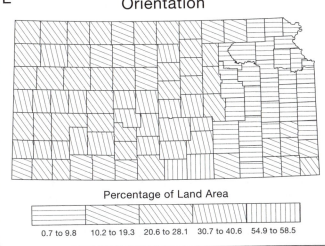

E Orientation

Percentage of Land Area

0.7 to 9.8   10.2 to 19.3   20.6 to 28.1   30.7 to 40.6   54.9 to 58.5

FIGURE 4.10 Visual variables for representing the percentage of wheat harvested in Kansas counties: (A) perspective height, (B) size, (C) spacing, (D) lightness, and (E) orientation. For color visual variables, see Color Plate 4.4.

	Nominal	Ordinal	Numerical
Spacing	P	M[c]	M[c]
Size	P	M	M
Perspective Height	P	M[a]	G[b]
Orientation	G	P	P
Shape	G	P	P
Arrangement	G	P	P
Lightness	P	G	M
Hue	G	G[d]	M[d]
Saturation	P	M	M

P = Poor   **M** = Marginally Effective   **G** = Good

[a] Since height differences are suggestive of numerical differences, use with caution for ordinal data.

[b] Hidden enumeration units and lack of a north orientation are problems.

[c] Not aesthetically pleasing.

[d] The particular hues selected must be carefully ordered, such as yellow, orange, red.

FIGURE 4.11 Effectiveness of visual variables for each level of measurement for areal phenomena. (After MacEachren 1994a, 33.)

Obviously, both variables are inappropriate for nominal data because different heights and sizes suggest quantitative rather than qualitative information.

Although other visual variables can be manipulated mathematically to create proportional (ratio) relationships, the resulting symbols cannot be interpreted easily in a ratio fashion. For example, consider the lightness variable shown in Figure 4.10D. It is easy to see that one shade is darker or lighter than another, but it is difficult to establish proportional relations (that one shade is twice as dark as another). Similar comments can be made for the visual variables spacing, saturation, and hue, with

the following caveats. First, note that we have given spacing only a moderate rating for ordinal information because, in our opinion, the symbols are not aesthetically pleasing, and there is the implication that low data values are qualitatively different from high data values. Second, we have given saturation only a moderate rating for ordinal information because it is our experience that people have a difficult time understanding what a "greater" saturation means.

We have rated hue as "good" for both nominal and ordinal data because some hues work well for nominal data, and other hues work better for ordinal data. For example, to display different soil types (alfisols, entisols, mollisols), red, green, and blue hues might be deemed appropriate (one of these hues does not inherently represent more than another). For ordinal and higher level data, logically ordered hues are necessary; for example, a yellow, orange, and red scheme (Color Plate 4.4B) is one possibility because orange is seen as a mixture of yellow and red (based on opponent process theory, a topic to be covered in Chapter 10).

The remaining visual variables (orientation, shape, and arrangement) are only appropriate for creating nominal differences. As an example, consider the orientation variable. You might try to create a logical progression of symbols by starting with horizontal lines for low values and then gradually changing the angle of the lines so that the highest class is represented by vertical lines, but an examination of Figure 4.10E reveals that this approach is not effective; the changing angle of the lines appears to create nominal differences, and the resulting map is "busy."

It should be noted that cartographers are not in complete agreement on the ratings displayed in Figure 4.11, so you might develop slightly different ratings. For example, one of our students rated the orientation variable "poor" (even for nominal data) because he felt it lacked aesthetic quality and that it was difficult to discriminate among different orientations.

SUMMARY

In this chapter, we have covered basic principles for symbolizing geographic phenomena. We have discovered that the spatial arrangement of the underlying phenomenon is an important consideration in selecting an appropriate symbology. For example, if the underlying phenomenon is **smooth** and **continuous** (e.g., yearly snowfall for Russia), then a **contour map** would be appropriate, but a **choropleth map** would be inappropriate.

Another important consideration in selecting symbology is the **level of measurement** of the data. Ideally, there should be a logical match between the level of measurement and the symbology (or **visual variable**) used to represent the data. For instance, if data are numerical (e.g., the magnitude of electrical generation at power plants in kilowatt hours), then the symbology should be capable of enabling a map reader to visualize numerical relations (e.g., a **proportional symbol map** would be appropriate) and in certain cases to obtain exact data values. Keep in mind, however, that we generally do not expect readers to acquire precise numerical information from maps; rather maps are primarily used to show spatial patterns.

Although the underlying phenomenon is an important consideration in selecting symbology, we have seen that map purpose can also play an important role. For example, if the mapmaker wishes to show "typical" values for enumeration units, then a choropleth map might be appropriate even when the underlying phenomenon is not coincident with enumeration units, as was the case with wheat harvested in Kansas. The mapmaker should realize, however, that in this instance a choropleth map might lead to incorrect perceptions of the underlying phenomenon.

We have also learned that for some mapping methods **data standardization** is important, that **raw totals** need to be adjusted to account for the area over which the data have been collected (typically an enumeration unit). The simplest form of adjustment is to divide raw totals by the areas of enumeration units (e.g., we could divide the number of people in counties by the areas of counties to create a map of population density). In subsequent chapters, we look at other forms of standardization.

FURTHER READING

Bertin, J. (1983) *Semiology of Graphics: Diagrams, Networks, Maps.* Madison, WI: University of Wisconsin Press. (Translated by W. J. Berg)

> Chapter 2 of this widely cited text focuses on visual variables. The text is a bit difficult to read as it has been translated from French to English.

Brewer, C. A. (1994a) "Color use guidelines for mapping and visualization." In *Visualization in Modern Cartography,* ed. by A. M. MacEachren and D. R. F. Taylor, pp. 123–147. Oxford: Pergamon.

> Pages 124–126 cover terminology for using color. We consider Brewer's work more fully in Chapter 13.

Burrough, P. A., and McDonnell, R. A. (1998) *Principles of Geographical Information Systems.* Oxford: Oxford University Press.

> Chapter 2 provides an alternative view of geographic phenomena by contrasting *entity* and *continuous field* approaches.

Chrisman, N. (2002) *Exploring Geographic Information Systems.* New York: Wiley.

> Discusses several extensions to the four standard levels of measurement (nominal, ordinal, interval, and ratio). For a theoretical discussion, see Chrisman (1998).

Forrest, D. (1999b) "Geographic information: Its nature, classification, and cartographic representation." *Cartographica* 36, no. 2:31–53.

> Presents a sophisticated approach for selecting symbology as a function of the phenomena, the spatial data, and the level of measurement.

MacEachren, A. M. (1994a) *Some Truth with Maps: A Primer on Symbolization and Design.* Washington, DC: Association of American Geographers.

> Spatial arrangement of geographic phenomena, levels of measurement, and visual variables are covered in pages 13–34.

5

Data Classification

OVERVIEW

*Data classification involves combining raw data into classes or groups, with each class represented by a unique symbol. For instance, data for a choropleth map might be grouped into five classes, with each class depicted by a different value of gray. The result of data classification is a **classed map**; in contrast, if each raw data value is depicted by a unique symbol, an **unclassed map** results.*

Although data classification is relevant to a wide variety of mapping techniques, it is commonly discussed in association with choropleth maps. Traditionally, cartographers argued for classed choropleth maps on two grounds: readers' inability to discriminate among many differing areal symbols, and the difficulty of creating unclassed maps using traditional photomechanical procedures. Today the latter constraint has been eliminated, as computer hardware is capable of producing unclassed maps; as a result, some cartographers now question whether data classification is a necessity.

*In this chapter, we consider two issues relevant to the classification of a single attribute: methods of classification and how spatial context can be used to simplify the appearance of choropleth maps. Section 5.1 considers six common methods of data classification: **equal intervals**, **quantiles**, **mean–standard deviation**, **maximum breaks**, **natural breaks**, and **optimal**. A common problem you will face is determining which of these methods should be used. Criteria that can assist in selecting a classification method include: (1) whether the method considers how data are distributed along the number line, (2) ease of understanding the method, (3) ease of computation, (4) ease of understanding the legend, (5) whether the method is acceptable for ordinal data, and (6) whether the method can assist in selecting an appropriate number of classes.*

Many cartographers have promoted the optimal method of classification because it does the best job of considering how data are distributed along the number line (by placing similar data values in the same class and dissimilar data values in different classes). The optimal method, however, does not score well on some of the other criteria previously listed; for example, the legend is difficult to understand and the method is unacceptable for ordinal data. Furthermore, we'll see in Chapter 18 that it is inappropriate for map comparison tasks. As a result, it is important that mapmakers learn about the advantages and disadvantages of other methods of classification.

One limitation of the classification methods described in section 5.1 is that they do not consider the spatial context of the data. If the purpose of classification is to simplify the appearance of the map, it can be argued that spatial context also should be considered. In light of this, section 5.2 covers two approaches for simplifying choropleth maps that do consider spatial context. The first uses the optimal classification method, but incorporates a spatial constraint by requiring that values falling in the same class be contiguous, whereas the second employs no classification approach, but rather smooths the raw data by changing values for enumeration units as a function of values of neighboring units.

*In the simplest view, data classification involves grouping observations that have similar scores on a single numeric attribute such as median income in California counties. It is also possible, however, to consider classifying observations on the basis of multiple attributes; for example, we might ask which counties have similar incomes and similar voting behavior. Such a process is termed **cluster analysis**. In section 5.3, we consider eight basic steps of a hierarchical cluster analysis.*

5.1 COMMON METHODS OF DATA CLASSIFICATION

In this section, we consider six common methods of data classification: equal intervals, quantiles, mean–standard deviation, maximum breaks, natural breaks, and optimal. These methods range from those that do not consider how data are distributed along the number line (e.g., equal intervals) to those that do (e.g., optimal). We point out advantages and disadvantages of each method, and develop a general set of criteria by which the methods can be evaluated. For simplicity, we assume that the intent is to visualize an attribute for a single point in time, as opposed to visualizing how an attribute changes over time (the latter involves comparing maps, which we consider in Chapter 18).

For illustrative purposes, we will work with two attributes: (1) the wheat data that we utilized in the preceding chapter (the percentage of land area from which wheat was harvested in Kansas in 1993), and (2) the percentage of foreign-born residents for counties in Florida in 1990. Raw data (sorted from low to high data values) and dispersion graphs for both attributes are given in Tables 5.1 and 5.2, and Figures 5.1 and 5.2, respectively. In examining the dispersion graphs, note that both attributes have a positive skew, with the foreign-born attribute having a distinctive outlier (Dade County, which includes the city of Miami).

Before attempting to classify data, it is essential to consider the kind of numerical data being mapped: bipolar, balanced, or unipolar. Remember from section 4.2 that bipolar data have either a natural or meaningful dividing point that can be used to partition the data. For example, a data set of "percent population change" has a natural dividing point of zero, which can be used to create two classes: values at or above zero and values below zero. Once bipolar data have been split in such a fashion, it might be appropriate to apply one of the methods discussed in this section to each subset of data.

For balanced and unipolar data, there generally is no natural dividing point, and thus the data can be classified directly using one of the methods discussed here. With such data, however, it might be desirable to create a meaningful dividing point prior to classifying. For example, we might compute a mean percentage value for wheat harvested in the 105 counties of Kansas and then split the data into values above and below the mean.

An important consideration in any method of classification is selecting an appropriate *number of classes*. To compare methods of classification in this section, we assume five classes, which are easy to discriminate on the gray-tone maps shown here.

Before selecting a classification method, it is essential to determine precisely how you wish to portray the data on the map. The precision you select will be a function of the initial data you have available, your impression of the quality of the data, and how easily you think readers can interpret the numeric values you provide. In our case, we felt that the data were of sufficient quality to report to the nearest tenth of a percent and that readers of this book would be comfortable with these values. The argument could certainly be made, however, that we should round to the nearest whole percent.

5.1.1 Equal Intervals

In the **equal intervals** (or **equal steps**) method of classification, each class occupies an equal interval along the number line. As a result, this method is identical to creating a grouped-frequency table (see section 3.3.1), except that cartographers commonly distinguish between the calculated class limits and the limits actually used for mapping. The steps for computation are as follows:

*Step 1: Determine the **class interval**, or width that each class occupies along the number line.* This is computed by dividing the range of the data by the number of classes. The results are as follows:

FOR WHEAT

$$\frac{\text{Range}}{\text{Number of classes}} = \frac{\text{High} - \text{Low}}{\text{Number of classes}}$$
$$= \frac{58.5 - 0.7}{5} = 11.56$$

FOR FOREIGN-BORN

$$\frac{\text{Range}}{\text{Number of classes}} = \frac{\text{High} - \text{Low}}{\text{Number of classes}}$$
$$= \frac{45.1 - 0.6}{5} = 8.90$$

Note that for both data sets no rounding is necessary; this differs from the murder rate data used to compute grouped-frequency tables in section 3.3.1.

Step 2: Determine the upper limit of each class. The upper limit for each class is computed by repeatedly adding the class interval to the lowest value in the data. (For the wheat data, adding the class interval 11.56 to 0.7 yields a value of 12.26.) The result is the right-hand set of numbers in the Calculated Limits column in Table 5.3.

Step 3: Determine the lower limit of each class. Lower limits for each class are specified so that they are just above the highest value in a lower valued class (for the

TABLE 5.1 Percentage of land area from which wheat was harvested in Kansas counties in 1993

Observation	County	% Wheat	Observation	County	% Wheat
1	Wyandotte	0.7	54	Ellsworth	21.1
2	Greenwood	1.5	55	Wallace	21.2
3	Jefferson	2.5	56	Jewell	21.6
4	Elk	2.8	57	Stevens	21.8
5	Miami	2.9	58	Smith	22.0
6	Lyon	2.9	59	Ottawa	22.4
7	Wabaunsee	2.9	60	Cheyenne	22.7
8	Chase	3.0	61	Lincoln	23.0
9	Pottawatomie	3.1	62	Ellis	23.1
10	Doniphan	3.5	63	Decatur	23.3
11	Bourbon	3.6	64	Marion	23.3
12	Johnson	3.7	65	Rawlins	23.3
13	Leavenworth	3.8	66	Cloud	23.4
14	Chautauqua	3.8	67	Clay	23.6
15	Franklin	3.9	68	Gove	23.7
16	Shawnee	4.1	69	Seward	24.0
17	Linn	4.1	70	Norton	24.7
18	Jackson	4.1	71	Hodgeman	24.8
19	Osage	4.2	72	Ness	24.9
20	Atchison	4.4	73	Kiowa	25.1
21	Riley	4.5	74	Hamilton	25.1
22	Douglas	5.2	75	Kearny	25.9
23	Woodson	5.8	76	Logan	26.1
24	Nemaha	6.7	77	Saline	26.4
25	Geary	6.8	78	Finney	27.0
26	Brown	7.3	79	Edwards	27.3
27	Coffey	7.5	80	Sherman	27.4
28	Butler	8.7	81	Dickinson	28.1
29	Anderson	9.8	82	Stanton	30.7
30	Morris	10.2	83	Stafford	31.5
31	Crawford	10.8	84	Pawnee	31.8
32	Allen	10.8	85	Ford	32.2
33	Neosho	12.2	86	Mitchell	32.7
34	Rooks	12.5	87	Wichita	33.1
35	Montgomery	13.1	88	Grant	33.5
36	Wilson	14.2	89	Rush	33.5
37	Clark	14.9	90	Lane	33.8
38	Washington	15.3	91	Thomas	34.3
39	Labette	15.7	92	Gray	34.3
40	Marshall	16.0	93	Haskell	34.4
41	Morton	16.7	94	Rice	34.4
42	Graham	16.8	95	Barton	34.6
43	Russell	17.0	96	Pratt	34.7
44	Cowley	17.3	97	Greeley	35.3
45	Phillips	18.1	98	Reno	35.9
46	Comanche	18.2	99	Harvey	36.2
47	Osborne	18.8	100	Scott	37.1
48	Meade	19.3	101	Kingman	39.1
49	Barber	20.6	102	McPherson	39.3
50	Cherokee	20.6	103	Sedgwick	40.6
51	Trego	20.8	104	Harper	54.9
52	Sheridan	21.0	105	Sumner	58.5
53	Republic	21.0			

TABLE 5.2 Percentage of foreign-born population in Florida counties in 1990

Observation	County	% Foreign-Born	Observation	County	% Foreign-Born
1	Madison	0.6	35	Leon	3.7
2	Calhoun	0.8	36	Lafayette	4.1
3	Dixie	0.8	37	Okaloosa	4.3
4	Baker	0.8	38	Glades	4.5
5	Taylor	0.9	39	Highlands	4.6
6	Jefferson	1.0	40	Citrus	4.9
7	Liberty	1.0	41	Lee	5.2
8	Bradford	1.0	42	Brevard	5.3
9	Wakulla	1.2	43	Manatee	5.4
10	Gadsden	1.2	44	DeSoto	5.5
11	Gilchrist	1.3	45	Hernando	5.5
12	Gulf	1.3	46	Volusia	5.8
13	Holmes	1.4	47	Pasco	5.9
14	Suwannee	1.5	48	Alachua	5.9
15	Hamilton	1.6	49	Sarasota	6.0
16	Walton	1.6	50	Indian River	6.1
17	Nassau	1.6	51	Okeechobee	6.3
18	Columbia	1.7	52	Seminole	6.3
19	Franklin	1.8	53	St. Lucie	6.3
20	Sumter	1.9	54	Charlotte	6.3
21	Jackson	2.2	55	Hardee	6.3
22	Putnam	2.2	56	Martin	6.8
23	Sana Rosa	2.2	57	Osceola	7.1
24	Levy	2.3	58	Pinellas	7.1
25	Union	2.4	59	Orange	7.5
26	Washington	2.4	60	Hillsborough	7.6
27	Escambia	2.7	61	Flagler	8.3
28	Clay	3.1	62	Monroe	10.1
29	Bay	3.4	63	Collier	10.5
30	Duval	3.5	64	Palm Beach	12.2
31	Lake	3.5	65	Hendry	14.6
32	St. Johns	3.6	66	Broward	15.8
33	Polk	3.6	67	Dade	45.1
34	Marion	3.6			

TABLE 5.3 Class limit computations for equal-intervals classification

Class	Calculated Limits	Legend Limits
	Kansas Wheat Data	
1	0.70 to 12.26	0.7 to 12.3
2	12.27 to 23.82	12.4 to 23.8
3	23.83 to 35.38	23.9 to 35.4
4	35.39 to 46.94	35.5 to 46.9
5	46.95 to 58.50	47.0 to 58.5
	Florida Foreign-Born Data	
1	0.6 to 9.5	0.6 to 9.5
2	9.6 to 18.4	9.6 to 18.4
3	18.5 to 27.3	18.5 to 27.3
4	27.4 to 36.2	27.4 to 36.2
5	36.3 to 45.1	36.3 to 45.1

wheat data, the lower limit of class 2 is 12.27, which is .01 more than the upper limit of class 1). Up to this point the equal-intervals method is the same as the grouped-frequency method.

Step 4: Specify the class limits actually shown in the legend. The class limits actually shown in the legend should reflect the rounded raw data on which the classification is based. Because we began with values rounded to the nearest tenth of a percent, we also should report class limits to the nearest tenth of a percent. As such, legend limits expressed to the nearest tenth of a percent are shown in Table 5.3.

Step 5: Determine which observations fall in each class. This involves simply comparing the raw values with the legend limits from step 4. Figures 5.1A and 5.2A present these observations in graphic form, and maps of the classified data appear in Figures 5.3A and 5.4A.

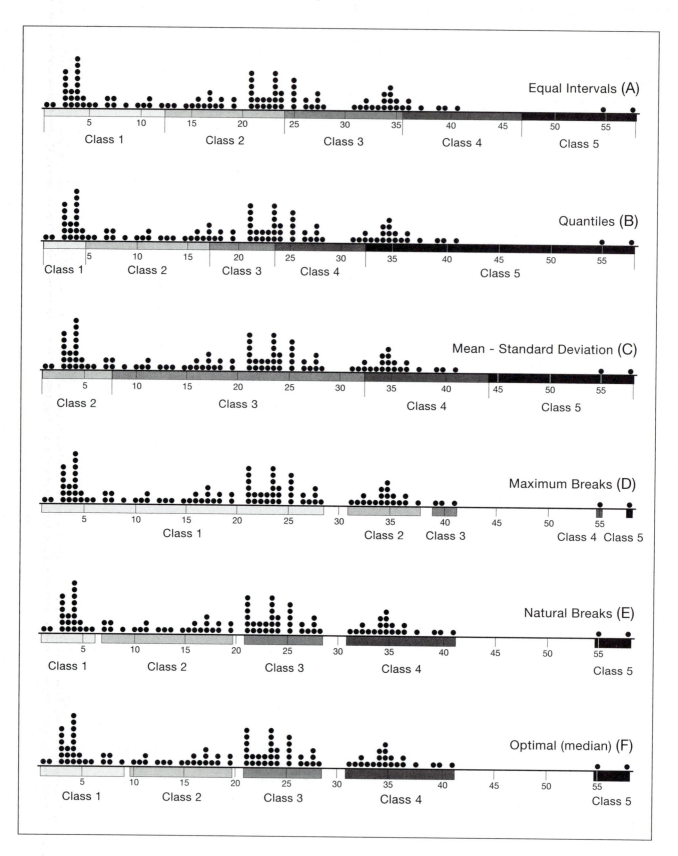

FIGURE 5.1 Dispersion graphs of the wheat data shown in Table 5.1 along with class breaks for various methods of data classification.

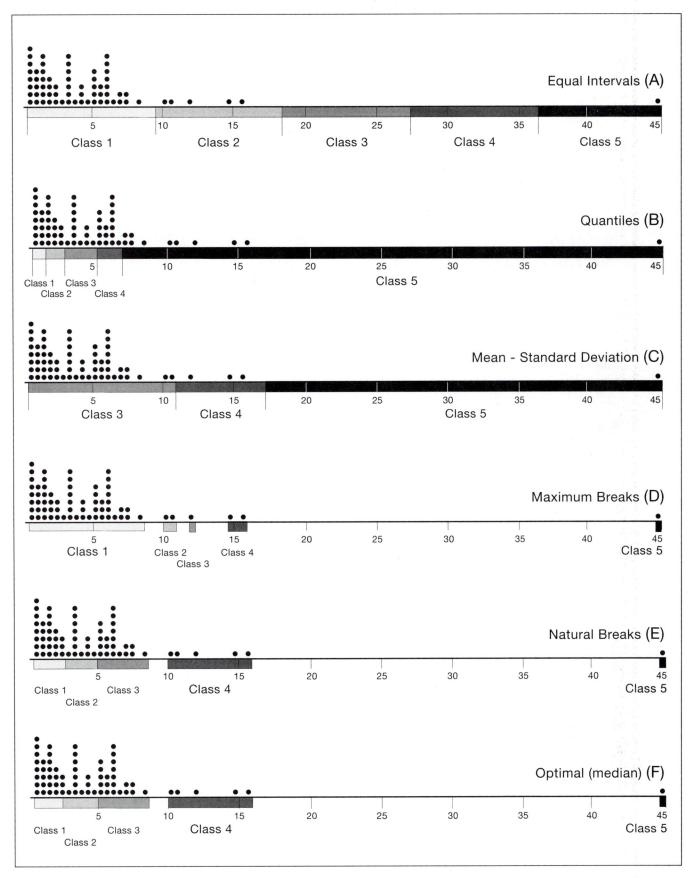

FIGURE 5.2 Dispersion graphs of the foreign-born data shown in Table 5.2 along with class breaks for various methods of data classification.

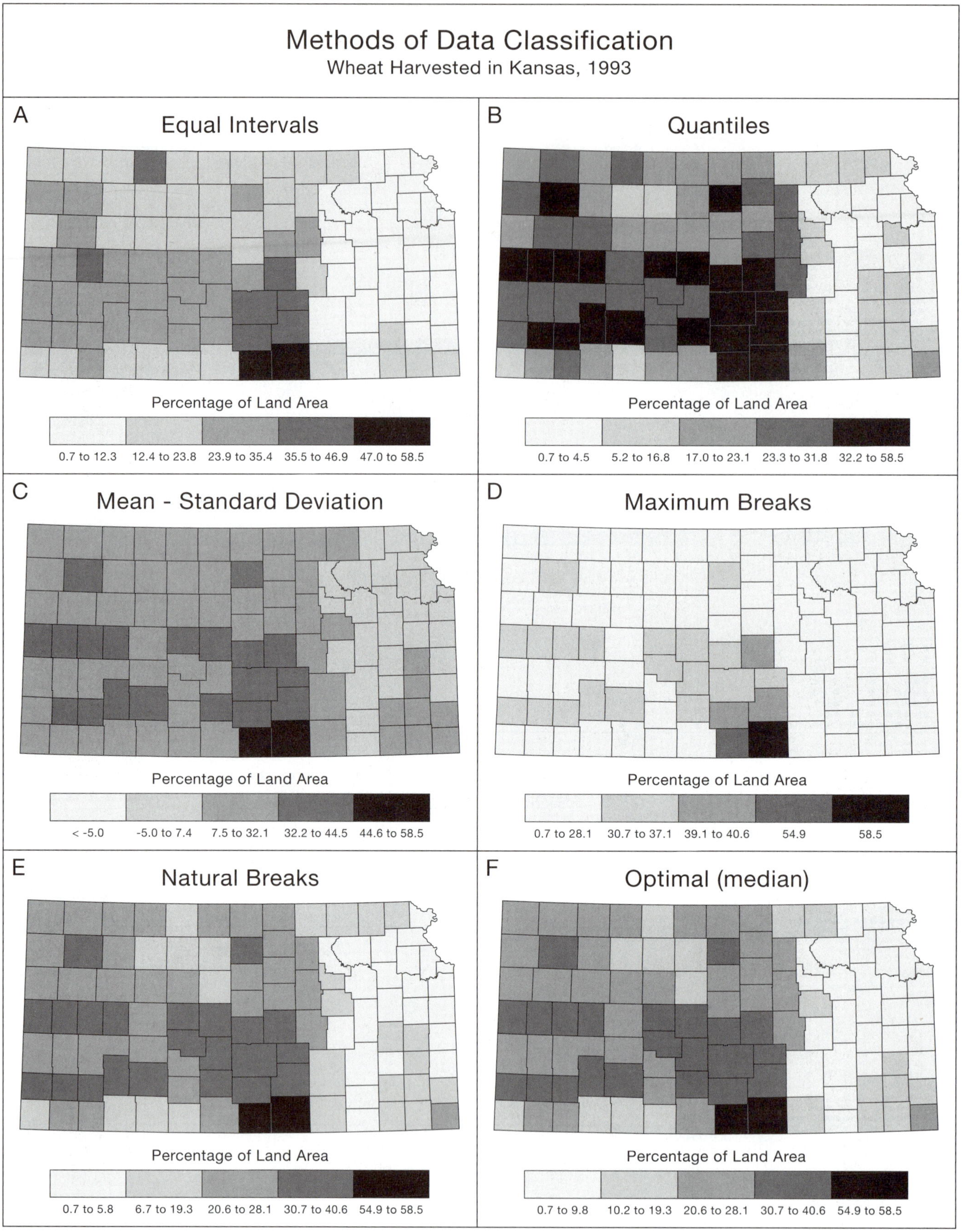

Methods of Data Classification
Wheat Harvested in Kansas, 1993

A Equal Intervals

Percentage of Land Area

| 0.7 to 12.3 | 12.4 to 23.8 | 23.9 to 35.4 | 35.5 to 46.9 | 47.0 to 58.5 |

B Quantiles

Percentage of Land Area

| 0.7 to 4.5 | 5.2 to 16.8 | 17.0 to 23.1 | 23.3 to 31.8 | 32.2 to 58.5 |

C Mean - Standard Deviation

Percentage of Land Area

| < -5.0 | -5.0 to 7.4 | 7.5 to 32.1 | 32.2 to 44.5 | 44.6 to 58.5 |

D Maximum Breaks

Percentage of Land Area

| 0.7 to 28.1 | 30.7 to 37.1 | 39.1 to 40.6 | 54.9 | 58.5 |

E Natural Breaks

Percentage of Land Area

| 0.7 to 5.8 | 6.7 to 19.3 | 20.6 to 28.1 | 30.7 to 40.6 | 54.9 to 58.5 |

F Optimal (median)

Percentage of Land Area

| 0.7 to 9.8 | 10.2 to 19.3 | 20.6 to 28.1 | 30.7 to 40.6 | 54.9 to 58.5 |

FIGURE 5.3 Choropleth maps illustrating various methods of data classification for the wheat data shown in Table 5.1. (A) Equal intervals, (B) quantiles, (C) mean–standard deviation, (D) maximum breaks, (E) natural breaks, and (F) optimal, based on medians.

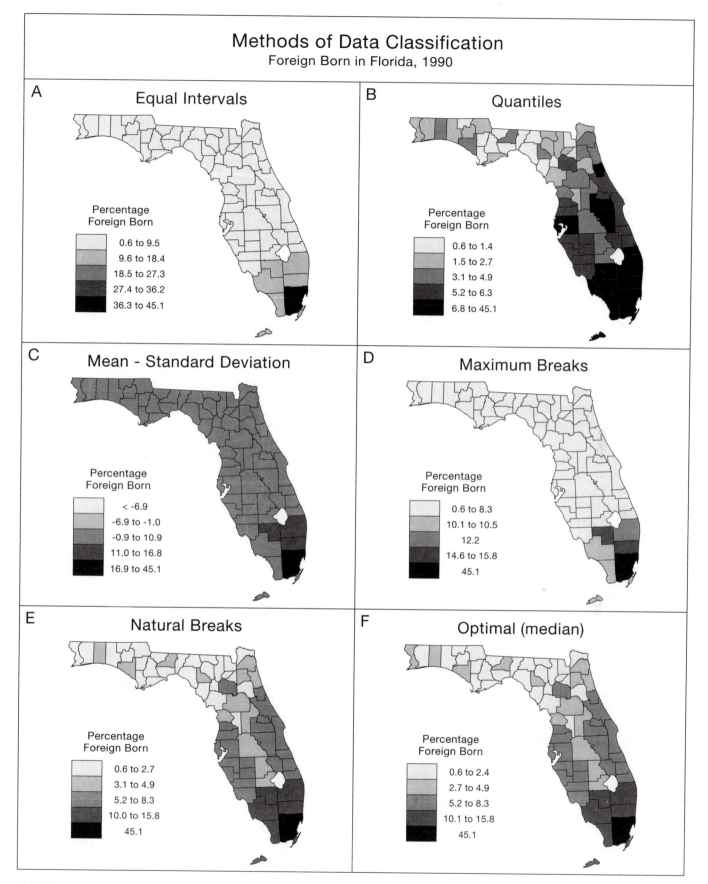

FIGURE 5.4 Choropleth maps illustrating various methods of data classification for the foreign-born data shown in Table 5.2. (A) Equal intervals, (B) quantiles, (C) mean–standard deviation, (D) maximum breaks, (E) natural breaks, and (F) optimal, based on medians.

An advantage of equal intervals is that these steps can be completed using a calculator, or even pencil and paper. As a result, this method was often favored before mapping software became available. A second advantage is that the resulting equal intervals will, in some cases, be easy for map users to interpret. For example, if you were making a five-class map of "percent urban population," and the data ranged from 0 to 100, the resulting classes would be convenient rounded multiples (0–20, 21–40, 41–60, 61–80, and 81–100). For the wheat and foreign-born data, however, note that the mapped limits do not readily reveal what the class interval might be.

A third advantage of equal intervals is that the legend limits contain no missing values (or gaps): For both data sets, the difference between the upper value in a class and the lower value in the next class is 0.1, which is the precision of the data. Gaps, which we will see occur in other classification methods (e.g., quantiles), might cause a reader to wonder why some of the data are missing from the legend. A related advantage for equal intervals is that the legend limits can be simplified so that only the lowest and highest values in the data and the upper limit of each class are shown (Figure 5.5). This approach should permit faster map interpretation, but it might also create confusion concerning the bounds of each class (e.g., the reader might wonder whether 12.3 falls in the first or second class).

The major disadvantage of equal intervals is that the class limits fail to consider how data are distributed along the number line (Figures 5.1A and 5.2A). For example, if you inspect the dispersion graph for the wheat data, you will note that the boundary between classes 4 and 5 falls in a blank area; as a result, the legend limits for classes 4 and 5 are not particularly meaningful. This problem is even more serious for the more sharply skewed foreign-born data, where the third and fourth classes have no members.

5.1.2 Quantiles

In the **quantiles** method of classification, data are rank-ordered and equal numbers of observations are placed in each class. Different names for this method are used, depending on the number of classes; for example, four- and five-class quantiles maps are referred to as *quartiles* and *quintiles*, respectively.

To compute the number of observations in a class, the total number of observations is divided by the number of classes.

WHEAT

$$\text{Number in class} = \frac{\text{Total observations}}{\text{Number of classes}} = 105/5 = 21$$

FOREIGN-BORN

$$\text{Number in class} = \frac{\text{Total observations}}{\text{Number of classes}} = 67/5 = 13.4$$

To determine which observations should be placed in each class, one simply progresses through the rank-ordered data until the desired number of members in a class is obtained. For example, for the wheat data, the first 21 observations in Table 5.1 would be considered members of the first class.

Identifying class membership for the foreign-born data is more complicated because the number of observations desired in each class, 13.4, is not an integer value. In such a situation, one should attempt to place approximately the same number of observations in each class. Thus, we placed 13 observations in the first class and 14 observations in the second class so that the total for the first two classes was 27 (approximately 13.4×2).

Because identical data values should not be placed in different classes, ties can complicate the quantiles method. This can be seen for the wheat data if a class break is attempted between the 63rd (21×3) and 64th numeric positions. Defining a break here doesn't make sense because the data value for both positions is 23.3. Because the value 23.3 also occurs at the 65th numeric position, we chose a class break between the 62nd and 63rd positions to minimize the number of tied values that would have to be moved to another class. (As an exercise, you should consider whether similar manipulations need to be made with the foreign-born data.)

Two approaches are possible for defining legend limits for the quantiles method. One is to specify the lowest and highest values of members in a class; for the wheat data, the 21 members of the first class range from 0.7 to 4.5, and so these limits would be shown in the legend. The other approach is to compute a class boundary as an average of the highest value in a class and the lowest value of the next class; using this approach for the wheat data, the upper limit for the first class would be $(4.5 + 5.2)/2$, or 4.9. We prefer the former approach because it more accurately reflects the data values falling in a class. An advantage of the latter approach, however, is

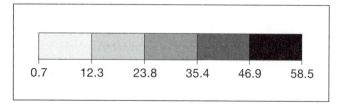

| 0.7 | 12.3 | 23.8 | 35.4 | 46.9 | 58.5 |

FIGURE 5.5 A legend for the equal intervals method that takes advantage of the continuous nature of class limits.

that only the lowest and highest values in the data and the upper limit of each class would need to be shown, as was done for the equal-intervals method in Figure 5.5.

As with equal intervals, an advantage of quantiles is that class limits can be computed manually. A second advantage is that because an equal number of observations fall in each class, the percentage of observations in each class will also be the same. Depending on the number of classes, this might simplify our discussion of the mapped data. For instance, on a five-class map we can refer to the upper or lower 20 percent of the data, whereas on a six-class map we would have to refer to the upper or lower 16.7 percent of the data, which is not a convenient round number. A related advantage is that the 50th percentile or median (a measure of central tendency in the data; see section 3.3.1) will be logically associated with the classes. In the case of an odd number of classes, the median will fall in the center of the middle class, whereas in the case of an even number of classes, the median will fall between the two middle classes.

A third advantage of quantiles is that because class assignment is based on rank order, quantiles are useful for ordinal-level data. For example, if the 50 states of the United States were ranked on "quality of life," the resulting ranks could be split into five equal groups: no numeric information would be necessary to create the classification. A fourth advantage is that if enumeration units are approximately the same size, each class will have approximately the same map area; in Chapter 18, we will see that this trait is useful when comparing maps.

The quantiles method shares the major disadvantage of equal intervals: It fails to consider how the data are distributed along the number line. For example, note that for the foreign-born data, the outlier for Dade County is included in the same class with values of considerably lower magnitude (Figure 5.2B). Another disadvantage of quantiles is that gaps result (for the wheat data, the highest value in class 1 is 4.5 and the lowest value in class 2 is 5.2, resulting in a gap of 0.7; Figure 5.3B). Gaps are problematic because the reader might wonder why they occur. However, they do permit the legend to reflect the range of data actually occurring on the map.

5.1.3 Mean–Standard Deviation

The **mean–standard deviation** method is one of several classification techniques that do consider how data are distributed along the number line. In this method, classes are formed by repeatedly adding or subtracting the standard deviation from the mean of the data, as shown in Table 5.4.* As with the equal intervals method, both

---

* We use mean and standard deviation formulas appropriate for a sample; because of the relatively large number of observations in each case, similar results would be obtained with population formulas.

TABLE 5.4 Class limit computations for mean–standard deviation classification

Class	Normal Distribution Limits	Calculated Limits	Legend Limits
\multicolumn{4}{c}{Kansas Wheat Data: $\bar{x} = 19.7$, $s = 12.4$}
1	$<\bar{x} - 2s$	<-5.0	<-5.0
2	$\bar{x} - 2s$ to $\bar{x} - 1s$	-5.0 to 7.4	-5.0 to 7.4
3	$\bar{x} - 1s$ to $\bar{x} + 1s$	7.4 to 32.1	7.5 to 32.1
4	$\bar{x} + 1s$ to $\bar{x} + 2s$	32.1 to 44.5	32.2 to 44.5
5	$>\bar{x} + 2s$	>44.5	44.6 to 58.5
\multicolumn{4}{c}{Florida Foreign-Born Data: $\bar{x} = 4.96$, $s = 5.94$}
1	$<\bar{x} - 2s$	<-6.9	<-6.9
2	$\bar{x} - 2s$ to $\bar{x} - 1s$	-6.9 to -1.0	-6.9 to -1.0
3	$\bar{x} - 1s$ to $\bar{x} + 1s$	-1.0 to 10.9	-1.1 to 10.9
4	$\bar{x} + 1s$ to $\bar{x} + 2s$	10.9 to 16.8	11.0 to 16.8
5	$>\bar{x} + 2s$	>16.8	16.9 to 45.1

calculated and legend limits can be computed. Calculated limits are computed by using the mean and standard deviation values listed in column 2 of Table 5.4 (Normal Distribution Limits). To create legend limits, calculated limits are adjusted so that identical values cannot fall in two different classes, and the limits of the lowest and highest classes reflect the lowest and highest values in the data.

A major disadvantage of the mean–standard deviation method is that it works well only with data that are normally distributed. This is particularly evident with the foreign-born data, in which the two lowest classes contain solely negative values and therefore have no members (Figure 5.4C). Even with the less severely skewed wheat data, there is still an empty class at the low end of the distribution (Figure 5.3C). One solution to this problem is to transform the data (as described in section 3.3.1), but this is inappropriate if the intention is to examine the *raw* data. Another disadvantage is that the mean–standard deviation method requires an understanding of some basic statistical concepts; a message on the map or in the text indicating that "classes were developed based on the mean and standard deviation" would not be meaningful if one had no statistical training.

A distinct advantage of the mean–standard deviation method, however, is that if the data are normally distributed (or near normal), the mean serves as a useful dividing point, enabling a contrast of values above and below it. This is most effectively accomplished if an even number of classes is used; for example, a six-class map could consist of the positive classes ($\bar{x}$ to $\bar{x} + s$, $\bar{x} + s$ to $\bar{x} + 2s$, and $>\bar{x} + 2s$) and the negative classes ($\bar{x}$ to $\bar{x} - s$, $\bar{x} - s$ to $\bar{x} - 2s$, and $<\bar{x} - 2s$). (For the five-class maps shown in Figures 5.3C and 5.4C, the two middle classes were combined.) Another advantage is that the legend contains no gaps that might confuse the reader.

5.1.4 Maximum Breaks

The mean–standard deviation method considers how data are distributed along the number line in a holistic sense—by trying to fit a normal distribution to the data. An alternative approach is to consider individual data values and group those that are similar (or, alternatively, avoid grouping those that are dissimilar). The **maximum breaks** method is a simplistic means for accomplishing this.* In this method, raw data are ordered from low to high, the differences between adjacent values are computed, and the largest of these differences serve as class breaks.

The foreign-born data provide a good illustration of how the maximum breaks method groups similar values (and avoids grouping dissimilar values): Note how the six highest data values are broken into four classes. In addition to considering how data are distributed along the number line, the maximum breaks method is easy to compute, simply involving subtracting adjacent values.

A disadvantage of maximum breaks is that by paying attention only to the largest breaks, the method seems to miss natural clusterings of data along the number line. For example, for the wheat data, maximum breaks places the two highest values (54.9 and 58.5) in separate classes, but the distance of these values from the rest of the data suggests that they should be in the same class.

5.1.5 Natural Breaks

The **natural breaks** method is one solution to the failure of maximum breaks to consider natural groupings of data. In natural breaks, graphs (e.g., the dispersion graph or histogram) are examined visually to determine logical breaks (or, alternatively, clusterings) in the data. Stated another way, the purpose of natural breaks is to minimize differences between data values in the same class and maximize differences between classes. Later we will see that this is also the objective of the optimal method, but with the optimal method, the classification is done using a mathematical measure of classification error, whereas with natural breaks, the classification is subjective.

To illustrate the computation of natural breaks, consider how we might divide the foreign-born data into five classes (you should examine the dispersion graph associated with the natural breaks method in Figure 5.2E). The highest value in the data (45.1) appears to be quite different from the rest of the data, so we will place it in a class by itself. Our next decision is how to handle the five values ranging from 10.1 to 15.8. We could divide

these into two separate classes (10.1–12.2 and 14.6–15.8), but then we would have to divide the remaining data into two classes. It seems easier to divide the remaining data into three classes, roughly corresponding to the three peaks in the dispersion graph, and so we group the values ranging from 10.1 to 15.8 in one class. We can see in this example that an obvious problem with natural breaks is that decisions on class limits are subjective, and therefore can vary among mapmakers.

5.1.6 Optimal

The **optimal** classification method is a solution to the limitations noted for maximum and natural breaks. The optimal method places similar data values in the same class by minimizing an objective measure of classification error. To illustrate, consider how a small hypothetical data set of nine values would be classified by the quantiles and optimal methods (Table 5.5). The quantiles method assigns the same number of observations to each class (three, in this case), and thus places similar values in different classes (e.g., 14 appears with 31 and 32 in class 2, even though 14 is more similar to 11, 12, and 13, the members of class 1). In contrast, the optimal method places similar values in the same class (the first class consists of 11, 12, 13, and 14, whereas the second class consists of 31, 32, and 33).

One measure of classification error commonly used in the optimal method is the sum of absolute deviations about class medians (ADCM). Computing this measure involves calculating the median of each class, the sum of absolute deviations of class members about each class median, and then adding the resulting sums of absolute deviations. For example, for quantiles, the median in the first class is 12 (remember, the median is simply the middle value in an ordered set), and the sum of absolute deviations for the class is $|11 - 12| + |12 - 12| + |13 - 12| = 2$. The sum of absolute deviations for all classes is $2 + 18 + 67 = 87$. In contrast, ADCM for the optimal method is $4 + 2 + 1 = 7$, which is obviously a smaller value and thus indicative of a better classification.

TABLE 5.5   Computing the sum of absolute deviations about class medians (ADCM)

Raw Data: 11, 12, 13, 14, 31, 32, 33, 99, 100

Quantiles Classification			Optimal Classification		
Class	Values	Error	Class	Values	Error
1	11, 12, 13	2	1	11, 12, 13, 14	4
2	14, 31, 32	18	2	31, 32, 33	2
3	33, 99, 100	67	3	99, 100	1
		ADCM = 87			ADCM = 7

---

* We have borrowed the term "maximum breaks" from the teaching of George Jenks.

The data for this hypothetical example were selected so that the results would be clear-cut. In the real world, the desired minimum-error classification is normally not obvious, and so researchers have developed computer-based algorithms for determining possible solutions. Here we consider two algorithms: the Jenks–Caspall and the Fisher–Jenks.

The Jenks–Caspall Algorithm

The **Jenks–Caspall algorithm**, developed by George Jenks and Fred Caspall (1971), is an empirical solution to the problem of determining optimal classes. It minimizes the sum of absolute deviations about class means (as opposed to medians). The algorithm begins with an arbitrary set of classes (say, the quantiles classes shown in Table 5.5), calculates a total map error analogous to ADCM (but involving the mean), and attempts to reduce this error by moving observations between adjacent classes.

Observations are moved using what Jenks and Caspall termed reiterative and forced cycling. In *reiterative cycling,* movements are accomplished by determining how close an observation is to the mean of another class; for example, for the quantiles data in Table 5.5, the value 14 is closer to the mean of class 1 (12) than is 13 to the mean of class 2 (25.7), so 14 would be moved to the first class. Movements based on the relation of observations to class means are repeated until no further reductions in total map error can be made.

In *forced cycling,* individual observations are moved into adjacent classes, regardless of the relation between the mean value of the class and the moved observation. After a movement, a test is made to determine whether any reduction in total map error has occurred. If error has been reduced, the new classification is considered an improvement and the movement process continues in the same direction. Forcing is done in both directions (from low to high classes and from high to low classes). At the conclusion of forcing, the reiterative procedure described earlier is repeated to see whether any further reductions in error are possible. Although this approach does not guarantee an optimal solution, Jenks and Caspall (1971, 236) indicated that they were "unable to generate, either purposefully or by accident, a better … representation in any set of data."

In addition to developing an automated algorithm for determining optimal classes, Jenks and Caspall (1971, 225) introduced three criteria for selecting a "best" classification. They introduced these criteria by posing three questions:

1. Which map provides the reader with the most accurate intensity values for specific places?

2. Which map creates the most accurate overview?

3. Which map contains boundaries that occur along major breaks in the statistical surface?

Corresponding to these questions, Jenks and Caspall discussed three kinds of error: tabular, overview, and boundary.

To understand these forms of error, it is helpful to consider a 3-D prism map, such as the one shown in Figure 4.2B, which has no error due to classification because each county is raised to a height proportional to the data. If the data were classed, error would arise as a result of counties in the same class being raised to the same height. The difference in *height* between corresponding prisms would constitute **tabular error**, and **overview error** would be the difference in *volume* between corresponding prisms. Tabular error is equivalent to the error measure described previously (sum of absolute deviations about the mean of each class), whereas overview error is weighted to account for the size of enumeration units.

Jenks and Caspall used the term **boundary error** to describe the error occurring along the boundary between two enumeration units on a classed map. They argued that the highest cliffs appearing on the unclassed prism map (such as Figure 4.2B) should ideally appear on the classed prism map. They computed a measure of boundary accuracy by dividing the sum of the n actual cliffs used on the classed map by the sum of the n largest cliffs occurring in the raw data.

Of these three kinds of error, cartographers (including Jenks himself) have focused on tabular error, primarily because of its simplicity. We have presented the three different kinds of error, however, to provide a broader perspective on the classification problem.

The Fisher–Jenks Algorithm

In contrast to the empirical approach used by Jenks and Caspall, the **Fisher–Jenks algorithm** has a mathematical foundation that guarantees an optimal solution. Walter Fisher (1958) was responsible for developing the mathematical foundation, and George Jenks (1977) introduced the idea to cartographers. Cartographers generally have chosen to recognize only Jenks for this contribution, so the reader might find the algorithm referred to as "Jenks's optimal method."

To understand the Fisher–Jenks algorithm, it is worthwhile to consider how an optimal solution might be computed using brute force. Imagine that you wanted to develop an optimal two-class map of the data 1, 3, 7, 11, and 22. With such a small data set, it is easy to list all possible two-class solutions and compute associated error measures (Table 5.6). If the process is so simple with a small data set, it would seem that for large data sets a computer could be used to determine an optimal solution by simply considering all possibilities. Unfortunately, for large data sets, the number of possible solutions becomes

TABLE 5.6 Computing ADCM for all potential two-class maps

Raw Data: 1, 3, 7, 11, 22

Solution 1			Solution 2		
Class	Values	Error	Class	Values	Error
1	1	0	1	1, 3	2
2	3, 7, 11, 22	23	2	7, 11, 22	15
	ADCM = 23			ADC M = 17	

Solution 3			Solution 4		
Class	Values	Error	Class	Values	Error
1	1, 3, 7	6	1	1, 3, 7, 11	14
2	11, 22	11	2	22	0
	ADCM = 17			ADCM = 14	

Solution 4 is optimal because it has the smallest total error (or ADCM).

prohibitively large; for example, Jenks and Caspall (1971, 232) calculated that for the 102 counties of Illinois there would be over 1 billion possible seven-class maps.

Rather than consider all solutions, the Fisher–Jenks algorithm takes advantage of the mathematical foundation provided by Fisher, which states that any optimal partition is simply the sum of optimal partitions of subsets of the data. We illustrate this concept by considering some initial steps for handling the data 1, 3, 7, 11, and 22 (Table 5.7). For computational simplicity, we use the median (and associated sum of absolute deviations); another version of the algorithm uses the mean (and associated sum of squared deviations about the mean).

Step 1 involves computing the sum of absolute deviations about the class median for any ordered subset of the raw data, ignoring how these subsets might fit into a particular classification. For example, the sum of absolute deviations for the first through third observations (the subset 1, 3, and 7) is $|1 - 3| + |3 - 3| + |7 - 3| = 6$. This result appears in row 1, column 3, of the matrix shown in step 1 of Table 5.7. The resulting sum of absolute deviations is termed the *diameter (D),* and is represented by *D(i, j),* where *i* and *j* identify the observations in step 1 (Hartigan 1975, Chapter 6); thus, *D(i, j)* for this example would be $D(1, 3) = 6$.

In step 2, the optimal solution for a two-class map of the complete data set is computed, along with optimal two-class solutions for subsets of the data. Together these are termed the optimal two partitions. Calculations for the optimal two-class map for the complete data set are shown in part (a) of step 2 (Table 5.7), and some of the results for subsets of the data are shown in part (b) of step 2. Although the subset calculations are not used in determining the optimal two-class map, they are used to determine optimal classifications for maps with a greater

TABLE 5.7 Initial steps in the Fisher–Jenks algorithm for optimal data classification

Raw Data: 1, 3, 7, 11, 22

Step 1. Compute the sum of absolute deviations about the class median for all ordered subsets of the data.

The following matrix shows the sum of absolute deviations about the median for the *i*th through the *j*th observation; for example, if $i = 1$ and $j = 3$, then the sum is $|1 - 3| + |3 - 3| + |7 - 3| = 6$. (Note that this result appears in the *first* row and the *third* column of the matrix.) The resulting sums of absolute deviations are commonly termed the *diameter (D)* and are represented by *D(i, j)*; for this example $D(1, 3) = 6$.

		jth observation				
		1	2	3	4	5
	1	0	2	6	14	29
*i*th	2		0	4	8	23
observation	3			0	4	15
	4				0	11
	5					0

Step 2. Compute all optimal two partitions.

a. The results for the optimal two-class map of the complete data set are as follows:

1 | 3 7 11 22
$D(1, 1) + D(2, 5) = 0 + 23 = 23$
1 3 | 7 11 22
$D(1, 2) + D(3, 5) = 2 + 15 = 17$
1 3 7 | 11 22
$D(1, 3) + D(4, 5) = 6 + 11 = 17$
1 3 7 11 | 22
$D(1, 4) + D(5, 5) = 14 + 0 = 14$

b. The following are some results for optimal two-class partitions of subsets of the data:

1 | 3 7 11
$D(1, 1) + D(2, 4) = 0 + 8 = 8$
1 3 | 7 11
$D(1, 2) + D(3, 4) = 2 + 4 = 6$
1 3 7 | 11
$D(1, 3) + D(4, 4) = 6 + 0 = 6$
1 | 3 7
$D(1, 1) + D(2, 3) = 0 + 4 = 4$
1 3 | 7
$D(1, 2) + D(3, 3) = 2 + 0 = 2$

Ideally, all optimal two partitions would be computed!

Step 3. Compute all optimal three-partitions (let Opt-2 represent an optimal two partition.)

a. The results for the optimal three-class map of the complete data set would be calculated. Some of the calculations follow. The question marks represent optimal two-class partitions not computed above.

1 | 3 7 11 22
$D(1, 1) + \text{Opt-2} = 0 + ?$
1 3 | 7 11 22
$D(1, 2) + \text{Opt-2} = 2 + ?$
1 3 7 | 11 22
$\text{Opt-2} + D(4, 5) = 2 + 11 = 13$
1 3 7 11 | 22
$\text{Opt-2} + D(5, 5) = 6 + 0 = 6$

b. Calculate all other optimal three-class partitions.

number of classes, such as a three-class map, for which some calculations are shown in step 3.

Advantages and Disadvantages of Optimal Classification

The obvious advantage of the optimal method is that it considers, in detail, how data are distributed along the number line. It is the "best" choice for classification when the intention is to place like values in the same class (and unlike values in different classes) based on the position of values along the number line.

Another advantage is that the optimal method can assist in determining the appropriate number of classes. When the median is used as the measure of central tendency, this is accomplished by computing the **goodness of absolute deviation fit** (**GADF**), which is defined as:

$$GADF = 1 - \frac{ADCM}{ADAM}$$

where ADCM is the sum of absolute deviations about class medians for a particular number of classes, and ADAM is the sum of absolute deviations about the median for the entire data set. An analogous measure can be computed when the mean is used as the measure of central tendency, and is known as the **goodness of variance fit** (**GVF**; Robinson et al. 1984, 363).

GADF ranges from 0 to 1, with 0 representing the lowest accuracy (a one-class map) and 1 the highest accuracy. If there are no ties in the data, then a GADF value of 1 will result only when each observation is a separate class (an n-class map will be required, where n is the number of classes). Ties can, however, considerably reduce the number of classes needed to achieve a GADF of 1.

For example, in the case of the foreign-born data, only 46 classes are needed (there are 67 data values).

It is important to note that an n-class map is equivalent to an "unclassed map." This might be confusing because the term *unclassed* suggests no classes, although there are actually n classes on an n-class map. "Unclassed" is commonly used to indicate that no classing or grouping has been applied to the data; for instance, it is not necessary to run the optimal program to create an unclassed map.

GADF calculations can assist in selecting an appropriate number of classes in two ways. One approach is to construct a graph of the number of classes against GADF values (Figure 5.6) and look for a point at which a curve fit to the data begins to flatten out. In the case of the foreign-born data, the curve appears to flatten out at about five classes. A flattening at this point indicates that a larger number of classes would not contribute substantially to a reduction in the classification error. (One should also bear in mind that a map with more classes would be more difficult to use, as a greater number of areal symbols would have to be differentiated.)

Another approach is to determine the number of classes for which the GADF first exceeds a certain value, say, .80 (the accuracy is 80 percent). For the foreign-born data, this approach yields a six-class map (Figure 5.6). Note, however, that if a more stringent value were used, say, .90, a nine-class map would be required. Admittedly, both of these approaches are subjective, but they are an improvement over choosing an arbitrary number of classes.

In addition to helping you determine an appropriate number of classes for the optimal method, the GADF technique could also be used to compute the accuracy of other classification methods, and thus determine whether those methods might be appropriate. For example, if you

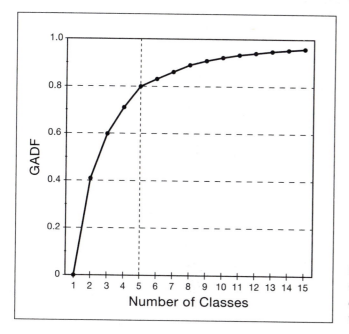

FIGURE 5.6 Graph of the number of classes plotted against GADF values. The curve appears to flatten out at about five classes, indicating that a greater number of classes would not substantially reduce the error on the map.

were to compute similar GADF values for optimal and quantiles, you might choose the quantiles method because it would be easier for the user to understand how the class limits were created. Alternatively, GADF values might be computed for various numbers of classes for quantiles to assist in determining an appropriate number of classes for that method.

Disadvantages of the optimal method include the difficulty of understanding the concept and the appearance of gaps in the legend. Another traditional disadvantage

was that software packages generally did not include the optimal method as an option, but this has changed (e.g., the popular package ArcView includes such an option, which it terms "natural breaks").

5.1.7 Criteria for Selecting a Classification Method

In discussing the common methods of classification, we have pointed out numerous criteria that might be used to judge their usefulness. Figure 5.7 summarizes these criteria

	Equal Interval	Quantiles	Mean SD	Maximum Breaks	Natural Breaks	Optimal
Considers distribution of data along a number line	P	P	G[a]	G	VG[b]	VG
Ease of understanding concept	VG	VG	VG	VG	VG	G[c]
Ease of computation	VG	VG	VG	VG	VG	G[d]
Ease of understanding legend	VG[e]	P	G	P	P	P
Legend values match range of data in a class	P	VG	P	VG	VG	VG
Acceptable for ordinal data	U	A	U	U	U	U
Assists in selecting number of classes	P	P	P	P	G	VG

P = Poor   G = Good   VG = Very Good   A = Acceptable   U = Unacceptable

[a] Rating would be poor if data are not normal.

[b] Although breaks are subjectively determined, the results are often similar to those obtained by the optimal method.

[c] Only a good rating is assigned because of the fairly complex nature of the algorithm.

[d] When the Fisher-Jenks algorithm is used, only about 1000 observations can be handled; this problem does not occur with the Jenks-Caspall algorithm.

[e] Only a good rating would be appropriate if round numbers are not used.

FIGURE 5.7 Criteria for selecting a method of classification.

and rates each classification method as very good, good, or poor on each measure (acceptable or unacceptable in the case of "acceptable for ordinal data"). One problem with any rating system is that it is a function of the computer environment the mapmaker has available and the knowledge of the map user. For Figure 5.7, we assume that computer software is available for creating all of the classification methods we have considered and that the map user is a college-level student with a basic foundation in introductory statistics.

Note that "ease of understanding legend" is a function of whether or not there are gaps in the legend: Remember that gaps between class limits can make the legend difficult to understand. The equal-intervals method receives a very good rating on this criterion because not only are there no gaps, but the rounded intervals can be very easy to understand (e.g., 0–25, 26–50, etc.). Some mapmakers might wish to avoid the problem of gaps by creating continuous legends for all classification methods (as in Figure 5.5). Remember, though, that this approach will not indicate the actual range of values falling in a class (the latter is dealt with in the criterion "legend values match range of data in a class").

An analysis of Figure 5.7 reveals that there is no single best method of classification. Although the optimal approach is often touted as the best method, it is best only in terms of grouping like values together (as a function of their position along a number line) and in selecting an appropriate number of classes. Clearly, there are several other criteria for which it is not the best. Ultimately, a mapmaker creating a display for presentation purposes must consider the purpose of the map and the knowledge of the intended audience before selecting a classification method.

A good illustration of the role of map purpose and intended audience is the effort of Cynthia Brewer and Trudy Suchan to create an atlas of the 2000 U.S. Census data (Brewer 2001). Rather than use one of the standard classification methods presented in this chapter, Brewer and Suchan used meaningful breaks (e.g., a percentage figure for the entire United States), rounded breaks, and breaks that were identical across a set of maps. Brewer (2001, 225) argued that the resulting map set is much more useful than if a "map-by-map optimization approach" were used.

5.2 USING SPATIAL CONTEXT TO SIMPLIFY CHOROPLETH MAPS

One limitation of the classification methods described in the preceding section is that they do not consider the spatial context of the data. If the purpose of classification is to simplify the appearance of the map, it can be argued that spatial context also should be considered. This section considers two approaches that do consider spatial context: one that begins with the optimal classification approach but incorporates a spatial constraint, and one that works solely with spatial context.

5.2.1 Spatial Constraint with Optimal Classification

To illustrate how a spatial constraint can be combined with optimal classification, consider Figure 5.8. The top portion of the figure lists raw data for 16 hypothetical enumeration units, and the bottom portion portrays two maps of these data. The map on the left is an optimal classification; note that breaks for this map occur between 6 and 9 and between 13 and 16. The GADF for this map is .76, and the complexity using the C_F face measure (see section 3.4.2) is 10/16 or .63.

A spatial constraint can be applied by starting with the optimal solution and then allowing data values to shift between classes so that the map pattern is simplified. One potential solution is shown in the right-hand map, where the values 9 and 16 have been shifted from the second and third classes to the first and second classes, respectively. The result is a slightly lower GADF (.70 as opposed to .76), but a considerably simpler complexity (.38 as opposed to .63).

Franky Declercq (1995) developed an algorithm that will find the least complex map for a particular number of classes, assuming that one is willing to shift a certain percentage of observations between classes. Obviously, one problem with this approach is that there is a trade-off between complexity and accuracy: As complexity decreases,

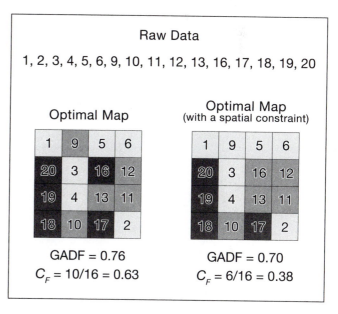

FIGURE 5.8 Applying a spatial constraint to the optimal classification approach.

accuracy will also decrease. Consequently, the mapmaker must use some subjectivity in determining what magnitude of GADF is acceptable for a particular application.

5.2.2 Spatial Context Only

Another approach for simplifying map pattern is to adjust the values of enumeration units as a function of values of surrounding units, an idea developed by Waldo Tobler (1973) and promoted by Adrian Herzog (1989). This notion is based on the concept of **random error**: If we repeatedly measure a value for an enumeration unit, we will likely get a different value each time (think of trying to calculate the population for a census tract for the 2000 Census—numerous variables could affect your result). This suggests that some change in the data we have collected is permissible (Clark and Hosking 1986, 14). Moreover, the spatially autocorrelated nature of geographic data (see section 3.4.2) provides a mechanism for making adjustments: Nearby units can assist in determining appropriate values for a particular enumeration unit because similar values are likely to be located near one another.

Although this approach could be used with a wide variety of data, it is easiest to implement for proportion data (e.g., the proportion of adults who smoke cigarettes) because the statistical theory for such data is well known. To illustrate, consider the hypothetical portion of a map shown in Figure 5.9, where it is assumed we wish to change the value for the central enumeration unit as a function of the surrounding units. The simplest formula for determining the value of the central unit would be to

average all four values, but such a formula would not consider two factors: (1) that we have presumably taken some care in collecting the data for the central unit (and thus would like to place greater weight on it), and (2) that those units having a longer common boundary with the central unit should have greater impact. Thus, a more appropriate formula is

$$V_e = W_c V_c + W_s \left(\sum_{i=1}^{n} \frac{L_i}{L_T} V_i \right)$$

where V_e = the estimated value for the central unit
V_c = the original value for the central unit
V_i = the original value for the ith surrounding unit
W_c = the weight for the central unit
W_s = the weight for the surrounding units
L_i = the length of the boundary between the ith unit and the central unit
L_T = the total length of the central unit boundary
n = the number of surrounding units

If we assume W_c and W_s values of .67 and .33, respectively, then for Figure 5.9 the formula would be calculated as follows:

$$V_e = .67(.15) + .33\left(\frac{1}{6}(.26) + \frac{2}{6}(.22) + \frac{3}{6}(.19) \right) = .17$$

Herzog indicated that a more complete algorithm should involve several other considerations. First, some limit should be placed on how much change is permitted to the central unit. For proportion data, Herzog suggested that the central value should not be moved outside of a 95 percent confidence interval around its original value (see Burt and Barber 1996, 272–274, for a discussion of confidence intervals for proportions). Second, surrounding values differing dramatically from the central value should not be used in the calculations. For proportion data, Herzog suggested a difference of proportions test (see Burt and Barber 1996, 322–324), with proportions differing significantly from the central unit not being used. Third, the process should be implemented in an iterative fashion, meaning that the entire map should be simplified several times; for example, this process could continue until all enumeration units reached the bounds of their confidence intervals or the largest change made in a unit was less than a small tolerance. Finally, the map resulting from the iterative process should be further smoothed by combining areas with nearly equal values.

Figure 5.10 illustrates the net effect of using Herzog's approach for the percentage of students in each commune of Switzerland who study at the University of

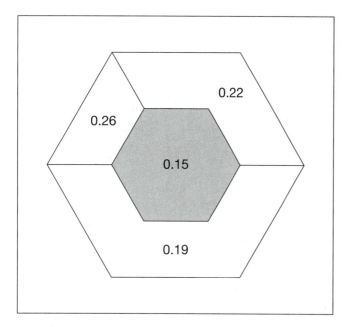

FIGURE 5.9 A hypothetical set of enumeration units for which a simplified value is desired for the central enumeration unit.

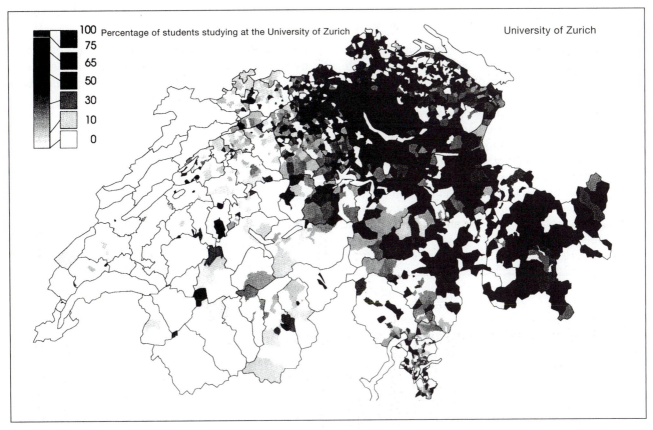

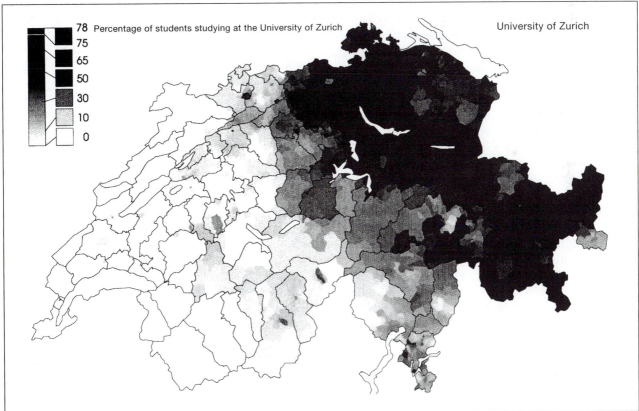

FIGURE 5.10 A comparison of an original unclassed map (*top*) and a simplified map (*bottom*) resulting from applying Herzog's method for simplifying choropleth maps. (An adaptation of Figure 1 from p. 214 of *Accuracy of Spatial Databases*, M. Goodchild and S. Gopal (eds.); courtesy of Taylor & Francis.)

Zurich: The top map is an original (unsimplified) unclassed map, and the bottom map is the simplified result. Although the bottom map is still unclassed, note that it is much simpler to interpret than the top map.

It should be noted that the methods we have examined for simplifying map appearance are relatively new and have not yet been incorporated in commercial software. Assuming that such methods do become readily available, it will be interesting to see to what extent they are used, and what impact this has on the use of more traditional classification methods that do not consider spatial context.

5.3 CLUSTER ANALYSIS

In the preceding sections of this chapter, we considered classification methods appropriate for a single numeric attribute. Here we consider classification methods appropriate for multiple numeric attributes. To illustrate, consider the data shown in Table 5.8 representing the value of various attributes for counties in New York State (excluding New York City) recorded by the U.S. Bureau of the Census for the 2000 Census. Here our interest is in whether there are certain counties that have similar scores on one or more of these attributes. For instance, you might wonder whether there are counties that have a high percentage of African Americans, a high infant mortality rate, and a decreasing population (a negative percent population change). You might tackle such questions by visually comparing separate maps of the attributes, but this would require that you mentally overlay the maps to derive various combinations of attributes for particular counties. More useful would be a method that combines the maps for you and tells you which counties are similar to one another and how they are similar—this is the purpose of the mathematical technique of **cluster analysis,** which we describe in this section.

Cluster analysis techniques can be divided into hierarchical and nonhierarchical methods. *Hierarchical* methods begin by assuming that each observation (county in the case of the New York data) is a separate cluster and then progressively combine clusters.* The process is hierarchical in the sense that once observations are combined in a cluster they remain in that cluster throughout the clustering procedure. *Nonhierarchical* methods presume a specified number of clusters and associated members, and then attempt to improve the classification by moving observations between clusters. Because hierarchical methods are a bit easier to understand and more common, we focus on them here.

5.3.1 Basic Steps in Hierarchical Cluster Analysis

To summarize the process of cluster analysis, we utilize an eight-step process that is a modification of six steps recommended by Charles Romesburg (1984).

Step 1. Collect an Appropriate Data Set

The first step is to collect a data set that you wish to cluster. Ultimately, we will cluster the data shown in Table 5.8, but for illustrative purposes we also will work with the small hypothetical data set for livestock and crop production listed in Figure 5.11A and plotted in graphical form in Figure 5.11B.

Romesburg (38) indicates that cluster analysis can be used for three purposes:

- To create a scientific question.
- To create a research hypothesis that answers a scientific question.
- To test a research hypothesis to decide whether it should be confirmed or disproved.

Because we were relatively unfamiliar with the census data for New York, we wanted to use it to create a scientific question—to explore the patterns that might arise. Those knowledgeable about the geography of New York might use the resulting patterns to look for other attributes to explain the results or consider other attributes that might be suitable for a cluster analysis.

Step 2. Standardize the Data Set

The second step in cluster analysis is to standardize the data set, if necessary. Two forms of standardization are possible. One is to account for enumeration units of varying size. Just as for choropleth maps (see section 4.4.1), this standardization is appropriate as larger enumeration units are apt to have larger values of a countable phenomenon.† For the New York State data, we standardized the raw number of African Americans living in each county by dividing by the total population of each county. For our small hypothetical agricultural data set, we will assume the data are already standardized.

A second form of standardization is to adjust for different units of measure on the attributes. For instance,

---

* Throughout this section we assume that we are clustering enumeration units, an areal phenomenon. It is possible, however, to cluster other forms of phenomena, such as points—thus, we have used the generic term *observation.*

† If data associated with points (as opposed to areas) are clustered, it might also be appropriate to standardize the data (see section 16.1 for appropriate methods).

TABLE 5.8 Cluster analysis data for New York State counties

County	% Population Change (1990–2000)	% Unemployed	% African American	Infant Mortality Rate (Deaths per 1000 Live Births)
Albany	0.6	2.8	11.1	6.1
Allegany	–1.1	6.7	0.7	5.4
Broome	–5.5	3.3	3.3	8.1
Cattaraugus	–0.3	6.4	1.1	9.6
Cayuga	–0.4	4.5	4.0	6.4
Chautauqua	–1.5	4.8	2.2	8.0
Chemung	–4.3	4.8	5.8	4.9
Chenango	–0.7	4.8	0.8	4.9
Clinton	–7.1	5.2	3.6	15.2
Columbia	0.2	2.9	4.5	6.0
Cortland	–0.7	5.9	0.9	3.6
Delaware	1.5	4.8	1.2	8.0
Dutchess	8.0	3.1	9.3	5.0
Erie	–1.9	4.8	13.0	8.2
Essex	4.6	6.6	2.8	5.2
Franklin	9.9	7.6	6.6	10.7
Fulton	1.6	5.8	1.8	4.6
Genesee	0.5	4.9	2.1	14.6
Greene	7.7	5.0	5.5	4.1
Hamilton	1.9	8.2	0.4	0.0
Herkimer	–2.1	5.0	0.5	8.6
Jefferson	0.7	8.2	5.8	5.2
Lewis	0.6	7.8	0.4	9.0
Livingston	3.1	4.6	3.0	8.5
Madison	0.4	4.4	1.3	4.8
Monroe	3.0	3.8	13.7	8.2
Montgomery	–4.4	5.8	1.2	3.4
Nassau	3.6	2.7	10.1	4.8
Niagara	–0.4	5.9	6.1	10.6
Oneida	–6.1	3.8	5.7	5.9
Onondaga	–2.3	3.5	9.4	6.7
Ontario	5.4	3.7	2.1	6.8
Orange	11.0	3.1	8.1	6.5
Orleans	5.6	5.3	7.3	5.4
Oswego	0.5	6.3	0.6	4.8
Otsego	2.1	4.7	1.7	5.1
Putnam	14.1	2.5	1.6	8.2
Rensselaer	–1.2	3.8	4.7	10.7
Rockland	8.0	3.0	11.0	4.3
Saratoga	10.7	3.2	1.4	4.8
Schenectady	–1.8	3.5	6.8	6.2
Schoharie	–0.8	4.8	1.3	14.9
Schuyler	3.0	5.5	1.5	4.6
Seneca	–1.0	4.9	2.3	5.4
Steuben	–0.4	4.9	1.4	5.9
St. Lawrence	0.0	8.0	2.4	6.8
Suffolk	7.4	3.2	6.9	6.2
Sullivan	6.8	5.0	8.5	4.8
Tioga	–1.1	3.3	0.5	4.7
Tompkins	2.6	2.7	3.6	8.2
Ulster	7.5	3.3	5.4	3.6
Warren	6.9	4.0	0.6	5.8
Washington	2.9	4.1	2.9	3.3
Wayne	5.2	4.4	3.2	10.3
Westchester	5.6	3.0	14.2	4.9
Wyoming	2.2	5.8	5.5	11.3
Yates	7.9	3.7	0.6	3.0

consider the case of median family income and percent voting in a presidential election. Median family income likely would be measured in thousands of dollars ranging from, say, 20,000 to 60,000, whereas percent voting in a presidential election is a percentage attribute ranging from, say, 30 to 50. If such data were clustered, the percentage values would have little impact on the cluster analysis; we would essentially be clustering on family income. The most common approach for standardizing for different units of measure is to compute z scores:

$$z_i = \frac{X_i - \overline{X}}{s}$$

where z_i is the resulting z score for the ith observation, X_i is a raw data value for the ith observation, $\overline{X}$ is the mean of an attribute, and s is the standard deviation of an attribute. Applying this formula to our hypothetical data in Figure 5.11A, we would find mean and standard deviation values of 14.80 and 9.65 for the livestock attribute, and a z score of $(5.00 - 14.80)/9.65 = 1.02$ for the first observation. Because we presume our hypothetical data are in the same units of measure, however, this standardization is not necessary and so we will work with the unstandardized data. We chose, however, to standardize the New York State data using z scores because three of the attributes were measured in percent and the other (infant mortality) was measured in number of deaths per 1,000 births.

Step 3. Compute Initial Resemblance Coefficients

The third step of cluster analysis is to compute an initial set of **resemblance coefficients** that express the similarity of each pair of observations. Although numerous coefficients are possible, we will use a *Euclidean distance* coefficient because it is commonly used and readily interpreted graphically. The Euclidean distance for two attributes is

$$d_{ij} = \sqrt{(X_i - X_j)^2 + (Y_i - Y_j)^2}$$

where d_{ij} is the Euclidean distance between the ith and jth observations, X_i and X_j are the values of observations i and j on attribute X, and Y_i and Y_j are the values of observations i and j on attribute Y. For instance, the Euclidean distance between the first two observations in Figure 5.11A is

$$d_{12} = \sqrt{(X_1 - X_2)^2 + (Y_1 - Y_2)^2}$$
$$= \sqrt{(5-5)^2 + (20-15)^2} = 5.00$$

Note that this value corresponds to the distance between points 1 and 2 in the graph in Figure 5.11B. Figure 5.11C provides a complete set of Euclidean distance coefficients for all pairs of observations listed in Figure 5.11A.

Step 4. Cluster the Data

The fourth step of cluster analysis is to actually cluster the data. In hierarchical cluster analysis, we begin by considering each observation as a separate cluster and then combine those clusters that are most similar to one another as expressed by the initial resemblance coefficients. In our case, observations (clusters) 1 and 2 have the shortest Euclidean distance (see Figure 5.11C) and so combine to form a new cluster in Figure 5.11D. For our purposes, we designate this as cluster 1-2.

When clusters are combined, we must recompute the resemblance coefficient between the newly formed cluster and all other existing clusters. Although numerous methods are possible for recomputing resemblance coefficients, we focus on one—the **unweighted pair–group method using arithmetic averages (UPGMA)**—because Romesburg (1984) indicated it has been commonly used. In UPGMA, revised resemblance coefficients are computed by averaging coefficients between observations in the newly formed cluster and observations in existing clusters. For the newly formed cluster 1-2, we would have

$$d_{3(1-2)} = \tfrac{1}{2}(d_{13} + d_{23}) = \tfrac{1}{2}(12.04 + 12.65) = 12.35$$

$$d_{4(1-2)} = \tfrac{1}{2}(d_{14} + d_{24}) = \tfrac{1}{2}(21.21 + 18.03) = 19.62$$

$$d_{5(1-2)} = \tfrac{1}{2}(d_{15} + d_{25}) = \tfrac{1}{2}(26.63 + 24.17) = 25.40$$

The results of these computations appear in the last row of the revised resemblance matrix shown in Figure 5.11E. Evaluating this revised resemblance matrix, we find that observations (clusters) 4 and 5 group next to form cluster 4-5 (see Figure 5.11F). Again, we compute revised resemblance coefficients by averaging coefficients between observations in the newly formed cluster and observations in existing clusters. For clusters 3 and 4-5, the result is

$$d_{3(4-5)} = \tfrac{1}{2}(d_{34} + d_{35}) = \tfrac{1}{2}(14.32 + 17.21) = 15.77$$

while for clusters 1-2 and 4-5, the result is

$$d_{(1-2)(4-5)} = \tfrac{1}{4}(d_{14} + d_{15} + d_{24} + d_{25})$$
$$= \tfrac{1}{4}(21.21 + 26.63 +$$
$$18.03 + 24.17) = 23.00$$

The resulting distances are shown in Figure 5.11G, where we see that the smallest Euclidean distance is 12.35 (between clusters 1-2 and 3). These clusters are combined in Figure 5.11H. Finally, we combine clusters 1-2-3 and 4-5 at an average Euclidean distance of 20.26 (the average of the lengths of the dashed lines shown in Figure 5.11H).

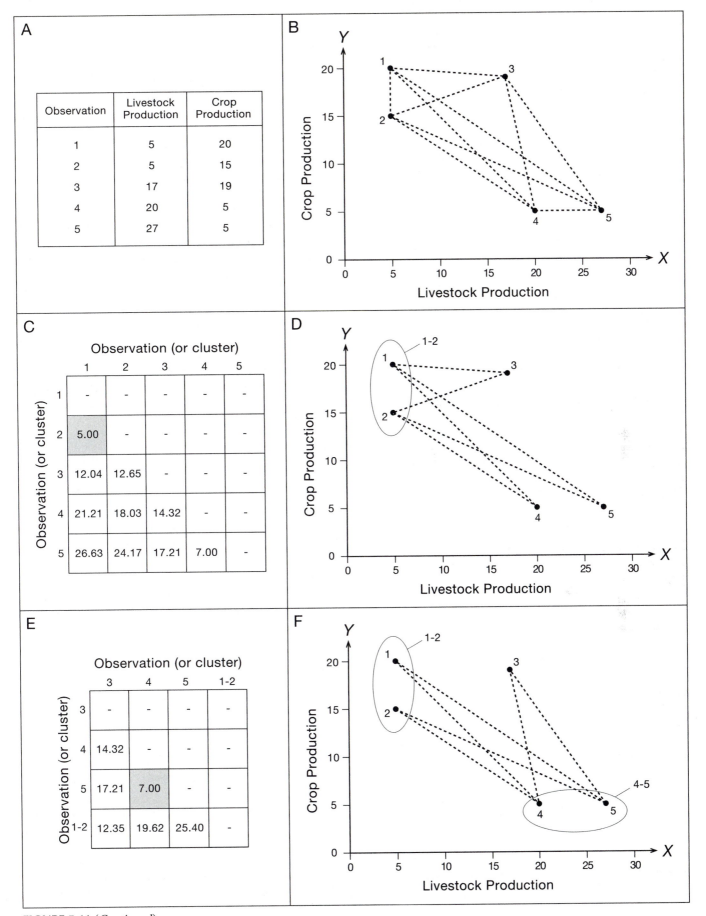

FIGURE 5.11 (*Continued*)

95

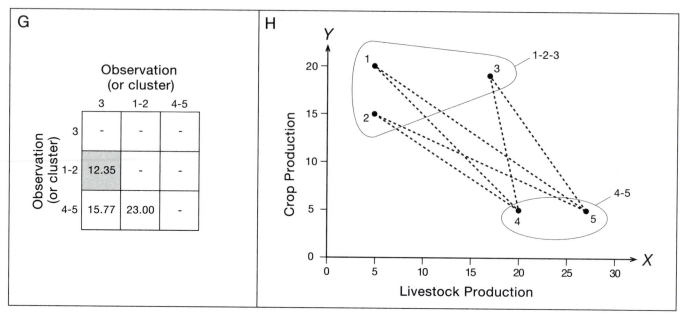

FIGURE 5.11   The basic cluster analysis process using the UPGMA method. (A) a hypothetical raw data set; (B) a graphical plot of each observation; (C) the initial resemblance matrix of Euclidean distance coefficients; (D) observations (or clusters) 1 and 2 combine because they have the smallest Euclidean distance coefficient; (E)–(H) the clustering process continues with clusters combining that have the smallest Euclidean distance coefficient.

The results of the clustering process normally are summarized in the form of a **dendrogram**, which is a treelike structure illustrating the resemblance coefficient values at which various clusters combined. For instance, for the hypothetical agricultural data, Figure 5.12 illustrates that clusters 1 and 2 combine at a Euclidean distance of 5, and clusters 4 and 5 combine at a distance of 7 (compare Figure 5.12 with Figure 5.11C and E). Figure 5.13A is a dendrogram of the results of UPGMA for the New York State data.

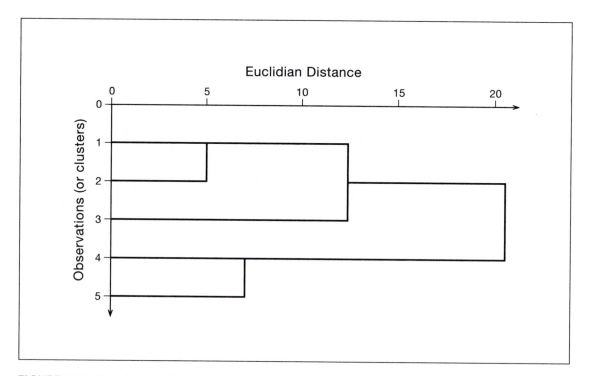

FIGURE 5.12   Dendrogram for the hypothetical data shown in Figure 5.11A using Euclidean distance as the resemblance coefficient.

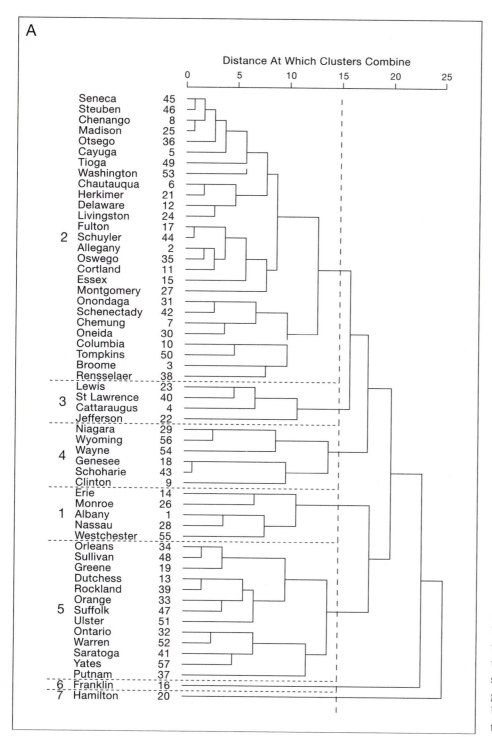

FIGURE 5.13 Dendrograms of (A) UPGMA and (B) Ward's method for the New York State data. Numbers on the left represent cluster numbers specified by the SPSS software used to generate the clusters; in the case of UPGMA, these numbers correspond to those shown in Table 5.9.

Step 5. Determine an Appropriate Number of Clusters
One key aspect of cluster analysis is determining an appropriate number of clusters. As we can see in the dendrograms, we begin with each observation as a separate cluster, and ultimately combine all observations to create a single cluster. Clearly the latter is not our desire as our goal is to create *groups* of observations, with each group being relatively homogeneous and different from another group. One common approach for selecting an appropriate number of clusters is to look for breaks in the dendrogram—places where combining two clusters would require that you make a fairly large jump in the resemblance coefficient. One simple way to do this is to cover the dendrogram with a piece of paper and then

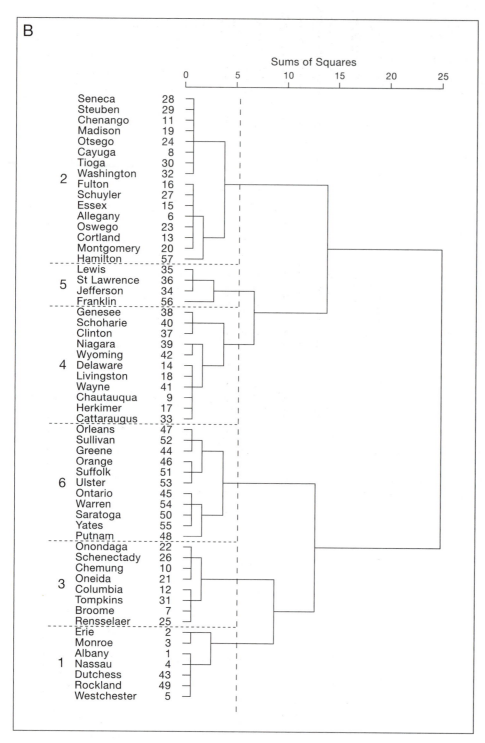

FIGURE 5.13 (*Continued*)

gradually slide the paper along the dendrogram (in our case from right to left). If you do this for the New York State data (Figure 5.13A), you will find a relatively large break for three clusters, but this produces the relatively uninteresting result of all but two observations falling in a single cluster. If you continue to slide the paper, you will find another relatively large break at seven clusters. Note that we have marked this solution in Figure 5.13A

with a line of long dashes and indicated the composition of clusters with lines of short dashes.

Step 6. Interpret the Clusters

One of the challenges of cluster analysis is interpreting the resulting clusters. One approach is to compute, for each attribute, the mean z score for all observations falling in each cluster; for instance, Table 5.9 shows the mean z

TABLE 5.9 Mean *z* scores for attributes on each cluster

Cluster	% Population Change	% Unemployed	% African American	Infant Mortality Rate
1	.02	−.86	2.19	−.08
2	−.56	−.03	−.40	−.23
3	−.40	1.95	−.50	.33
4	−.47	.31	−.17	2.07
5	1.37	−.67	.26	−.48
6	1.74	1.95	.63	1.36
7	−.04	2.35	−1.03	−2.26

scores for the New York State data for the UPGMA method. We can name a cluster by using attributes with mean values that are relatively high or low (values near ±1.0 represent moderately high or low scores, whereas values near ±2.0 represent distinctly high and low scores). Thus, cluster 1 has moderately low unemployment and a high percentage of African Americans, whereas cluster 6 has a rapidly increasing population, high unemployment, and moderately high infant mortality. (Keep in mind that these statements are being made relative to other counties in New York State. A high value here might be low relative to some other area of the country.)

Step 7. Map the Resulting Clusters

Once you have selected an appropriate set of clusters, you will want to map the results (Color Plate 5.1A) . In symbolizing clusters, you should bear in mind that clusters are likely to be qualitatively different, and so symbols used to represent the clusters should reflect this qualitative difference. For the New York State data, we chose to use qualitatively different hues developed by Cynthia Brewer and Mark Harrower (*http://www.ColorBrewer.org*), we discuss their program ColorBrewer more fully in Chapter 13.

Step 8. Determine Whether Clusters Are Really Appropriate

One should be cautious in interpreting the results of a cluster analysis, as it is possible to cluster any data set, even a set of random numbers, for multiple attributes. One approach used to judge the effectiveness of a cluster analysis is the **cophenetic correlation coefficient**, which measures the correlation between raw resemblance coefficients (Euclidean distance values in the case of UPGMA) and resemblance coefficients derived from the dendrogram (normally referred to as the *cophenetic* coefficients). For our hypothetical data, these coefficients are depicted in Figure 5.14 as the *resemblance* and *cophenetic* matrices, respectively. Values for the resemblance matrix are taken from Figure 5.11C, and those for the cophenetic matrix are determined by considering the Euclidean distance at which observations combine in the dendrogram shown in Figure 5.12. For instance, cluster 1-2-3 does not combine with cluster 4-5 until a distance of 20.26, and so cophenetic values for observations 1 and 4, 2 and 4, and 3 and 4 are all 20.26. Once the two matrices are determined, they are converted to two linear lists of *X* and *Y* values, as shown in Figure 5.14. A correlation coefficient *r* (see section 3.3.2) is then computed between *X* and *Y*. For the matrices shown in Figure 5.14, *r* is equal to

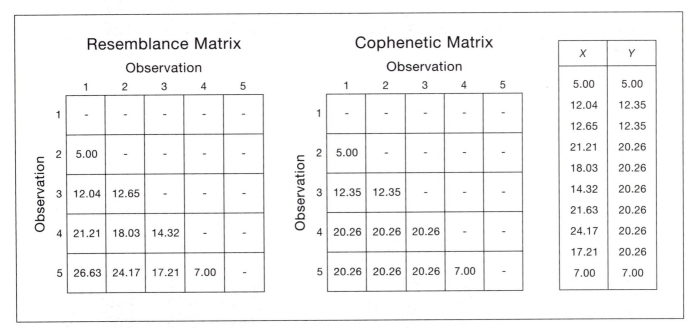

FIGURE 5.14 Computation of the cophenetic correlation involves relating corresponding values in the resemblance and cophenetic matrices.

.91, which suggests that the clusters are reflecting the true relationships in the data, as the maximum possible r is 1.0.* Unfortunately, we could not compute a cophenetic value for the New York State data, as the SPSS software we used does not provide this capability.

5.3.2 Ward's Method

A slight variation of the eight-step process described in the preceding section is necessary for **Ward's method**, which is another commonly used hierarchical method. In Ward's method, clusters are combined that lead to the smallest increase in sums of squares (SS) about the means of clusters—in this sense, only one resemblance coefficient is possible. At the beginning of

---

* For more sophisticated approaches to evaluating cluster solutions, see Aldenderfer and Blashfield (1984, 62–74).

the clustering process, the Total SS is 0 because each observation is a separate cluster (the mean score for an attribute within a cluster is simply equal to the value of the attribute). To determine which clusters need to be combined, we need to calculate what the increase in SS will be for each possible combination. To illustrate, let's presume that we wish to cluster the five hypothetical observations shown in Table 5.10A. When going from five to four clusters, there are 10 possible ways of combining clusters, just as there were 10 Euclidean distance values computed in Figure 5.11C. Computations for one of these combinations (where observations 1 and 2 are grouped) are shown in Table 5.10B, and computations for another (where observations 2 and 3 are combined) are shown in Table 5.10C; the SS for eight other combinations also would need to be computed. It turns out that Table 5.10B produces the smallest increase in SS, and so this is the appropriate combination.

TABLE 5.10   Some basic computations involved in Ward's cluster analysis method

(A) A hypothetical data set to be clustered

		Observation (or cluster)				
		1	2	3	4	5
Attribute	1	5	5	17	20	27
	2	20	15	19	5	5

(B) Sums of squares (SS), presuming that observations 1 and 2 combine to form a cluster

Cluster (1-2)

		Observation			
		1	2	$\overline{X}$	SS
Attribute	1	5	5	5	$(5-5)^2 + (5-5)^2$
	2	20	15	17.5	$+(20-17.5)^2 + (15-17.5)^2$
					12.5

Cluster 3

		Observation		
		3	$\overline{X}$	SS
Attribute	1	17	17	$(17-17)^2$
	2	19	19	$+(19-19)^2$
				0

Cluster 4

		Observation		
		4	$\overline{X}$	SS
Attribute	1	20	20	$(20-20)^2$
	2	5	5	$+(5-5)^2$
				0

Cluster 5

		Observation		
		5	$\overline{X}$	SS
Attribute	1	27	27	$(27-27)^2$
	2	5	5	$+(5-5)^2$
				0

Total SS = 12.5 + 0 + 0 + 0 = 12.5

(C) Sums of squares (SS), presuming that observations 2 and 3 combine to form a cluster

Cluster 1

		Observation		
		1	$\overline{X}$	SS
Attribute	1	5	5	$(5-5)^2$
	2	20	20	$+(20-20)^2$
				0

Cluster (2-3)

		Observation			
		2	3	$\overline{X}$	SS
Attribute	1	5	17	11	$(5-11)^2 + (17-11)^2$
	2	15	19	17	$+(15-17)^2 + (19-17)^2$
					80

Cluster 4

		Observation		
		4	$\overline{X}$	SS
Attribute	1	20	20	$(20-20)^2$
	2	5	5	$+(5-5)^2$
				0

Cluster 5

		Observation		
		5	$\overline{X}$	SS
Attribute	1	27	27	$(27-27)^2$
	2	5	5	$+(5-5)^2$
				0

Total SS = 0 + 80 + 0 + 0 = 80

The approach that we have outlined here would be used at each subsequent step in the clustering process, with those clusters being combined that lead to the smallest increase in SS. This approach is analogous to the variant of the Fisher–Jenks algorithm that minimizes the sum of squared deviations about the mean (see section 5.1.6). In Ward's method, however, note that we are dealing with multiple attributes and we do not necessarily achieve an optimal result at each step (in terms of having the minimum possible SS).

Figure 5.13B and Color Plate 5.1B depict, respectively, a dendrogram and map of the results for Ward's method for the New York State data. Because the dendrogram for Ward's method has a different shape than the UPGMA method, it does not make sense to use the same number of clusters as for the UPGMA method—for Ward's method a relatively large break in the dendrogram appears at six rather than seven clusters. Also note that individual observations did not form separate clusters, as was the case for the UPGMA method. It is not uncommon to have different clustering strategies produce somewhat different results. Generally, geographers have tended to favor Ward's method, but we have utilized both techniques here to illustrate the range of clustering approaches.

SUMMARY

In this chapter, we initially examined several common methods of data classification that combine raw data values into groups (or classes). Methods that do consider how data are distributed along the number line (e.g., **natural breaks** and **optimal**) are desirable because they place similar data values in the same class (and dissimilar data values in different classes). Methods that do not consider the distribution of data along the number line (e.g., **equal intervals** and **quantiles**) might, however, also be desirable because they satisfy other criteria. For example, equal intervals is desirable because it is easy to understand, easy to compute, and has an easily understood legend (at least when rounded percentage data are used, such as 0–100 percent urban). In Chapter 18, we will see that certain methods (e.g., quantiles) might also be useful in map comparison.

One limitation of traditional classification methods (e.g., equal intervals and optimal) is that they fail to consider the spatial context of the data. If the purpose of classification is to simplify the appearance of the map, it can be argued that spatial context should also be considered. In this chapter, we looked at two approaches that consider spatial context when simplifying a choropleth map. The first used the optimal classification method but incorporated a spatial constraint by requiring that values falling in the same class be contiguous, whereas the second employed no classification approach, but rather smoothed the raw data by changing values for enumeration units as a function of values of neighboring units. It will be interesting to see to what extent commercial software vendors adopt such novel models.

In this chapter, we also examined cluster analysis methods, which are appropriate when we wish to combine multiple numeric attributes on a single map; for example, we might want to know which census tracts are similar in terms of voting behavior, income, and attendance in private schools. In the eight-step clustering process, we saw that there are numerous issues that must be tackled in cluster analysis, including standardizing the data, selecting a clustering method (we considered two hierarchical methods—UPGMA and Ward's), selecting an appropriate number of clusters, interpreting those clusters, and symbolizing the clusters. We should caution you that we have only touched the surface of cluster analysis, as entire books are written on this topic.

FURTHER READING

Brewer, C. (2001) "Reflections on mapping Census 2000." *Cartography and Geographic Information Science* 28, no. 4:213–235.

> Describes how meaningful breaks, rounded breaks, and breaks that were identical across a set of maps were used to design an atlas of 2000 U.S. Census data.

Chang, K. (1982) "Multi-component quantitative mapping." *The Cartographic Journal* 19, no. 2:95–103.

> Discusses a range of approaches for mapping multiple attributes—cluster analysis is one of the methods covered.

Coulson, M. R. C. (1987) "In the matter of class intervals for choropleth maps: With particular reference to the work of George F. Jenks." *Cartographica* 24, no. 2:16–39.

> Summarizes data classification methods, with an emphasis on Jenks's work.

Cromley, R. G. (1996) "A comparison of optimal classification strategies for choroplethic displays of spatially aggregated data." *International Journal of Geographical Information Systems* 10, no. 4:405–424.

> Illustrates a variety of criteria for optimally classifying data by treating classification as an integer programming problem; also see Cromley and Mrozinski (1997; 1999).

Evans, I. S. (1977) "The selection of class intervals." *Transactions, Institute of British Geographers (New Series)* 2, no. 1:98–124.

> A classic article on methods of data classification.

Herzog, A. (1989) "Modeling reliability on statistical surfaces by polygon filtering." In *Accuracy of Spatial Databases*, ed. by M. Goodchild and S. Gopal, pp. 209–218. London: Taylor & Francis.

> A detailed discussion of Herzog's method for simplifying patterns on choropleth maps.

Jenks, G. F., and Caspall, F. C. (1971) "Error on choroplethic maps: Definition, measurement, reduction." *Annals, Association of American Geographers* 61, no. 2:217–244.

The classic article on optimal data classification.

Lindberg, M. B. (1990) "Fisher: A Turbo Pascal unit for optimal partitions." *Computers & Geosciences* 16, no. 5:717–732.

Describes a computer implementation of the Fisher–Jenks optimal method of data classification.

Maxwell, B. A. (2000) "Visualizing geographic classifications using color." *The Cartographic Journal* 37, no. 2:93–99.

Presents a method for selecting colors to depict clusters of data associated with narrow coastal zones.

Monmonier, M. S. (1982) "Flat laxity, optimization, and rounding in the selection of class intervals," *Cartographica* 19, no. 1:16–27.

Argues that round-number breaks should be used with optimal data classification to create a more readable map.

Romesburg, H. C. (1984) *Cluster Analysis for Researchers.* Belmont, CA: Lifetime Learning Publications.

Provides an overview of methods for cluster analysis; also see Aldenderfer and Blashfield (1984) and Everitt et al. (2001).

Rowles, R. A. (1991) Regions and Regional Patterns on Choropleth Maps. Unpublished PhD dissertation, University of Kentucky, Lexington, KY.

Compares regions on an optimally classified choropleth map with those formed by a cluster analysis method.

6

Scale and Generalization

OVERVIEW

*This chapter presents basic concepts in scale and generalization. Scale is a fundamental geographic principle, although there is often confusion about the exact meaning of **geographic scale**, **cartographic scale**, and **data resolution**. Geographers think about scale as area covered, where large-scale studies cover a large region such as a state. Cartographers think about scale mathematically and use the standard **representative fraction (RF)** to express the relationship between map and Earth distance. For instance, most national map series are established at a specific scale, such as the French 1: 25,000 BD Topo, and the USGS's 1:24,000 series. Data resolution refers to the granularity of the data, such as the pixel size of a remote sensing image. Directly related to the concept of scale is the idea of generalization, and modifying the information content so that it is appropriate at a given scale. It would not be possible, for instance, to depict the street network for the entire United States with the country mapped at a scale that would fit one page—only major highways could be depicted.*

*Section 6.1 introduces the concept of geographic and cartographic scale and covers how scale controls the amount of map space and thus the appropriate information content. The concepts of cartographic and geographic scale and representative fraction are explained. Section 6.2 provides some basic definitions of **generalization**, including a discussion of some fundamental generalization operations. Section 6.3 discusses several conceptual models of the generalization process. One of the more complete models divides the process into why, when, and how components of generalization. Section 6.4 describes the many operations that have been designed for the generalization process and provides frameworks for their organization. In particular, extensive discussions of the **simplification** and **smoothing** operations are included. Section 6.5 illustrates a variety of* generalization methods applied to several scales of a TIGER database for the Tampa-St. Petersburg area of Florida.

6.1 GEOGRAPHIC AND CARTOGRAPHIC SCALE

Scale is a fundamental concept in all of science and is of particular concern to geographers, cartographers, and others interested in geospatial data. Astronomers work at a spatial scale of light years, physicists work at the atomic spatial scale in mapping the Brownian motion of atoms, and geographers work at spatial scales from the human to the global. Within the fields of geography and cartography, the terms **geographic scale** and **cartographic scale** are often confused. Geographers and other social scientists use the term scale to mean the extent of the study area, such as a neighborhood, city, region, or state. Here, large scale indicates a large area—such as a state—whereas small scale represents a smaller entity—such as a neighborhood. Climatologists, for instance, talk about large-scale global circulation in relation to the entire Earth; in contrast, urban geographers talk about small-scale gentrification of a part of a city. Alternatively, cartographic scale is based on a strict mathematical principle: the **representative fraction (RF)**. The RF, which expresses the relationship between map and Earth distances, has become the standard measure for map scale in cartography. The basic format of the RF is quite simple, where RF is expressed as a ratio of map units to earth units (with the map units standardized to 1). For example, an RF of 1:25,000 indicates that one unit on the map is equivalent to 25,000 units on the surface of the Earth. The elegance of the RF is that the measure is unitless—with our example the 1:25,000 could represent inches, feet, or meters. Of course, in the same way that $\frac{1}{2}$ is a larger fraction than $\frac{1}{4}$, 1:25,000 is a larger scale than 1:50,000. Related to this concept, a scale of 1:25,000 depicts relatively little area but in much greater detail, whereas a scale of 1:250,000 depicts a larger area in less detail. Thus, it is the cartographic scale that determines the mapped space and level of geographic detail possible. It is important to realize that with cartographic scale the larger the representative fraction, the more detail (information content) is possible; the smaller the representative fraction, the less detail is possible. At the extreme, architects work at very large scales, perhaps 1:100, where individual rooms and furniture can be depicted, whereas a standard globe might be constructed at a scale of 1:30,000,000, allowing for only the most basic of geographic detail to be provided. As noted in Chapter 11, there are design issues that have to be considered when representing scales on maps, and a variety of methods for representing scale, including the RF, the verbal statement, and the graphical bar scale.

The term **data resolution**, which is related to scale, indicates the granularity of the data that is used in mapping. If mapping population characteristics of a city—an urban scale—the data can be acquired at a variety of resolutions, including census blocks, block groups, tracts, and even minor civil divisions (MCDs). Each level of resolution represents a different "grain" of the data. Likewise, when mapping biophysical data using remote sensing imagery, a variety of spatial resolutions are possible based on the sensor. Common grains are 79 meters (Landsat Multi-Spectral Scanner), 30 meters (Landsat Thematic Mapper), 20 meters (SPOT HRV multispectral), and 1 meter (Ikonos panchromatic). Low resolution refers to coarser grains (counties) and high resolution refers to finer grains (blocks). Cartographers must be careful to understand the relationship among geographic scale, cartographic scale, and data resolution, and how these influence the information content of the map.

6.1.1 Multiple-Scale Databases

Increasingly, cartographers and other geographic information scientists require the creation of multiscale/multiresolution databases from the same digital data set. This assumes that one can generate, from a master database, additional versions at a variety of scales. The need for such multiple-scale databases is a result of the requirements of the user. For instance, when mapping census data at the county level a user might wish to have significant detail in the boundaries. Alternatively, when using the same boundary files at the state level, less detail is needed. Because the generation of digital spatial data is extremely expensive and time-consuming, one master version of the database is often created and

smaller scale versions are generated from this master scale. Further details are provided later.

6.2 DEFINITIONS OF GENERALIZATION

6.2.1 Definitions of Generalization in the Manual Domain

Generalization is the process of reducing the information content of maps due to scale change, map purpose, intended audience, and/or technical constraints. For instance, when reducing a 1:24,000 topographic map (large scale) to 1:250,000 (small scale), some of the geographical features must be either eliminated or modified because the amount of map space is significantly reduced. Of course, all maps are to some degree generalizations, as it is impossible to represent all features from the real world on a map, no matter what the scale. A quote from Lewis Carroll's (1893) *Sylvie and Bruno Concluded* nicely illustrates this concept:

> "That's another thing we've learned from your Nation," said Mein Herr, "map making." "But we've carried it much further than you. What do you consider the largest map that would be really useful?"
>
> "About six inches to the mile."
>
> "Only about six inches!" exclaimed Mein Herr. "We very soon got to six yards to the mile. Then we tried a hundred yards to the mile. And then came the grandest idea of all! We actually made a map of the country on a scale of a mile to the mile!"
>
> "Have you used it much?" I enquired.
>
> "It has never been spread out yet," said Mein Herr. "The farmers objected: they said it would cover the whole country, and shut out the sunlight! So now we use the country itself, as its own map, and I assure you it does nearly as well."

Cartographers have written on the topic of cartographic generalization since the early part of the 20th century. Max Eckert, the seminal German cartographer and author of *Die Kartenwissenschaft*, wrote in 1908, "In generalizing lies the difficulty of scientific map making, for it no longer allows the cartographer to rely merely on objective facts but requires him to interpret them subjectively" (p. 347). Other cartographers also have struggled with the intrinsic subjectivity of the generalization process as they have attempted to understand and define cartographic generalization. For instance, in 1942 John K. Wright argued that, "Not all cartographers are above attempting to make their maps seem more accurate than they actually are by drawing rivers, coasts, form lines, and so on with an intricacy of detail derived largely from the imagination" (p. 528). Wright identified two major components of the generalization process: simplification—the reduction of raw information that is too intricate—and amplification—the enhancement of information that is too sparse. This idea that generalization could be broken down into a logical set of processes, such as simplification and amplification, has become a common theme in generalization research.

Erwin Raisz (1962), for example, identified three major components of generalization: combination, omission, and simplification. Arthur Robinson and his colleagues (1978) identified four components: selection, simplification, classification, and symbolization. In Robinson et al.'s model, selection was considered a preprocessing step to generalization itself. Selection allowed for the identification of certain features and feature classes whereas generalization applied the various operations, such as simplification. This is detailed in their model, as discussed later.

6.2.2 Definitions of Generalization in the Digital Domain

In a digital environment, Robert McMaster and Stuart Shea (1992) noted that "the generalization process supports a variety of tasks, including: digital data storage reduction; scale manipulation; and statistical classification and symbolization. *Digital generalization* can be defined as the process of deriving, from a data source, a symbolically or digitally-encoded cartographic data set through the application of spatial and attribute transformations" (p. 3). McMaster and Shea listed the objectives of digital generalization as: (1) the reduction in scope and amount, type, and cartographic portrayal of mapped or encoded data consistent with the chosen map purpose and intended audience; and (2) the maintenance of graphical clarity at the target scale. The theoretical "problem" of generalization in the digital domain is straightforward: the identification of areas to be generalized and the application of appropriate operations, as discussed later.

6.3 MODELS OF GENERALIZATION

To better understand the complexity of generalization, researchers have attempted to design conceptual models of the process. Some efforts have focused on fundamental operations and the relationship among them, whereas others have created complex models.

6.3.1 Robinson et al.'s Model

Arthur Robinson and his colleagues (1978) developed one of the first formal models or frameworks to better understand the generalization process. They separated the process into two major steps: selection (a preprocessing step) and the actual process of generalization, which involves the geometric and statistical manipulation

of objects. *Selection* involves the identification of objects to retain in (or eliminate from) the database. For instance, in developing content for a thematic map, often a minimal amount of base material is selected, such as major roads, political boundaries, or urbanized areas. Detailed base information, such as place names and hydrologic networks, are often eliminated because this base information is not deemed critical. On the other hand, considerable base information is often selected for detailed topographic maps, as this information is deemed critical. *Generalization* involves the processes of simplification, classification, and symbolization. Simplification is the elimination of unnecessary detail in a feature, classification involves the categorization of objects, and symbolization is the graphic encoding. Simplification is discussed further shortly, and both classification and symbolization are covered in other chapters in this book. In the 1970s, Joel Morrison (1974) formalized this series of steps (selection, simplification, classification, and symbolization) in the form of a set theory model. By applying set theory, he attempted to show how the basic data content was transformed through the generalization process. He defined each of the operations in terms of their set properties, including "one-to-one" (injective) and "onto" (surjective). For instance, if information is lost in the generalization process, then the relationship cannot be one-to-one. Although somewhat beyond the scope of this introductory book, the basic idea was to clarify the processes of generalization through formal mathematics.

6.3.2 Kilpeläinen's Model

Although the European literature contains numerous conceptual frameworks, few have had as significant an influence on American workers as the models of Tiina Kilpeläinen (1997) and Kurt Brassel and Robert Weibel (1988). Kilpeläinen developed alternative frameworks for the representation of multiscale databases. Assuming a master cartographic database, called the Digital Landscape Model (DLM), she proposed a series of methods for generating smaller scale Digital Cartographic Models (DCMs). The master DLM is the largest scale, most accurate database possible, whereas secondary DLMs are generated for smaller scale applications (Figure 6.1). The DLM is only a computer representation and cannot be visualized. DCMs, on the other hand, are the actual graphical representations, derived through a generalization or symbolization of the DLM. In her model, each DCM is generated directly from the initial master database or from the previous version. A separate DLM is created for each scale or resolution, and the DCM is directly generated from each DLM. The master DLM is used to generate smaller scale DLMs, which are then

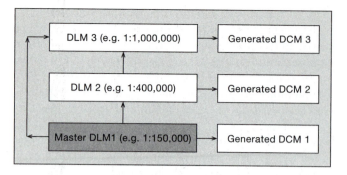

FIGURE 6.1 Kilpeläinen's notion of the Digital Landscape Model (DLM) and Digital Cartographic Model (DCM).

used to generate a DCM at that level. The assumption is that DCMs are generated on an as-needed basis.

6.3.3 Brassel and Weibel's Model

Brassel and Weibel (1988) have worked extensively at The University of Zurich in developing methods for terrain generalization. Their research has two primary objectives: to design a strategy for terrain generalization that is adaptive to different terrain types, scales, and map purposes, and to implement this strategy in an automated environment as fully as possible. Toward these ends, they have developed a model of terrain generalization that consists of five major stages: structure recognition, process recognition, process modeling, process execution, and data display and evaluation of results. In structure recognition, the specific cartographic objects—as well as their spatial relations and measures of importance—are selected from the source data. Process recognition identifies the necessary generalization operators and parameters by determining "specifically how the source data are to be transformed, which types of conflicts have to be identified and resolved, and which types of objects and structures have to be carried into the target database" (Weibel 1992, 134). Process modeling then compiles the rules and procedures—the exact algorithmic instructions—from a process library: a digital organization of these rules.

The final stages of Brassel and Weibel's model involve process execution, in which the rules and procedures are applied to create the generalization, data display, and finally, evaluation. Specifically, their model includes three different generalization methods: a global filtering, a selective (iterative) filtering, and a heuristic approach based on the generalization of the terrain's structure lines. For a given generalization problem that is constrained by the terrain character, map objective, scale, graphic limits, and data quality, the appropriate technique is selected through structure and process recognition procedures. The authors also depict the application of specific generalization operations, including selection, simplification, combination, and displacement, to illustrate the application of these operations to digital terrain models.

6.3.4 McMaster and Shea's Model

In an attempt to create a comprehensive conceptual model of the generalization process, McMaster and Shea (1992) identified three significant components: the theoretical objectives, or *why to generalize*; the cartometric evaluation, or *when to generalize*; and the fundamental operations, or *how to generalize* (Figure 6.2).

Why Generalization Is Needed: The Theoretical Objectives of Generalization

The theoretical or conceptual elements of generalization include reducing complexity, maintaining spatial accuracy, maintaining attribute accuracy, maintaining aesthetic quality, maintaining a logical hierarchy, and consistently applying the rules of generalization. Reducing complexity is perhaps the most significant goal of generalization. The question for the cartographer is relatively straightforward: How does one take a map at a scale of, perhaps, 1:24,000 and reduce it to 1:100,000? More important, the question is how the cartographer reduces the information content so that it is appropriate for the scale. Obviously, the complexity of detail that is provided at a scale of 1:24,000 cannot be represented at 1:100,000; some features must be eliminated and some detail must be modified. For centuries, through considerable experience, cartographers developed a sense of what constituted

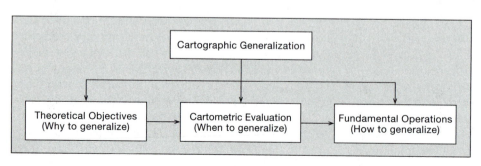

FIGURE 6.2 An overview of McMaster and Shea's model of generalization.

appropriate information content. Figure 6.3 nicely illustrates this point. This figure depicts the very same feature—a portion of Ponta Creek in Kemper and Lauderdale Counties, Mississippi—at four different scales: 1:24,000, 1:50,000, 1:100,000, and 1:250,000. These features, digitized by Philippe Thibault (2002) in his doctoral dissertation, effectively show the significantly different information content as one reduces scale from 1:24,000 to 1:250,000. In the top portion of the illustration, the general form of the line stays the same, although the fine-level detail is lost. The bottom part of the illustration depicts an enlargement of the smaller scale features to match the feature at 1:24,000. Note, for instance, the enlargement of the 1:250,000 scale line by 1,041.67 percent to match Ponta Creek at 1:24,000. In this case, the cartographer has manually generalized Ponta Creek through a series of transformations including simplification, smoothing, and enhancement (as described later) as a holistic process, unlike current computer approaches that require a separation of these often linked processes. The set of decisions required to generalize cartographic features based on their inherent complexity is difficult if not impossible to quantify, although as described next, several attempts have been made over the past decade.

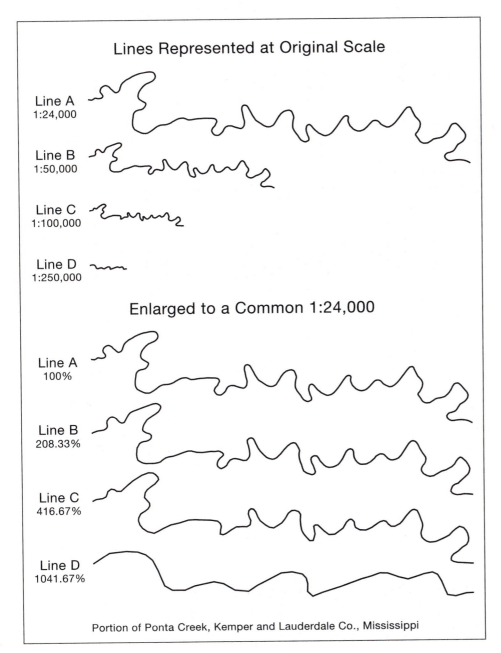

Lines Represented at Original Scale

Line A
1:24,000

Line B
1:50,000

Line C
1:100,000

Line D
1:250,000

Enlarged to a Common 1:24,000

Line A
100%

Line B
208.33%

Line C
416.67%

Line D
1041.67%

Portion of Ponta Creek, Kemper and Lauderdale Co., Mississippi

FIGURE 6.3 Depiction of Ponta Creek (in Mississippi) at four different scales. (Courtesy of Philippe Thibault.)

Clearly, there is a direct and strong relationship among scale, information content, and generalization. John Hudson (1992) explained the effect of scale by indicating what might be depicted on a map 5 by 7 inches:

- A house at a scale of 1:100
- A city block at a scale of 1:1,000
- An urban neighborhood at a scale of 1:10,000
- A small city at a scale of 1:100,000
- A large metropolitan area at a scale of 1:1,000,000
- Several states at a scale of 1:10,000,000
- Most of a hemisphere at a scale of 1:100,000,000
- The entire world with plenty of room to spare at a scale of 1:1,000,000,000

He explained that these examples, which range from largest ($1{:}10^2$) to smallest ($1{:}10^9$), span eight orders of magnitude and a logical geographical spectrum of scales. Geographers work at a variety of scales, from the very large—the neighborhood—to the very small—the world. Generalization is a key activity in changing the information content so that it is appropriate for these different scales. However, a rough guideline that cartographers use is that scale change should not exceed 10×. Thus if you have a scale of 1:25,000, it should only be used for generalization up to 1:250,000. Beyond 1:250,000, the original data are "stretched" beyond their original fitness for use.

Two additional theoretical objectives important in generalization are maintaining the spatial and attribute accuracy of features. Spatial accuracy deals primarily with the geometric shifts that necessarily take place in generalization. For instance, in simplification coordinate pairs are deleted from the data set. By necessity, this shifts the geometric location of the features, creating "error." The same problem occurs with feature displacement, where two features are pulled apart to prevent a graphical collision. A goal in the process is to minimize this shifting and maintain as much spatial accuracy as possible. Attribute accuracy deals with the subject being mapped, or the statistical information such as population density or land use. For instance, classification, a key component of generalization, often degrades the original "accuracy" of the data through data aggregation.

When Generalization Is Required

In a digital cartographic environment, it is necessary to identify those specific conditions where generalization will be required. Although many such conditions can be identified, six of the fundamental conditions include:

1. Congestion
2. Coalescence
3. Conflict
4. Complication
5. Inconsistency
6. Imperceptibility

As explained by McMaster and Shea (1992), *congestion* refers to the problem when, under scale reduction, too many objects are compressed into too small a space, resulting in overcrowding due to high feature density (Figure 6.4). Significant congestion results in decreased communication, for instance, where too many buildings are in close proximity. *Coalescence* refers to the condition where features graphically collide due to scale change. In these situations, features actually touch. This condition thus requires the implementation of the displacement operation, as discussed shortly. The condition of *conflict* results when, due to generalization, an inconsistency between or among features occurs. For instance, if generalization of a coastline eliminated a bay with a city located on it, either the city or the coastline would have to be moved to ensure that the urban area remained on the coast. Such spatial conflicts are difficult to both detect and correct. The condition of *complication* is dependent on the specific conditions that exist in a defined space. An example is a digital line that changes in complexity from one part to the next, such as a coastline that progresses from very smooth to very crenulated, like Maine's coastline. Barbara Buttenfield (1991) demonstrated the use of line-geometry-based structure signatures as a means for controlling the tolerance values, based on complexity, in the generalization process. Later, details are provided on other techniques for detecting changes in linear complexity.

Despite the fact that many problems in generalization require the development and implementation of mathematical, statistical, or geometric measures, little work on generalization measurement has been reported. Two basic types of measures can be identified: procedural and quality assessment. Additionally, some measures are used to assess individual features, whereas others are utilized in a more global manner.

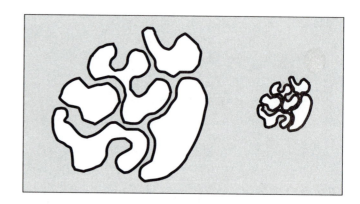

FIGURE 6.4   How a change in scale can create congestion.

Procedural measures are those needed to invoke and control the process of generalization. Such measures might include those to: (1) select a simplification algorithm, given a certain feature class; (2) modify a tolerance value along a feature as the complexity changes; (3) assess the density of a set of polygons being considered for agglomeration; (4) determine whether a feature might undergo a type change (e.g., area to point) due to scale modification; and (5) compute the curvature of a line segment to invoke a smoothing operation. *Quality assessment measures* evaluate both individual operations, such as the effect of simplification, and the overall quality of the generalization (e.g., poor, average, excellent). Several studies have reported on mathematical and geometric measures, including Buttenfield (1991), McMaster (1986; 1987) and Plazanet (1995), yet no comprehensive framework of the existing and potential measures and their characteristics has been developed.

One general classification of measures, as presented by McMaster and Shea (1992), includes the following classes: density, distribution, length and sinuosity, shape, distance, and Gestalt.

- *Density measures* are used to evaluate multifeature relationships, and can include such metrics as the number of point, line, or area features per unit area; average density of point, line, or area features; and the number and location of centroids of point, line, or area features. An example of a density measure for urban blocks might be the number of blocks within a 500-meter radius. The lakes region of northern Minnesota, the Thousand Islands in the St. Lawrence Seaway, and the deltaic region of the Mississippi River are all examples of geometries where a high degree of complexity might need simplification. European researchers apply density measures to complex building configurations in cities to delete or fuse structures together. A complicating factor here, of course, is the actual configuration and number of buildings in an urban environment. Jones et al. (1995) detailed a series of measures for such built structures based on their data structure.

- *Distribution measures* are used to assess the overall spatial configuration of the map features. For example, we might measure the dispersion, randomness, and clustering of point features. Linear features can be assessed by their overall pattern—an example would be the calculation of the distribution of a stream network based on the number of first-, second-, and third-order streams, or whether the pattern is dendritic or trellis. In a similar way, areal features can be evaluated by their intrinsic distribution, such as the spatial configuration of a series of islands.

- *Length and sinuosity measures* are often applied to single linear or areal boundary features such as the

calculation of stream network lengths. Some sample length measures include total number of coordinate pairs, total length, and the average number of coordinates or standard deviation of coordinates per inch. Sinuosity measures can include total angular change, average angular change per inch, average angular change per angle, sum of positive or negative angles, and total number of positive and negative turns (McMaster 1986). One common sinuosity measure involves calculating the individual angularity between segments, often noted as either positive or negative (Figure 6.5). Another sinuosity measure accumulates these into curvilinear segments, defined as continuous runs, or turns, of positive or negative angles. Yet another measure, as described in more detail later, computes the trend line along the curves to create a medial trend line. Additionally, specific measures for feature classes have been designed in various domains of knowledge, such as common morphometric measurements compiled from physical geography, hydrology, and geology.

- *Shape measures* have been commonly applied in the geographic literature for measuring the form of objects, and are useful in the determination of whether an area feature can be represented at a new scale (Christ 1976). In general, the most important components of shape are the overall elongation of the polygon and the efficiency or sinuosity of its boundary, but many metrics can be used: geometry of point, line, or area features; perimeter of area features; centroid of line or area features, X and Y variances of area features, covariance of X and Y area features, and the standard deviation of X and Y area features (Bachi 1973). One of the best-known shape measures was developed by Boyce and Clarke (1964). Called the radial line method, it calculates the lengths of a set of radials (the number of radials is user defined) from the centroid of a polygon to the edges of the boundary. These accumulated lengths are then compared to the set of lengths that would be expected on the most regularized form—the circle. The larger the index

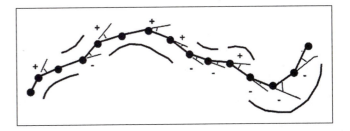

FIGURE 6.5 A common sinuosity measure involves calculating the individual angularity between segments making up a line.

value, the more the shape varies from a circle. Such a method could be applied to a series of polygons to assess basic complexity and need for generalization. Other commonly used measures compute relationships between the area and perimeter of polygons (Muehrcke et al. 2001, 358–359).

- *Distance measures* involve computing the distance between the basic geometric forms—points, lines, and areas. Distances between each of these forms can be assessed by examining the shortest perpendicular distance or shortest Euclidean distance between each form. In the case of two geometric lines, two different distance calculations exist: (1) line-to-line and (2) line buffer-to-line buffer. Figure 6.6, for instance, shows a simple straight line and the line's buffer, which is equidistant from the line itself. Such buffers are commonly used in GISs to measure the proximity of features. Distance measures related to buffers are crucial for many fundamental operations of generalization; for instance, under scale reduction the features or their respective buffers might be in conflict.

- *Gestalt measures* are based in the use of Gestalt theory, which helps to indicate perceptual characteristics of the feature distributions through an isomorphism—that is, the structural relationship that exists between a stimulus pattern and the expression it conveys (Arnheim 1974). Common examples of this relationship include closure, continuation, proximity, and similarity (Wertheimer 1958). Although the existence of these Gestalt characteristics is well documented, few techniques have been developed that would accurately serve to identify them.

6.4 THE FUNDAMENTAL OPERATIONS OF GENERALIZATION

6.4.1 A Framework for the Fundamental Operations

In the McMaster and Shea model discussed earlier, the third major component involves the fundamental operations or how to generalize. Most of the research in generalization assumes that the process can be broken down into a series of logical operations that can be classified according to the type of geometry of the feature. For instance, a simplification operation is designed for linear

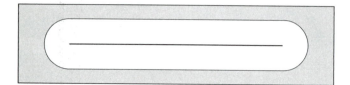

FIGURE 6.6   A line and its buffer, which is equidistant from the line.

features, whereas an amalgamation operator works on areal features. Table 6.1 provides a framework for the operations of generalization, dividing the process into those activities for vector- and raster-mode processing. The types of generalization operations for vector and raster processing are fundamentally different. Vector-based operators require more complicated strategies because they operate on strings of *x*–*y* coordinate pairs and require complex searching strategies. In raster-based generalization, it is much easier to determine the proximity relationships that are often the basis for determining conflict among the features. Next, a more detailed discussion of individual vector-based operations is provided, and Figure 6.7 provides graphic depictions of some key operations.

6.4.2 Vector-Based Operations

Simplification

Simplification is the most commonly used generalization operator. The concept is relatively straightforward, because it involves at its most basic level a "weeding" of unnecessary coordinate data. The goal is to retain as much of the geometry of the feature as possible, while eliminating the maximum number of coordinates. Most simplification routines utilize complex geometrical criteria (distance and angular measurements) in selecting significant or critical points. A general classification of simplification methods consists of five approaches: independent point routines, local processing routines,

TABLE 6.1   A framework for generalization operations (After McMaster and Monmonier 1989, and McMaster 1989b.)

Raster-mode generalization	Vector-mode generalization
Structural generalization	Point feature generalization
Simple structural reduction	Aggregation
Resampling	Displacement
Numerical generalization	Line feature generalization
Low-pass filters	Simplification
High-pass filters	Smoothing
Compass gradient masks	Displacement
Vegetation indices	Merging
	Enhancement
Numerical categorization	Areal feature generalization
Minimum-distance to means	Amalgamation
Parallelopiped	Collapse
Maximum-likelihood classification	Displacement
Categorical generalization	Volume feature generalization
Merging (of categories)	Smoothing
Aggregation (of cells)	Enhancement
Nonweighted	Simplification
Category-weighted	
Neighborhood-weighted	
Attribute change	
	Holistic generalization
	Refinement

Spatial Operator	Original Map	Generalized Map
Simplification Selectively reducing the number of points required to represent an object	15 points to represent line	13 points to represent line
Smoothing Reducing angularity of angles between lines		
Aggregation Grouping point locations and representing them as areal objects	Sample points	Sample areas
Amalgamation Grouping of individual areal features into a larger element	Individual small lakes	Small lakes clustered
Collapse Replacing an object's physical details with a symbol representing the object	Airport City boundary School	Airport Presence of city School
Merging Grouping of line features	All railroad yard rail lines	Representation of railroad yard
Refinement Selecting specific portions of an object to represent the entire object	All streams in watershed	Only major streams in watershed
Exaggeration To amplify a specific portion of an object	Bay Inlet	Bay Inlet
Enhancement To elevate the message imparted by the object	Roads cross	Roads cross; one bridges the other
Displacement Separating objects	Stream Road	Stream Road

FIGURE 6.7 Fundamental operations of generalization. (Courtesy of Philippe Thibault.)

constrained extended local processing routines, unconstrained extended local processing routines, and global methods. **Independent point routines** select coordinates based on their position along the line, nothing more. For instance, a typical nth point routine might select every third point to quickly weed coordinate data. Although computationally efficient, these algorithms are crude in that they do not account for the true geomorphological significance of a feature. **Local processing routines** utilize immediate neighboring points in assessing the significance of the point. Assuming a point to be simplified x_n, y_n, these routines evaluate its significance based on the relationship to the immediate neighboring points, x_{n-1}, y_{n-1}, and x_{n+1}, y_{n+1}. This relationship is normally determined by either a distance or angular criterion, or both. **Constrained extended local processing routines** search beyond the immediate neighbors and evaluate larger sections of lines, again normally determined by distance and angular criteria. Certain algorithms search around a larger number of points, perhaps two, three, or four in either direction, whereas others use more complex criteria. **Unconstrained extended local processing routines** also search around larger sections of a line, but the search is terminated by the geomorphological complexity of the line, not by algorithmic criterion. Finally, **global algorithms** process the entire line feature at once and do not constrain the search to subsections. The most commonly used simplification algorithm—the Douglas–Peucker—takes a global approach and processes a line "holistically." Details of the Douglas–Peucker algorithm can be found in McMaster (1987) and McMaster and Shea (1992), and comparisons of the algorithms can be found in McMaster (1987). Table 6.2 provides details on algorithms that can be found in each of the five categories.

The effect of the Douglas–Peucker algorithm can be seen in Figure 6.8, where the algorithm is applied to Hennepin County, Minnesota, using a 350-meter tolerance value. The original spatial data, taken from the United States Bureau of the Census TIGER files, is in light gray, whereas the generalized feature is in black. Note that many of the original points have been eliminated, thus simplifying the feature. Unfortunately, the effects of this approach—as with most generalization processes—are not consistent, as the algorithm behaves differently depending on the geometric or geomorphological significance of the feature. In areas that are more "natural," such as streams and rivers, the simplification produces a

TABLE 6.2   A classification of algorithms used to simplify cartographic features (After McMaster and Shea, 1992, *Generalization in Digital Cartography*. p. 73, copyright Association of American Geographers.)

Category 1:	Independent point algorithms	
	Do not account for the mathematical relationships with the neighboring pairs, operate independent of topology	
	Examples:	Nth point routine
		Random selection of points
Category 2:	Local processing routines	
	Utilize the characteristics of the immediate neighboring points in determining significance	
	Examples:	Distance between points
		Angular change between points
		Jenks's algorithm (distance and angular change)
Category 3:	Constrained extended local processing routines	
	Search continues beyond immediate coordinate neighbors and evaluates sections of a line	
	Extent of search depends on distance, angular, or number of points criterion	
	Examples:	Lang algorithm
		Opheim algorithm
		Johannsen algorithm
		Deveau algorithm
		Roberge algorithm
		Visvalingam algorithm
Category 4:	Extended local processing routines	
	Search continues beyond immediate coordinate neighbors and evaluates sections of a line	
	Extent of the search is constrained by geomorphological complexity of the line, not by algorithmic criterion	
	Example:	Reumann–Witkam algorithm
Category 5:	Global routines	
	Considers the entire line, or specified line segment; iteratively selects critical points	

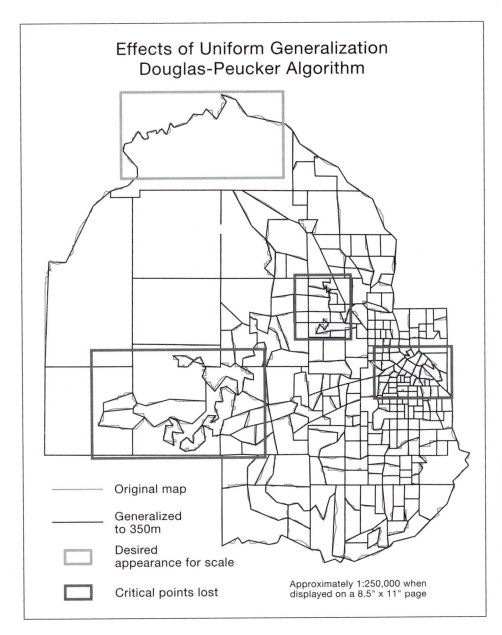

FIGURE 6.8 Overview of effects of Douglas–Peucker simplification in Hennepin County, Minnesota. (After McMaster, R. B., and Sheppard, E. (2004) "Introduction: Scale and Geographic Inquiry," p. 9. In *Scale and Geographic Inquiry: Nature, Society and Method,* ed. by E. Sheppard and R. B. McMaster. Courtesy of Blackwell Publishing.)

relatively good approximation. However, in urban areas, where the census geography follows the rectangular street network, the results are less satisfactory. In many cases, logical right angles are simplified to diagonals.

Figure 6.9 shows an enlargement of several parts of Figure 6.8. At the top, it is clear that the algorithm works well on the northern boundary of Hennepin County, which follows the Mississippi River. The bottom three enlargements depict where essential critical points have been lost, resulting in simplifications that are not deemed acceptable. A significant problem in the generalization process involves the identification of appropriate tolerance values for simplification. Unfortunately, this is mostly an experimental process, where the user has to test and

retest values until an acceptable solution is empirically found. As explained previously, cartographers often turn to measurements to ascertain the complexity of a feature and to assist in establishing appropriate tolerance values.

Figure 6.10 depicts the calculation of one specific measurement, the *trendline*, for the Hennepin County data set. The trendlines for a digitized curve are based on a calculation of angularity, or where the lines change direction. Where a curve changes direction, for example, from left to right, a mathematical inflection point is defined (theoretically, the point of no curvature). The connection of these inflection points, which indicates the general "trend" of the line, is called the trendline. The complexity of a feature can be approximated by looking

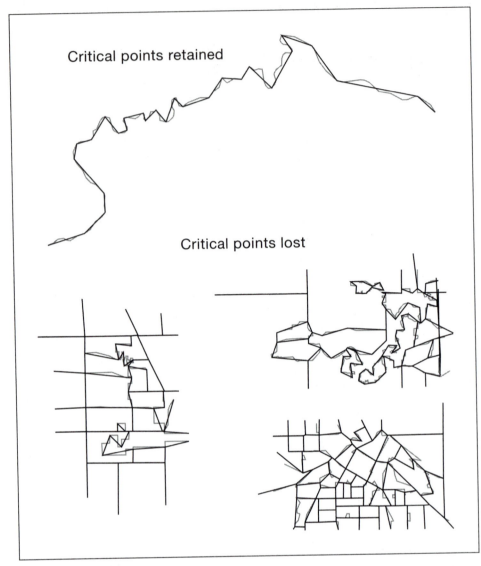

Critical points retained

Critical points lost

FIGURE 6.9   Enlargements of areas outlined in Figure 6.8. (Courtesy of National Historical Geographic Information System.)

at the trendlines for an entire feature or for the entire data set. A simple measure of complexity derived from the trendline is the trendline/total length of a line, or the sinuosity of a feature. Along relatively straight line segments, with little curvalinearity, the trendline will be very close to the curve and the trend line/total length ratio will be nearly the same (e.g., the relatively straight line near the middle of Figure 6.10). However, a highly complex curve, such as the northern border of Hennepin County, will deviate significantly from the trendline, and the length of the trendline will be greater. Thus, the greater the difference between the actual digitized curve and the trendline, the more complex the feature.

Smoothing

Although often assumed to be identical to simplification, **smoothing** is a much different process. The smoothing operation shifts the position of points to improve the appearance of the feature (Figure 6.7). Smoothing algo-

rithms relocate points in an attempt to plane away small perturbations and capture only the most significant trends of the line (McMaster and Shea 1992). As with simplification, there are many approaches for the process—a simple classification is provided in Table 6.3.

Research has shown that a careful integration of simplification and smoothing routines can produce a simplified, yet aesthetically acceptable, result (McMaster 1989a).

Aggregation

As depicted in Figure 6.7, **aggregation** involves the joining together of multiple point features, such as a cluster of buildings. This process involves grouping point locations and representing them as areal units. The critical problem in this operation is determining both the density of points needed to identify a cluster to be aggregated and the boundary around the cluster. The most common approach is to create a Delaunay triangulation of points and use measures of distance along the

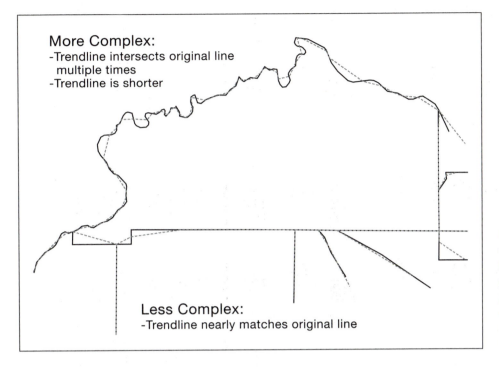

More Complex:
-Trendline intersects original line multiple times
-Trendline is shorter

Less Complex:
-Trendline nearly matches original line

FIGURE 6.10 Comparison of the original unsimplified line and a trendline, an indicator of the degree of simplification. The image is of the northern portion of Hennepin County shown in Figure 6.8. (Courtesy of National Historical Geographic Information System.)

TABLE 6.3 A classification of algorithms used to smooth cartographic features (After McMaster and Shea 1992, *Generalization in Digital Cartography*, pp. 86–87, copyright Association of American Geographers.)

Category 1:	**Weighted averaging**
	Calculates an average value based on the positions of existing points and neighbors, with only the end points remaining the same; maintains the same number of points as the original line; algorithms can be easily adopted for different smoothing conditions by adjusting tolerance values (e.g., number of points used in smoothing); all algorithms use local or extended processors
	Examples: Three-point moving average
	Five-point moving average
	Other moving average methods
	Distance-weighted averaging
	Slide averaging
Category 2:	**Epsilon filtering**
	Algorithm uses certain geometrical relationships between the points and a user-defined tolerance to smooth the cartographic line; endpoints are retained, but the absolute number of points generated for the smoothed line is algorithm dependent; approaches are local, extended local, and global
	Examples: Epsilon filtering
	Brophy algorithm
Category 3:	**Mathematical approximation**
	Develop a mathematical function or series of mathematical functions that describe the geometrical nature of the line; number of points on the smoothed line is variable and is controlled by the user; retention of the endpoints and of the points on the original line is dependent on the choice of algorithms and tolerances; function parameters can be stored and used to later regenerate the line at the required point density; approaches are local, extended local, and global
	Examples: Local processing: cubic spline
	Extended local processing: b-spline
	Global processing: bezier curve

Delaunay edges to calculate density and boundary (Jones et al., 1995).

Amalgamation

Amalgamation is the process of fusing together nearby polygons, and is needed for both noncontinuous and continuous areal data (Figure 6.7). A noncontinuous example is a series of small islands in close proximity with size and detail that cannot be depicted at the smaller scale. A continuous example is with census tract data, where several tracts with similar statistical attributes can be joined together. Amalgamation is a very difficult problem in urban environments where a series of complex buildings might need to be joined.

Collapse

The **collapse** operation involves the conversion of geometry. For instance, it might be that a complex urban area is collapsed to a point due to scale change and resymbolized with a geometric form, such as a circle. A complex set of buildings may be replaced with a simple rectangle—which might also involve amalgamation.

Merging

Merging is the operation of fusing together groups of line features, such as parallel railway lines, or edges of a river or stream (Figure 6.7). This is a form of collapse, where an areal feature is converted to a line. A simple solution is to average the two or multiple sides of a feature, and use this average to calculate the new feature's position.

Refinement

Refinement is another form of resymbolization, much like collapse (Figure 6.7). However, refinement is an operation that involves reducing a multiple set of features such as roads, buildings, and other types of urban structures to a simplified representation. The concept with refinement is that such complex geometries are resymbolized to a simpler form, a "typification" of the objects. The example of refinement shown in Figure 6.7 is a selection of a stream network to depict the "essence" of the distribution in a simplified form.

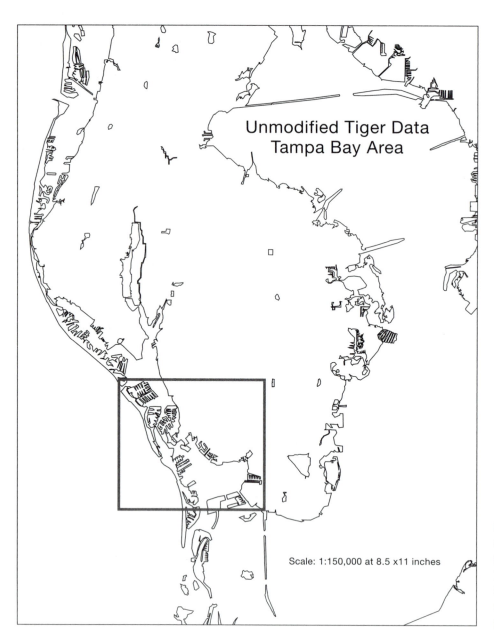

Unmodified Tiger Data
Tampa Bay Area

Scale: 1:150,000 at 8.5 x11 inches

FIGURE 6.11 Unsimplified representation of the Tampa Bay-St. Petersburg, Florida, region at a scale of 1:150,000 when displayed on a 8.5″ × 11″ page. (Courtesy of National Historical Geographic Information System.)

Exaggeration

Exaggeration is one of the more commonly applied generalization operations. Often it is necessary to amplify a specific part of an object to maintain clarity in scale reduction. The example in Figure 6.7 depicts the exaggeration of the mouth of a bay that would close under scale reduction.

Enhancement

Enhancement involves a symbolization change to emphasize the importance of a particular object. For instance, the delineation of a bridge under an existing road is often portrayed as a series of cased lines that assist in emphasizing that feature over another.

Displacement

Displacement is perhaps the most difficult of the generalization operations, as it requires complex measurement (Figure 6.7). The problem might be illustrated with a series of cultural features in close proximity to a complex coastline. Assume, for example, that a highway and railroad follow a coastline in close proximity, with a series of smaller islands offshore. In the process of scale reduction, all features would tend to coalesce. The operation of displacement would pull these features apart to prevent this coalescence. What is critical in the displacement operation is the calculation of a displacement hierarchy because one feature will likely have to be shifted away from another (Nickerson and Freeman 1986; Monmonier and McMaster 1990). A description of the mathematics involved in displacement can be found in McMaster and Shea (1992).

6.5 AN EXAMPLE OF GENERALIZATION

Figure 6.11 depicts the raw TIGER vector data for the Tampa-St. Petersburg area of Florida. These data, encoded at a scale of 1:150,000, show the complexity of both the natural and human-created coastline along the Florida coast. Figure 6.12 depicts a simplification of the inset

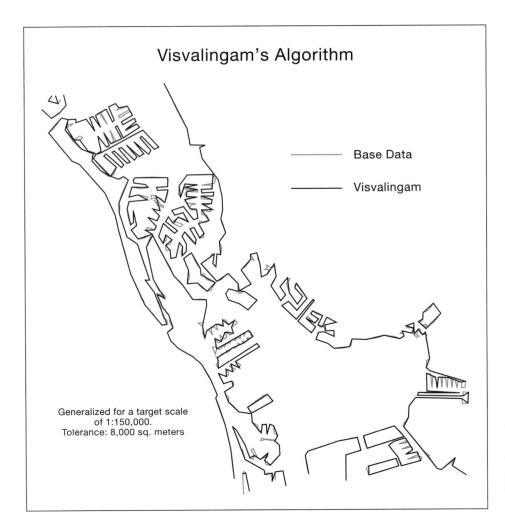

Visvalingam's Algorithm

——— Base Data

——— Visvalingam

Generalized for a target scale
of 1:150,000.
Tolerance: 8,000 sq. meters

FIGURE 6.12 Simplified representation of the inset region shown in Figure 6.11; here simplification is accomplished using Visvalingam's algorithm. (Courtesy of National Historical Geographic Information System.)

area in Figure 6.11 using the Visvalingam algorithm (Visvalingam and Williamson 1995), which uses an areal tolerance (in this case 8,000 square meters) to select critical points and is considered to be robust in maintaining the original character of the line. This is a somewhat novel approach in that most simplification algorithms use a linear distance to determine the proximity between the original feature and the simplified version. An areal tolerance measures the "area" of change as points are eliminated and the two features are displaced. When this area is too large, based on the user-defined tolerance, no further points are eliminated. Note in particular the performance of the simplification algorithm along the complicated "canaled" coastline, where it becomes difficult to retain the rectangular nature of this human-created landscape.

Although considerable developments in automated generalization have taken place over the last 30 years, it is still difficult to solve generalization problems with off-the-shelf software due to the limited capability of the algorithms and complexity of the databases. At the

National Historical Geographic Information System (NHGIS) project at the University of Minnesota (*http://www.nhgis.org/*), work is currently underway to design generalization software for specific problems, such as this coastline example. One algorithm, developed by Kai Chi Leung and programmed by Ryan Koehnen, is designed to retain the critical right angle geometry of such landscapes, while also reducing the number of canals. Figure 6.13 shows the same inset as before, but with both the Visvalingam and Leung–Koehnen algorithms applied. Note that many of the smaller canals have been generalized, and are now aggregated into larger units that will more easily be reduced through scale change.

A major goal of the NHGIS project is to provide multiple scale versions of census data to enable users to select the scale most appropriate for their use. To illustrate, consider Figure 6.14, which again depicts the Tampa-St. Petersburg area of Florida, but is intended to appear at a scale of 1:400,000; an enlargement of the central portion

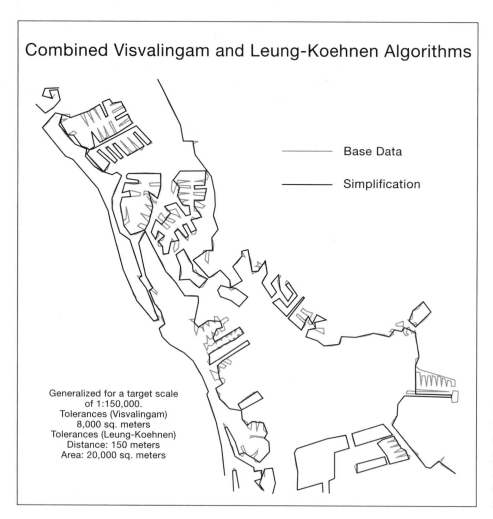

Combined Visvalingam and Leung-Koehnen Algorithms

Base Data

Simplification

Generalized for a target scale
of 1:150,000.
Tolerances (Visvalingam)
8,000 sq. meters
Tolerances (Leung-Koehnen)
Distance: 150 meters
Area: 20,000 sq. meters

FIGURE 6.13 Simplified representation of the inset region shown in Figure 6.11; here simplification is accomplished using both the Visvalingam and Leung–Koehnen algorithms. (Courtesy of National Historical Geographic Information System.)

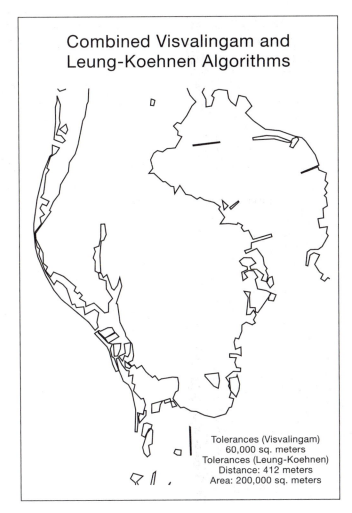

Combined Visvalingam and Leung-Koehnen Algorithms

Tolerances (Visvalingam)
60,000 sq. meters
Tolerances (Leung-Koehnen)
Distance: 412 meters
Area: 200,000 sq. meters

FIGURE 6.14 Generalized representation of the Tampa Bay-St. Petersburg, Florida, region shown in Figure 6.11. (Courtesy of National Historical Geographic Information System.)

significant parts of the coastline retained. For such problems it is often necessary to look beyond standard approaches and create custom-designed solutions.

SUMMARY

Scale is a fundamental process in geography and cartography. It requires the cartographer to select the appropriate information content given the map purpose and intended audience. The set of processes used to manipulate (change the scale of) the spatial information are collectively known as **generalization**. Cartographers have tackled the problem of generalization for centuries in the manual domain, but conversion to the digital world has created many new challenges. This chapter has provided a discussion of the various forms of scale, including **cartographic scale**, which is depicted with the representative fraction, such as 1:24,000. It should be noted, however, that geographers and other spatial scientists often conceptualize scale very differently, such as human geographers' views on the social construction or political construction of scale.

We reviewed major definitions and models of the generalization process. Generalization models include those by Robinson and his colleagues, Morrison, Weibel and Brassel, and McMaster and Shea. We have provided details of the McMaster and Shea model, which was designed for the generalization process in a digital environment. A critical part of the generalization process involves the identification and implementation of the fundamental operations, such as line **simplification**. For each of the operations, multiple approaches or computer algorithms have been designed, such as the Douglas and Peucker simplification routine. Fuller details of most algorithms can be found in the cartographic, geographic, and computer science literature.

Finally, we provided an example of generalization using ongoing work at the National Historic Geographic Information System (NHGIS), housed at the University of Minnesota. Here we saw that the complexity of the real world often requires the creation of custom-designed solutions.

of Figure 6.14 including both the generalization and the original base data is shown in Figure 6.15. Both Figures 6.14 and 6.15 have been generalized using the Visvalingam and Leung–Koehnen algorithms. Note the extreme level of generalization here, with only the

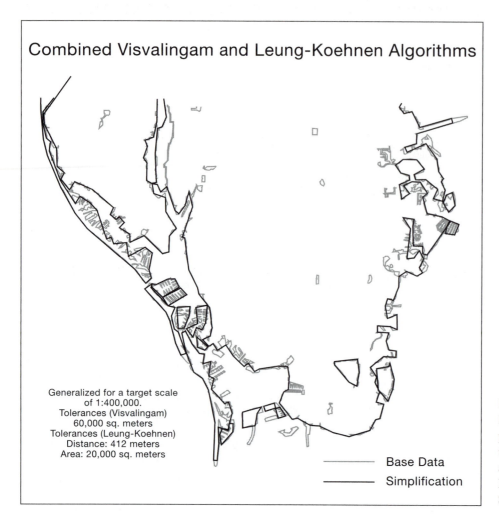

Combined Visvalingam and Leung-Koehnen Algorithms

Generalized for a target scale
of 1:400,000.
Tolerances (Visvalingam)
60,000 sq. meters
Tolerances (Leung-Koehnen)
Distance: 412 meters
Area: 20,000 sq. meters

———— Base Data

———— Simplification

FIGURE 6.15 An enlargement of the lower left portion of Figure 6.14; simplification is accomplished using both the Visvalingam and Leung–Koehnen algorithms. (Courtesy of National Historical Geographic Information System.)

FURTHER READING

Buttenfield, B. P. l. (1990) "NCGIA Research Initiative 3: Multiple Representations." *http://www.ncgia.ucsb.edu/research/initiatives.html#i3.*

This NCGIA working paper summarizes the results of Initiative 3 on multiple representations held in 1989 and includes a comprehensive bibliography of earlier work on the subject.

Goodchild, M. F., and Quattrochi, D. A. (1997) "Scale, multiscaling, remote sensing, and GIS." In *Scale in Remote Sensing and GIS*, ed. by D. A. Quattrochi and M. F. Goodchild, pp. 1–11. New York: Lewis.

This edited volume explores scale from a variety of perspectives, including spatial and temporal statistical analysis, multiple scaled data for the analysis of biophysical phenomena, and landscape ecology.

Jensen, J. R. (1996) *Introductory Digital Image Processing: A Remote Sensing Perspective.* 2nd ed. Upper Saddle River, NJ: Prentice Hall.

Provides a detailed description of many image processing techniques, including those considered to be part of raster-based generalization.

João, E. M. (1998) *Causes and Consequences of Map Generalization.* Bristol, PA: Taylor and Francis.

Focuses on the quantitative effects of map generalization. The study uses a series of European maps and also provides a methodology for studying the effects of generalization.

McMaster, R. B., and Shea, K. S. (1992) *Generalization in Digital Cartography.* Resource Publications in Geography. Washington, DC: Association of American Geographers.

This monograph reviews the history of generalization, including many of the basic models, and provides details of the fundamental operations.

Sheppard, E., and McMaster, R. B. (eds.) (2004) *Scale and Geographic Inquiry: Nature, Society, and Method.* Oxford: Blackwell.

This edited volume reviews recent research in scale from the human-social, biophysical, and methodological perspectives.

Weibel, R. (1995) *Cartography and Geographic Information Systems* 22, no. 4 (special issue on "Map Generalization").

This special issue contains a series of papers on both the data models for generalization and generalization algorithms. Several of the papers focus on road generalization.

7

The Earth and Its Coordinate System

OVERVIEW

This chapter provides fundamental material related to the shape and size of the Earth and the nature of its coordinate system. Knowledge of these concepts is essential for understanding the material on map projections covered in Chapters 8 and 9.

Section 7.1 introduces the basic characteristics of the Earth's coordinate system, which is comprised of an imaginary network of lines called the **graticule**. The graticule is further composed of lines of **latitude** and **longitude** that criss-cross the Earth's surface, allowing point locations to be uniquely described in terms of a **sexagesimal system** (degrees, minutes, and seconds). The arrangement of this coordinate system is based on the Earth's position relative to the sun as well as its alignment with celestial objects such as the North Star. This alignment establishes the **Equator** as the reference line for latitude values, which start at the Equator and run 90° to the North Pole and 90° to the South Pole. On the other hand, the **Prime Meridian**, passing through Greenwich, England, serves as the reference line for longitude values that run 180° east and 180° west of this line.

Section 7.2 discusses the historical solution to the importance of accurately determining latitude and longitude. Determining longitude proved to be particularly difficult because there is no convenient reference line dividing the Earth into Eastern and Western Hemispheres (the location of the Prime Meridian is arbitrary). Two events were critical to providing accurate longitude locations. First, in the 1700s John Harrison invented the **chronometer**—a time piece using spring mechanisms rather than a pendulum for movement, an important fact when sailing at sea because pendulum clocks were unusable at sea due to the constant rocking motion. The second event was the

International Meridian Conference, which established the meridian running through Greenwich, England, as the Prime Meridian that would be the starting point for longitude values dividing the world into the Western and Eastern Hemispheres.

Section 7.3 explains the investigations of **geodesy**, the science of understanding and explaining the Earth's size and shape. Approximately 2000 years ago, the Greek Eratosthenes was able to calculate the circumference of the Earth as 40,500 kilometers (25,170 miles), a figure not far from our present-day value of 40,075 kilometers (24,906 miles). Determining the correct shape of the Earth has proved more problematic, as it wasn't until about 300 years ago that Sir Isaac Newton suggested that the Earth is an **oblate spheroid** or **ellipsoid**, bulging at the Equator due to **centrifugal force**. This ellipsoid concept extends to the **reference ellipsoid** as a solid body that more accurately defines the Earth's shape than a simple spherical assumption. Advances in satellite technology have led to the concept that the Earth is not a smooth mathematically definable surface (as described by the reference ellipsoid), but due to differences in gravitational forces can be modeled as a **geoid**—a shape that the Earth would take on if the seas were allowed to flow over land adjusting to the gravitational differences across its surface and creating a single water body.

7.1 BASIC CHARACTERISTICS OF THE EARTH'S GRATICULE

Addresses, zip codes, and telephone area codes are just some of the many ways that places on the Earth's surface can be located. Although useful, such approaches are limited in that they have no uniformity (e.g., each country has its own form of postal code system), they locate areas rather than specific points, and they do not completely cover the Earth (no area codes exist over the world's oceans). To overcome these limitations, a system composed of latitude and longitude known as the **graticule** was developed and is a critical foundation for understanding the concepts on map projections that appear in Chapter 8.

To understand the graticule, we start by recalling that the Earth rotates about its axis, called the **axis of rotation**, which is aligned with the **North Star** (*Polaris*) and passes through the **North** and **South Poles** (Figure 7.1). Perpendicular to this axis and positioned halfway between the poles, we can envision an imaginary plane passing through the Earth's center. The trace of this plane on the Earth's circumference creates a circle called the **Equator**, which divides the Earth into the **Northern** and **Southern Hemispheres**.

It is interesting to note that the alignment of the Earth's axis of rotation with the North Star is not exact and is purely coincidental. In fact, the North Star is actually a small distance away from the Earth's axis of rotation. Moreover, the distance between them is changing from year to year due in part to the Earth's change in orbital path and axial tilt. Because of this constant change, the North Star alignment will someday no longer provide utility as a pole star.

7.1.1 Latitude

Describing a point on the Earth's surface requires that a location's latitude and longitude be known with respect to an origin. In the case of **latitude**, the Equator serves as a convenient origin because it divides the Earth into two equal halves. Figure 7.2A illustrates lines of latitude, shown as thin gray lines, with the Equator represented by the thicker gray line. Because lines of latitude are parallel to each other, they are often called **parallels**. Latitude values are reported in angular measurements of **degrees**, **minutes**, and **seconds**. Similar in concept to units of time, this **sexagesimal system** has a base unit of 60 (in the decimal system the base unit is 10) where each degree is divided into 60 minutes and each minute is divided again into 60 seconds. In this system, the ° symbol denotes the number of degrees, a single quote (') indicates minutes, and double quotes (") specify the number of seconds. There are 90° of latitude north and south of the Equator (designated as 0°) for a total of 180° from pole to pole. It is customary to apply the terms North and South to designate latitude locations above or below the Equator. In some cases, plus (+) and minus (−) signs are attached as a prefix to the degree values indicating latitude locations above and below the Equator, respectively. Thus, the latitude of the Washington Monument can be specified as 38° 53' 22" N or +38° 53' 22".

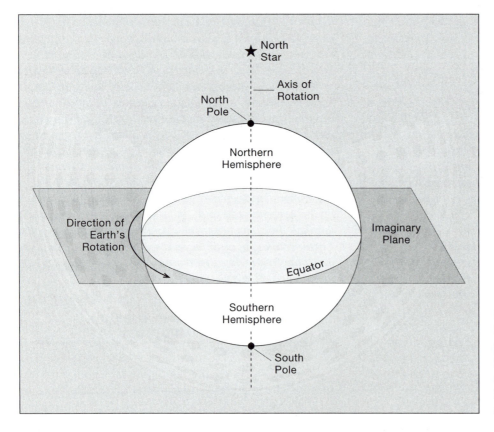

FIGURE 7.1 The Equator is formed by the intersection of an imaginary plane perpendicular to and bisecting the Earth's axis of rotation, which is aligned with the North Star. The Equator divides the Earth into Northern and Southern Hemispheres.

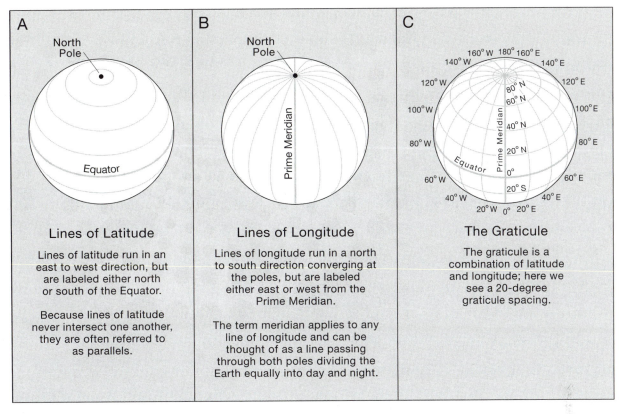

FIGURE 7.2   Latitude and longitude on the Earth's surface.

7.1.2 Longitude

Longitude, using the same sexagesimal system as latitude, measures location east or west of an origin called the **Prime Meridian**, which has an associated value of 0° longitude coinciding with the Royal Observatory in Greenwich, England. In Figure 7.2B, lines of longitude (also called **meridians**) are shown as thin gray lines, with the Prime Meridian represented by the thicker gray line. Lines of longitude run north–south from pole to pole, but are measured east or west of the Prime Meridian. There are 180° of longitude east and west of the Prime Meridian for a total of 360° of longitude. As with latitude, plus and minus signs are attached as a prefix to indicate location east or west of the Prime Meridian, respectively.* Thus, the longitude of the Washington Monument can be specified as 77° 02' 7"W or −77° 02' 7". The 180th meridian is directly opposite the Prime Meridian and partially coincides with the **International Dateline** (the line dividing days of the Earth's rotation). Figure 7.2C shows the graticule—lines of latitude and longitude in combination with a spacing of 20°.

To utilize latitude and longitude when locating positions on the Earth's surface two quantities must be known: the angular distance from the Equator to a given location (the latitude) and the angular distance from the Prime Meridian to a given location (the longitude). Figure 7.3 illustrates how such angles are determined. In this case, assume the point in question is positioned at 60° North latitude and 45° East longitude. The latitude value of 60° is the angle formed between (1) a line passing through the point on the Earth's surface and the center of the Earth, and (2) a plane passing through the Equator and the center of the Earth (the plane of the Equator). Latitude is usually symbolized by the Greek letter ϕ,[†] as shown in Figure 7.3A. Similarly, Figure 7.3B shows that a longitude value of 45° is the angle formed between (1) a plane passing through the point and the center of the Earth, and (2) a plane passing through the Prime Meridian. Longitude is usually symbolized by the Greek letter λ. The conventional manner of referencing a location on the Earth's surface is to state the latitude

---

* The "+ = east" and "− = west" convention is not universal. Different countries and mapping organizations sometimes assign "+ = west" and use a complete 360° system of longitude rather than dividing the world into two 180° halves.

[†] The adoption of ϕ is not universal as a symbol for latitude. The Greek letters ψ and φ are also used to reference latitude.

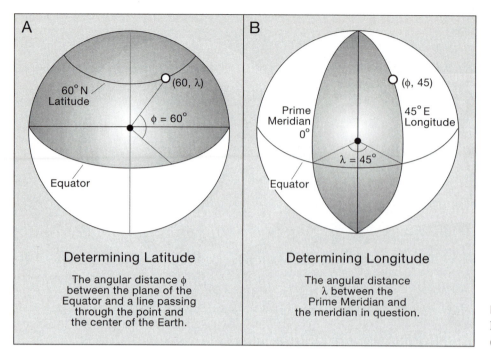

FIGURE 7.3   Determining the angular values of (A) latitude and (B) longitude.

value first and the directional counterpart (i.e., North or South), then the longitude value along with its directional counterpart (i.e., East or West). For example, the U.S. population center for the 2000 Census is located in Phelps County, Missouri, at 37° 41' 49" North latitude and 91° 48' 34" West longitude.

7.1.3 Distance and Directions on the Earth's Spherical Surface

The arrangement of the parallels and meridians on the Earth's surface sets up two important relationships: distances and directions. These relationships are important to cartographers because various types of maps and the projections that serve as their frameworks are used for accurately representing distances and directions.

The shortest distance between any two points is a straight line. On the Earth's curved surface, however, the shortest distance between two points is an *arc* of a **great circle**. A great circle results from the trace of the intersection of any plane and the Earth's surface as long as the plane passes through the Earth's center. Because lines of longitude are traces of planes that intersect the Earth's surface and its center, all meridians are great circles. It is important to note, however that not all great circles coincide with meridians. The Equator is also a great circle because its plane intersects the Earth's center, but no other parallels intersect the Earth's center. Rather, all other parallels form circles on the Earth's surface called **small circles**. However, not all small circles coincide with a parallel. In general terms, a small circle results when a plane passes through the Earth's curved surface, but does

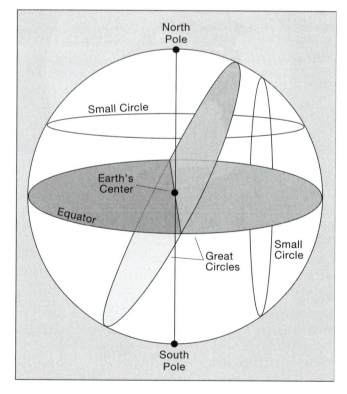

FIGURE 7.4   Examples of great and small circles on the Earth's surface.

not intersect the Earth's center. Figure 7.4 illustrates great and small circles on the Earth's surface.

Great circles also establish an important relationship regarding measurement of directions on the Earth's surface. In Figure 7.5 an arc of a great circle is shown as a

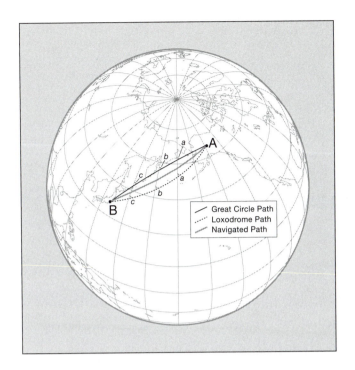

FIGURE 7.5 A great circle arc from point A to B creates a constantly changing angle (or azimuth) along each meridian the great circle arc intersects. The angle at *a* is different from the angles at *b* and *c*. The loxodrome from point A to point B crosses each meridian at the same angle at *a'*, *b'*, and *c'*. The navigated course (thick gray line) results from a series of short loxodromes that approximate the great circle path.

solid line between points A (Fairbanks, Alaska) and B (Tokyo, Japan). As the great circle arc crosses the Earth's surface it also intersects each sequential meridian between A and B. At *a*, *b*, and *c* the angle made between each meridian and the great circle arc is called an **azimuth** (or direction).* Along a great circle arc, the azimuth at each meridian intersection constantly changes, which is problematic for navigators. For instance, airplanes ideally fly along great circles, the more direct route, but their flight paths must be constantly adjusted because the azimuths along the great circle arc are also constantly changing. From a navigation standpoint, having to constantly alter a course is not very practical.

Lines on the Earth's surface that have constant direction can also be described. **Loxodromes** (or **rhumb lines**) are special lines that intersect all meridians at the same angle. A simple example will illustrate this concept. All small circles that coincide with parallels as well as the Equator intersect meridians at right angles and therefore are loxodromes. A loxodrome can be constructed that is not aligned with a parallel or meridian; for instance, Figure 7.6 shows that a line constructed on the Earth's surface from A to B so that it intersects all meridians at the same constant angle will, if continued beyond point B at the same angle, spiral toward the North Pole. This constant angle relationship is due to the convergence of the meridians at the poles. In a more practical example, Figure 7.5 shows the loxodrome path between points A

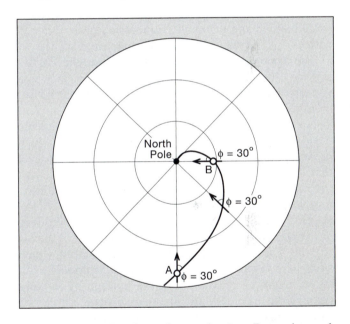

FIGURE 7.6 A line drawn from point A to B crossing each meridian at a constant angle is called a loxodrome. If extended, this line will continue to spiral toward the North Pole.

and B as a dashed line. Note that as the loxodrome path intersects each meridian, the angle at *a'* is the same as the angle at *b'* and *c'*. Loxodromes are a preferred course for navigation due to the minimum number of alterations to the course. Although navigating along a great circle route across the Earth's surface is the most direct route, it is not practical for navigators to constantly change their course. In most instances, the great circle route will be approximated by a series of short loxodromes that

---

* Azimuths are customarily measured in a clockwise fashion starting with geographic north as the origin and passing through 360°. This definition of azimuth is not universally followed. For instance, azimuths can be measured from magnetic north rather than geographic north.

necessarily lengthen the travel time, but facilitate the navigation process.

7.2 A BRIEF HISTORY OF LATITUDE AND LONGITUDE

The concept of latitude can be traced back to early Greek civilization where Dicaerchus (a student of Aristotle) placed the first line of latitude on a map of the known world running from the Pillars of Hercules (Strait of Gibraltar) through the Mediterranean Sea. Other lines of latitude were identified and later placed on maps according to direct observation of nature. The seasonal progression of the sun, for example, provided the latitude values for the Equator, Tropic of Cancer (at 23° 30' 0" N), and Tropic of Capricorn (at 23° 30' 0" S). Expanding on the concept of latitude, Strabo (1st century BCE) provided in his *Geography* a detailed description of the world. His published world map in 18 CE illustrated the *oikumene* (inhabitable Earth) that was divided into climate regions called *climata*. The *climata* zones included the frigid, temperate, and torrid, which were equally placed above and below the Equator and were aligned with various parallels that ran through notable cities.*

The desire to accurately fix a location on the Earth's surface with respect to latitude came easily to early navigators, but required the development of various instruments, such as the *astrolabe*. Developed by the ancient Greeks, the astrolabe's chief utility was to measure the angular height of the sun or star above the horizon, which provided one's latitude. The astrolabe relied on the fact that the North Star remained at a fixed position above the Earth's surface. By aligning the instrument with the horizon and the North Star, the angle (or latitude) of the position could be read from the instrument. The modern day **sextant** (Figure 7.7) is a direct descendant of the astrolabe. The sextant makes use of two mirrors, which allow light to pass through, and a moveable arm. The observer aligns the sextant with the horizon by looking through the eyepiece and mirror A. Mirror B is attached to a moveable arm, which, once the sextant is aligned with the horizon, is positioned so that the sun's rays are reflected off mirror B and pass onto mirror A. The resulting image cast through the eyepiece is one object (the sun) superimposed on the other (the horizon). The angle between the two objects is then read off the scale, giving the latitude of the observer.

The North Star facilitates determining latitude position in the Northern Hemisphere because its location in the sky is fixed and can easily be found because the Big Dipper (*Ursa Major*) constellation points to it. In the

* For more information, see *http://www.henry-davis.com/MAPS/AncientWebPages/115mono.html.*

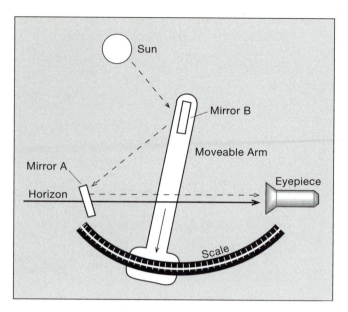

FIGURE 7.7   The sextant, a device for measuring latitude.

Southern Hemisphere, however, there is no South Star.[†] Therefore, an alternative method must be used to find the South Pole and the Southern Cross constellation is the key. Figure 7.8 illustrates the Southern Cross and its relationship to the South Pole. Simply extend a line from the top of the Southern Cross through its bottom 4.5 times its length and this will reach the South Celestial Pole; this process works regardless of the time of night or season.

Unlike the Equator for latitude, there is no natural origin for longitude. This created considerable difficulty

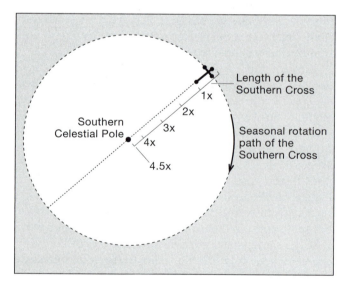

FIGURE 7.8   Determining the location of the South Pole using the Southern Cross constellation.

[†] A star (*Sigma Octantis*) is almost as close to the South Pole as the North Star is to the North Pole, but this star is barely visible.

in determining longitude positions. Moreover, there is no fixed celestial body (e.g., the North Star) that provides information for determining longitude. Those who pursued this problem knew that, similar to latitude, a fixed entity needed to be known and then the value of that entity at a specific location could be compared to the fixed value to arrive at a longitude estimate.

As a way to solve this problem, attention turned to the concept of time. It was known that the Earth makes one complete revolution about its axis in a 24-hour period. Thus, each location on the Earth's surface makes a complete 360° rotation during this 24-hour time period. Through simple division, each hour of rotation corresponds to 15° of longitude change. Knowing this, the task was then to compare the current local time with some agreed-on fixed time. The difference in time could easily be converted into degrees of longitude locating the position in question. For example, if a ship carried a clock set to London time it could compare local noon at its current position with the time on the clock set to London time and simply calculate the time difference and thus determine its longitude relative to London. Although simple in theory, it proved extremely difficult to develop a timepiece (known as a **chronometer**) that would keep accurate time, especially while at sea. At the time, pendulum clocks were in common use on land, but were unusable on ships due to the constant rocking and the gravitational differences experienced across the oceans.

In 1714 England established the Board of Longitude to promote the development of a practical and useful method for finding longitude at sea. The problem was so pressing that the Board offered a monetary award of £20,000 to anyone who could make a timepiece capable of longitude accuracy within one-half a degree (2.5 minutes of time) on a journey from England to the West Indies. Over the six weeks of the voyage, the clock would be allowed a daily error of less than three seconds, something barely possible for the best clocks on land. John Harrison stepped up to the challenge, developing four different timepieces over nearly 30 years, finally arriving at a design he called H4 that satisfied the Board of Longitude's requirements on time accuracy. Interestingly, Harrison did not receive the full monetary award until after an act of the British Parliament rewarded him the prize at the age of 79.

Another problem that plagued those desiring precise longitude was deciding which line of longitude should serve as the origin—similar to what the Equator offers latitude. Prior to the mid-1800s, there was no universally agreed-on meridian that was assigned 0° longitude. Rather, it was customary for each country to place the origin of longitude in that nation's capital. This practice, although demonstrating a sense of national pride, proved incredibly confusing to those who utilized maps of various countries. In an attempt to set an internationally

recognized Prime Meridian, 41 delegates from 25 nations met at the International Meridian Conference in Washington, DC, in 1884. By the end of the conference, Greenwich had won the prize of 0° longitude by an overwhelmingly favorable vote. There were two main reasons for choosing Greenwich. First, the United States had already chosen Greenwich as the basis for its own national time zone system. Second, in the late 19th century almost 75 percent of the world's commerce depended on sea charts, which used Greenwich as the Prime Meridian. Today, most nations adopt the meridian running through the London Observatory at Greenwich, England, as the Prime Meridian.*

One final point about longitude is that the meridian concept also played an important role in the development of the meter as a standard of measure. When, in 1791, the French Académie Royale des Sciences was instructed to create a new system of units, it was decided that this new system should be set on a base 10 system and that the fundamental measuring units should be based on values that were unchanging. To this end, l'Académie Royale des Sciences calculated the distance of an imaginary line that began at the North Pole and ended at the Equator running through Paris. They then divided this line into exactly 10 million identical pieces. The length of one of these pieces became the base unit for the new system of measurement—the meter. This was the beginning of the metric system.

7.3 DETERMINING THE EARTH'S SIZE AND SHAPE

One of the more important historical scientific pursuits has been determining the Earth's size and shape—a pursuit that is still ongoing in the field of **geodesy**. You might think that after 2000 years of scientific inquiry, scientists would have a firm grasp on the Earth's size and shape. To some extent they do, but the Earth's interior and exterior constantly change, making it difficult to pin down its exact size and shape.

7.3.1 The Earth's Size

There is a rich history on the efforts involved in deriving estimates of the Earth's size. One early investigator was Eratosthenes (276–195 BCE) who worked at the Alexandria library in Egypt. To determine the Earth's circumference, Eratosthenes applied the following formula for calculating the circumference of a circle (C):

$$C = 360° \cdot d/\phi$$

---

* The Royal Observatory has an interesting Web page full of information about the history of the Prime Meridian (see *http://www.rog.nmm.ac.uk/*).

where *d* is the distance between two locations on the Earth's surface, and *φ* is the angular (or latitude) difference separating the two locations. Eratosthenes selected Syene (present-day Aswan, Egypt) and a city to the north, Alexandria, Egypt, as the two locations because these cities were aligned along a meridian, allowing the distance to be computed more accurately.

Modern scholars are unsure of how Eratosthenes arrived at an estimate of distance; some think that the distance was an educated guess, whereas others think he based it on the number of days it took for a camel to walk between the two cities! In any case, he arrived at a value of approximately 810 kilometers (500 miles). To obtain the angular or latitude difference between these cities, Eratosthenes knew that at noon on a day that has the greatest amount of daylight in the Northern Hemisphere (the Summer Solstice) the sun was directly overhead a local well in Syene. On that same day and time, he measured an angle cast by a shadow in Alexandria and calculated it to be 7° 12'. Through a simple geometric relationship, the angle he measured on the Earth's surface at Alexandria was the same as the one that would be measured at the Earth's center (Figure 7.9). Substituting these values into the equation, he arrived at an estimated circumference of 40,500 kilometers (25,170 miles):

C = 360° · 810 kilometers/7°12' = 40,500 kilometers

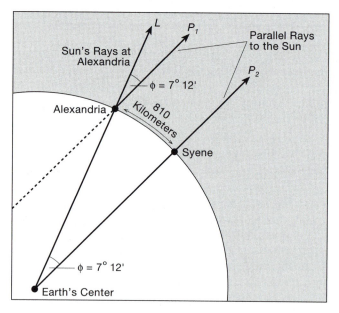

FIGURE 7.9   Method employed by Eratosthenes to estimate the Earth's size. Through geometry, a line *L* that intersects another line P_1 creates an angle *φ*. If that line continues and intersects another line P_2 parallel to P_1, the angle formed at the second intersection is the same as the angle at the first intersection.

This result is very close to contemporary measurements of the Earth's size—40,075 kilometers (24,906 miles).

Although a remarkable feat for his time, there is considerable doubt placed on the accuracy of Eratosthenes's measurement, as his method introduced many unaccounted errors. We have already mentioned the problem of determining distance, but there were other problems. For instance, Syene and Alexandria are not along the same meridian—Aswan is located at 32° 53' 56" E and Alexandria is located at 29° 55' 09" E—an east–west angular difference of 2° 58' 47". We also know that Syene is located at 24° 5' 15" North latitude and not along the Tropic of Cancer, which is at 23° 30' 0" North latitude, and thus the sun's rays were not perfectly vertical. Moreover, there is no agreement on the exact conversion between modern-day miles or kilometers and the units Eratosthenes used.* Regardless of his results, or modern-day interpretations thereof, his measurement should be regarded as one of the first to scientifically investigate the Earth's size.

7.3.2 The Earth's Shape

Similar to investigations of the Earth's size, there has been a constant quest to arrive at a definitive description of the Earth's shape. Recorded evidence of these investigations comes from various Greek scholars, such as Pythagoras (6th century BCE), who relied on philosophy to arrive at the spherical nature of the Earth's surface. Pythagoras argued that the Earth must be a sphere because the sphere was considered the perfect shape and the Greeks inhabited a perfect world. Other scholars, such as Aristotle (4th century BCE) and Archimedes (3rd century BCE), made estimates of the Earth's size through direct observation. For instance, Aristotle noticed that as ships approached from or sailed toward the horizon, the hull disappeared first rather than becoming smaller and smaller.

From our everyday experience, as individuals have for over 2000 years, we see evidence suggesting the Earth is a sphere. One simple way to see this shape is to note that during a lunar eclipse, the edge of the Earth's shadow is a circular arc. A more involved proof is the relative position of the North Star with respect to locations on the Earth's surface. At the Equator, the apparent position of the North Star is very low with respect to the horizon,[†] but as one progresses toward the North Pole, the

---

* The units in which Eratosthenes reported distance were stadia, which originated in ancient Greece. From *How Many? A Dictionary of Units of Measurements* (see *http://www.unc.edu/~rowlett/units/dictL.html*).

[†] Although there are many different definitions of horizons, the meaning here is the general term, which means the apparent or visible junction of the Earth and sky as seen from any position on Earth's surface.

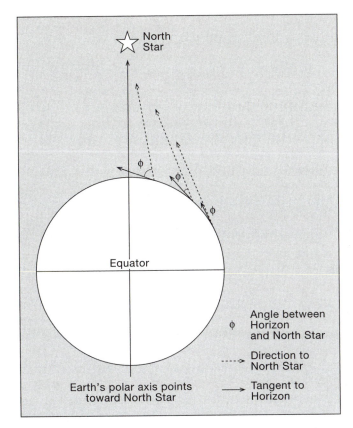

FIGURE 7.10 Positions on the Earth's surface suggesting a spherical surface. Note that the position of the North Star is relative and not to scale.

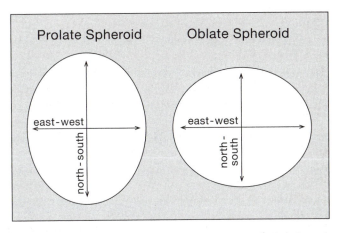

FIGURE 7.11 The prolate spheroid, shown on the left, has an east–west axis that is compressed or flattened relative to the north–south axis. The oblate spheroid, on the right, shows the opposite relationship. Both figures are greatly exaggerated to show the differences.

apparent position with respect to the horizon rises, until at the North Pole, the North Star is directly overhead (see Figure 7.10). This phenomenon can only be explained if the Earth has a spherical surface—a fact that early navigators utilized for locating their latitude in the Northern Hemisphere.

The Prolate versus Oblate Spheroid Controversy

The notion of the Earth as a perfect sphere existed for more than a 1000 years after the early Greeks first proposed the idea. The idea was not challenged until the late 1600s, when the Cassini family first undertook a comprehensive, large-scale survey of France through a newly developed technique called **triangulation**.* During their survey, the Cassinis encountered the fact that one degree of latitude in the northern part of France was not the

same length as one degree of latitude in the southern part of France. They concluded that the Earth was not a perfect sphere, but rather a **prolate spheroid**—an ellipse that bulges in the north–south direction and is compressed in the east–west direction (Figure 7.11).

About the same time period, other pieces of evidence regarding the Earth's gravitational forces were being investigated that would later be paramount in providing a more exact description of its shape. For instance, Jean Richer (1630–1696) examined the periods or motions of pendulums at different locations on the Earth. While on expedition to Cayenne, French Guyana, he found that a pendulum beat more slowly than it did in Paris. From this Richer theorized that gravity must be weaker at Cayenne, and therefore Cayenne was further from the Earth's center than Paris, which contradicted the Cassinis' prolate spheroid assumption.

In concert with Richer and other gravity investigations, Sir Isaac Newton (1642–1727) was promoting his laws of gravitational motion. His 1687 treatise stated that every object in the universe attracts every other object with a force that is equal to the product of their masses divided by the square of the distance between them. He applied his gravitational concept to the Earth's shape and suggested that the Earth's rotating mass creates a **centrifugal force** essentially pushing mass away from its center. He substantiated his assumption by noting that different latitudes on the Earth's surface travel at different speeds. For example, in Figure 7.12 we can see that each latitudinal position completes one full rotation per day, but that positions at the Equator have much further to travel and therefore must travel faster. Thus, according to Newton's law, the centrifugal force must be stronger at the Equatorial regions, which he proposed

---

* Triangulation is a method of surveying where two points having known locations are established and the distance between them is measured. This line is called the baseline. Next, using one end of the baseline, an angle is measured to a distant point. At the other end of the baseline, an angle is measured to the same point closing the triangle. Using the angles of this triangle, the length of any unknown side might be computed. Other triangles are established from this initial triangle creating a triangulated network.

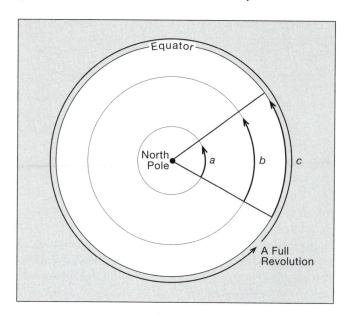

FIGURE 7.12 The notion of centrifugal force is supported by differing speeds of different latitude—positions near the Equator (*c*) must travel faster than those near the poles (*a*).

creates an **oblate spheroid** where the Equator bulges, but the poles are flattened (Figure 7.11).

Whether the Earth was a prolate or an oblate spheroid became the focus of a heated debate among the scientific community in the early 1700s. To resolve the debate, l'Académie Royale des Sciences in Paris organized two expeditions—one to the equatorial region of Ecuador and another to Lapland (the border between Sweden and Finland) to measure the length of one degree of latitude on the Earth's surface. The expeditions took more than nine years to complete, but finally reported that one degree of latitude at the Equator was 111.321 kilometers, whereas at Lapland the same distance was 111.900 kilometers. Thus, these findings proved Newton correct—the Earth is an oblate spheroid.

The fact that the Earth was described as an oblate spheroid provided scientists a new figure to describe the Earth's shape in simple mathematical terms. Basically, an oblate spheroid can be described as an ellipse, which has a semimajor axis—*a*—and a semiminor axis—*b* (Figure 7.13). The degree to which the ellipse deviates from a circle is called the **flattening constant** *f* and is computed as:

$$f = (a - b)/a$$

A two-dimensional ellipse can be rotated about its semiminor axis, arriving at a 3-D figure called a **reference ellipsoid**, which can be used to describe the Earth's size and shape. On a reference ellipsoid, the semimajor axis coincides with the *equatorial radius*, which is the distance from the Earth's center to a point on the Equator, and the semiminor axis coincides with the *polar radius*,

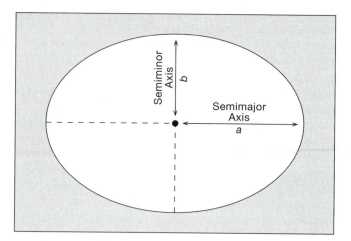

FIGURE 7.13 The reference ellipsoid and associated parameters.

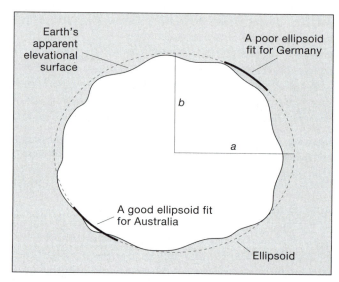

FIGURE 7.14 A comparison between the irregular Earth's surface and a smooth reference ellipsoid surface. (This illustration is greatly exaggerated to show the effects.)

which is the distance from the Earth's center to the North Pole.

Armed with the idea of a reference ellipsoid, various countries undertook national surveys to create ellipsoids that best fit local areas. Figure 7.14 illustrates this concept, and Table 7.1 lists several prominent historical reference ellipsoids. In total, dozens of reference ellipsoids have been defined, and currently there are more than 30 in use by various countries around the world.

Reference Ellipsoid and the Graticule

Although defined in general terms at the beginning of this chapter, the same location can take on different latitude values, depending on the shape of the Earth's surface used

TABLE 7.1 Several common ellipsoids developed for national mapping programs (From NIMA at *http://earth-info.nima.mil/ GandG/tm83581/8358010.gif.*)

Ellipsoid Name	Date of Inception	Semimajor Axis (meters)	Semiminor Axis (meters)	Flattening	Where Used
Airy	1830	6,377,563.44	6,356,256.91	1/299.33	Great Britain
Everest	1830	6,377,276.35	6,356,075.42	1/300.8	India
Clarke	1866	6,378,206.4	6,356,583.8	1/294.99	North America
International	1924	6,378,388.0	6,356,911.95	1/297.0	Select areas
Krassovsky	1940	6,378,245.0	6,356,863.03	1/298.3	Soviet Union
WGS	1972	6,378,135.0	6,356,750.5	1/298.26	U.S. Defense Dept.
GRS	1980	6,378,137.0	6,356,752.31	1/298.26	World
WGS	1984	6,378,137.0	6,356,752.31	1/298.26	World

and the map purpose involved. Figure 7.15A shows geodetic and geocentric latitude on a reference ellipsoid. **Geodetic latitude** computed on a reference ellipsoid is measured by an angle that results when a perpendicular

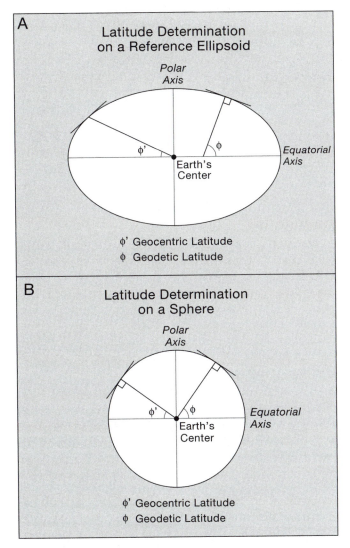

FIGURE 7.15 A comparison between geocentric and geodetic latitudes on (A) a reference ellipsoid and (B) a sphere.

line at the surface is drawn toward the Earth's center. In every instance, except when the location in question is directly over the Equator or one of the poles, the perpendicular line will never intersect the Earth's center. Rather, the line will pass through the equatorial plane at some other location. The angle measured at the intersection of the line and the equatorial plane is called the geodetic latitude. On the other hand, **geocentric latitude** computed on a reference ellipsoid is measured by an angle that results when a line at the Earth's surface is drawn intersecting the plane of the Equator at the Earth's center. Given the same point on the reference ellipsoid, the geocentric and geodetic latitude will be off by a small amount. This amount is significant when creating accurate maps (e.g., large-scale topographic maps) of local areas. For instance, surveyors commonly use geodetic latitude when referencing their survey positions because their surveys are tied to a specific reference ellipsoid for the country in which they are surveying. Geocentric latitudes are commonly used for global phenomena, such as satellite ground tracking maps.

On a spherical surface, geodetic and geocentric latitude are computed using the same methods previously outlined. Note, however, in Figure 7.15B that geodetic and geocentric latitude computations produce the same result on a spherical surface—both latitude computations measure the angle that results when a line at the Earth's surface is drawn intersecting the plane of the Equator at the Earth's center. In other words, geodetic and geocentric latitude computations produce the same latitude value when using a spherical model of the Earth. When using a spherical model for the Earth, it is common practice to translate geodetic latitude values computed on the ellipse to latitude as measured on the sphere.*

---

* Technically speaking, one would specify an auxiliary latitude, which is a special kind of latitude that takes the place of geodetic latitude and is used when cartographers project the ellipsoid onto a sphere. The particular auxiliary latitude chosen, however, depends on the kind of map to be created. For example, authalic latitudes can be computed to create a sphere having the same surface area as the Earth—desirable in thematic mapping applications. Snyder (1987) explained five types of auxiliary latitudes and provided a mathematical discussion for their derivation.

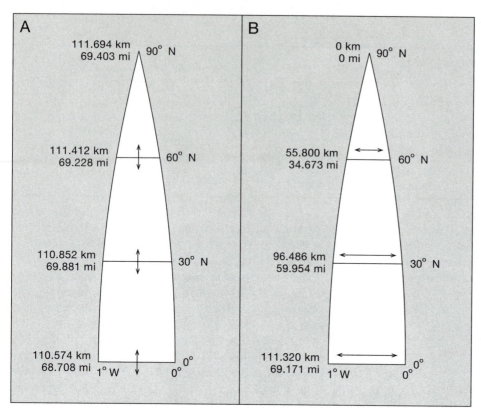

FIGURE 7.16 (A) The different lengths of one degree of latitude as measured along the meridian centered at 0° 30' west longitude, on the GRS80 reference ellipsoid. (B) The different lengths of one degree of longitude as one travels from the Equator to the North Pole as computed on the GRS80 reference ellipsoid.

Because the ellipsoid is not perfectly spherical, the length of one degree of geodetic latitude is not constant—the flattening at the poles means that one degree of latitude here is longer than one degree at the Equator. It is customary to measure one degree of latitude beginning at the equator and moving toward a pole, with the measurement centered on a multiple of, for example, 10° (Figure 7.16A). Table 7.2 lists lengths of one degree of latitude for the GRS80 reference ellipsoid. Appendix A lists the lengths of one degree of latitude for the GRS80 reference ellipsoid in one-degree increments.

TABLE 7.2 The length of one degree of latitude as measured along a meridian based on the GRS80 reference ellipsoid

Latitude	Kilometers	Statute Miles
0°	110.574	68.708
10°	110.608	68.729
20°	110.704	68.789
30°	110.852	68.881
40°	111.034	68.994
50°	111.229	69.115
60°	111.412	69.228
70°	111.562	69.321
80°	111.660	69.382
90°	111.694	69.403

Note. Values were computed using INVERSE, a free program available from the National Geodetic Survey at *http://www.ngs.noaa.gov/cgi-bin/Inv_Fwd/ inverse.prl*. Values are rounded to the nearest thousandths place.

On the sphere, distance between successive degrees of latitude is constant and is derived by first calculating the Earth's circumference and then dividing the result by 360°. So, assuming the Earth to have a circumference of 40,075 kilometers (24,906 miles), then the length of one degree of latitude is approximately 111.2 kilometers (69.1 miles).

Due to the convergence of meridians at the poles, the length of one degree of longitude becomes progressively smaller as one travels from the Equator to the poles, regardless of whether we assume a spherical Earth. This notion is illustrated in Figure 7.16B, and Table 7.3 lists lengths of one degree of longitude for the GRS80 reference ellipsoid. Appendix A lists the lengths of one degree of latitude for the GRS80 reference ellipsoid in one-degree increments.

The Geoid

Up to the late 1950s, suitable reference ellipsoids were developed on the Earth's surface utilizing surveying instruments that required visual contact between the instruments. As such, these surveys were very localized in nature, and thus no single reference ellipsoid could describe the entire Earth in very precise terms. In addition, each new reference ellipsoid had its own center that made it difficult for other countries to adopt existing reference ellipsoids. Satellite measurements have changed this, as models describing the Earth's size and shape are now computed based on the Earth's center of mass for the whole

TABLE 7.3 Lengths of one degree of longitude as measured along a specific parallel

Longitude	Length in Kilometers	Statute Miles
0°	111.320	69.171
10°	109.639	68.127
20°	104.647	65.025
30°	96.486	59.954
40°	85.394	53.061
50°	71.696	44.550
60°	55.800	34.673
70°	38.186	23.728
80°	19.394	12.051
90°	0	0

Note. Values were computed using INVERSE, a free program available from the National Geodetic Survey at *http://www.ngs.noaa.gov/cgi-bin/Inv_Fwd/inverse.prl.* Values are rounded to the nearest thousandth place.

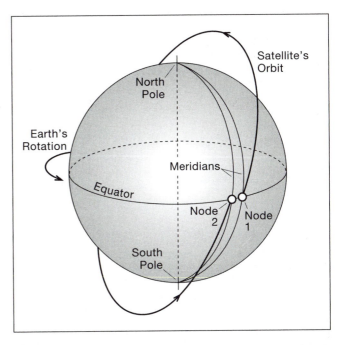

FIGURE 7.17 The precession of the nodes. Due to Earth's elliptical nature, the orbital path of a satellite is in the opposite direction of Earth's gravity, causing its path to predictably cross a different meridian with each orbit (after Maling 1992).

planet. This new idea allows a single reference ellipsoid to be used that gives a better fit to the entire Earth than previously possible.

The first United States satellite measurements of the Earth's shape were from the National Aeronautical and Space Administration's (NASA's) Vanguard rocket program in the late 1950s. If the Earth was perfectly spherical, satellites would operate in an orbital path around the Earth and that path would coincide with a plane. Deviations from this plane due to variations in the Earth's gravitational forces resulted in a predictable motion called *precession of the nodes.* Figure 7.17 illustrates this concept. The amount of change that occurred with each orbital path was measured very precisely, providing an indication of the gravitational forces acting on the Earth's surface and thus a measure of the Earth's shape. In fact, the Vanguard rocket program showed a slight bulge in the Northern Hemisphere's polar areas and a slight depression around the Southern Polar areas giving rise to the concept of a *pear-shaped* Earth. The North Pole was reported to be about 10 meters (32.8 feet) further from the Equator and the South Pole was found to be resting approximately 30 meters (98.4 feet) closer to the Equator than an elliptical Earth model suggested (Maling 1992 14).

As satellite measurements have continued, a more descriptive and informative term has been applied to describe the Earth's shape—the **geoid** (meaning Earthlike). Understanding the geoid concept is somewhat difficult because it is not directly observable. In general terms, the geoid is defined as a two-dimensional curved surface the Earth would take on if the oceans were allowed to flow freely over the continents without currents, tides, waves, and so on, creating a single undisturbed water body. This water body would then be free to adjust to differences in the gravitational forces (i.e., centrifugal force caused by the Earth's rotation) as well as the uneven distribution of the Earth's mass that exists at its surface. Thus, this new surface would have *undulations*

(i.e., peaks and valleys) that reflect the influence of gravity. One can think of this "new" surface as coinciding with mean sea level, which varies in height some 100 meters (328 feet) worldwide.

Unlike the reference ellipsoid, the geoid is impossible to define in simple mathematical terms. To compute a geoidal surface requires, among other parameters, knowing the specific nature of gravitational forces that exist across the Earth's surface. These forces, called gravitational anomalies, are not evenly distributed, nor are they equal in their intensity, as the Earth's crust contains different rock densities (e.g., metamorphic and igneous rocks are more likely to contain the iron-rich mineral magnetite and thus influence gravity to a greater extent than sedimentary rocks), but they can be measured and mapped. At every location on the Earth's surface a line can be placed so that it is perpendicular to the local gravitational anomaly. As shown by Figure 7.18, the line perpendicular to the geoid will not coincide with a line that is perpendicular to the chosen reference ellipsoid's surface. A geoidal surface (or equipotential* surface) can be created from these perpendicular gravitational measurements. In general terms, the Earth's crust is thicker over continents where the geoid typically rises but falls over the oceans where the crust is thinner. However, even the smooth undulation in the geoid surface has minor

---

* The term equipotential surface is defined as a surface having the same potential of gravity everywhere.

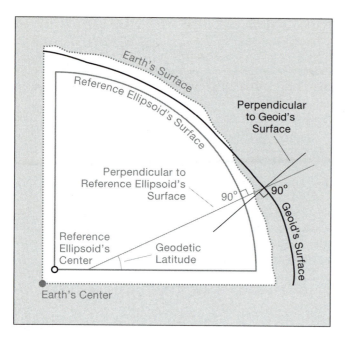

FIGURE 7.18 The relationship among surfaces representing the Earth, reference ellipsoid, and geoid. Note that when determining geodetic latitude, a line perpendicular to the reference ellipsoid is not perpendicular to the geoid.

highs and lows that can rise as much as 60 meters (196.8 feet) in some areas. Color Plate 7.1 shows one geoid model developed by the U.S. Coast and Geodetic Survey for the conterminous United States.

Geodetic Datum

The relationship between the geoid and a chosen reference ellipsoid is important for a variety of mapping applications, especially large-scale mapping projects such as aeronautical charting. Because the geoid is a complex undulating surface and the reference ellipsoid is a smooth, regular, mathematically defined surface, the two typically do not coincide, except at one point. This single point defines the origin of a datum. By definition, a datum is any quantity that serves as a base value (a value of zero) by which other values are referenced. A **geodetic datum**, then, is a model that describes the location, direction, and scale relationships with respect to an origin on the Earth's surface. With respect to geodesy, there are two types of datums: horizontal and vertical. A **horizontal datum** specifies locations in terms of geodetic latitude and longitude as computed from a chosen reference ellipsoid relative to an origin, which coincides with the Equator for latitude values and the Prime Meridian for longitude values. The horizontal datum provides a precise description of the parameters for the reference ellipsoid so that the semimajor axis (equatorial radius) value and flattening constant for the Earth are known, providing an accurate

means to establish latitude and longitude. A **vertical datum** provides a base value that enables elevations to be determined, and as mentioned earlier is tied to the concept of mean sea level, which relates to the geoid. Although these two datum concepts have been historically separate because the surfaces to which they refer are different (i.e., reference ellipsoid and geoid), they are commonly tied together today because they are conceptually related. Choosing an appropriate datum is necessary, for example, in surveying, to accurately provide latitude and longitude based on a specific horizontal datum and a derived elevation value determined through the geoid.

Historically, the United States used the Clarke Ellipsoid of 1866 as the reference ellipsoid and the geoid whose origin is located at Meades Ranch in Kansas (located near the geographic center of the United States) for its large-scale topographic mapping program. Taken together, the Clarke 1866 reference ellipsoid and the geoid origin of Meades Ranch comprise the **North American Datum of 1927 (NAD27)**. However, since the advent of **satellite geodesy**, large errors have been encountered when comparing the results of today's highly accurate survey techniques with NAD27 values. To remedy this discrepancy, a new reference ellipsoid, the Geodetic Reference System (GRS) was recommended by the International Association of Geodesy and subsequently adopted by the National Geodetic Survey,* creating the **North American Datum of 1983 (NAD83)**. Table 7.4 contrasts NAD27 and NAD83 on several key parameters. NAD83 utilizes the Geodetic Reference System of 1980 as the reference ellipsoid, and the geoid is now centered at the Earth's center of mass. As a result of changing from NAD27 to NAD83, most locations within North America have new latitude and longitude values. This readjustment has resulted in the publication of new coordinate data for approximately 250,000 geodetic control survey points throughout the United States yielding coordinate shifts that exceed 400 meters (1,312.3 feet) in some parts of the country.†

Geodetic Datums and Thematic Cartography

Although the knowledge of reference ellipsoids, geoids, geodetic datums, and geodesy are certainly worthwhile topics of investigation, they are limited in significance to large- and medium-scale mapping applications that involve accurate measurements of distances, directions, or areas. For example, a ship's navigator uses a nautical chart

---

* See the National Geodetic Survey (NGS) home page (*http://www.ngs. noaa.gov/*) for further information on this agency and its activities.

† The USGS and NGS have collaborated to provide these tables for the determination of the datum shifts as they relate to the use of mapping and charting products at scales of 1:10,000 and smaller (U.S. Geological Survey 1989).

TABLE 7.4 A listing of parameter values comparing the reference ellipsoids of NAD27 and NAD83

Datum Parameters	NAD27	NAD83
Reference ellipsoid name	Clarke 1866	Geodetic Reference System 80
Semimajor axis (a)	6,378,206.4 meters	6,378,137.0 meters
Semiminor axis (b)	6,356,583.8 meters	6,356,752.3 meters
Flattening	1/294.9786982	1/298.2572221
Intended use	North America	Worldwide

Note: Values taken from Snyder (1987).

to accurately plot a ship's intended course so that its precise location can be known at any instant. In the realm of thematic mapping, the focus is on examining spatial patterns and distributions and so knowledge of precise direction, distance, and area is not critical.

To illustrate the limited applicability of datums to thematic mapping, consider a small-scale thematic mapping exercise that requires symbolizing data for each country of the world. Assume for this problem that the model of the Earth is the GRS80 reference ellipsoid as opposed to a spherical model of the Earth. In this case, the actual difference in length between the two reference ellipsoids' semiaxes is approximately 21.3 kilometers (13.2 miles), or the circumference associated with the polar axis is about 135 kilometers (83.3 miles) smaller than that for the Equatorial axis (40,075–39,940 = 135 km) and the flattening is very close to 1/298. If you were to draw a page-size map of the world with the GRS80 parameters, the Equator would be represented by a line 12.75 cm (7.9 inches) long. At this size, the difference between the Equatorial and polar radii of the GRS80 reference ellipsoid parameters would be about 0.21 mm. Because 0.21 mm is also the approximate width of a line that can be used to represent fine detail on maps, the flattening would be hidden by the width of the pen used to draw world outlines. So, the selection of a more accurate model of the Earth using the GRS80 reference ellipsoid would not produce a visual difference in the final appearance of the map as opposed to the spherical model. This is an important conclusion because it permits the assumption that the Earth can be regarded as spherical for thematic applications. This assumption drastically simplifies the mathematics involved when creating a map projection.

SUMMARY

In this chapter, we have examined the basic concepts behind the size and shape of the Earth and its geographic coordinate system. We learned that the **graticule** serves as the framework for the imaginary network of **latitude**

and **longitude** lines, allowing any point location on the Earth's surface to be uniquely identified. The **Equator** serves as the dividing line for the **Northern** and **Southern Hemispheres** and references all latitude locations north or south of that line. On the other hand, the **Prime Meridian** divides the Earth into Eastern and Western Hemispheres and is the reference line for all longitude locations east or west of that line. Although the graticule is important for referencing spatial activity that takes place on the Earth's surface, a more important concept, especially with regard to map making, is the changes to the graticule's appearance that occur as a result of the map projection, a topic addressed in Chapters 8 and 9.

The Earth's size and shape are important considerations in determining locational accuracy. Historically, beginning with the work of Eratosthenes, the Earth's shape evolved from a simple spherical assumption to a more complex shape called an **oblate spheroid**. This figure of the Earth was developed after Newton proposed his gravitational laws in which a rotating body produces **centrifugal force**, causing a bulging of the Earth at the Equator. When creating a model of the Earth, geodesists frequently utilize a **reference ellipsoid**, which is a smooth, mathematically definable figure that flattens at the poles and bulges at the Equator.

Numerous reference ellipsoids have been developed, each defined by different parameters for the semimajor and semiminor axes, with the objective of trying to accurately map specific regions of the Earth. In recent years, the reference ellipsoid concept has been updated to reflect discoveries of gravitational variation across the Earth's surface. This variation has been modeled as a **geoid**, the two-dimensional curved surface the Earth would take on if the oceans were allowed to flow freely over the continents. A **datum** takes the geoid and a specific reference ellipsoid in combination to produce a reference for horizontal locations defined by latitude and longitude (established by the reference ellipsoid) and vertical elevations (defined by the geoid). The **North American Datum of 1927 (NAD27)** served as the basis on which the United States topographic mapping program was established, but has been updated to **NAD83** to reflect the contributions of gravitational measurements made by satellites.

The discussions of the graticule and the Earth's size and shape within this chapter are important to all facets of cartography. The graticule serves as the framework around which locations on the Earth's surface are recorded and subsequently analyzed with maps. This framework is expanded on in Chapter 8, where we discuss the characteristics of the graticule in detail as an important preliminary to understanding map projections. The Earth's size and shape are equally important to mapping. Although the reference ellipsoid, geoid, and datum concepts are more appropriate to large-scale topographic mapping applications, every cartographer needs to be aware of their importance, especially given the increased reliance on and utilization of digital spatial data.

FURTHER READING

Bowditch, N. (1995) *The American Practical Navigator*. Bethesda, MD: National Imagery and Mapping Agency.

> Chapter 2 provides an overview of geodesy, datums, and map projections.

Brown, L. A. (1990) *The Story of Maps*. Dover, DE: Dover Publications.

> A historical account focusing on maps and those who made them from the earliest Greeks through the European Renaissance.

Defense Mapping Agency. (1981) *Glossary of Mapping, Charting, and Geodetic Terms*. 4th ed. Washington, DC: U.S. Department of Defense.

> A glossary of cartographic terms relating to geodesy, datums, and map projections.

Iliffe, J. C. (2000) *Datums and Map Projections*. Boca Raton, FL: CRC Press.

> Chapters 1 through 6 cover aspects of the geoid, datums, and positioning on the Earth. Although much of the first half of the text is readable for the introductory student, a grasp of mathematics is essential to make full use of this text.

Maling, D. H. (1992) *Coordinate Systems and Map Projections*. New York: Pergamon.

> A readable book for the introductory student; the focus is on all aspects of map projections. Chapters 1 through 4 cover the basics of the Earth's shape and size.

National Oceanic and Atmospheric Administration. (1989) *North American Datum of 1983*. Washington, DC: U.S. Department of Commerce.

> A comprehensive overview of the North American Datum of 1983.

National Ocean Service. (1983) *Geodesy for the Layman*. 5th ed. Washington, DC: National Oceanic and Atmospheric Administration. Available at *http://earth-info.nima.mil/GandG/geolay/toc.htm*

> A classic nonmathematical treatment of geodesy. Many illustrations and simple explanations provide the novice with a clear understanding of geodesy.

Robinson, A. H., Morrison, J. L., Muehrcke, P. C., Kimerling, A. J., and Guptill, S. C. (1995) *Elements of Cartography*. 6th ed. New York: Wiley.

> Chapter 4 presents an overview of basic geodesy.

Smith, J. R. (1997) *Introduction to Geodesy: The History and Concepts of Modern Geodesy*. New York: Wiley.

> Provides readers with a basic overview and understanding of geodesy without the complex mathematical formulas and equations necessary to actually practice geodesy.

Snyder, J. P. (1987) *Map Projections: A Working Manual*. Washington, DC: United States Geological Survey.

> A comprehensive overview of map projections including worked examples of mathematical equations for many common map projections.

Sorbel, D. (1995) *Longitude: The True Story of a Lone Genius Who Solved the Greatest Scientific Problem of His Time*. New York: Viking Penguin.

> Presents an overview of basic science, cultural history, and personality conflicts that encompassed John Harrison during his efforts to solve the question of longitude determination.

United States Geological Survey. (1989) *North American Datum of 1983, Map Data Conversion Tables*. Denver, CO: United States Geological Survey.

> These conversion tables, produced cooperatively by the U.S. Geological Survey and the National Oceanic and Atmospheric Administration, are designed to allow users of maps and charts based on the NAD27 to convert scaled or digitized information to NAD83.

Wilford, J. N. (1982) *The Mapmakers: The Story of the Great Pioneers in Cartography from Antiquity to the Space Age*. New York: Vintage Books.

> Presents a history of cartography from antiquity to the space age, focusing on the impacts of technology on cartography.

Woodward, D., and Lewis, M. (1998) *The History of Cartography: Cartography in the Traditional African, American, Arctic, Australian, and Pacific Societies*. Vol. 2. Chicago: University of Chicago Press.

> Takes a detailed and critical look at the history of cartography, its technologies, and advances made in the African, American, Arctic, Australian, and Pacific societies. Also see Woodward and Harley (1987; 1994) and Harley and Woodward (1992).

8

Elements of Map Projections

OVERVIEW

The process of transforming the Earth onto a flat surface is accomplished through a **map projection**. *One of the more complex processes in cartography, the map projection plays an important role in the map's appearance (e.g., shapes of landmasses and the arrangement of the graticule) as well as the kinds of map uses that are possible (e.g., accurately measuring distances as opposed to visualizing the spatial variation of a data set symbolized by dots throughout various countries).*

In this chapter, we consider the fundamental elements of map projections. In section 8.1, the general concept of the map projection is presented. The projection process begins conceptually with the **reference globe** *(a model of the Earth at some chosen scale) and a* **developable surface** *(a mathematically definable surface onto which the land masses and graticule are projected from the reference globe). Traditionally, the cartographer set the scale of the reference globe and then chose one of the developable surfaces (* **cone**, **plane**, *or* **cylinder***) onto which the landmass and graticule were projected. Once projected, the developable surface was "unrolled," producing a map. Today, computers and mathematical equations have replaced this laborious task with software capable of producing hundreds of projections. Although the reference globe and developable surface are no longer relied on when creating projections, they help in conceptually understanding the projection process and are used throughout the chapter.*

Computers make creating projections a rather trivial process compared to what it was just a few years ago, but understanding the mathematical process involved in creating a projection remains important if projection software is to be utilized effectively. Section 8.2 reviews some general projection equations, explaining the parameters necessary to create a projection, and illustrating some different projections that result.

In addition to understanding the mathematical equations associated with projections, the cartographer needs to comprehend the various characteristics that a given projection possesses. Section 8.3 describes the characteristics of **class**, **case**, *and* **aspect** *of a projection. The projection's class refers to the developable surface used in creating the projection. Thus, the cone produces the* **conic class**, *the plane produces the* **planar class**, *and the cylinder produces the* **cylindrical class** *of map projections. The developable surface can be positioned so that it touches the reference globe along one point or line creating the* **tangent case**, *or two lines creating the* **secant case**. *The decision to use the tangent or secant case depends, in part, on the level of accuracy desired for the map. The developable surface can also be centered over any location on the reference globe. If centered over one of the poles, a* **polar aspect** *results; if the center is somewhere along the Equator, an* **equatorial aspect** *results; or, if the center is not at either pole or along the Equator, an* **oblique aspect** *results.*

Regardless of how carefully the cartographer constructs a projection, **distortion** *is an inevitable consequence. Section 8.4 explains distortion and illustrates its consequences. One way to understand distortion is to realize that, unlike on the Earth, scale varies considerably across a projection and thus a map's surface. This variation in scale can be analyzed across a projection through the* **scale factor**. *This numeric assessment explains how much departure there is at a given location from the scale that is found at that same location on the Earth's surface. The scale factor is also a useful underlying concept for* **Tissot's indicatrix**, *which provides a visual means of showing how* *distortion varies at point locations across the projection. In addition, the indicatrix provides a quantitative analysis of distortion that describes the amount and type of distortion that occurs at points across the map. When the indicatrix is mapped across a projection, certain distortion patterns become apparent and can be classified according to the projection's class. Examining these distortion patterns provides insight on how suitable a particular projection's class, case, and aspect are for minimizing distortion over a specific geographic area.*

Although no projection can avoid some form of distortion, most projections minimize distortion in either areas, angles, distances, or direction. In section 8.5, we consider classes of projections that have been named on this basis. ***Equivalent projections*** *preserve areal relationships, whereas* ***conformal*** *projections preserve angular relations. Preserving distance and directional relationships are accomplished on* ***equidistant*** *and* ***azimuthal*** *projections, respectively.* ***Compromise projections*** *preserve no specific property, but rather strike a balance among various projection properties.*

8.1 THE MAP PROJECTION CONCEPT

When trying to create a model of the Earth, a globe is an obvious possibility. Globes are particularly suitable for representing the Earth because they preserve areal, angular, distance, and directional relationships. Unfortunately, globes of any size are rather inconvenient to carry around and store, and detailed representations of even large countries would require rather sizable globes. Using portions of a globe to reproduce a specific country would reduce some of the bulkiness, but measuring, for example, distances across a globe's curved surface is difficult.

Maps have four distinct advantages over globes. First, the two-dimensional nature of maps makes many cartometric activities (e.g., measuring areas, plotting a course, calculating direction, etc.) much easier than on the curved surface of a globe. Second, maps can show considerable detail for a given landmass. Consider the United States Geological Survey's (USGS) 1:24,000 topographic map series. This series maps portions of the United States in greater detail than could be found on even the largest globe. Third, most maps are of a dimension that makes them easy to work with and very portable—most 1:24,000 sheets are roughly 56 cm × 66 cm (22 in. × 26 in.). Fourth, maps are less costly to produce and purchase. A single USGS 1:24,000 map sheet costs about $6.00, whereas a 30 cm (12 in.) diameter globe is priced around $50.00.

Although maps have numerous advantages over globes, they suffer from distortion, which is a natural consequence that results when the Earth's two-dimensional curved surface is projected to a map—when a **map projection** is created. The map projection is one of the most

important concepts in cartography, as the way in which the Earth's two-dimensional curved surface is projected directly impacts the appearance of the *graticule* (lines of latitude and longitude) and landmasses, and ultimately the kinds of uses for which a map can be applied.

8.2 THE REFERENCE GLOBE AND DEVELOPABLE SURFACES

The techniques employed to create map projections have changed considerably over the past 30 years. Prior to the advent of computers, map projections were laboriously created through manual drafting techniques. Two key concepts from that era include the **reference globe** and **developable surface**. These conceptual aids were used to assist the mapmaker in envisioning how the Earth's two-dimensional curved surface would be projected to create a map. Although computers have replaced manual drafting techniques, these concepts are still useful in explaining how a map projection is created.

When the cartographer begins creating a map projection, an initial step involves conceptually reducing the Earth's size to a smaller imaginary globe. The final size of this globe is set according to the principal map scale, which is usually dictated by the map purpose. Figure 8.1 shows the relationship between the Earth and the reference globe, which is a model of the Earth at a reduced scale that is used to project the graticule and landmasses onto the map. The reference globe thus serves as a *model* of the Earth sized to the principal scale of the final map.

A developable surface is a simple mathematical surface that can be flattened to form a plane without compressing or stretching any part of the original surface. There are three developable surfaces: **cylinders**, **cones**, and **planes**. To project the graticule from the reference globe to a map, first, a developable surface is conceptually placed over the reference globe touching it at one

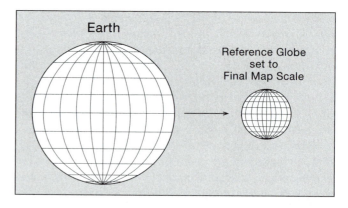

FIGURE 8.1   The relationship between the Earth and the reference globe. Note that the sizes of the Earth and reference globe are not to scale.

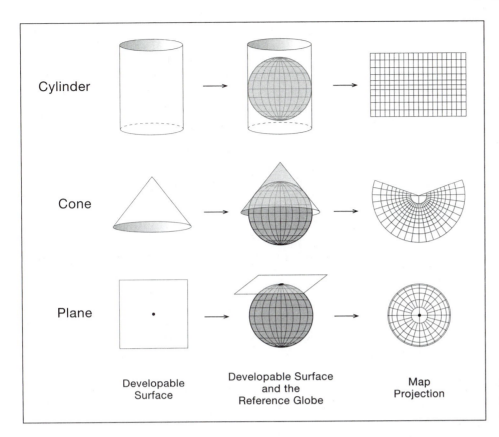

Cylinder

Cone

Plane

Developable
Surface

Developable Surface
and the
Reference Globe

Map
Projection

FIGURE 8.2   How developable surfaces and the reference globe are used to create map projections.

point or along either a parallel or meridian (Figure 8.2); second, the graticule and geographic landmasses are projected onto the developable surface. Afterward, the developable surface is "unrolled," revealing the graticule and landmasses.

Figure 8.3 provides a detailed view of how the reference globe and developable surface can be used to create a map projection. Conceptually speaking, a projection is created by first shining an imaginary light on the reference globe. The light source, called the **point of projection**, can be located in one of several positions with respect to the reference globe. For example, in Figure 8.3 the point of projection changes from the center of the reference globe (Figure 8.3A) to a position on the side of the reference globe opposite where the developable surface touches the reference globe, to a point at infinity (Figure 8.3C). The light casts an image of the graticule and landmasses onto the developable surface (in this case a plane), which touches the reference globe at one point, creating a projection. Note how changing the location for the point of projection influences the spacing of the graticule.

8.3 THE MATHEMATICS OF MAP PROJECTIONS

The reference globe and developable surfaces are conceptual aids that help illustrate the projection process,

but they are not used to create projections today. Rather, the field of mathematics is utilized to create projections, and so it is important to understand some of the basic mathematical manipulations used to project the Earth onto a map.

For illustrative purposes, assume you want to transform a point on the Earth's curved surface to a flat map and the point in question is defined by the latitude and longitude values of 40° N and 60° W, as shown in Figure 8.4. Our goal is to project this coordinate pair to a corresponding set of *x* and *y* **Cartesian coordinates**.

At a minimum, all map projections must involve at least two mathematical equations: one defining the *x* value and another defining the *y* value. A simple map projection involving two equations is as follows:

$$x = R * (\lambda - \lambda o)$$
$$y = R * \phi$$

where λ is the longitude value, ϕ is the latitude value, λo is the value of the **central meridian** (the location of the projection's east–west center), and R is the radius of the reference globe. The *x* and *y* coordinate values are commonly referred to as **plotting coordinates** and specify the longitude and latitude values in a Cartesian coordinate system.

To compute the *x* and *y* values using these equations, four steps are necessary. First, the longitude value for the

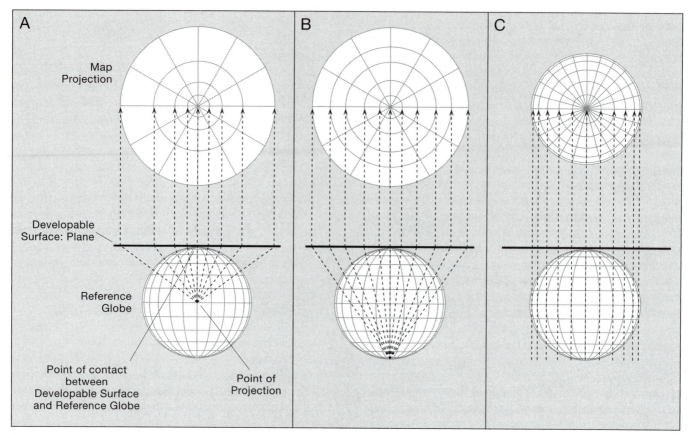

FIGURE 8.3   A detailed view of how the reference globe and developable surface can be used to create a map projection. From A to C, the point of projection changes from the center of the reference globe to a point on the reference globe opposite where the developable surface touches the reference globe to a point at infinity.

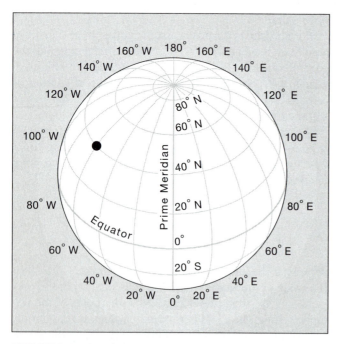

FIGURE 8.4   A point on Earth's curved surface described by the graticule. In this case, the point is located at 60° W longitude and 40° N latitude.

central meridian must be selected. Any longitude value can be selected, but if 0° is chosen, the central meridian will coincide with the Prime Meridian. Second, a value of R must be specified, indicating the radius of the reference globe (ultimately setting the final scale of the map). For example, if a world map was to be created with a principal scale of 1:30,000,000, then a reference globe with a radius (R) of 21.24 cm (8.36 in.) would be used. To simplify the computations that follow, we assume an R value of 1.0.

Third, all latitude and longitude degree values must be converted into radians. This conversion is especially necessary when using map projection software or computer programming languages, as neither can compute trigonometric functions specified in degree values. Converting degree values to radians involves multiplying the degree measurement by the constant $\pi/180$, where π is approximately 3.1415. For instance, 90° in radians equals $90° \times (\pi/180) = 1.5707$ or $\pi/2$, and 180° equals π radians. Because longitude values on the Earth range from $-180°$ to $+180°$ and all x values are projected to longitude degree values, then x values range from $-\pi$ to $+\pi$. On the other hand, because latitude values on the Earth range from $-90°$ to $+90°$ and all y values are projected to latitude values, then y values range from $-\pi/2$ to $+\pi/2$. For instance, 60° W converts to -1.047 radians and 40° N converts to 0.698 radians.

The fourth step in computing x and y values involves inserting longitude and latitude radian values into the equations. Using our values of -1.047 and 0.698, we have:

$$x = 1.0 * (-1.047 - 0) = -1.047$$
$$y = 1.0 * 0.698 = +0.698$$

If all remaining point locations on the Earth's curved surface defined by latitude and longitude were computed (e.g., every 15°) the plate carrée projection would be generated (as shown on the right side of Figure 8.5). The plate carrée projection is one of the oldest projections, having been developed by the ancient Greeks. Note that the lines of latitude and longitude for the plate carrée

are spaced at equal intervals. Furthermore, although all lines of longitude are equal in length (as they should be), they do not converge at the poles, as they ideally should.

We can expand on the simple pair of equations just introduced by adding trigonometric functions. For example, consider what happens if we compute the sine of the latitude—we would have the following:

$$x = R * (\lambda - \lambda o)$$
$$y = R * \sin \phi$$

This mathematical transformation produces the Lambert cylindrical projection developed in 1772 by Johann H. Lambert (Figure 8.6). Note that the meridians are equally spaced as they are on the plate carrée projection in Figure 8.5, but that the spacing of the parallels decreases as their distance from the Equator increases. As an

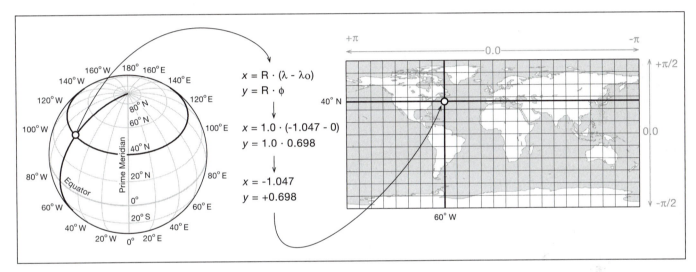

FIGURE 8.5   The mathematics of a map projection transforms a point on the Earth's curved surface represented by a latitude and longitude pair to x and y coordinate pairs on a map. In this case, the plate carrée projection results. The ranges of x and y coordinate values are shown in gray along the sides of the map projection.

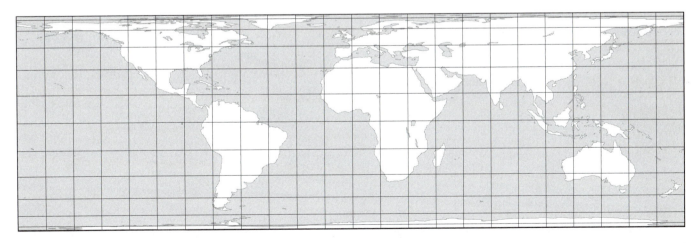

FIGURE 8.6   The Lambert cylindrical projection with a 15° graticule spacing.

alternative, we can change just the spacing of the *x* values by introducing the cosine function to our simple equations. This time, we take the cosine of the latitude value:

$$x = R * (\lambda - \lambda o) * \cos \phi$$
$$y = R * \phi$$

The sinusoidal pseudocylindrical projection results, as shown in Figure 8.7. The term *pseudocylindrical* is applied here because this projection shares only some of the visual characteristics of the graticule with the plate carrée cylindrical projection, namely that the lines of latitude are parallel and in this case, equally spaced. However, the meridians of this projection are curved lines converging at the North and South Poles.

Now, combining the cosine and sine functions together yields the following equations:

$$x = R * (\lambda - \lambda o) * \cos \phi$$
$$y = R * \sin \phi$$

This set of mathematical equations produces the polycylindrical projection shown in Figure 8.8. In this case,

some of the graticule characteristics of the previous projections should be apparent. For instance, the meridians are curved as in the sinusoidal projection and the parallels are parallel as in the other two cylindrical projections.

8.4 MAP PROJECTION CHARACTERISTICS

Through manipulation of the mathematical equations, numerous projections are possible, where each projection has specific characteristics, making it useful for a particular mapping purpose. This section discusses the various characteristics of map projections based on class, case, and aspect.

8.4.1 Class

Earlier, we saw that the developable surface concept was useful in illustrating the manner in which the graticule and landmasses on the reference globe are projected to the map. This concept is also a constructive way to describe the overall shape and appearance of the graticule after the projection process is complete, referred to as

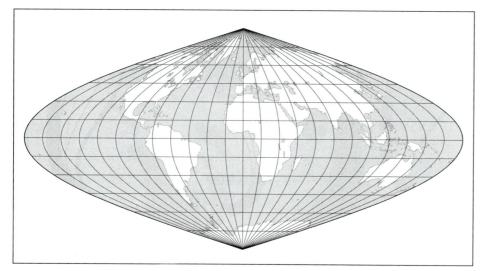

FIGURE 8.7   The sinusoidal projection with a 15° graticule spacing.

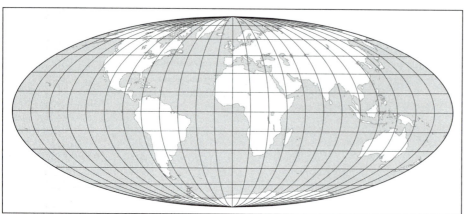

FIGURE 8.8   A polycylindrical projection with a 15° graticule spacing.

the projection's **class**. The three common map projection classes are **cylindrical, conic**, and **planar**.

The cylindrical class of projections results from wrapping the developable surface of a cylinder around the reference globe, projecting the landmasses and graticule onto the cylinder, and then unrolling the cylinder. On cylindrical projections, lines of longitude typically appear as straight, equally spaced, parallel lines, whereas lines of latitude appear as straight parallel lines that intersect the lines of longitude at right angles (Figure 8.9A and B). The spacing of the parallels distinguishes one cylindrical projection from another.

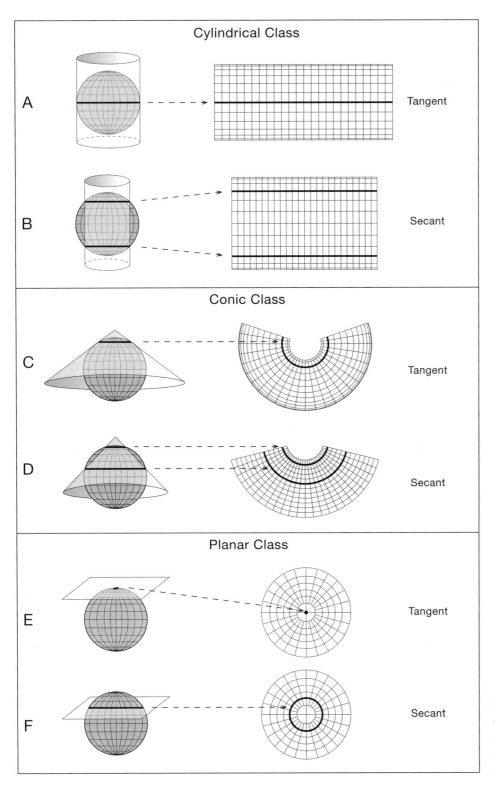

FIGURE 8.9 A comparison of the different classes of projections: (A–B) cylindrical, (C–D) conic, and (E–F) planar. Also shown are the tangent and secant cases for each class.

The conic class of projections results from wrapping the developable surface of a cone around the reference globe, projecting the landmasses and graticule onto the cone, and then unrolling the cone. On conic projections, lines of longitude typically appear as straight lines of equal length radiating from a central point (usually one of the poles), whereas lines of latitude appear as concentric circular arcs centered about one of the poles (Figure 8.9C and D). The overall shape of most conic projections can be described as a pie wedge, where a pie would be a full circle. Note that in the right-hand portion of Figure 8.9C, the pie wedge is slightly more than one-half, whereas in Figure 8.9D, the pie wedge is slightly less than one-half. The angular extent of the pie wedge and the spacing of the parallels distinguish one conic projection from another.

The planar class of projections results from positioning the developable surface of a plane next to the reference globe and projecting the landmasses and graticule onto the plane.* On planar projections, lines of longitude

_____

* Some texts refer to the planar class of projections as azimuthal. We have chosen to use planar because it is easier to relate to the developable surface concept (you can see that a plane is used).

typically appear as straight, equally spaced, parallel lines that radiate from the center (when the center of the projection is one of the poles), and lines of latitude appear as equally spaced concentric circles centered about a point, for example, one of the poles (Figure 8.9E and F). The spacing of the parallels distinguishes one planar projection from another.

There are many variations of the three developable surfaces that create other interesting projections. Like the cylindrical projections, the **pseudocylindrical** class of projections shows parallels as straight, nonintersecting lines. In addition, pseudocylindrical projections are visually distinguished by curved meridians that are equally spaced along every parallel and converge at the poles, which are represented by points or lines. Figure 8.10A illustrates the Craster parabolic pseudocylindrical projection with the meridians converging to points, and Figure 8.10B shows the Eckert III pseudocylindrical projection with meridians converging to lines representing the poles.

The **polyconic** class of projections conceptually employs multiple cones rather than one cone. Polyconic projections display curved lines (nonconcentric circular arcs) representing the parallels in a fashion similar to the conic projections, but the meridians curve toward the central

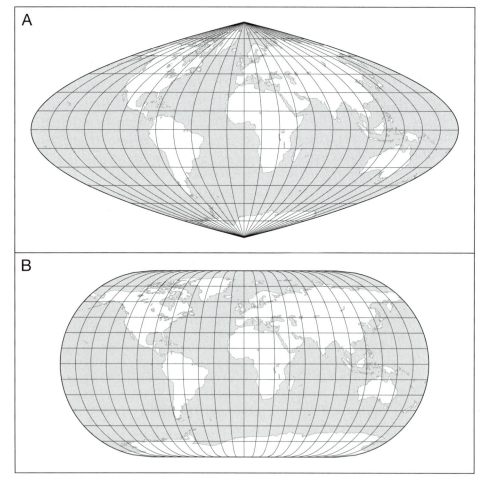

FIGURE 8.10 A comparison of the (A) Craster parabolic and (B) Eckert III pseudocylindrical projections. Each projection has a 15° graticule spacing.

meridian. Figure 8.11 illustrates the rectangular polyconic projection. There are many other variations of the basic developable surface concept. However, unlike the cylinder, cone, and plane, each variation requires a mathematical rather than a visual approach to explain how each projection is developed.

8.4.2 Case

The **case** of a projection relates to how the developable surface is positioned with respect to the reference globe and is either **tangent** or **secant**. A brief example illustrates the tangent idea. Conceptually speaking, imagine a ball rolling across the floor. At any given time, there is exactly one point in common between the ball and the floor. This point of contact is called the point of tangency. If all points of contact were connected together (either on the ball or on the floor), a line of tangency would result.

In the **tangent case** of a map projection, the reference globe touches the developable surface along only one line or at one point. Figure 8.9A, C, and E illustrates the tangent case for the cylindrical, conic, and planar map projections, respectively. Note that due to the way in which the developable surface is positioned with respect to the reference globe, the tangent case of the planar class utilizes a tangent point, rather than a line. The **secant case** of a projection occurs when the developable surface passes through the reference globe. In the secant case of the cylindrical and conic map projections (Figure 8.9B and D), there are two secant lines, whereas in the case of the planar projection there is one secant line (Figure 8.9F).*

---

* In the strictest sense, a secant case for the planar class is not possible, but is included here for conceptual completeness.

For a given projection class, the mapmaker usually has a variety of choices for the desired specific line(s) or point of tangency. For instance, with the conic class, any line of latitude can be selected as the tangent line. In the secant case of the conic class, any two lines of latitude can be selected. If these lines are nonequally spaced on opposite sides of the Equator, equally spaced on the same side of the Equator, or nonequally spaced on the same side of the Equator, the parallels and meridians take on the familiar cone shape. If however, the two lines are equidistant from and on opposite sides of the Equator, the parallels and meridians appear as straight parallel lines.

Secant lines and points of tangency each have the same scale as the principal scale of the reference globe. Thus, secant lines are called **standard lines**† and points of tangency are called **standard points**. All other lines and points will have either a larger or smaller scale than the principal map scale of the reference globe. Figure 8.12 illustrates the concept of a standard line and its impact on scale variation across a map. In the figure, a portion of the reference globe is represented by the dashed line, and the developable surface is represented by the solid line of gray values. Note that the developable surface cuts the reference globe creating two standard lines that have the same scale as the reference globe. The area between the standard lines projected to the developable surface (shown by a light gray solid line) has a compressed scale, whereas the areas beyond the standard lines (shown by a darker gray solid line) have an exaggerated scale. Figure 8.13 shows an interesting comparison of the effect of selecting tangent and secant cases for the conic class

---

† In many cases, standard lines coincide with lines of latitude and are referred to as standard parallels. Note that *standard line* is a general term—a standard line can also coincide with small circles or meridians.

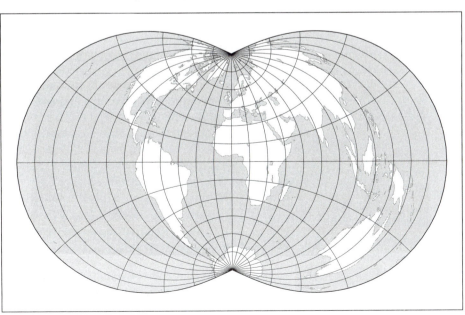

FIGURE 8.11 The rectangular polyconic projection with a 15° graticule spacing.

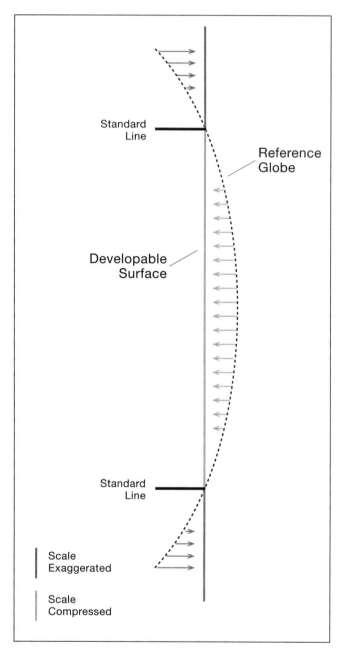

FIGURE 8.12   The effect of standard lines on scale variation on a map.

of projections. Figure 8.13A shows the Euler conic projection, developed by Leonhard Euler in 1777, displaying the tangent case with only one standard line at 25° N. Figure 8.13B shows the secant case with two standard lines located at 25° N and 50° N. Note that for both figures the meridians converge toward the poles and the parallels form curves concave toward the North Pole. Figure 8.13C shows the secant case with standard lines at 25° S and 50° S. Note that when placing the standard lines in the southern latitudes the parallels are curves

that concave toward the South Pole whereas the meridians converge to the South Pole.

The choice of the tangent or secant case, as well as their placement with respect to the reference globe, impacts the shape of the landmasses and the arrangement of the graticule. For instance, in Figure 8.13A, the convergence of the meridians at the North Pole is not as great as the meridian convergence in Figure 8.13B. As a result, landmasses in the northern latitudes are not as stretched in an east–west direction in Figure 8.13B. Figure 8.13A and B also illustrates that the placement of the standard lines in the northern latitudes greatly distorts the landmasses in the extreme Southern Hemisphere (e.g., Antarctica is stretched out the length of the map). In contrast, when both standard lines are placed in the Southern Hemisphere (as in Figure 8.13C), the landmasses there are not as distorted, but the landmasses in the Northern Hemisphere are greatly distorted.

8.4.3 Aspect

The aspect of a projection deals with the placement of the projection's center with respect to the Earth's surface. In general terms, a projection can have one of three aspects: **Equatorial, oblique**, and **polar**. An equatorial aspect is centered somewhere along the Equator, a polar aspect is centered about one of the poles, and an oblique aspect is centered somewhere between a pole and the Equator. The aspect of the projection can be defined more precisely in terms of the placement of the central meridian and the **latitude of origin**. Figure 8.14 shows four different centers for the equidistant cylindrical projection. For these maps, the central meridian is located at –96° W and the latitude of origin begins at 0° and moves northward every 30° of latitude until the projection is centered over the North Pole.

8.5 DISTORTION ON MAP PROJECTIONS

Creating a map projection has the consequence of introducing **distortion** in the final map. In a general sense, distortion involves altering the size of the Earth's landmasses and arrangement of the Earth's graticule when they are projected to the two-dimensional flat map. There are numerous ways in which distortion can be examined and analyzed on a projection, many of which are covered in a thorough review by Karen Mulcahy and Keith Clarke (2001).

8.5.1 A Visual Look at Distortion

One way to examine distortion on a map projection is to visually compare the sizes of landmasses and the arrangement of the graticule on the spherical Earth to how the

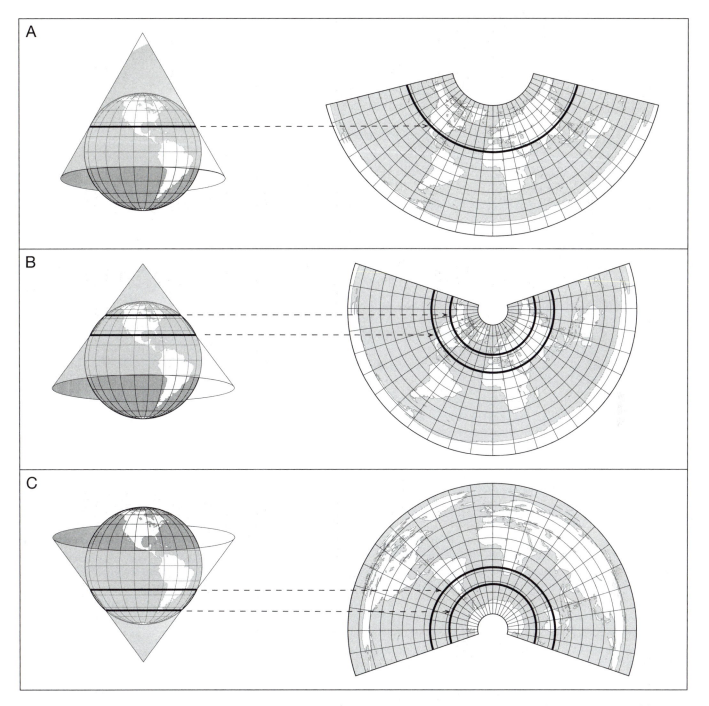

FIGURE 8.13 The Euler conic projection on which the location of the standard lines is represented by solid black line(s). A has the standard line tangent to 25° N. B has standard lines secant at 25° N and 50° N. C has standard lines secant at –25° S and –50° S.

landmasses and graticule arrangement appears on the map projection. To illustrate, Figure 8.15A shows an orthographic projection presenting a view over the North Atlantic Ocean similar to what one would see if looking at the Earth from space. Note the size of Greenland relative to the United States in this figure. Now, compare the size of Greenland to the United States on the Mercator projection in Figure 8.15B; obviously, Greenland is now relatively much larger than the United States. As a point of reference, Greenland has a land area of 2,166,086 km^2, which is smaller than the United States at 9,158,960 km^2, but on the Mercator projection, Greenland is represented as larger than the United States. In general, you see that landmasses in the upper latitudes on the Mercator projection are considerably exaggerated—a limitation when using this projection for maps of the world.

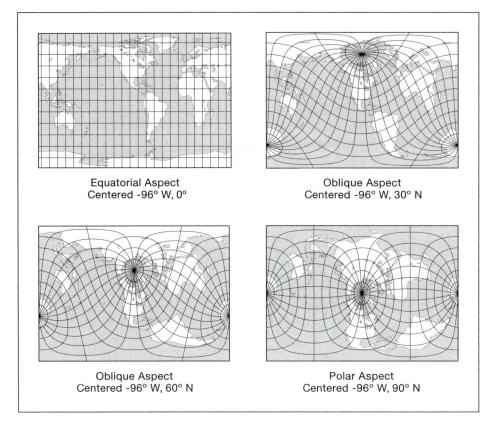

Equatorial Aspect
Centered -96° W, 0°

Oblique Aspect
Centered -96° W, 30° N

Oblique Aspect
Centered -96° W, 60° N

Polar Aspect
Centered -96° W, 90° N

FIGURE 8.14   The different aspects of a cylindrical projection.

Depending on the way in which a projection is developed, distortion can also cause the Earth's graticule to take on a number of different appearances. For instance, on the Earth's surface, all meridians converge to the poles. This relationship is preserved on the orthographic projection, but on the Mercator projection in Figure 8.15B all meridians are equally spaced straight lines. As a result of this nonconvergence, the graticule on the Mercator projection is stretched as compared to its representation on the orthographic projection.

A visual inspection of a map projection is useful in providing a general overview of the distortion present on the map, but a more sophisticated approach is needed to quantitatively analyze the amount and kind of distortion throughout a projection. Moreover, once the quantitative analysis is complete, there remains the need to visualize the distribution of distortion on a projection. There have been several approaches to quantitative distortion analysis that have been developed, but our discussion focuses on one of the more common—Tissot's indicatrix.

8.5.2 Scale Factor

To understand the details of Tissot's indicatrix, we begin with an introduction of the **scale factor** (**SF**), which is a numerical assessment of how the map scale at a specific location on the map compares with the map scale at the standard point or along the standard line(s). Figure 8.9 showed that the scale on the map is the same as the reference globe only where the developable surface comes in contact with the reference globe at the standard point or along standard line(s). In all other locations, depending on the way the projection is mathematically created, the scale experiences either exaggeration or compression.

The SF at any given location is computed using the following formula:

$$Scale\ factor\ =\ \frac{Local\ scale}{Principal\ scale}$$

where *local scale* is the scale computed at a specific location and *principal scale* is the scale computed at the standard point or along the standard line(s). Using this formula, we can compute how much deviation exists between the local scale and the principal scale at any location on a given projection. To illustrate, assume that we are working with the quartic authalic projection (suitable for world maps) shown in Figure 8.16, and that this projection has a principal scale of 1:250,743,970 along the Equator. Now, assume you wish to compute the SF at some point along the parallel at 30° N, say between 0° and 15° W (shown on Figure 8.16 as *a*). You would first calculate the local map scale between 0° and 15° W, which can be found using the following steps. Measuring along the parallel at 30° N between 0° and 15° W you arrive at a *map distance* of 0.60 cm (0.236 in.). To compute the corresponding Earth

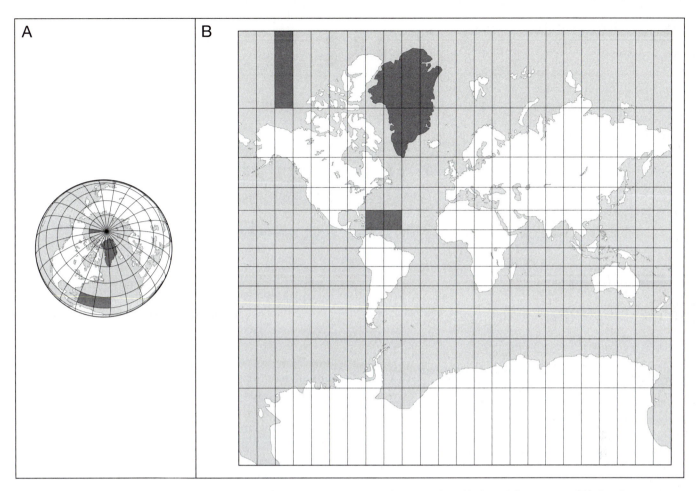

FIGURE 8.15 (A) The orthographic projection gives the appearance of the Earth as if viewed from space. (B) On the Mercator projection, note the exaggerated size of the landmasses, especially in the upper latitudes. Also, the meridians do not converge at the poles and the spacing of the lines of latitude increases from the Equator to the poles. Although it might not appear so, both projections are shown at the same scale.

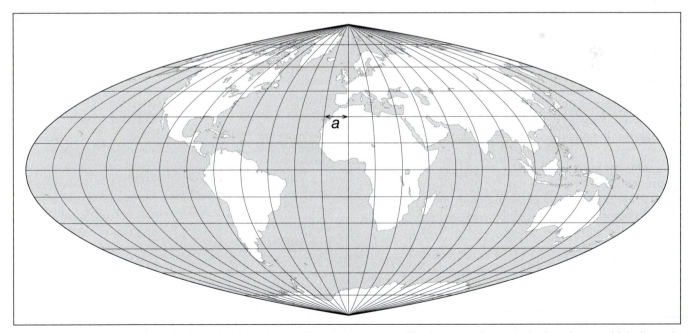

FIGURE 8.16 The quartic authalic projection with a 15° graticule spacing. The letter *a* references the location at which the scale factor is computed.

distance for that same distance, look in Appendix A and find the distance for one degree of longitude for 30°, which is 96,486.28 meters. Because the Earth distance between 0° and 15° is 15°, we need to multiply 96,486.28 by 15°. Doing so, we arrive at an Earth distance of 1,447,294.2 meters, or 1447.2942 km.

Next, substituting the 0.60 cm and 1447.2942 km into the basic scale equation, we obtain the following local scale:

Map scale = Earth Distance/Map distance

Map scale = 1447.2942 km/0.60 cm

Before solving, the units in the equation must be the same, so convert km to cm:

1447.2942 km * 100,000 cm = 144,729,420 cm

Therefore,

Map scale = 144,729,420 cm/0.60 cm = 1:241,215,700

The local scale along the 30° parallel is thus computed as 1:241,215,700, which is considerably smaller than the principal scale along the Equator of 1:250,743,970. Knowing the principal scale of the map projection (1:250,743,970) and the local scale along the 30° N latitude (1:241,215,700), we can compute the SF at this location. Substituting the different scale values into the SF equation, we arrive at the following:

$$SF = \frac{1:241,215,700}{1:250,743,970} = 0.962$$

This value suggests that the scale has been compressed, meaning that the distances on the Earth's curved surface along 30° N latitude have been *shrunk* with respect to its true size. On the other hand, if a local scale was computed to be 1:275,000,000 and the principal scale was 1:250,743,970, then the SF would be 1.09, and an *exaggeration* of scale would result. Obviously, if we had a location on the same projection with a local scale of exactly 1:250,743,970, the SF would be 1.0, indicating that there has been *no* change in scale at that location.

8.5.3 Tissot's Indicatrix

Nicolas Tissot, a French mathematician, was one of the pioneers in analyzing distortion found on map projections. Tissot developed the **indicatrix**, which provides a graphical representation of distortion at various points across a projection. The indicatrix, as found on the reference globe's surface where all spatial relationships are preserved, results in no distortion, and is called the *unit circle*.* In other words, at each and every point on the reference globe, an infinitely small circle exists that can be described as a unit circle having a radius (or SF) of 1.0 as shown in Figure 8.17. Even though the indicatrix is displayed as a unit circle of some areal extent, it is important to remember that the appearance of the indicatrix on the reference globe is restricted to measurements made around infinitely small points and is not considered valid when discussing landmasses or water bodies of any great extent.

_____
* In mathematics, a unit circle radius equals 1 unit.

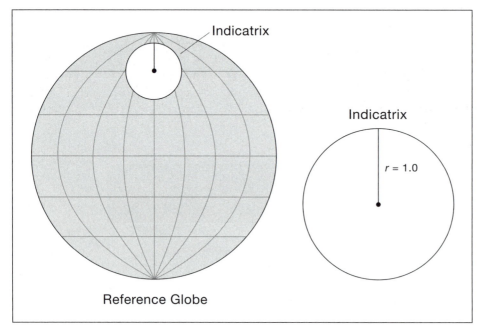

FIGURE 8.17 The indicatrix and its characteristics as shown on the reference globe. The radius of the indicatrix on the reference globe is 1.0.

After projection, the indicatrix has two characteristics that allow the amount and kind of distortion to be analyzed. First, the indicatrix is defined by two specific radii called the semimajor (a) and semiminor (b) axes. By convention and through mathematical manipulation, the semimajor axis is aligned in the direction of the maximum SF, and the semiminor axis is aligned in the direction of the minimum SF. These two axes are always perpendicular to one another. Second, on the indicatrix there exist two lines called l and m that intersect forming an angle called 2ω. The specific changes in the SFs along a and b as well as the alteration of 2ω after projection provide a useful means to analyze distortion.

There are four general instances of the indicatrix that result after a projection is created. In the first instance, there is no change in the SFs along a and b, so the indicatrix is still a unit circle (Figure 8.18A), indicating no distortion. The second instance is where the SFs along a and b change unequally, creating an ellipse, but the area of this ellipse is equal to the area of the unit circle. In this instance, there is an increase in the SF along a and the SF along b decreases proportionately (Figure 8.18B). This leads to a change in the angle 2ω, and so the projection is said to possess *angular distortion*. Note, however, that the area of the unit circle is preserved, as found on the reference globe, and so there is no areal distortion. In the third instance, the SFs along a and b change in the same manner (e.g., $a = 2.0$ and $b = 2.0$). Here, the indicatrix is still a circle, but it is larger (or smaller) in size than the unit circle found on the reference globe (Figure 8.18C), indicating *areal distortion*. Because the angle 2ω has been preserved, there is no angular distortion. The fourth instance is where a and b change unequally, producing an ellipse, and the area of the ellipse does not equal the unit circle (Figure 8.18D). In this case, there is both areal and angular distortion.

When the indicatrix is plotted on a projection (e.g., every 15°), the type and amount of distortion can be visualized across the projection. For example, Figure 8.19 shows the indicatrix for the Winkel Tripel projection. In this case, note that at approximately 44° 28' 25" N/S is the only location where the indicatrix is a circle and the SFs along both a and b are 1.0. At all other locations the indicatrix takes on an elliptical shape, suggesting that there is angular distortion throughout the projection. In a similar light, the ellipses do not appear to have the same areal extent, indicating that areal relations are not preserved either in the Winkel Tripel projection.

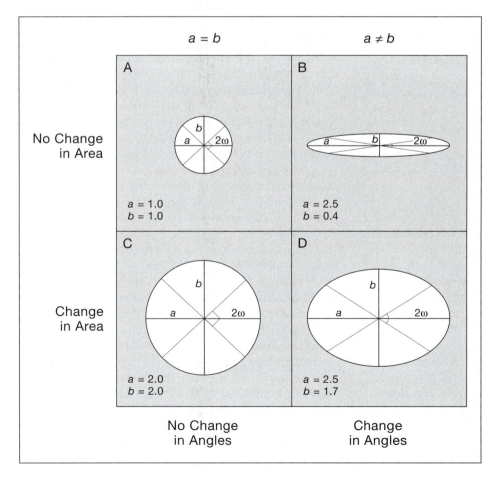

FIGURE 8.18 The possible configurations of the indicatrix after projection. The indicatrix in (A) is the unit circle as found on the reference globe, and thus there is no distortion in scale factors, areal, or angular relations. In (B), the scale factors are not equal; the area of the original indicatrix has been preserved, and note the change in 2ω, suggesting angular distortion. The scale factors in (C) are greater than 1.0, but equal to one another and thus the area is exaggerated, but there is no angular distortion. In (D), there is both angular and areal distortion.

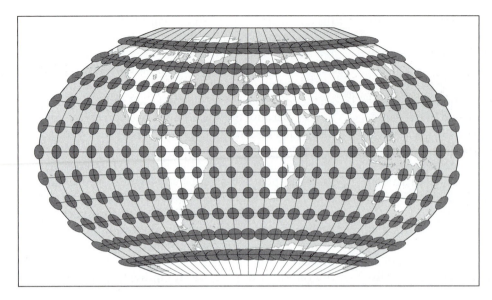

FIGURE 8.19 The indicatrix on the Winkel Tripel projection. Note that the indicatrix is a circle only along the central meridian at approximately 44° 28' 25" N/S. The indicatrix takes on a variety of elliptical shapes and sizes throughout the remainder of the projection, suggesting that this projection preserves neither areas nor angles. The indicatrix is plotted every 15°.

8.5.4 Distortion Patterns

Figure 8.9 showed that the developable surface contacts the reference globe at either a single standard point or along one or two standard lines. Where the standard point or the standard line(s) contact the reference globe, no distortion in scale is present—the SF at these locations is 1.0. However, we learned that at other locations throughout the map projection scale varies. This scale variation can be mapped, producing characteristic *distortion patterns* that are unique to the specific class and case of projection, as shown in Figure 8.20, where darker grays indicate greater distortion.

8.6 PROJECTION PROPERTIES

A map projection is said to possess a specific property when it preserves one of the spatial relationships (areas, angles, distances, and directions) found on the Earth's surface. The preservation of a particular property is achieved by controlling the SFs throughout the projection. For instance, projections that preserve either areas or angles throughout the projection are called **equivalent** (equal area) and **conformal**, respectively. Projections also are capable of preserving several *special* properties. When all distances from a particular location are correct, then the projection is said to be **equidistant. Azimuthal projections**, which were introduced earlier as planar projections, preserve directions or azimuths from one central point to all others. **Minimum error projections** possess no specific property, but by mathematically optimizing SFs they achieve lower overall distortion across a projection than can be achieved when one property is

preserved. Next, we consider each of the projection properties in detail.

8.6.1 Preserving Areas

Equivalent projections preserve landmasses in their true proportions, as found on the Earth's surface. To illustrate equivalency, consider the landmasses in Figure 8.21, in which we see four samples of Greenland that have been taken from different projections (each projection has the same approximate principal scale). Figure 8.21A through 8.21C are from equivalent projections, whereas Figure 8.21D is from a nonequivalent projection. To preserve areal relations in Figure 8.21A through C, the shapes of landmasses have been distorted (e.g., in Figure 8.21A, the northern portion of Greenland is expanded in an east-to-west direction compared to its representation on the Earth).

To ensure that areas are preserved on equivalent projections, the SFs must be controlled so that each indicatrix contains the same area. For instance, Figure 8.22 shows the indicatrix on the Albers equivalent conic projection, which possesses two standard lines at 30° N and 45° N latitude. On this projection the indicatrix appears as a circle along each standard line. At all other locations, the indicatrix takes on an elliptical shape, suggesting that the scale factors a and b change across this projection. In the upper latitudes, the ellipses display a contorted appearance, elongated in an east–west direction and compressed in a north–south direction. Recall that for equivalence, b must decrease proportionately as a increases. The difference in lengths of a and b indicates angular distortion at these locations. Regardless of the appearance of the ellipses, all are of the same size, ensuring that areas are preserved across the projection.

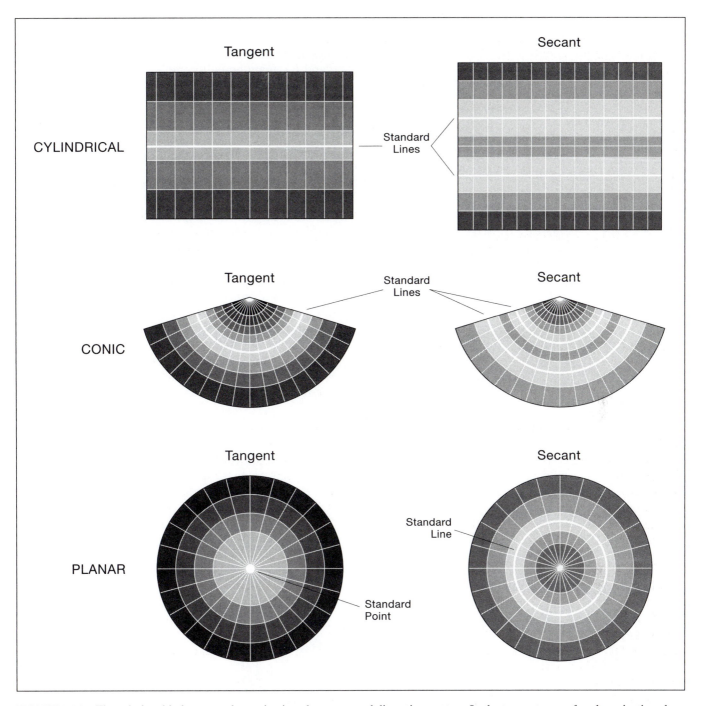

FIGURE 8.20 The relationship between the projection class, case, and distortion pattern. In the tangent case of each projection class, the standard line or point is the location where the scale factor is 1.0, shown by the thicker white line or point. In the secant case of each projection class, the standard lines (note that the planar class only has one standard line) have a scale factor equal to 1.0. Darker gray shading indicates increasing distortion.

8.6.2 Preserving Angles

Conformal projections preserve angular relationships around a point by uniformly preserving scale relations about that point in all directions. This concept requires some explanation because the term *conformal* has been misunderstood and misused by many. It often has been interchanged with *orthomorphic*, meaning "correct shape," and this has conveyed the assumption that conformal projections preserve *shapes* of entire landmasses (both large and small). Unfortunately, conformal maps do not preserve shapes of landmasses per se. Rather, preservation of shapes is found only at infinitely small points.

Conformal projections preserve angular relations by ensuring that SFs change along *a* and *b* at the same rate. As

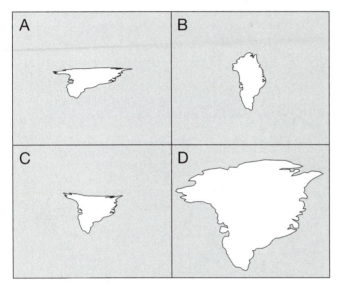

FIGURE 8.21   Greenland depicted on three different equivalent projections—(A) the Eckert IV pseudocylindrical, (B) the Lambert azimuthal equivalent, and (C) the Albers equivalent conic—and one nonequivalent projection, (D) the Miller cylindrical projection.

Figure 8.23 shows, the indicatrix throughout the Lambert conformal conic projection is displayed as a circle, but *a* and *b* necessarily change uniformly in size. At all locations, *a* and *b* remain equal to one another, permitting 2ω to maintain the same angle as found on the reference globe.

8.6.3 Preserving Distances

In simple terms, projections that maintain the principal scale from two points on the map to any other point on the map are said to be **equidistant**. For instance, if the two points are the poles,* then all meridians are straight lines that have the same principal scale; an example is the plate carrée projection (Figure 8.5). Recall that equivalent and conformal projections achieve their properties by varying the SFs during the transformations. With these projections, scale cannot be preserved along both *a* and *b*. Scale on a projection can be preserved only along a single line from one point to any other point (usually placed at the center of the projection), between two points, or along standard line(s). For instance, the azimuthal equidistant projection (Figure 8.24A) portrays correct scale from the projection's center along a straight line to any other point on the map. Examples of other

---

* In technical terms, any two points that are exactly opposite each other on the Earth's surface are called antipodes. For example, a location at 39° N and –83° W would have its antipode located at –39° S (the opposite hemisphere) and 97° E (180°–83°).

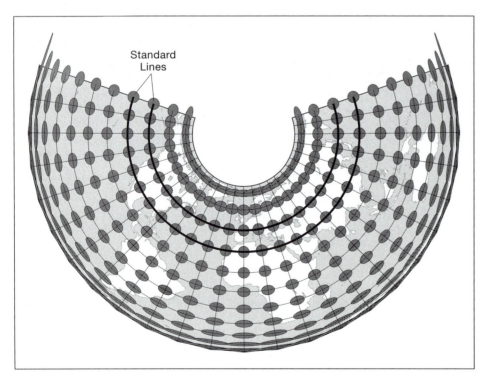

FIGURE 8.22   Indicatricies for the Albers equivalent conic projection, which has standard lines at 30° N and 45° N latitude (shown by the thicker black lines). Indicatricies are circles along the standard lines; for all other locations, indicatricies are elliptical in shape, but retain the same area as indicatricies along the standard lines.

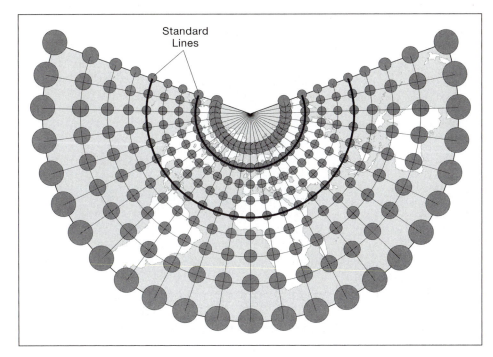

Standard
Lines

FIGURE 8.23 Indicatricies for the Lambert conformal conic projection, which has standard lines at 15° N and 60° N latitude. All indicatricies are circles (indicating no angular distortion), but the indicatricies change in size when moving away from the standard lines (indicating areal distortion).

equidistant projections include the equirectangular cylindrical and Euler equidistant conic (Figures 8.24B and 8.24C), where scale is correct along all meridians; and the doubly equidistant azimuthal (Figure 8.24D), which preserves scale along a straight line from either of two central points—in this case between Washington, DC, and London, England.

In some cases, a mapmaker might wish to focus on a certain area of the map, and show the relationship of that area to the larger geographical context. In other words, the central area to be emphasized is shown at a larger scale and the peripheral area is shown at a smaller scale. Such a **variable scale map projection** presents an interesting view of the Earth and, as the title suggests, the SFs change considerably across the projection but can be controlled to yield interesting and useful results. For example, Figure 8.25 shows the Hagerstrand logarithmic projection centered at 96° W, 40° N. On this projection, notice that the center shows the U.S. Midwest enlarged compared to the rest of the landmass, which becomes increasingly smaller toward the map edge.

8.6.4 Preserving Directions

Directions or azimuths can be preserved on **azimuthal projections**. On this kind of projection, directions are preserved from the center of the map to any other point on the map. When measuring azimuths from the center of an azimuthal projection, all straight lines drawn or measured to distant points also represent great circle routes. As such, azimuthal projections have been used extensively for navigation and pinpointing locations.

Azimuthal projections are among the oldest known projections, some having been developed by the ancient Greeks. Figure 8.26 shows five common azimuthal projections: (A) Lambert equivalent, (B) stereographic conformal, (C) equidistant, (D) orthographic, and (E) gnomonic. Each azimuthal projection is conceptually developed in a similar fashion by placing a plane tangent to one point on the reference globe. As such, the SF on these azimuthal projections is only true at the center point and greater than 1.0 at all other locations. It is useful to note that the azimuthal projections, in addition to preserving direction, each preserve another property: The stereographic is conformal, the Lambert is equal area, the azimuthal equidistant is equidistant only along the meridians (here the SF is set to 1.0 so that one can measure distance from the center of the projection to any other point), and the gnomonic shows all great circle routes as straight lines. The orthographic displays a spherical appearance suggesting the roundness of the Earth, which is a correct representation of the Earth if viewed from a distant point in space.

8.6.5 Compromise Projections

In both equivalent and conformal mapping, the sizes and shapes of landmasses are often visually disturbing, sometimes to the point of being unrecognizable. A solution to this problem is **compromise projections**, which manipulate the SFs so that the extreme angular and areal distortion found on equivalent and conformal projections is not present. Thus, a compromise projection strikes a balance between the distortion in area present on conformal

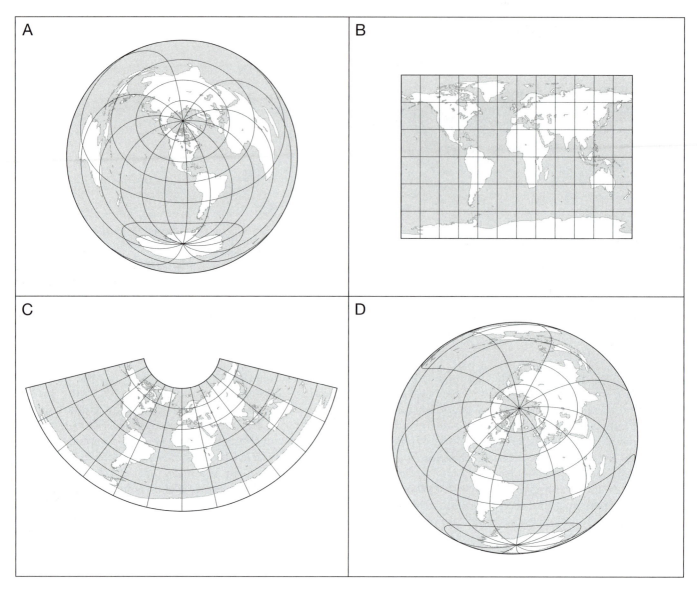

FIGURE 8.24 Four equidistant projections: (A) the azimuthal equidistant, (B) the equirectangular cylindrical, (C) the Euler equidistant conic, and (D) the two-point (or doubly) equidistant.

projections and the angular distortion that is common on a purely equivalent projection. With compromise projections, even though no projection property is completely preserved, the combined areal and angular distortion is usually less than if a single property was preserved and generally gives a better visual representation of landmasses.

Probably the most notable compromise projection is the Robinson. Its popularity increased dramatically when the National Geographic Society in 1988 made the decision to replace the Van der Grinten I projection (Figure 8.27A) with the Robinson (Figure 8.27B) as the projection for their world maps. Even though the Van der Grinten I is a compromise projection, the National Geographic Society believed the Robinson

presented a better visual impression of the shape of the landmasses and so it was employed up to 1998 when the National Geographic Society replaced it with the Winkle Tripel (Figure 8.27C), another compromise projection.

A variation on the compromise projection concept is the **minimum error** projection. As previously discussed, each projection property has the objective of preserving one of the spatial relationships (area, angles, directions, and distances) on the final map. Although this is often advantageous for specific mapping applications, in some cases, the preservation of one property often leaves considerable areas, angles, or distances exaggerated in other parts of the map. In some mapping applications, such as general reference maps, there is no need to preserve a

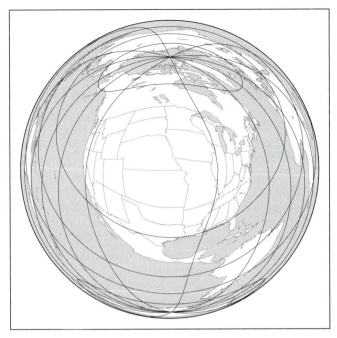

FIGURE 8.25 The Hagerstrand logarithmic projection with a 30° graticule spacing. Here, the appearance of the map is determined by taking the logarithmic distance from the center to all other map locations.

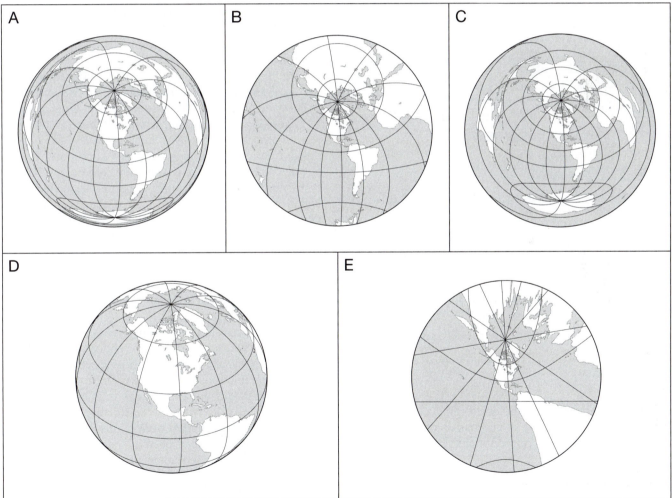

FIGURE 8.26 Five common azimuthal projections each centered at 96° W and 40° N: (A) Lambert equivalent, (B) stereographic conformal, (C) equidistant (scale is preserved from the center to any point on the map), (D) orthographic, and (E) gnomonic (allows all great circles to be drawn as straight lines).

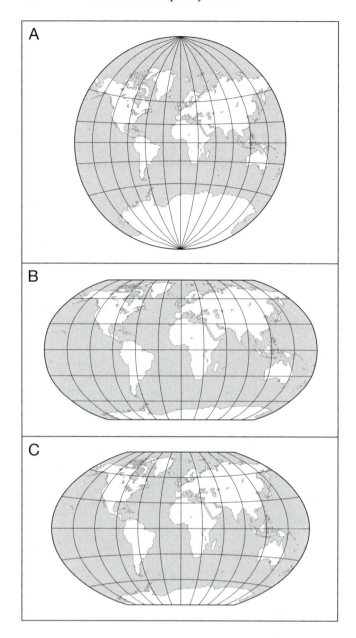

FIGURE 8.27   Three compromise projections: (A) Van der Grinten I, (B) Robinson, and (C) Winkel Tripel.

specific property. Rather, the desire might be to lessen the consequences of having landmasses modified from how they appear on the reference globe.

SUMMARY

In this chapter we examined the fundamental elements of map projections. We began with a discussion of the reference globe and developable surface as conceptual aids in understanding how the Earth's curved two-dimensional surface is projected onto a map. Today, computer software and mathematics are heavily involved in developing projections, and so we illustrated how mathematical equations could be used to produce a map projection.

Using software to create a projection is not difficult, but does require the user to understand the importance of the basic projection characteristics of **class**, **case**, and **aspect**. The projection's class is directly related to the developable surface used in creating a projection—**cones**, **planes**, and **cylinders** produce **conic**, **planar**, and **cylindrical** projections. The case of a projection relates to the way in which the developable surface is positioned with respect to the reference globe. When the developable surface touches the reference globe, the **tangent case** results, where only one line (or point) of contact is shared between the two surfaces. When the developable surface passes through the reference globe, the **secant case** results, where contact is along either one or two lines. The projection's **aspect** relates to the way in which the developable surface can be positioned anywhere on the reference globe. Changing the aspect allows the projection to be centered at one of the poles (a **polar aspect**), along the Equator (an **equatorial aspect**), or somewhere in between a pole and the Equator (an **oblique aspect**).

Distortion is an inevitable consequence of the projection process and involves the alteration of scale about a point on a map compared to that same point on the Earth. An important consideration in determining the distortion at each point on the map is the **scale factor**, which is the ratio of the local scale to the principal scale. Closely associated with the scale factor is **Tissot's indicatrix**, which is one of the more common graphical devices used to visualize distortion across the map. If the indicatrix is mapped across the projection, definite *distortion patterns* emerge that in part reflect the nature of the developable surface that was conceptually used to create the projection.

Important spatial relationships on the Earth include the ability to measure distances, directions, areas, and angles. Each projection has certain properties that allow these spatial relationships to be preserved on a map. Preserving areal and angular relations is made possible through **equivalent** and **conformal** projections. **Equidistant** and **azimuthal** projections permit distances and directions to be measured in specific ways across a projection. One consequence of preserving a specific property is that the other properties are often distorted (e.g., on conformal projections, areal relations are distorted). Therefore, **compromise projections** have been developed that do not preserve any specific property.

FURTHER READING

Bugayevskiy, L. M., and Snyder, J. P. (1995) *Map Projections: A Reference Manual*. London: Taylor & Francis.

Provides a mathematical treatment of the projection process.

Canters, F., and Decleir, H. (1989) *The World in Perspective: A Directory of World Map Projections*. New York: Wiley.

Includes a useful discussion of the principles of the map projection process, mathematical derivation of map distortion, and illustrations and a historical overview of many world map projections.

Chamberlin, W. (1947) *The Round Earth on Flat Paper: Map Projections Used by Cartographers*. Washington, DC: National Geographic Society.

A broad introduction to fundamental map projection elements, an explanation of the importance of map projections in cartography, and many useful illustrations.

Deetz, C. H., and Adams, O. A. (1945) *Element of Map Projections with Applications to Map and Chart Construction*. 5th ed. Washington, DC: U.S. Department of Commerce Coast and Geodetic Survey.

Provides a historical perspective on the methods used to manually construct projections.

Iliffe, J. C. (2000) *Datums and Map Projections*. Boca Raton, FL: CRC Press.

Chapters 7 through 11 review some fundamental aspects and mathematical computations of the map projection process.

Maling, D. H. (1989) *Measurements from Maps: Principles and Methods of Cartometry*. Oxford: Pergamon.

Reviews the nature of measurements taken from, and consequently the distortion that results, as scale varies across a map.

Maling, D. H. (1992) *Coordinate Systems and Map Projections*. 2nd ed. Oxford: Pergamon.

Covers a broad range of topics on map projections, including chapters devoted to mathematical principles, Tissot's indicatrix, and quantitative analysis.

McDonnell, P. W. (1991). *Introduction to Map Projections*. 2nd ed. Rancho Cordova, CA: Landmark Enterprises.

Explains, using spherical trigonometry, the various equations for numerous map projections; involves less complex mathematics than the Bugayevskiy and Snyder or Yang, Snyder, and Tobler texts.

Robinson, A. H., Morrison, J. L., Muehrcke, P. C., Kimerling, A. J., and Guptill, S. C. (1995) *Elements of Cartography*. 6th ed. New York: Wiley.

Chapter 5 presents a basic nontechnical discussion on map projections.

Snyder, J. P. (1987) *Map Projections: A Working Manual*. Washington, DC: U.S. Geological Survey.

A useful overview of basic map projection concepts, numerous worked examples of projection equations (both spherical and ellipsoidal forms), and a brief history of each projection.

Snyder, J. P. (1993) *Flattening the Earth: Two Thousand Years of Map Projections*. Chicago: University of Chicago Press.

A concise and readable overview of the history of map projections, their development, key individuals, and mathematical evolution as a key factor in the evolution of increasingly sophisticated map projections.

Snyder, J. P., and Steward, H. (1988) *Bibliography of Map Projections*. Washington, DC: U.S. Geological Survey.

Includes 2551 entries of various texts and articles relating to map projections.

Snyder, J. P., and Voxland P. M. (1989) *An Album of Map Projections*. Washington, DC: U.S. Geological Survey.

Illustrates numerous projections suitable for world maps. Also provides a useful overview of distortion analysis, including a discussion of Tissot's indicatrix.

Tissot, N. A. (1881) *Memoire sur la Representation des Surfaces et les Projections des Cartes Geographiques*. Paris: Gautier-Villars.

Discusses the development of his indicatrix.

Yang, Q., Snyder, J. P., and Tobler, W. R. (2000) *Map Projection Transformation: Principles and Applications*. London: Taylor & Sons.

Presents a thorough treatment of the mathematics of the map projection process.

9

Selecting an Appropriate Map Projection

OVERVIEW

To assist you in applying the concepts presented in Chapters 7 and 8, this chapter presents an overview of methods for selecting an appropriate projection for a variety of thematic maps. We begin by reviewing several common approaches to selecting map projections. We focus on a **projection selection guideline** developed by John Snyder because it offers the greatest utility to beginning map designers. Specifically, Snyder's guideline utilizes three separate tables that initially recommend projections based on the extent of the geographic area to be mapped: (1) world, (2) hemisphere, and (3) continent, ocean, or smaller region. Each table is further organized according to additional considerations made during the projection selection process. For instance, the recommended projections for the world category are arranged according to the desired property (e.g., conformal) and any special characteristic (e.g., the projection's center).

To explore the utility of Snyder's selection guideline, in section 9.2 we discuss how an appropriate projection can be selected for four sample data sets. For our first data set, we imagine creating a world map of literacy rates by country. In this case, we choose to create a choropleth map and thus need to select a projection that preserves areal relations. Here we introduce the advantages of using an **interrupted** projection for world maps, on which the graticule has been "cut," creating **lobes** that encompass specific geographic areas (e.g., South America). On an interrupted projection, lobes showing specific geographic areas defined by longitude bounds are created that have lower distortion than if those same areas were shown on a noninterrupted world map.

The second data set focuses on creating a map showing population distribution of Russia at the oblast and kray level. In this case, we combine proportional symbol and dot maps so that we can depict both urban and rural pop-

ulation, respectively. Here we need to select a projection that preserves areal relations so that the map user can properly compare dot distributions in different geographic areas; thus, we focus on the selection of various parameters for conic projections: standard parallels, central meridian, and central latitude.

The third data set examines migration from Europe and Asia to the United States. This map's objective is twofold. First, we want to show the general migration route over which immigrants traveled; second, we want the map user to see the spherical nature of the Earth over which the immigrants traveled. To meet the first requirement, we choose graduated flow lines as the symbology. The map user can thus visualize the number of immigrants from the different Asian and European countries, as well as the general routes taken. To meet the second requirement, it is important for the map to communicate the spatial proximity between the United States, Europe, and Asia; thus, we focus on planar map projections that present a hemisphere.

The fourth data set focuses on a small geographic area: Kansas. Specifically, the data set entails the paths of F4 and F5 tornados across Kansas during the past 50 years. In this map, we symbolize tornado paths as arrows—the length of the arrow indicates the distance the tornado traveled across the ground and the direction of the arrow indicates the general direction the tornado took. Selecting projections for this small geographic area highlights the fact that regardless of the projection property chosen (conformal vs. equidistant), small geographic areas are equally well represented by most projections.

The chapter concludes with a discussion of five key objectives that should be emphasized when selecting a map projection for a thematic data set. First, in most cases, the mapmaker should select a projection with the lowest *distortion. Second, amounts of distortion can be kept small by aligning the geographic area (or data set) under consideration with the standard line(s) or positioning the map's center with the standard point. Third, as the amount of geographic area under consideration increases (e.g., mapping North America compared to Kansas), distortion becomes a more important consideration. Fourth, just because a projection has seen considerable exposure (e.g., it has been used in prominent atlases) does not mean that the projection is suitable for your specific application. Fifth, an often overlooked aspect of the map projection is its influence on the overall map design—a topic that is not well studied by cartographers.*

9.1 POTENTIAL SELECTION GUIDELINES

Selecting an appropriate map projection is one of the more involved tasks in cartography. In essence, when selecting a projection, a match must be made between the map's purpose and various projection properties and characteristics. This task is difficult because of the many variables involved when creating a map, such as the map scale, amount of the Earth to be mapped, level of generalization, and thematic symbology used. Similarly, any given projection has numerous characteristics (class, case, and aspect) and associated properties (equivalent, conformal, equidistant, and azimuthal). Rarely will a single projection have all the characteristics and properties necessary to satisfy all variables involved in the mapmaking process.

In an attempt to recommend projections for specific map purposes, various **projection selection guidelines** have been developed. Although the format of each guideline differs, the general purpose is to provide the mapmaker with one or more projections that can reasonably be applied to a particular purpose. For example, a simple selection guideline discussed by Frederick Pearson (1984) related the choice of projection to the range of latitude depicted on the map. Equatorial regions lying 30° either side of the equator are mapped with cylindrical projections; midlatitude regions between 30° and 65° are mapped with conic projections; and polar areas above 65° are mapped with planar projections. This guideline's logic rests on the fact that the location of the standard point or line(s) of a projection class lies within the geographic area recommended for that class, hence distortion is low throughout the region of interest. For instance, those areas lying along the Equator are matched with cylindrical projections that have one standard line coinciding with the Equator or two standard lines equally spaced on either side of the Equator. Although Pearson's guideline provides a starting point for selecting projections, it generally is not very useful, as it does not consider the map purpose.

Arthur Robinson and his colleagues (1995) described another simple guideline for selecting projections based on the relationship between projection properties and the intended map purpose. For instance, they recommended conformal projections for analyzing, measuring, or recording angular relationships, as in navigation, piloting, and surveying; suitable projections include the Mercator, transverse Mercator, Lambert conformal conic, and stereographic. They recommended equivalent projections for geographic comparison across a map, a common purpose of thematic maps. For example, a dot map's primary goal is to visually compare different dot distributions across geographic areas, and this comparison is greatly facilitated if the geographic areas are represented in their correct proportions. Equivalent projections that they recommended include the cylindrical equivalent, Lambert's azimuthal, Albers and Lambert conic, and most equivalent pseudo-cylindrical projections. Robinson et al. also noted that when recording and tracking direction of movement, planar projections are useful, including the orthographic (useful for showing the Earth as if viewing it from space), azimuthal equidistant (for showing correct directions and distances from the center to any other point), and the gnomonic (all great circles are represented by straight lines). Robinson et al. also mentioned that some projections have been designed for specific applications. One such example is the space oblique Mercator, which is used to record the ground track of satellites orbiting above the Earth's surface.

9.1.1 Snyder's Hierarchical Selection Guideline

Although the preceding approaches are conceptually useful, we feel that neither guideline provides the level of detail or the flexibility needed to adequately select an appropriate projection. Thus, we focus on a guideline developed by John Snyder (1987) that permits you to select a map projection for many mapping purposes. Snyder presented a hierarchical list of suggested projections that is organized according to the region of the world to be mapped, projection property (e.g., equivalent, conformal, equidistance, azimuthal), and characteristic (e.g., class, case, and aspect). The advantage of a hierarchical approach is that the mapmaker can begin with a broad question (e.g., size of geographic region to be mapped) and logically proceed to more detailed questions about the mapping situation (e.g., an oblique, polar, or equatorial aspect). In so doing, the mapmaker is led down a particular path through the hierarchy until an appropriate projection is recommended.

Snyder's hierarchy begins with a division of geographic areas into three categories: (1) world, (2) hemisphere, and (3) continent, ocean, or smaller region. Snyder also drew a distinction between the appropriate models of the Earth for each geographic area (see section 7.3.2). For instance, for maps of the world, hemisphere, continents, and oceans, a spherical model of the Earth is reasonable. With these large geographic extents, the map uses do not call for highly accurate maps such as large-scale topographic ones. For smaller regions, an ellipsoidal model should be used, as smaller regions often form the base of maps on which highly accurate measurements are made (e.g., navigational charts).

World Map Projections

Table 9.1 presents Snyder's map projection guideline when creating a map of the entire world. The mapmaker begins by choosing from among the following projection properties: conformal, equivalent, equidistant, straight rhumb lines, and compromise distortion. If a conformal projection is desired, then the mapmaker has several projections from which to choose according to the scale variation across the projection. For instance, when the map purpose requires that a constant scale be shown along the equator, a meridian, or an oblique great circle, then the Mercator, transverse Mercator, and oblique Mercator projections should be chosen, respectively. If constant scale is not a requirement, then the Lagrange, August, and Eisenlohr projections are recommended (Figure 9.1). The Lagrange conformal projection shows the world within a circle—an advantage over the stereographic conformal projection, which commonly only

shows one hemisphere. The August and Eisenlohr projections are both conformal epicycloidal* projections and are useful because they are conformal throughout—on many conformal projections, the property of conformality is not maintained at the poles. A shortfall of the epicycloidal conformal projections is the rapidly increasing scale distortion from the projection's center. For instance, note that Antarctica's appearance in the conformal projections in Figure 9.1 is greatly exaggerated (in reality, South America's area extent is roughly 19 million km^2, whereas Antarctica is only 14 million km^2).

If an equivalent projection is desired, there are three general categories: noninterrupted, interrupted, and oblique aspect. An **interrupted projection** shows the graticule as "cut" along specific meridians creating one or more **lobes** as shown by the Eckert V pseudocylindrical projection in Figure 9.2. Each lobe has its own central meridian that creates an area of lower distortion than if the projection was not interrupted. The number of lobes and placement of each cut is determined by the mapmaker. In Figure 9.2A, two hemispheres are created (the cut is placed along the Prime Meridian and 180°). In contrast, in Figure 9.2B, the cuts are placed along the Atlantic and Indian Oceans so that the continents have

---

* An epicycloidal projection is conceptually developed as a curve traced by a point on a circle's circumference of radius a that rolls along the outside of a fixed circle having a radius of $2a$.

TABLE 9.1   Snyder's map projection guideline showing projections for mapping the world

Region Mapped	Property	Characteristic	Named Projection
World	Conformal	Constant scale along Equator	Mercator
		Constant scale along a meridian	Transverse Mercator
		Constant scale along an oblique great circle	Oblique Mercator
		No constant scale anywhere on the map	Lagrange
			August
			Eisenlohr
	Equivalent	Noninterrupted	Mollweide
			Eckert IV & VI
			McBryde or McBryde–Thomas
			Boggs Eumorphic
			Sinusoidal
			Other miscellaneous pseudocylindricals
			Hammer (a modified azimuthal)
		Interrupted	Any of the above except Hammer
			Goode's Homolosine
		Oblique aspect	Briesemeister
			Oblique Mollweide
	Equidistant	Centered on a pole	Polar azimuthal equidistant
		Centered on a city	Oblique azimuthal equidistant
	Straight rhumb lines		Mercator
	Compromise distortion		Miller cylindrical
			Robinson pseudocylindrical

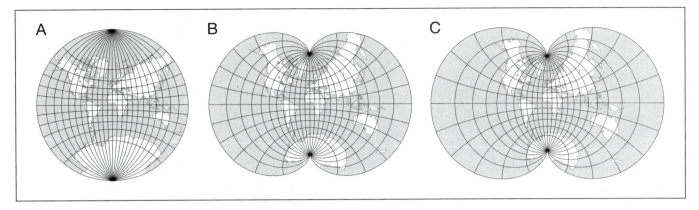

FIGURE 9.1 Different conformal projections of the world. The Lagrange (A) presents the world in a circle, whereas the August (B) and Eisenlohr (C) are epicycloids.

lower distortion. In Figure 9.2C, the cuts are again placed along the oceans, but an additional cut is made in the southern Pacific Ocean. Although the cuts in Figure 9.2B and Figure 9.2C are positioned over water to provide continuity of the landmasses, depending on the map's purpose, the mapmaker might choose to place the cuts over land, preserving the continuity of the oceans.

The oblique aspect (a projection with a center that is neither the pole nor the Equator) is warranted when geographic areas not aligned along the Equator or centered at the pole must be brought to the center of the projection. For example, Figure 9.3 shows the Briesemeister modified azimuthal projection,* which is equivalent and is centered over northwestern Europe at 10° E and 45° N, thus bringing northern Europe, the Norwegian Sea, and Scandinavia to the map's center.

In situations where equidistance is required, Snyder suggested the azimuthal equidistant projection. The location of the map's geographic focal point dictates the aspect of the projection (maps centered about a pole require a polar aspect, whereas maps centered about the Equator require an equatorial aspect; all other locations use an oblique aspect). Snyder also included projections in his guidelines that require straight rhumb lines or compromise distortion. Straight rhumb lines are a special property that only a few projections possess; for instance, on the Mercator projection all rhumb lines (lines of constant compass bearing) appear as straight lines; this is obviously needed in certain mapping applications such as navigation. Some projections do not have a specific property and were described as compromise projections in Chapter 8. These projections typically have lower overall distortion than projections that preserve a single property. In many cases, compromise projections are developed specifically

to match the shape of landmasses to their appearance on a globe, as in the Robinson projection (Figure 8.27B).

Map Projections for a Hemisphere

Table 9.2 presents Snyder's projection guidelines for maps of a hemisphere. Planar projections are suited for mapping hemispheres as they have a standard point placed at the center of the projection and can be easily recentered over any area of importance. Planar projections are also interesting in that many of them combine the azimuthal property (showing directions correctly from the projection's center to all other points) with an additional property (e.g., the stereographic projection is both conformal and azimuthal), whereas most projections only possess one useful property (e.g., the Albers conic projection is exclusively equivalent). When showing a hemisphere, the following properties are appropriate: conformal, equivalent, equidistant, and global look. For each property, note that a specific planar projection is recommended.

Map Projections for a Continent, Ocean, or Smaller Region

Table 9.3 presents Snyder's guidelines for maps of a continent, ocean, or smaller region. Note that this hierarchy includes a subdivision that distinguishes between the predominant directional extent of the landmass: east–west, north–south, oblique (not aligned in an east–west or

---

* The Briesemeister projection, always centered at 10° E and 45° N, is a special oblique case of the Hammer modified azimuthal projection.

TABLE 9.2 Snyder's projection selection guideline showing planar projections for mapping a hemisphere

Region Mapped	Property	Named Projection
Hemisphere	Conformal	Stereographic conformal
	Equivalent	Lambert azimuthal equivalent
	Equidistant	Azimuthal equidistant
	Global look	Orthographic

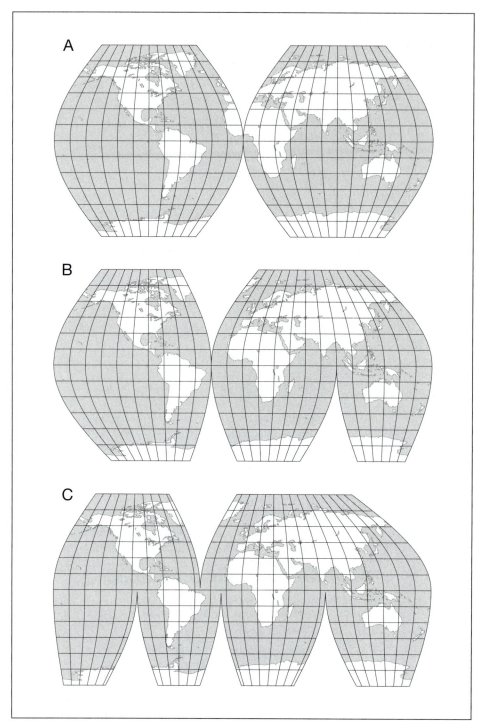

A

B

C

FIGURE 9.2   The Eckert V pseudo-cylindrical projection showing different placement of cuts.

north–south direction), or equal extent in all directions. This subdivision matches the position of the standard line(s) or point to the directional extent of the landmass, ultimately leading to a reduction of distortion for the geographic area under consideration. For instance, a map of Canada, which has considerable east–west extent, will have lower distortion when the standard lines can be positioned to coincide with the east–west extent. In contrast,

Antarctica, a landmass that is generally equal extent in all directions, will be more appropriately mapped with a single standard point, as distortion increases radially with distance from this point.

After examining directional extent of the landmass, you should consider the location of the landmass to be mapped. For instance, given a landmass of predominant east–west extent on the Earth's surface, the region can

TABLE 9.3 A portion of Snyder's map projection guideline showing projections for mapping a continent, ocean, or smaller region

Region Mapped	Directional Extent	Location	Property	Named Projection
Continent, ocean, or smaller region	East–West	Along the Equator	Conformal	Mercator
			Equivalent	Cylindrical equivalent
		Away from the Equator	Conformal	Lambert conformal conic
			Equivalent	Albers equivalent conic
	North–South	Aligned anywhere along a meridian	Conformal	Transverse Mercator
			Equivalent	Transverse cylindrical equivalent
	Oblique	Anywhere	Conformal	Oblique Mercator
			Equivalent	Oblique cylindrical equivalent
	Equal extent	Polar, Equatorial, or oblique	Conformal	Stereographic
			Equivalent	Lambert azimuthal equivalent

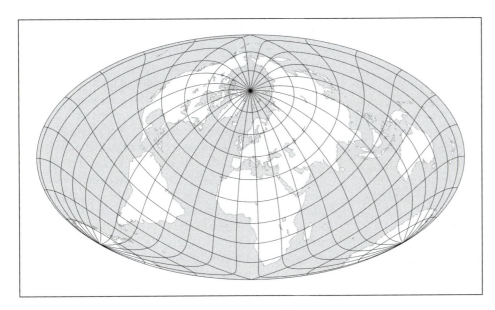

FIGURE 9.3 The Briesemeister modified azimuthal projection centered at 10° E and 45° N.

be located along the Equator or away from the Equator. If the landmass is located along the Equator, one of the cylindrical projections that have the Equator as a standard line is suitable (Mercator or cylindrical equivalent). If the landmass' east–west extent is located away from the Equator (e.g., the midlatitudes), then a conic projection with one or two standard lines is a suitable choice (Lambert conformal conic or Albers equivalent conic). For landmasses having a predominant north–south extent, a transverse Mercator or transverse cylindrical equivalent projection with a standard line corresponding to a meridian centered on the landmass is recommended. For landmasses with extents that are not aligned north–south or east–west, Snyder recommended an oblique Mercator or oblique cylindrical equivalent projection with a standard line aligned to the predominant extent of the landmass. If a landmass is equal in its predominant extent, then a planar projection having a standard point is recommended (stereographic or Lambert azimuthal equivalent). The last step is to examine which projection property (conformal or equivalent) is appropriate—topographic maps usually require conformal projections, whereas thematic maps usually require equivalent projections.

9.2 EXAMPLES OF SELECTING PROJECTIONS

In this section, we utilize Snyder's guideline as a foundation for selecting projections for four fictitious mapping situations. For each mapping situation we discuss the logic of Snyder's guideline and why the projections he recommended are appropriate. Additional insights into selecting projections not covered by Snyder's guidelines are also explored.

9.2.1 A World Map of Literacy Rates

For our first example, imagine that we wish to create a world map illustrating the distribution of literacy rates by country. We begin our selection process by considering the data, symbolization method, intended audience, and overall purpose for the map. Literacy rate is normally computed as the percentage of all people who can read. Although the underlying phenomenon of literacy rates is arguably spatially continuous in nature, we presume that we wish to examine the phenomenon at the level of countries, for which data are available (in the CIA Factbook*). Given our desire to portray data at the country level, choropleth symbolization is suitable. We further presume that the purpose is to create a map that allows for a comparison of literacy rates by country around the world—noting the concentrations of high and low literacy rates, and permitting us to examine the spatial pattern. We also presume that we wish to highlight those areas of the world with low literacy rates.

With our purpose in mind, we now consider Snyder's projection guideline. Because our focus is on the world, we start with Table 9.1. Our first step is to select a desired projection *property*. Given our choice of choropleth symbolization, a critical goal is to preserve the relative areas of each enumeration unit as they would appear on a globe. If this is not done, then areas appearing larger than they should will result in greater visual emphasis on the associated data. This sort of thinking leads us to choose an equivalent projection such as the Mollweide shown in Figure 9.4A. For comparison, Figure 9.4B shows a Mercator conformal projection, which is inappropriate. Note how the Mercator projection places emphasis on the high literacy rate category in the upper latitudes, suggesting that it dominates the world.

Once the property of equivalence is decided, the mapmaker has to choose the *characteristic* of the projection: noninterrupted, interrupted, or oblique aspect. As mentioned earlier, an interrupted projection reduces distortion across a geographic area by creating one or more lobes. Two factors to consider when deciding on

the appropriateness of an interrupted projection are the intended audience and where the data exist on the map. Recall that interrupted projections remove the continuity of the oceans or landmasses, which could make the map difficult to comprehend. If the audience is assumed to have poor geographic knowledge (e.g., elementary school children), an interrupted projection would be inappropriate. On the other hand, if the audience is assumed to have good geographic knowledge (e.g., policymakers), then an interrupted projection would be acceptable. In deciding whether an interrupted projection is appropriate, a second factor is the distribution of the data across the map. In the literacy rate example, the data are concentrated over the land and not over the oceans. Because the data are concentrated on land, the continuity of the oceans can be sacrificed to create an interrupted projection. Therefore, the cuts can be placed over the oceans, with each region of interest placed in a different lobe. In our case, South America, Africa, Australia, North America, and Eurasia correspond to the different regions. Figure 9.4C shows the interrupted Mollweide projection. Comparing Figure 9.4C with the noninterrupted Mollweide projection in Figure 9.4A, we see that the cuts along the oceans create a discontinuity that might make the map difficult to interpret for those with limited geographic knowledge.

Having examined the possibility of an interrupted or noninterrupted projection, another consideration is whether or not to specify an oblique aspect. Most world maps employ an equatorial aspect, where the map is centered at the Equator along a chosen central meridian—usually selected so that either the Atlantic or Pacific Ocean is central to the map. In cases where the geographic area of interest is not along the Equator, the projection's center must be moved to allow the geographic area of interest to be brought to the center of the map—an advantage of the oblique aspect. Recall that on most projections lower amounts of distortion are usually found at a projection's center. This is especially true on world maps where, for example, polar areas suffer the greatest distortion when a point somewhere along the Equator is the projection's center. To determine whether an oblique projection is necessary, we need to again examine the geographic distribution of the data and determine if a particular geographic area of the Earth has high or low literacy rates that are important and if so, where they are. Based on our original statement, we wish to focus on those areas with low literacy rates in an attempt to highlight their situation. In reviewing the literacy data, Central Africa and portions of Central Asia have the lowest literacy rates and should be brought to the center of the map. These geographic areas are situated along the Equator and thus an oblique aspect is not called for. However, we should consider recentering the projection over

---

* The CIA World Factbook is available at *http://www.odci.gov/cia/publications/factbook/*.

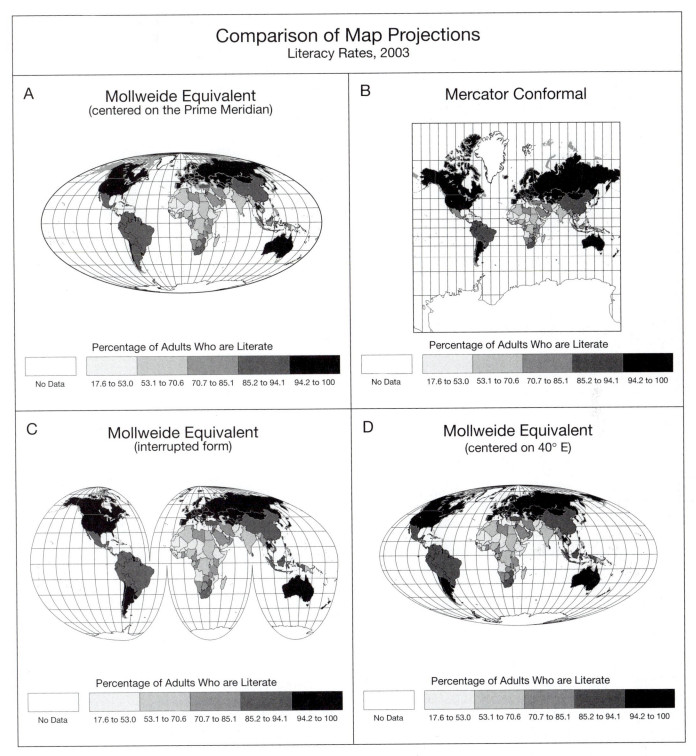

FIGURE 9.4 World literacy rates mapped on the (A) Mollweide equivalent projection, (B) Mercator conformal projection, (C) an interrupted Mollweide projection, and (D) a Mollweide projection that has been recentered along the Equator at 40° E.

the area of interest. In this case, the area of low literacy rates primarily falls near 40° E and so a central meridian is chosen to coincide with 40° E, as shown on the Mollweide projection in Figure 9.4D.

An additional thought on selecting an appropriate projection not directly addressed by Snyder is the idea of the poles being represented by a point or a line. In Figure 9.5, the quartic authalic and Eckert IV equivalent

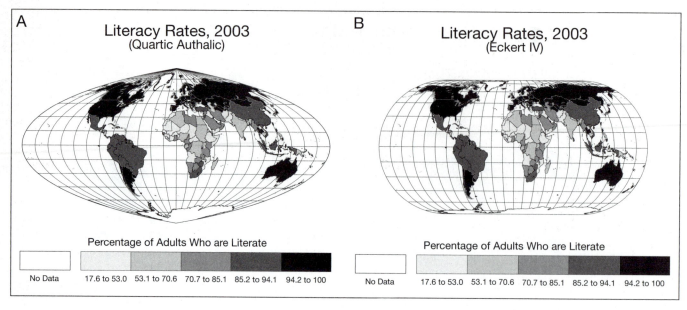

FIGURE 9.5   The literacy rate data shown on the (A) quartic authalic and (B) Eckert IV equivalent pseudocylindrical projections. The quartic authalic represents the poles as points and the Eckert IV represents the poles as lines.

projections are shown. Notice that the quartic authalic projection (Figure 9.5A) represents the poles as points, whereas the Eckert IV projection (Figure 9.5B) represents the poles as lines. On the quartic authalic projection, landmasses in the upper latitudes are compressed and difficult to see due to the convergence of the meridians to a point (especially at smaller scales). On the other hand, on the Eckert IV projection, the landmasses in the upper latitudes are stretched in an east–west direction due to the nonconvergence of the meridians, making recognition of the landmasses more apparent, but distorting their overall shapes.

To help determine whether the poles should be represented by a point or a line, we can examine the map's geographic focus. Because our area of focus is Africa and Central Asia and not the upper latitude areas, it might make sense to use a projection where the poles are represented by a point. According to Snyder's guideline, the Mollweide, Boggs, and Sinusoidal projections all represent the poles as a point. On the Mollweide projection the meridians curve to the poles more gently than the Boggs or Sinusoidal and the landmasses are not as compressed in the upper latitudes, thus making it a suitable choice for the literacy rate data.

9.2.2 A Map of Russian Population Distribution

For our second example, we consider mapping the population distribution in Russia for 1993. As in the previous example, we begin the process of selecting a projection by considering the data, symbolization method, intended audience, and overall purpose of the map. We assume that this map is to be included in a textbook for high school students who are taking an introductory course in geography. In this context, the map will serve a supporting role in the discussion of Russia and its cultural geography. Population can be thought of as discrete entities (individual people), but in some cases can be assumed to be more continuous in nature (e.g., a visit to New York City can give the impression that people are everywhere). Thus, in conceptual terms population distribution is highly variable. In Russia, there are very dense urban centers and vast expanses of emptiness. To capture this variation, a combined proportional symbol–dot method seems appropriate, as it facilitates the observance of high population concentrations (urban areas) and expanses of low population (e.g., in central Siberia).

Starting with Snyder's guideline, we turn toward the continent, ocean, or smaller region selection guideline (Table 9.3). The first selection criterion is to determine the directional extent of the landmass. Russia clearly has a greater east–west than north–south extent (Figure 9.6): The east–west extent is approximately 5,163 miles (ranging from 27° E along the Baltic border to 170° W in the Bering Sea), whereas its north–south extent is 2,764 miles (ranging from 40° N along the Azerbaijan border to 80° N in the Arctic Ocean). Moreover, Russia is positioned away from the equator—60° N is the approximate central latitude of the country's north–south extent. Based on this discussion and according to Table 9.3, we see that

FIGURE 9.6 An orthographic projection shows the considerable east–west extent of Russia.

Russia has a considerable east–west directional extent and is located away from the Equator in the midlatitudes, calling for a conic projection.

To select a specific conic projection, we need to consider the appropriate projection property. Snyder's guideline (Table 9.3) offers conformal or equivalent properties for the conic projections. Because we are creating a combined graduated symbol–dot map, an equivalent projection is necessary to ensure that map users will correctly interpret the relationship between the phenomenon represented by the dots and the geographic area in which it is contained. To demonstrate the importance of an equivalent projection, examine Figure 9.7, which

shows a fictitious dot map for some enumeration unit. In Figure 9.7A the enumeration unit is shown on an equivalent projection, whereas in Figure 9.7B the same enumeration unit is projected on a conformal projection. Although the number of dots does not change (each map has 15 dots), there obviously is an apparent difference in the concentration of the dots over the given area.

Snyder recommends the Albers equivalent conic projection for thematic maps of landmasses that have a predominant east–west extent and are located away from the Equator because most conic projections have one or two standard lines that coincide with a line of latitude that can be placed to minimize the distortion of the mapped area. Although there are no universally accepted rules on the placement of standard parallels on conic projections, the overall goal is to select latitudes that will result in the lowest distortion over the mapped area. One approach is to place one standard parallel at one-sixth the latitudinal distance from the southern latitude limit of the map and the other standard parallel at one-sixth the latitudinal distance from the northern latitude limit of the map.* For example, Russia's southern latitude limit is approximately 41° 43′N (found between the border of Russia and Azerbaijan), and the northern latitude limit is approximately 81° 53′N (along the northern border of Ostrov Komsomolec—an island in the Arctic Ocean). This latitude range is close to 40° 10′. One-sixth this latitudinal range of 40° 10′ is 6.7° or 6° 42′. Given the latitudinal range and the general placement guideline presented earlier, the two standard parallels would be placed at 48° 25′ N (i.e., 41° 43′ + 6° 42′) and 75° 11′ N (i.e., 81° 53′ – 6° 42′). You should realize that the approach for standard parallel placement discussed here is only a rule of thumb, and more sophisticated mathematical approaches to selecting standard parallels have been developed. For details, see Snyder (1993) and Maling (1992).

In addition to the selection and placement of suitable standard parallels, most conic projections have other parameters that can be modified. One parameter is the central meridian, which should be chosen so that it is at the center of the longitudinal range. The longitude limits of Russia are approximately 163° (ranging from 27° E, Ostorov Gogland—a small island in the Gulf of Finland—to 170° W, the boundary between Russia and Alaska). Given this range, the central meridian for Russia would fall at approximately 108° 30′E. Another parameter for most conic projections is the central latitude, which is the latitude that falls halfway between the northern and

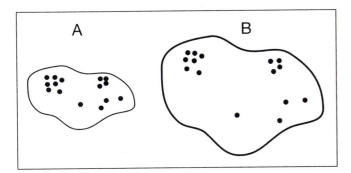

FIGURE 9.7 A fictitious data set represented through dot symbolization on (A) an equivalent projection, and (B) a conformal projection.

---

* Although the "one-sixth" rule will not yield the *lowest* distortion over the mapped area per se, the procedure will result in low distortion without the complex mathematics associated with finding the standard parallels that result in the lowest distortion over the mapped area.

southern latitudinal limits.* Assuming the northern and southern latitude values of 41° 43′N and 81° 53′N, respectively, the central latitude would be placed at 61° 48′N. Supplying these values for the Albers conic projection parameters, a combined graduated symbol–dot map of the population distribution of Russia is shown in Figure 9.8.

It is interesting to note that the Albers equivalent projection was introduced by Heinrich Albers in 1805, but was not frequently employed until the 1900s when it became the primary projection used by the United States Geological Survey for maps of the conterminous United States.[†] The standard parallels for these maps were placed at 29° 30′N and 45° 30′N and resulted in a maximum

scale error throughout the map of no more than $1\frac{1}{4}$ percent (Deetz and Adams 1945). Other countries, such as Russia, also have utilized a single conic projection for atlas mapping of their entire country. For example, the *Bol'shoy Sovetskiy Atlas Mira* relied heavily on the Kavrayskiy IV equidistant conic projection for atlases of the former Soviet Union (Snyder 1993). An interesting comparison of the graticule spacing of the parallels between the Albers equivalent conic and the Kavrayskiy IV equidistant conic is shown in Figure 9.9. On the Albers equivalent projection (Figure 9.9A), the spacing of the parallels decreases from the standard lines toward the polar regions (a characteristic appearance of the lines of latitude on equivalent projections), whereas the Kavrayskiy IV equidistant projection (Figure 9.9B) shows the parallels as equally spaced along a meridian (a characteristic appearance of the lines of latitude on equidistant projections).

9.2.3 A Map of Migration to the United States

For our third example, we focus on an appropriate projection for a series of maps (a small multiple) showing

---

* Note that although the central latitude definition might seem obvious, it is a parameter that is available when defining or changing projection parameters in many Geographic Information System (GIS) software systems as the central latitude is used to center the projection on the page.

[†] *The National Atlas*, published by the United States Geological Survey in 1970, contains numerous maps drawn on the Albers equivalent projection.

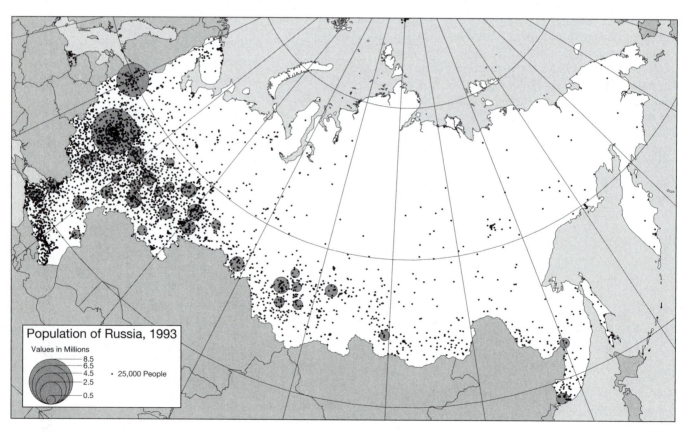

FIGURE 9.8   A combined proportional symbol–dot map showing the population distribution across Russia on an Albers equivalent projection. Data obtained from United Nations Educational Scientific and Cultural Organization (UNESCO), 1987 through United Nations Environment Program (UNEP)/Global Resource Information Database (GRID)-Geneva at *http://www.grid.unep.ch/index.php*.

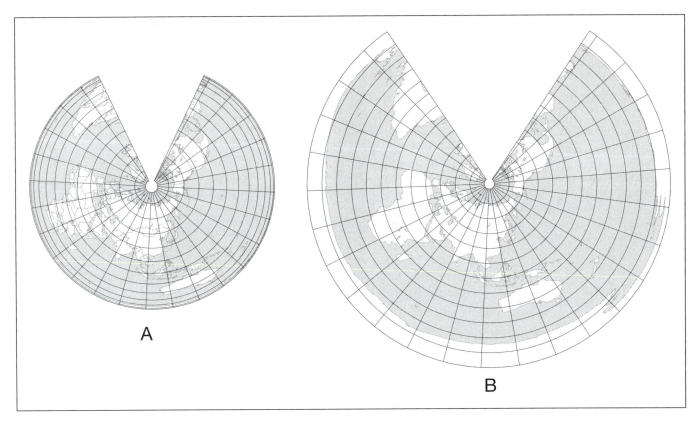

FIGURE 9.9 The (A) Albers equivalent and (B) Albers Kavrayskiy IV equidistant conic projections. Both projections are at the same scale and have standard parallels at 47° N and 62° N.

migration to the United States. We assume the map will be included as part of an atlas describing a history of U.S. ethnicity and will be published by the U.S. Census Bureau for the general public. Although the atlas will have maps showing migration from all populated continents, this specific small multiple is designed to show the number of individuals who migrated to the United States from European and Asian countries during each 10-year period between 1960 and 2000. A goal is not to simply show the number of immigrants per country, but to provide a general sense of the route by which the migration took place. Thus, we argue that the flow map is an appropriate symbolization method.

Turning to Snyder's selection guideline, we have a problem in that the mapped area does not neatly fit any of his predefined geographic categories. The closest geographic category that our mapped area fits into is a hemisphere. In this case, given the source areas for the immigrants (European and Asian countries) and their ultimate destination (United States), the Northern Hemisphere is the logical geographic area to be mapped. Because the United States should figure prominently in the center of the map and it is located neither along the Equator nor at a pole, the projection calls for an oblique

aspect. Moreover, the longitudinal range of our data encompasses approximately 180°—a hemisphere.

Therefore, we can proceed with Snyder's scenario that considers projections for a hemisphere (in our case, the Northern Hemisphere). Examining Table 9.2, we now must determine which projection property is desirable for our data. Conformality preserves angles at infinitesimally small points throughout the mapped area, but this property is not required for our data. Equivalent projections preserve areal relations throughout the mapped area. In the previous two examples (world literacy rates and Russian population distribution) there was a justifiable need for equivalent projections. However, the enumeration units of the migration data are not directly associated with the flow line symbol (i.e., unlike a choropleth map, the enumeration units on a flow map are not the symbol); thus, there is no direct need to preserve areal relations. An equidistant projection preserves distances from one point to all other points, but also is not directly related to our map purpose. Global look projections (the orthographic being among them) give the appearance of the Earth as if looking on it from outer space, which often produces an eye-catching map. However, global look projections are typically limited to showing a hemisphere and have considerable distortion near the map edges.

Although no specific property appears to be an obvious choice for a projection for the migration data, a mapping situation such as this provides an opportunity to be creative and make use of the flexibility that planar projections provide. One planar projection appropriate for the migration data is the azimuthal equidistant. Although discounted earlier on the basis that our data do not require directions or distances to be measured correctly, the projection does have interesting qualities that make it suitable for the migration data. For instance, the azimuthal equidistant projection has lower overall scale distortion than projections that preserve a single property, such as equivalence. As a result, equidistant projections tend to minimize the distortion of shapes of land masses. For this reason, Maling (1992, 109) indicated that "equidistant map projections are often used in atlas maps, strategic planning maps and similar representations of large parts of the Earth's surface." Another interesting quality of the azimuthal equidistant projection is that it displays the entire world (in its default configuration), whereas most planar projections are only capable of showing a single hemisphere. Although the azimuthal equidistant projection can show the entire world (Figure 9.10A), it can easily be cropped to focus on specific geographic areas, as shown in Figure 9.10B.

Another useful planar projection that would allow the migration routes to be effectively shown, as well as provide an interesting perspective view of the Earth, is the vertical perspective azimuthal projection. With this projection, the amount of land that is shown can be manipulated—note that Figure 9.11A shows much more land within the border of the projection than Figure 9.11B, which focuses on North and South America. The vertical perspective azimuthal projection also can be used to zoom in on a portion of the Earth's surface, which is similar to using a camera to zoom in on an object. For instance, Figure 9.12 illustrates a zoom for North and Central America; notice the dramatic effect this projection has in emphasizing the United States as the surrounding area fades into the background. Obviously, with a projection such as this, the scale changes dramatically from the center outward. Applying the vertical perspective azimuthal projection to the migration data, we see an interestingly designed map in Figure 9.13. Here, the map's focus is the United States, which is clearly the convergence point for the flow lines. The shapes of the landmasses are shown close to their appearance as found on the globe so as to not confuse potential map users about the geographic area. The projection also provides a dramatic appearance to the data, capturing the map user's attention.

The azimuthal equidistant and vertical perspective azimuthal projections both are capable of representing the geographic area of interest for our purposes. However, there are subtle differences between the two projections. When comparing Figures 9.10A and 9.10B, note that in Figure 9.10B we cropped a considerable portion of the world's land masses. In its uncropped appearance,

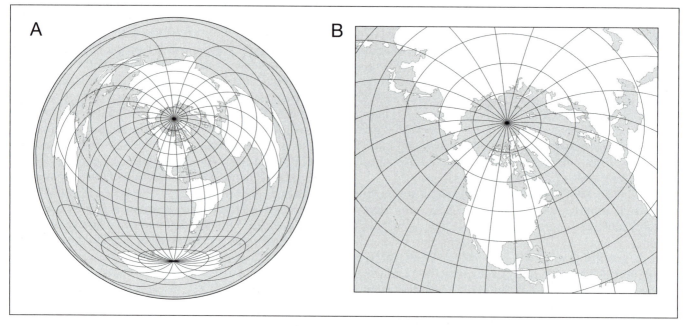

FIGURE 9.10   (A) The azimuthal equidistant planar projection showing the entire world and (B) the same projection cropped to focus on the migration data.

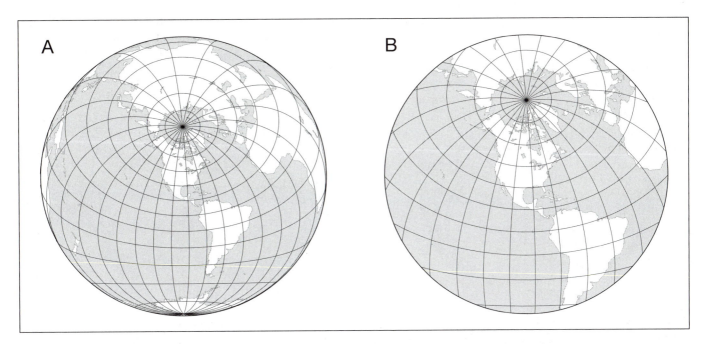

FIGURE 9.11 Two perspective views of the Earth through the vertical perspective azimuthal projection.

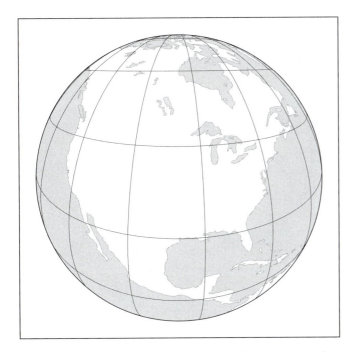

FIGURE 9.12 The vertical perspective azimuthal projection zoomed into North and Central America.

perspective azimuthal projection is developed, a hemisphere can be shown (Figure 9.11A) without the extreme consequences of distortion found on the azimuthal equidistant projection. If the region to be mapped is smaller than a hemisphere, the vertical azimuthal perspective projection can also be used by zooming in on the region of interest—note that Figure 9.11B is zoomed in on the Americas. Therefore, by using the vertical perspective azimuthal projection, the data and geographic area of importance can be shown as well as placing the data and geographic area of importance in spatial context, showing the surrounding landmasses without the considerable distortion found on the azimuthal equidistant projection. The choice between the two projections comes down to a concern for distortion and aesthetic appeal. Using the azimuthal equidistant projection, the mapmaker should crop the map by including only those geographic areas necessary for the map, thus minimizing distortion. If the mapmaker wishes to present more of a hemispheric look, then the vertical perspective azimuthal projection should be used.

9.2.4 A Map of Tornado Paths Across Kansas (1950–2000)

In the fourth projection selection example, we focus on the paths that category F4 and F5* tornadoes took across

the azimuthal equidistant projection (Figure 9.10A) shows the entire world. A consequence of displaying this much geographic area is that toward the map edges, a considerable amount of distortion is present—notice how the southwestern corner of Australia is stretched. On the other hand, due to the way in which the vertical

---

* The Fujita scale is used to indicate the intensity of a tornado where F0 is the least intense and F5 is the most intense according to the damage inflicted by the tornado.

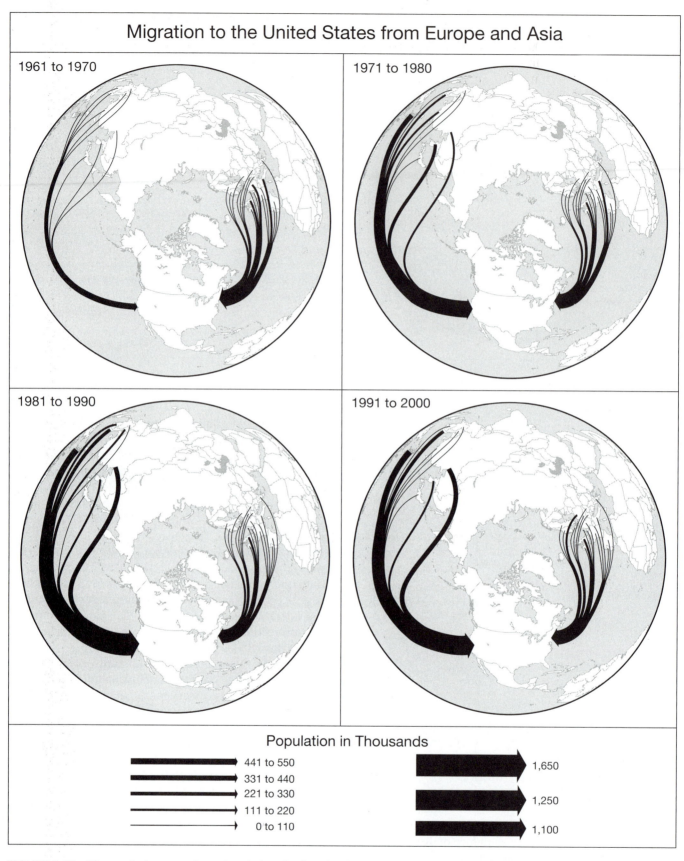

FIGURE 9.13 The vertical perspective azimuthal projection showing the migration data during different time periods.

Kansas from 1950 to 2000. As before, we begin by considering the data, symbolization method, intended audience, and overall map purpose. Assume that this map is to be created for weather experts who are interested in learning about counties in Kansas that historically have been hit by tornadoes. Specifically, they are interested in using this map to visualize the precise track and distance along the ground each tornado took. This map requirement focuses on a geographic area to be mapped that is comparatively smaller than the other map examples we have considered.

Next, we consider which of Snyder's categories to use for our mapping situation. Due to its small geographic area, Kansas does not exclusively match one of Snyder's geographic categories. To select a projection for a small area (e.g., a state) requires examining that area's directional extent and location to match these characteristics to a specific projection's standard point or line(s), thus keeping distortion low. Because we are dealing with a smaller region for this data set, Table 9.3, which handles selecting a projection for a continent, ocean, or smaller region as the geographic area mapped, will provide suitable guidance for selecting a projection for Kansas.

The first step is to determine the directional extent of the region to be mapped. In this case, Kansas has a considerably greater east–west than north–south extent. In fact, Kansas is about twice as long in an east–west extent (about 400 miles vs. about 200 miles). In the second step, we determine the positional characteristic of Kansas. Lying between 37° N and 40° N, Kansas is positioned a considerable distance from the Equator. For the third step, we decide on the appropriate projection property. Notice that for our east–west extending landmass that is located away from the Equator, Snyder's table lists conic

projections that are either conformal or equivalent. Because the data focus on the tracings of tornado paths across Kansas, angles and distances would be the appropriate projection properties to consider. We can eliminate equivalent projections because this property is not appropriate—we are not interested in preserving areal relations. Recall that conformal projections preserve accurate angles about individual point locations (e.g., where a tornado touches down). However, conformal projections do not preserve scale and therefore distances shown on the projection would not be the same as found on the Earth. On the other hand, recall that equidistant projections allow for correct distances to be measured as they would be found on the Earth, but do not preserve angles. However, equidistant projections restrict the way in which distances are preserved (e.g., all distances are correct from the center of the map to any other point). Thus, as in the previous example, there is no single suitable map projection for this particular mapping situation. A conformal or equidistant conic projection might be sufficient, but there are trade-offs associated with each projection property and resulting distortion.

To help solve this dilemma between the conformal and equidistant property, using various distortion measures, we will compare the differences in the amount of distortion between the Lambert conformal conic and equidistant conic projections. To begin our distortion analysis, examine Figure 9.14 where we see the Lambert conformal conic (Figure 9.14A) and the equidistant conic (Figure 9.14B) projections for Kansas. The north–south latitudinal limits range from approximately 35° N to 42° N, and the east–west longitudinal limits range from about 90° W to 104° W. Each projection shares the same conic parameters of a central meridian placed at 98° W,

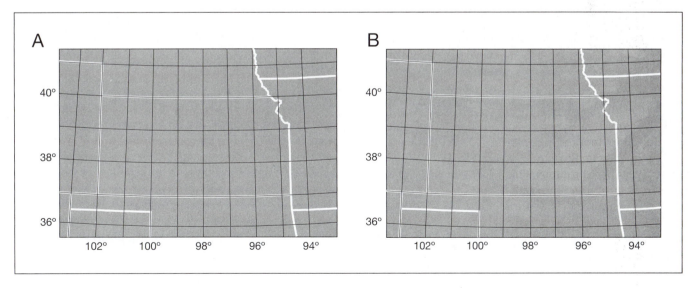

FIGURE 9.14 The (A) Lambert conformal conic and (B) equidistant conic projections of Kansas.

a central latitude at 38° 30′ N, and two standard parallels—one at 40° 50′ N and the other at 36° 10′ N.* Although the projections differ in their respective properties, there appears to be nothing that visually distinguishes the mapped appearance of Kansas on the two projections.

Now, examine Figure 9.15, which shows Tissot's indicatrix for the same mapped area. Recall that the indicatrix appears as a circle of different sizes on conformal projections (i.e., circles are increasingly larger the greater their distance from the standard point or lines),

but it appears as ellipses on an equidistant projection. Figure 9.15A does indeed show the indicatrix as a series of circles on the Lambert conformal conic projection, but Figure 9.15B also portrays the indicatrix as a series of circles on the equidistant projection, suggesting that for this data set, the choice of projection property does not have a significant impact on the display of the data. We can determine whether the geometric shapes in Figure 9.15 are actually circles or ellipses by examining Table 9.4, which shows various distortion values for the Lambert conformal and equidistant conic projections. Beginning on the left side, the first two columns in the table list every degree of latitude from 33° N to 45° N along 98° W longitude. Also listed for each projection are a, the scale factor for the semimajor axis; b, the scale

_____

* The location of these standard parallels was set at one-sixth the distance from the limiting parallels.

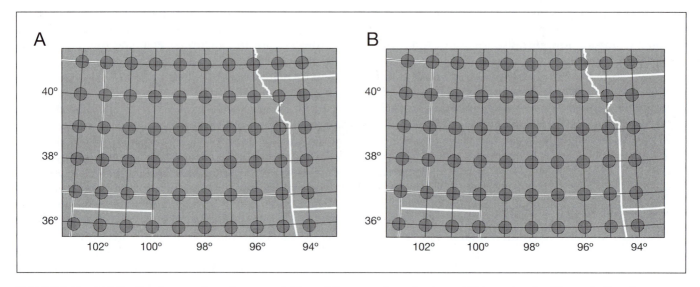

FIGURE 9.15 (A) The Lambert conformal conic and (B) equidistant conic projections of Kansas showing Tissot's indicatrix at every 1° spacing.

TABLE 9.4 Values for a, b, area distortion, and angular distortion for the Lambert conformal and equidistant conic projections between latitudes 33° and 45°

	Lambert Conformal Conic				Equidistant Conic			
Latitude	a	b	Area Distortion	Angular Distortion	a	b	Area Distortion	Angular Distortion
33	1.004	1.004	1.007	0.0°	1.004	1.000	1.004	0.21°
34	1.002	1.002	1.004	0.0°	1.002	1.000	1.002	0.12°
35	1.000	1.000	1.000	0.0°	1.001	1.000	1.001	0.06°
36	1.000	1.000	1.000	0.0°	1.000	1.000	1.000	0.0°
37	1.000	1.000	1.000	0.0°	1.000	1.000	1.000	0.03°
38	0.999	0.999	0.998	0.0°	1.000	0.999	0.999	0.05°
39	0.999	0.999	0.998	0.0°	1.000	0.999	0.999	0.05°
40	1.000	1.000	0.999	0.0°	1.000	1.000	1.000	0.02°
41	1.000	1.000	1.000	0.0°	1.001	1.000	1.000	0.00°
42	1.001	1.001	1.002	0.0°	1.002	1.000	1.001	0.06°
43	1.002	1.002	1.005	0.0°	1.003	1.000	1.002	0.14°
44	1.004	1.004	1.008	0.0°	1.005	1.000	1.004	0.23°
45	1.006	1.006	1.012	0.0°	1.007	1.000	1.006	0.34°

factor for the semiminor axis; area distortion; and angular distortion values.

In examining the table, note that the scale factors for *a* and *b* on the Lambert conformal conic projection at any given latitude are both equal, a characteristic of conformal projections. The scale factors equal 1.0 at the location of the standard parallels. On this projection, the areal distortion is not severe—ranging from 1.012 (a very slight exaggeration of areas) to 0.998 (a very slight compression of areas). There is no angular distortion on this conformal projection—only values of 0.0° are reported in the Angular Distortion column. For the equidistant conic projection, the values for *a* and *b* are not equal at any parallel except at the standard lines. The amount of areal distortion is slightly less (ranging from 1.006–0.999) than on the Lambert conformal conic projection (1.012–0.998). There is also no more than 0.34° (one-third) of a degree of angular distortion on the equidistant conic projection over this geographic area, which is rather low.

After reviewing the distortion values, we can see that neither projection preserves scale throughout the geographic area of interest, which means that regardless of which projection is selected, there will be some distortion present in any distance measurement taken from the map, but the measurement error will be negligible. Because neither the Lambert conformal nor the equidistant projection has a substantial amount of scale error across the mapped area, computing distances on either projection will result in the same approximate result. In terms

of angular distortion, the Lambert conformal conic projection preserves angular relations, which facilitate measuring the direction of each tornado's path.

Based on the preceding discussion, we selected the Lambert conformal conic projection to map the tornado data (Figure 9.16). We should stress again that given the small geographic area of interest and the tornado data set, the choice of projection is rather inconsequential. However, we chose conformality over equidistance because on conformal projections, angular relations are preserved at every point and the errors from any distance measurement on this projection are negligible. Note in Figure 9.16 that the data are represented as flow lines where the origin of the flow line indicates the location where each tornado touched down. The flow line extent is the distance over which the tornado traveled, and the arrowhead indicates the direction of the path. The point of the arrowhead indicates where the tornado dissipated. Examining the pattern of tornado paths, we see the general trend is from the southwest to northeast, and that the south central portion of the state appears to have a concentration of the most intense tornado activity for the time period.

9.2.5 Discussion

The four projection examples each provided interesting opportunities to see how map projections can be selected. In summary, there are three key objectives from our examples that we would like to emphasize. First, in

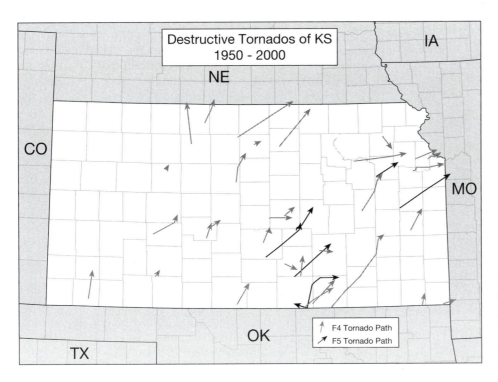

FIGURE 9.16 The Lambert conformal conic projection showing the paths of F4 and F5 tornadoes across Kansas from 1950 to 2000. Data from the National Climate Data Center found at *http://www.ncdc.noaa.gov/oa/climate/climatedata.html.*

most cases, it is the objective of the mapmaker to select a projection for which distortion is the lowest. Of course, the kinds of distortion that are to be considered are dependent on the purpose to which the map is to be put (e.g., for many thematic activities, equivalent projections are suitable). Second, amounts of distortion can be kept small by aligning the geographic area (or data set) under consideration with the standard line(s) or positioning the map's center with the standard point. Third, as the extent of geographic area under consideration increases (e.g., from country to hemisphere), distortion becomes a more important consideration. For example, we noted in the map of Kansas tornado paths that the scale variation was very small on either projection. Had we, however, used the Lambert conformal or equidistant conic projection to map the entire world, we would have noted extreme scale distortion toward the edge of each projection, especially on the Lambert conformal conic. Thus, when mapping smaller geographic areas such as a state, variation in distortion values is less of a concern than for larger geographic areas.

In addition to those key objectives learned from our examples, there are two important considerations to keep in mind when selecting map projections. First, just because a projection has seen considerable usage (e.g., in prominent atlases) does not mean that the projection is suitable for your specific application. In many cases, a projection's usage was dictated by its availability and not its suitability. One example is the overuse of the Mercator projection as the basis for thematic maps (e.g., showing world population distribution), an obvious misuse of the Mercator projection. Second, an often overlooked aspect of map projections is their influence on overall map design. Although this idea is more difficult to grasp, there are clear enhancements that projections can make with regard to map design. For example, Richard Edes Harrison in his *Look at the World* atlas used the orthographic projection to present some powerful images of World War II in illustrating the spatial relationships between the United States and the rest of the world. For instance, Color Plate 9.1A shows the spatial proximity of North America to Europe, and Color Plate 9.1B illustrates the proximity of South America and Antarctica. Harrison's use of the orthographic and other azimuthal projections throughout *Look at the World* was a significant departure from other atlases that relied on the Mercator projection. By using azimuthal projections, especially the orthographic, Harrison attempted to show how spatially connected the continents are. For instance, in Color Plate 9.1A, the orthographic projection illustrates how close North America is to Europe, which is contrary to the image given by the Mercator projection.

SUMMARY

In this chapter, we examined several commonly referenced map projection selection guidelines, noting that they are limited in helping novice cartographers through the often confusing process of selecting a projection. In contrast, we feel that Snyder's guideline is well organized and presents a logical hierarchy for selecting an appropriate projection. Snyder's guideline begins with a focus on the geographic area to be mapped: world; hemisphere; and continent, ocean, or smaller region. Once the geographic area to be mapped has been selected, a closer examination of projection properties and characteristics takes place. For example, under the world category, the specific projection properties (conformal) and specific characteristics (e.g., constant scale along the Equator) are examined. Once the property and characteristic are decided, the guideline offers a specific projection.

We relied on Snyder's guideline to help in the selection of appropriate map projections for four different thematic data sets. The first data set focused on world literacy rates. In selecting a projection for this data set, we focused on the world category, chose to preserve areal relationships (an equivalent projection to ensure that areas were preserved in their correct proportions and no area visually dominated the map), and an interrupted projection (to reduce distortion by placing cuts over the oceans where data were not present). The result was an interrupted Mollweide projection. In addition, discussion was raised regarding the impression of selecting a projection with poles that were represented by points or lines, a topic not directly addressed by Snyder's guideline.

In the second data set, we looked at the population distribution of Russia. In this instance, we felt it desirable to create a combined proportional symbol–dot map that would present the spatial variation of Russia's population—some areas have high population densities whereas others are vast areas of emptiness. Because the geographic area of interest was Russia, we utilized Snyder's guidelines for a continent, ocean, or smaller region. Noting that Russia has a considerable east–west extent, we selected a conic projection. Moreover, we chose an equidistant conic projection to enable the user to properly compare areas of differing densities. This selection also allowed us to discuss how to select several parameters associated with conic projections: standard parallels, central meridians, and the central latitude.

The third data set focused on migration patterns from Europe and Asia to the United States. The objective for this map was to show the routes by which immigration took place, as well as the spatial proximity between source of migration and the United States—essentially encompassing a hemisphere. To help us select a projection for this data set, we utilized Snyder's table listing various planar projections suitable for representing a hemisphere. This table lists several planar projections according to specific projection properties. We noted in our discussion that the planar projections offer a considerable range of design options not found with other map projection classes. In the end, we selected the vertical perspective azimuthal projection to highlight the migration paths and the spatial proximity of Europe and Asia to the United States, as well as the spherical nature of the Earth.

The last data set involved a historical look at the most destructive tornado paths across Kansas. Here, we utilized Snyder's selection guideline for a continent, ocean, or smaller region. Using his guideline, we selected a conic projection. However, the small size of the geographic area prompted us to investigate the nature of the distortion pattern across the mapped area. We were curious to see what the impact would be of using projections with different properties—equidistant versus conformal, which are two properties useful for our tornado data. In the distortion analysis, it became clear that neither projection offered any substantial benefit because such a small geographic area was under consideration.

The chapter concluded with a brief overview of five key criteria that you should consider when selecting map projections. First, in most cases, it is the objective of the mapmaker to select a projection for which distortion is the lowest. Second, amounts of distortion can be kept small by aligning the geographic area (or data set) under consideration with the standard line(s) or positioning the map's center with the standard point. Third, as the amount of geographic area under consideration increases, distortion becomes a more important consideration. Fourth, just because a projection has seen considerable exposure (e.g., use in prominent atlases) does not mean that the projection is suitable for your specific application. Fifth, an often overlooked aspect of the map projection is its influence on the overall map design—a topic that is not well studied by cartographers.

FURTHER READING

American Cartographic Association. (1986) *Which Map Is Best: Projections for World Maps*. Bethesda, MD: American Congress on Surveying and Mapping.

Presents a general overview of map projections.

American Cartographic Association. (1988) *Choosing a World Map: Attributes, Distortions, Classes, Aspects*. Bethesda, MD: American Congress on Surveying and Mapping.

Presents a general overview of map projections with special attention to choosing world map projections.

American Cartographic Association. (1991) *Matching the Map Projection to the Need*. Bethesda, MD: American Congress on Surveying and Mapping.

Discusses several map projections and how their properties and characteristics are appropriate for specific data sets.

Bugayevskiy, L., and Snyder, J. P. (1995) *Map Projections: A Reference Manual*. London: Taylor & Francis.

Chapter 7 takes a mathematical look at selecting projections.

Canters, F. (2002) *Small-Scale Map Projection Design*. New York: Taylor & Francis.

Chapter 6 focuses on automated methods for selecting projections for small-scale maps.

Canters, F., and Decleir, H. (1989) *The World in Perspective: A Directory of World Map Projections*. West Sussex, England: Wiley.

Part 1 examines selecting map projections from the standpoint of distortion analysis.

Deetz, C. H., and Adams, O. A. (1945) *Element of Map Projections with Applications to Map and Chart Construction*. 5th ed. Washington, DC: U.S. Department of Commerce Coast and Geodetic Survey.

Briefly describes selecting projections for a variety of purposes.

De Genst, W., and Canters, F. (1996) "Development and implementation of a procedure for automated map projection selection." *Cartography and Geographic Information Systems* 23, no. 3:145–171.

Discusses the development of an automated approach for selecting map projections.

Harrison, R. E. (1944) *Look at the World*. New York: Knopf.

Presents numerous hemispheric maps based on the orthographic azimuthal projection.

Hsu, M. (1981) "The role of projections in modern map design." *Cartographica* 18, no. 2:151–186.

Reviews selecting projections based on map design issues.

Jankowski, P., and Nyerges, T. (1989) "Design Considerations for MaPKBS-Map Projection Knowledge-Based System." *The American Cartographer* 16, no. 2:85–95.

Presents MaPKBS, an expert system for the selection of a suitable map projection.

Maling, D. (1992) *Coordinate Systems and Map Projections*. 2nd ed. Oxford, England: Pergamon.

Chapters 11 and 12 cover various methods used to select map projections.

Nyerges, T., and Jankowski, P. (1989) "A knowledge base for map projection selection." *The American Cartographer* 16, no.1:29–38.

Describes the development of a knowledge base for selecting map projections.

Pearson, F. (1984) *Map Projection Methods*. Blacksburg, VA: Sigma Scientific.

Chapter 10 presents a simple map projection selection guideline.

Peters, A. (1983) *The New Cartography*. New York: Friendship Press.

Discusses development of the Peters projection, which is designed to better represent the Earth's equatorial regions than other cylindrical projections such as the Mercator. The claims that Peters makes in this text caused quite a stir among cartographers.

Robinson, A. (1974) "A new projection: Its development and characteristics." *International Yearbook of Cartography* 14: 145–155.

Discusses the development of the Robinson projection.

Robinson, A. H., Morrison, J. L., Muehrcke, P. C., Kimerling, A. J., and Guptill, S. C. (1995) *Elements of Cartography*. 6th ed. New York: Wiley.

Chapter 5 briefly describes some commonly used map projections.

Snyder, J. P. (1993) *Flattening the Earth: Two Thousand Years of Map Projections*. Chicago: University of Chicago Press.

Useful discussion throughout the text on the selection and application of map projections.

Snyder, J. P. (1994) *Map Projections: A Working Manual*. Washington, DC: U.S. Government Printing Office.

Provides some limited discussion on Snyder's rationale for his selection guideline.

10

Principles of Color

OVERVIEW

*The increasing use of color on maps means that mapmakers have a greater need to understand the proper use of color. To assist you in developing this understanding, this chapter covers some basic principles of color. How color is processed by the human visual system is the topic of section 10.1. Here we learn about the nature of **visible light**, the structure of the eye, theories of color perception (focusing on **opponent-process theory**), **simultaneous contrast** (how color perception is influenced by its surroundings), **color vision impairment**, and the visual processing that takes place beyond the eye. Some of this material (e.g., details of the structure of the eye) might seem far removed from cartography, but developing rules for sound map design requires knowledge of how our visual system processes information.*

*Section 10.2 describes hardware considerations relevant to the production of soft-copy color maps (those produced using a computer screen). We will cover hardware considerations relative to hard-copy (printed) maps in Chapter 12, which deals with map production and dissemination. In the realm of soft-copy maps, this chapter focuses on how **cathode ray tubes (CRTs)** are able to produce millions of colors by using the **additive colors red, green, and blue (RGB)**.*

*Specifying appropriate colors is a common problem faced by cartographers. For example, you might like to create a smooth progression of colors extending from a light desaturated green to a dark saturated green. Section 10.3 covers numerous **color models** that have been developed for specifying colors. Some of these models are hardware-oriented (RGB and CMYK) and therefore of limited use to the mapmaker. Others are user-oriented (**HSV, Munsell,** and Tektronix's **HVC**) and thus permit color specification in terms mapmakers are apt to be familiar with, such as hue, lightness, and saturation. This section also considers the **CIE***

color model, which, in theory, allows the mapmaker to reproduce colors specified by others (e.g., a friend tells you that a particular color progression is very effective, and you want to be sure that you use the same progression).

10.1 HOW COLOR IS PROCESSED BY THE HUMAN VISUAL SYSTEM

10.1.1 Visible Light and the Electromagnetic Spectrum

We see maps as **visible light**, whether it is reflected from a paper map or emitted from a computer screen. Visible light is a type of **electromagnetic energy**, which is a waveform having both electrical and magnetic components (Figure 10.1).* The distance between two wave crests is known as the **wavelength of light**. Because visible wavelengths are small, they are typically expressed in nanometers (nm), which are 1 billionth of a meter each. Visible wavelengths range from 380 to 760 nm. Figure 10.2 relates visible light to other forms of electromagnetic energy that humans deal with; the complete continuum of wavelengths is called the **electromagnetic spectrum**.

We have all seen or read about how a prism splits sunlight into the color spectrum (red, orange, yellow, green, blue, indigo, and violet). This phenomenon occurs because the visible portion of sunlight consists of a broad range of wavelengths, rather than being concentrated at a particular wavelength. Different colors arise in a prism as a function of how much each wavelength is bent, with shorter wavelengths (e.g., blue) bent more than longer wavelengths (e.g., red). Note that the colors in the visible portion of the spectrum in Figure 10.2 are arranged from short

---

* Light also consists of photons (packets of energy), which behave as particles when light strikes a surface (Birren 1983, 20).

to long wavelength (from violet to red), and thus match the colors we might see using a prism.

10.1.2 Structure of the Eye

The basic features of the eye of concern to cartographers are shown in Figure 10.3. After passing through the

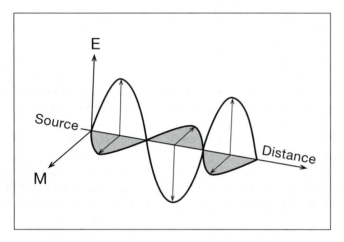

FIGURE 10.1   Electromagnetic energy is a waveform having both electrical (E) and magnetic (M) fields. Wavelength is the distance between two crests.

cornea (a protective outer covering) and the **pupil** (the dark area in the center of our eye), light reaches the **lens**, which focuses it on the **retina**. Changing the shape of the lens, an automatic process known as **accommodation**, focuses images. As we age, our lens becomes more rigid, and our ability to accommodate thus weakens. Generally, around the age of 45, our ability to accommodate becomes so weak that corrective lenses (glasses or contacts) are necessary. The **fovea** is the portion of the retina where our visual acuity is the greatest. The **optic nerve** carries information from the retina to the brain.

A term used to describe the size of an image projected onto the retina is **visual angle**, the angle formed by lines projected from the top and bottom of an image through the center of the lens of the eye (ϕ_1 in Figure 10.4). (Note that ϕ_1 is identical to ϕ_2; thus the degrees of coverage in the visual field correspond to degrees of coverage on the retina.)

An enlargement of the retina is shown in Figure 10.5. Note that it consists of three major layers of nerve cells (rods and cones, bipolar cells, and ganglion cells), along with two kinds of connecting cells (*horizontal* and *amacrine cells*), which enable cells within the major layers to communicate with one another. **Rods and cones** are specialized nerve cells that contain light-sensitive chemicals called **visual pigments**, which generate an electrical

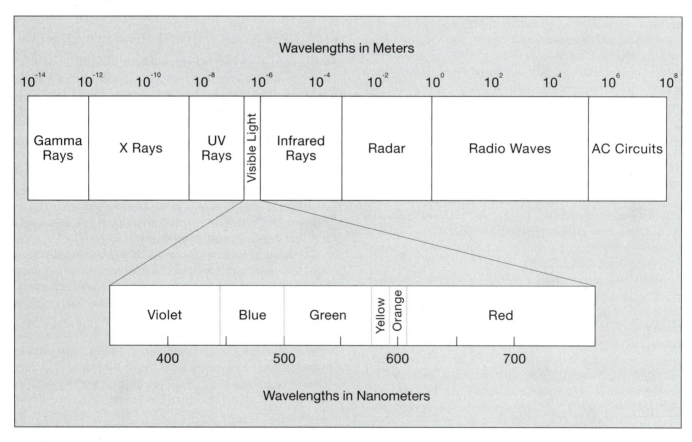

FIGURE 10.2   Relation of visible light to other forms of electromagnetic energy.

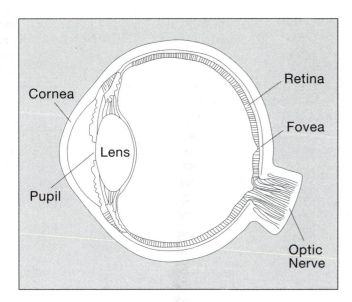

FIGURE 10.3 Basic features of the eye relevant to cartography.

response to light. The concentration of cones is greatest at the fovea, and the highest concentration of rods is about 20° on either side of the fovea. Overall there are about 120 million rods and 6 million cones.

Cones function in relatively bright light and are responsible for color vision, whereas rods function in dim light and play no role in color vision. The cones are of primary interest to cartographers because most maps are viewed in relatively bright light (an exception would be maps viewed in the dim light of an aircraft cockpit). Physiological examination of cones taken from the eye of a person with normal color vision reveals three distinct kinds based on the wavelength to which they are most sensitive: short (blue), medium (green), and long (red)

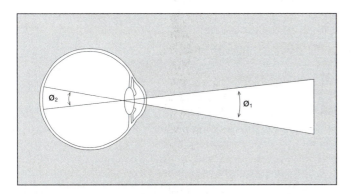

FIGURE 10.4 The projection of an image onto the back of the eye. A visual angle is computed by measuring the angle ϕ_1 formed by lines projected from the top and bottom of the image passing through the lens of the eye. Note that ϕ_1 is identical to ϕ_2, and thus the degrees of coverage in the visual field will correspond to degrees of coverage on the retina.

(Bowmaker and Dartnall 1980).* Alan MacEachren (1995, 56) noted that the distribution and sensitivity of these three kinds of cones vary in the retina: Although blue cones cover the largest area, they are least sensitive, thus making blue inappropriate for small map features.

The major function of the **bipolar** and **ganglion cells** (Figure 10.5) is to merge the input arriving from the rods and cones. Although there are about 126 million rods and cones, there are only about 1 million ganglion cells. Considerable convergence must take place between the rods and cones and the ganglion cells; each single ganglion cell corresponds to a group of rods or cones, or what is termed a **receptive field**. These receptive fields are circular in form and overlap one another.

10.1.3 Theories of Color Perception

Psychology textbooks (e.g., Goldstein 2002) generally consider two major theories of color perception: trichromatic and opponent process. The **trichromatic theory**, developed by Thomas Young (1801) and championed by Hermann von Helmholtz (1852), presumes that color perception is a function of the relative stimulation of the three types of cones (blue, green, and red). If only one type of cone is stimulated, that color is perceived (e.g., a red light would stimulate primarily red cones, and thus red would be perceived). Other perceived colors would be a function of the relative ratios of stimulation (a yellow light would stimulate green and red cones, and so yellow would be perceived).

The **opponent-process theory**, originally developed by Ewald Hering (1878), states that color perception is based on a lightness–darkness channel and two opponent color channels: red–green and blue–yellow. Colors within each opponent color channel are presumed to work in opposition to one another, meaning that we do not perceive mixtures of red and green or blue and yellow; rather, we see mixtures of pairs from each channel (red–blue, red–yellow, green–blue, and green–yellow).

For many years, proponents of the two theories of color perception hotly debated their merits, presuming that only one theory could be correct. It is now apparent, however, that both can help explain the way we see color. The trichromatic theory is correct in the sense that our color vision system is based on three types of cones and that information from these cones combines to produce the perception of color. The manner, however, in which information from the cones combines is based on opponent-process theory.

There is both psychophysical and physiological evidence in support of opponent-process theory. The psychophysical

---

* Hubel (1988, 163–164) indicated that technically the terms *violet, green,* and *yellowish-red* would probably be more appropriate.

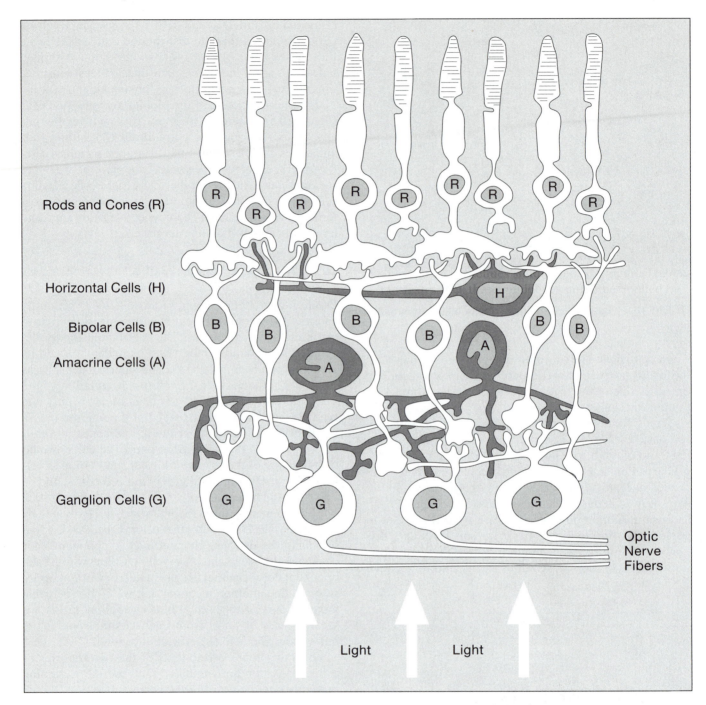

Rods and Cones (R)

Horizontal Cells (H)

Bipolar Cells (B)

Amacrine Cells (A)

Ganglion Cells (G)

Optic
Nerve
Fibers

Light   Light

FIGURE 10.5   Major layers of cells found in the retina. (Adapted from J. E. Dowling and B. B. Boycott, 1966, "Organization of the primate retina: electron microscopy," *Proceedings of the Royal Society of London*, 166, Series B, pages 80–111, Figure 23; courtesy of John E. Dowling)

evidence comes from the seminal work of Leo Hurvich and Dorothea Jameson (1957), which showed that a color of an opposing pair could be eliminated by adding light for the other color in the pair; for example, when yellow light is added to blue light, the blue eventually disappears. The physiological evidence is based on an analysis of how electrical signals pass through cells in the nervous system. In this regard, an important concept is that nerve cells

fire at a constant rate even when they are not stimulated. Firing above this constant rate is termed *excitation*, and firing below it is termed *inhibition*. By studying electrical activity in cells, physiologists have noted linkages between the blue, green, and red cones and the bipolar and ganglion cells; for example, a red light might excite red cones, which in turn excite bipolar and ganglion cells (Derrington et al. 1983; De Valois and Jacobs 1984).

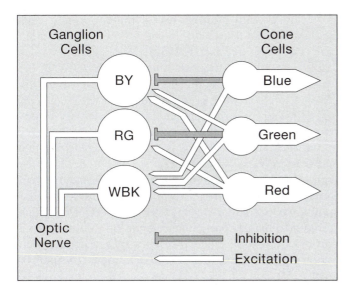

FIGURE 10.6 A model of how color information reaching the cones can be converted to opponent processes. (After Eastman, 1986. First published in *The American Cartographer*, 13(4), p. 326. Reprinted with permission from the American Congress on Surveying and Mapping.)

Although experts in human vision are reasonably sure that certain colors are in opposition to one another and that excited and inhibited nerve cells play a role, the precise linkage between the cones and bipolar and ganglion cells is unknown. One model that has been suggested is shown in Figure 10.6. In this model the blue–yellow channel is excited by green and red cones and inhibited by blue cones; the red–green channel is excited by red cones and inhibited by green cones; and red, green, and blue cones stimulate the lightness–darkness channel.

10.1.4 Simultaneous Contrast

One problem sometimes encountered when reading maps is that the perceived color of an area might be affected by the color of the surrounding area, a problem known as **simultaneous contrast**, or **induction** (Brewer 1992). This concept is illustrated for lightness in Figure 10.7. Here the gray tones in the central boxes are physically identical, but the one on the left appears lighter. This occurs because a gray tone surrounded by black shifts toward a lighter tone, whereas the same tone surrounded by white shifts toward a darker tone. Note that in this case the shifts are toward the opposite side of the lightness–darkness channel in the opponent-process model.

When different hues are used, the apparent color of an area will tend to shift toward the opponent color of the surrounding color. For example, Color Plate V of Hurvich (1981) illustrates an example in which a gray tone is surrounded by either green or blue. When the surround is green, the gray tone appears reddish; in contrast, when the surround is blue, the gray tone appears yellowish.

Simultaneous contrast is believed to be due to the receptive fields mentioned earlier. Receptive fields are not uniform; rather, distinctive centers and surrounds characterize them, with visual information in the surround having an impact on the information found in the center. For a detailed discussion of how simultaneous contrast operates, see Hurvich (1981).

10.1.5 Color Vision Impairment

Up to this point, we have assumed map readers with normal color vision. Actually, a substantial number of people have some form of color vision impairment. The highest percentages are found in the United States and Europe (approximately 4 percent, primarily males), and the lowest incidence (about 2 percent overall) appears in the Arctic and the equatorial rainforests of Brazil, Africa, and New Guinea (Birren 1983).

The color vision–impaired can be split into two broad groups: **anomalous trichromats** and **dichromats**. These groups are distinguished on the basis of the number of colors that must be combined to match any given color; anomalous trichromats require three colors, whereas dichromats use two. For both groups, the most common

FIGURE 10.7 An illustration of simultaneous contrast for a black-and-white image. The central gray strips are physically identical, but the one surrounded by black appears lighter.

problem is distinguishing between red and green; for anomalous trichromats, there is some difficulty, whereas for dichromats the colors cannot be distinguished.

Two hypotheses have been proposed for color vision impairment: (1) a change in the colors to which cone cells are sensitive and (2) changes in one of the opponent-process channels (normally the red–green one). Assuming that changes in the cones cells are the cause, the two major groups have been divided into subgroups: protanomalous and deuteranomalous for anomalous trichromats and protanopes and deuteranopes for dichromats. The two subgroups differ on the basis of the types of cones affected; for example, protanopes and deuteranopes are presumed to be missing red and blue cones, respectively. In Chapter 13 we consider how color schemes for maps can be adjusted to account for color vision impairment.

10.1.6 Beyond the Eye

It is important to realize that the eye is part of the larger visual processing system shown in Figure 10.8. Note first that information leaving the eyes via the optic nerves crosses over at the *optic chiasm;* up to this point information from each eye is separate, but pathways beyond this point contain information from both eyes. After passing through the optic chiasm, each pathway enters the *lateral geniculate nucleus (LGN)*. Physiological experiments with animals reveal that opponent cells similar to those found within the retina are also found here (De Valois and Jacobs 1984).

Interpretation of the visual information begins in the **primary visual cortex**, the first place where all of the information from both eyes is handled. As with the LGN, our knowledge of processing in this area is largely a function of physiological experiments with animals. Probably the most significant of these is the work of David Hubel and Torsten Wiesel, who received the 1981 Nobel Prize for their efforts. They found three kinds of specialized cells in the primary visual cortex: *simple cells*, which respond best to lines of particular orientation; *complex cells*, which respond to bars of particular orientation that move in a particular direction; and *end-stopped cells*, which respond to moving lines of a specific length or to moving corners or angles. Not only did they discover these different kinds of cells, but they also mapped out where they occur within the primary visual cortex (Goldstein 2002).

Although such findings are certainly significant, specialists still have not provided an explanation of how the brain handles a complex real-world situation, such as a map. As a result, we concur with MacEachren (1995, 64), who argued that "from a cartographic point of view … we are interested in … how the brain processes visual signals not because this knowledge is likely to tell us how maps work, but because these processes put limits on what symbolization and design variations might work."

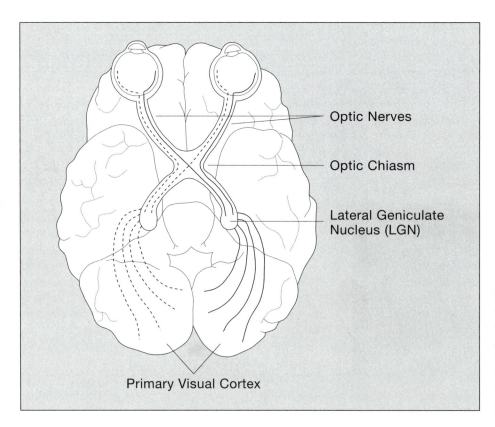

Optic Nerves

Optic Chiasm

Lateral Geniculate Nucleus (LGN)

Primary Visual Cortex

FIGURE 10.8 An overview of the visual processing system viewed from underneath the brain. (From *Sensation and Perception 3rd edition* by Goldstein. © 1989. Reprinted with permission of Wadsworth, a division of Thomson Learning: *www.thomsonrights.com*. Fax 800-730-2215.

10.2 HARDWARE CONSIDERATIONS IN PRODUCING SOFT-COPY COLOR MAPS

This section considers some hardware aspects of producing color maps in soft-copy form. The term **graphic display** is commonly used to describe the computer screen (and associated color board) on which a map is displayed in soft-copy form; examples include cathode ray tubes (CRTs), liquid crystal displays (LCDs), plasma displays, and electroluminescent displays. We focus on CRTs and LCDs because they are the most commonly used.

10.2.1 Vector versus Raster Graphics

Images on graphic displays can be generated using two basic hardware approaches: vector and raster. In the **vector** approach images are created much like we would draw a map by traditional pen-and-ink methods: The hardware moves to one location and draws to the next location. In contrast, in the **raster** approach the image is composed of **pixels** (or **picture elements**), which are created by scanning from left to right and from top to bottom (Figure 10.9). Prior to about 1980, the vector approach was more common, but today virtually all graphic displays use a raster approach.

10.2.2 CRTs

Images on CRT screens are created by firing electrons from an **electron gun** at *phosphors*, which emit light when they are struck. Monochrome CRTs contain a single

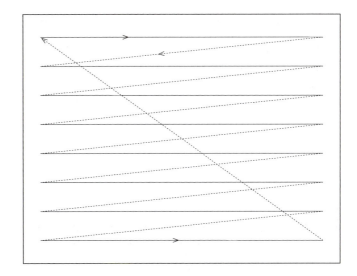

FIGURE 10.9   The order of refresh on a CRT.

electron gun, whereas color CRTs contain three guns, normally designated as R (red), G (green), and B (blue). The names for the guns have nothing to do with the type of electrons they fire, but are a function of which type of phosphor the electrons strike on the screen. In Figure 10.10;* we see two common arrangements of electron guns and phosphors (delta and in-line). Note that a shadow mask or aperture grill (termed a metal mask in the figure) is

---

* Foley/Feiner/Hughes/Van Dam, COMPUTER GRAPHICS: PRINCIPLES AND PRACTICE, Fig. 4.14 (p. 159) 4.15 (p. 160), © 1990 Addison Wesley Publishing Co. Reprint by Permission of Pearson Education, Inc.

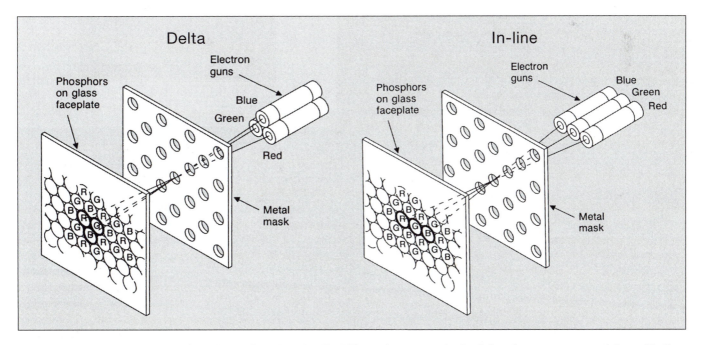

FIGURE 10.10   Cross-sectional view of a portion of a color CRT illustrating two methods of phosphor arrangement: delta and in-line.

positioned so that each electron gun can hit only one type of phosphor.

Different colors on a CRT screen result from the principle of **additive color**: The colored phosphors are visually added (or combined) to produce other colors. This principle is normally demonstrated with overlapping colored circles (as in Color Plate 10.1), but phosphors on a CRT screen do not actually overlap. Rather, we see a mixture of color because our eye cannot resolve the very fine detail of individual phosphors; the concept is analogous to pointillism techniques used in 19th-century painting.

Together three phosphors compose a pixel. One measure of the **resolution** of a monitor is the number of *addressable pixels*, normally specified as the number of pixels displayable horizontally and vertically.* Common resolutions range from 640×480 up to 1280×1024. Problems with lower resolution systems include *jaggies* or **aliasing** (a staircase appearance to diagonal lines), an inability to smoothly vary the size of small symbols, and difficulty in creating crisp text.† In a later section, we will see that to compare the resolution of monitors and printers, it is useful to calculate the number of pixels per inch. For example, for a 17-inch monitor with 1024×768 displayable pixels, there are approximately 75 pixels per inch.

Because phosphors on a CRT screen have a low *persistence* (stay lit only briefly), the screen must be **refreshed** constantly. Refresh takes place by scanning

---

* See Peddie (1994, 7–10) for a variety of definitions for resolutions.

† Aliasing can be handled using antialiasing routines (Foley et al. 1996, 132–142), but such routines are not common in mapping and design software.

across the screen from left to right and top to bottom (Figure 10.9). Two types of refresh are possible: interlaced and noninterlaced. In the *interlaced* method, every odd-numbered scan line is refreshed in the first 1/60 of a second, and every even-numbered scan line in the next 1/60 of a second, for a total refresh rate of 1/30 of a second. If information on adjacent scan lines is similar, then this approach is acceptable, but if the information is different (as on a map with horizontal political borders 1 pixel thick) the screen will appear to flicker. Although flicker is undesirable, interlacing is sometimes used on high-resolution systems (e.g., 1280×1024) because so many pixels must be addressed. Fortunately, the trend is toward *noninterlaced* systems in which the entire screen is refreshed from top to bottom in 1/60 of a second or less.

The **frame buffer** is an area of memory that stores a digital representation of colors appearing on the screen. We consider the frame buffer in detail because it determines the number of colors available for cartographic applications. Consider first a monochrome system in which only black-and-white pixels are possible. In such a system, we can think of the frame buffer as an area of memory in which each bit corresponds to a pixel on the screen; bits can have values of either 0 or 1, corresponding to either lit or unlit pixels (Figure 10.11). The resulting layer of bits is termed a *bit-plane*.

Now consider a monochrome system in which shades of gray are possible. In this case there is a layer of bit-planes, with the lightness of a pixel a function of whether the bits associated with that pixel are set to 0 or 1 (Figure 10.12). Because each bit-plane corresponds to a power of 2, the lightness value for any pixel is the sum of the products of the frame buffer values times the

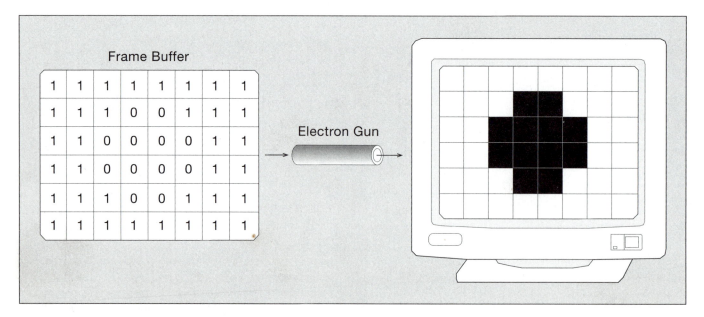

FIGURE 10.11   A monochrome system in which only black and white pixels are possible.

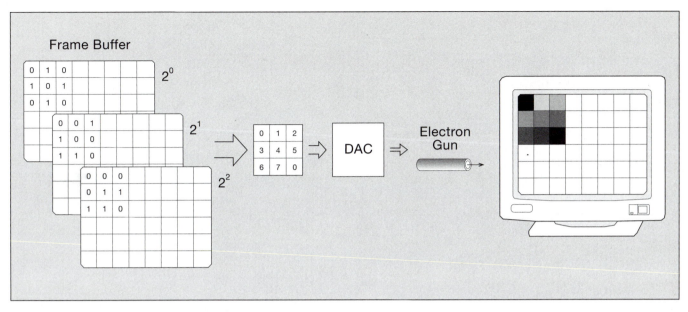

FIGURE 10.12   A monochrome system in which shades of gray are possible.

corresponding power of 2 (for the upper right pixel in Figure 10.12, the result is $0 \times 2^0 + 1 \times 2^1 + 0 \times 2^2 = 2$. In general, if n is the number of bit-planes, then 2^n different values will be possible (in Figure 10.12, n is 3, so $2^n = 8$ possible values), and the values will range from 0 to $2^n - 1$ (0–7 in Figure 10.12).

Also note in Figure 10.12 that a **digital-to-analog converter (DAC)** is placed between the frame buffer and the electron gun. The purpose of the DAC is to convert the digital information stored in the frame buffer into the analog signal produced by the electron gun.

Now let's consider color display systems. In sophisticated color display systems, a set of bit-planes is assigned to each color gun (Figure 10.13). Values for individual color guns are computed in a manner identical to the monochrome system just described. Thus, in Figure 10.13 each gun can fire with eight intensities. The total number of colors possible for a pixel is the product of the number of intensities for each color gun, or $2^n \times 2^n \times 2^n$; in Figure 10.13 the result is $8 \times 8 \times 8 = 512$. Typically, sophisticated color display systems have 8 bit-planes per color gun, meaning they can, in theory, display $2^8 \times 2^8 \times 2^8 = 16.8$ million colors.

Although 16.8 million colors might seem excessive, three points must be kept in mind. First, we can distinguish a much larger number of colors than is commonly recognized; Goldstein (2002, 189) suggested an upper limit of 1 million. Second, the differences between light intensities for a color gun are not constant throughout the range of possible digital values. For example, a change in intensity between 0 and 1 might not be the same as between 200 and 201; some changes might be so small as

to be indistinguishable. Third, the actual intensities used do not account for the fact that our eye–brain system discriminates among lighter shades better than darker ones.

Unfortunately, display systems in which a set of bit-planes is assigned to each color gun are beset by two problems: (1) large memory requirements, and (2) slow speed of changing a geographic region of uniform color (as might be desired in cartographic animation). The solution to such problems is **color lookup tables**, in which values in the frame buffer serve as indexes to a table that provides the actual values sent to the color guns. Figure 10.14 shows a simplified example for a table with 8 rows (traditionally, tables have consisted of 256 rows, but 1024 rows are now common). Note that the bright red pixels on the screen could be changed to a moderate red simply by replacing the 255 for "Red" at lookup position 2 in the table with a value of 127.

Dithering is another approach that conserves frame buffer memory and permits the display of additional colors. In dithering, new colors are created by presuming that readers will perceptually merge colors displayed in adjacent pixels. For example, Color Plate 10.2 shows a dithered orange (A) and the individual yellow and red pixels (B) used to create that orange. One problem with dithering is that the resulting colors often exhibit pattern (a qualitative characteristic), which is inappropriate for use in a quantitative series.

10.2.3 LCDs

LCDs consist of a complex sandwich of a light source, glass plates, polarizing film, liquid crystals, a source of

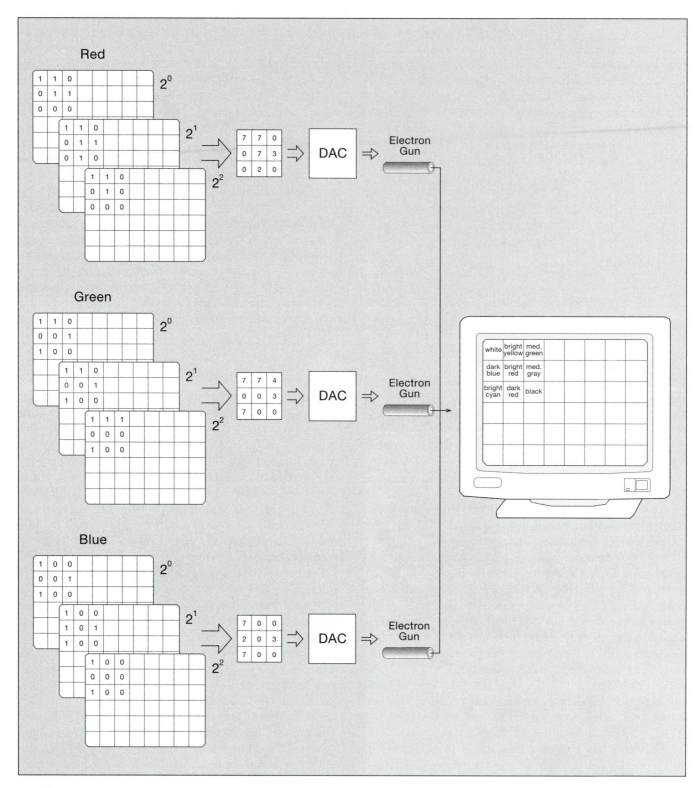

FIGURE 10.13 A color system in which several bit-planes are allotted to each color gun.

electrical power (e.g., transistors), and color filters (Figure 10.15). The basic principle is that the liquid crystals found at each pixel location initially are all in the same orientation. When polarized light passes through the crystals, the crystals direct the light so that it passes through another polarized filter. We see the result as a

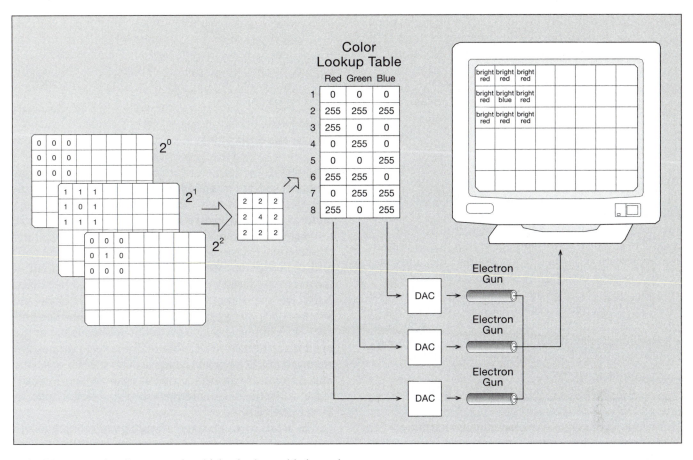

Color Lookup Table

	Red	Green	Blue
1	0	0	0
2	255	255	255
3	255	0	0
4	0	255	0
5	0	0	255
6	255	255	0
7	0	255	255
8	255	0	255

FIGURE 10.14   A color system in which a lookup table is used.

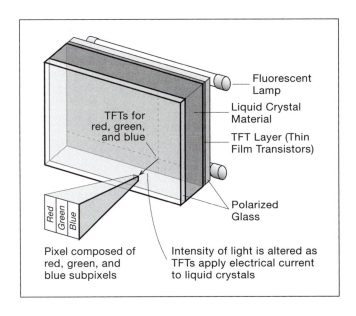

FIGURE 10.15   Diagram of a liquid crystal display (LCD) system. (After Morgenstern and Seff, 2002, and Marshall Brain's HowStuffWorks (*http://www.howstuffworks.com/lcd.htm*)).

bright light. If, however, an electrical charge is applied to a liquid crystal, then the crystal is bent and some of the light does not pass through—the intensity of the light that we see is a function of how strong a charge is applied. Color is created by using colored filters that are associated with separate liquid crystals at each pixel location. Just as in CRTs, red, green, and blue filters are used.*

LCDs have been used in laptop computers and computer projection devices since the mid-1990s. Today they are also challenging the CRT in its traditional area of dominance—the desktop computer. Advantages of LCDs compared to the CRT include their light weight, small depth (or small footprint), absence of flicker, low power consumption, and the absence of potentially harmful x-rays and low-frequency magnetic fields. Disadvantages of LCDs include a smaller range of available color, optimal performance at only one resolution (say

*For more on how LCDs work, see Marshall Brain's HowStuffWorks (*http://www.howstuffworks.com/lcd.htm*). Much of what we said about the frame buffer for the CRT applies to the LCD, although LCDs do not require a DAC.

1280 × 1024), difficulty of viewing from a wide angle, degradation in performance due to changes in ambient temperature, and the greater expense (especially for large-format displays). As technology continues to evolve, these disadvantages might disappear, and the LCD might replace the CRT. Of course, advancements in CRTs must also be considered; for example, in recent years a flat-screen CRT has been developed (Morgenstern and Seff 2002).

One limitation of both LCDs and CRTs is that they cannot handle large map displays (e.g., an entire USGS topographic sheet). The largest CRT and LCD screens generally do not exceed 21 inches (along a diagonal). One alternative to displaying an entire large-format map is, of course, to use *pan*, *scroll*, and *zoom* functions, but there are times when you would rather examine the entire map at once or compare a variety of maps simultaneously; for example, imagine comparing 50 maps showing changes in wheat production for Kansas on a yearly basis over a 50-year period. Although large, low-cost map displays are not available today, it will be interesting to see what becomes available as technology continues to evolve.

10.3 MODELS FOR SPECIFYING COLOR

This section considers six **color models** that have been used for specifying colors appearing on maps: **RGB**, **CMYK**, **HSV**, **Munsell**, **HVC**, and **CIE**. The RGB and CMYK models are hardware-oriented because they are based on hardware specifications of red, green, and blue color guns, and cyan, magenta, yellow, and black ink, respectively. In contrast, the HSV, Munsell, and HVC models are user-oriented because they are based on how we

perceive colors (using attributes such as hue, lightness, and saturation). The CIE system is neither hardware- nor user-oriented; however, it is "optimal" in the sense that if you provide someone with the CIE coordinates of a color you created, that person should be able to create exactly the same color.

10.3.1 The RGB Model

In the RGB model, colors are specified based on the intensity of red, green, and blue color guns. The range of intensities for color gun values can be represented as a cube, as shown in Figure 10.16. In Figure 10.16A, gray tones (or completely desaturated colors) are found along the diagonal extending from "White" to "Black." In general, lighter colors are found around the White point of the cube, and darker colors are found around the Black point of the cube. As you move away from the White–Black line, you move toward more saturated colors; for example, at the "Red" point you would be at the maximum saturation of red. Finally, note that hues are arranged in a hexagonal fashion around the White–Black line. The latter can be seen most easily if you look directly down the diagonal from White to Black, as seen in Figure 10.16B.

The RGB model has the advantage of relating nicely to the method of color production on graphic displays, but it has two major disadvantages. One is that common notions of hue, saturation, and lightness are not inherent to the model; although we discussed these notions earlier, they are not used to specify colors. Another disadvantage is that equal steps in the RGB color space do not correspond to equal visual steps; for example, a color

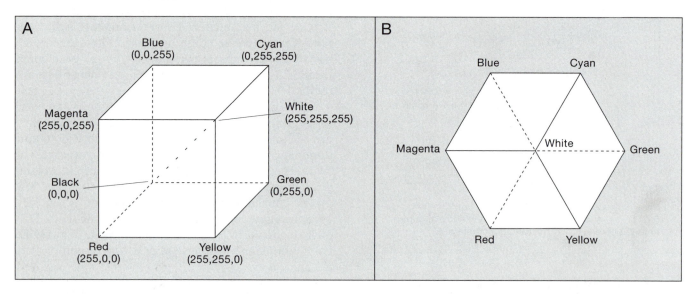

FIGURE 10.16 (A) Schematic form of an RGB color cube for specifying color. (It is assumed that the red, green, and blue color guns each have a maximum value of 255.) (B) The cube viewed looking directly down the diagonal from White to Black.

gun value of 125, 0, 0 will not appear to fall midway between values of 0, 0, 0 and 250, 0, 0. Typically, you will find that an incremental change in low RGB values represents a smaller visual difference than the same incremental change in high RGB values. In spite of these disadvantages, RGB values frequently are used as an option for specifying color in software packages, presumably because of their long history and the consequent familiarity that many users have with them.

10.3.2 The CMYK Model

Because printed maps are based on reflected (as opposed to emitted) light, they create color using a subtractive (as opposed to an additive) process. The three basic **subtractive primary colors** are cyan, magenta, and yellow (Color Plate 10.1); black ink, however, is utilized when a true black is desired. Together cyan, magenta, yellow, and black compose the CMYK color model. If we think of cyan, magenta, and yellow as analogous to the red, green, and blue color guns, then it is also possible to conceive of the CMY portion of CMYK as a cube: A certain percentage of cyan, magenta, and yellow would correspond to a particular point in the cube. Black would need to be added to create true shades of gray within the cube. Given the analogy to the RGB cube, it makes sense that CMYK will share the same disadvantages: lack of relation to common color terminology and equally spaced colors in the model will not correspond to equal visual steps. In Chapter 12, we consider in more detail how CMYK colors are used to create colors for printed maps.

10.3.3 The HSV Model

In contrast to RGB and CMYK, the HSV model is more intuitive from a map design standpoint because it allows users to work directly with hue, saturation, and value (lightness). Color space in HSV is represented as a hexcone, as shown in Figure 10.17. The logic of the hexcone can be seen if you compare it with the color cube for RGB shown in Figure 10.16B; note that the hexagonal structure of the hues in the cube are retained in the hexagonal structure at the base of the hexcone. Value changes occur as you move from the apex of the cone to its base, whereas saturation changes occur as you move from the center to the edge of the cone.

The intuitive notions of hue, saturation, and value in HSV have led to its common use in software. Although HSV is commonly used, it also has disadvantages. One is that different hues having the same value (V) in HSV will not all have the same perceived value. As an example, consider the base of the cone, where the highest value green and red are found. If you create such colors on your monitor, green will appear lighter than red. In a

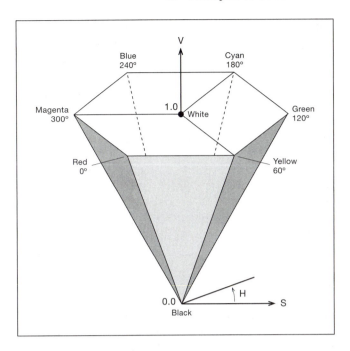

FIGURE 10.17 The HSV (hue, saturation, and value) color system represented as a hexcone.

similar fashion, different hues having identical saturations (S) will not have the same perceived saturations (Brewer 1994b). HSV also shares a disadvantage noted for RGB: Selecting a color midway between two colors will not result in a color that is perceived to be midway between those colors.

10.3.4 The Munsell Model

The Munsell color model is a user-oriented system that was developed prior to the advent of computers. Munsell colors are specified using the terms *hue, value* (for lightness), and *chroma* (for saturation). The general structure of the model (Figures 10.18 and 10.19 and Color Plate 10.3) is similar to HSV (hues are arranged in a circular fashion around the center, chroma increases as one moves outward from the center, and value increases from bottom to top). Note, however, that in contrast to HSV, the Munsell model is asymmetrical; for example, if you were to hold the model in your hands, you would note that the lightest green would be higher on the model than the lightest red. The asymmetry occurs because the model is perceptually based (the lightest possible green does appear brighter than the lightest possible red).

Ten major Munsell hues are recognized, and these are split into five principal (represented by a single letter, such as Y for yellow) and five intermediate (represented by two letters, such as YR) hues (Figures 10.18 and 10.19). Each major hue is also split into 10 subhues (consider the 10 subhues shown for 5R in Figure 10.19). Values

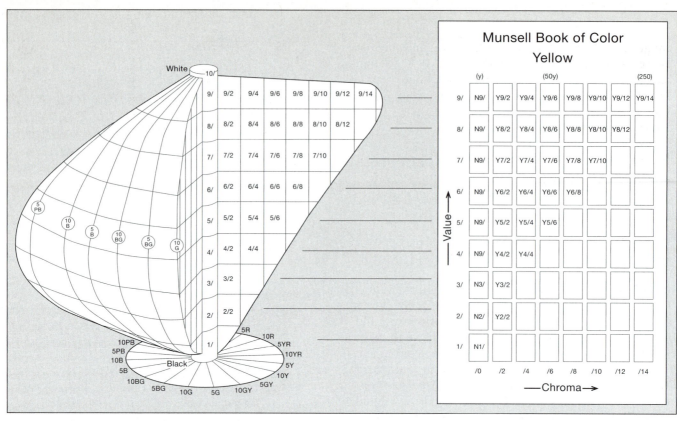

FIGURE 10.18 Three-dimensional representation of the Munsell color solid. A vertical slice through yellow is shown in detail. (Courtesy of Leo M. Hurvich.)

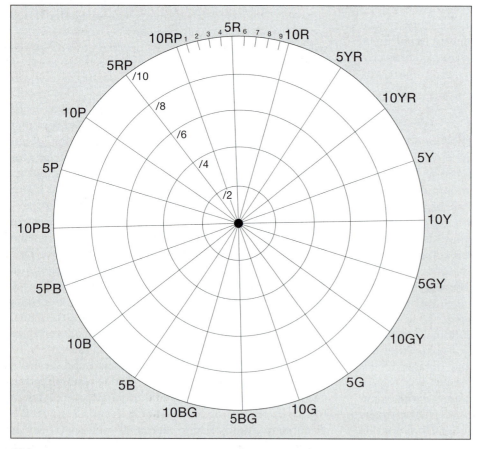

FIGURE 10.19 A horizontal slice through the Munsell color solid shown in Figure 10.18.

range from 0 to 10 (darkest to lightest), and chromas range from 0 to 16 (least to most saturated). Due to the asymmetry of the model, not all values and chromas occur for each hue. Munsell colors are represented symbolically as H V/C; thus, 5R 5/14 is a red of moderate value and high saturation (a crimson).

An important characteristic of the Munsell model is that equal steps in the model represent equal perceptual steps. Thus, a color that is numerically midway between two other colors should appear to be perceptually midway between those colors. For example, color 5R 5/5 should appear midway between 5R 2/2 and 5R 8/8. Remember that this is not a characteristic of the other color models described thus far.

10.3.5 The HVC Model

The HVC (hue, value, and chroma) color model developed by Tektronix was an attempt to duplicate the Munsell system on computer graphic displays. The similarity of HVC and Munsell is seen in the use of the same three terms (hue, value, and chroma), and the irregular shape of the color space (compare Figures 10.18 and 10.20). The two models differ, however, in color notation. Munsell uses the H V/C notation, whereas in HVC, hue is specified in degrees counterclockwise from 0° (red), value varies vertically from 0 to 100, and chroma varies from 0 to 100 from the central axis to the edge of the model (Figure 10.20). Although the literature suggests that HVC is effective (e.g., Taylor et al., 1991), software associated with the model is no longer distributed. Apparently, many designers outside the field of cartography either do not

require or do not see the advantage of specifying colors that are equally spaced in the visual sense, and so Tektronix has chosen not to support the software. We have included the HVC model here because it is illustrative of the type of model that developers should consider including in cartographic software.

10.3.6 The CIE Model

CIE is an abbreviation for the French *Commission International de l'Eclairage* (International Commission on Illumination). In theory, careful color specification in the CIE model means that anyone in the world should be able to recognize and reproduce a desired color. CIE colors can be specified in several ways (Yxy, L*u*v*, L*a*b*), but in all cases a combination of three numbers is used. We consider the Yxy model (commonly referred to as the 1931 CIE model) first because it forms the basis of other CIE methods. In the Yxy model, the x and y coordinates define a two-dimensional space within which hue and saturation vary (Figure 10.21). Note that hues are arranged around a central *white point* (or *equal-energy point*) and that saturation increases as one moves outward from the white point. The Y portion of the model provides the third dimension—the lightness or darkness component (Color Plate 10.4).

The structure of the Yxy model is similar to both HVC and Munsell (all have hues arranged in a circular fashion, desaturated colors in the middle, and a vertical lightness axis), but note that in CIE hues and saturations are not related in a simplistic fashion to the x and y axes. The reason for this can be found in the manner in which CIE was established. CIE was developed using the notion that most colors can be defined by a mixture of three colors (roughly speaking, we can call these red, green, and blue). The appropriate combination of three colors needed to match selected colors was determined using human observers (the average response of the observers was termed the *standard observer*). To understand the matching process, imagine that you were asked to view a screen on which a single circle was projected using a standard light source. In the top portion of the circle a test color appeared, and in the bottom portion you manipulated the three colors to produce a color identical to the test one. If you repeated this task for many test colors, you would discover that various amounts of the three colors would be required to make appropriate matches.

Results of actual CIE matching experiments are shown in Figure 10.22. The three curves correspond to the three colors combined in the experiments. The x axis represents the wavelength of the test color, whereas the y axis represents the relative magnitudes of the three colors needed to match the test color. For example, a test color at 530 nm would require 0.005 of blue, 0.203 of

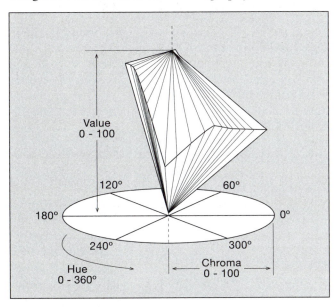

FIGURE 10.20   The HVC color system developed by Tektronix. (Taylor et al., 1991. First published in *Information Display*, 7 (4/5), p. 21. Courtesy of Tektronix, Inc.)

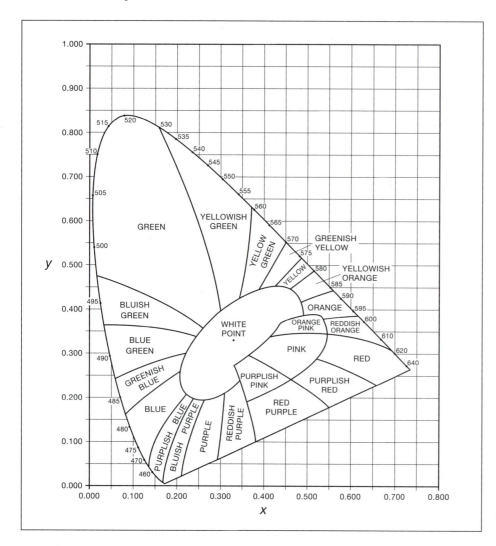

FIGURE 10.21 The *Yxy* 1931 CIE system. Hues are arranged in a continuum around a central white point. Saturation is at a maximum on the edge of the horseshoe and at a minimum at the white point. Numerical values on the edge of the horseshoe represent wavelengths in nanometers. Because differing lightnesses cannot be shown by this diagram, a three-dimensional diagram is required, as in Color Plate 10.4. (After Kelly, K. L. (1943) "Color designations for lights," *Journal of the Optical Society of America*, 33, no. 11:627–632.)

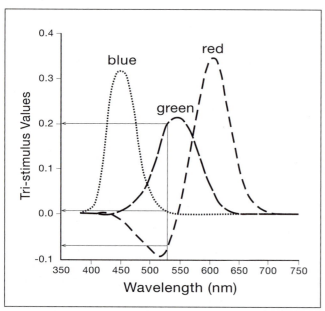

FIGURE 10.22 Curves representing the relative proportion of CIE primaries needed to match test colors at various wavelengths.

green, and −0.071 of red (Wyszecki and Stiles 1982, 750); these are known as *tristimulus values*. The negative value for red is necessary because in some cases you would find that you could not match the test color with any combination of the three colors; to achieve a match, you would have to mix one of the three with the test color, and this is recorded as a minus value in the graph. To avoid having to work with negative values, the developers of the CIE model transformed the results shown in Figure 10.22 to purely positive values, which are commonly referred to as *X*, *Y*, and *Z*. To get coordinates for the *Yxy* system, the *X*, *Y*, and *Z* values were converted to proportional values:

$$x = X/(X + Y + Z)$$
$$y = Y/(X + Y + Z)$$
$$z = Z/(X + Y + Z)$$

Because these proportions add to 1, it is not necessary to plot *z* in the *Yxy* system (*z* would be 1 − (*x* + *y*)). One problem with plotting proportional values is the elimination of information about lightness or darkness. (We get the same proportions when X, Y, and Z are all 10 units as we do when they are all 20 units.) This problem was handled in CIE by arbitrarily assigning the lightness information to Y and plotting this as the third dimension, as shown in Color Plate 10.4 (Hurvich 1981, 284).

It should also be noted that the CIE Yxy coordinate values can be adjusted to account for the lighting conditions under which the colors are viewed. For example, you might want to consider the potential effects of natural sunlight versus fluorescent room light. This is a capability that is not generally included in other color models.

One problem with the 1931 CIE diagram is that colors are not equally spaced in a visual sense, just as was true for RGB, CMYK, and HSV. Fortunately, two **perceptually uniform color models** (or simply **uniform color models**) have been developed based on CIE: L*u*v* (CIELUV) and L*a*b* (CIELAB). CIELUV is appropriate for graphic displays, and CIELAB is appropriate for printed material.

10.3.7 Discussion

Given the variety of color models presented here, it is natural to ask which models mapmakers need to be familiar with. At present, the RGB, CMYK, and HSV models are most frequently used in mapping software. This is problematic because the RGB and CMYK models do not relate easily to our common notions of hue, saturation, and lightness, and all three models do not produce equally spaced colors (in the visual sense). The Munsell model is not generally found in mapping software, but it is useful because it relates well to our notion of hue, saturation, and lightness, and it stresses the notion of creating equally spaced colors. The HVC model is useful for illustrating how the Munsell model can be included in software. The CIE model, although thus far not commonly used in mapping software, is important to understand because you might see reference to CIE colors when reading the results of cartographic research; for example, recent articles by Olson and Brewer (1997) and Brewer et al. (1997) reported colors in CIE coordinates.

SUMMARY

In this chapter, we examined several basic principles of color that will assist you in understanding the use of color elsewhere in this book. In considering how color is processed by the human visual system, we found that there are three types of **cones** in the **retina** of the eye— red, green, and blue—each sensitive to certain wavelengths of light; for example, red cones are sensitive to long-wavelength red light. Our perception of color is not simply a function of which cones are stimulated; we must also consider the notion that cones excite or inhibit other nerve cells (e.g., the **bipolar** and **ganglion** cells). The net result is that we perceive color according to the **opponent-process theory**, which states that three channels are involved in color perception: a lightness–darkness channel and two opponent-color channels (red–green and blue–yellow). Perceived colors are a function of a mix of colors in the opponent channels; for example, we see orange as a mixture of red and yellow.

It is important to keep in mind that not everyone sees color in the same way. About 4 percent of the population in the United States and Europe (mostly males) has some form of **color vision impairment**; a common impairment is the inability to discriminate between red and green colors. Another characteristic of color perception is **simultaneous contrast**—that our perception of a color might be a function of the colors that surround it. Fortunately, as we will see in Chapter 13, special color schemes have been developed to handle the problems of color vision impairment and simultaneous contrast.

Maps produced on a computer screen (soft-copy maps) and maps printed on paper (hard-copy maps) utilize fundamentally different processes. Soft-copy maps use an *additive* process involving red, green and blue, whereas hard-copy maps utilize a *subtractive* process involving cyan, magenta, yellow, and black. The emphasis of this chapter was on soft-copy maps; Chapter 12 covers hard-copy production and dissemination.

Today, CRTs are the dominant technology for the production of soft-copy maps, but LCDs are becoming quite common, and might eventually displace CRTs as the dominant technology. An important concept in producing color on CRTs is the **frame buffer**, an area of memory that stores a digital representation of colors appearing on the screen; in sophisticated color display systems, the frame buffer is divided into a series of *bit-planes* for each of the **electron guns** (red, green, and blue). Sometimes **color lookup tables** are utilized to make rapid changes in the colors of regions (as in animation); in this case, values in the frame buffer serve as indexes to a lookup table that provides the actual values sent to the color guns.

We considered a variety of **color models** that can be used to specify colors displayed on maps. Some of these are hardware-oriented (**RGB** and **CMYK**), and thus of limited use to the mapmaker. Others are user-oriented (**HSV**, **Munsell**, and Tektronix's **HVC**), and thus permit color specification in terms mapmakers and map users are apt to be familiar with, such as hue, lightness, and saturation. We also examined the **CIE** model, which, in theory, allows a mapmaker to reproduce colors specified by others. At present, the RGB, CMYK, and HSV models are most frequently used in mapping software. This is problematic because the RGB and CMYK models do not relate easily to our common notions of hue, saturation, and lightness, and all three models do not produce equally spaced colors (in the visual sense).

FURTHER READING

Birren, F. (1983) *Colour*. London: Marshall Editions.

A general text on color and its varied uses.

Computer Graphics 31, no. 2, 1997 (entire issue).

The bulk of this issue deals with novel graphic displays such as HDTV and virtual reality (the latter is considered in Chapter 24 of this text).

Fairchild, M. D. (1998) *Color Appearance Models*. Reading, MA: Addison-Wesley.

Reviews principles of color vision and describes numerous color appearance models.

Foley, J. D., van Dam, A., Feiner, S. K., and Hughes, J. F. (1996) *Computer Graphics: Principles and Practice*. 2nd ed. Reading, MA: Addison-Wesley.

A standard reference on computer graphics. Chapter 4 covers hardware issues, and Chapter 13 covers issues related to color models; also see Chapter 1 and pp. 602–656 of Rogers (1998).

Garo, L. A. B. (1998) "Color theory." *http://www.uncc.edu/lagaro/cwg/color/index.html*.

An online resource covering principles of color in cartography.

Goldstein, E. B. (2002) *Sensation and Perception*. 6th ed. Pacific Grove, CA: Wadsworth.

A standard text on perception; Chapter 6 is a useful supplement to section 10.1 of this text.

Hubel, D. H. (1988) *Eye, Brain, and Vision*. New York: Scientific American Library.

A treatise on how the eye–brain system functions.

Hunt, R. W. G. (1987a) *Measuring Colour*. Chichester, England: Ellis Horwood.

A thorough treatment of color models.

Hunt, R. W. G. (1987b) *The Reproduction of Colour in Photography, Printing & Television*. Tolworth, England: Fountain Press.

A good reference text on color vision.

Hurvich, L. M. (1981) *Color Vision*. Sunderland, MA: Sinauer Associates.

Covers the details of how we perceive color, with an emphasis on opponent-process theory.

Itten, J. (1973) *The Art of Color: The Subjective Experience and Objective Rationale of Color*. New York: van Nostrand Reinhold.

Discusses principles of color from an artist's perspective.

Peddie, J. (1994) *High-Resolution Graphics Display Systems*. New York: Windcrest/McGraw-Hill.

An extensive review of graphic display systems.

Travis, D. (1991) *Effective Color Displays: Theory and Practice*. London: Academic Press.

Chapters 1 to 3 deal with computer hardware, the human visual system, and color models, respectively.

Wyszecki, G., and Stiles, W. S. (1982) *Color Science: Concepts and Methods, Quantitative Data and Formulae*. 2nd ed. New York: Wiley.

A detailed reference on technical aspects of color.

11

Elements of Cartographic Design

OVERVIEW

This chapter introduces common map elements that are employed in the creation of thematic maps, presents typography and its application in cartography, and describes the process of cartographic design. The general goal is to provide you with guidance in the creation of effective, attractive maps that efficiently communicate geographic information. Specific goals include guidance in choosing which map elements to employ and how to implement them, how to implement effective typography, and to provide assistance in the process of designing high-quality maps. Section 11.1 begins with a clarification of widely misunderstood terms associated with the alignment and centering of map features. A firm understanding of these terms is essential in understanding the remainder of the chapter.

Section 11.2 introduces common **map elements**. It is the cartographer's job to select appropriate map elements, and to implement them properly according to the needs of the **map user**, who represents the map's intended audience. Detailed descriptions of map elements are provided, together with rules and guidelines related to their implementation. The elements are introduced in the order we recommend they be placed when constructing a map. The **frame line** and **neat line** help to organize the content of a map and to define its extent. The **mapped area** is the region of Earth being represented; it consists of **thematic symbols** and **base information**. The **inset** is a smaller map included within the context of a larger map. The **title** and **subtitle** convey the map's theme, plus additional information such as the region or date. The **legend** defines all of the symbols that are not self-explanatory, and normally includes a **legend heading** that further explains the map's theme. The **data source** informs the map user of where the thematic data were obtained. The **scale** provides an indication of how much reduction has taken place, or allows the map user to measure distances. The scale can take the form of a **representative fraction**, a **verbal scale**, or a **bar scale**. **Orientation** refers to the indication of direction, and is represented by a **north arrow** or **graticule**.

Section 11.3 is dedicated to **typography**, the art or process of specifying, arranging, and designing type. **Type** refers to the words that appear on maps. Type is organized according to **type family**, **type style**, **typeface**, **type size**, **font**, and also according to whether **serifs** are present. Type can be modified for specific purposes by altering **letter spacing**, **word spacing**, **kerning**, and **leading**. General and specific guidelines are provided for the use of type in cartography. Also included in section 11.3 is a description of **labeling software** that employs aspects of **expert systems** to place type labels automatically, according to rules and guidelines specified by cartographers.

Section 11.4 describes **cartographic design** as a process in which the cartographer conceptualizes and visualizes the map to be created, according to the needs of the intended map user. Aspects of cartographic design have been derived from the results of **map design research**, and have been influenced by principles of graphic design and **Gestalt principles** of perceptual organization. The design process is distilled to eight general procedures, including the establishment of an **intellectual hierarchy**, the creation of rough **sketch maps**, and the creation and evaluation of **rough drafts**. The design process culminates with an evaluation of the map by members of the intended audience. Effective cartographic design is dependent, in part, on the appropriate choice and implementation of map elements. Other aspects of cartographic design are (1) **visual hierarchy**—a graphical representation of the intellectual *hierarchy, conveyed by adjusting the **visual weight** of map elements; (2) **contrast**—the visual differences between map elements that distinguish them, and imply their relative importance; (3) **figure-ground**—the emphasis of selected features by making them appear closer to the map user, sometimes through the use of **screening**; and (4) **balance**—the harmonious organization of map elements and empty space that incorporates the concept of **available space**.*

This chapter was not written with specific software applications in mind. Graphic design applications such as Macromedia FreeHand and Adobe Illustrator typically provide the greatest control over graphics and type, but recent advances in the design capabilities of GISs have narrowed the gap, allowing you to produce high-quality maps in a user-friendly GIS environment. The examples in this chapter reflect simple thematic maps, although the principles set forth can also be applied to more complex thematic maps and general reference maps.

11.1 ALIGNMENT AND CENTERING

Before we begin, it is important to clarify terms associated with the alignment and centering of map features. Alignment and centering of map features is achieved either through visual approximation or measurement. Terms such as *visually centered* refer to the placement of map features through visual approximation, so that they *look* centered within an area. Alignment and centering is also achieved through measurement, resulting in map features that are *precisely* aligned or centered. If the term "visually centered" appears, it refers to visual approximation; all other references to alignment and centering in this chapter are based on measurement, as described next.

Terms such as "horizontally aligned to right," and "vertically centered" are used repeatedly in this chapter to describe spatial relationships between map features. These terms are widely misunderstood and warrant clarification. Map features such as symbols and type are often arranged in relation to imaginary lines, as illustrated in Figure 11.1. Horizontal alignment and centering (Figure 11.1A) involve the side-to-side movement of map features in relation to a *vertical* line (the square, circle, star, and "Black Mountain" text are moved in relation to the dashed line). Vertical alignment and centering (Figure 11.1B) involve the top-to-bottom movement of map features in relation to a *horizontal* line. Be aware that the terms horizontal and vertical are often confused in this context. (The terms are even misused in the manual of one prominent graphic design software application.) For example, the square, circle, star, and "Black Mountain" text in Figure 11.1A are *vertically distributed* (arranged from top to bottom), but alignment and centering are achieved through side-to-side (horizontal) movement. In Figure 11.1C, the alignment and centering controls of two

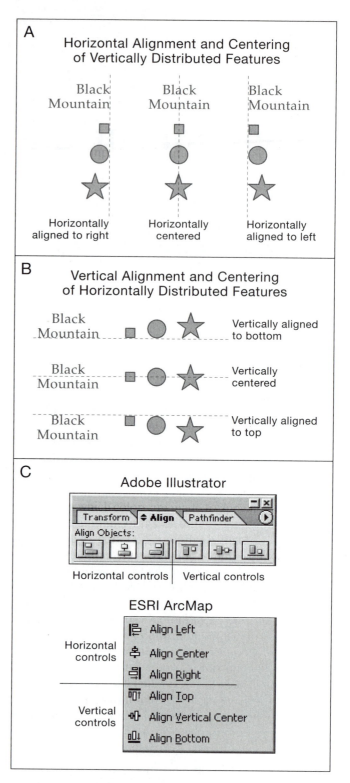

FIGURE 11.1   (A) Options for horizontal alignment and centering. (B) Options for vertical alignment and centering. (C) Alignment and centering controls in Illustrator and ArcMap. Illustrator screenshot used by permission © 2003 Adobe Systems Incorporated. All rights reserved. Adobe and Illustrator are registered trademarks of Adobe Systems Incorporated in the United States and/or other countries. ArcMap screenshot used by permission from ESRI.

popular graphic design and GIS applications are presented. Although the terminology differs slightly from application to application, the concepts just presented hold true. A firm grasp of these concepts and terms is crucial in fully understanding this chapter, and will help you when creating maps.

11.2 MAP ELEMENTS

The various purposes a map can serve, together with the wide variety of mapping techniques available, can lead to the perception that every map is completely unique. Despite the great variety of maps in the world, it is important to recognize that most are created from a common set of **map elements**. These map elements represent the building blocks of cartographic communication: the transmission of geographic information through the use of maps. The following list presents the most common map elements, each of which is described in detail beginning in section 11.2.1.

1. Frame line and neat line
2. Mapped area
3. Inset
4. Title and subtitle
5. Legend
6. Data source
7. Scale
8. Orientation

These eight map elements are listed in a logical progression representing the order in which we recommend they be placed when constructing a map. For example, the frame line establishes the size and shape of the initial map space, and should be placed first; the mapped area and inset are among the largest map elements, and should be placed after the initial space has been defined by the frame line; the title, subtitle, and legend are intermediate in size, and should be placed in the space that remains after placing the mapped area and inset; smaller map elements such as the data source, scale, and orientation (typically a north arrow), should be placed last in the space that remains. The concept of *available space* is referenced repeatedly in this chapter, and is covered in detail in section 11.4.5.

Cartographic design, described in detail later in this chapter, requires that you choose which map elements to include on a particular map. In addition to choosing appropriate map elements, you need to decide on the most appropriate implementation of each element. The content, style, size, and position of each map element must be carefully considered, even before map construction begins.

Proper choice and implementation of map elements is governed in large part by the purpose of the map and its intended audience. The **map user** represents the intended audience; virtually every decision you make should be made in reference to the needs of the map user. The proper choice and implementation of map elements result in a minimization of "map noise," which refers to unnecessary or inappropriate symbolization, design, and typography that interfere with the map user's ability to interpret the map.

The rules and guidelines associated with map elements presented here are based on convention, research, common sense, and to a lesser degree, the opinions of the authors based on many years of experience. We believe it is important to build a foundation of knowledge and experience by following these specific rules and guidelines. After a foundation has been built, we encourage you to consider different points of view, and to "break the rules" in creative ways. Always be prepared to explain or defend your reasoning when choosing and implementing map elements.

11.2.1 Frame Line and Neat Line

The frame line and neat line help to organize the map's contents and to define its extent. A **frame line** encloses all other map elements (Figure 11.2A); it is similar to a picture frame because it focuses the map user's attention on everything within it. The frame line should be the first map element placed because it defines the initial available space that all other map elements will be placed within. A **neat line** might also appear *inside* a frame line. The neat line is used to crop (clip, or limit the extent of) the mapped area, as it does to the San Francisco Bay area in Figure 11.2B. A frame line should be used in most situations; a neat line is used when the mapped area needs to be cropped. In certain cases, a frame line can also act as a neat line, enclosing all map elements *and* cropping the mapped area (Figure 11.2C).

The style of the frame line and neat line should be subtle. A single, thin, black line should be used; slightly thicker lines are appropriate when working with larger formats, such as wall maps and posters. Many large-format maps (e.g., wall maps of the world) employ ornate frame lines, which often detract attention from more important map elements, such as the mapped area. Overly thick or ornate lines should be avoided for the same reason that you would avoid choosing a gaudy picture frame for a fine painting or photograph (Figure 11.2D).

The size (width and height) and position of the frame line depend on the desired map size and page dimensions. The size and position of the neat line are normally dictated by the frame line, the mapped area, and the other map elements.

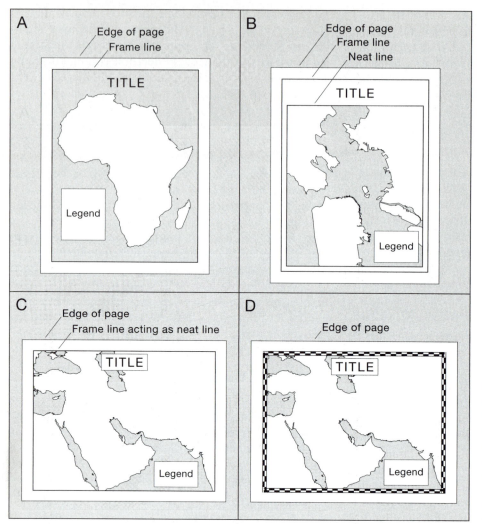

FIGURE 11.2 (A) The frame line encloses all map elements. (B) The neat line crops the mapped area. (C) A frame line can also act as a neat line. (D) Thick or ornate lines should be avoided. (Note that the edge of the page is illustrated to put the frame and neat lines into context. The edge of the page is not illustrated elsewhere in this chapter.)

11.2.2 Mapped Area

The **mapped area** is the region of Earth being represented, as illustrated by Kenya in Figure 11.3. It consists of **thematic symbols** that directly represent the map's theme (we say little about thematic symbols in this chapter, as we focus on them in Chapters 13–19). The mapped area can also include **base information** that provides a geographic frame of reference for the theme. Base information includes, but is not limited to, boundaries, transportation routes, landmarks, and place names. In Figure 11.3, the shaded polygons are thematic symbols representing the percentage of religious or private secondary schools within each district of Kenya. The railroads represent base information that provides a geographic frame of reference for people who might be familiar with them, but might not be familiar with the district boundaries. In this case, the base information also suggests a possible relationship between the railroads and the districts having the highest percentage of secondary schools. Base information that illuminates the map's theme in this manner is of particular

value. As another example, consider Figure 11.7. Four types of base information are included here: interstate highways, city boundaries, city names, and the state line. Without this base information, the thematic symbols (precipitation contours) would be virtually meaningless—the map user would be unable to associate them with a particular geographic region.

In some cases, the mapped area appears as a "floating" geographic region, disconnected from neighboring regions, as illustrated by Kenya in Figure 11.3. Use of a floating mapped area often eases the placement of other map elements, because of the empty space that results from the omission of surrounding geography. However, this also removes the region of interest from its geographic context, making the map more abstract, and possibly confusing the map user. In other cases, a neat line is used to crop the mapped area, as illustrated by the Kansas City region in Figure 11.7. A mapped area that is cropped represents the region of interest within its geographic context, but can make the placement of other map elements more difficult, due to a lack of empty space.

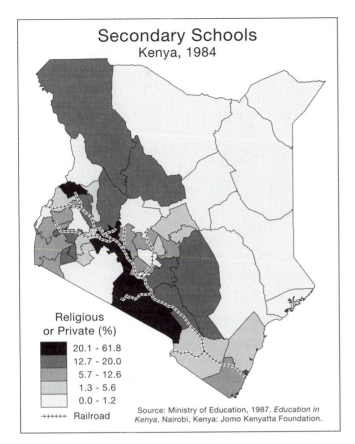

FIGURE 11.3 The mapped area is composed of thematic symbols (shaded polygons in this example) and base information (railroads in this example).

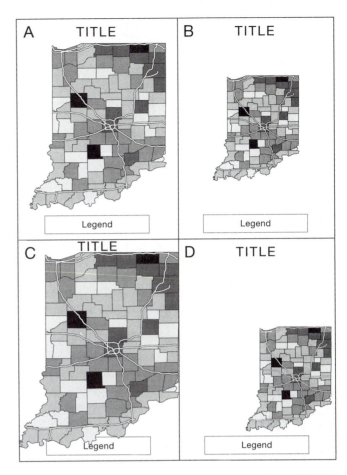

FIGURE 11.4 Sizing and position of the mapped area. (A) Appropriate sizing and position. (B) Insufficient size, but appropriate position. (C) Excessive size, but appropriate position. (D) Insufficient size and inappropriate position. (Note the inclusion of interstate highways, which act as base information.)

Consideration of the map's purpose, the geographic region of interest, and other map elements will determine whether the mapped area should be floating or cropped.

The size of the mapped area is dependent on several factors, including the page size, margins, and space needed for other map elements. A general guideline is to make the mapped area as large as possible within the available space without being "too close" to the frame line, leaving ample room for the remaining map elements (Figure 11.4A). Maximum size is important because the mapped area—thematic symbols in particular—is instrumental in communicating the map's information. Figure 11.4B represents a mapped area that does not take full advantage of the available space, whereas Figure 11.4C illustrates a mapped area that is too large: It touches the frame line and leaves inadequate space for the title and legend.

Position of the mapped area is dependent on several factors, including the shape of the geographic region, page dimensions, and the other map elements. If possible, the mapped area should be visually centered, both horizontally (side to side) and vertically (top to bottom) within the available space, as defined by the frame line. The mapped areas in Figures 11.4A, 11.4B, and 11.4C are properly centered; the mapped area in Figure 11.4D is off-center both horizontally and vertically. A properly centered mapped area can lend a sense of balance to a map, but irregularly shaped geographic regions, together with other factors, often make centering impossible. The country of Chile is a classic example of a geographic region that is commonly positioned off-center to accommodate additional map elements.

11.2.3 Inset

An **inset** is a smaller map included within the context of a larger map. Insets can serve several purposes: (1) to show the primary mapped area in relation to a larger, more recognizable area (a locator inset), as illustrated in Figure 11.5A; (2) to enlarge important or congested areas (Figure 11.5B); (3) to show topics that are related to the map's theme, or different dates of a common

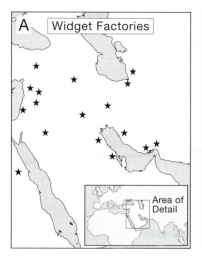

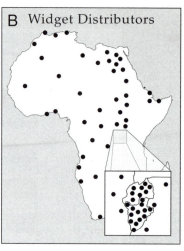

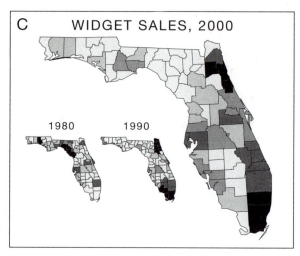

FIGURE 11.5   Insets serve several purposes: (A) locating the primary mapped area; (B) enlarging an important or congested area; and (C) showing related themes or dates. An additional purpose is showing related areas (Figure 11.14).

theme, represented by smaller versions of the primary mapped area, as illustrated by the 1980 and 1990 insets in Figure 11.5C; and (4) to show areas that are related to the primary mapped area that are in a different location, or cannot be represented at the scale of the primary mapped area, as illustrated by Alaska and Hawaii in Figure 11.14.

The style of the inset can vary. In Figure 11.5A, the inset is relatively subtle; its only purpose is to help orient the map user. In Figures 11.5B and 11.5C, the insets take on a central focus of the map and attract attention along with the primary mapped area. The size and position of the inset are equally variable, depending on the purpose of the inset, the size of the map, and the other map elements.

11.2.4 Title and Subtitle

Most thematic maps require a **title**, although a title is sometimes omitted when a map is used as a figure in a written document, assuming that the theme is clearly expressed in the figure caption. A well-crafted title can draw attention to a map, however, and thus we recommend using a title in virtually all situations—even when the theme is reflected in the figure caption. The title of a thematic map should be a succinct description of the map's theme (Figure 11.6A), whereas the title of a general-reference map is normally a statement of the region being represented. Unnecessary words should be omitted from the title, but care should be exercised to avoid abbreviations that might not be understood by the map user. For example, the title "A Map of the Population Density of New Hampshire Counties in 2002" is excessively wordy; "NH Pop. '02" is abbreviated to the point of being cryptic; whereas "Population Density" or "Population Density in New Hampshire, 2002" clearly expresses the theme

---

A

Long Term Debt
AVERAGE AGE, 2001
Museums of Modern Art

B

Population Density
New Hampsire, 2002

Number of Chickens
Harper County

BIRTH RATE INCREASE
1950 - 2000

Subtitle horizontally centered below title

FIGURE 11.6   (A) Titles. The title of a thematic map succinctly describes the map's theme. (B) Titles with subtitles. The subtitle is used to further explain the title.

in few words. Notice in the last example that the theme, geographic region, and year all appear in the title, and that the theme is aptly stated *before* the region and year. Use of the word *map* in a title is a statement of the obvious, and should be avoided.

The **subtitle**, if employed, is used to further explain the title. The name of the region and the data collection date are common components of the subtitle (Figure 11.6B). "New Hampshire, 2002" would be an appropriate subtitle for the "Population Density" title of the previous example. The name of the region is often omitted

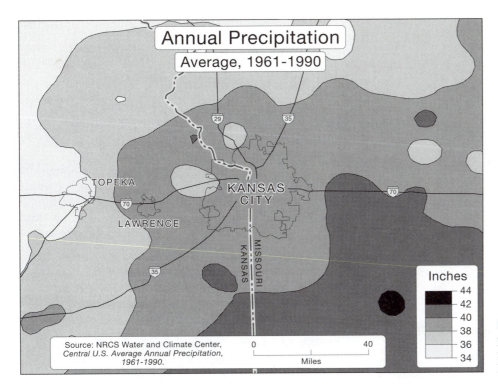

FIGURE 11.7 The title and subtitle placed at top center. Note how bounding boxes are used to mask the underlying mapped area.

when the cartographer feels that the map user will easily identify the region. For example, most residents of the United States will recognize the shape of the United States, and most readers of an article focusing on Japan will recognize the shape of Japan.

In the name of legibility, the style of the title and subtitle should be plain. Avoid italics and ornate type styles. Although the use of bold type can emphasize the title and subtitle, it is normally not required if appropriate type sizes are chosen. Use a subtle bounding box around the title and subtitle only if it is necessary to mask the underlying mapped area to improve legibility, as illustrated in Figure 11.7.

The title should generally be the largest type on a map; the subtitle should be visibly smaller, and is normally the second largest type on a map. Both the title and subtitle should be limited to one line each in most cases. If possible, the title should be placed toward the top of the map, where the map user is accustomed to seeing titles, and be horizontally centered within the frame line (Figure 11.7). A case can also be made for keeping the title directly above the legend, as illustrated in Figure 11.8, to relieve the map user from jumping back and forth between the title and legend (Monmonier 1993). Many large-format wall maps make use of this technique, as distances between the title and legend can be great. The subtitle, if used, should be located directly below the title, and should be horizontally centered with it, as illustrated in Figure 11.6B. The horizontal centering of multiple lines of type creates a "self-balancing" block of type, and is used frequently in cartography.

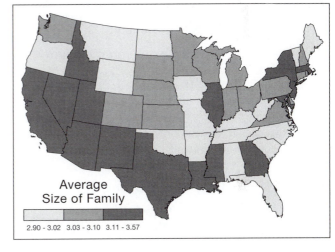

FIGURE 11.8 The title placed directly above the legend.

11.2.5 Legend

The **legend** is the map element that defines all of the thematic symbols on a map. Symbols that are self-explanatory or not directly related to the map's theme are normally omitted from simple thematic map legends. In contrast, legends for general reference maps, such as Forest Visitor maps issued by the U.S. Forest Service, often define all symbols found on the map—even if self-explanatory. General guidelines for legend design are provided in this section; legend design and content for particular thematic map types are described in Part II of this text.

The style of the legend should be clear and straight-forward. Use a subtle bounding box around the legend only if it is necessary to mask the underlying mapped area, as illustrated in Figure 11.7. Special care should be exercised to ensure that symbols in the legend are *identical* to those found in the mapped area. This includes size, color, and orientation if possible. A well-designed legend is self-explanatory, and does not need to be identified with "Legend" or "Key" labels.

Representative symbols should be placed on the left and defined to the right, as illustrated in Figure 11.9A. This arrangement is customary in English dictionaries, which are read from left to right; it allows the map user to view the symbol first, and then its definition. In most cases, representative symbols should be vertically distributed (placed in sequence from top to bottom), and horizontally centered with one another as illustrated in Figure 11.9B.

Symbols should be vertically centered with their definitions* (Figure 11.9B). Textual definitions such as "Community Garden," and definitions consisting of individual numbers such as "25,000," should be horizontally aligned to left (Figure 11.9B). In most cases, definitions consisting of ranges of numbers should be horizontally aligned with one another on the separator between numbers (Figure 11.9C). Numbers are normally separated by a hyphen, which is compact, or the word "to," which can prevent the appearance of consecutive hyphens when negative numbers are represented. For clarity, spaces should be included to the right and left of each separator.† Definitions containing numbers of 1,000 or greater should incorporate commas, and decimal numbers smaller than one should incorporate a leading zero, as illustrated by "1,021.8" and "0.4" in Figure 11.9C. Space considerations might require that the legend be oriented in a horizontal fashion, with the definitions horizontally centered below the symbols they represent, as illustrated in Figure 11.9D.

Notice in Figure 11.9 that all the point and line symbols are separated from one another, whereas the areal symbols (the rectangles, or "boxes") are either separated from one another (Figure 11.9A) or are connected, and share a common boundary (Figure 11.9C). Areal symbols within the mapped area are almost always connected, and share common boundaries, suggesting that legend boxes should also be connected. However, the type of data represented by the boxes also influences whether they should be connected or separated. Legend boxes should be connected when representing a range of numeric values associated with an attribute (e.g., per capita income), because connected boxes help to emphasize the idea that a gradation of values (associated with a single attribute) is being represented. Legend boxes should be separated when representing unique (qualitative) categories of data (e.g., sheep, goats, and pigs), because separated boxes help reinforce the idea that distinctly different entities are being represented. Areal symbols are sometimes represented in legends by irregular, amorphous polygons, as illustrated in Figure 11.9E. Irregular

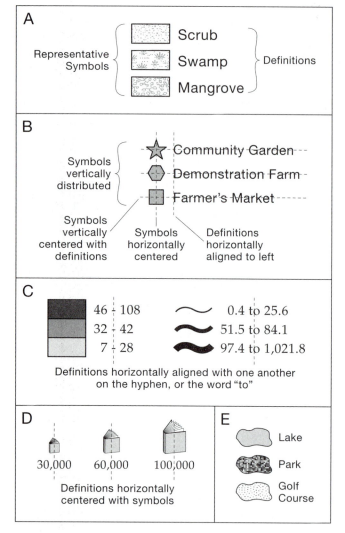

* The legend in Figure 11.7 is an exception. Here, definitions represent the values of the boundaries *between* symbols.

† An en dash can also be used as a separator. An *en* is a unit of measure equal to half the point size of the type being used. En dashes are half the width of *em* dashes, and are used to separate ranges of numbers in written documents, without the use of spaces to the right and left of the dash (Romano 1997, 223). For clarity in map legends, we recommend the use of the hyphen (which is shorter than the en dash) or the word "to," with spaces on both sides of either one.

FIGURE 11.9   (A) The legend is composed of representative symbols and definitions. (B) Distribution and alignment of symbols and definitions. (C) Horizontal alignment of definitions composed of numeric ranges. (D) A horizontally oriented legend. (E) Irregular polygons used to represent areal data.

polygons are most appropriate when representing unique categories of areal data that are amorphously shaped, such as lakes and golf courses. They should be separated because they represent unique categories of data. Symbols that represent a singular feature on a map (e.g., one symbol represents one well) should be singular in the legend, not plural (Figure 11.10A).

A **legend heading** is often included, and is used to further explain the map's theme. The unit of measurement (for quantitative data) and the enumeration unit are common components of the legend heading. For example, a map with the title "College Graduates" might have a legend heading such as the one illustrated in Figure 11.10B. The legend heading should normally be placed above the legend, and be horizontally centered with it; multiple lines of type should be horizontally centered with one anoth-

er (Figure 11.10B). The horizontal centering of the legend heading, legend, and related map elements such as the data source creates a group of objects that is self-balancing; it is neither "left-heavy" nor "right-heavy," and often contributes to an overall sense of balance in a map. Balance is further described in section 11.4.5. The same objects are sometimes horizontally aligned to the left (Figure 11.10C) or to the right (Figure 11.10D), but the self-balancing effect is lost.

A category that represents an absence of areal data can be represented on the mapped area in a neutral color such as white, or with a subtle texture, which is not used in any other category. A simple note below the legend can be used to inform the map user of the "no data" category, as illustrated in Figure 11.10B. A potential problem with this method is that very light gray areas might be confused with the white no-data category, and textures might draw undue attention. Another option is to place a small dot within no-data areas; the dot must then be identified as representing no data in the legend. One problem with this method is that a point symbol is used to represent an areal feature, possibly confusing the map user.

Legend symbols are often organized into groups, based on a particular criterion. For example, Figure 11.11 represents three possible methods of grouping the symbols that might appear on a map having to do with natural gas pipeline facilities. In Figure 11.11A, two groups are formed according to whether the symbols represent natural or cultural features. In Figure 11.11B, three groups are formed according to the general geometric form of the symbols: point, line, or area. In Figure 11.11C, two groups are formed according to whether the symbols are thematic in nature and directly related to the map's theme (the upper group), or represent base information (the lower group). This last approach is probably the most appropriate for thematic maps that include base information in the legend.

The legend should be large enough to allow the map user to employ it easily, but should not be so large as to occupy vast areas of space, or to challenge thematic symbols in the mapped area. The legend heading should be smaller than the subtitle, and the definitions should be smaller than the legend heading, as illustrated in Figure 11.10B.

Position of the legend is dependent on available space, as defined by the other map elements; if possible, the legend should be visually centered within the available space. For example, in Figure 11.12 space is available to the left and right of South America. Each section of the legend, which has been split in two because of the large number of categories, is visually centered within its available space. The currency map in Figure 11.12 is qualitative in nature, and thus the polygons would best be represented using colors of unique hue, as illustrated

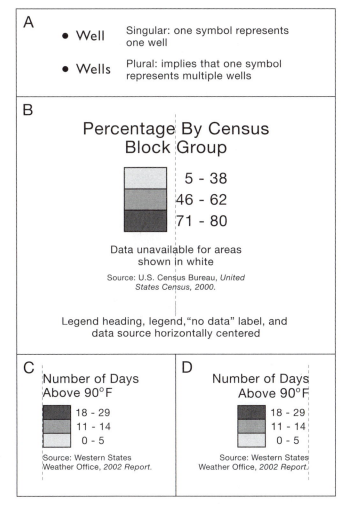

FIGURE 11.10 (A) Singular versus plural definitions. (B) Typical choropleth legend with legend heading and "no data" label. The data source appears at the bottom. (C) Legend heading, legend, and data source horizontally aligned to left. (D) The same objects horizontally aligned to right.

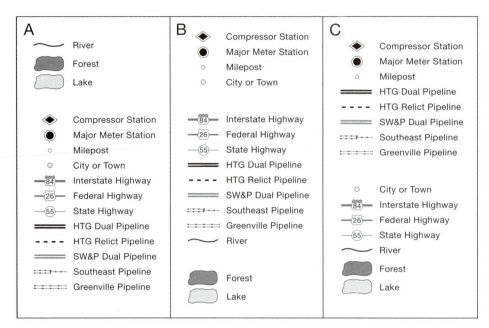

FIGURE 11.11   Grouping of legend symbols according to different criteria: (A) natural and cultural groups; (B) point, line, and polygon groups; and (C) thematic symbol and base information groups.

FIGURE 11.12   The legend is centered within the available space if possible. In this example, the legend has been split into two sections, each centered within its available space. (Compare with color version shown in Color Plate 11.1.)

in Color Plate 11.1. Line and dot patterns should be avoided if color reproduction is feasible.

11.2.6 Data Source

The **data source** allows the map user to determine where the thematic data were obtained. For example, in Figure 11.12, the currency information was acquired from the *National Trade Data Bank*, a publication of the U.S. Department of Commerce. Sources of base information such as roads or administrative boundaries are normally omitted from the data source on thematic maps. The data source should be formatted in a manner similar to that of a standard bibliographic reference, but is often more concise and less formal. Evaluation of the intended audience's needs will dictate whether to include a fully detailed data source or a simplified version. Many map users have no intention of tracking down a data source, but are likely to want a general idea of where the data were obtained. Most data sources in this text are simplified versions.

The style of the data source should be plain and subtle, as illustrated in Figure 11.13. The word "Source:" should preface the data source to avoid ambiguity; the data source indicates where the data came from, not map authorship. If necessary, a separate block of text can be used to indicate map authorship. If the data were obtained from a publication, the name of the publication should be italicized (as in a bibliographic reference); all other type should be normal. If the data were obtained from a preexisting map, special methods of citation

```
Source: United States Central Intelligence Agency. Map
          File #505103 (547149) 2-82.

Source: The International Bank for Reconstruction and
          Development/The World Bank.
       World Development Report, 2000: Poverty.

Source: Field Survey by Nigel Tufnel and
       David St. Hubbins, December, 2003.


     Multiple lines are horizontally centered
```

FIGURE 11.13 Data sources. The data source indicates where thematic data were obtained. It is among the smallest type on a map.

should be employed, based on the type of source map used (Clark et al. 1992). Multiple lines of type in a data source should be horizontally centered with one another (Figure 11.13). The data source should be among the smallest type on a map; its purpose is to inform the curious, not to attract attention. Optimally, the data source is placed directly below the legend to which it refers, and is horizontally centered with the legend heading and legend, as illustrated in Figure 11.10B.

Related to the data source is information regarding the map projection used. The type of projection can become increasingly important to the map user as larger geographic regions are represented and distortion of the mapped area increases. Include a separate block of text indicating the type of projection (and information related to the projection) if you feel it will help the map user to interpret the map's information accurately. Like the data source, blocks of text related to map authorship and map projections should be among the smallest on the map.

11.2.7 Scale

The **scale** either indicates the amount of reduction that has taken place on a given map or allows the map user to measure distances. Calculation of map scale is described in Chapter 6.

As discussed in Chapter 6, the **representative fraction** (e.g., 1:24,000) is a ratio of map distance to Earth distance, which indicates the extent to which a geographic region has been reduced from its actual size. For example, on a map with a scale of 1:24,000, one unit of distance on the map (e.g., 1 mile) represents 24,000 of the same units on Earth, regardless of the unit of measure used. The representative fraction becomes invalid if the map on which it appears is enlarged or reduced, and cannot easily be used to measure distances. For these reasons, the representative fraction does not appear on most thematic maps.

Although not actually spoken (it is a block of text), the **verbal scale** reads like a spoken description of the relationship between map distance and Earth distance. "One inch to the mile" means that 1 inch of distance on the map represents 1 mile on Earth. Like the representative fraction, the verbal scale becomes invalid if the map on which it appears is enlarged or reduced. Distances on a map can be determined with accuracy using the verbal scale, but only in conjunction with a ruler. Rough measurements can be taken by estimating the length of an inch.

The **bar scale**, or scale bar, resembles a ruler that can easily be used to measure distances on a map. Its ability to indicate distance, together with its ability to withstand enlargement and reduction of a map, make it the preferred format for inclusion on a thematic map. You should normally include a bar scale on a thematic map if distance information can enhance the map user's understanding of the theme. The map in Figure 11.12 lacks a bar scale because distance information lends no insight into the type of currency each country uses. However, the map in Figure 11.14 includes bar scales because distance information can potentially enhance the map user's understanding of the theme. For example, general measurements of the state of Colorado taken with the bar scale yield an area of approximately 93,750 square miles. Taking into consideration the minimum number of possible farms in Colorado's category (28,000), it can be estimated that there is at least one farm per every four square miles, on average. This information is directly related to the map's theme and would be difficult to acquire without a bar scale.

Bar scales lose much of their utility when used on maps in which the scale changes greatly from one location to another, such as on a map of the entire world. A bar scale should be employed with caution in such cases, because it will only allow the map user to measure distances accurately along certain lines (e.g., standard parallels or principal meridians). A variable bar scale is useful with certain map projections (e.g., Miller Cylindrical), as it reflects changes in scale in relation to latitude (Figure 11.15A). A basic understanding of map projections and associated scale changes is essential in providing the map user with a bar scale that is actually useful and accurate.

The maximum distance value represented in a bar scale should always be round and easy to work with. Compare the maximum values illustrated in Figure 11.15B with those in Figure 11.15D. A maximum value of 400 miles is preferred to 437 miles (quick, what's one quarter of 437?). Decimal values such as 94.3 are difficult to work with and should be avoided in favor of integers. Some bar scales include an "extension scale," which consists of numbers and tic marks positioned to the left of zero, as

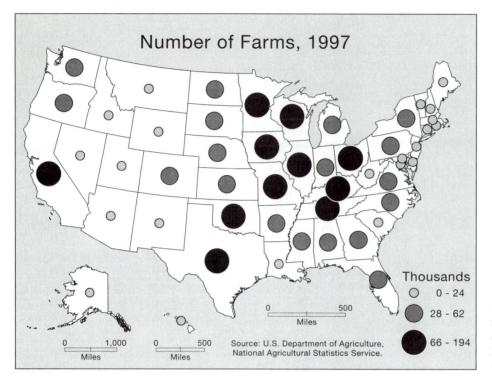

FIGURE 11.14 A bar scale (or bar scales, in this example) should be included if distance information can enhance the map user's understanding of the theme.

illustrated in Figure 11.15C. The extension scale can be useful when employing a particular method of measuring map distances,* but is also a source of confusion for many map users, who expect numbers seen to the left of zero to be negative (as on a number line). Potential confusion, together with the fact that small intermediate distance tics can be included to the right of zero, argue against the use of extension scales on thematic maps.

Like the data source, the bar scale should be subtle; its purpose is to inform the curious, not to attract attention. Bulky, complex, and sloppy designs (Figure 11.15B) should be avoided in favor of slender, simple, and precise designs (Figure 11.15D). Line weights should be fine and type should be among the smallest on a map. Include intermediate distance tics, but only as many as necessary for making general measurements.

The bar scale should be long enough to be useful but not so long as to be cumbersome. In Figure 11.15E, the 50-mile bar scale is too short to be useful, and the 1,000-mile scale is cumbersome—it's actually wider than the mapped area (Iceland). The 500-mile bar scale is a good solution because it is long enough to be useful but doesn't take up too much space. The optimal length is directly tied to the size of the mapped area and to the quest

---

* Distance on a map is transferred as tic marks to a scrap of paper, which is held up to the bar scale. The right tic mark on the scrap is aligned with a value on the bar scale located to the right of zero, so that the left tic mark on the scrap falls within the extension scale. Total distance is calculated by adding distances found to the right and left of zero.

for a maximum distance value that is round and easy to work with. If possible, the bar scale should be placed below the mapped area, where the map user is accustomed to finding it, as illustrated in Figure 11.14.

Notice in Figure 11.14 that both Alaska and Hawaii (the insets) have their own bar scales: If a bar scale is used for the primary mapped area, an individual bar scale is normally included with an inset, unless distances on the inset can be inferred from the primary mapped area and its bar scale.

11.2.8 Orientation

Orientation refers to the indication of direction on a map. Orientation can be indicated with a **north arrow**, or through the inclusion of a **graticule** (a system of grid lines, normally representing longitude and latitude). The north arrow and graticule are normally omitted from thematic maps, unless there is a good reason to include one or the other. One case in which you *would* include an indication of orientation is if the map is not oriented with geographic or "true" north at the top. This can occur when the shape of the mapped area, together with the page dimensions, require the cartographer to rotate the mapped area, as illustrated by California in Figure 11.16A. Map users expect to find north at the top of a map; they should be notified if it isn't. A second case in which you would include an indication of orientation is when the map is intended for use in navigation, surveying, orienteering, or any other function in which the determination of direction is crucial.

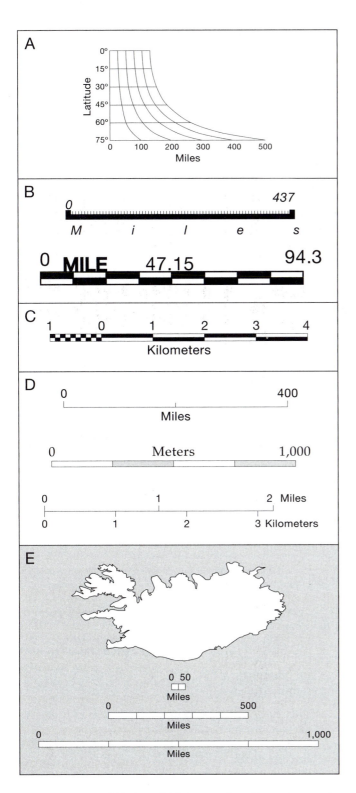

FIGURE 11.15 (A) A variable bar scale reflects changes in scale in relation to latitude. (B) Bulky, complex, poorly designed bar scales. (C) Bar scale incorporating an "extension scale" to the left of zero. (D) Slender, simple, well-designed bar scales. (E) The 500-mile bar scale represents the most appropriate length in this example.

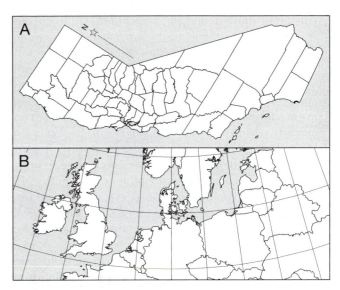

FIGURE 11.16 (A) Use of a north arrow on a map not oriented with north at the top. (B) Meridians of a graticule indicating direction of north.

The graticule indicates direction through the orientation of grid lines, typically meridians that run north–south, as illustrated in Figure 11.16B. Notice that the meridians in this example are not parallel to one another—this indicates that the direction of north is variable, depending on which part of the map one focuses on. Just as caution should be exercised when employing bar scales on maps with variable scale, caution should be exercised when placing north arrows on maps in which the direction of north is variable. For example, it would be inappropriate to include a north arrow on the map in Figure 11.16B, because it would only indicate north for one particular location, not the entire mapped area. In addition to orientation, the graticule can also provide positional information. This can be important when a map's theme is in some way related to latitude or longitude. For example, a graticule, or at least marginal tic marks, is often included on thematic maps of natural vegetation, as this attribute is strongly influenced by latitude. Labels indicating the values of grid lines should be included when employing a graticule in this fashion. For example, a meridian might be labeled "60° E," and a parallel labeled "20° S."

The style of the north arrow and graticule should be simple and subtle. Like the data source and bar scale, these map elements should not attract attention. Line weights should be fine and type should be among the smallest on a map. Bulky and complex north arrows should be avoided (Figure 11.17A), and only north should be indicated (if necessary, the map user can infer the other cardinal directions). Subtle and simple north arrows are illustrated

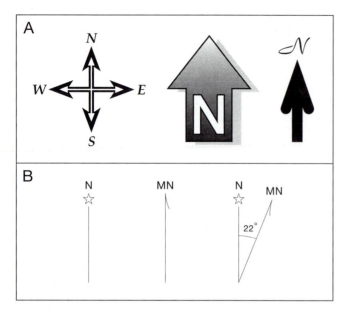

FIGURE 11.17   (A) Bulky and complex north arrows. (B) Subtle and simple north arrows for geographic north (N), magnetic north (MN), and both combined (a compound north arrow).

in Figure 11.17B, where a five-pointed star (representing Polaris, the North Star) indicates geographic north, and an arrowhead (representing a compass needle) represents magnetic north. The arrowhead is not the most logical choice for indicating geographic north because compass needles point to magnetic, not geographic, north. Regardless of this logical inconsistency, the arrowhead is commonly used instead of the five-pointed star to indicate geographic north. Compound north arrows indicate both geographic and magnetic north, and include the declination (in degrees) between the two. Compound north arrows should be reserved for use on maps intended for navigation with a compass.

The north arrow should be relatively small; it should be large enough to find and use, but should not attract attention. The north arrow should be placed in an out-of-the-way location, preferably near the bar scale.

11.3 TYPOGRAPHY

Type, or text, refers to the words that appear on maps. **Typography** is the art or process of specifying, arranging, and designing type. Several of the map elements described in the previous section are partly composed of type—the legend, for example—and others, such as the data source, are composed entirely of type. Type can be considered a special sort of symbol, or even a map element in its own right. Type that is well designed and smartly applied can make a map easier to understand

and more attractive. The rules and guidelines for the use of type in cartography are derived from general rules of typography, but have been modified over time to reflect the specific purposes of mapmaking. Fortunately, these rules and guidelines are relatively well defined.

11.3.1 Characteristics of Type

Type is commonly organized according to characteristics such as type family, type style, and type size. **Type family** refers to a group of type designs that reflect common design characteristics and share a common base name—Palatino, for example (Figure 11.18A). Within a type family, type is differentiated by **type style**: Roman (normal), bold, and italic are common type styles[*] (Figure 11.18B). Additional type styles include condensed, expanded, light, and extra bold. Type of a particular family and style is referred to as a **typeface**, such as Palatino Roman or Helvetica Bold (Figure 11.18C). A particular typeface is further differentiated by **type size** (Figure 11.18D). Type size (height) is measured in points (one point equals 1/72"). Although type sizes are described in points, the actual height of a given character cannot be inferred from its point size; the point size refers to the height of the metal block on which type was created prior to the development of digital type. A **font** is a set of all alphanumeric and special characters of a particular type family, type style, and type size, such as Palatino Italic, 12 point (Romano 1997). Many typographical terms are

---

[*] When applying type styles, it is best to use a member of a type family that has been specifically designed with that style (e.g., Bookman Bold). Many software applications allow roman type to be crudely modified into italic or bold, resulting in type that is unsuitable for high-quality printing.

A Type Family	B Type Style (Based on Palatino)
Palatino Helvetica Bookman Gill Sans	Roman **Bold** *Italic*
C Typeface	D Type Size (Based on Palatino Roman)
Palatino Roman **Helvetica Bold** *Bookman Italic* Gill Sans Condensed	Six point Ten point Fourteen point

FIGURE 11.18   (A) Type family. (B) Type style. (C) Typeface (a particular type family and type style). (D) Type size.

misunderstood and misused. For example, the term *font* is commonly (and inappropriately) used to refer to a type family or typeface.

Uppercase and lowercase letters are used in cartography, but lowercase letters have proven to be easier to read. This is because lowercase letters are less blocky, and they provide more detail that helps differentiate one letter from another. The majority of type on a map should be set in title case, as illustrated in Figure 11.19A. **Title case** is composed of lowercase letters with the first letter of each word set in uppercase. Conjunctions and other "linking words" (in, on, or, of, per, by, for, with, the, and, over, etc.) are set in lowercase. Title case is appropriate for use in titles, subtitles, legend headings, legend definitions, labels for point and line features, and so on. **Sentence case** is composed of lowercase letters with the first letter of each sentence set in uppercase, and is appropriate when formal sentences are used, such as in textual explanations or descriptions appearing on a map. Words set in all uppercase are sometimes used as short titles and as labels for areal features, as described later.

Serifs are short extensions at the ends of major letter strokes, as illustrated in Figure 11.19B. Type families with serifs are termed serifed; type families without are termed sans serif (without serifs). Serifed type is preferred in the context of written documents because the serifs provide a horizontal guideline that helps to tie subsequent letters together, reducing eye fatigue. In cartographic applications, where type is used primarily as short labels and descriptions, both serifed and sans serif type is used; one has not proved to be more effective than the other. In certain situations, a serifed type family can be used for one category of features (e.g., cultural) and a sans serif type family can be used for another category of features (e.g., natural).

Not surprisingly, **letter spacing** refers to the space between each letter in a word, and **word spacing** refers to the space between words. Minimal letter and word spacing (Figure 11.20A) results in compact type that is often easier to place on complex maps (words occupy less space from beginning to end). Slightly increased letter and word spacing results in type that appears less "cramped" and is easier to read. You should employ slightly increased spacing if possible. The blocky nature of all-uppercase type normally requires greater letter and word spacing than lowercase type to prevent it from looking cramped. Exaggerated letter and word spacing is often employed in conjunction with all-uppercase type when labeling areal features, as described later. Letter and word spacing should be kept consistent within individual blocks of text, and among labels that are otherwise similar. For example, each line of type in a three-line data source should have the same letter and word spacing, as should every label that identifies a restaurant on a particular map.

Kerning refers to the variation of space between two adjacent letters, as illustrated in Figure 11.20B. Different combinations of adjacent letters require different amounts of kerning if spaces between letters are to be visually consistent. For example, the letter pair WA requires the removal of space between letters to look consistent with the space between MN. Digital type includes preset "kerning pairs" that automatically set the space between various letter pairs. Kerning can also be performed

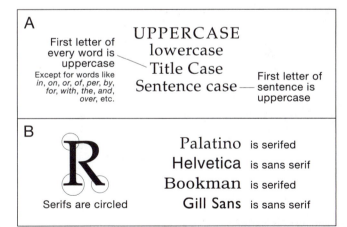

FIGURE 11.19 (A) Uppercase, lowercase, title case, and sentence case. The majority of type on a map should be set in title case. (B) Serifs are extensions at the ends of letter strokes.

FIGURE 11.20 (A) Letter spacing and word spacing can be altered for different situations. (B) Kerning adjusts the space between two individual, adjacent letters. (C) Leading is the space between lines of type, from baseline to baseline.

manually, after letter and word spacing has been specified, to adjust the space between particular letter pairs that still appear to be too close together or too far apart. Kerning is measured according to a unit called an *em*, which is equal to the point size of the type being used.

Leading (pronounced like *heading*), or line spacing, refers to the vertical space between lines of type, according to their *baselines*, and is altered to place lines of type closer together or further apart (Figure 11.20C). Leading should be great enough to allow multiple lines of type to be read easily, but not so great as to result in wasted space between lines. (An exception to this guideline is presented in section 11.3.6.)

11.3.2 General Typographic Guidelines

The following is a list of general guidelines for the use of type in cartography.

1. Avoid the use of decorative type families, and use bold and italic styles sparingly. Script, cursive, and otherwise fancy and ornate styles are unnecessarily difficult to read. They should be avoided in favor of more practical type families, such as those used as examples in this chapter. The overuse of bold styles can overshadow other map elements, and is normally not required if appropriate type sizes are chosen. If possible, italic type should be reserved for two applications: to label hydrographic (water) features, and to identify publications in the data source. Italics are appropriate for hydrographic features because their slanted form resembles the flow of water. (It is also conventional to use the color cyan for hydrographic labels and features.) The use of italics for publications is standard bibliographic practice. The wide variety of features on general reference maps might require that you use bold and italic styles outside of these guidelines.

2. Avoid using more than two type families on a given map. Simpler maps can, and should, be limited to one type family. Variations in style and size within a family normally offer sufficient variety for all but the most complex maps. For the sake of consistency, map elements such as the title, subtitle, legend heading, legend definitions, data source, and scale should all employ the same typeface. If two type families are required (e.g., to label a wide variety of map features), choose families that are distinctly different—one serifed and one sans serif, for example. As mentioned earlier, a serifed family can be used for one category of features (e.g., cultural) and a sans serif family can be used for another category of features (e.g., natural).

3. Choose a realistic lower limit for type size; all type needs to be readable by the intended audience. Factors for consideration include the age and visual acuity of the map user, map reproduction method, anticipated lighting conditions, and the map user's physical proximity to the map. Mark Monmonier (1993) recommended a lower size limit of 7 points for lowercase type, but this value is conservative. Type as small as 3.2 points (usually uppercase, sans serif) is commonly used on congested street maps where space is at a premium (Jonathan Lawton, California State Automobile Association, personal communication, 2002). Readability is ultimately tied to the typeface used, crispness of reproduction, and other factors. The only way to ensure the readability of small type is to provide a sample to members of the intended audience.

4. Generally speaking, type size should correspond with the size or importance of map features. For example, type representing the names of large cities should be noticeably larger than type used to represent small cities. Type size is also partially dictated by the relative importance of map elements, as described in sections 11.2 and 11.4. Because map users are not sensitive to slight differences in type size, avoid differences of less than two points if possible (Shortridge 1979).

5. Critically evaluate and apply type specifications such as type family, type style, type size, letter spacing, word spacing, kerning, and leading. Do not passively accept the default settings provided by software applications. Instead, consider the purpose of each block of type in the context of the map, and apply type specifications accordingly.

6. All type should be spell-checked. Special attention should be focused on the most current spelling of place names, which change over time and are often controversial. Also, be aware that certain older place names are considered to be offensive or derogatory by today's standards.

11.3.3 Specific Guidelines: All Features

The following is a list of specific guidelines for the placement of type associated with point, linear, and areal features.

1. Orient type horizontally, as illustrated in Figure 11.21A. One exception is when labeling a map that includes a graticule with curved parallels, in which case the type should be oriented with the parallels (Figure 11.21B). Another exception is when labeling diagonal or curved linear and areal features, in which case the type should reflect the orientation of the features, as illustrated in Figure 11.21C.

2. Avoid overprinting and, when unavoidable, minimize its effects. **Overprinting** is a phenomenon that occurs when a block of type is placed on top of another

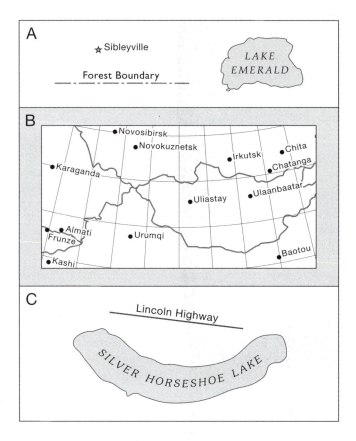

FIGURE 11.21   (A) Type placed horizontally. (B) Type oriented with a graticule. (C) Diagonal and curved type.

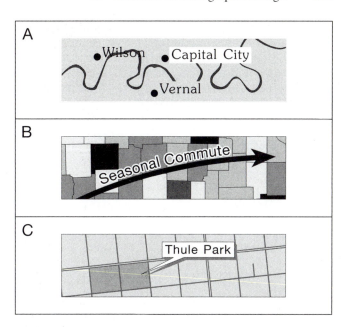

FIGURE 11.22   (A) Masks placed underneath type. (B) A halo around type. (C) A callout combines a mask with a leader line.

graphic object (e.g., a river), obscuring the type and making it difficult to read, as illustrated by "Wilson" in Figure 11.22A. The effects of overprinting can be minimized through the use of either a mask, halo, or callout. A **mask** is a polygon (e.g., a white rectangle) that is placed underneath type, but above the underlying graphics, as illustrated by "Capital City" in Figure 11.22A. As seen in this example, masks can sometimes obscure too much of the underlying graphics, and should be used with caution. Masks can also be specified with the same color as the background area, allowing them to blend in better, as illustrated by "Vernal" in Figure 11.22A. A **halo** is an extended outline of letters in a type label (Figure 11.22B). Haloes cover less of the underlying graphics than masks, while still allowing the type to be read. **Callouts** are a combination of mask and leader line (Figure 11.22C). Callouts are effective, but should be used with caution because they are visually dominant and can overshadow other map elements.

3. Ensure that all type labels are placed so that they are clearly associated with the features they represent. In pursuit of this goal, it is often useful to place larger

type labels first, followed by intermediate and then smaller labels (Imhof 1975).

11.3.4 Specific Guidelines: Point Features

The following is a list of specific guidelines for the placement of type associated with point features.

1. When labeling point features, select positions that avoid the overprinting of underlying graphics according to the sequence of preferred locations illustrated in Figure 11.23A. This sequence is based on the work of Pinhas Yoeli (1972), but is modified by the authors according to the idea that, if possible, the symbol should be placed on the left and defined to the right (as in a legend). Notice, also, that the least preferred locations for a label are directly to the right and left of the symbol. This results in an "unfavorable optical coincidence" (Imhof 1975, 132), in which the point symbol might be misinterpreted as a type character in the label.

2. Do not allow other map features to come between a point symbol and its label (Figure 11.23B). Emphasize the association between the label and symbol by placing the label close to the symbol, even if it means choosing a less preferred location.

3. If the sequence of preferred locations does not provide a suitable option, consider using a mask, halo, or callout. Another option is to use a simple leader line as illustrated in Figure 11.23C. Leader lines should be

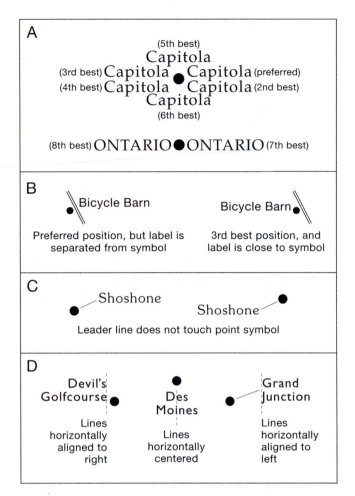

FIGURE 11.23 (A) Sequence of preferred locations for labeling point features. (B) A road coming between a point symbol and its label. (C) Leader lines are used if necessary. (D) Multiple-line labels are horizontally aligned or centered to imply association with the symbol.

very thin (e.g., 0.25 point), not include an arrowhead, and point to the center of the point symbol without actually touching it.

4. Multiple-line labels should be placed according to the sequence of preferred locations, and individual lines of type should be horizontally aligned or centered to emphasize the association between the label and point symbol (Figure 11.23D).

5. Point symbols on land that are close to coastlines should be labeled entirely on land if possible. Point symbols that touch coastlines should be labeled either entirely on land or entirely on water, depending on which option offers greater legibility (Wood 2000). Avoid overprinting the coastline with type (Figure 11.24).

6. These guidelines should be followed as closely as possible. In practice, however, it is often impossible to ad-

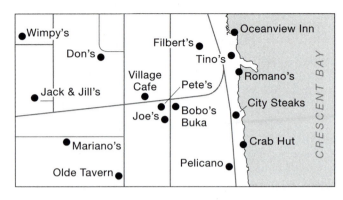

FIGURE 11.24 Guidelines for labeling point features applied simultaneously.

here to all guidelines simultaneously. Figure 11.24 illustrates how these guidelines might be applied on a map of restaurant locations in a coastal region.

7. Do not exaggerate letter or word spacing for point features. Exaggerated spacing weakens the association between a point symbol and its label, and tends to emphasize the *areal* extent of features (point features have no areal extent).

11.3.5 Specific Guidelines: Linear Features

The following is a list of specific guidelines for the placement of type associated with linear features.

1. When labeling linear features, place type above the features, close to, but not touching them, as illustrated in Figure 11.25A. Descenders, such as the

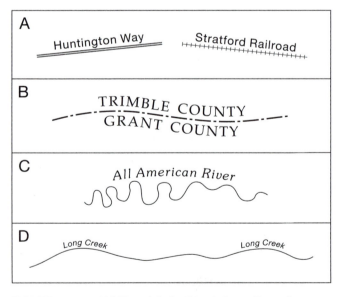

FIGURE 11.25 (A) Type labels placed above linear features. (B) Labels on both sides of a boundary. (C) Label following the general trend of a complex curve. (D) Long feature labeled twice.

lower extensions of "g" and "y" in "Huntington Way" should just clear the line symbol. Type is placed above linear features because it appears to rest on the feature instead of hanging below it, and because the bottom edge of lowercase type is normally less ragged than the upper edge, resulting in a more harmonious relationship between label and symbol. One exception is when labeling areas on both sides of a boundary, in which case type appears above *and* below the line, centered with one another (Figure 11.25B).

2. When labeling linear features that have complex curves, follow the *general* trend of the feature (Figure 11.25C), as type that curves too much is difficult to read.

3. Very long linear features can be labeled more than once (Figure 11.25D). The use of multiple labels is preferred to the exaggeration of letter and word spacing to emphasize linear extent.

4. Labels for linear features should be placed upright, not upside down. Correctly placed labels read from left to right, whereas incorrectly placed labels read from right to left, as illustrated in Figure 11.26. By convention, type that is absolutely vertical should be readable from the right side of the page. These rules actually apply to all type that appears on a map, but are most closely associated with linear features.

11.3.6 Specific Guidelines: Areal Features

The following is a list of specific guidelines for the placement of type associated with areal features.

1. When labeling areal features that are large enough to fully contain a label, visually center the label within the feature, as illustrated in Figure 11.27A. Don't allow the label to crowd the areal symbol—allow a space of at least one and one half times the type size (1.5 ems) between the ends of the label and the boundary of the feature (Imhof 1975). As when labeling linear features, follow the *general* trend of areal features that have complex curves.

2. Consider using all-uppercase type when labeling areal features (Figure 11.27A). The blocky nature of uppercase type can help to emphasize areal extent.

3. Exaggerated letter and word spacing can be used to emphasize areal extent, and is most effective when applied to all-uppercase type (Figure 11.27B); lowercase type tends to look disjointed when exaggerated spacing is applied (Figure 11.27C). Caution should be exercised when exaggerating letter and word spacing, as individual letters can become so far apart that the map user would have trouble seeing the label as anything other than individual letters. An extreme upper limit of four times the type size (4 ems) should be observed for letter spacing. Maximum word spacing can exceed

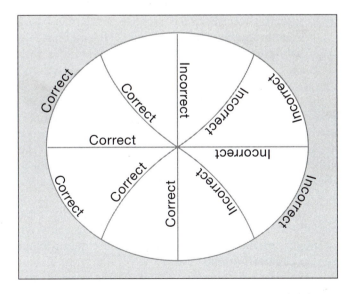

FIGURE 11.26 Type should be placed upright, and should read from left to right. Vertical type should be readable from the right side of a page.

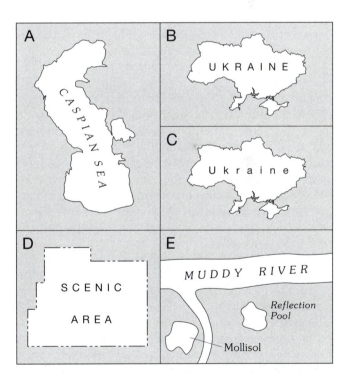

FIGURE 11.27 (A) A type label visually centered within an areal feature. (B) Uppercase type with exaggerated letter and word spacing, emphasizing the areal extent of the feature. (C) Lowercase type is not as well suited to exaggerated spacing. (D) Exaggerated leading. (E) A river labeled as an areal feature, a small areal feature (reflection pool) labeled as a point feature, and a leader line used to help identify an areal feature.

this value, but care should be exercised to ensure that the relationship between words is clear.

4. Leading can also be exaggerated to emphasize areal extent, as illustrated in Figure 11.27D. Leading should not be so great that the relationship between lines of type is lost.

5. Features typically thought of as being linear (e.g., rivers) that are represented at such a large scale that they appear as areas should be labeled as areal features, as illustrated in Figure 11.27E.

6. Areal features that are too small to contain a label should be labeled as if they were point symbols, as illustrated by the reflection pool in Figure 11.27E.

7. If necessary, leader lines can be used with areal features. Leader lines should be very thin (e.g., 0.25 point), not include an arrowhead, and just enter the areal symbol, as illustrated by the Mollisol in Figure 11.27E.

11.3.7 Automated Type Placement

Throughout this section, we have assumed that the cartographer is making all decisions regarding type placement. When using general-purpose graphic design software (e.g., FreeHand or Illustrator), this is almost certainly the case. The reader must bear in mind, however, that specialized **labeling software** has been developed for automatically positioning type, often within the context of a GIS. Labeling software focuses primarily on the placement of type associated with map features (e.g., streams) as opposed to positioning textual map elements, such as the title. Development of this software has been the focus of both computer scientists and cartographers, who have created sophisticated algorithms (including heuristics, or "rules of thumb") based on established rules and guidelines of cartographic type placement. Most labeling software applications incorporate aspects of cartographic **expert systems**, which make decisions based on rules and guidelines obtained from cartographic experts (Zoraster 1991).

Labeling software is designed to approach or achieve "optimal" placement of type labels that avoids both the overprinting of underlying map elements and conflicts among type labels. Two general approaches have emerged: (1) placement of each label in its preferred position, followed by an iterative reorganization of labels to avoid or minimize conflict among labels (Freeman 1995); and (2) casual, sub-optimal placement of labels, followed by an iterative reorganization of labels until the combined placement of all labels approaches an optimal state (Edmondson et al. 1996; Pinto and Freeman 1996).

The biggest potential advantage of labeling software is its ability to save time, as manual type placement remains one of the most time-consuming aspects of map construction, particularly when labeling linear and areal features (Barrault 2001). One problem associated with labeling software is the fact that optimal solutions are often computer-intensive when a considerable amount of type must be positioned. Another problem is that the wide variety of maps, together with variations in map scale and complexity, make it difficult to achieve satisfactory results in all situations. The finished product normally requires some interactive editing to arrive at a solution that is visually acceptable. As more sophisticated algorithms are developed and the speed of computers increases, it appears inevitable that type placement will become a more fully automated process.

11.4 CARTOGRAPHIC DESIGN

Cartographic design is a partly mental, partly physical process in which maps are conceived and created. The design process appears formally as Step 4 in the map communication model that was introduced in Chapter 1 (Figure 11.28), but it encompasses aspects of all five steps, from imagining the real-world distribution to evaluating the resulting map. The design process does not end until the final map has been completed. The word *design* also describes the appearance of a map; a map can have a particular design, but design in this sense is only the end result of the design *process*. Successful cartographic design results in maps that effectively communicate geographic information. Current theories suggest that maps do not necessarily communicate *knowledge*, but rather, stimulate and suggest it through the transmission of information (Montello 2002).

Cartographic design involves the conceptualization and visualization of the map to be created, and is driven by two goals: (1) to serve the purpose of the map based on its intended audience and use, and (2) to communicate the map's information in the most efficient manner, with simplicity and clarity. Edward Tufte (1990, 53) echoed this second goal, eloquently stating that "Confusion and clutter are failures of design, not attributes of information." The physical act of placing, modifying, and arranging map elements is often referred to as the separate activity of map construction, or "layout." Because of the holistic nature of the design process, map construction is considered here to be largely integrated with the cartographic design process.

Cartographic design is directed in large part by rules, guidelines, and conventions, but is relatively unstructured. A single, optimal solution to a given mapping problem generally does not exist; rather, several acceptable solutions are usually possible. "Good design is simply the best solution among many, given a set of constraints imposed by the problem" (Dent 1999, 241). If map design were easy and straightforward, this chapter would be

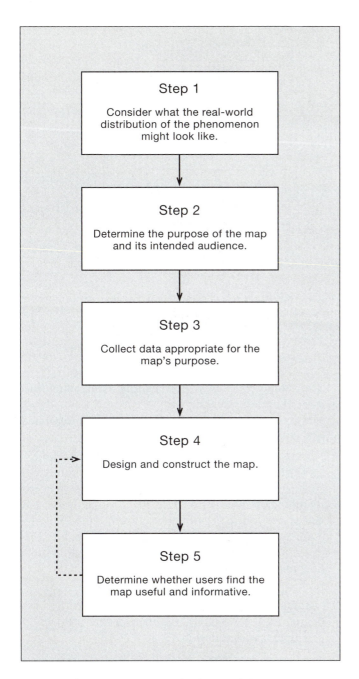

Step 1

Consider what the real-world distribution of the phenomenon might look like.

Step 2

Determine the purpose of the map and its intended audience.

Step 3

Collect data appropriate for the map's purpose.

Step 4

Design and construct the map.

Step 5

Determine whether users find the map useful and informative.

FIGURE 11.28 Map communication model.

unnecessary, as cartographic design expert systems would be used to design most maps. Cartographic expert systems, however, are currently limited to narrow aspects of cartographic design, and lack the ability to completely and consistently design top-quality maps. As was the case with map elements, we believe it is important to build a foundation of cartographic design skills by following the specific rules and guidelines presented here, and then to consider alternative approaches. Always be prepared to explain or defend your design decisions.

Many aspects of cartographic design have been guided by the results of **map design research**. Arthur Robinson (1952) sparked enthusiasm for this research with *The Look of Maps*, in which he emphasized the importance of a map's function over its form and called for objective experimentation with regard to map design. Much of this research has focused solely on determining which mapping techniques are most effective, without regard for why they are effective (a behaviorist view). In contrast, *cognitive map design research* focuses on why certain techniques are effective by applying knowledge structures to the ways that people perceive maps. This research has been driven by the idea that understanding cognitive processes, and applying their principles to map design, can result in more effective maps (Montello 2002).

Although most map design research represents a scientific approach to understanding how maps work, the "art" of maps also plays an important role in cartographic communication. The artistic aspect of maps is guided less by experiment and more by intuition and critical examination (MacEachren 1995). It is difficult to anticipate the map user's sensitivity to the artistic aspects of a map. However, it seems likely that a map that has been created with an artistic synthesis of contrast, balance, color, and so on has a greater chance of communicating information than a map that has been created in the absence of an artistic sensibility.

The link between cartographic design and graphic design is strong. *Graphic design* has been described as "problem solving on a flat two-dimensional surface ... to communicate a specific message" (Arntson 2003, 2). Both cartographic design and graphic design emphasize the communication of information through graphical means—the primary difference being that graphic design is mainly oriented toward advertisements and packaging, as opposed to maps. Both incorporate Gestalt principles of perceptual organization. **Gestalt principles** attempt to describe the manner in which humans see the individual components of a graphical image and organize them into an integrated whole. Principles such as similarity, proximity, continuation, and closure are not directly addressed in this chapter but represent the theoretical underpinning for many of the design guidelines presented (MacEachren 1995). The Gestalt principle of figure-ground is described in section 11.4.4.

Many examples of well-designed maps appear in this chapter. It is important for you to build a mental inventory of designs and design possibilities from nonmap sources as well. Borden Dent (1999) referred to this mental inventory as an *image pool*, which can be built by critically viewing as much art, graphic design, and—of course—maps as possible. Use the maps in every chapter of this textbook to help build your image pool, as the vast majority have been well designed.

11.4.1 The Design Process

The design process can be distilled into the following list of general procedures. This list is an expanded version of Step 4 of the map communication model (Figure 11.28); it assumes that Steps 1 through 3 have been completed. Be aware that the procedures in this list are iterative, and need to be repeated until the map has been completed. They will sometimes need to be executed simultaneously, or out of the prescribed order.

1. Determine how the map will be reproduced. Reproduction considerations, such as the printing method to be used, will impact almost every aspect of the design process, and need to be resolved first. See Chapter 12 for a discussion of map reproduction.

2. Select a scale and map projection that is appropriate for the map's theme. See Chapter 6 for a discussion of scale and generalization. See Chapter 9 for a discussion on choosing an appropriate projection.

3. Determine the most appropriate methods for data symbolization and classification. See Chapter 4 for an introduction to symbolization and Chapter 5 for a discussion of data classification. See Chapters 13 through 19 for discussions of specific mapping techniques.

4. Select which map elements to employ, and decide how each will be implemented. See section 11.2 for a discussion of map elements. You must also decide how to implement type, as discussed in section 11.3.

5. Establish a ranking of symbols and map elements according to their relative importance. This ranking is referred to as an **intellectual hierarchy**, or "scale of concepts" (Monmonier 1993), and usually takes the form of a list. The intellectual hierarchy often varies depending on the type of map and its purpose, but the following is a general hierarchy (from most to least important):

   - Thematic symbols and type labels that are directly related to the theme
   - Title, subtitle, and legend
   - Base information such as boundaries, roads, place names, and so on
   - Data source and notes
   - Scale, frame and neat lines, north arrow, and so on

6. Create one or more sketch maps. A **sketch map**, also called a thumbnail sketch, is a rough, generalized hand drawing that represents your developing idea of what the final map will look like, as illustrated in Figure 11.29. The sketch map should include all selected map elements, and should reflect the intellectual hierarchy established in the previous step. Methods for graphically emphasizing and deemphasizing map elements are described in sections 11.4.3 and 11.4.4.

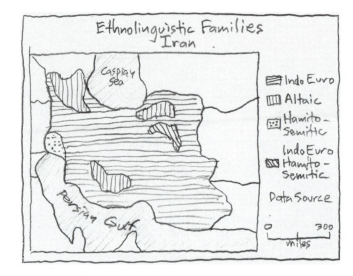

FIGURE 11.29   A rough sketch map. The map resulting from this sketch is illustrated in Color Plate 11.2.

You should experiment with various arrangements of map elements, striving for balance between them. Methods for establishing good balance are discussed in section 11.4.5.

7. Construct the map in your chosen software application. Place, modify, and arrange map elements according to your sketch map, and in the order that was recommended in section 11.2. Print **rough drafts** that will allow you to reevaluate and refine the evolving map. Color Plate 11.2 illustrates the end result of this procedure, based on the sketch map of ethnolinguistic families (Figure 11.29).

8. If possible, allow members of the intended audience to evaluate the map's effectiveness (Step 5 in the map communication model), and incorporate useful suggestions into your design.

11.4.2 Visual Hierarchy

Visual hierarchy refers to the graphical representation of the intellectual hierarchy described earlier in which symbols and map elements were ranked according to their relative importance. When implementing the visual hierarchy, thematic symbols are graphically emphasized, and base information is deemphasized. Similarly, more important map elements such as the title and legend are graphically emphasized, and less important elements such as the data source and bar scale are deemphasized. An effective visual hierarchy attracts the map user's eyes to the most important map elements first, and to the less important elements later.

Visual hierarchy is implemented by applying *contrast* to map features, as described in section 11.4.3. The **visual weight** of map features refers to the relative amount of at-tention that they attract, and can be manipulated to emphasize or deemphasize features. The map in Figure 11.30A reflects an inverted (incorrect) visual hierarchy based on

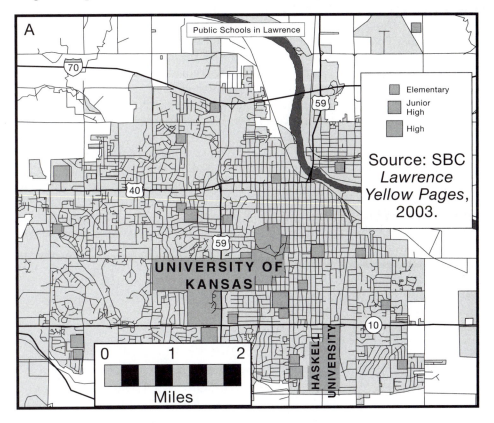

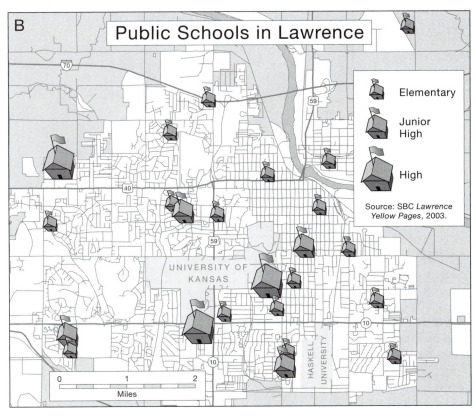

FIGURE 11.30 (A) An inverted (in-correct) visual hierarchy. (B) A correct-ly applied visual hierarchy, appropriately reflecting the intellectual hierarchy.

the general intellectual hierarchy listed earlier. Thematic symbols (symbols representing elementary, junior high, and high schools) have been deemphasized in favor of base information such as roads. More important map elements such as the title and legend have also been deemphasized, whereas less important map elements have been given far too much visual weight. Try looking at this map while squinting, and identify the map features that stand out. The bar scale, data source, universities, river, and roads should certainly *not* be the most noticeable features on a map focusing on public elementary, junior high, and high schools. The map in Figure 11.30B has a visual hierarchy that appropriately reflects the intellectual hierarchy listed earlier. Thematic symbols are visually dominant, as are the title and legend; they are dominant because of their greater visual weight, achieved through the manipulation of contrast (described next). Base information (universities, river, roads, etc.) is subdued, as is the data source and bar scale. An effective visual hierarchy results in maps that clearly reflect the relative importance of symbols and map elements. A map with an appropriate visual hierarchy is easier to interpret, and is more attractive.

11.4.3 Contrast

Contrast refers to visual differences between map features that allow us to distinguish one from another. Contrast adds interest to a map by providing graphical variety; it can be used simply to differentiate features, or to imply their relative importance. Several techniques can be used to create contrast, including manipulation of the visual variables presented in Chapter 4: spacing, size, perspective height, orientation, shape, arrangement, and all aspects of color.

The map in Figure 11.31A lacks appropriate contrast in four respects: type size, lightness and size of thematic symbols (circles), size of lines (line width), and difference between the mapped area and the background. Type on this map is insufficiently differentiated by increments of 0.5 point. Remember that in section 11.3.2, a minimum difference of 2 points was recommended. The title (10 pt.), legend heading (9.5 pt.), legend definitions (9 pt.), data source (8.5 pt.), and bar scale type (8 pt.) are almost impossible to differentiate by size. This lack of contrast contributes to a monotonous

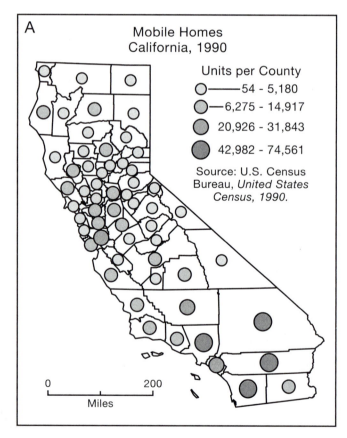

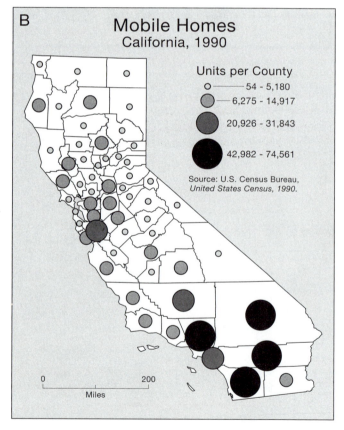

FIGURE 11.31   (A) Insufficient contrast in type size, lightness and size of thematic symbols (circles), line width, and difference between the mapped area and the background. (B) Sufficient contrast in all respects.

design in which the relative importance of map elements is unclear at best, and misleading at worst. The data source in this map appears almost as large as the title, but is far less important. A lack of contrast between the lightness of the circles, together with a lack of contrast in circle size, contributes to a dull design in which it is difficult to differentiate between classes. (See Chapter 13 for a discussion of contrast in gray tones, and Chapter 16 for a discussion of contrast in the size of point symbols.) The lines in Figure 11.31A also lack contrast; the frame line, circle outlines, county boundaries, bar scale, and leader lines in the legend are all one point wide. Again, this lack of contrast contributes to a monotonous design in which the relative importance of map features is unclear. Bar scales and leader lines are less important than thematic symbols or base information (county boundaries), and should appear so. Finally, the map in Figure 11.31A lacks sufficient contrast between the mapped area and the background; the mapped area appears to blend in with the background.

The map in Figure 11.31B exhibits sufficient contrast in each of the four respects just described. It is easier to interpret, more visually stimulating, and more attractive. Notice the significant difference the gray background makes—it enhances the mapped area, and makes it appear to be more important than the background. This technique creates a special type of contrast called figure-ground, which deserves special attention.

11.4.4 Figure-Ground

Figure-ground refers to methods of accentuating one object over another, based on the perception that one object stands in front of another and appears to be closer to the map user. Map design research has failed to produce guidelines for figure-ground that will assuredly work in every situation (MacEachren and Mistrick 1992), but we have found the following guidelines work well in most cases. When accentuating points or lines, the figure-ground relationship is established by making the points or lines darker than their surroundings. In Figure 11.32A, a lack of contrast results in a situation where no features advance easily as figures, and none recede as ground. In Figure 11.32B, the base information has been lightened, allowing the points and trails to emerge as figures. **Screening** is a term that describes the lightening of graphics to reduce their visual weight, as illustrated by the administrative boundaries in Figure 11.32B, which have been reduced from black to gray (20 percent black).

When accentuating an area, the figure-ground relationship is established by making the area lighter than its surroundings, as illustrated by the study area in Figure 11.32C. The figure-ground relationship accentuating Western Europe in Figure 11.33A is similar, but is actually a special case—one in which land is the figure, and water is the ground: the classic *land–water contrast* problem. Making an area lighter than its surroundings effectively emphasizes the area, but it is not always appropriate. For example, in situations when the mapped area is dense with areal thematic symbols, and when color is limited to shades of gray (Figure 11.33B), it would be inappropriate to apply a shade of gray to the surrounding area because the areal thematic symbols provide enough contrast between the mapped area and its surroundings, and the application of gray to the surrounding area might cause it to be confused with the thematic symbols (which are also gray). Notice in Figure 11.33C that there are actually three levels, or layers, of information represented. The circles act as figures in relation to the countries (which act as ground)—the circles are dominant, and appear closer to the map user. The countries, although subordinate to the circles, act as

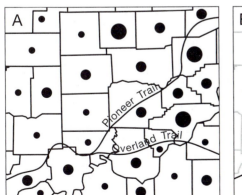 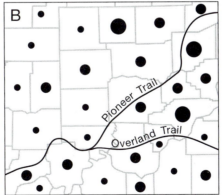 

FIGURE 11.32   (A) Ambiguous figure-ground relationship. (B) Darker points and trails emerge as figures; lighter administrative boundaries recede as ground. (C) Lighter study area emerges as figure.

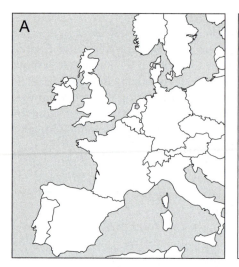
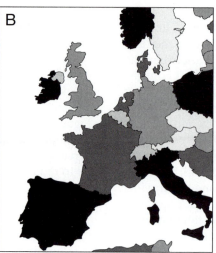
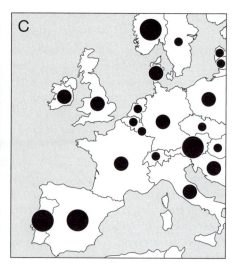

FIGURE 11.33 (A) Land–water contrast as a special case of the figure-ground relationship. (B) Situation in which the application of gray to the surrounding area (water) would be inappropriate. (C) Three levels of information established through figure-ground relationships.

figures in relation to the water (which acts as ground)—the countries appear to be further away than the circles, yet closer than the water.

Figure 11.34 represents two alternative methods of establishing a figure-ground relationship that accentuates areas. In Figure 11.34A, the mapped area is established as the figure (although less strongly than in previous examples) because it appears to stand in front of the graticule, which acts as ground. Stylized effects such as the vignette (boundary shading) surrounding the mapped area in Figure 11.34B can enhance the figure-ground relationship, but should be used with caution, as they can become visually dominant.

11.4.5 Balance

Balance refers to the organization of map elements and empty space, resulting in visual harmony and equilibrium.

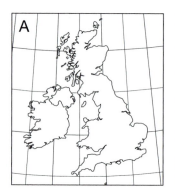

FIGURE 11.34 Figure-ground established using (A) a graticule and (B) a vignette.

The map elements in a well-designed map tend to complement one another, whereas those in a poorly designed map appear to compete for space, resulting in visual disharmony. Before attempting to achieve balance, you need to identify the initial **available space**—the area the map will occupy—as defined by the frame line, and illustrated in Figure 11.35A.

Once the initial available space is defined, the placement of larger map elements such as the mapped area and inset can be considered (the inset is excluded from this example). The mapped area should be as large as possible within the available space, while leaving ample room for the remaining map elements. The mapped area should also be visually centered within the available space if possible. The concept of visually centering objects has been discussed, but warrants repetition: Visually centering objects is accomplished through visual approximation, not precise measurement; what matters most is that an object *looks* centered within the available space, as Africa does in Figure 11.35B. With the mapped area placed, the title, which is intermediate in size, can be placed at top center (Figure 11.35B). Notice that the title appears to be visually centered (vertically) within the space above the mapped area, as well as being horizontally centered within the frame line.

After the mapped area and title have been placed, the evolving map needs to be reevaluated for available space (Figure 11.35B); locations must be identified for the remaining map elements, based on guidelines for each map element as presented in section 11.2. Like the title, the legend is normally intermediate in size—larger than the data source, bar scale, and north arrow—and should be placed next. Visually center the legend within a larger

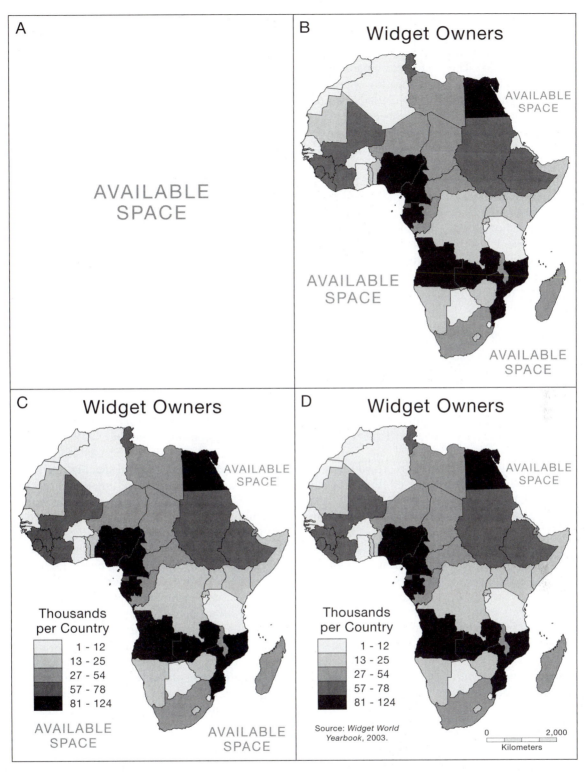

FIGURE 11.35 Establishment of balance by visually centering map elements within available spaces. (A) The frame line defines the initial available space. (B) The mapped area and title are placed. (C) The legend is placed. (D) The data source and bar scale are placed.

area of available space (Figure 11.35C), reevaluate the remaining available space, and then place the smaller map elements such as the data source and bar scale, as illustrated in Figure 11.35D. These smaller map elements should also be visually centered within the available spaces. Notice in Figure 11.35D that available space remains, even after all map elements have been placed. It is not necessary to fill all available spaces, but an effort should be made to occupy the larger areas with well-placed map elements.

The addition of each map element alters the preexisting balance and available space. Map elements will probably need to be rearranged several times to achieve good balance. Useful questions for the cartographer to ask are whether a map looks left-heavy, right-heavy, top-heavy, bottom-heavy, or whether certain areas appear cramped or barren. Top-heavy designs are of particular concern because they tend to make people uncomfortable; humans are intrinsically aware of gravity, and tend to feel more comfortable with objects that are closer to the ground, or the bottom of a page (Arntson 2003). If the answer to any of the previous questions is yes, then the map elements should be rearranged with the goal of visually centering them within the available spaces. Consideration should also be given to the different visual weights of map elements. Don't place too many "heavy" objects in the same area, but rather, try to balance heavier objects with lighter ones.

Instead of allowing map elements to crowd one another, try to use them to balance one another. For example, the bar scale in Color Plate 11.3A crowds the legend, competing for its space, whereas the bar scale in Color Plate 11.3B (along with the data source) helps to counterbalance the legend. The map in Color Plate 11.3A is poorly balanced in many respects, but most of the problems are rooted in the fact that the map elements were not visually centered within the available spaces. Color Plate 11.3B represents a well-balanced design in which all map elements exist in harmony and equilibrium. Map elements were placed according to the same sequence described for Figure 11.35. Certain individuals are intrinsically better than others at judging balance, but experience can improve one's skills.

SUMMARY

In this chapter, we have introduced **map elements** as the building blocks of cartographic communication, and have described their appropriate selection and usage in response to needs of the **map user**. The **frame line** and **neat line** should be subtle, and used to define a map's extent. **Thematic symbols** directly represent a map's theme, and should stand out. They are often used in conjunction with

base information in the **mapped area**. The **inset** is a smaller map used in the context of a larger map, and can serve several purposes. The topic of a thematic map is clearly expressed with a prominent, legible **title**, and further explained in the **subtitle**. Map symbols that are not self-explanatory are defined in the **legend** of a thematic map; representative symbols and their definitions are arranged in various ways, according to the type of map and other factors. The map user is told where thematic data were obtained through the use of a subtly designed **data source** that is similar to a bibliographic reference but often simpler. The **bar scale** was identified as the most practical tool to allow the map user to take general measurements from a thematic map. A subtle, easy-to-use bar scale should be used if it will help illuminate a map's theme. A subtle **north arrow** or **graticule** also can be employed if the map is not oriented with north at the top, or if the map user will require directional information.

We have also presented aspects of **typography**, and have identified it as being central to the utility and attractiveness of a map. The cartographer needs to ensure that type is legible, and select appropriately from **type families**, **type styles**, and **type sizes**. Lowercase type is more legible than uppercase, and type set in **title case** (a combination of the two) is appropriate for most cartographic applications. **Serifed** and **sans serif** type are both used in cartography, as neither has proven to be more effective. Blocks of type can occupy more or less space by altering **letter spacing**, **word spacing**, and **leading**; spaces between letter pairs can be fine-tuned through **kerning**. General guidelines for the use of type were described, including the need to avoid decorative type styles and minimize the use of bold and italic; the need to limit a map to two type families; the selection of a minimum type size that will be readable by the intended audience; the sizing of type to correspond with the relative importance of map features; the need to critically evaluate and specify all aspects of type; and the importance of spell-checking type that appears on a map. Specific guidelines have been provided for the use of type associated with point, linear, and areal features, and an overview of **automated type placement** was given.

We learned that **cartographic design** is a holistic process, both mental and physical, that results in the creation of clear, efficient maps for the communication of geographic information. Cartographic design is strongly guided by the needs of the map user and the purpose of the map. It is guided by rules, guidelines, and conventions, but there is often no optimal result of the design process; the cartographer decides on the best solution from among many. Cartographic design has been guided by the results of **map design research**, and is strongly influenced by graphic design. The design process was distilled into eight general procedures, which were incorporated into the map communication model introduced in Chapter 1: (1) determine how the map will be reproduced; (2) select an appropriate scale and projection; (3) decide on symbolization and classification methods; (4) select appropriate map elements and determine how to implement them; (5) establish an intellectual hierarchy that represents the relative importance of symbols and map elements; (6) create rough sketch maps that represent the evolving map design; (7) construct the map in your chosen software application based on your **sketch maps**, and print **rough drafts** that will allow you to reevaluate the evolving map; and (8) allow members of the intended audience to evaluate the map's effectiveness, and integrate useful suggestions into your design.

The **visual hierarchy** was described as being the graphical representation of the intellectual hierarchy, achieved through the manipulation of contrast. Contrast refers to visual differences between map features that allow us to tell one from another, and sometimes allow certain map features to appear more or less important. We learned that **figure-ground** is a special application of contrast that allows certain map features to be emphasized by appearing closer to the map user. Finally, **balance** was described as the organization of map elements and empty space, resulting in a map that appears to be in a state of visual harmony and equilibrium. Good balance can be obtained, in part, through the establishment, identification, and use of **available space**.

FURTHER READING

Arntson, A. E. (2003) *Graphic Design Basics*. 4th ed. Fort Worth, TX: Harcourt Brace College Publishers.

An introduction to principles of graphic design.

Barrault, M. (2001) "A methodology for the placement and evaluation of area map labels." *Computers, Environment, and Urban Systems* 25:33–52.

Discusses automated type placement with an emphasis on evaluation of type placement and measures of fitness.

Clark, S. M., Larsgaard, M. L., and Teague, C. M. (1992) *Cartographic Citations, A Style Guide*. Chicago: American Library Association, Map and Geography Round Table, MAGERT Circular No. 1.

A guide to cartographic citation.

Dent, B. D. (1999) *Cartography: Thematic Map Design*. 5th ed. Boston: McGraw-Hill.

A classic cartography textbook with chapters on cartographic design and typography.

Edmondson, S., Christensen, J., Marks, J., and Shieber, S. M. (1996) "A general cartographic labelling algorithm." *Cartographica* 33, no. 4:13–23.

An example of automated suboptimal positioning of type labels, which are subsequently reorganized to approach an optimal state.

Forrest, D. (1999a) "Developing rules for map design: A functional specification for a cartographic design expert system." *Cartographica* 36, no. 3:31–52.

A detailed description of how expert systems might be used to design maps.

Freeman, H. (1995) "On the automated labeling of maps." In *Shape, Structure, and Pattern Recognition*, ed. by D. Dori and A. Bruckstein, pp. 432–442. Singapore: World Scientific.

A seminal paper on automated text placement, in the context of soil survey maps.

Imhof, E. (1975) "Positioning names on maps." *The American Cartographer* 2, no. 2:128–144.

A classic reference on guidelines for positioning type on maps.

MacEachren, A. M. (1995) *How Maps Work: Representation, Visualization, and Design*. New York: Guilford.

Includes comprehensive discussions of map design research and Gestalt principles of perceptual organization.

Monmonier, M. S. (1993) *Mapping It Out: Expository Cartography for the Humanities and Social Sciences*. Chicago: University of Chicago Press.

Introduces practical aspects of cartographic design for noncartographers.

Montello, D. R. (2002) "Cognitive map-design research in the twentieth century." *Cartography and Geographic Information Systems* 29, no. 3:283–304.

An overview of the origins, rise, fall, and rebirth of cognitive map design research.

Pinto, I., and Freeman, H. (1996) "The feedback approach to cartographic areal text placement." In *Advances in Structural and Syntactical Pattern Recognition*, ed. by P. Perner, P. Wang, and A. Rosenfeld, pp. 341–350. Berlin: Springer-Verlag.

An example of automated suboptimal positioning of type labels, which are subsequently reorganized to approach an optimal state. See also Freeman (1995).

Robinson, A. H. (1952) *The Look of Maps*. Madison, WI: University of Wisconsin Press.

The book that gave birth to map design research.

Robinson, A. H., Morrison, J. L., Muehrcke, P. C., Kimerling, A. J., and Guptill, S. C. (1995) *Elements of Cartography*. 6th ed. New York: Wiley.

A classic cartography textbook with chapters on cartographic design and typography.

Tufte, E. R. (1990) *Envisioning Information*. Cheshire, CT: Graphics Press.

Focuses on the visual display of information, with many cartographic examples.

Wood, C. H., and Keller, C. P., eds. (1996) *Cartographic Design: Theoretical and Practical Perspectives*. Chichester, England: Wiley.

Includes essays focusing on the modern role and character of cartographic design.

Wood, C. H. (2000) "A descriptive and illustrated guide for type placement on small scale maps." *The Cartographic Journal* 37, no. 1:5–18.

An extensive collection of typographical guidelines for use in cartography.

Zoraster, S. (1991) "Expert systems and the map label placement problem." *Cartographica* 28, no. 1:1–9.

Explains the role of expert systems in the automated positioning of type.

12

Map Reproduction

OVERVIEW

This chapter presents topics related to print and nonprint *map reproduction*, and introduces *map dissemination* with regard to nonprint reproduction. In section 12.2, we encourage you to plan ahead for map reproduction to avoid problems associated with specific map reproduction methods. In section 12.3, we introduce *map editing*—a critical activity that can save you time and money during map reproduction.

Section 12.4 describes the role of *raster image processing* in print reproduction. The process of printing a digital map is distilled to four steps: The map is created in *application software*, converted into *page description data*, interpreted by a *raster image processor*, and printed on a printing device. In section 12.5, *halftone* and *stochastic screening* methods are described as ways to create *tints* from base colors.

We introduce aspects of color printing in section 12.6. *Process colors* are mixed on the page by combining cyan, magenta, yellow, and black (CMYK colors). *Spot colors* are premixed and thus allow for exact color matches based on printed color swatches. *High-fidelity process colors* allow for increased color variety, vibrancy, and color-matching capabilities. *Continuous tone* printing allows for smooth color transitions without the use of screening. *Color management systems* identify and correct for differences in color that are introduced by electronic devices.

In section 12.7, we present options for low-volume print reproduction, including *monochromatic* and *color laser printers*. Color copy machines can be used as color laser printers through the use of a *print controller*. *Ink-jet printers* range from small, inexpensive units to expensive, large-format, high-quality devices. Additional low-volume printing methods include *thermal-wax transfer*, *dye-*

sublimation, and *xerography*. Low-volume reproduction methods do not normally offer volume discounts.

Offset lithography is identified in section 12.8 as the dominant method for high-volume print reproduction, characterized by excellent print quality, high speed, and volume discounts. *Prepress* is described as a phase of high-volume reproduction centered on the *service bureau*, in which preparations are made for offset printing. *Film negatives (color separations)* and *printing plates* are produced for each base color on a map, and *proofs* are created for use in map editing. Portable document formats are preferred for the delivery of digital maps to the service bureau. Proofing methods range from low-end techniques such as *on-screen display* to high-end techniques such as *separation-based proofs*, which are created from color separations. The *offset lithographic printing press* transfers images from the printing plate to the print medium via a series of rolling cylinders. *Registration* refers to the proper alignment of colors produced by an offset press; *trapping* is used to correct for improper registration. *Computer-to-plate* and *direct-to-press* printing methods streamline the offset printing process by eliminating the film negative. These methods can be faster and less expensive than traditional separation-based offset lithography, but eliminate the possibility of creating separation-based proofs.

In section 12.9, we describe television- and computer-based nonprint reproduction and dissemination. Nonprint reproduction involves the use of *videocassettes*, *DVD* recorders, and file duplication performed by computers. Television-based map dissemination has several limitations, many of which are addressed by *digital television*. Computer-based maps can be disseminated on various media and via *local area networks*, although it

229

*is difficult to anticipate how a map will appear on different systems. The **Internet** and **World Wide Web** offer a wide range of map dissemination options. **FTP** can be used to distribute digital maps in a rudimentary fashion over the Internet, and **HTTP** and **HTML** can be employed for Web-based dissemination. The **Java** and **JavaScript** programming languages allow for the creation of custom mapping applications, and sophisticated map dissemination environments can be created through the use of **map servers**.*

12.1 REPRODUCTION VERSUS DISSEMINATION

Map reproduction involves printing a map in small or large quantities (print reproduction), or the electronic duplication of a map in digital form (nonprint reproduction). **Map dissemination** refers to the distribution of reproduced maps in physical or electronic form. The term dissemination is also used to describe the combination of electronic duplication *and* electronic distribution, in which maps are duplicated and distributed simultaneously. For example, when a map is downloaded from the Internet, it is reproduced *and* disseminated at the same time, whereas when maps are printed, reproduction and dissemination are distinctly different processes, occurring at different times. Emphasis is given to print reproduction in this chapter because of the breadth and depth of the topic; nonprint reproduction is simple in comparison. Electronic dissemination is discussed in relation to nonprint reproduction toward the end of the chapter (section 12.9).

12.2 PLANNING AHEAD

Reproduction methods can be quite specific—each provides certain benefits but also imposes strict limitations. The reproduction method you choose will influence almost every aspect of the cartographic design process. If you design a map without a specific reproduction method in mind, you could very well end up with a map that cannot be practically or economically reproduced. For example, if you design a color map using the RGB color model and then attempt to print to a CMYK device, you will encounter problems with color accuracy that will take time (and probably money) to resolve. Another example involves the reproduction of folded maps. Many large-format maps are designed to be folded in specific patterns so that, for example, the title and cover art appear on the front, and the legend appears on the back when folded. If you design a map with a special fold pattern in mind without ensuring that an appropriate folding mechanism is available, you will be forced to redesign your map. If you hire a second party to reproduce a map, choose carefully,

and make sure to establish a good relationship with this party before making major design decisions.

The following is a list of questions that need to be answered in the early stages of the design process. If you take the time to answer these sorts of questions carefully, the map reproduction process is less likely to present problems and roadblocks that will prevent you from successfully completing a mapping project.

1. Who is the intended audience, and what is the purpose of the map? As in the design process, the answer to this question influences almost every aspect of the reproduction process.

2. What is your budget? Different reproduction methods vary greatly in cost.

3. When is your deadline? Make sure to allow enough time for reproduction considering the speed of the method you have chosen. Consider that problems often arise that delay the reproduction of a map.

4. Will the map be printed, or displayed on a television or computer screen? Significant differences between print and display methods require you to consider aspects such as size, resolution, and the color model to be used.

5. If the map is to be printed, what material will be used? Paper? Cloth? Plastic? Different materials and different grades of material absorb ink and toner differently.

6. If the map is to be displayed, what will the file size(s) be? If the map is to be disseminated via the Internet, the user's connection speed and computer processing power need to be considered.

7. Will the map be limited to black, white, and gray tones, or will it be full color? How many colors will be used? Printers and display devices have strict limitations regarding color, and the cost of print reproduction generally increases when more colors are used.

8. What size will the map be? Printers and display devices have strict limitations on size.

9. How many copies are required? Certain reproduction methods offer a significant decrease in the cost per unit as the number of copies increases; other methods don't.

10. Will the map be folded? What will the fold pattern be? Folding mechanisms are limited to certain types of patterns.

11. What level of print or display quality is acceptable? Different reproduction methods provide various levels of quality.

12. Will you copyright the map? Will the map infringe on an existing copyright?

12.3 MAP EDITING

A map must be carefully edited before being reproduced (Figure 12.1). **Map editing** is the critical evaluation and correction of every aspect of a map; it begins the first time the cartographer views the map in its early stages, and culminates with a final edit that occurs just before reproduction begins. Editing could easily be treated as an aspect of cartographic design, but is presented here because of its critical role in map reproduction. Because reproduction methods consume time and money, you simply can't afford to reproduce a map that contains errors. The following is a list of general questions that should be addressed when editing:

1. Map design: Does the design serve the purpose of the map based on its intended audience and use? Does it communicate the map's information in the most efficient manner, with simplicity and clarity?

2. Completeness: Are any features, map elements, or type labels missing?

3. Accuracy: Are features, map elements, and type labels correctly placed? Are words spelled correctly? Are numeric values correct?

Overfamiliarity and fatigue are problems faced by cartographers who edit their own maps. The cartographer spends so much time working on a map that he or she loses a certain degree of objectivity, and often misses errors that are obvious to others. One solution to this problem is to have a separate individual (or individuals) edit the map. Other tips for editing include the following:

1. Edit with "fresh eyes," at the beginning of a work session.

2. Edit large maps in sections.

3. View maps upside-down or sideways—this forces you to see in new ways, possibly illuminating errors.

4. If possible, edit after several days of separation from a map.

12.4 RASTER IMAGE PROCESSING FOR PRINT REPRODUCTION

Virtually all modern print reproduction methods involve **raster image processing**: the conversion of a digital map into a *raster image* that can be processed directly by a raster-based printing device.* The resulting raster image is representative of the **raster data model** (a general category of computer data) consisting of rows and columns of square pixels, as illustrated in Figure 12.2A. In contrast to the raster model is the **vector data model**, which is composed of mathematically defined points, together with the lines that connect them. The vector data model closely resembles what is drawn by hand, and allows the representation of discrete points, lines, and polygons, as illustrated in Figure 12.2B. High-quality type is also represented using the vector data model. The majority of thematic maps are created using the vector data model because of its ability to represent clearly defined features such as wells, roads, administrative polygons, and so on. The raster data model is commonly used when representing continuous surfaces such as terrain, or when representing the results of raster-based GIS analysis or

---

* Certain devices, such as pen plotters, do not require that data be converted into a raster image for printing, but these devices are less widely used today.

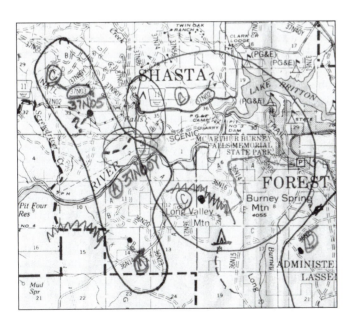

FIGURE 12.1 A Forest Visitor Map in the process of being edited. Carefully executed corrections, additions, and deletions save time and money during reproduction.

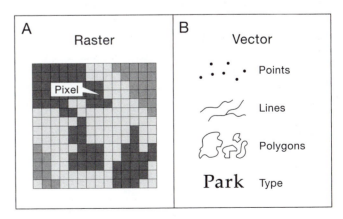

FIGURE 12.2 (A) Raster and (B) vector data models.

satellite image classification. Many maps incorporate both raster and vector data models. Regardless of the data models used in the creation of a map, the map is normally converted into a raster image for printing.

12.4.1 Printing the Digital Map

Several steps are involved in printing a digital map, as illustrated in Figure 12.3. First, the cartographer creates the map in digital form using **application software**. A description of the wide range of application software capable of producing maps is beyond the scope of this text, but general categories include graphic design, GIS, and remote sensing applications; examples include FreeHand, ArcGIS, and Imagine, respectively. The digital map file produced by application software is converted, via the **printer driver**, into **page description data** that consist of a set of printing instructions, describing every graphical and textual component of the map. The page description data are written in a **page description language** (PDL) such as Adobe's **PostScript**, or Hewlett Packard's **Printer Control Language** (PCL). PDLs are *device independent* to a certain extent, allowing the user to print from, or to, almost any device equipped to interpret a particular PDL. PostScript has proven to be more device independent and better suited to high-end printing than other PDLs, making it the de jure standard page description language—PostScript is the only PDL to be recognized

by the International Organization for Standardization (ISO) (Adobe Systems Incorporated 1997).

After the page description data have been created by the printer driver, they are interpreted by a **raster image processor** (RIP). A RIP is software, or a combination of software and hardware, that interprets page description data and converts them into a raster image that can be processed directly by a printing device. For example, a PostScript-compatible RIP interprets PostScript page description data, and converts them into a raster image for printing. RIPs exist inside printing devices such as laser printers, inside dedicated computing devices such as the ColorSpan RIPStation (a stand-alone RIP), or as software applications that operate inside computers, such as ESRI's ArcPress. Many less expensive printing devices are incompatible with PDLs, and don't possess a RIP. These devices depend on RIP software inside the computer to create the raster image, which is subsequently sent to the printer.

The RIP produces raster images of a specific resolution that normally coincides with the maximum resolution supported by the printing device that will be used. Resolution refers to the number of pixels per inch, or dots per inch (dpi), that a device can print. After the RIP has created the raster image, the image is processed by the printing device, which in turn, creates the printed map.

12.5 SCREENING FOR PRINT REPRODUCTION

Screening is a technique used in most print reproduction methods in which colors are made to appear lighter by reducing the amount of ink or toner* applied to the *print medium* (e.g., paper). Screening is also referred to as dithering, but the term screening is more commonly used in the printing industry. Screening is used to create tints of a particular *base color*, and to represent continuous tone surfaces. A **tint** is a lighter version of a base color: Gray is a tint of the base color black, and pink is a tint of the base color red. A continuous tone surface is composed of tints that continuously vary in lightness within a given area, as illustrated by the terrain beneath the regional park label in Figure 12.4.

A standard (monochromatic) laser printer is only capable of printing black because it uses black toner. Through the application of screening, however, the same laser printer is capable of producing what *appear* to be gray tones, even though black is the only color used (Figure 12.4). Screening can produce what appear to be lighter versions of *any* base color, not just black. In the screening process, ink or toner of a particular color, say dark green, is applied to the print medium in

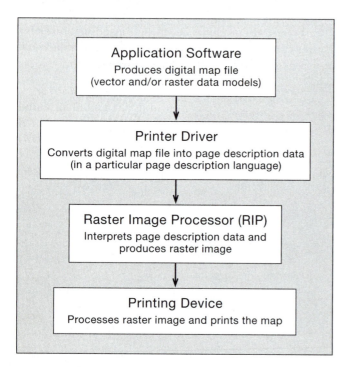

FIGURE 12.3   General steps involved in printing a digital map.

---

* Toner is powder, normally black, that is used to form images in laser printers and copy machines.

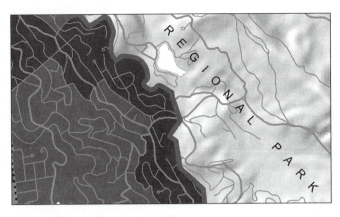

FIGURE 12.4 Black is the only color used in this map; the illusion of gray tones is produced through screening.

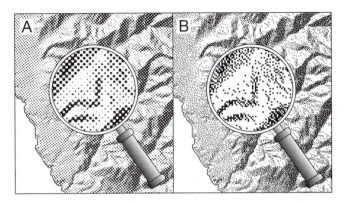

FIGURE 12.5 (A) Halftone screening, produced by equally spaced dots of variable size. (B) Stochastic screening, produced by pseudorandomly spaced dots of uniform size.

a pattern of individual dots, allowing the color of the underlying print medium (normally white) to show through. If the pattern is fine enough, the human eye is capable of combining the color of the individual dark green dots with the white of the underlying print medium, allowing the map user to perceive light green instead of dark green; the pattern itself typically goes unnoticed. Most color printing devices work on this principle; each color that the device can produce can be lightened through screening, allowing for the creation of far more colors than the number of inks or toners would imply.

12.5.1 Halftone and Stochastic Screening

Several methods of screening have been developed, but two have emerged as the most widely used: halftone and stochastic.* **Halftone screening** is used in almost all print reproduction methods, with the exception of ink-jet printing, which normally employs stochastic screening. Halftone screening involves the application of ink or toner in a pattern of equally spaced dots of *variable* size, as illustrated in Figure 12.5A. The size of each dot determines the degree of lightness that is achieved; the smaller the dot, the lighter the result, as more of the underlying print medium shows through. Halftone dots are composed of pixels produced by a RIP, with the pixel size normally corresponding with the maximum resolution of the printing device to be used (Figure 12.6). As a result, halftone dots are always coarser than the maximum resolution of the printing device. Smaller pixels (Figure 12.6A) allow for the creation of finer, better defined halftone dots, and allow a larger number of tints to be derived from a base color.[†]

---

* Certain screening methods represent a combination of halftone and stochastic approaches (Hyatt 2000).

[†] The number of tints that can be derived from a base color can be calculated using this formula: (printer resolution/screen frequency)^2 + 1.

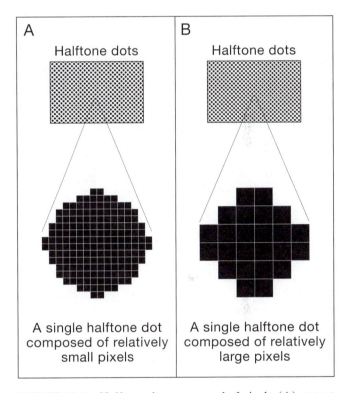

FIGURE 12.6 Halftone dots composed of pixels: (A) output from a higher resolution device; (B) output from a lower resolution device.

Halftone screening is sometimes referred to as *amplitude modulation (AM)* screening, because variations of lightness are achieved through the alteration of the size, or amplitude, of each dot. Halftone dots increase in size until a 50 percent tint is surpassed, at which point all color is removed from the dots and is applied to the area surrounding the dots, essentially creating white dots on a color background (70 percent and 90 percent in Figure 12.7A). As tints greater than 50 percent

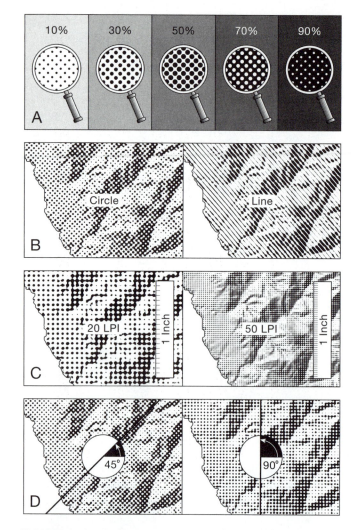

FIGURE 12.7   Halftone screening parameters: (A) tint percentage, (B) cell type, (C) screen frequency, measured in lines per inch (lpi), (D) screen angle, measured in degrees, counterclockwise from horizontal.

Halftone screening has been practiced and refined for about 200 years, whereas stochastic screening is a relatively new method. While the halftone method is well-defined, reliable, and relatively forgiving, the stochastic method is still being refined, and is relatively unforgiving of inaccuracies in the reproduction process. Very small dots are difficult to print correctly, and require that the entire reproduction process be tightly controlled. The stochastic method, however, offers several improvements over halftone screening. For example, it can result in cleaner images with greater contrast because individual dots can be placed closer together, and because there is less overprinting of inks when colors are mixed on the page (Romano 1997). It can also eliminate the possibility of undesirable moiré patterns (described later) that are inherent with halftone screening. Although stochastic screening has been predicted to eclipse halftone screening, this has not yet occurred—both methods are widely used.

12.5.2 Halftone Screening Parameters

Halftone screening is controlled by four parameters: tint percentage, cell type, screen frequency, and screen angle. The **tint percentage** controls the degree to which the appearance of an ink or toner is lightened, as illustrated in Figure 12.7A. Tint is specified in percentage, with lower values resulting in lighter colors. The **cell type** refers to the shape of each individual mark of the halftone pattern; the circle (dot) is the most widely used shape, but the line and other shapes are also used (Figure 12.7B).

    Screen frequency, or screen ruling, refers to the spacing of halftone dots within a given area, or more specifically, the spacing of the *lines* that the dots are arranged in (Figure 12.7C). The screen frequency parameter controls how coarse or fine the halftone pattern is. Lower frequencies produce a coarse halftone pattern in which individual dots are clearly visible, and are typically employed only when a lower resolution printing device is used. Higher frequencies produce a fine halftone pattern in which it is more difficult to discern individual dots. This pattern is easier for the human eye to interpret as representing a solid, but lighter, color. Screen frequency is measured in lines per inch (lpi). Lpi is related to, but should not be confused with, dpi (dots per inch). Dpi refers to the maximum resolution of a printing device, independent of the lpi setting of a halftone screen. Halftones with higher lpi settings consist of smaller dots that are most accurately produced by high-resolution printing devices.

    The **screen angle** parameter controls the angle at which the lines of halftone dots are oriented, as illustrated in Figure 12.7D. Screen angle becomes important when multiple halftone patterns are printed in the same area. For example, imagine one halftone pattern that represents a 30 percent tint of black, and another that represents a 30

are specified, the background area expands, eventually filling the areas formerly occupied by dots.

    Stochastic screening involves the application of ink or toner in a pattern of very small, pseudorandomly* spaced dots of *uniform* size, as illustrated in Figure 12.5B. The density of dots within a given area determines the degree of lightness that is achieved; dots spaced further apart produce lighter results. Stochastic screening is also referred to as *frequency modulation (FM)* screening, because variations of lightness are achieved through the alteration of the *spacing*, or frequency, of each dot. The term *stochastic* refers to the pseudorandom approach used to place dots within a given area.

---

* Computing devices are incapable of generating truly random numbers.

percent tint of cyan. To mix the two colors, both halftone patterns need to be printed in the same area, allowing black and cyan to "mix" on the page. If both screen angles are identical (and if all other halftone parameters are equal), then the black halftone dots will cover and block out the cyan dots. To prevent this, the screen angle for the black halftone would need to be offset from the cyan halftone, preventing the direct overlapping of dots, as illustrated in Color Plate 12.1. The concept of mixing colors on the page is described further in section 12.6.1. Screen angles need to be precisely specified to prevent the creation of a **moiré pattern** (Figure 12.8), which is an unwanted print artifact resulting from the interplay of dots in overlying halftones (Adams and Dolin 2002). Most software applications and printing devices have been programmed to use appropriate screen angles that prevent the creation of moiré patterns.

12.5.3 Stochastic Screening Parameters

Stochastic screening is typically controlled by one parameter only: tint percentage. As with halftone screening, the tint percentage controls the degree to which the appearance of an ink or toner is lightened. Tint is specified in percentage, with lower values resulting in lighter colors. The additional parameters associated with halftone screening do not apply to stochastic screens. The cell type is always a very small dot, usually representing the smallest pixel size a particular printing device can produce. Because these dots are placed pseudorandomly within a given area, screen frequency and screen angle become meaningless. The irregular spacing of dots eliminates the potential problem of moiré patterns inherent with halftone screening.

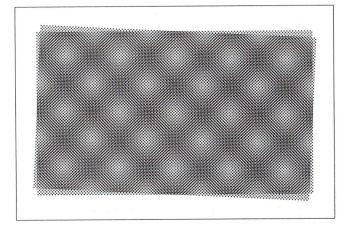

FIGURE 12.8 Moiré pattern produced from the incorrect specification of halftone screen angles.

12.6 ASPECTS OF COLOR PRINTING

12.6.1 Process Colors

In contrast with CRTs, which generate color through the addition of red, green, and blue light (the additive primary colors—see Chapter 10), most printing devices generate color using inks and toners based on the **subtractive primary colors**: *cyan* (C), *magenta* (M), and *yellow* (Y). In theory, a mixture of pure C, M, and Y (without screening) will result in pure black. In reality, this mixture produces a "muddy" dark brown that lacks crisp detail, and gray tones produced by mixing C, M, and Y are not pure. As a result, black (K) is normally used in conjunction with C, M, and Y, creating CMYK, or the **process colors**. Most color printing devices mix the process colors on the page by applying them, in sequence, to the same area. Inks and toners based on the process colors are semi-opaque, or translucent, allowing them to combine on the page; new colors are created where the process colors overlap. This, together with screening techniques, allows for the creation of a wide variety of colors, and is referred to as **four-color process printing**. When process colors are mixed on the page, tints of each base color are represented by halftone patterns, each with a unique screen angle to avoid moiré patterns. This results in a special pattern of inks called a **rosette** (Color Plate 12.2). The human eye interprets fine rosettes (ones that result from halftones with high screen frequencies) as representing a solid color. Four-color process printing is an efficient method of producing many colors from four base colors. However, the actual number of colors that are possible is fewer than the number possible using the additive primary colors (RGB). Specifically, vibrant colors such as orange, red, green, and blue are difficult to reproduce using the process colors (Viggiano and Hoagland 1998).

12.6.2 Spot Colors

An alternative to mixing process colors on the page is to use solid, or **spot colors**. Spot colors take the form of opaque inks that are *premixed* before they reach the printing device. As with process colors, tints can be created from spot colors through screening. Exact color matches are easier to achieve using spot colors because they do not rely on the printing device for mixing—a situation that often results in color inconsistencies from device to device. The *Pantone Matching System* (PMS) is the most universally accepted spot-color system, but TOYO and other systems are also popular. Spot colors can be selected on-screen, as illustrated in Figure 12.9, or (more appropriately) from printed color swatches. Spot inks are mixed according to specific formulas, such as those found in the *Pantone Color Formula Guide*

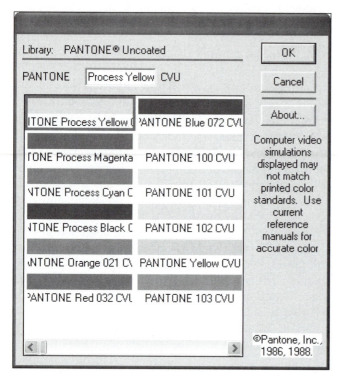

FIGURE 12.9 On-screen selector for Pantone spot colors within Macromedia FreeHand (reproduced in gray tones). Notice the advisory regarding accuracy of on-screen colors. Used by permission from Macromedia, Inc.

(Adams and Dolin 2002). A disadvantage of using spot colors is the cost of reproduction when a map has many colors. Unlike the process colors, which can produce a wide variety of colors through mixing, spot colors require that separate inks be used for each color, increasing the cost of reproduction.

12.6.3 High-Fidelity Process Colors

New methods of color printing have been developed in response to the disadvantages of process and spot colors, specifically, the limited variety and vibrancy of process colors, and the expense of spot colors when used in large numbers. **High-fidelity process colors** are based on the traditional process colors (CMYK), but include two or three additional colors. This allows for dramatic increases in the variety and vibrancy of colors, and also allows for accuracy in color matching that rivals the Pantone Matching System.

Various approaches to high-fidelity color have been taken. Early attempts included the incorporation of red, green, and blue inks with the process colors (Kenny 2000). This approach is still practiced but requires a total of seven colors, and has been largely eclipsed by six-color approaches, which have been developed by a wide variety of companies including Epson, Roland, Hewlett Packard, Lexmark, and Pantone. Most of these systems incorporate black, modified versions of C, M, and Y, and two additional colors, usually orange and green, although some systems include light cyan and light magenta. Six-color systems produce colors with more variety, vibrancy, and contrast than traditional process-color systems, but are usually more expensive.

Pantone has emerged with the de jure standard in high-fidelity color with *Hexachrome*, a six-color system that incorporates black, more saturated versions of C, M, and Y, plus orange and green (Pantone Incorporated 2002). The Hexachrome system is capable of accurately reproducing over 90 percent of the spot colors in the Pantone Matching System, as opposed to traditional process-color systems, which can only reproduce about 40 percent (Kenny 2000). Special software is required to work with Hexachrome, but it is either incorporated into existing software or can be accessed through the use of *plug-ins*—software modules that work in conjunction with software applications.

12.6.4 Continuous Tone Color Printing

Some printing devices are capable of producing **continuous tone** output: full-color prints that are achieved without the use of screening techniques. Tints are created through subtle variations in the volume or density of ink or toner applied to the page. The term *contone* refers either to true continuous tone printing or to a set of hybrid techniques that incorporate a limited set of continuous tone methods with screening, producing results that approach true continuous tone output. Until recently, contone techniques have been best suited to printing images of continuous tone such as photographs, as opposed to maps, which are composed primarily of discontinuous tones and fine detail; their use has been limited primarily to the creation of intermediate-quality proofs, as described in section 12.8.3. Recent improvements in the resolution and "apparent resolution" of continuous tone devices have made them more suitable for cartographic applications—especially those involving continuous tone surfaces, such as terrain—but the technology remains relatively expensive.

12.6.5 Color Management Systems

A common problem confronting cartographers is matching the colors on a graphic display with those produced by a printing device. Without your deliberate intervention, on-screen colors will rarely, if ever, truly match colors that are printed. This is partly due to differences between the color models employed by graphic displays and most printing devices (RGB vs. CMYK). It is also partly due to differences between software applications and display and printing technologies. Each device in the workflow, including scanners, graphic displays, and printers, introduces subtle variations in color. Even devices that are supposedly identical—two identical printers, for example—can produce different results due to variations in calibration, operating conditions, and so on. All of these factors contribute to differing color **gamuts** (ranges of colors produced) among devices. **Color management systems** (CMSs) provide consistency and predictability of color by identifying differences in gamuts, and by correcting the variations in color introduced by each device (Agfa Corporation 1997).

CMSs are centered on the **color profile**, which is an electronic file that describes the manner in which a particular device introduces color variations. Before a profile is created, the device—a color printer, for example—needs to be warmed up to normal operating temperature and calibrated so that it operates within the manufacturer's specifications. A sample page is printed, which consists of a series of strictly defined colors. Colors on the printed sample page are measured using a *colorimeter* or *spectrophotometer* (both devices precisely measure color). The measured values are then sent to the CMS, which creates the device's color profile by comparing the colors on the sample page with the colors as they were originally defined. Once color profiles have been created for every device in the workflow, the CMS is able to correct for variations in color, resulting in printed colors that are *as similar as possible* to those seen on the graphic display, captured by the scanner, and so on. It is impossible to match all colors in every situation, however, because the color gamuts of electronic devices can be substantially different.

Early CMSs were proprietary and worked only with a particular manufacturer's equipment. Early systems were also designed to work at the application level, as opposed to the operating system level, which is the trend today. Color management has become partly standardized with the formation of the **International Color Consortium** (ICC). The ICC has developed a standard for a vendor-neutral, cross-platform color profile called the *ICC color profile*. The ICC has also established architectures for the *Color Management Module* (CMM), which actually performs color corrections, and the *Color Management Solution*, which facilitates the color management process at the operating system level (Adams and Dolin 2002). Apple's ColorSync (Macintosh) and Microsoft's ICM2 (Windows) are examples of ICC Color Management Solutions.

12.7 LOW-VOLUME PRINT REPRODUCTION

There is no discrete boundary between the numbers of maps created in low-volume versus high-volume print reproduction, but for convenience, we will set this boundary at 200 maps. Low-volume print reproduction methods do not normally offer a significant decrease in the cost per unit as the number of copies increases—the last copy costs about the same as the first. Most low-volume reproduction is performed "in house" on printing devices that are commonly owned by individuals or small businesses. In certain cases, especially when a map incorporates color, a second party is hired for reproduction.

12.7.1 Laser Printing

Monochromatic laser printers have proven to be practical and reliable for low-volume reproduction of small-format maps consisting of black, white, and gray tones (Figure 12.10). The term *monochromatic* refers to the fact that these devices are capable of printing only one color, normally black. Gray tones are simulated through screening, normally of the halftone type. Maximum resolution for standard laser printers ranges from 600 dpi to 1200 dpi. Maximum page sizes typically range from 8.5″ × 11″ to 11″ × 17″. Laser printing involves the application of a latent (invisible) image onto a metal drum by *electronically etching* it with a light source such as a laser, or a light-emitting diode (LED), thereby selectively altering the electrical charges on the drum. The drum is then dusted with positively charged toner, which sticks only to the portions of the drum that have been electronically etched (negatively charged). The toner is then transferred to the print medium, and is fused to it by means of heat and pressure. Distinct advantages of laser printers are speed, cost of operation, and relatively sharp depiction of detail.

Color laser printers work on the same basic principles, except that the process is repeated four times, once for each of the process colors. The image is either transferred

FIGURE 12.10 Hewlett Packard monochromatic laser printer. Used by permission from Hewlett Packard, Inc.

to the page one color at a time or applied to an intermediate transfer surface such as a belt, which subsequently transfers all colors to the page simultaneously (Bojorquez 1995). Tints are normally created through screening, typically of the halftone type, but more expensive units can produce continuous or near-continuous tones. Color laser printers are significantly more expensive than monochromatic laser printers and are more expensive to operate, but image quality is generally quite good. The process also results in images that are relatively resistant to streaking and fading.

Color copy machines are technologically similar to color laser printers and are sometimes used as printing devices. These devices have not traditionally been designed to accept print jobs from computers, but combinations of hardware and software have been developed that connect a computer with a copy machine, allowing users to treat the copier like any other printing device. The necessary hardware and software is referred to as a **print controller**, and is essentially a stand-alone RIP (Figure 12.11). This arrangement yields results very similar to those of a color laser printer, and is a practical solution for organizations that already own or lease a color copier.

12.7.2 Ink-Jet Printing

The quality of ink-jet printing has advanced dramatically over the past decade; equally dramatic has been the drop in prices for these devices. Like laser printers, maximum resolution for standard **ink-jet printers** ranges from 600 dpi to 1200 dpi. Page sizes for personal use devices typically range from 8.5″ × 11″ to 11″ × 17″, but large-format ink-jet devices are capable of printing page widths exceeding four feet, and in lengths limited only by the

FIGURE 12.12   An Epson large-format ink-jet printer employing six base colors. Used by permission from Epson America, Inc.

length of paper stored on a roll. Ink-jet printers form images by squirting ink onto the print medium. Most ink-jet printers are four-color devices containing inks for the process colors, but some, such as the six-color Epson illustrated in Figure 12.12, employ more. Colors are mixed on the page, allowing for the creation of a wide variety of colors. Tints are normally created through screening, typically of the stochastic type, but more expensive units can produce continuous or near-continuous tones. Ink-jet output has traditionally been more susceptible to streaking and fading than laser printer output.

12.7.3 Additional Printing Methods

Thermal-wax transfer printers apply a wax-based ink to the print medium. The ink is melted by a thermal print head and is transferred to the page, where it resolidifies when the print head is turned off. The ink is stored on *transfer rolls,* each consisting of four page-sized sections of ink, each section representing one of the four process colors. Each of the process colors is applied to the page in sequence as a mechanism draws each section of the transfer roll into contact with the page. Tints are normally created through screening, typically of the halftone type. Printing quality can be higher than for ink-jet printers, but the initial purchase cost is somewhat greater, and best results are obtained only with special print media. In addition, ink costs can be relatively high because the transfer roll is discarded after each page is printed, regardless of how much ink was actually used. On the plus side, thermal-wax transfer printers produce excellent transparencies.

Dye-sublimation printers (or *thermal-dye printers*) are similar to thermal-wax transfer printers in that they use heat to move colorant from a transfer roll or ribbon

FIGURE 12.11   An EFI Fiery ® print controller and raster image processor (RIP). Photography: Stan Musilek. Used by permission © Electronics for Imaging, Inc. 2002.

to the print medium. They differ in several respects, however. Dye-sublimation printers use a special dye as opposed to a wax-based ink. Another difference is that dye-sublimation printers use a sublimation process (converting solid ink to gas) as opposed to simply melting the ink. Advantages of this process are that 256 different tints can be created from each of the process colors (continuous tone is possible), and the resulting image cannot be scratched easily. Dye-sublimation printers produce high-quality continuous-tone images, but are less well suited to the creation of crisp, finely detailed lines or small type—common characteristics of maps. The initial purchase price is high, and expensive specialized paper is required. As with thermal-wax transfer printers, ink costs can be relatively high because the transfer roll is discarded after each page is printed, regardless of how much ink was actually used.

The use of color copy machines as printing devices was described previously. Copy machines can also be used to reproduce preprinted maps, either in black and white or in color. Copy machines produce images through **xerography**, which is related to laser printing. Monochromatic copying is relatively inexpensive, but the cost for color copies is significantly higher. A primary drawback to the use of copy machines is the loss in image quality inherent in the copying process.

12.8 HIGH-VOLUME PRINT REPRODUCTION

When large numbers of maps are required (e.g., more than 200), issues of cost and time become critical; methods for low-volume reproduction become too costly or time-consuming, especially when full-color output is required. In contrast with low-volume reproduction methods, which are plentiful and varied, high-volume reproduction is dominated by a single method: offset lithography. *Lithography* is a printing process in which ink is made to stick only to certain areas of a printing surface (through chemical means), and is subsequently transferred to a print medium (e.g., paper). **Offset lithography** is a form of lithography in which ink is transferred to an *intermediate printing surface* before being transferred to the print medium. Virtually all mass-produced maps are the result of offset lithography, which is characterized by excellent print quality, high printing speed, and a significant decrease in the cost per unit as the number of copies increases—the last copy costs significantly less than the first. The cost per unit decrease occurs because the majority of expenses are incurred in preparation for printing; the cost of paper and ink for additional copies is minimal in comparison with the costs of prepress and press setup, activities that occur before printing begins.

12.8.1 The Prepress Phase

The **prepress** phase of high-volume map reproduction consists of various technologies and procedures that make offset lithographic printing possible. The *press* in prepress refers to the **offset lithographic printing press** (Figure 12.13), which is further described in section 12.8.4. The abbreviated terms *offset press* and *offset printing* are often used in regard to this device. Many aspects of prepress revolve around the **service bureau**, a business that specializes in the creation of products such as film negatives, printing plates, and proofs, and that provides print-related services. The service bureau often plays a key role in the reproduction process, intermediate between the cartographer and the press operator (the person who operates the offset press).

The **film negative** is a representation of a map, or one color of a map, on clear plastic film. It is composed of black, gray tones produced through screening, and clear areas. It represents the opposite of what will ultimately be printed: Black on film results in white when printed; a 70 percent tint on film results in a 30 percent tint when printed. The film negative plays two important roles in prepress: It is used in the creation of printing plates and in the creation of high-quality proofs. The service bureau creates the film negative from a digital map file. The file is processed according to the steps described earlier and is illustrated in Figure 12.3. The film negative is printed on a device called an **imagesetter**, which is essentially a very high-resolution, large-format laser printer that prints on clear film. A typical resolution of 2,400 dpi allows for the creation of extremely well-defined linework and fine halftone screen frequencies. A single film negative is

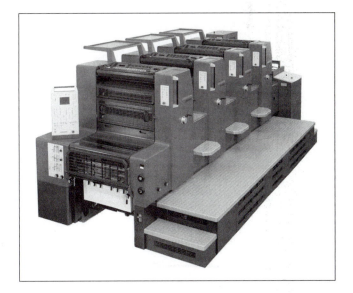

FIGURE 12.13 A Heidelberg four-color offset lithographic printing press. Used by permission from Heidelberg USA.

produced for each base color that a map consists of. A map that consists of one base color, say black, requires only one film negative (Figure 12.14A); a map that will be reproduced using the four-color process printing method requires four film negatives—one each for cyan, magenta, yellow, and black (Figure 12.14B). Each negative produced for a multicolor map is called a **color separation**, because it represents just one of the base colors (process or spot) of which a map is composed.

The **printing plate** is a sheet of aluminum (or polyester) that is ultimately mounted to a roller on an offset lithographic printing press. It receives a positive, latent (invisible) image from a film negative through photographic means. The photosensitive surface of the printing plate is *photographically etched* when exposed to light that passes through an overlying film negative. One printing plate is created for each film negative that is produced, as illustrated in Figure 12.14. After it is mounted on the offset press, the printing plate is the first representation of a digital map to come into contact with ink, which is transferred to the print medium shortly afterward. Plates are produced by service bureaus or by press operators, depending on the capabilities of each.

The **proof** is a representation of what the final, reproduced map will look like. It is an essential component of the prepress phase, and is used in conjunction with editing to ensure that your map will be reproduced just as you intend. Cruder, less expensive proofs are created repeatedly during the map design process (rough drafts), but higher quality (and more expensive) proofing methods are required in the prepress phase. Proofs are created using various technologies, and are produced by

individuals, service bureaus, and press operators. Proofing methods are described in detail in section 12.8.3.

12.8.2 File Formats for Prepress

A digital map can be delivered to the service bureau in a variety of formats. For example, it can be delivered in the native format of the application software that created it (an Illustrator file, an ArcMap file, etc.), but this can be a risky option for several reasons. First, the service bureau might not have the exact software application or fonts that you used. Second, related data such as linked images or GIS datasets might not transfer well to the service bureau's computer, making it difficult or impossible for them to properly open the file. Finally, application files are easily editable, allowing the possibility that the service bureau could edit your file, either intentionally or by accident. Delivering a digital map in the application's native file format *can* yield good results, but only if the service bureau can assure you that these issues will not pose problems.

Another option is to deliver the digital map as a file that contains page description data (e.g., a PostScript file). Page description files can be written by choosing the Print to File option when printing your map. This can be a practical option because related data and fonts get *embedded* into page description data, and because the service bureau has the ability to download page description data directly to RIPs and printing devices without the software application that was used to create the map. Downsides to this approach include the fact that page description files can be very large, they need to contain specific information regarding the printing device that will be used (information you might not have), and their contents cannot be directly viewed on screen.

A more attractive option is to deliver the digital map in a *portable* document format such as **Encapsulated PostScript** (EPS) or **Portable Document Format** (PDF). An EPS file is a subset of the PostScript page description language that allows digital maps and other documents to be transported between software applications and between different types of computer. It consists of PostScript code for high-resolution printing and an optional low-resolution raster image for on-screen display. Because they are written in PostScript, EPS files can be very large, but they don't require the specific printing device information required by page description files. EPS files also have the ability to embed related data and fonts. A PDF file is similar to an EPS file in that it is related to PostScript but is "smarter" and more efficient. In addition to being able to embed related data and fonts, PDF files have the ability to embed, or encapsulate, features such as hyperlinks, movies, and keywords for searching and indexing. Zooming and panning capabilities for

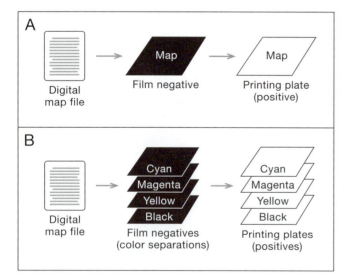

FIGURE 12.14   (A) Film negative and printing plate for a map employing one base color. (B) Film negatives and printing plates for a full-color map to be printed using process colors.

on-screen viewing are also provided. PDF files can be viewed using Adobe Acrobat Reader (a free download), and can be created from virtually any application via Adobe's Acrobat Distiller. The Distiller compresses and optimizes digital maps for printing, Web display, and so on. Although the PDF format was originally intended for on-screen display, it is quickly becoming the format of choice for the delivery of maps and other documents to service bureaus for high-end printing (Cardin et al. 2001).

Portable document files are created by converting digital maps from their native file format into EPS or PDF files. This is normally accomplished either by selecting the Export command within the application that was used to create the map, or by "printing" the map to a PDF file using Acrobat Distiller. Be aware that on occasion, certain graphical elements of your map (e.g., custom line patterns) will be altered or lost during the conversion process. If you choose to deliver your digital map to the service bureau in a portable document format, it is especially important that you proof your map based on output from the EPS or PDF file so that any changes resulting from the file conversion can be identified.

12.8.3 Proofing Methods

Proofing methods range from low-cost, low-quality techniques to high-cost, high-quality techniques, as illustrated in Figure 12.15. The proofing method that most closely resembles the ultimate reproduction method is usually the most reliable. Proofs are useful in both low-volume and high-volume reproduction, but the emphasis here is on proofing for high-volume reproduction using an offset press. In print reproduction (as opposed to display) the correction of errors during editing is generally cheapest when working with crude proofing methods, and becomes increasingly expensive with more sophisticated proofing methods.

On-screen display (or soft proofing) is the least expensive proofing method (assuming that a computer was used to create the map); it involves the viewing of a digital map on a graphic display. Maps are typically displayed using application software, document-viewing software such as Acrobat Reader, which displays PDF files, or in Web browsers. On-screen display is the primary proofing method for nonprint reproduction but is the crudest of methods for print reproduction. On-screen display allows you to perform basic editing and spell checking, but should not be entirely trusted when editing for print reproduction for three reasons:

1. Because computer screens are often smaller than the map being edited, the entire map cannot be seen at full size.
2. Because computer screen resolution is far lower than the resolution of printing devices, fine detail will not be visible on screen unless the cartographer is zoomed in on a portion of the map.
3. On-screen colors rarely, if ever, truly match colors that are printed. CMSs can improve the accuracy of on-screen colors, but accurate color proofing is still beyond the scope of on-screen display.

Despite the drawbacks of on-screen proofing, the PDF file format provides various features that enhance on-screen proofing. When viewed in Acrobat, PDF files can be zoomed and panned, searched, spell checked, and commented on. Acrobat also provides proofing simulation tools that allow the user to see what the map might look like when printed on paper. Because of these features, PDF files are often accepted as contract proofs (described later) in noncritical printing situations (*WhatTheyThink.com* 2002).

Creation of **monochromatic composites** is the second least expensive proofing method. This involves the printing of a map in black, white, and gray tones, typically on a

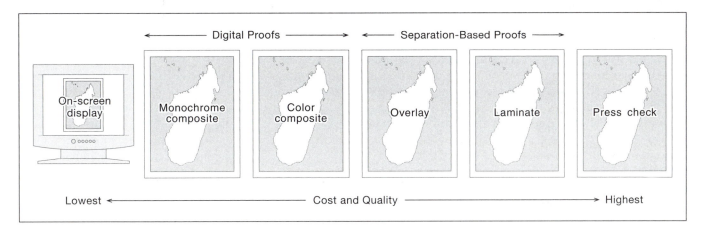

FIGURE 12.15   Proofing methods for print reproduction. The most reliable method is that which most closely resembles the intended reproduction method.

laser printer. Monochromatic composites are classified as **digital proofs**, because they are created without film negatives. The term *composite* refers to the fact that all color information is applied to the page at virtually the same time, and in this case, is represented as gray tones. Monochromatic proofs created by laser printers are sufficiently detailed for general editing tasks, but color cannot be discerned, making monochromatic proofing unsuitable for color editing. Laser prints are typically limited to page sizes of 11″ × 17″ or smaller, but this limitation can be compensated for by printing larger maps in sections, or tiles, that can be assembled using scissors and tape.

Color composites represent a middle ground between low- and high-quality proofing methods and are also classified as digital proofs. Various printing devices can produce color composites, including color laser, ink-jet, thermal-wax transfer, and dye-sublimation printers, although ink-jet devices are probably the most widely used. Ink-jet printers range from inexpensive letter-size units to expensive, large-format devices developed primarily for producing color composite proofs. Inexpensive ink-jet printers cannot reproduce fine detail as well as a monochromatic laser printer, but normally have the ability to represent color. However, the color accuracy of inexpensive ink-jet printers is not great enough to warrant using them for critically evaluating colors. High-end, large-format ink-jet printers, such as the *Iris* line of proofing devices, allow large maps to be printed on one sheet, and when used in conjunction with a CMS can produce colors that are far more accurate than those produced by inexpensive, small-format ink-jet printers. Color laser printers produce high-quality images but are usually limited to page sizes of 11″ × 17″ or smaller. As with monochromatic composites, color composites for larger maps can be produced on smaller format devices, printed in sections, and assembled using scissors and tape.

The highest quality, most expensive proofs available (before the offset press is run) are created from color separations. **Separation-based proofs** are created from color separations (film negatives), one for each base color. Because they are produced from the same film negatives that will be used to create printing plates, separation-based proofs provide a level of quality unattainable with lower priced options. They are of such high quality that they can act as *contract proofs*: contractually binding documents. If the cartographer is satisfied with a separation-based proof, and if the print operator assures the cartographer that the final, printed map will look *very similar* to the proof, then the proof can act as part of the contract between the two parties.

There are two categories of separation-based proofs: overlay and laminate. **Overlay proofs** consist of a stack of transparent film sheets that are bound on one edge like a book. Examples include DuPont's Cromacheck and Imation's Color Key products. Each film sheet contains a positive image of one base color and its tints, taken directly from its corresponding color separation. The sheets are registered to overlay correctly, and when viewed together, the colors in each sheet combine to produce an image that is very similar to what the printed map will look like. Individual sheets can be lifted, allowing you to isolate problems with particular color separations (Color Plate 12.3). **Laminate proofs** also consist of overlying transparent film sheets, but the individual sheets are melded into a single sheet (individual sheets cannot be lifted), which is often mounted to a backing material similar to the intended print medium. Examples include DuPont's Cromalin and Imation's Matchprint products. Laminate proofs are considered to be superior to overlays in image quality and are more expensive (Agfa Corporation 1999). Although separation-based proofs represent the most reliable prepress proofs, color composites (digital proofs) are improving rapidly and will likely emerge as practical alternatives to separation-based proofs.

The ultimate (and most expensive) proof is referred to as the **press check**. At the press check, printing plates are mounted on the offset press, the press is inked and calibrated, and sample prints are made. If the cartographer is satisfied with the samples, the press run can begin. As the printing press runs, the press operator periodically compares the prints with the approved samples and makes adjustments to the press if necessary. Cartographic errors discovered during the press check will be extremely expensive to correct, as prepress and press setup activities will need to be repeated. Because of this, the press check should be viewed as an opportunity to identify issues that the press operator can address, rather than an opportunity to identify cartographic errors. For example, the press operator is capable of producing slight variations in the lightness of particular colors by altering the density of ink that is applied to the page without increasing the cost of the print job.

12.8.4 Offset Lithographic Printing

Once the prepress phase is complete, the offset lithographic printing process can begin. Offset presses are mechanical printing devices that incorporate aspects of photographic and digital technologies. Presses are categorized, in part, according to the number of base colors they can produce. The simplest presses are limited to one base color and its tints, whereas others are capable of printing ten or more base colors and their tints. Tints are typically created through halftone screening. The offset press contains one or more **printing units** (Figure 12.16), composed of many cylinders that ultimately transfer ink to the print medium. Each printing unit is capable of

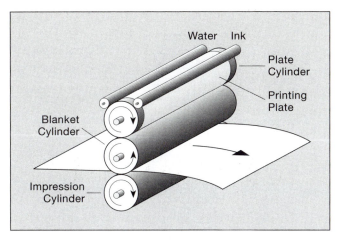

FIGURE 12.16 A printing unit (simplified) of an offset lithographic printing press.

printing one base color. There are three primary cylinders in each printing unit: the *plate cylinder*, to which a printing plate is mounted; the *blanket cylinder*, which receives an image from a printing plate and subsequently transfers it to the print medium; and the *impression cylinder*, which helps move the print medium through the press. There are also several smaller cylinders that come into contact with the plate cylinder; these are used to apply ink and water to the printing plate. For simplicity, they are represented in Figure 12.16 as one cylinder for water and one for ink.

Offset presses are also categorized as being either *sheet-fed* or *web-fed*. Sheet-fed presses use pre-cut print media (individual sheets), and are typically capable of producing between 10,000 and 12,000 prints per hour. Web-fed presses use uncut print media stored on a roll, and are up to four times faster, producing almost 50,000 prints per hour (Fleming 2002). Web-fed presses commonly have integrated folding mechanisms and are used for printing newspapers, magazines, and so forth. Both categories of offset press are blazingly fast in comparison with methods for low-volume reproduction. Most offset presses print on one side of the print medium at a time although some, called *perfectors*, are capable of printing on both sides simultaneously.

As stated earlier, the printing plate is the first representation of a digital map to come into contact with ink. The ink used in offset printing is oil-based—a key characteristic in determining the manner in which ink adheres to the printing plate. After a latent image is transferred to a printing plate during prepress, the plate is chemically treated in a process that converts image areas (areas that have been photographically etched) into *oleophilic* (oil-liking) surfaces, and nonimage areas into *oleophobic* (oil-fearing) surfaces. As the offset press runs, water and ink are applied to the printing plate by the series of smaller cylinders near

the top of the printing unit; ink adheres only to the image areas and is washed away from nonimage areas.

Once ink is applied to the printing plate, the image on the plate could, in theory, be transferred directly to the print medium. If this were to occur, however, the resulting image would be reversed, as illustrated by the image on the blanket cylinder in Figure 12.17. To print a normal, nonreversed image, it is necessary to transfer the image from the printing plate onto the blanket cylinder, which subsequently transfers the image onto the print medium. The *offset* in offset lithography refers to this process—the printing plate is offset from the print medium by the blanket cylinder. The rubber coating on the blanket cylinder allows it to perform two additional duties. First, it acts as a buffer between the easily worn printing plate and the print medium. Second, it applies ink to porous print media more effectively than the rigid printing plate would be able to. As ink is transferred to the blanket cylinder and then to the print medium, it is forced to spread out due to pressures exerted between cylinders. This effect is termed **dot gain**. It tends to deteriorate fine detail and make tints slightly darker, as halftone dots are increased in size. Dot gain is sometimes compensated for by reducing the size of halftone dots when film negatives and printing plates are produced (Romano 1997).

The procedure just described accounts for the application of a single base color to the print medium. Multicolor print jobs are produced in the same manner, except that multiple printing units are employed, one for each base color, as illustrated in Figure 12.18. As the print medium passes through the press, it receives a different color ink from each successive print unit. The print units are spaced at a particular distance from one another to allow each successive color to print on the same region of the print medium. This allows all base colors to overlay correctly and mix on the page.

Registration refers to the alignment of colors in multicolor printing. When using the process colors, for example, a map feature composed of a mixture of cyan, magenta, and yellow *should* have crisp, well-defined

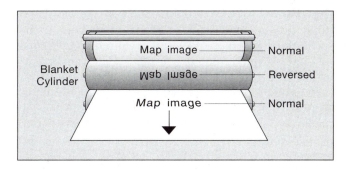

FIGURE 12.17 A blanket cylinder transferring the map image from the printing plate to the print medium.

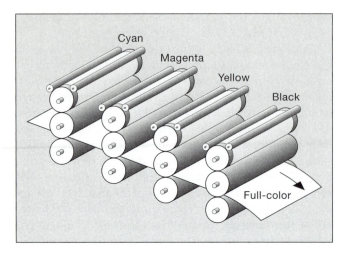

FIGURE 12.18 Multicolor printing achieved through the use of multiple printing units. In this example, process colors are mixed on the page, resulting in full-color output.

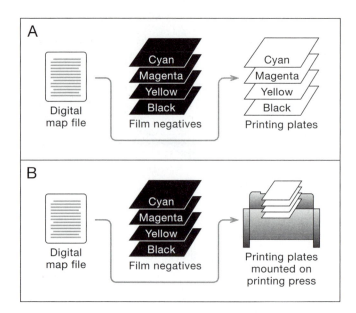

FIGURE 12.19 (A) Computer-to-plate printing. (B) Direct-to-press printing. (Note that in both cases the film negative is bypassed.)

edges, indicating that all three inks have been applied to the exact same area of the page. In reality, misalignment often prevents the perfect placement of colors, and so misregistration occurs. Misregistration can result from poorly produced film negatives or printing plates, or from an imprecise or poorly calibrated printing press. **Trapping** refers to a series of techniques used to minimize the effects of misregistration. It involves the manipulation of ink placement to improve the appearance of areas where inks overlap incorrectly, or where gaps occur (where inks should print, but don't). The trapping process is becoming largely automated through the use of special software employed by service bureaus, relieving the cartographer of what used to be a tedious and often difficult procedure (Agfa Corporation 1999).

12.8.5 Computer-to-Plate and Direct-to-Press

The procedures described in the previous section represent the current standard for offset lithographic printing. However, newer methods are increasing in popularity, deemphasizing or abandoning photographic technology in favor of digital technology. One such method is termed **computer-to-plate** (CTP), also known as direct-to-plate. This method is similar to traditional offset lithography, except that the film negative component is bypassed entirely; the printing plate is digitally imaged by a **platesetter**, directly from a digital map file (Figure 12.19A). The printing plate is normally coated with a photosensitive material and is photographically etched when selectively exposed to laser light within the platesetter, although some platesetters use thermal technology for imaging. CTP methods offer lower costs,

shorter prepress times, and fewer registration problems. They can also improve print quality because one generation of image transfer is removed from the printing process. A primary disadvantage of CTP printing is related to proofing. High-quality, separation-based proofs are created directly from film negatives; without negatives, alternative, lower quality forms of proofing must be employed. Another disadvantage is the initial cost of CTP equipment.

Direct-to-press methods take CTP one step further. Direct-to-press methods are based on either offset lithographic technology (direct imaging presses) or color laser printing technology (digital toner-based printing), and are currently tailored primarily to lower volume, fast-turnaround reproduction needs. A separate platesetter is not involved. With offset-oriented devices, the printing plate is imaged while it is mounted on the printing press (Figure 12.19B); if color laser printer technology is involved, the device acts much like a color laser printer, as described in section 12.7.1. Direct-to-press methods further streamline the prepress phase of reproduction, and virtually eliminate registration problems because the plates are imaged after they are mounted on the press. Initial equipment costs are high, and proofing remains a problem with offset-oriented devices. However, laser-oriented devices allow the printing plate to be reimaged on the fly, allowing inexpensive, last-minute changes to be made (Adams and Dolin 2002). The overall print quality of laser-based devices is very good, but doesn't match the quality of offset-based devices (Agfa Corporation 2000).

12.9 NONPRINT REPRODUCTION AND DISSEMINATION

In this section, we focus on methods of duplicating and distributing static and dynamic maps that will *not* be printed. This discussion is not focused on software tools used to create electronic maps or on mapping applications themselves, but rather on methods for reproducing and disseminating digital maps.

12.9.1 Television-Based

Television was the first medium used to disseminate maps widely in nonprint form. Most common has been the use of maps (particularly animated) for the weather segment of newscasts; even small stations spend thousands of dollars for such technology. Although commonly used in weather reporting, maps are less frequently used in other areas, such as national and international news. One reason for the lack of maps is that the home television screen imposes a number of constraints on image viewing. Because a television is a form of CRT display, some of these constraints are identical to those found in CRTs associated with personal computers. For example, the relatively low resolution of television (525 scan lines in the United States) means that small type and symbols are not possible and features can appear jagged. This problem has become less of an issue with the advent of **digital television** (DTV). DTV is a relatively new standard for high-definition television (HDTV) that provides higher resolution images and better sound quality. Digital television is currently being broadcast, and DTV-capable televisions are available but still relatively expensive. *Set-top boxes* are available that convert DTV signals into the NTSC format that can be viewed on traditional television sets, but this lower cost solution produces results that are inferior to those of true digital television (Lim 1998). Regardless of the technology employed, televisions produce low-resolution images in comparison with printed output. The display of color maps on televisions is also problematic, because the traditional (NTSC) television broadcast signal causes some colors (e.g., highly saturated reds and pale pastels) to transmit incorrectly (Caldwell 1981). Other constraints on the display of maps on television are related to the television medium itself. Particularly important is the limited time viewers have to examine maps, with the average news map appearing for less than 15 seconds (Caldwell 1979). Another constraint is that there is often little time to prepare a map for a late-breaking news story.

An important mode of map reproduction and dissemination (particularly for dynamic maps) associated with television has been the **videocassette**. Maps can be easily reproduced by copying from cassette to cassette, or from a computer, and can be viewed by anyone with a videocassette player. The **Digital Versatile Disc** (DVD) is gaining popularity as a high-quality medium for the reproduction and dissemination of maps for display on television sets.

12.9.2 Computer-Based

The reproduction of maps in digital form is simple in comparison with print reproduction: It consists primarily of digital file duplication, as performed by a computer's operating system. Reproduced digital maps can be stored on media such as the CD, DVD, and Zip disk, which are also commonly used for the dissemination of electronic maps. The **local area network** (LAN) offers an efficient method for electronic map dissemination within a limited area, such as within a business. Static digital maps can be disseminated in a wide variety of file formats, including the native format of the application software that created the map, a page description file, a portable document format such as EPS or PDF, or in one of many raster file formats. Raster formats commonly used include JPEG, GIF, TIFF, and BMP, although almost any raster format can be used if the map user has software that will display it. Dynamic maps can be disseminated as *executable files* that encapsulate a map that was created in a programming language such as Visual Basic, or that was created in an application such as Macromedia Director and saved as a *Projector* file (Figure 12.20). The Projector is an executable file that can include animation, audio, interactivity, and other features.

An important consideration in computer-based dissemination is that the characteristics of graphic displays vary, and thus the cartographer cannot anticipate exactly what a map will look like on the map user's computer. Display color and resolution can differ from system to system. The problem of color accuracy will be partly overcome as more people employ CMSs, but differences in resolution will likely remain an obstacle to the consistent representation of maps on computers.

12.9.3 Internet- and Web-Based

The **Internet** plays an important role in map dissemination, as it is capable of distributing static and interactive maps to large numbers of users almost instantaneously. Whereas the reproduction of digital maps is relatively simple, electronic map dissemination can be quite complex; a comprehensive discussion is beyond the scope of this text. The following is an overview of some of the technologies, protocols, and file formats that make Internet and Web mapping possible. The Internet allows for the transfer of digital maps via basic protocols such as **File Transfer Protocol** (FTP). Using FTP software, such as

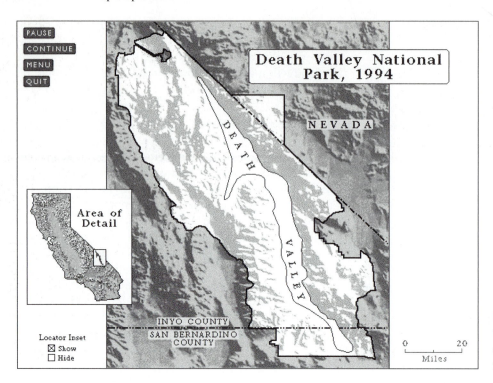

FIGURE 12.20 Screen shot of an interactive dynamic map that was disseminated as a Projector executable file.

WS_FTP, the cartographer can copy a digital map to a *file server*—a computer connected to the Internet and dedicated to the sharing of data and other software—where it can be accessed by a user, who downloads it to his or her computer using similar FTP software. This is an example of simultaneous reproduction and dissemination that was mentioned at the beginning of this chapter; the file is duplicated and distributed at the same time. Digital maps are often compressed to reduce their file size, allowing for faster dissemination. Several file formats, including JPEG and PDF, are automatically compressed when they are created. Other file formats need to be compressed using a dedicated compression utility, such as WinZip.

The Internet allows for basic map dissemination, but the World Wide Web offers far greater flexibility. The **World Wide Web** is a subset of the Internet, based on the **Hypertext Transfer Protocol** (HTTP)—a set of methods defining the manner in which documents are sent and retrieved. The primary file format for the Web is the **Hypertext Markup Language** (HTML). HTML is a relatively simple, text-based programming language that is used to define the content and appearance of Web pages. Web browsers such as Netscape and Internet Explorer are capable of interpreting HTML documents and displaying their contents (text and related images). At its core, HTML is only capable of displaying static maps in two raster formats (GIF and JPEG), and interactivity is limited to the use of hyperlinks that, when selected, open a different image, or direct the user to a different Web location (Kraak and Brown 2001).

To increase the functionality of Web browsers beyond what is possible with HTML, specialized software applications called plug-ins are used that allow Web browsers to interpret additional file formats. For example, you can disseminate a static digital map stored in PDF format from a Web site by placing the PDF file on a file server, and by providing a hyperlink to the map on your Web page. Assuming that the map user has the Acrobat Reader plug-in installed, the user can simply click on the hyperlink, causing the map to be reproduced, transferred, and displayed in their Web browser. A dynamic map can be disseminated in the same way, except that a different file format and plug-in will be used. For example, a dynamic map can be created in Macromedia's Flash format and stored on a file server. The user can download the map and display it in a Web browser, as long as the Flash plug-in is installed on the user's computer.

The functionality of the Web can be further expanded to disseminate maps through the use of Java and JavaScript. **Java** is a platform-independent, object-oriented programming language that allows the user to interact with maps and mapping applications via the Web. To use Java applications, or *applets*, the Web browser must have access to the *Java Virtual Machine* (JVM), which interprets Java code—JVM is included with both major Web browsers. Executable code and related data are downloaded from a Web site (the server), stored on the user's computer (the client), and processing is performed locally by the user's central processing unit (CPU). This arrangement allows Java applets to perform

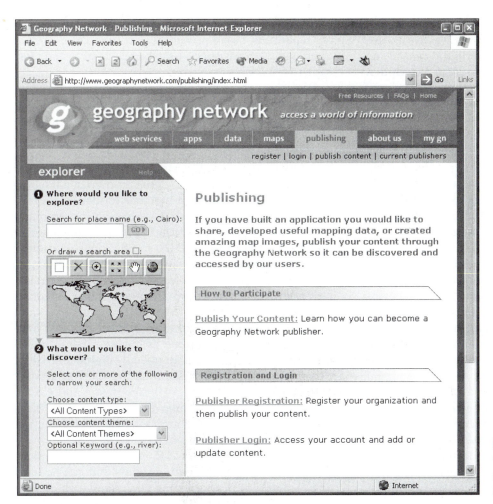

FIGURE 12.21 The Geography Network Web site, where digital maps and mapping applications can be published. Source: ESRI.

quickly once they have been downloaded (Kähkönon et al. 1999). **JavaScript** is similar to Java but is simpler and less robust. Unlike Java, which is interpreted by the JVM, JavaScript is embedded directly into HTML code, and is interpreted and displayed by the Web browser itself without the need for the JVM. The *Common Gateway Interface* (CGI) is an additional protocol that allows users to interact with mapping software and cartographic databases located on remote computers.

The most sophisticated Web-based map dissemination environments are composed of a combination of the technologies just described (Kraak and Brown 2001). For example, ESRI's ArcIMS (Internet Map Server) incorporates Java components and applets such as the Java Viewer, a plug-in-based map browser that can be customized using JavaScript. ArcIMS is designed to allow the user to interact with data stored locally, or on remote servers via CGI. Dedicated map servers such as this allow the user to create, query, and manipulate maps, and to view and interact with the results, all within a Web browser (Environmental Systems Research Institute Incorporated 2002). Examples of sophisticated map dissemination environments can be found on

the Geography Network (*www.geographynetwork.com*), a Web site where maps and mapping applications can be "published" (Figure 12.21).

It is important to note that HTML is being replaced gradually by the **eXtensible Markup Language** (XML). XML represents an approach to programming Web content that is better structured, more flexible, and more robust than HTML. The **Scalable Vector Graphics** (SVG) file format, which is based on XML, is an open, object-oriented standard for Web-based vector graphics. SVG is the most promising new standard for high-quality static and dynamic vector-based maps on the Web (Neumann and Winter 2001).

SUMMARY

In this chapter, we have presented topics related to print and nonprint **map reproduction** and electronic map **dissemination**. We have encouraged you to plan ahead for map reproduction, and have stressed the importance of **map editing**, both of which can help you avoid difficulties when reproducing a map.

Raster image processing was described as being central to printing digital maps, with the printing process distilled to four steps: creation of the digital map using **application software**, conversion into **page description data**, interpretation by a **RIP** and creation of a raster image, and printing of the image on a printing device. **Halftone** and **stochastic screening** methods were presented as ways to create **tints** from base colors. Aspects of color printing were described, including **process colors** that mix on the page, **spot colors** that are premixed, **high-fidelity process colors** that increase color variety and vibrancy, **continuous tone printing** that allows smooth color transitions without screening, and **color management systems** that identify and correct for variations in color introduced by electronic devices.

Methods for low-volume print reproduction were described, including monochromatic and color **laser printing**, **ink-jet printing**, and the use of color copy machines through the use of a **print controller**. **Thermal-wax transfer**, **dye-sublimation**, and **xerographic** techniques were also presented. Methods for low-volume reproduction do not normally offer volume discounts.

Offset lithography was identified as the dominant method for high-volume print reproduction, characterized by high quality, high speed, and volume discounts. **Prepress** involves the creation of **film negatives** or **color separations**, each of which represents one base color, **printing plates** that will be mounted to the offset press, and **proofs** that facilitate map editing. Prepress is centered on the **service bureau**. We learned that portable document formats such as **EPS** and **PDF** are preferred file formats for the delivery of digital maps to the service bureau.

Proofing methods were described on a scale ranging from low-end to high-end. The **on-screen** method is the cheapest and crudest, followed by **monochromatic composites** normally produced on laser printers; **color composites** produced on color laser printers, ink-jet printers, and so on; **separation-based proofs** created from color separations; and the **press check**—the highest quality and most expensive proof.

We learned that **offset lithographic printing presses** contain **printing units** composed of rolling cylinders, and that images are transferred from the printing plate to the blanket cylinder and then to the print medium. **Dot gain** was described as the spreading of ink due to pressures between cylinders on the press. **Registration** was described as the alignment of base colors on the print medium, and **trapping** was identified as a series of methods for correcting misregistration. **Computer-to-plate** was described as a streamlined approach to offset printing in which the film negative is eliminated; **direct-to-press** is similar, but the printing plates are imaged after they are mounted on the press.

Nonprint reproduction was introduced as being simple in comparison with print reproduction. It involves the use of **videocassette** and **DVD** recorders and digital file duplication on a computer. Television-based map dissemination was described as having several limitations, many of which are addressed by **digital television**. Computer-based map dissemination can be accomplished via several types of storage media and **LANs**. **Internet**- and **Web**-based map dissemination were identified as examples of simultaneous map reproduction and dissemination. Simple methods of dissemination involve **FTP**, **HTTP**, and **HTML**. We learned that plug-ins expand the variety of file formats that can be used online, and that programming languages such as **Java** and **JavaScript** allow for the creation and dissemination of digital maps. These technologies, together with **map servers**, allow sophisticated map dissemination environments to be published on the web. **XML** and **SVG** were identified as promising new standards for Web-based map dissemination.

FURTHER READING

Adams, J. M., and Dolin, P. A. (2002) *Printing Technology.* 5th ed. Albany, NY: Delmar, Thompson Learning.

> A comprehensive overview of printing technologies, including chapters on prepress, offset lithography, and digital printing.

Adobe Systems Incorporated. (1997) *The Adobe PostScript Printing Primer.* San Jose, CA: Adobe Systems Incorporated.

> An introduction to the PostScript page description language.

Agfa Corporation. (1997) *The Secrets of Color Management.* Agfa-Gevaert N.V., Septestraat 27, B-2640 Mortsel, Belgium.

> A primer on color management systems.

Agfa Corporation. (1999) *From Design to Distribution in the Digital Age.* Agfa-Gevaert N.V., Septestraat 27, B-2640 Mortsel, Belgium.

> A primer on the digital publishing process.

Agfa Corporation. (2000) *An Introduction to Digital Color Printing.* Agfa-Gevaert N.V., Septestraat 27, B-2640 Mortsel, Belgium.

> A primer on digital printing workflows.

Bruno, M. H., ed. (2000) *Pocket Pal: A Graphic Arts Production Handbook.* Memphis, TN: International Paper Company.

> A popular guide to graphic arts production.

Caldwell, P. S. (1979) *Television News Maps: An Examination of Their Utilization, Content, and Design*. Unpublished PhD dissertation, University of California at Los Angeles, Los Angeles, CA.

A dated but extensive treatment of maps in television. Also see Caldwell (1981).

Cardin, J., Castellanos, A., and Romano, F. (2001) *PDF Printing and Publishing: The Next Revolution After Gutenberg.* Agfa Corporation. Agfa-Gevaert N.V., Septestraat 27, B-2640 Mortsel, Belgium.

An in-depth description of the Portable Document Format and how it fits into modern, digital reproduction workflows.

Cartwright, W., and Stevenson, J. (2000) "A toolbox for publishing maps on the World Wide Web." *Cartography* 29, no. 2:83–95.

A description of tools for Web-based map dissemination.

Fleming, P. D. (2002) *http://www.wmich.edu/~ppse/Offset/*.

A comprehensive, Web-based overview of offset lithography.

Gartner, G., ed. (2002) *Maps and the Internet 2002*. Vienna, Austria: Geowissenschaftliche Mitteilungen, Band 60, TU Wien, Institute of Cartography and Geomedia Technique.

Papers presented at the 2002 Annual Meeting of the International Cartographic Association Commission on Maps and the Internet.

Hyatt, J. (2000) "Stochastic screening in cartographic applications." *Cartouche* 40:14–19.

An overview of screening methods and their use in cartography.

Kraak, M-J., and Brown, A., eds. (2001) *Web Cartography: Developments and Prospects.* London: Taylor & Francis.

Includes chapters related to Web-based map dissemination.

Lim, J. S. (1998) "Digital television: Here at last." *Scientific American* 278, no. 5:78–83.

An introduction to digital television and related standards.

Monmonier, M. S. (1993) *Mapping It Out: Expository Cartography for the Humanities and Social Sciences*. Chicago: University of Chicago Press.

Chapter 5 includes a discussion of copyright issues related to map reproduction.

Neumann, A., and Winter, A. M. (2001) "Time for SVG—Towards high quality interactive web-maps." *Proceedings, 20th International Cartographic Conference*, International Cartographic Association, Beijing, China, CD-ROM.

Describes the capabilities and shortcomings of SVG in the context of Web-based mapping technologies.

13

Choropleth Mapping

OVERVIEW

This chapter covers the choropleth map, which is arguably the most commonly used (and abused) method of thematic mapping. Section 13.1 considers the nature of data appropriate for choropleth mapping. Ideally, choropleth maps should be used for phenomena with spatial variation that coincides with the boundaries of enumeration units. In practice, this seldom occurs, and so data values depicted for enumeration units must be viewed as "typical," as opposed to being uniform throughout enumeration units. In Chapter 4, we stressed the need to standardize data for choropleth mapping, and standardized by dividing the raw data by the area of enumeration units. Here we consider additional methods for standardizing data; for example, areas of enumeration units can be accounted for indirectly by taking the ratio of two raw totals that do not involve area, such as a ratio of males to females.

Section 13.2 considers whether data for a choropleth map should be classed and, if so, which classification method should be used. Classed maps are commonly used because they are considered simpler than unclassed maps, and so we initially make the assumption that you wish to create a classed map. Because we examined methods of data classification extensively in Chapter 5, we only touch on classification methods in this chapter.

Section 13.3 considers numerous factors involved in selecting appropriate color schemes for choropleth maps. Some key factors include the kind of data (bipolar, polar, or balanced), the ease of assigning names to colors (when using multiple hues, users should be able to easily assign names to the hues), ease of interpretation for the color-vision-impaired, and avoiding problems of simultaneous contrast. These and other factors have been utilized by Cindy Brewer and her colleagues at Penn State University to develop a broad set of color schemes for choropleth maps.

Section 13.4 describes various approaches for specifying particular colors of a scheme. For example, you might decide that varying lightnesses of blue are appropriate for a unipolar data set, but how should you select the shades of blue making up the scheme? We will consider approaches for soft-copy maps, black-and-white printed maps, and colored printed maps. We will find that the Web site for ColorBrewer is especially useful for providing detailed color specifications.

In section 13.5, we consider some issues pertaining to designing legends for choropleth maps. One key is to utilize readers' prior cognitive framework. For instance, in designing a vertical legend, it makes sense to display high values at the top of the legend, using the notion that people associate "up" with "higher." On the other hand, to enhance the overall map design, a horizontally arranged legend might be appropriate.

In section 13.6, we return to the issue of classed versus unclassed maps. We argue that an unclassed map is appropriate when you wish to illustrate numerical relations among data values, or when you wish to explore data (where classed and unclassed maps are both alternatives). We also consider various experimental studies of classed versus unclassed maps. Limitations of such studies include a lack of direct comparison between perceived regions on classed and unclassed maps, a failure to measure the time required to process the map, and viewing maps in contrived experimental situations.

13.1 SELECTING APPROPRIATE DATA

Ideally, the choropleth technique is most appropriate for a phenomenon that is uniformly distributed within each enumeration unit, changing only at enumeration unit boundaries. For instance, state sales tax rates are appropriate because they are constant throughout each state, changing only at state boundaries. Because this ideal is seldom achieved in the real world, you should be cautious in using the choropleth map. If you wish to focus on "typical" values for enumeration units, you can utilize the choropleth map, but you should realize that error might be present in the resulting map. Instead of using the choropleth map because you have data readily available for enumeration units, you should consider whether some of the other symbolization methods discussed in this book might be more appropriate. (See section 4.4 for further elaboration of this issue.)

Of particular concern in choropleth mapping is the size and shape of the enumeration units. Ideally, the method works best and is more accurate when there is not significant variation in the size and shape of the units. The problem can be understood by considering the county-level map of population density shown in Figure 1.4. Imagine trying to compute a measure of population density for one of the large counties in California. Because people are unlikely to be uniformly spread throughout the county, the shade shown is a poor representation for the county. In contrast, a very small county, such as one of the five counties of New York City, would be well represented because the population density is relatively uniform. The net result is that there is considerable variation across the map in terms of our ability to properly represent population density. Also note that the largest counties provide the biggest visual impact, but also have

potentially the largest error of representation. Thus, the choropleth technique is more appropriately used when enumeration units are similar in size, such as for the counties of Iowa, remembering that ideally a phenomenon should be relatively uniform throughout enumeration units.*

Another important issue in selecting data for choropleth maps is ensuring that raw-total data have been adjusted to account for varying sizes of enumeration units. In Chapter 4 (see Figure 4.8), we standardized by dividing an area-based raw total (acres of wheat harvested for Kansas counties) by the areas of those counties. Because the numerator and denominator were both in the same unit of measurement (acres), a proportion (or percentage) resulted. We now consider additional approaches for standardizing data. One approach is to divide an area-based raw total by some other area-based raw total. For example, for acres of wheat harvested, we might divide by acres of wheat planted, with the resulting ratio providing a measure of success of the wheat crop (Figure 13.1A). As an alternative, we might divide acres of wheat harvested by acres harvested for all major crops, producing a map illustrating the relative importance of wheat to the agricultural economy of each county.

A second approach for standardizing data is to create a density measure by dividing a raw total not involving area by either the areas of enumeration units or some area-based raw total. For example, if bushels of wheat

---

* For studies of the effect of size and shape of enumeration units, see Coulson (1978) and MacEachren (1985).

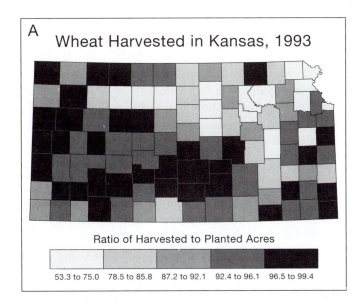

A Wheat Harvested in Kansas, 1993

Ratio of Harvested to Planted Acres

| 53.3 to 75.0 | 78.5 to 85.8 | 87.2 to 92.1 | 92.4 to 96.1 | 96.5 to 99.4 |

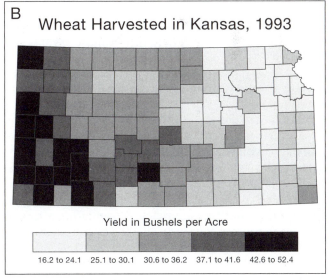

B Wheat Harvested in Kansas, 1993

Yield in Bushels per Acre

| 16.2 to 24.1 | 25.1 to 30.1 | 30.6 to 36.2 | 37.1 to 41.6 | 42.6 to 52.4 |

FIGURE 13.1 Examples of approaches for standardizing data. (A) Dividing acres of wheat harvested by acres of wheat planted; both are area-based raw totals and thus a proportion (or percentage) results. (B) Dividing bushels of wheat harvested by acres of wheat harvested; the result is a density measure (bushels per acre). (Data source: Kansas Agricultural Statistics 1994.)

produced is divided by the area of the county (in acres), then the result is bushels of wheat per acre. This approach might not be meaningful, however, if the raw-total data occur only within a portion of the enumeration unit. For example, with the wheat data, it makes more sense to divide the bushels produced by the acres of wheat harvested, yielding bushels of wheat produced per acre of wheat (Figure 13.1B). A third approach for standardizing data is to compute the ratio of two raw totals, neither of which involves area. For example, we might divide the value of wheat harvested (in dollars) by the value of all crops harvested (in dollars). The resulting proportion would indicate the relative value of wheat in each county. Although area is not included in the formula, this approach indirectly standardizes for area because larger areas tend to have larger values for both attributes.

When computing the ratio of two raw totals not involving area, it is common to express the result as a *rate*. Although we are all familiar with rates, we often aren't aware of how they are computed. To illustrate, consider how cancer death rates are computed for counties. First, we establish a simple proportional relationship as follows:

$$\frac{\text{Cancer deaths for county}}{\text{Population for county}} = \frac{\text{Number of cancer deaths}}{100{,}000 \text{ people}}$$

We then solve for the number of cancer deaths by rewriting the equation as follows:

$$\frac{\text{Number of}}{\text{cancer deaths}} = \frac{\text{Cancer deaths for county}}{\text{Population for county}} \times 100{,}000$$

The resulting number of cancer deaths is termed the cancer death rate. More generally, the formula for rates is:

$$\text{Rate} = \frac{\text{Magnitude for category of interest}}{\text{Maximum possible magnitude}} \times \frac{\text{Units}}{\text{of the}}_{\text{rate}}$$

A fourth standardization approach is to compute a summary numerical measure (e.g., mean or standard deviation) for each enumeration unit. For example, we could compute the average size of farms in each county by dividing the acreage of all farms by the number of farms. Note that this approach accounts for a larger acreage in a larger county by dividing by a greater number of farms.

In general, note that all of the standardization approaches discussed thus far involve ratios. Thus, one might suggest that simply computing ratios is the key to standardization. A simple example illustrates that this is not the case. Imagine that for each county you computed the mean number of acres of wheat harvested over a 10-year period. Clearly, the values would be ratios (the numerator would be the sum of acreages over 10 years and the denominator would be 10), but the data would not be standardized because the denominator in each case would be 10. No adjustment would have been made to account for the fact that larger counties tend to have larger numbers of acres harvested each year.

It is important to recognize that the various standardization approaches lead to quite different maps of the phenomenon being investigated. For example, Figure 4.8B (representing the proportion of land area from which wheat was harvested) revealed a peak in the south central part of the state, with a tendency for higher values in the western two-thirds of the state. In contrast, Figure 13.1A (illustrating the percentage of planted wheat actually harvested) reveals a more random pattern, although we note that low values extend across the northeastern and north central sections of the state. The map of wheat yield (Figure 13.1B) produces yet another pattern, with progressively higher values as one moves westward across the state. Clearly, different methods of standardization produce different views of the spatial pattern of wheat harvested in Kansas. It is your responsibility as the mapmaker to think carefully about which standardization procedure is most appropriate, given your data and the message that you wish to communicate.

13.2 DATA CLASSIFICATION

One important decision with respect to data classification is whether to class the data. A potential advantage of an unclassed map is that it can more accurately portray the data; for example, on a gray-tone map the intensity of a shade can be related directly to the magnitude of individual data values. In contrast, a potential advantage of a classed map is that the limited number of categories (on a five-class map, there will be five categories) makes the map easy to process. In deciding whether to class your data, you must also consider whether you wish to present a map to others or explore a data set on your own. When presenting a map to others, you might be limited to showing one map (as in a printed publication), whereas when exploring data, you can visualize the data in many ways (both classed and unclassed maps can be used). For the bulk of this chapter, we assume that you wish to create a map with a limited number of categories, and that you wish to present the map to others, making the classed map an obvious choice. In section 13.6 we return to the issue of classed versus unclassed maps, and consider some research that has been done on this topic. We should also point out that the bulk of choropleth maps produced today are classed, even though unclassed maps can be created easily. Finally, many common mapping packages do not provide an option for mapping unclassed data, which might make your decision quite straightforward.

Once you have decided that classification is appropriate, you need to consider the range of classification methods described in Chapter 5. Remember that before selecting a method, you need to consider the kind of data you have collected (bipolar, balanced, or unipolar) and what level of rounding you wish (see section 5.1 for further discussion). For our purposes, we assume that we wish to map the percentage of the adult population in U.S. states that did not graduate from high school (Table 13.1). This data set is unipolar, and so we chose not to divide the data into two groups, as we would likely do with bipolar data. Because the data are reported to the nearest tenth of a percent and we felt that readers of this book would be comfortable with this level of precision, we chose to retain it for the classification.

Next, you need to consider the actual methods of classification, such as equal intervals, quantiles, mean–standard deviation, and optimal. Because we were creating a map for this book and knew that readers would be familiar with the optimal method (from Chapter 5), we chose it. If we were creating the map for a daily newspaper, we might select a different method; for instance, if we wished to focus on percentiles in the data (say the top or bottom 20 percent), then we would use quantiles. The last decision is to select an appropriate number of classes. An advantage of the optimal method is that it can assist in selecting an appropriate number of classes. Although we considered plotting the number of classes against GADF to determine the number of classes, we chose five classes because we wished to map this data set with both color and gray-tone maps, and we felt the gray-tone maps would be most easily interpreted if no more than five classes of data were shown.

13.3 FACTORS FOR SELECTING A COLOR SCHEME

In this section, we consider numerous factors that can be utilized in selecting a color scheme, focusing largely on the recent work of Cynthia Brewer and her colleagues: Brewer (1994a; 1996), Brewer et al. (1997), Brewer (2001), and Harrower and Brewer (2003).

13.3.1 Kind of Data

In her early work, Brewer (1994a) suggested that the *kind of data* (unipolar, bipolar, balanced) should play an important role in selecting a color scheme. For unipolar data, she recommended that the sequential steps in the data should be represented by sequential steps in lightness; Brewer termed this a **sequential scheme**. The most obvious example would be tones of gray, although in general

TABLE 13.1 Percentage of the adult population, in U.S. states, that did not graduate from high school (based on 1990 data)

State	% Not Graduating
Alabama	33.1
Arizona	21.3
Arkansas	33.7
California	23.8
Colorado	15.6
Connecticut	20.8
Delaware	22.5
Florida	25.6
Georgia	29.1
Idaho	20.3
Illinois	23.8
Indiana	24.4
Iowa	19.9
Kansas	18.7
Kentucky	35.4
Louisianna	31.7
Maine	21.2
Maryland	21.6
Massachusetts	20.0
Michigan	23.2
Minnesota	17.6
Mississippi	35.7
Missouri	26.1
Montana	19.0
Nebraska	18.2
Nevada	21.2
New Hampshire	17.8
New Jersey	23.3
New Mexico	24.9
New York	25.2
North Carolina	30.0
North Dakota	23.3
Ohio	24.3
Oklahoma	25.4
Oregon	18.5
Pennsylvania	25.3
Rhode Island	28.0
South Carolina	31.7
South Dakota	22.9
Tennessee	32.9
Texas	27.9
Utah	14.9
Vermont	19.2
Virginia	24.8
Washington	16.2
West Virginia	34.0
Wisconsin	21.4
Wyoming	17.0

a sequential scheme can be achieved by holding hue and saturation constant, and varying lightness (e.g., progressing from a light orange to a dark orange). Concurring with earlier cartographers' recommendations (e.g., Robinson et al. 1984), Brewer advocated using light colors for low data values and dark colors for high data values, respectively. She argued that if the opposite is done, as is sometimes the case on graphic displays, a clear legend is essential. These recommendations are supported by studies by McGranaghan (1989; 1996).

Although Brewer argued that for sequential schemes lightness differences should predominate, she stressed that visual contrast could be enhanced if saturation differences also were used. This can be seen in Color Plate 13.1 in which a pure lightness scheme is compared with a combined lightness–saturation scheme. In Color Plate 13.1A only lightness changes—from a light to a dark green; in Color Plate 13.1B lightness again changes from light to dark green, but saturation also changes, increasing from the first to the third class and then decreasing for the latter two classes. Although in this case the increase and decrease in saturation does not correspond logically to the continual increase in the data, Brewer (1994a, 137) indicated this is acceptable "if high saturation colors do not overemphasize unimportant categories."

Brewer suggested that hue differences could also be used for sequential schemes, but that such differences should be subordinate to lightness differences. Color Plate 13.2A illustrates the middle five classes of a yellow to green to purple scheme that she advocated. Brewer did not recommend a greater range of hues, although she noted that "sequential schemes may be constructed that use the entire color circle" (as in Color Plate 13.2B). In our opinion, the latter implies qualitative differences and therefore should not be used for numerical data.

For bipolar data, Brewer (1994a, 139) recommended a **diverging scheme**, in which two hues diverge from a common light hue or neutral gray. An example converging on a light hue is a dark green–greenish yellow–yellow–orange–dark orange scheme, and an example converging on a neutral is a dark red–light red–gray–light blue–dark blue scheme (Color Plate 13.3 illustrates the latter). Brewer argued that the greatest flexibility is achieved if diverging schemes are thought of as two sequential schemes running end to end, with the lightest tone toward the middle.

Brewer discussed balanced data in association with bivariate color schemes because balanced data technically involve two attributes. We consider balanced data here because the associated legend appears only slightly more complicated than a legend for a univariate map (on a five-class map, percent English might be labeled above the legend boxes and percent French below the legend boxes). Brewer indicated that although balanced schemes can be created by overlaying two sequential schemes—she mixed magenta and cyan to create a purple—she recommended using standard sequential schemes to emphasize the high end of the data, or alternatively a diverging scheme to emphasize the midpoint of the balanced data.

Brewer's (1994a) early work was based on "personal experience, cartographic convention and the writings and graphics of others" (124). In more recent work, Brewer and her colleagues (1997) performed experiments with people to determine the effectiveness of various color schemes; in particular, they applied diverging, sequential, and spectral schemes to unipolar data. The logic of applying a diverging scheme to unipolar data is that if a quantiles classification is used, the median will fall in the middle class and serve as a logical dividing point in the data.* A **spectral scheme** is based on the electromagnetic spectrum (red, orange, yellow, green, blue, indigo, and violet). Traditionally, cartographers opposed using spectral schemes because yellow is inherently a light color, and thus seems out of place if shown in the middle of a spectral scheme. Brewer and her colleagues, however, argued that a satisfactory scheme could be developed by progressing from *dark* blue to *bright* yellow to *dark* red. Interestingly, Brewer and her colleagues found that the diverging, sequential, and spectral schemes yielded very similar visualizations of data, in part because they took great care in developing these schemes. Color Plate 13.4 illustrates some of their schemes applied to the high school education data.

More recently, Harrower and Brewer (2003) developed the program ColorBrewer (*http://www.ColorBrewer. org*) to assist those who know little about map design in selecting appropriate color schemes. ColorBrewer provides three basic schemes: sequential, diverging, and qualitative. Sequential schemes "are suited to ordered data that progress from low to high," whereas diverging schemes "put equal emphasis on mid-range critical values and extremes at both ends of the data range." Qualitative schemes are not relevant to this chapter as they are intended for qualitative (or nominal) data. A total of 18 sequential and nine diverging schemes are possible, with sequential schemes ranging from three to nine classes, and diverging schemes from three to 11 classes. Using a hypothetical spatial pattern, ColorBrewer allows users to see whether individual colors can be easily discriminated on a map (see Color Plate 13.5). The program also allows users to see how the colors will look when overlaid with other information (e.g., a road network), and provides an indication of whether the color scheme will work in various situations (e.g., projected on a screen or photocopied; see the lower left of Color Plate 13.5). We consider ColorBrewer again in section 13.4 when we deal with the details of color specification.

---

* For the optimal classification of the high school education data, the median also happens to fall in the middle class.

13.3.2 Color Naming

In addition to working with *kind of data*, Brewer and her colleagues have utilized several other factors in developing their color schemes. One important factor for diverging schemes is selecting colors that are not confused with one another, and thus are readily "named" as distinct colors. For instance, if red and pink were assigned to opposite ends of a diverging scheme, a user might have difficulty associating a particular color with the proper end of the diverging scheme. In contrast, blue and red would not be confused, as blue colors look quite different than red colors and thus are readily named as distinct colors. The second column of Table 13.2 indicates which color pairs are easily named (an "OK" appears in the table).

TABLE 13.2 Color pairs appropriate for diverging color schemes

Easily Named Colors	Confusions		
	Color Naming	Color Vision Impairment	Simultaneous Contrast
Pink-Red	*	*	OK
Pink-Orange	*	*	OK
Pink-Brown	*	*	OK
Pink-Yellow	OK	*	OK
Pink-Green	OK	*	*
Pink-Blue	OK	*	OK
Pink-Purple	*	*	OK
Pink-Gray	*	*	OK
Red-Orange	*	*	OK
Red-Brown	*	*	OK
Red-Yellow	OK	*	OK
Red-Green	OK	*	*
Red-Blue	OK	OK	OK
Red-Purple	*	OK	OK
Red-Gray	OK	*	OK
Orange-Brown	*	*	OK
Orange-Yellow	*	*	OK
Orange-Green	OK	*	OK
Orange-Blue	OK	OK	OK
Orange-Purple	OK	OK	OK
Orange-Gray	OK	*	OK
Brown-Yellow	*	*	OK
Brown-Green	*	*	OK
Brown-Blue	OK	OK	*
Brown-Purple	*	OK	OK
Brown-Gray	*	*	OK
Yellow-Green	*	*	OK
Yellow-Blue	OK	OK	*
Yellow-Purple	OK	OK	OK
Yellow-Gray	*	OK	OK
Green-Blue	*	*	OK
Green-Purple	OK	*	*
Green-Gray	*	*	OK
Blue-Purple	*	*	OK
Blue-Gray	*	OK	OK
Purple-Gray	*	*	OK

Source: Based on Brewer 1996, p. 81.

13.3.3 Color Vision Impairment

The various forms of color vision impairment described in section 10.1.5 can be represented in the CIE color model by a series of *confusion lines*, or lines along which colors are confused with one another. For example, Figure 13.2 illustrates confusion lines for protanopes and deuteranopes (two subgroups of dichromats, people who cannot distinguish red from green). Colors running along the confusion lines will be confused by these groups, whereas colors running roughly perpendicular to the lines should be distinguishable.

Using CIE diagrams like the one shown in Figure 13.2, Olson and Brewer (1997) developed sets of confusing and accommodating color schemes. As an example, Figure 13.3 portrays their results for a diverging color scheme representing gains or losses in the number of manufacturing jobs. For the confusable rendition, gains are represented by a yellow-green to dark green scheme, whereas losses are represented by a yellow-orange to red scheme. Note that these colors all fall along the same set of confusion lines in the CIE diagram. In contrast, the accommodating rendition ranges from light blue to dark blue and light orange-red to red, and is oriented perpendicular to the confusion lines. Not surprisingly, Olson

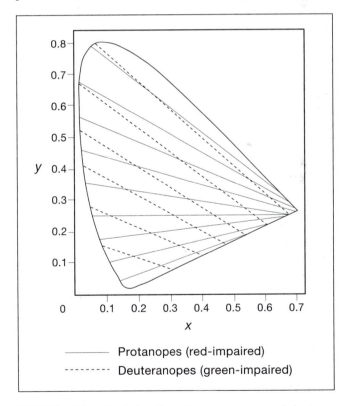

FIGURE 13.2 Confusion lines for protanopes and deuteranopes drawn on an *Yxy* 1931 CIE diagram. Colors of similar lightness will be difficult to discriminate if they are placed along the same confusion line. (After Olson and Brewer 1997, p. 108.)

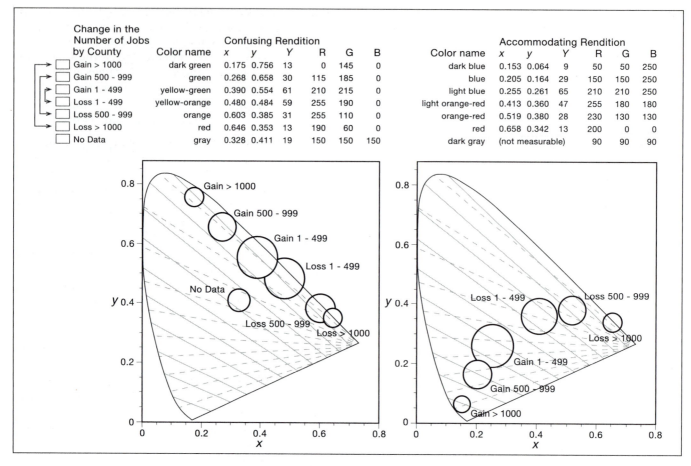

FIGURE 13.3 Confusing and accommodating diverging color schemes for those with red–green vision impairment. Note that colors for the confusing scheme are arranged along similar confusion lines, whereas colors for the accommodating scheme cross the confusion lines. (After Olson, J. M. and Brewer, C. A. (1997). "An evaluation of color selections to accommodate map users with color-vision impairments." *Annals, Association of American Geographers*, 87, no.1, p. 131. Courtesy of Blackwell Publishing.)

and Brewer found that color-impaired readers interpreted accommodating schemes more easily.

The third column of Table 13.2 lists diverging color pairs that are acceptable for the color-vision-impaired. If you are showing a diverging scheme to the general public, you should consider using these color pairs. If the maps are to be shown in an interactive environment, then you might wish to provide an option for viewers to indicate whether they are color-vision-impaired. Those without impairment would then be able to take advantage of the broader set of easily named colors shown in column 1 of the table.

Although the multiple hues present in some sequential color schemes could, in theory, create problems for the color-vision-impaired, lightness differences normally enable proper differentiation of classes (e.g., a dark red to orange to yellow). Standard spectral schemes (that progress through red, orange, yellow, green, and blue)

are inappropriate for the color-impaired, but careful modification of the standard scheme can yield suitable colors—for instance, Brewer (1997) recommended a red, orange, yellow, blue-green, blue, and purple-blue as one of several options.

13.3.4 Simultaneous Contrast

As discussed in section 10.1.4, simultaneous contrast causes the color of an area to shift toward the opponent color of a surrounding color (e.g., a gray tone surrounded by green will appear reddish). The fourth column of Table 13.2 lists diverging color pairs that are not affected by simultaneous contrast. Although not many color pairs are affected by simultaneous contrast, note that many of those affected do coincide with easily named color pairs. The fact that a color pair is unacceptable from the standpoint of simultaneous contrast does not necessarily prevent it from being used on a particular map. Rather, you need to examine map patterns to see whether situations arise in which simultaneous contrast is likely to be a problem (the program ColorBrewer facilitates this process).

13.3.5 Map Use Tasks

Map use tasks refers to whether a map is used to obtain specific or general information, and whether this information is acquired while looking at the map or recalled from memory. To examine the role color schemes play in map use tasks, Janet Mersey (1990) had map readers interpret a spatial distribution using a variety of color schemes. In Color Plate 13.6 we use the high school graduation data to illustrate the various color schemes that Mersey tested (she utilized another unipolar data set). For specific acquisition tasks, Mersey found that unordered hues (Color Plate 13.6A) worked best. This is not surprising given that shades for ordered schemes are difficult to discriminate and this discrimination is complicated by simultaneous contrast. For general acquisition tasks, Mersey found that ordered schemes (Color Plates 13.6E and 13.6F) performed better, thus supporting the traditional thinking on using shades of a single hue. Mersey focused just on general tasks in her discussion of results for memory, finding that lightness-based schemes outperformed hue-based schemes, with the hue–lightness scheme (Color Plate 13.6D) the best overall.

Although Mersey's study illustrates that certain map use tasks are more effectively accomplished with particular color schemes, you should realize that it is often difficult to predict how a map will be used. It is probably more useful to utilize color schemes that permit a broad range of map use tasks. In Mersey's case, she found that the hue–lightness scheme worked best, scoring highest on four of the 10 tasks tested, and a close second or third on the remaining tasks. In more recent research, Brewer and her colleagues (1997) found that a broad range of carefully selected color schemes worked well for a wide variety of map use tasks.

13.3.6 Color Associations

The fact that certain colors are often associated with particular phenomena might allow you to pick a logical color scheme (e.g., in the United States people associate the color green with money). You should be aware, however, that these associations are cognitive and cultural in nature, and thus could change over time and be inconsistent among map users. A good illustration of the potential for change over time is the use of blue and red for cool and warm temperatures. In an early study, Cuff (1973) concluded these associations were not effective,

but a more recent study by Bemis and Bates (1989) found the opposite. Bemis and Bates suggested that the different results might be a function of the more common use of this scheme since the time of Cuff's study and thus users' greater familiarity with it.

13.3.7 Aesthetics

The aesthetics of a color scheme are an important consideration, regardless of how effective that scheme might be otherwise. As an example, consider the study of Slocum and Egbert (1993) in which participants compared traditional static maps (the map is viewed all at once) with sequenced maps (the map is built while the reader views it). Their study was split into two major parts: a formal one in which participants viewed one of the two types of maps (static or sequenced) and performed various map use tasks, and an informal one in which participants viewed both types of maps and were asked to comment on them. Slocum and Egbert used a yellow–orange–red hue–lightness scheme similar to Mersey's in the formal portion, and a lightness-based blue scheme in the informal portion. Although the major purpose of the informal portion was to have participants compare the method of presentation (static or sequenced), they often commented on the color schemes, indicating their preference for the blue scheme over the supposedly more appropriate yellow–orange–red scheme.

Interestingly, studies of color preference outside the discipline of cartography have also found the color blue appealing. For example, using people's rating of Munsell color chips, Guilford and Smith (1959) found that colors were preferred in the order blue, green, purple, red, and yellow. More recently, McManus et al. (1981) found that people's preferences were more variable than Guilford and Smith implied but that blue and yellow were still the most and least preferred, respectively. One limitation of such studies is that they tend to focus on particular groups of people. For example, McManus et al. studied only "undergraduate members of the University of Cambridge" (653). What might the results have been if the study were done in other areas of the world? Because color preference varies among individuals, and there are likely to be differences among cultures, a key point to remember is that color schemes *you* find attractive might not be attractive to others.

13.3.8 Age of the Intended Audience

The age (and presumably experience) of the intended audience is an important consideration in selecting colors. For example, young children would probably not be familiar with the blue–red scheme commonly used for representing temperature data. Research by Trifonoff (1994; 1995) suggests that young children have a strong preference for color over black-and-white maps: In comparing four methods of symbolization (gray tones, proportional circles, proportional circles coded redundantly with gray tones, and shades of red), children overwhelmingly chose the colored symbolization (shades of red) as most desirable. Although this finding might be true for only young children, Brewer and her colleagues (1997) found their participants (college students) had a strong preference for color maps, as a black-and-white scheme was decidedly less preferred.

13.3.9 Presentation versus Data Exploration

From Chapter 1, recall that maps can be used in two basic ways: to *present* information to others (for *communication* purposes) and to *explore* data. In selecting color schemes, we have made the assumption that the former is our goal. If exploration is instead our goal, then the previously recommended schemes can still be used, but a greater variety of schemes also might be useful, even what might be considered "inappropriate" schemes. For example, John Berton (1990, 112) argued:

> The continuous blending of color is not conducive to highlighting artifacts in the interior of the data set. While it is usually critical that a palette not "double-back" on itself... certain kinds of banding effects which violate the traditional order of the color wheel can provide excellent markers for transitional areas in the data.

In section 21.3.10 we discuss the Transform software, which provides a range of unusual color sequences (one of these is illustrated in Color Plate 21.19).

13.3.10 Economic Limitations and Client Requirements

In an ideal world, there are no economic limitations, and thus the cartographer has sole responsibility for selecting colors. In the real world, of course, this usually is not the case. One obvious economic limitation is the expense of color reproduction in book or journal form: Although color might communicate information more effectively than black-and-white, it might not be feasible from an economic standpoint. As an example, academic journals often require authors to pay for color reproduction, which can cost $500 or more per page.

If you are employed in a cartographic production laboratory, you generally do not design a map completely on your own, but rather you must respond to client requirements or desires. Although you might suggest an ideal, or optimal, set of colors, clients might reject these because they find other colors more pleasing or have traditionally used other colors. For example, the director of a cartographic production laboratory told us "the school district made us make a boundary map for attendance areas in hot pink and blue to 'match their old map.'" Although mapmakers can try to dissuade clients from choosing such schemes, they must bear in mind that clients are keeping the laboratory in business.

13.4 DETAILS OF COLOR SPECIFICATION

In the previous section we considered numerous factors involved in selecting color schemes. Our emphasis was on the general nature of color schemes, as opposed to detailed color specifications. In this section, we consider such details. Because we need to create color schemes on a variety of soft-copy (on-screen) and hard-copy (printing) devices, we have split this section into three major parts: methods for soft-copy display, methods for black-and-white printed maps, and methods for colored printed maps.

13.4.1 Methods for Soft-Copy Display

It is logical to discuss methods for soft-copy display first because we normally design our maps on screen. Methods for soft-copy display include (1) color ramping, (2) the CIE color model, and (3) the ColorBrewer software.

Color Ramping

A simple and traditional approach for soft-copy color specification is **color ramping**, in which users select endpoints for a desired scheme from a color palette, and the computer automatically interpolates values between the endpoints on the basis of RGB values. As an example, imagine that a user selects endpoints of white (RBG values of 255, 255, 255) and black (RBG values of 0, 0, 0), and wishes to produce a five-class map. Simple interpolation between these endpoints yields RGB values of: 191, 191, 191; 128, 128, 128; and 64, 64, 64. The problem with this approach is that a simple arithmetic increase in a color gun does not correspond to an arithmetic increase in perceived lightness, and thus the resulting shades will not *appear* equally spaced. Although interpolation algorithms might be modified to account for this discrepancy, they generally are not. As a result, color ramping must be used with caution.

CIE Color Model

A more sophisticated approach for specifying colors is to utilize the CIE color model. This can be accomplished in two ways. One is to use equivalencies that have been established between the perceptually uniform Munsell color space and the 1931 *Yxy* CIE colors (see Chapter 10 for a discussion of the Munsell and CIE color models). For example, a Munsell 2.5R 9/6 is equivalent to the following *Yxy* CIE coordinates: .7866, .3665, and .3183.[†] The resulting CIE colors can be converted to RGB values using equations provided by Travis (1991, 93–97). A second approach is to specify endpoints for a color scheme in RGB form, convert these to the 1976 CIE L*u*v* perceptually uniform color space, interpolate colors in the uniform color space, and then convert the interpolated colors back to RGB. Equations for this process are provided in Appendix B.

One limitation of both of these approaches is that they require you to establish the gamma function for each color gun on your monitor. **Gamma function** refers to the relation between the voltage of a color gun and the associated luminance of the CRT display. Equations for converting between CIE and RGB assume this relation is linear, but generally it is nonlinear, having the form $L = D^\gamma$, where D is the digital value sent to a color gun, L is the luminance measured on the display, and γ is an exponent describing the nonlinearity. (The term *gamma function* comes from the use of γ as the exponent.) Travis (154–160) described how the gamma function can be linearized; the basic result is a lookup table equating standard RGB values (those that would normally be used in the absence of any correction) with gamma-corrected RGB values.

ColorBrewer

As we discussed in the preceding section, ColorBrewer provides a broad set of color schemes appropriate for choropleth maps. Rather than using a perceptually uniform color space, Brewer developed these schemes using "her knowledge of the relationships between CMYK color mixture and perceptually ordered color spaces, such as Munsell" (Harrower and Brewer 2003, 30). Because a typical user would not have Brewer's knowledge, he or she presumably would have difficulty creating these schemes from scratch, but this problem is irrelevant as ColorBrewer provides the specifications for color schemes in a variety of formats, including CMYK, RGB, hexadecimal, Lab, and ArcView 3.x. Thus, users do not have to interpolate colors themselves (as with color ramping) nor do they have to deal with the cumbersome CIE and gamma function concepts.

13.4.2 Methods for Black-and-White Printed Maps

Because of the high cost traditionally associated with producing colored printed maps, considerable research has dealt with the selection of appropriate areal symbols for black-and-white choropleth maps (see Kimerling 1985

for a comparison of various studies). Such research has resulted in **gray curves** or **gray scales**, which express the relation between percent area inked and perceived blackness. Separate gray curves have been developed for smooth and coarse (or "textured") shades, respectively.

Smooth Shades

For smooth shades, or those in which the pattern of marks making up the shade is not apparent,[◊] the **Munsell curve** is generally considered appropriate. In examining this curve (Figure 13.4), note that at the light end, a small difference in percent area inked equates to a large difference in perceived blackness, and at the dark end, a large difference in percent area inked equates to a small difference in perceived blackness. The Munsell curve can be utilized in a four-step process:

Step 1. Pick the smallest and largest perceived blackness that you desire. Let's assume that you select values of 12 and 100 as initial choices. (A value of 12 equates to 6 percent area inked and so will produce a gray tone that can be seen as a figure against the white background typically used on printed maps. A value of 100 will produce a solid black.)

[◊] Castner and Robinson (1969) found that shades having 75 or more marks per inch were perceived as smooth gray tones (the pattern of marks was not apparent).

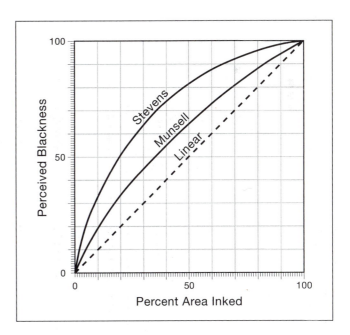

FIGURE 13.4 Munsell and Stevens gray curves for converting desired perceived blackness to percent area inked. The curves are intended for fine-toned black-and-white areal symbols. (After Kimerling 1985, 137.)

[†] For a complete set of equivalent colors, see Travis (1991, 214–270).

Step 2. Determine the contrast between each pair of perceived blackness values, assuming the values are equally spaced from one another. To accomplish this, divide the range of perceived blackness values by the number of classes minus 1. Assuming five classes, and the preceding perceived blackness values, the result is:

$$\frac{PB_{max} - PB_{min}}{NC - 1} = \frac{100 - 12}{4} = \frac{88}{4} = 22$$

where PB_{max} and PB_{min} are the perceived maximum and minimum blackness, and NC is the number of classes.

Step 3. Interpolate the intermediate perceived blackness values. This involves simply adding the contrast value derived in step 2 to each perceived blackness value, beginning with the lowest. For perceived blackness values of 12 and 100, and a contrast value of 22, we have: 12 + 22 = 34; 34 + 22 = 56; 56 + 22 = 78; and 78 + 22 = 100.

Step 4. Determine the percent area inked corresponding to each perceived blackness value. Using Figure 13.4, this can be accomplished by drawing a horizontal line from the perceived blackness value to the Munsell curve, and from this point drawing a vertical line to the proper percent area inked value. For perceived blackness values of 12, 34, 56, 78, and 100, you should find percent area inked values of approximately 6, 20, 41, 67, and 100.

Although the Munsell curve is the most widely accepted gray curve for smooth shades, the **Stevens curve** (Figure 13.4) has been argued as more appropriate for unclassed maps (Kimerling 1985, 141; MacEachren 1994a, 105). Note that the Stevens curve is displaced above the Munsell curve; thus, for light gray tones, a small difference in percent area inked leads to a very large difference in perceived blackness, whereas for dark tones, a very large difference in percent area inked is required to achieve a small difference in perceived blackness.

To understand why the Munsell and Stevens curves might be appropriate for classed and unclassed maps, respectively, it is necessary to consider the methods by which these curves were constructed. The Munsell curve was constructed using **partitioning**, in which a user places a set of areal shades between white and black such that the resulting shades appear equally spaced. For example, in a study that essentially duplicated the Munsell curve, Kimerling (1975) had people place seven gray tones (chosen from a set of 37) between white and black. It can be argued that the tones resulting from such an approach will have maximum contrast with one another and thus be appropriate for a classed choropleth map.

In contrast, the Stevens curve was created using **magnitude estimation**, in which a user estimates the lightness or darkness of one shade relative to another. Stevens and Galanter (1957) described the approach as follows:

Typically a particular gray was shown to [the observer] and he was told to call it by some number. It was then removed and the stimuli were presented twice each in irregular order. [The observer] was told to assign numbers proportional to the apparent lightness. (p. 398)

Some cartographers argue that this process is similar to how readers compare individual areas on an unclassed map. Readers do not, of course, use a number system, but they might conceive of ratios of gray tones (that a given gray tone is, say, five times as dark as another). It seems more likely, however, that readers consider only ordinal relations (one tone is considered darker or lighter than another or "much" darker or lighter than another). If this is the case, the Stevens curve would seem to be inappropriate for maps.

One of the problems with applying either the Munsell or Stevens curve to unclassed maps is that the four-step process just described is tedious with a large number of classes. To solve this problem, we digitized the curves and used curvilinear regression to fit the digitized data (Davis 2002, 207–214). The results for Munsell and Stevens, respectively, were:

$$PA = .287029 + .329157PB + .008753PB^2$$
$$-6.24654 \times 10^{-7}PB^4 + 4.16010$$
$$\times 10^{-11}PB^6$$

$$PA = .465370 + .013841PB^2 - 1.26098$$
$$\times 10^{-4}PB^3 + 8.38227 \times 10^{-11}PB^6$$

where PB represents a desired perceived blackness and PA represents the percent area ink necessary to achieve that level of blackness. The equations produced an excellent fit in the case of Munsell (the maximum error between predicted and digitized values for percent area inked was 0.3 percent) and a reasonably good fit in the case of Stevens (the maximum error was 3.4 percent).

Although these equations might appear complex, they can be incorporated into a spreadsheet in the following three-step procedure:

Step 1. Determine the proportion of the data range represented by each data value. To accomplish this, we subtract the minimum of the data from each raw data value, and divide the result by the range of the data.

$$Z_i = \frac{X_i - X_{min}}{X_{max} - X_{min}}$$

where X_i is a raw data value, X_{min} is the minimum of the raw data, X_{max} is the maximum of the raw data, and Z_i is the proportion of the data range. The resulting Z values will range from 0 to 1.

Step 2. Convert the values calculated in step 1 to perceived blackness. This involves computing

$$PB_i = Z_i(PB_{max} - PB_{min}) + PB_{min}$$

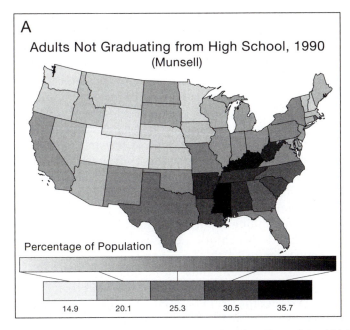

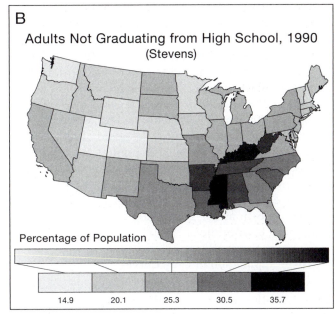

FIGURE 13.5 Unclassed maps illustrating the effect of the (A) Munsell and (B) Stevens gray curves.

where Z_i, PB_{min}, and PB_{max} are defined as before, and PB_i is the perceived blackness for an individual data value. (To achieve the 6 to 100 percent area inked used in the classed example, you would use PB_{min} and PB_{max} values of 13 and 100 for Munsell, and 22 and 100.8 for Stevens.)*

Step 3. Insert the perceived blackness values into whichever equation (Munsell or Stevens) you wish to use. Figure 13.5 provides a comparison of the effect of the Munsell and Stevens curves for the unclassed high school education data set. Note that when the entire distribution is considered, the Munsell curve produces a slightly darker map. On a more detailed level, the Munsell curve enables greater discrimination for low values, whereas the Stevens curve permits greater discrimination for high values.

Coarse Shades

For coarse shades, or those in which the pattern of marks making up the shade is apparent, the **Williams curve** (Figure 13.6) is usually recommended (see, e.g., Robinson et al. 1995, 394–396). In a manual environment, the Williams curve was applied to preprinted coarse shades (**Zip-a-Tone** was a common example). Although coarse shades can be created using the computer, smooth shades are more commonly used today, probably because they are considered more aesthetically pleasing. Note that the Williams curve is similar to the Munsell and Stevens

curves near the origin, but that it crosses over a hypothetical 45° line connecting the lower left and upper right corners of the diagram (compare Figures 13.4 and 13.6). To calculate percent-area-inked values for the Williams curve, a four-step procedure similar to that described for classed maps would be appropriate. It is important to realize that the Williams curve was intended for symbols

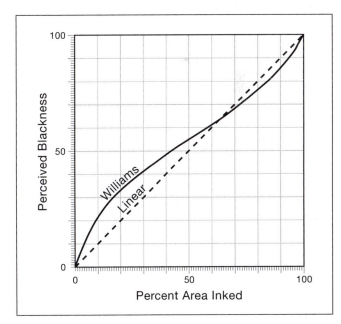

FIGURE 13.6 The Williams gray curve, which is recommended for converting perceived blackness to percent area inked when coarse areal shades are used. (After Kimerling 1985, 139.)

* The value of 13 does not match the value of 12 used earlier because the equation is an approximation of the curve shown in the graph and the graph must be visually interpreted.

having a relatively constant number of marks per inch. As a result, it should not be used for a symbol in which the number of marks per inch varies considerably, such as Tobler's (1973) crossed-line shading (as in Figure 4.10C). Rather, because people tend to view crossed-line and gray-tone shades having a similar percent area inked as equivalent, the Munsell or Stevens scales should be used for crossed-line shading (Slocum and McMaster 1986). More recently, Leonard and Buttenfield (1989) developed a gray scale for 300 dpi laser printers. We have chosen not to illustrate their curve because it was developed using percent reflectance and was not converted to a percent ink equivalent.* Their paper does, however, offer useful suggestions for generating maximum contrast maps using laser printers.

Practical Application of Gray Curves

Regardless of which gray curve is used, you should be aware that the results in a real-world production environment are likely to differ from those suggested by the curves; that is, if you specify a particular percent area inked using software, there is no guarantee that the hardware will actually produce the same percent area inked. (This should be obvious to anyone who has used laser printers for word processing: Once a toner cartridge has been used heavily, words will not appear as solid black.) The only real solution to this problem is to analyze available hardware carefully to ensure that percent-area-inked specifications match software specifications, and that black marks are actually produced.

Even if one has taken a great deal of care in producing maps within one's own shop, consideration must also be given to what will happen if the map is reproduced (e.g., using a fax machine, a photocopy machine, or offset lithography). Most of us are familiar with how degraded an image becomes as a result of using fax or copy machines. In the case of offset lithography, the basic problem is that individual marks making up an areal symbol might be larger on the printed map (than suggested by numerical specifications in a digital file) because of "overinking or too much pressure between the paper and blanket cylinder" (Monmonier 1980, 25). Although this problem might be handled by applying a correction factor similar to that described earlier for gray scales, it is more realistic to find a high-quality printer, have a good working relationship with that printer, and thus try to avoid (or at least limit) this problem.

Handling Solid Black Areal Symbols

One of the issues involved in using black-and-white areal symbols is that when a solid black (100 percent area

inked) is used for the highest class, boundaries of contiguous enumeration units in the highest class disappear (Figure 13.7A). One solution to this problem is to reduce the percent area inked for the highest class (and, of course, change the percent area inked for intermediate classes to maintain maximum contrast); for instance, in Figure 13.7B we utilized an 80 percent area inked for the highest class. The problem with this solution is that the contrast between classes is reduced, thus making it more difficult to determine which class an enumeration unit is a member of. A second solution is to shade all boundaries with a gray tone that appears to be midway between the two darkest shades (Figure 13.7C). This differentiates areas falling in the highest class but weakens the boundary between other classes, producing a less distinctive "figure." A third solution is to decrease the percent area inked only for boundaries between enumeration units falling in the highest class; Figure 13.7D shows the result of doing this with an 80 percent-area-inked shade. We thought that the latter approach was effective in differentiating areas in the highest class, and so chose it for maps in this book.

13.4.3 Methods for Colored Printed Maps

For colored printed maps, we recommend using the CMYK values presented in ColorBrewer, as Brewer and her colleagues (2003) took considerable care to ensure that the colors specified in ColorBrewer would look appropriate not only on screen, but also in print. To see what the color schemes are apt to look like, you should examine Brewer et al.'s (2003) paper, where a high-quality printed version of the color schemes appears.

13.5 LEGEND DESIGN

General principles of legend design were presented in section 11.2.5. Here we consider some finer points of legend design that pertain to choropleth maps. Although the design of the legend might seem a minor point, a map intended for presentation could fail to communicate if the legend is poorly designed. One issue in legend design is whether a horizontal or vertical legend should be used (Figure 13.8). An argument for a horizontal legend is that its orientation matches the traditional number line; as such, values should increase from left to right (Figure 13.8A would be appropriate, whereas Figure 13.8B would be inappropriate). For a vertical legend, high values can be shown either at the top or bottom (Figure 13.8C and 13.8D). The logic of showing high values at the top is that people associate "up" with "higher"—think of climbing a mountain: High elevations are at the top. One problem, however, with having high values at the top is that we normally read from left to right and from top to

---

* See MacEachren (1994a, 103–105) for a discussion and graph of the Leonard–Buttenfield curve.

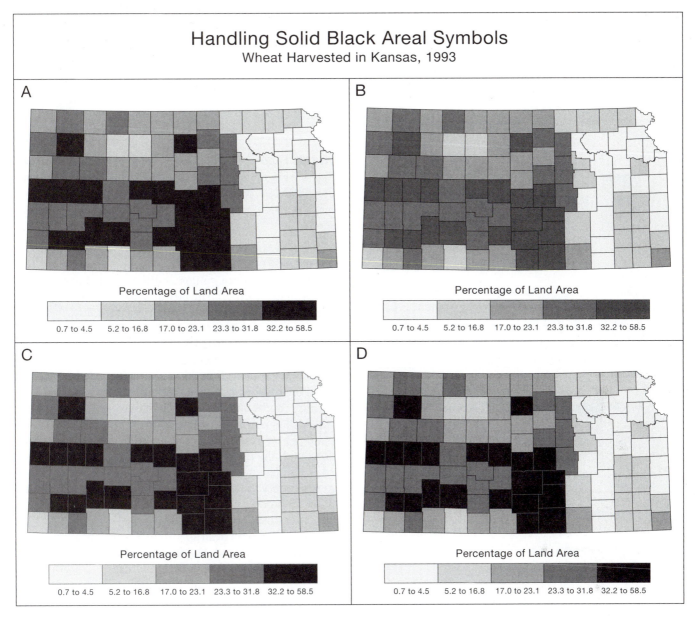

Handling Solid Black Areal Symbols
Wheat Harvested in Kansas, 1993

A

Percentage of Land Area

0.7 to 4.5   5.2 to 16.8   17.0 to 23.1   23.3 to 31.8   32.2 to 58.5

B

Percentage of Land Area

0.7 to 4.5   5.2 to 16.8   17.0 to 23.1   23.3 to 31.8   32.2 to 58.5

C

Percentage of Land Area

0.7 to 4.5   5.2 to 16.8   17.0 to 23.1   23.3 to 31.8   32.2 to 58.5

D

Percentage of Land Area

0.7 to 4.5   5.2 to 16.8   17.0 to 23.1   23.3 to 31.8   32.2 to 58.5

FIGURE 13.7 Alternatives for handling solid black areal symbols: (A) solid black is used for the highest class and for all boundaries, (B) the percent area inked for the highest class is reduced (and intermediate shades are adjusted to maintain maximum contrast), (C) boundaries are depicted with a shade perceived to be midway between the shades for the two highest classes, and (D) a gray tone is used only for boundaries between contiguous enumeration units in the highest class.

bottom; as such, it might seem awkward to find values increasing from left to right, but decreasing from top to bottom. The decision to use a horizontal or vertical legend likely will depend on available map space; for instance, we chose a horizontal design for most choropleth maps in this book because the design seemed to fit well with the geographic shapes and column widths we were dealing with.

Numeric values should be placed either at the bottom of legend boxes (for a horizontal legend) or to the right of boxes (for a vertical legend); thus, the design in Figure 13.9A should be avoided. The logic is that we read from left to right and from top to bottom; we normally see an areal symbol on the map and want to know the associated range of values. Legend boxes should be placed directly next to one another, as opposed to being separated by spaces (avoid Figure 13.9B). The logic is that enumeration units are contiguous to one another, so legend boxes also should be contiguous. (See section 11.2.5 for some exceptions to this rule.)

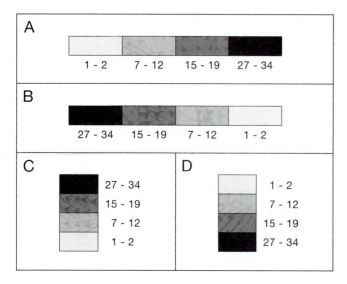

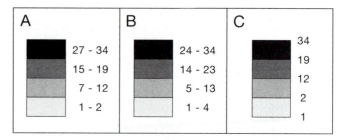

FIGURE 13.10   Methods for indicating class limits: (A) the actual range of data falling in each class is shown, (B) class limits are expanded to eliminate gaps, (C) only the minimum and maximum data values and the upper limit of each class are shown.

FIGURE 13.8   Some horizontal and vertical legends. (A), (C), and (D) are appropriate, whereas (B) is inappropriate because the legend limits decrease from left to right (unlike a number line).

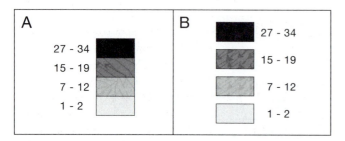

FIGURE 13.9   Inappropriate legend designs: (A) the numeric values for classes precede the legend boxes, (B) the legend boxes are not contiguous.

Several approaches are possible for specifying class limits, including: (1) indicating the range of data actually falling in each class, which produces numeric gaps between classes (e.g., the gap between "1-2" and "7-12" in Figure 13.10A); (2) eliminating these gaps by expanding classes (Figure 13.10B), and (3) indicating the minimum

and maximum data values and the upper limit of each class (Figure 13.10C). The advantage of the first approach is that the reader will know precisely the values falling in each class.* The latter two approaches avoid the problem of gaps, but they do not show the range of data actually occurring in each class on the map. An advantage of the last approach is that the reader will have fewer numbers to work with, but there might be confusion concerning the bounds of each class (e.g., in Figure 13.10C which class is "19" in?). As explained in Chapter 11, a hyphen can be used to separate numeric values, as we have done with the figures here. If negative values are included in the legend, then the word "to" is frequently used to separate values. Because some of the data sets in this book included negative values, we generally have used "to" for consistency.

It might also be desirable to integrate a graphical display with the legend. For instance, Figure 13.11 shows two approaches for integrating a dispersion graph. In Figure 13.11A, the legend boxes are sized to reflect the range of the data in each class, whereas in Figure 13.11B the legend boxes are equal in size. A problem with the

---

*Note that this would not be true on an equal interval map, where the legend intervals normally do not match the range of data in each class.

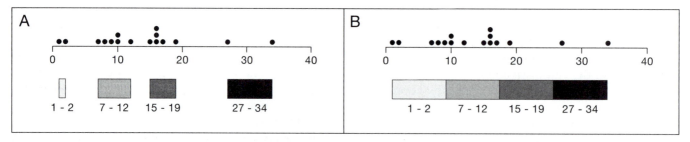

FIGURE 13.11   How a graphical display can be integrated with the legend: (A) legend boxes are scaled to reflect the range of data in each class, (B) legend boxes are constant.

former approach is that the size of the small legend box for the first class might make it difficult to match the shade with a shade on the map.

13.6 CLASSED VERSUS UNCLASSED MAPPING

Prior to the early 1970s, classed choropleth maps were the norm.* Cartographers argued that classed maps were essential because of the limited ability of the human eye to discriminate shades for areal symbols; also, practically speaking, classed maps were the only option because of the time and effort required to produce unclassed maps using traditional photomechanical procedures. In 1973, the latter constraint was eliminated when Waldo Tobler introduced a method for creating unclassed maps using a line plotter (Figure 4.10C could have been produced using such a device). Today, unclassed maps can be created using a wide variety of hardware devices.

The development of unclassed mapping led to a hotly contested debate on its merits and demerits (Dobson 1973; 1980a; 1980b; Muller 1980a; Peterson 1980), and numerous experimental studies (Peterson 1979; Muller 1979; 1980b; MacEachren 1982b; Gilmartin and Shelton 1989; Mersey 1990; Mak and Coulson 1991). Although the results of the experimental studies can be helpful in selecting between classed and unclassed maps, we argue that two criteria should be considered first: (1) whether the cartographer wishes to maintain the numerical relations among data values, and (2) whether the map is intended for presentation or exploration. In this section, we consider these two criteria, and then appraise the results of some of the experimental studies.

13.6.1 Maintaining Numerical Data Relations

To illustrate the notion of maintaining numerical data relations, consider the classed and unclassed maps for foreign-owned agricultural land and high school graduation shown in Figure 13.12. The optimal method was used to create the classed maps because the optimal method does the best job of minimizing classification error, and so will produce classed maps that are most similar to the unclassed maps. Shades for the classed maps were selected using a conventional maximum-contrast approach in which tones are perceptually equally spaced from one another. In contrast, shades on the unclassed maps were made directly proportional to the values falling in each enumeration unit, thus maintaining the numerical relations among the data (see section 13.4.2 for the equations).

---

* Interestingly, the first choropleth map (produced by Charles Dupin in 1827) was unclassed, but this rapidly gave way to the classed approach (Robinson 1982, 199).

Clearly, the maps within each pair look different, with the difference most distinct for the agricultural maps. The difference for the agricultural maps is a function of the severe skew in the data (examine Figure 13.13 and note the concentration of data on the left, with Maine a notable outlier at 14.1). The unclassed agricultural map is a spatial expression of this skew, as we see numerous low-valued gray tones and one black tone (Maine). In contrast, on the classed map we see several different tones of gray, with Maine not appearing quite so distinct. Although unclassed maps do a better job of portraying the actual data relations, a disadvantage is that for skewed distributions the ordinal relations in much of the data might be hidden. For example, in the case of foreign-owned agricultural land, it is difficult to determine that states bordering the Great Lakes have lower values than Deep South states. When the data are classed, these differences become more obvious.

13.6.2 Presentation versus Data Exploration

When a map is intended for presentation, normally only one view of the data can be shown, so one must make a choice between classed and unclassed maps. If, however, the intention is to explore data, then numerous options are possible. One is to compare a variety of classification approaches visually; for example, Color Plate 13.7, taken from the program ExploreMap (discussed fully in Chapter 21), compares four methods of classification. With the appropriate software, such a comparison might also include an unclassed map. Another option in data exploration is to apply unclassed shading to only a portion of a data set. For example, for the agricultural data, Maine could be assigned a unique symbol, with the remainder of the data displayed using the unclassed method (Figure 13.14). Note how this approach makes it easier to contrast the pattern of gray tones in states bordering the Great Lakes with the pattern of gray tones in the Deep South.

The approach we have taken in applying unclassed shading to a portion of a data set is similar to Brassel and Utano's (1979) **quasi-continuous tone method**, in which the lightest and darkest shades were reserved for classes of outlying low and high data values and the remaining shades were assigned in an unclassed manner to intermediate data. Rather than use the darkest shade for the outlier (Maine), we used a unique symbol so that it would be clearly distinguished from other values. If color were used, it might be useful to portray outlying values with gray tones and apply a continuous color scheme (e.g., shades of red) to the remaining data. In an interactive environment it is, of course, possible to make such changes dynamically, focusing on any subset of the data desired; we consider this notion in more detail in Chapter 21.

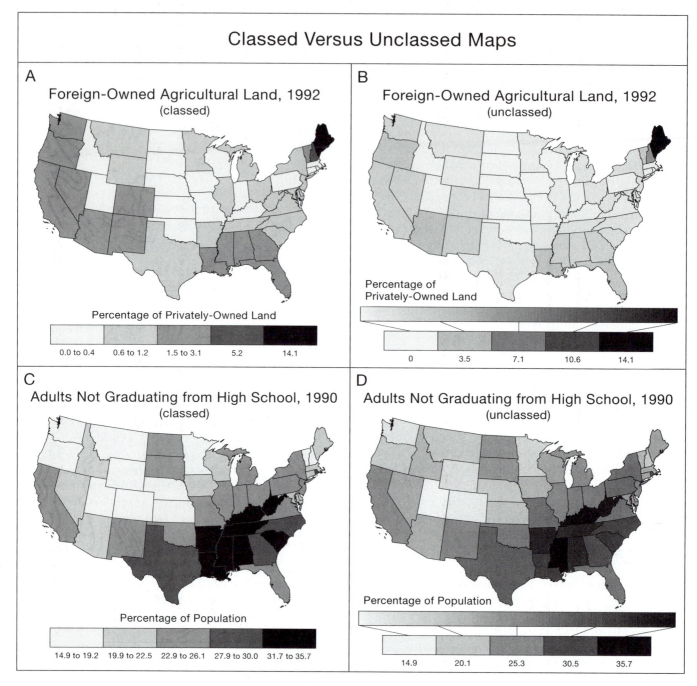

FIGURE 13.12   Optimally classed (A) and (C) and unclassed (B and D) maps of two attributes: foreign-owned agricultural land and high school education.

13.6.3 Summarizing the Results of Experimental Studies

This section summarizes the results of some experimental studies that might assist in determining whether classed or unclassed maps are appropriate. The summary is based on studies by Muller (1979), MacEachren (1982b), Gilmartin and Shelton (1989), Mersey (1990), and Mak and Coulson (1991). Studies by Peterson (1979) and Muller (1980b) also dealt with classed and unclassed maps, but these studies are not considered here because they dealt with map comparison, a topic to be covered in Chapter 18. One problem in summarizing the studies is that two of them (MacEachren and Mersey) did not use unclassed maps, but rather varied the number of classes (from 3 to 11 and 3 to 9, respectively); thus, in the following discussion it is useful conceptually to substitute

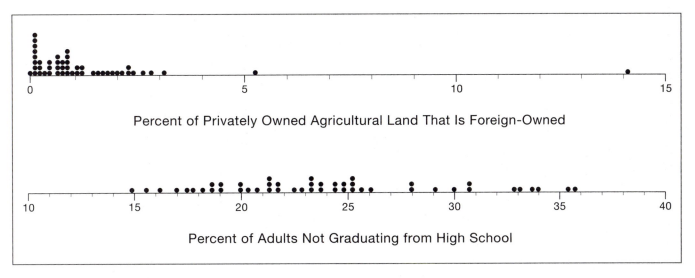

FIGURE 13.13   Dispersion graphs for the attributes shown in Figure 13.12.

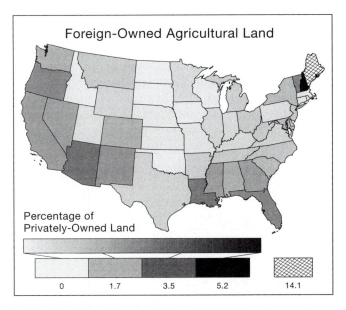

FIGURE 13.14   An unclassed map in which continuous-tone shading is applied to a portion of the data set (in this case, Maine is separated from the remaining data and represented by a cross-hatched shade). This is sometimes referred to as quasi-continuous tone mapping.

"maps with few classes" for classed maps and "maps with many classes" for unclassed maps. The results of the studies can be summarized conveniently under the types of information that readers acquire and recall from maps: specific and general.

Specific Information

For the *acquisition* of specific information, classed maps generally have been more effective. This result might seem surprising because unclassed maps are usually touted as being more accurate (because data are not grouped into classes). The high accuracy of unclassed maps is, however, mathematical, not perceptual. Visually matching a shade on an unclassed map with a shade in the legend is difficult because there are so many shades and their appearance is affected by simultaneous contrast. Note, however, that for some data sets unclassed maps can be useful for visualizing ordinal relations. For example, consider ascertaining the ordinal relation of states on the unclassed education map shown in Figure 13.12: Montana appears to have a higher value than Wyoming, which in turn is higher than Colorado. On the classed map, such ordering is impossible to determine when the data fall in the same class (as for these three states). Although ordinal relations can be obtained from unclassed maps, the task can be difficult if enumeration units are far apart (and thus likely appear in different contexts), or if a distribution is highly skewed (as with the foreign-owned agricultural land data). It also must be realized that the ability to acquire specific information will depend on the number of classes. For example, on a two-class map the error resulting from classification clearly will be greater than the error resulting from incorrect estimation on an unclassed map. Although experimental studies do not indicate an ideal number of classes, it appears that the error resulting from classification will be unacceptably large on maps with four or fewer classes.

For the *recall* of specific information, results are inconclusive. MacEachren found that maps with fewer classes were more effective, whereas Mersey found that although a three-class map was most effective, five- and seven-class maps did not elicit as accurate a response as a nine-class map.

General Information

For *acquisition* of general information, studies generally have revealed no significant difference for classed and unclassed maps. Only Mak and Coulson found any significant difference, and this occurred only when individual regions were analyzed; they concluded that "for more complex classed maps (six to eight classes) classification may have a distinct advantage over unclassed maps" (121). For *recall* of general information, MacEachren found no significant relationship between the number of classes and subjects' ability to recall data, and Mersey found that a greater number of classes decreased recall effectiveness. Based on his findings, MacEachren suggested that classification might be unnecessary for general tasks. In contrast, Mersey noted:

> A large number of map symbols, even on a map with regular surface trends, may create a notion in the mind of the user that the distribution is more complex than in fact it is. Using many categories to provide the user with more detailed specific information may serve as "noise" when the same user attempts to reconstruct the thematic distribution from memory. (p. 125)

Discussion

One problem with interpreting the results of these studies is that usually there was no direct comparison between classed and unclassed maps. For example, Mak and Coulson asked subjects to divide each map into five ordinal-level regions (ranging from "low" to "high"). Rather than comparing the resulting regions directly on classed and unclassed maps, they analyzed the consistency for each type of map (classed and unclassed). As a result, it is possible that the consistency was similar for both but the locations of regions differed.

Another problem is that only one of the studies (Gilmartin and Shelton) measured processing time. Measuring processing time is important because classed and unclassed maps could provide the same information, but one map could take less time to process. (The more rapidly processed map clearly would be more desirable.) If classed maps are processed more rapidly, as traditional classification proponents such as Dobson argue, and the information acquired (or recalled) is identical, an argument can be made for classing the data. If, however, the information acquired or recalled varies for classed and unclassed maps, faster processing is a moot point: In this case, the unclassed map might be argued for on the grounds that it provides a more correct portrayal of the spatial distribution.

The studies that we have considered focused on a *quantitative* assessment of how effectively users could acquire or recall information from classed and unclassed maps. A different approach is to assess users' *attitudes* about classed and unclassed maps using *qualitative* methods. Kevin Spradlin (2000) accomplished this by having individuals utilize classed and unclassed maps in role-playing scenarios—the individuals then rated the mapping techniques using word pairs (e.g., "easy to use" vs. "difficult to use") and participated in open-ended interviews in which they provided their thoughts and opinions on the mapping techniques. Generally, Spradlin found that classed maps were preferred, although he noted that this might have been a function of some of the tasks performed in the role-playing scenarios.

SUMMARY

Important issues involved in choropleth mapping include making certain that the choropleth map is a suitable option for a particular data set, deciding whether or not a data set should be classed, selecting a method of classification, choosing a color scheme, and designing a readable legend. You should use the choropleth map only when you wish the reader to associate "typical" values with enumeration units. If the spatial variation of the underlying phenomenon does not coincide with enumeration unit boundaries, and you wish readers to visualize the phenomenon correctly, then another mapping technique (e.g., the isopleth or dot map) is more appropriate. Ideally, the choropleth map should be used only when the size and shape of enumeration units are similar; if the units differ greatly in size and/or shape, then the resulting pattern should be treated with caution. There are numerous methods for standardizing data to account for differing sizes of enumeration units, including dividing two area-based raw totals (e.g., dividing the number divorced by the number married), calculating a rate (e.g., the number of suicides per 100,000 people), and calculating a summary numerical measure (e.g., the median value of homes).

Now that unclassed maps can be constructed using modern computer hardware, the cartographer is faced with the question of whether a classed or unclassed map should be created. In this chapter, we saw that two criteria can be used to assist in selecting between classed and unclassed maps. The first is whether or not you wish to portray correct numerical relationships among the data. If the intent is to maintain correct data relations, then unclassed maps are appropriate because they provide a spatial expression of numerical relations in the data. A second criterion is whether the intention is to present or explore data. For presentation purposes, it generally is only possible to show one map of a distribution, and thus one must make a choice between classed and unclassed maps. If, however, the intent is to explore data, then several visualizations of the data are possible: a classed map, an unclassed map, or a focus on a subset of the data.

Experimental studies for evaluating the effectiveness of classed and unclassed maps have revealed mixed results. For example, classed maps have been more effective for acquiring specific information, but neither technique has performed consistently better in terms of acquiring general information. The experimental studies have been limited by a failure to directly compare classed and unclassed maps, by not considering the time necessary for readers to process the map, and by not fully considering users' thoughts and opinions; thus, it appears that more research is needed to fully compare the effectiveness of classed and unclassed maps.

If you decide that a classed map is appropriate, you first should consider the kind of data you have collected: Bipolar data often have a logical dividing point that can be used to initially split the data into two classes. You should also consider the level of precision you desire in your final map *before* submitting the data to a data classification method. Choosing an appropriate method of data classification is a function of several factors, such as whether the method considers the distribution of the data along the number line, the ease of understanding the legend, and whether it can assist in selecting the number of classes. These and other factors are dealt with in Chapter 5.

Numerous factors should be considered in selecting a color scheme, including the kind of data (a **diverging scheme** is clearly appropriate for bipolar data, but can also be used for unipolar and balanced data), the ease of naming colors (this is important for creating an acceptable diverging scheme), ease of interpretation for the color-vision-impaired, and avoiding problems of simultaneous contrast. The Web site for ColorBrewer provides a useful approach for specifying *particular* colors in a scheme. For black-and-white printed maps, we have seen how **gray curves** can be used to select a set of gray tones that appear to be equally spaced from one another.

Legends for choropleth maps should be designed so that readers can readily interpret the map. Either a horizontal or vertical legend can be used, depending on available map space. Numeric values should progress in a logical fashion (for instance, in a horizontal legend values should increase from left to right, as on a number line). The legend should be constructed with the presumption that readers will enter the legend to determine the numeric value for an areal symbol—thus, in a vertical legend areal symbols should appear to the left of numeric values (we read from left to right and should encounter the symbols first). Legend boxes should be placed contiguous to one another because on the map enumeration units are found contiguous to one another.

FURTHER READING

Brassel, K. E., and Utano, J. J. (1979) "Design strategies for continuous-tone area mapping." *The American Cartographer* 6, no. 1:39–50.

Describes various methods for symbolizing unclassed maps.

Brewer, C. A. (1994a) "Color use guidelines for mapping and visualization." In *Visualization in Modern Cartography*, ed. by A. M. MacEachren and D. R. F. Taylor, pp. 123–147. Oxford: Pergamon.

Discusses color schemes appropriate for various kinds of data (unipolar, bipolar, and balanced).

Brewer, C. A. (1997) "Spectral schemes: Controversial color use on maps." *Cartography and Geographic Information Science* 24, no. 4:203–220.

Presents recommendations for using spectral color schemes.

Cromley, R. G. (1995) "Classed versus unclassed choropleth maps: A question of how many classes." *Cartographica* 32, no. 4:15–27.

Describes a method for optimally classifying data by treating classification as an integer programming problem; argues that this approach can assist in choosing an appropriate number of classes.

Dobson, M. W. (1980a) "Perception of continuously shaded maps." *Annals, Association of American Geographers* 70, no. 1:106–107.

An example of some arguments against unclassed maps; counterarguments for unclassed maps can be found in Muller (1979; 1980a).

Eastman, J. R. (1986) "Opponent process theory and syntax for qualitative relationships in quantitative series." *American Cartographer* 13, no. 4:324–333.

Introduces the notion that opponent process theory might be useful in selecting color schemes, and presents an experiment designed to evaluate the theory.

Kennedy, S. (1994) "Unclassed choropleth maps revisited: Some guidelines for the construction of unclassed and classed choropleth maps." *Cartographica* 31, no. 1:16–25.

Summarizes the debate concerning classed and unclassed choropleth maps, and provides some guidelines for constructing both kinds of maps.

Kimerling, A. J. (1975) "A cartographic study of equal value gray scales for use with screened gray areas." *American Cartographer* 2, no. 2:119–127.

An example of an experimental study for developing a gray curve.

Kimerling, A. J. (1985) "The comparison of equal-value gray scales." *American Cartographer* 12, no. 2:132–142.

Compares various gray curves that have been developed.

Mak, K., and Coulson, M. R. C. (1991) "Map-user response to computer-generated choropleth maps: Comparative experiments in classification and symbolization." *Cartography and Geographic Information Systems* 18, no. 2:109–124.

An example of a study that compared readers' ability to interpret classed and unclassed maps.

Mersey, J. E. (1990) "Colour and thematic map design: The role of colour scheme and map complexity in choropleth map communication." *Cartographica* 27, no. 3:1–157.

A study of the role of map use tasks in selecting color schemes.

Muller, Jean-C. (1979) "Perception of continuously shaded maps." *Annals, Association of American Geographers* 69, no. 2:240–249.

Describes a method for creating unclassed maps, along with a study of readers' ability to interpret classed and unclassed maps.

Olson, J. M., and Brewer, C. A. (1997) "An evaluation of color selections to accommodate map users with color-vision impairments." *Annals, Association of American Geographers* 87, no. 1:103–134.

An experimental study of color schemes for the color-vision-impaired; includes RGB and *Yxy* values for suitable schemes.

Peterson, M. P. (1992) "Creating unclassed choropleth maps with PostScript." *Cartographic Perspectives* no. 12:4–6.

Describes an approach based on PostScript that permits 9,999 different gray tones to be created for unclassed choropleth maps. This is in contrast to programs such as Freehand that only permit 99 gray tones. Maps in this book were created using Freehand.

Spradlin, K. L. (2000) *An Evaluation of User Attitudes Toward Classed and Unclassed Choropleth Maps*. Unpublished M.A. thesis, University of Kansas, Lawrence, KS.

A qualitative study of users' thoughts about classed and unclassed maps.

Tobler, W. R. (1973) "Choropleth maps without class intervals?" *Geographical Analysis* 5, no. 3:262–265.

The paper that initiated the debate concerning classed and unclassed maps.

Travis, D. (1991) *Effective Color Displays: Theory and Practice*. London: Academic Press.

Contains a variety of information related to color, including formulas for converting between CIE and RGB (pp. 89–97), discussion of the gamma function (pp. 154–160), and specifications for matching Munsell and CIE colors (Appendix 3).

14

Isarithmic Mapping

OVERVIEW

Isarithmic maps (the most common form being the *contour map*) depict smooth continuous phenomena, such as rainfall, barometric pressure, depth to bedrock, and, of course, the earth's topography. After the choropleth map, the isarithmic map is probably the most widely used thematic mapping technique, and is certainly one of the oldest, dating to the 18th century. Section 14.1 considers the kind of data appropriate for isarithmic mapping. As with the proportional symbol map (see Chapter 16), two kinds of point data are used in isarithmic mapping: true and conceptual. *True point data* can actually be measured at a point location (e.g., temperatures recorded at weather stations). In contrast, *conceptual point data* are collected over either an area or volume (e.g., murder rates for census tracts), but the data are considered located at points (e.g., the centroid of a census tract) for the purpose of symbolization. The terms *isometric map* and *isopleth map* are used to describe maps resulting from using true and conceptual point data, respectively. Like the choropleth map, the isopleth map requires that standardized data be used to account for the area over which conceptual data are collected.

True or conceptual points at which data are collected are termed *control points*. A fundamental problem in isarithmic mapping is *interpolating* unknown values between known values of irregularly spaced control points. Section 14.2 considers *manual interpolation*, or interpolation by "eye." Although manual interpolation is uncommon today, it often serves as a yardstick against which automated interpolation methods are compared.

Section 14.3 discusses several common automated interpolation methods for true point data: triangulation, inverse-distance, and kriging. *Triangulation* fits a set of triangles to control points and then interpolates along the edges of these triangles. *Inverse-distance* lays an equally spaced grid of points on top of control points, estimates values at each grid point as a function of their distance from control points, and then interpolates between grid points. *Kriging* considers the spatial autocorrelation in the data, both between a grid point and surrounding control points, and among control points. Although the mathematical basis underlying kriging is complex, learning about it is useful because the resulting maps are often more accurate than those produced by other interpolation methods (kriging is sometimes said to produce an *optimal interpolation*).

In section 14.4, we consider six criteria that can be utilized to select an appropriate interpolation method for true point data: (1) correctness of estimated data at control points (does the method honor the raw data?), (2) correctness of estimated data at noncontrol points (how well does the method predict unknown points?), (3) ability to handle discontinuities (e.g., geologic faults), (4) execution time, (5) time spent selecting interpolation parameters, and (6) ease of understanding. Although automated interpolation methods can save time (by avoiding manual interpolation), they do have their limitations. In section 14.5, we'll look at some of these limitations (e.g., "jagged lines" and "spurious details").

Section 14.6 covers *pycnophylactic interpolation*, a technique appropriate for conceptual point data. Pycnophylactic interpolation begins by raising each enumeration unit to a height proportional to the value of its associated control point. This 3-D surface is then gradually smoothed, keeping the volume within each enumeration unit constant. This is accomplished using a cell-based smoothing process analogous to generalizing procedures used in image processing.

In the last section of the chapter, we consider some issues associated with symbolizing data resulting from these *interpolation approaches. Basic symbolization techniques include **contour lines**, **hypsometric tints** (shading the areas between contour lines), **continuous-tone maps** (an unclassed isarithmic map), and **fishnet** symbolization (a netlike structure that simulates the 3-D character of a smooth continuous surface). An intriguing possibility is to view a spectral color scheme with special glasses that enhance the **color stereoscopic effect** (long-wavelength colors will appear nearer than short-wavelength colors). In Chapter 15, we look at symbolization methods that are especially appropriate for topography.*

14.1 SELECTING APPROPRIATE DATA

In considering the kind of data appropriate for isarithmic mapping, it is important to remember that the underlying phenomenon is presumed to be *continuous* and *smooth*. Thus, the phenomenon is presumed to exist throughout the geographic region of interest and change gradually between individual point locations (as opposed to abruptly). For instance, think of the mean yearly snowfall for the state of Michigan. There is potentially a value for every location, and the change between locations is relatively gradual. Phenomena that are largely continuous and smooth but that have some discontinuities can be handled with the isarithmic approach, but you must specify the discontinuities, and have software available that will handle them. For instance, a geologist might be interested in creating an isarithmic map showing the height of the Dakota sandstone in Kansas based on a series of sample wells. If any faults in the bedrock were known, these would have to be specified; otherwise, a smooth transition between elevation values would be presumed along the fault zone.

Data for isarithmic mapping are sampled from an underlying phenomenon at point locations termed **control points**. As with the proportional symbol map, two forms of data can be collected at control points: true and conceptual. With **true point data**, values are actually measured at a point location; for instance, weather data, such as snowfall and insolation, are measured at individual weather stations. In contrast, **conceptual point data** are collected over an area (or volume), but the data are presumed to occur at point locations; for instance, we might collect population data for counties, and presume that the population is located at the centroid of counties. Maps produced using true and conceptual point data are termed **isometric** and **isopleth maps**, respectively. You should be especially cautious when working with conceptual point data, as the assumption of continuity might not be met. For instance, population-related data might be characterized by sharp discontinuities that would be more appropriately mapped by the dot or dasymetric method.

As with the choropleth map, the isopleth map requires that data be standardized to account for the area over which conceptual point data are collected. Thus, it would make no sense to map the raw number of people in counties, as a larger county would tend to have a higher population total. Methods of standardization for isopleth maps are identical to those we have discussed for choropleth maps (see Chapter 13).

14.2 MANUAL INTERPOLATION

A key problem in isarithmic mapping is that data are available only at irregularly spaced control points, but the phenomenon is presumed to exist throughout the region of interest. Therefore, it is necessary to determine the values of intermediate points through the process of **interpolation**. **Manual interpolation**, or interpolation by "eye," was the only approach available prior to the development of computer-based methods, which did not become popular until the 1970s. Although not commonly practiced today, it is important that you be aware of manual interpolation because it is frequently used as a yardstick against which automated methods are compared. Manual interpolation is accomplished by mentally connecting neighboring control points with straight lines, and then linearly interpolating along these lines to create contours (or isolines) of equal value. For instance, in Figure 14.1 you can see how a contour line of 30 would be drawn through a set of four control points. *Linear* interpolation refers to the idea that the contour line is positioned proportionally between the control points (e.g., in Figure 14.1 the contour line of 30 is 4 units from 26

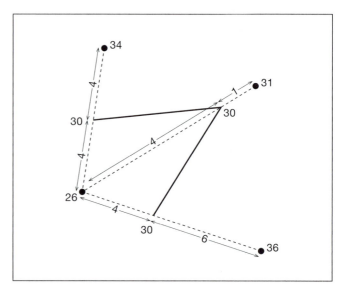

FIGURE 14.1   Manual interpolation involves a linear interpolation between control points.

and 1 unit from 31, and so is positioned 4/5 of the distance between 26 and 31).

A strict definition of linear interpolation produces the angular contour line shown in Figure 14.1. Because real-world phenomena generally do not exhibit such angularity, contour lines are smoothed by considering the trend of control point values within the region. This is illustrated in Figure 14.3A for a larger data set. In Figure 14.3A, also note that when neighboring control points are connected, a set of triangles is formed. These triangles are also utilized in *triangulation*, one of several automated interpolation methods that we consider. When control points are positioned near the corners of a square, it might not be clear where to draw the contours. For instance, in Figure 14.2A and 14.2B we see a situation in which the area in the middle of the four control points can be either below or above the 25-contour line. This problem is handled by averaging the four control points to create an additional control point (Figure 14.2C).

14.3 AUTOMATED INTERPOLATION FOR TRUE POINT DATA

In this section, we consider three automated interpolation methods appropriate for true point data: triangulation, inverse-distance, and kriging.

14.3.1 Triangulation

Triangulation is logical to address first because it emulates how contour maps are made manually. A key step in triangulation is connecting neighboring points to form a set of triangles that are analogous to those employed in manual contouring. John Davis (2002, 375) indicated that one of the challenges for those developing the triangulation approach was determining a "best" set of triangles,

as "simply entering the [control] points in a different sequence could result in contour lines with a conspicuously different appearance." Fortunately, this problem has disappeared, as **Delaunay triangles** now provide a unique solution regardless of the order in which control points are specified. Delaunay triangles are closely associated with **Thiessen polygons**, which are formed by drawing boundaries between control points such that all hypothetical points within a polygon are closer to that polygon's control point than to any other control point. For example, in Figure 14.3B, hypothetical point S is closer to control point C than to any other control point, so it is part of the Thiessen polygon associated with control point C. Delaunay triangles are created by connecting control points of neighboring Thiessen polygons. For example, in Figure 14.3B, triangle ABC is formed because the Thiessen polygons associated with control points A, B, and C are all neighbors of one another.

Once Delaunay triangles are formed, contour lines are created by interpolating along the edges of triangles in a fashion similar to manual interpolation. Delaunay triangles are desirable for this purpose because the longest side of any triangle is minimized, and thus the distance over which interpolation must take place is minimized. As with manual interpolation, a strict linear interpolation along triangle edges leads to angular contour lines, as shown in Figure 14.3C. Smoothing of these lines is obviously necessary if the underlying phenomenon is to be properly represented. Although details of the smoothing procedure are beyond the scope of this text, the process is the 3-D equivalent of splining, which is illustrated later for the inverse-distance method (Davis, 2002, 378); the result is shown in Figure 14.3D.*

---

* Interested readers with a mathematical background should consult McCullagh 1988, 757–761 for details on smoothing methods for both triangulation and inverse-distance methods.

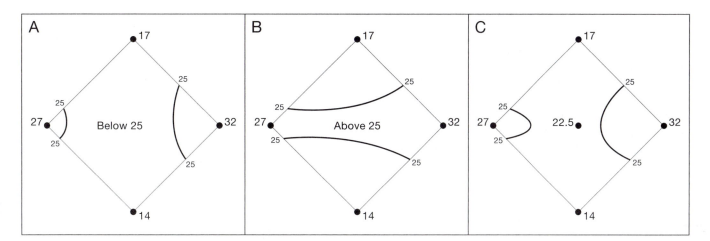

FIGURE 14.2   When the control points to be manually interpolated form a square, a fifth control point is created by averaging the four original control points.

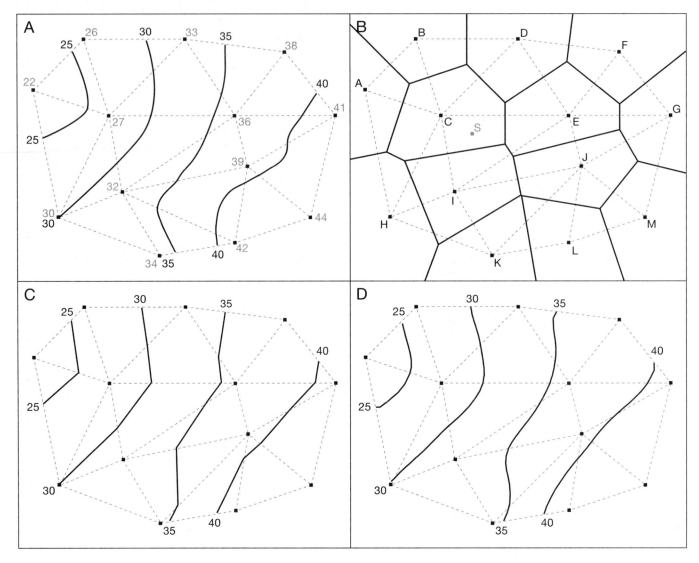

FIGURE 14.3   Hypothetical data showing (A) manual contouring, (B) Thiessen polygons and Delaunay triangles, (C) simple linear interpolation along Delaunay triangle edges, and (D) smoothed interpolation along Delaunay triangle edges.

To assist in comparing triangulation with the inverse-distance and kriging approaches, Figure 14.4 portrays contour maps of all three methods. These maps are based on precipitation values for 376 weather stations within South Carolina, North Carolina, and Georgia. More specifically, the data cover a four-day period associated with Hurricane Hugo in 1989. Although only South Carolina is shown in Figure 14.4, data for adjacent states were included so that the contour map would be accurate along the state boundary of South Carolina.

14.3.2 Inverse-Distance

Inverse-distance interpolation involves three steps: (1) laying a grid on top of the control points, (2) estimating values at each grid point (or grid node) as a function of distance to control points, and (3) interpolating between the grid points to actually create a contour line. Because a grid is used, the inverse-distance method is sometimes termed **gridding**. There is no analogy to this method in manual contouring—a grid is used because "it is much easier to draw contour lines through an array of regularly spaced grid nodes than it is to draw them through the irregular pattern of the original points" (Davis 2002, 380).

The term *inverse-distance* is used because control points are weighted as an inverse function of their distance from grid points: Control points near a grid point are weighted more than control points far away. The basic formula for estimating a value at a grid point is

$$\hat{Z} = \frac{\sum_{i=1}^{n} Z_i/d_i^k}{\sum_{i=1}^{n} 1/d_i^k}$$

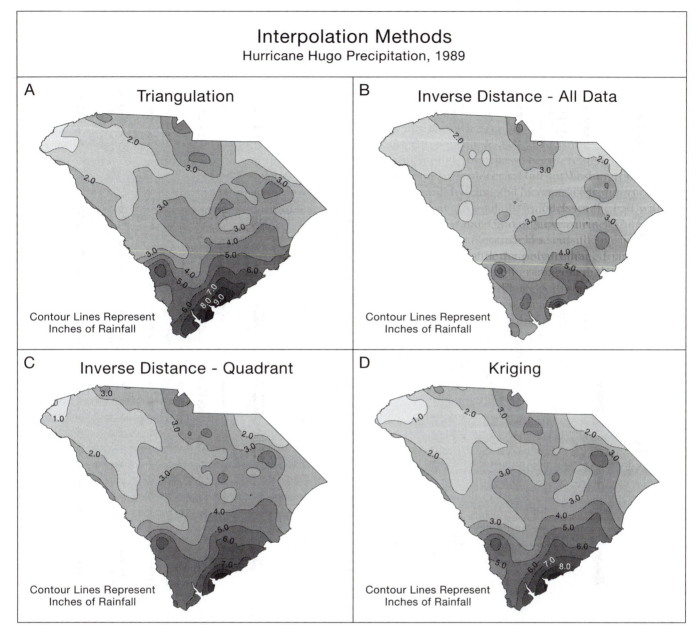

FIGURE 14.4 Contour maps of precipitation data associated with Hurricane Hugo: (A) triangulation, (B) inverse-distance using all data, (C) inverse-distance using a quadrant approach, and (D) kriging. (Data source: Southeast Regional Climate Center; Telnet to 198.202.229.3.)

where $\hat{Z}$ = estimated value at the grid point

$\quad Z_i$ = data values at control points

$\quad d_i$ = euclidean distances from each control point to a grid point

$\quad k$ = power to which distance is raised

$\quad n$ = number of control points used to estimate a grid point

To illustrate, Figure 14.5 depicts a portion of a hypothetical grid laid on top of four control points, along with calculations for the distances from each control point to the central grid point. If we assume $k = 1$ (for simplicity in

manual computation), then the central grid point value is:

$$\hat{Z} = \frac{(Z_1/d_1^1) + (Z_2/d_2^1) + (Z_3/d_3^1) + (Z_4/d_4^1)}{(1/d_1^1) + (1/d_2^1) + (1/d_3^1) + (1/d_4^1)}$$

$$= \frac{(40/2.24) + (60/1.00) + (50/1.00) + (40/1.41)}{(1/2.24) + (1/1.00) + (1/1.00) + (1/1.41)}$$

$$= 49.5$$

Normally, however, k is set to 2, which cancels the square root computation in distance calculations, and thus saves computer time (Isaaks and Srivastava 1989, 259).

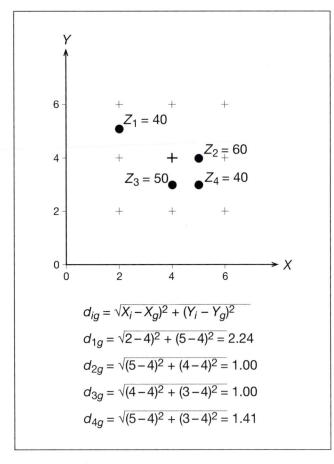

$$d_{ig} = \sqrt{X_i - X_g)^2 + (Y_i - Y_g)^2}$$

$$d_{1g} = \sqrt{2 - 4)^2 + (5 - 4)^2} = 2.24$$

$$d_{2g} = \sqrt{(5 - 4)^2 + (4 - 4)^2} = 1.00$$

$$d_{3g} = \sqrt{(4 - 4)^2 + (3 - 4)^2} = 1.00$$

$$d_{4g} = \sqrt{(5 - 4)^2 + (3 - 4)^2} = 1.41$$

FIGURE 14.5   Computation of distances for inverse-distance contouring. Four control points are overlaid with a hypothetical grid; distance calculations are shown from each control point to the central grid point.

With real-world data, the number of grid and control points obviously will be greater than this simple hypothetical case. A larger number of control points leads one to wonder which control points should be used to estimate each grid point. One approach is to use *all* control points; in this case, *n* in the formula for $\hat{Z}$ is simply the number of control points. This approach seems unrealistic, however, because control points far away from a grid point are unlikely to be correlated with that grid point. This sort of thinking, and the need to minimize computations, led software developers to create various strategies for selecting an appropriate subset of control points.

To illustrate the range of strategies, we consider four available within an early version of Surfer, a common software package for creating isarithmic maps: All Data, Simple, Quadrant, and Octant.* The All Data strategy uses *all* control points in the calculation of each grid point. The Simple strategy requires that control points fall within an ellipse (typically a circle) of fixed size; normally, only a subset of these points (say, the eight nearest) is used to make an estimate. Quadrant and Octant strategies also require that control points fall within an ellipse, but the ellipse is divided into four and eight sectors, respectively, with a specified number of control points used within each sector. Surfer also includes options for the minimum number of points that must fall within the ellipse and the maximum number of empty sectors permitted.

To better understand these strategies, the Simple, Quadrant, and Octant strategies are illustrated with hypothetical data in Figure 14.6. For sake of comparison,

---

* These options are possible in the present version of Surfer, but not using a single command.

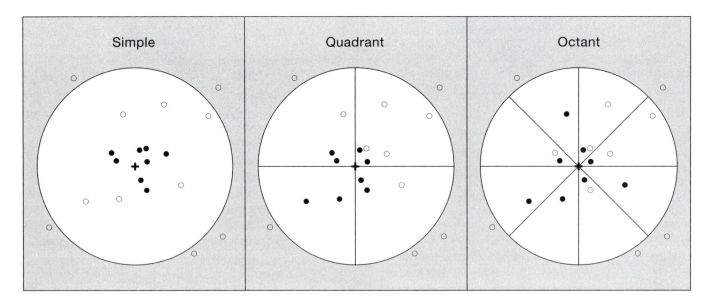

FIGURE 14.6   Three search strategies used within Surfer, a common contouring package. The cross in the middle represents a grid point location to be estimated, and the large circle represents the search radius within which control points must be located. Actual control points selected in each case are shown in black.

assume that each strategy requires a total of eight points to make an estimate. Note that the Simple strategy takes the nearest eight points regardless of the direction of the points; Quadrant, the nearest two points within each of four sectors; and Octant, the nearest point in each of eight sectors. For these hypothetical data, it appears that either the Quadrant or Octant strategy is preferable to the Simple strategy because the latter does not use any control points southwest of the grid point.

To further illustrate the effect of the strategies, consider the contour maps of the Hurricane Hugo data using the All Data and Quadrant strategies (Figure 14.4B and 14.4C). For the Quadrant strategy, we specified that two control points be used in each sector, permitted a grid point to be calculated using as few as six control points, and allowed a maximum of one sector to have no control points. We chose these parameters, in part, to demonstrate how the All Data and Quadrant strategies can lead to quite different-looking maps. Because precipitation data were not available for the ocean surface, the latter two parameters also ensured that grid points would be calculated near the land–ocean interface.

For the Quadrant strategy, we also specified differing ellipse radii (2° latitude and 2.4° longitude) because the control points were specified in degrees latitude and longitude, and at this latitude 1° of latitude is not equal to 1° of longitude.* One could also modify the radii to account for the possibility that the phenomenon being mapped is more highly related (autocorrelated) along one axis than another. For example, the user's guide for Surfer (Golden Software 2002, 109) suggests that for temperature data in the upper Midwest of the United States a longer axis in the east–west direction might be appropriate because temperatures are more similar in that direction. For the Hugo precipitation data, one might consider orienting the ellipse along the direction of the storm track, assuming that it was constant.

Comparing the All Data and Quadrant maps in Figure 14.4, we see that they have similar overall patterns, but that the All Data map exhibits more gradual changes; for example, if you compare the north central portion of the maps, you will note that the distance between the 2.0 and 3.0 contour lines is greater on the All Data map. This result is not surprising given that all control points are included in grid calculations, and thus we would expect estimated values for grid points closer to the average of all control point values. It is also interesting to note that the Quadrant map is very similar to the triangulated map. One wonders whether this would be true with other data sets; we will deal with this question in more detail in section 14.4.

In discussing inverse-distance interpolation, we have focused on grid point calculations because of the considerable number of options involved. Remember, however, that after grid point calculation, an interpolation must also take place between grid points. As with triangulation, a simple linear interpolation leads to angular contour lines; angularity can be avoided either by increasing the number of grid points (which increases computations and memory requirements) or by smoothing the contours. Smoothing is accomplished by **splining**, which involves fitting a mathematical function to points defining a contour line (Davis 2002, 228–234); this concept is illustrated graphically in Figure 14.7.

One problem with inverse-distance interpolation is that it cannot account for trends in the data. For example, a visual examination of z values for control points might suggest a value for a grid point higher than any of the control points, but the inverse-distance method cannot produce such a value. This problem exists because the formula for inverse-distance is a weighted *average* of z values for control points: An average of values cannot be lesser or greater than any input values.[†]

14.3.3 Kriging

The term **kriging** comes from Daniel Krige, who developed the method for geological mining applications (Cressie 1990). Kriging is similar to inverse-distance interpolation in that a grid is overlaid on control points, and values are estimated at each grid point as a function of distance to control points. Rather than simply considering distances to control points independently of one

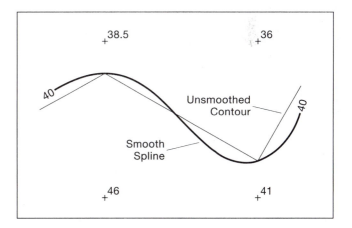

FIGURE 14.7 Fitting a smooth curve (a spline) to an angular contour line. (After Sampson 1978, 50; courtesy of Kansas Geological Survey.)

---

* Differing ellipse radii would not be necessary if the control points were specified in terms of an equidistant projection.

[†] One solution to this problem is to utilize the technique of *trend surfaces* (Davis 2002, 384–385).

another, however, kriging utilizes the spatial autocorrelation in the data, both between the grid point and the surrounding control points and among the control points themselves. Understanding how spatial autocorrelation is used in kriging requires an understanding of semivariance and the semivariogram.

14.3.4 Semivariance and the Semivariogram

Consider the simplified graphic shown in Figure 14.8A, which portrays attribute data for five hypothetical equally spaced control points. Normally, of course, control points are not equally spaced, but it is easier to understand semivariance if initially we assume that they are. **Semivariance** is defined as

$$\gamma_h = \frac{\sum\limits_{i=1}^{n-h}(Z_i - Z_{i+h})^2}{2(n-h)}$$

where Z_i = values of the attribute at control points
  h = multiple of the distance between control points
  n = number of sample points

In Figure 14.8A, the distance between control points is 10; h can take on values of 1, 2, 3, and 4 (distances of 10, 20, 30, and 40, respectively, are possible); and n is 5. Note that the largest possible multiple is 4 because the maximum possible distance between the control points is 40.

To illustrate computations for semivariance, we can insert the data from Figure 14.8A in the preceding formula for $h = 1$.

$$\gamma_1 = \frac{\sum\limits_{i=1}^{5-1}(Z_i - Z_{i+1})^2}{2(5-1)}$$

$$= \frac{(Z_1 - Z_2)^2 + (Z_2 - Z_3)^2 + (Z_3 - Z_4)^2 + (Z_4 - Z_5)^2}{8}$$

$$= \frac{(20 - 30)^2 + (30 - 35)^2 + (35 - 40)^2 + (40 - 45)^2}{8}$$

$$= 21.88$$

A summary of these calculations for all values of h is shown in Figure 14.8B. You should compute the column for $h = 2$ to ensure that you understand the method of computation.

Real-world computation of semivariance requires two modifications to this approach. First, because control points are not normally equally spaced, a tolerance must be allowed in both distance and direction between control points. For example, consider the control points shown in Figure 14.9. If we assume the distance between

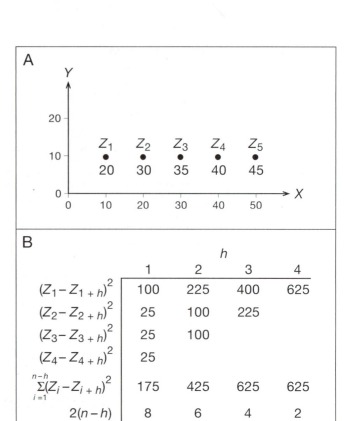

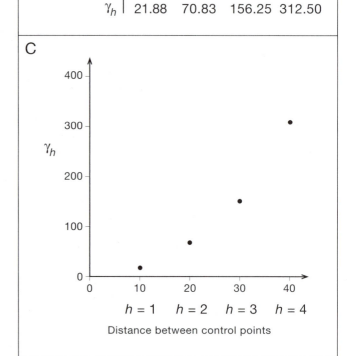

FIGURE 14.8   Computation of semivariance using equally spaced hypothetical control points: (A) the equally spaced points, (B) semivariance computations, and (C) semivariogram.

control points is 5 meters, note that no control point occurs exactly 5 meters east of point A. However, if we permit a distance tolerance of 1 meter and an angular tolerance of 20°, then four control points are east of point

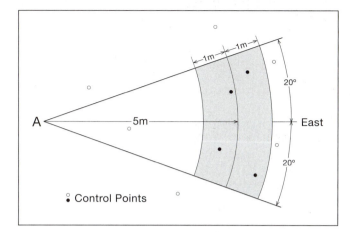

FIGURE 14.9 Determining which control points will be used in semivariance computations. A fixed distance yields no points directly 5 meters east of point A. A tolerance of 1 meter and 20° however, yields four control points (From APPLIED GEO-STATISTICS by E. H. Isaaks and R. Mohan Shrivastava. Copyright © 1989 by Oxford University Press, Inc. Used by permission of Oxford University Press, Inc.)

A. The second modification is to calculate semivariance in a variety of directions; thus, in addition to computing along the x axis (east–west direction), computations should be made in north–south, northwest–southeast, and northeast–southwest directions. For now, we assume that such computations are combined to create a single semivariance value.

The **semivariogram** is a graphical expression of how semivariance changes as h increases. For example, Figure 14.8C is a semivariogram for the hypothetical data shown in Figure 14.8A. Clearly, as h increases (i.e., as control points become more distant), the semivariance also increases. This basic feature is characteristic of most data and should not surprise us: We expect nearby geographical data to be more similar than distant geographical data. The behavior of semivariograms with larger data sets is characterized by the idealized curve shown in Figure 14.10. Note that the semivariance increases as it did in Figure 14.8C, but eventually reaches a plateau. The value for semivariance at which this occurs is known as the *sill*, and the distance at which it occurs is known as the *range*. The plateau normally indicates that the data are no longer similar to nearby values, but rather that the semivariance has approached the variance in the entire data set.

When using the semivariogram in kriging, it is necessary to make an estimate of the semivariance for some arbitrary distance between points. For example, in Figure 14.8C we might wish to estimate the semivariance for a distance of 26. This is normally accomplished by fitting a curve (or model) to the set of points comprising the semivariogram, a process known as *modeling the semivariogram*. Once an equation for the model is determined, a value for distance can be inserted in the equation, and a value for

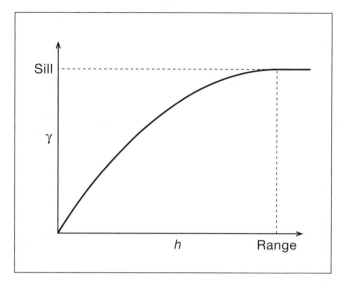

FIGURE 14.10 Idealized semivariogram illustrating a flattening in the semivariance values. The distance at which this occurs is known as the range and the associated semivariance value is the sill.

semivariance computed. The simplest model is a straight-line (linear) one; other common models include the spherical and exponential (Figure 14.11).

Kriging Computations

There are two major forms of kriging: ordinary and universal. We focus on **ordinary kriging** in which the mean of the data is assumed constant throughout geographic

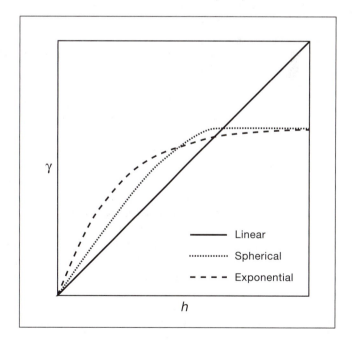

FIGURE 14.11 Models commonly used to summarize the points comprising the semivariogram: linear, spherical, and exponential. (After Olea 1994, 31.)

space (there is no trend or drift in the data).* To simplify the discussion, we assume that only three control
points are used to estimate a grid point; later we relax
this constraint.

In a fashion similar to inverse-distance, kriging uses a
weighted average to compute a value at a grid point. For
three control points, the equation is

$$\hat{Z} = w_1 Z_1 + w_2 Z_2 + w_3 Z_3$$

where $\hat{Z}$ = estimated value at a grid point
$Z_1, Z_2,$ and Z_3 = data values at the control points
$w_1, w_2,$ and w_3 = weights associated with each
control point

The w_i are analogous to the $1/d$ values used in inverse-
distance computations; the formula appears simpler for
kriging because the w_i are constrained to sum to 1.0.

In kriging, the weights (w_i) are chosen to minimize the
difference between the estimated value at a grid point
and the true (or actual) value at that grid point. This is
analogous to the situation in regression analysis in which
we minimize the difference between the estimated value
of a dependent attribute and its true value (see section
3.3.2). In kriging, minimization is achieved by solving for
the w_i in the following simultaneous equations:

$$w_1 \gamma(h_{11}) + w_2 \gamma(h_{12}) + w_3 \gamma(h_{13}) = \gamma(h_{1g})$$

$$w_1 \gamma(h_{12}) + w_2 \gamma(h_{22}) + w_3 \gamma(h_{23}) = \gamma(h_{2g})$$

$$w_1 \gamma(h_{13}) + w_2 \gamma(h_{23}) + w_3 \gamma(h_{33}) = \gamma(h_{3g})$$

where $\gamma(h_{ij})$ = the semivariance associated with the distance between control points i and j, and $\gamma(h_{ig})$ is the semivariance associated with the distance between the ith
control point and a grid point (Davis 1986, 385).[†] Note that
these equations consider not only the distance from control
points to the grid point (used in calculating $\gamma(h_{ig})$), but also
the distances between the control points themselves (used
in calculating $\gamma(h_{ij})$); contrast this approach with the
inverse-distance method, which considers only distances
between the grid point and control points.

To illustrate the nature of these equations, assume that
we are given three hypothetical control point values and
wish to estimate a value for a grid point (Figure 14.12).
Figure 14.12B lists the X and Y coordinates for all points
and 14.12C lists the distances between pairs of points.

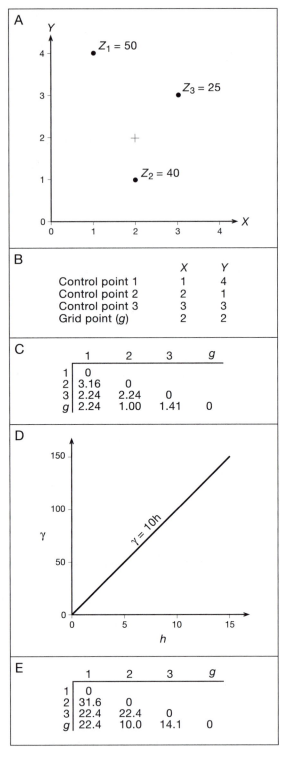

FIGURE 14.12 Semivariance computations for a grid point
and three control points: (A) graph of the grid point and control points; (B) X and Y coordinates for the points; (C) distances between points; (D) linear model of a semivariogram
for a data set from which the points were presumed to have
been taken; (E) semivariances associated with distances between points. (After Davis 1986, 387.)

---

* See Davis (2002, 428–437) for a discussion of the more complicated
method of **universal kriging**, which does account for the trend in the
data.

[†] We reference Davis's (1986) second edition here because we feel it is
more appropriate for the introductory student. Please be aware, however, that more recent information on kriging is available in Davis's
(2002) third edition.

Also assume that Figure 14.12D is an appropriate linear model for a semivariogram associated with a larger data set from which the three points were taken: The model ($\gamma = 10h$) indicates that the semivariance is simply 10 times the distance (when h is 5, γ is 50).

Determining a semivariance for use in the simultaneous equations requires plugging distance values into the model. For example, for control points 1 and 2, the distance is 3.16 and the semivariance is 31.6. Repeating this process for each pair of points produces the semivariance values shown in Figure 14.12E. Inserting these values into the simultaneous equations, we have

$$w_1 0.00 + w_2 31.6 + w_3 22.4 = 22.4$$
$$w_1 31.6 + w_2 0.00 + w_3 22.4 = 10.0$$
$$w_1 22.4 + w_2 22.4 + w_3 0.00 = 14.1$$

Davis (1986, 385–388) provided details of how these equations can be solved using matrix algebra. We did so using the statistical package SPSS, and found the following values for w: $w_1 = 0.15$, $w_2 = 0.55$, and $w_3 = 0.30$. Inserting these and the values for the control points into the basic weighted-average formula for kriging, we have

$$\hat{Z} = 0.15(50) + 0.55(40) + 0.30(25)$$
$$= 37$$

Using kriging in real-world situations (e.g., with the Hurricane Hugo data) differs from this simplified example in two important respects: (1) more than three control points normally are used to estimate each grid point, and (2) more than one semivariogram might be appropriate. For the number of control points, one normally specifies a search strategy similar to that used for the inverse-distance method. For example, Edward Isaaks and Mohan Srivastava (1989, Chapter 14) suggested using three points in each quadrant, for a total of 12 points overall, and a search radius slightly larger than the average spacing between control points. For the Hurricane Hugo data, we followed Isaaks and Srivastava's recommendations (using the same ellipse radii as for the Quadrant map). Additionally, we permitted a grid point to be calculated using as few as six control points and a maximum of one sector to have no control points; remember that these parameters were necessary to have grid points estimated along the land–water boundary. The resulting map (Figure 14.4D) appears very similar to both the Quadrant and Triangulated maps. Again, one wonders whether the similarity among these maps is a function of this particular data set. Or do these various interpolation methods yield similar results for a wide variety of data?

More than one semivariogram might be appropriate for a particular data set because the nature of spatial autocorrelation can vary as a function of the direction of semivariance calculation (e.g., east–west vs. north–south directions), as in the upper Midwest temperature example discussed when considering the inverse-distance method.* For the Hurricane Hugo precipitation data, two semivariograms might have been useful if we wished to consider the direction of hurricane movement. For example, if over the four-day period of rainfall, the hurricane moved in a constant direction, we might have used one semivariogram parallel to the movement and one perpendicular to it. Because we did not have data regarding hurricane movement, we chose a single semivariogram. We chose a linear model based on a visual examination of a sample semivariogram.

Although kriging is admittedly more complex than other methods of interpolation, it can produce a more accurate map; as such, it is often said to produce an **optimal interpolation**. It must be stressed that this is only true if one has properly specified the semivariogram(s) and associated semivariogram models. This is important to recognize because software for kriging might provide simplified defaults (e.g., a single semivariogram and linear model) that will not produce an optimal kriged map. Another advantage of kriging is that it provides a measure of the error associated with each estimate, known as the *standard error of the estimate*. This error measure can be used to establish a *confidence interval* for the true value at each grid location, assuming a normal distribution of errors. For example, if the kriged estimate at a grid point is 2.3 inches of rainfall and the corresponding standard error is 0.5 inches, then a 95 percent confidence interval for the true value at that grid point would be 2.3 $\pm$ 2(0.5). This is equivalent to saying we are 95 percent certain that the true value is within the range 1.3 to 3.3.[†]

14.4 CRITERIA FOR SELECTING AN INTERPOLATION METHOD FOR TRUE POINT DATA

So far in the chapter, we have covered three methods for interpolating true point data: triangulation, inverse-distance, and kriging. Given the similarity of the maps resulting from applying these three methods to the Hurricane Hugo data, how does one decide which method should be used? This section considers six criteria that can be considered in selecting an interpolation method: (1) correctness of estimated data at control points (does the method honor the raw data?), (2) correctness of estimated data at noncontrol points (how well does the method predict unknown points?), (3) ability to handle discontinuities (e.g., geologic faults), (4) execution time, (5) time spent selecting interpolation parameters,

---

* It might also be necessary to fit different models (equations) to each semivariogram (Isaaks and Srivastava 1989).

[†] For a more sophisticated approach for providing a measure of error associated with kriging, see Chainey and Stuart 1998.

and (6) ease of understanding. For the present discussion, we assume that the goal is to create an isarithmic map depicting *values* of a smooth continuous phenomenon. If the intention is to map various derivative measures of elevation (e.g., slope and drainage), then triangulation is a natural choice (Clarke 1995, 144–148).

14.4.1 Correctness of Estimated Data at Control Points

One advantage often attributed to triangulation is that estimated values for control points will be identical to the raw values originally measured at those control points; this is known as **honoring the control point data**. For example, for the triangulated map shown in Figure 14.3D, the contour line for 30 passes directly through control point H, which has a raw value of 30. In contrast, the inverse-distance and kriging methods do not, in general, honor control points. An exception would be when control and grid points coincide: In this case, the distance between control point and grid point will be zero, and the value for the grid point can simply be made equivalent to the control point value. Although honoring control point data is worthwhile, using it as the key criterion for evaluating interpolation is risky for two reasons. First, the process presumes that data are measured without error; it is common knowledge that physical recording devices (e.g., a rain gauge) are imperfect, and their readings are also be subject to human error. Rather than exactly honoring control points, it makes sense that such estimates should be within a small tolerance of the raw data. A second problem with honoring control points is that the process does not necessarily guarantee correctness at noncontrol point locations. For example, in an evaluation of the inverse-distance method, Davis (1975) found that options providing good estimates at control points did poorly at noncontrol points.

14.4.2 Correctness of Estimated Data at Noncontrol Points

Two basic approaches have been used for evaluating correctness of data at noncontrol points: cross-validation and data splitting. **Cross-validation** involves removing a control point from the data to be interpolated, using other control points to estimate a value at the location of the removed point, and then computing the *residual*, the difference between the known and estimated control point values; this process is repeated for each control point in turn. If cross-validation is done for a variety of interpolation methods, the resulting sets of residuals can be compared. In **data splitting**, control points are split into two groups, one to create the contour map, and one to evaluate it. Residuals are computed for each of the control points used in the evaluation stage. One problem

with data splitting is that it is impractical with small data sets because it makes sense to use as many control points as possible to create the contour map. Ideally, you should use either cross-validation or data splitting to evaluate the accuracy of any contour maps that you create. Because this can be time consuming, it is useful to consider the results of those who have already done this sort of analysis. The results of two such studies are summarized next.*

Isaaks and Srivastava (1989, 249–277, 313–321, 338–349) evaluated triangulation, inverse-distance, kriging, and other methods for a set of digital elevation values in the Walker Lake area near the California–Nevada border. One important conclusion of their study was that the "'best' [interpolation method] depends on the yardstick we choose" (272). For example, triangulation produced a smaller standard deviation of residuals than an inverse-distance method (using all control points within a specified search ellipse), but the inverse-distance method minimized the size of the largest residual. Overall, their study revealed that kriging was best (318–321), although it was only slightly more effective than an inverse-distance approach in which four control points were required in each quadrant surrounding a grid point.

Franky Declercq (1996) evaluated triangulation, inverse-distance, and kriging for two types of data: one relatively smooth and one more abrupt in nature. Smooth data consisted of mean annual hours of sunshine for the European Union, and abrupt data consisted of soil erodability (*K* values) in northern Belgium. One of Declercq's major conclusions was that there was little difference in inverse-distance and kriging methods. He argued that the number of control points used to make an estimate is more important than the general interpolation method: He recommended few control points (four to eight) for smooth data, and many (16 to 24) for more abrupt data. He also recommended using either a quadrant- or octant-sector approach. Lastly, he recommended not using triangulation approaches because they produce "highly inaccurate values and erratic images ... in poorly informed regions or with abruptly changing data" (p. 143).

14.4.3 Handling Discontinuities

In addition to honoring the data, another advantage of triangulation is its ability to handle discontinuities, such as geologic faults or cliffs associated with a topographic surface. McCullagh (1988, 763) indicated that "If the locations of these special lines are entered as a logically connected set of points, the triangulation process ... will automatically relate them to the rest of the data."

* For other studies that have compared interpolation methods, see Mulugeta (1996).

He further indicated that although grids can handle discontinuities, recognizing a fault in gridded data is almost impossible (p. 766).

14.4.4 Execution Time

The amount of computer time it takes to create a contour map can be an important consideration for large data sets. In general, a simple linear interpolated triangulation is faster than either inverse-distance or kriging because of the numerous computations that must be done at individual grid points with the latter methods. Kriging is especially time consuming because of the simultaneous equations that must be solved ($n + 1$ simultaneous equations must be solved for each grid point, where n is the number of control points). *Smoothed* triangulation methods, however, can take substantially longer than either inverse-distance or kriging methods (McCullagh 1988). The study by Declercq provides some comparative figures. Using a 50-megahertz 486-based microcomputer, Declercq computed grid point estimates for a 121×361 grid in less than five minutes using either linear triangulation or inverse-distance, but almost an hour was required for some kriging computations. A smoothed triangulation approach (using a quadratic-spline interpolation) was clearly the slowest, requiring almost 12 hours. Given the similarity of contour maps resulting from inverse-distance and kriging, the faster execution time of inverse-distance would seem to favor it when many grid points must be computed. The usefulness of such studies, however, is problematic, as processor speeds, storage capacity, and the amount of RAM increases exponentially each year. In the end, a thorough understanding of both interpolation routines and of the specific nature of the data being analyzed is critical.

14.4.5 Time Spent Selecting Parameters and Ease of Understanding

Time spent selecting parameters refers to how long it takes to make decisions such as the appropriate power for distance weighting, the number of control points to use in estimating a grid point, which model to use for the semivariogram, or which smoothing method to use for triangulation. One can avoid this time by simply taking program defaults, but such an approach is risky. Kriging is the most difficult in this regard because of the need to consider whether one or more semivariograms are appropriate and what sort of model should be used to fit each semivariogram. The time spent selecting parameters will obviously become longer if one wishes to evaluate correctness using cross-validation or data splitting. This time can be eliminated by using the evaluations of others, such as those described previously, although more studies appear necessary to cover the broad range of data types that geoscientists are apt to deal with. Ease of understanding refers to the ease of comprehending the conceptual basis underlying the interpolation method. Clearly, kriging scores poorly on this criterion, as the kriging literature (both published articles and software user guides) is often complex.

14.5 LIMITATIONS OF AUTOMATED INTERPOLATION APPROACHES

Although automated interpolation is desirable because it eliminates the effort involved in manual interpolation, a study by Gebeyehu Mulugeta (1996) illustrated some potential limitations of automated interpolation. Mulugeta compared automated and manual interpolation of two data sets: precipitation totals for a single storm event (June 28, 1982) in a five-county area of Michigan, and bedrock-surface altitudes for a township in Michigan. These data were automatically interpolated via kriging and manually interpolated by experts (climatologists and meteorologists for the precipitation data and geomorphologists for the bedrock-surface altitude data). A quantitative analysis using data splitting revealed that the computer-interpolated precipitation map was significantly more correct than the manually interpolated one, and that there was no significant difference in computer and manually interpolated bedrock-surface maps. Together these results suggested that the computer-interpolated maps were at least as good as, if not better than, the manual ones.

Comments elicited from experts, however, raised serious questions about the appropriateness of automated interpolation. Criticisms noted by the experts can be understood by comparing the computer-interpolated map of the storm event with a couple of the manually interpolated maps created by the experts (Figure 14.13). The most significant problem noted by experts was that "isolated peaks and depressions [on the computer-interpolated map were] overly smoothed and too circular." (Consider A1–A5 in Figure 14.13A.) The experts would have preferred "elongated bands ... characteristic ... of storm-precipitation surfaces" (p. 335). A second problem was that the computer-interpolated map portrayed "features ... not warranted by the data points" (335). An example was the peak at B4 in the middle of the computer-interpolated map, which "many experts showed ... as a saddle between two elongated 2[-inch] isarithms [Figure 14.13B] or enclosed ... in a 1.5[-inch] isarithm [Figure 14.13C]" (335). Other problems noted by experts on the computer-interpolated map included "jagged lines" and "spurious details" (D1–D4 and F1–F4 in Figure 14.13A).

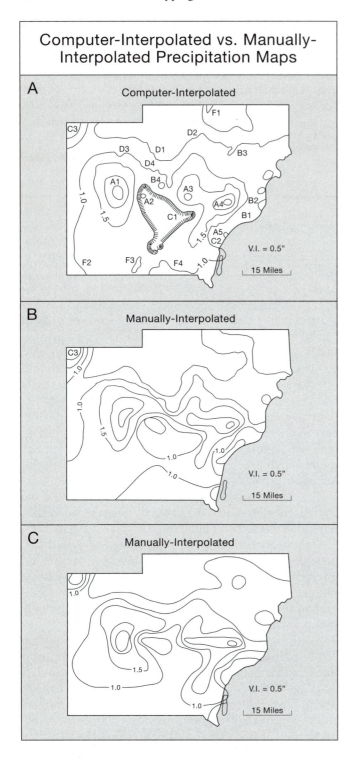

FIGURE 14.13 A comparison of (A) a computer-interpolated contour map and (B–C) manually interpolated contour maps. (After Mulugeta, G. (1996). "Manual and automated interpolation of climatic and geomorphic statistical surfaces: An evaluation." *Annals, Association of American Geographers*, 86, no. 2, p. 335. Courtesy of Blackwell Publishing.)

Mulugeta indicated that, ideally, mapmakers should edit automated contour maps to alleviate such problems. Unfortunately, if the mapmaker does not have expertise in the domain being mapped, this task will be difficult; for example, those unfamiliar with storm precipitation surfaces would not likely see the need for "elongated bands." Other problems, such as "jagged lines" and "spurious details," are capable of being edited by one without domain expertise, and thus we encourage you to correct such problems within common design programs such as Freehand.

14.6 TOBLER'S PYCNOPHYLACTIC APPROACH: AN INTERPOLATION METHOD FOR CONCEPTUAL POINT DATA

Although the interpolation methods discussed in section 14.3 were developed to handle true point data, they also have been commonly used for conceptual point data (and thus for isopleth mapping). For example, we used kriging to create the isopleth map of wheat harvested in Kansas counties described in Chapter 4 (the map is repeated in Figure 14.15B). We accomplished this by assigning the percentage of wheat harvested to the centroids of each county, which served as control points for interpolation.

Waldo Tobler's (1979) pycnophylactic (or volume-preserving) method is a more sophisticated approach for handling conceptual point data. To visualize this method, consider the standardized data associated with enumeration units as a clay model in which each enumeration unit is raised to a height proportional to the data (as in Figure 4.2B). The objective of the pycnophylactic method is to "sculpt this surface until it is perfectly smooth, but without allowing any clay to move from one [enumeration unit] to another and without removing or adding any clay" (520).* Relating this concept to the other interpolation approaches we have considered, we can think of volume preservation as a form of *honoring* the data associated with each *enumeration unit*.

To illustrate the pycnophylactic method in more detail, we use a simplified algorithm developed by Nina Lam (1983, 148–149). For this purpose, presume that we are given raw counts (RC_i) for the three hypothetical enumeration units shown in Figure 14.14A. These raw counts might be the number of people or acres of wheat harvested in each enumeration unit. In step 1 of Lam's algorithm, a set of square cells is overlaid on top of the enumeration units, and it is determined which cells fall in each enumeration unit (Figure 14.14B). A cell is considered part of an enumeration unit if its center falls within that unit; this can be determined by a so-called

---

* See *http://www.ncgia.ucsb.edu/pubs/gdp/pop/pycno.html* for a visual representation of pycnophylactic interpolation.

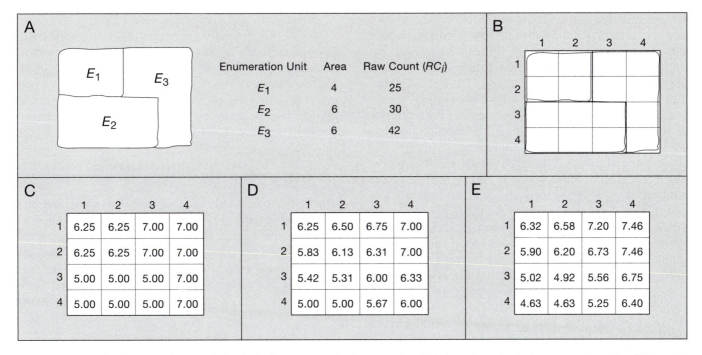

FIGURE 14.14   Basic steps of pycnophylactic (volume-preserving) contouring: (A) three hypothetical enumeration units; (B) square cells overlaid on the enumeration units; (C) initial density values for each cell (computed by dividing the raw count for each enumeration unit by the number of cells in that unit); (D) smoothed cell values (achieved by averaging neighboring cells); (E) smoothed values adjusted so that the sum within an enumeration unit equates to the original total sum for that enumeration unit (the volume is preserved). (After Lam 1983, 148–149.)

point-in-polygon test (Clarke 1995, 207–209). Note that four cells fall within enumeration unit E_1. In step 2 of the algorithm, a raw count for each cell is determined by dividing the raw count for each enumeration unit by the number of cells in that unit; for example, cells within enumeration unit 1 receive a value of 25/4, or 6.25 (Figure 14.14C). Note that this step essentially standardizes the data by computing a density measure.

Steps 3 to 5 of Lam's algorithm are executed in an iterative fashion. The steps are as follows:

Step 3. Each cell is computed as the average of its nondiagonal neighbors. For example, cell (2,2) becomes

$$\frac{6.25 + 6.25 + 7.00 + 5.00}{4} = 6.13$$

The results for all cells for the first iteration are shown in Figure 14.14D. This step accomplishes the smoothing portion of the algorithm.

Step 4. The cell counts within each enumeration unit at the end of step 3 (Figure 14.14D) are added to obtain a total smoothed count value, SC_i, for each enumeration unit. For example, the total for enumeration unit 1 is

$$SC_1 = 6.25 + 6.50 + 5.83 + 6.13 = 24.71$$

The results for enumeration units 2 and 3 are 32.40 and 39.39, respectively.

Step 5. All cell values are multiplied by the ratio RC_i/SC_i. For cell (2,2), the result is

$$6.13 \times (RC_1/SC_1) = 6.13 \times (25/24.71) = 6.20$$

The results for all cells for the first iteration are shown in Figure 14.14E.

Note that if the counts in each cell of an enumeration unit in Figure 14.14E are added, the resulting sum is equal to the original raw count for that unit; for example, for enumeration unit 1, we have

$$6.32 + 6.58 + 5.90 + 6.20 = 25.00$$

As a result, steps 4 and 5 enforce the pycnophylactic (volume-preserving) constraint. Remember that steps 3 to 5 are executed in an iterative fashion. For the second iteration, the results shown in E would be placed where C currently is depicted in Figure 14.14, and steps 3 to 5 would be executed again. Iteration continues until there is no significant difference between the raw and smooth counts for each enumeration unit or there is no significant change in the cell values compared with the last iteration.

One issue stressed by Tobler and not dealt with in this algorithm is how the boundary of the study area is handled. His computer program for pycnophylactic interpolation, PYCNO, provides two options: one in which zeros are presumed to occur outside the bounds of the region, and one in which a constant gradient of change is presumed across the boundary. The former would be appropriate if the region is surrounded by water, as when mapping population along the coast of the United States. The latter would be appropriate when mapping a phenomenon that is presumed to have similar characteristics in surrounding enumeration units, as when contouring wheat harvest data for the state of Kansas.

To contrast pycnophylactic interpolation with methods intended for true point data, Figure 14.15 portrays contour maps of the percentage of wheat harvested in counties of Kansas, using both the pycnophylactic and kriging methods. Although the two maps appear similar, there are some notable differences. One is that the magnitude of the highest contour on the pycnophylactic map is 70, whereas the highest contour on the kriged map is only 50. A higher valued contour appears on the pycnophylactic map because of its volume-preserving character. To understand this, imagine each of the counties in the form of the clay model described by Tobler earlier. In such a model, you would find that Sumner County is highest, with a raw standardized data value of 58.5 percent. To show a smooth transition to lower valued surrounding counties, Sumner's edges must be beveled off, but to retain the same volume, its center portion must be built up, thus resulting in a value above 58.5 in its interior. In contrast, the kriging method we used cannot produce a value higher than any control point because each grid point is a weighted average of surrounding control points.*

---

* Universal kriging can estimate values beyond the range of the data, but it would not necessarily honor the data associated with the enumeration unit.

Although the pycnophylactic method is arguably more appropriate for isopleth mapping than point-based interpolation methods, it must be emphasized that the method should only be used with continuous phenomenon. If the phenomenon is not continuous, nothing is gained by using the pycnophylactic approach. Instead, it might be more appropriate to use another mapping method, such as the dot map introduced in Chapter 17.

14.7 SYMBOLIZATION

The last step in isarithmic mapping is to symbolize the interpolated data. When mapping the Earth's topography, a variety of specialized symbolization techniques are possible—we consider these in Chapter 15. In this chapter, we consider symbolization approaches that are appropriate for either topographic or nontopographic phenomena.

14.7.1 Some Basic Symbolization Approaches

A range of basic symbolization approaches is shown in Figure 14.16. **Contour lines** (Figure 14.16A) frequently have been used to depict smooth continuous phenomena, particularly when maps were produced manually because little production effort was involved. An obvious problem with contours, however, is that visualizing the surface requires careful examination of the position and values of individual contour lines—the surface does not simply pop out at you.

The addition of **hypsometric tints** (or shaded areas) between contour lines (Figure 14.16B) enhances the ability to visualize a 3-D surface because light and dark tints

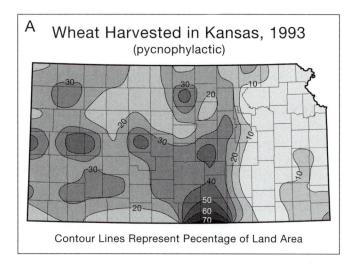

A   Wheat Harvested in Kansas, 1993
(pycnophylactic)

Contour Lines Represent Pecentage of Land Area

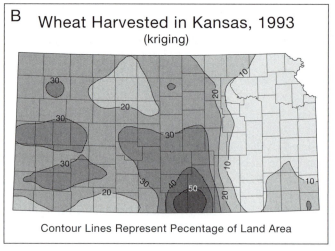

B   Wheat Harvested in Kansas, 1993
(kriging)

Contour Lines Represent Pecentage of Land Area

FIGURE 14.15   Comparison of interpolation methods for data collected for enumeration units: (A) the pycnophylactic approach, which expressly deals with the fact that the data were collected from enumeration units; (B) the kriging approach, which treats the areal data as conceptual point locations.

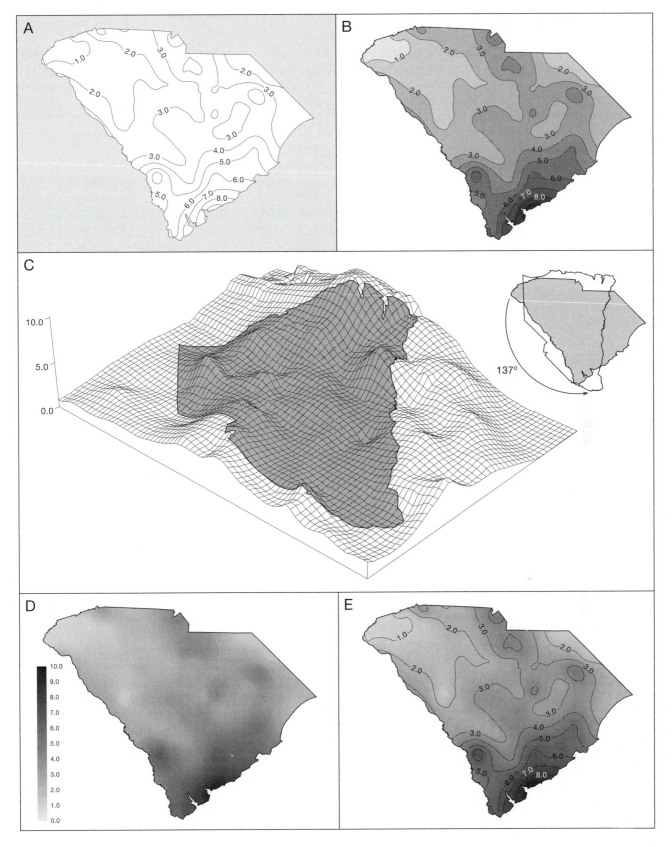

FIGURE 14.16 Comparison of several methods for symbolizing smooth continuous phenomena: (A) using only contour lines; (B) combining contour lines and hypsometric tints; (C) a fishnet symbolization; (D) continuous tones analogous to those used on unclassed choropleth maps; (E) continuous tones combined with contour lines.

can be associated with low and high values, respectively. This visualization can be further enhanced by using one of the color schemes that we recommended for choropleth maps (see Chapter 13) instead of gray tones. Note that a broader range of color schemes is possible on isarithmic maps because symbols of similar value must occur adjacent to one another (since the order of symbols on the map matches the order in the legend, the legend needs to be consulted less frequently). Hypsometric tints can also be shown without contour lines, as is common on TV weather maps.*

One problem with hypsometric tints is that the limited number of tones suggests a stepped surface, rather than the smooth one that occurs in reality. This problem can be ameliorated by creating a **continuous-tone map** (Kumler and Groop 1990), in which each point on the surface is shaded with a gray tone (or color) proportional to the value of the surface at that point (Figure 14.16D). This approach is analogous to unclassed choropleth mapping (see section 13.6), in which enumeration units are shaded with a gray tone proportional to the data value in that unit. One problem with interpreting a continuous-tone map is that it is difficult to associate numbers in the legend with particular locations, but this problem can be solved by overlaying continuous tones with traditional contour lines (Figure 14.16E).

Another approach that assists in interpreting smooth continuous phenomena is **fishnet** symbolization (Figure 14.16C). With this approach, not only does the surface change gradually, but we can actually "see" that certain points are higher or lower than others. Although a fishnet symbolization is useful, it has the same disadvantages as a prism map (discussed in section 4.5), including the blockage of low points by high ones, and that rotation might produce a view unfamiliar to readers who normally see maps with north at the top.

When choosing among the various symbolization methods shown in Figure 14.16, it is useful to consider a study by Mark Kumler and Richard Groop (1990) that evaluated black-and-white maps similar to those shown in Figure 14.16, plus several continuous-tone color maps. People were asked to complete the following tasks while looking at the maps: "locat[e] surface extrema, interpret . . . slope directions between points on the surface, estimat[e] relative surface values, and estimat[e] absolute values at points on the surface" (p. 282). Based on the accuracy with which people completed these tasks, Kumler and Groop found that continuous-tone maps were most effective. Also, when asked to pick their favorite approach, the majority of people selected one of the continuous-tone maps (a full-spectral scheme). Although Kumler and Groop's study suggests continuous-tone maps are particularly ef-

fective, the study was limited because it did not include noncontinuous-tone color maps. An interesting follow-up to their study would be a comparison of continuous-tone and noncontinuous-tone color maps.

The 3-D look of fishnet symbolization can also be achieved by using stereo pairs and anaglyphs, both of which permit stereo views (Clarke 1995, 273–274). With **stereo pairs**, two maps of the surface are viewed with a stereoscope, which you might be familiar with from examining 3-D views of aerial photographs. With **anaglyphs**, two images are created, one in green or blue, and one in red; these images are viewed with anaglyphic glasses, which use colored lenses to produce a 3-D view. In their book *Infinite Perspectives: Two Thousand Years of Three-Dimensional Mapmaking*, Brian and Jeffrey Ambroziak (1999) described a specialized anaglyph approach—the Ambroziak Infinite Perspective Projection (AIPP). In contrast to traditional anaglyphs, in which the image is viewed from only a single vantage point, the AIPP permits viewing from multiple positions. Additionally, AIPP is "self-scaling" (vertical exaggeration increases as one moves away from the image), permits both features that rise above and descend below the image plane (with no gaps of missing pixels), and permits the inclusion of additional data that are not part of the main anaglyph image (Ambroziak and Ambroziak, 1999, 86–87). Images viewed using the AIPP produce a striking 3-D appearance—to see this, we encourage you to examine the images found in the Ambroziaks' book.

14.7.2 Color Stereoscopic Effect

The **color stereoscopic effect** refers to the notion that colors from the red portion of the electromagnetic spectrum appear slightly nearer to the viewer than colors from the blue portion, primarily because the lens of the eye refracts light as a function of wavelength (see Travis 1991, 135–139, for a more detailed explanation). This effect can be used as an argument for utilizing a spectral color scheme (red, orange, yellow, green, blue, indigo, and violet). However, as we indicated in Chapter 13, cartographers normally recommend that spectral schemes be used with care, as yellow (an inherently light color) appears in the middle of the scheme.

The problem with yellow can be overcome if the color stereoscopic effect is enhanced. J. Ronald Eyton has discussed several approaches for enhancing the effect. In his early work, Eyton (1990) focused on how spectral colors should be generated and how they might be combined with contours and hill shading to enhance the color stereoscopic effect. He indicated that, ideally, the subtractive CMYK process should not be used to create spectral colors because the process is not purely subtractive.

---

* For a discussion of the use of color on weather maps, see Monmonier (1999a).

He stated "subtractive ink dots . . . adjacent to each other produce desaturated colors that are formed additively" (21). As one solution to this problem, Eyton created maps on a computer graphics display and then produced hard copies by recording the images onto film. As another solution, he printed maps using **fluorescent inks**, which produced "brilliant, intense color" (23–24). Maps resulting from his use of fluorescent ink can be found in Plates 1 to 4 of his paper.

In spite of taking these careful approaches, Eyton found little color stereoscopic effect when spectral colors were used alone. (He found that the effect could be enhanced with a reading magnifier, but the desired order of the colors was distorted, with green appearing above yellow and orange, instead of below them). Eyton did find, however, that when contour lines were added, users perceived a distinct color stereoscopic effect. Although Eyton could find no explanation for this enhancement in the literature, he hypothesized that the rapid change in slope associated with closer contours might provide depth cues.

Eyton achieved a particularly dramatic color stereoscopic effect when he combined spectral colors with a hill-shaded display. (**Hill shading** is a process in which terrain is shaded as a function of its orientation and slope relative to a presumed light source, typically from the northwest; see Chapter 15 for more details.) Eyton indicated that many viewers found a combination of spectral colors, contours, and hill shading to be most effective for enhancing the color stereoscopic effect.

In later work, Eyton (1994) described how the color stereoscopic effect could be enhanced via special viewing glasses. Originally developed by Steenblik (1987), these glasses are now available through *http://www.chromatek. com/*. The glasses can produce dramatic 3-D images when spectral colors are combined with contours or hill shading. We encourage you to acquire a set of glasses and examine Color Plate 14.1, which was created using a spectral scheme developed by Eyton (1990, 24–26).

14.7.3 Spatial Frequency and the Use of Color

Bernice Rogowitz and Lloyd Treinish (1996) suggested that the *spatial frequency* of the data (whether the data change rapidly or slowly over space) should determine whether a lightness- or saturation-based color scheme is used.* In particular, based on the human spatial vision literature, they argued that a lightness-based scheme (they termed this a "luminance-varying" scheme) should be used for high-frequency spatial data, and a saturation-

based scheme should be used for low-frequency spatial data. We illustrate this notion using the images shown in Color Plate 14.2, which depict the amount of ozone in the upper troposphere and lower stratosphere in the spring in the Southern Hemisphere. For all images, the amount of ozone is depicted by the height of the surface—this allows us to see the nature of the spatial distribution regardless of which color scheme is used. In Color Plate 14.2A, a traditional spectral scheme is used—concurring with cartographers, Rogowitz and Treinish argued against using such a scheme. In Color Plate 14.2B, a lightness scheme is used, which Rogowitz and Treinish argued will allow you to see the high-frequency artifacts of atmospheric circulation. In Color Plate 14.2C, a saturation-based scheme captures the broad depressed region (of ozone depletion) in the middle (a low-frequency characteristic). Finally, in Color Plate 14.2D, they argued that a combined lightness-saturation scheme captures both high-and low-frequency features.

SUMMARY

In this chapter, we have focused on **isarithmic maps**, which are used to depict smooth continuous phenomena. Two basic forms of data are utilized on isarithmic maps: true point data and conceptual point data. **True point data** can actually be measured at point locations (e.g., the hours of sunshine received over the course of the year), whereas **conceptual point data** are collected over areas (or volumes), but conceived as being located at points (e.g., birth rates for census tracts within an urban area). The terms **isometric map** and **isopleth map** are used to describe maps resulting from using true and conceptual point data, respectively.

A basic problem in isarithmic mapping is **interpolating** values between the known **control point** locations associated with either true or conceptual point data. We covered three interpolation methods appropriate for true point data: triangulation, inverse-distance, and kriging. **Triangulation** fits a set of triangles to control points and then interpolates along the edges of these triangles, in a fashion analogous to manual contouring. **Delaunay triangles** are commonly used because the longest side of any triangle is minimized, and thus the distance over which interpolation must take place is minimized. Triangulation is advantageous in that it honors the original control points, handles discontinuities well, and is an obvious choice for elevation data when information such as slope and drainage are also desired. A disadvantage of triangulation is that *smoothed* contours (a nonlinear interpolation) can require a lengthy execution time.

Inverse-distance interpolation lays an equally spaced grid of points on top of control points, estimates values at each grid point as a function of their distance from

---

* See Lloyd Treinish's home page (*http://www.research.ibm.com/people/ 1/lloydt/*) for other related papers.

control points, and then interpolates between grid points. The term **inverse-distance** comes from the fact that nearby control points are weighted more than distant points in calculating a grid point. Because points far away from a grid point are apt to be uncorrelated with the grid point, it makes sense to use a limited set of control points in calculating a grid point. Rather than simply considering the nearest control points, it also makes sense to consider the direction of the control points from the grid point (e.g., an octant search can be used). An advantage of inverse-distance is that large data sets can be efficiently interpolated.

Kriging is similar to inverse-distance in that a grid is overlaid on control points and values are estimated at each grid point as a function of distance to control points. Rather than considering distances to control points independently of one another, however, kriging considers the spatial autocorrelation in the data, both between the grid point and the surrounding control points, and among the control points themselves. The spatial autocorrelation in the data is evaluated by creating a **semivariogram**, which illustrates changes in the **semivariance** (variation in the data associated with control points) with increasing distance between control points. Like inverse-distance, kriging calculates an estimate at a grid point by using a weighted average of nearby control points. The weights, however, are not simply a function of the distance of the control points from the grid point, but rather consider the nature of the semivariogram computed for the control points. An advantage of kriging is that it can produce an **optimal interpolation** (if the semivariograms and associated **semivariogram models** are properly specified). Disadvantages of kriging are that execution times can be lengthy for large data sets, time is required to select the appropriate semivariograms and associated models, and the method is difficult to understand.

Ideally, you should evaluate the accuracy of interpolation methods by examining **residuals** (the difference between known and estimated control point values). This can be accomplished using either cross-validation or data splitting. In **cross-validation**, a control point is removed from the data, and other control points are used to interpolate a value at the location of the eliminated control point. If this process is repeated for each control point, then a set of residuals can be computed for each control point. In **data splitting**, control points are split into two groups, one to create the contour map and one to evaluate it. Residuals are computed for each of the control points used in the evaluation stage. Because the process of evaluating residuals can be time consuming, an alternative for evaluating the accuracy of interpolation methods is to use the results of past studies such as Declercq's.

Although automated interpolation methods for true point data clearly avoid the effort of manual interpolation, they are not without problems. Experts note that automated interpolation methods do not consider the nature of the underlying phenomenon; for example, an automated interpolation of precipitation data is not likely to consider the "elongated bands . . . characteristic . . . of storm-precipitation surfaces." Automated interpolation methods also tend to produce "jagged lines" and "spurious details" not warranted by the data. One solution to these problems is to graphically edit contour maps using design software such as Freehand.

This chapter also considered one method for handling conceptual point data: Tobler's **pycnophylactic interpolation**. The pycnophylactic method begins by assuming that each enumeration unit is raised to a height proportional to the value of its associated control point. This 3-D surface is then gradually smoothed, keeping the volume within each individual enumeration unit constant; thus, volume added somewhere within an enumeration unit must be balanced by taking volume from somewhere else in that unit. The smoothing is accomplished using a cell-based generalization process analogous to procedures used in image processing. Although pycnophylactic interpolation is appropriate for conceptual point data (and thus isopleth maps), it should be used only if the assumption of a smooth continuous phenomenon seems reasonable.

Methods for symbolizing interpolated data include **contour lines**, **hypsometric tints** (shading the areas between contour lines), **continuous-tone maps** (a form of unclassed map), and **fishnet** symbolization (a netlike structure that simulates the 3-D character of a smooth continuous surface). An intriguing possibility is to view a spectral color scheme with special glasses that enhance the **color stereoscopic effect** (long-wavelength colors will appear nearer than short-wavelength colors). In Chapter 15, we look at additional symbolization methods that are especially appropriate for topography.

FURTHER READING

Bucher, F. (1999) "Using extended exploratory data analysis for the selection of an appropriate interpolation model." In *Geographic Information Research: Trans-Atlantic Perspectives*. ed. by M. Craglia and H. Onsrud, pp. 391–403. London: Taylor & Francis.

   Proposes an expert system for selecting an appropriate method of interpolation.

Burrough, P. A., and McDonnell, R. A. (1998) *Principles of Geographical Information Systems*. Oxford: Oxford University Press.

   Chapters 5 and 6 cover numerous methods of interpolation.

Cressie, N. (1990) "The origins of kriging." *Mathematical Geology* 22, no. 3:239–252.

   Discusses the origins of kriging within a variety of disciplines.

Cressie, N. A. C. (1993) *Statistics for Spatial Data*. Rev. ed. New York: Wiley.

An advanced treatment of statistical methods for handling spatial data; includes an extensive section on kriging.

Davis, J. C. (2002) *Statistics and Data Analysis in Geology*. 3rd ed. New York: Wiley.

Triangulation, inverse-distance, and kriging methods are discussed on pages 370–397, 254–265, and 416–443. For kriging, the less statistically inclined student might wish to consult Davis's (1986) second edition.

Declercq, F. A. N. (1996) "Interpolation methods for scattered sample data: Accuracy, spatial patterns, processing time." *Cartography and Geographic Information Systems* 23, no. 3:128–144.

Evaluates the accuracy of various interpolation methods using a variety of approaches.

DeLucia, A. A., and Hiller, D. W. (1982) "Natural legend design for thematic maps." *The Cartographic Journal* 19, no. 1:46–52.

Examines a novel approach for designing legends for isarithmic maps; for a critique of this paper, see Paslawski (1983).

Isaaks, E. H., and Srivastava, R. M. (1989) *Applied Geostatistics*. New York: Oxford University Press.

A widely referenced text on interpolation methods; the focus is on kriging.

Journal of Geographic Information and Decision Analysis 2, no. 2, 1998 (entire issue).

This special issue deals with advanced issues in interpolation.

Kumler, M. P. (1994) "An intensive comparison of triangulated irregular networks (TINs) and digital elevation models (DEMs)." *Cartographica* 31, no. 2:1–99.

Triangulation and an equally spaced grid are two common approaches for representing elevation (topographic) data. This study compares the effectiveness of these two approaches.

Lam, N. S. (1983) "Spatial interpolation methods: A review." *Cartography and Geographic Information Systems* 10, no. 2:129–149.

An overview of a variety of interpolation methods, many of which are not covered in this chapter.

McCullagh, M. J. (1988) "Terrain and surface modelling systems: Theory and practice." *Photogrammetric Record* 12, no. 72:747–779.

Describes and compares grid-based (using inverse-distance or kriging) and triangulation interpolation methods.

Miller, E. J. (1997) "Towards a 4D GIS: Four-dimensional interpolation utilizing kriging." In *Innovations in GIS 4*, ed. by Z. Kemp, pp. 181–197. Bristol, PA: Taylor & Francis.

Our present chapter deals with interpolation methods appropriate for 21/2-D phenomena in which one *z* value is associated with each *x* and *y* location. Miller's work considers interpolation methods for true 3-D phenomena in which multiple *z* values are associated with each *x* and *y* location. Miller also considers spatial data with a temporal component. Also see Miller (1999).

Mitas, L., and Mitasova, H. (1999) "Spatial interpolation." In *Geographical Information Systems*, Volume 1, ed. by P. A. Longley, M. F. Goodchild, D. J. MaGuire, and D. W. Rhind, pp. 481–492. New York: Wiley.

A review of various methods of spatial interpolation.

Monmonier, M. (1999a) *Air Apparent: How Meteorologists Learned to Map, Predict, and Dramatize Weather*. Chicago: The University of Chicago Press.

Pages 223–226 discuss the use of color on weather maps; also see Monmonier (1999b).

Mulugeta, G. (1996) "Manual and automated interpolation of climatic and geomorphic statistical surfaces: An evaluation." *Annals, Association of American Geographers* 86, no. 2:324–342.

Compares the accuracy of automated and manual methods for interpolation.

Olea, R. A. (1999) *Geostatistics for Engineers and Earth Scientists*. Boston: Kluwer Academic.

Provides a mathematical treatment of kriging; a full understanding requires a solid mathematical background, but the book contains many useful thoughts on kriging.

Robeson, S. M. (1997) "Spherical methods for spatial interpolation: Review and evaluation." *Cartography and Geographic Information Systems* 24, no. 1:3–20.

Reviews and evaluates spherical methods for interpolation, which account for curvature of the Earth and are essential when interpolating global-scale phenomena. Also see Raskin et al. (1997).

Tobler, W. R. (1979) "Smooth pycnophylactic interpolation for geographical regions." *Journal of the American Statistical Association* 74, no. 367:519–536.

A detailed discussion of the pycnophylactic interpolation method; understanding this paper requires a solid background in calculus.

Tobler, W., Deichmann, U., Gottsegen, J., and Maloy, K. (1997) "World population in a grid of spherical quadrilaterals." *International Journal of Population Geography* 3:203–225.

Describes the development of a subnational database of population for the world based on pycnophylactic interpolation.

Veve, T. D. (1994) *An Assessment of Interpolation Techniques for the Estimation of Precipitation in a Tropical Mountainous Region*. Unpublished M.A. thesis, Pennsylvania State University, University Park, PA.

Compares several interpolation methods, including inverse-distance, kriging, and co-kriging. The latter permits the inclusion of ancillary attributes in the kriging process (e.g., elevation in the case of interpolating precipitation in a mountainous area).

Ware, J. M., and Jones, C. B. (1997) "A multiresolution data storage scheme for 3-D GIS." In *Innovations in GIS 4*, ed. by Z. Kemp, pp. 9–24. Bristol, PA: Taylor & Francis.

Describes a sophisticated triangulation-based approach for representing topographic and geologic data.

Webster, R., and Oliver, M. A. (2001) *Geostatistics for Environmental Scientists*. Chichester, England: Wiley.

A textbook on kriging for the application-minded environmental scientist.

15

Symbolizing Topography

OVERVIEW

Whereas section 14.7 dealt with approaches for symbolizing a broad range of smooth continuous phenomena, in this chapter we consider approaches for symbolizing a particular type of smooth continuous phenomenon—the Earth's topography. By topography, we mean the Earth's elevation (both above and below sea level) and associated features found on the Earth—its landscape. Specialized techniques have been developed for symbolizing topography because of its importance in everyday life and in the burgeoning area of GIS.

Section 15.1 briefly considers the kinds of data normally used in symbolizing topography. Fundamental data for depicting topography consist of elevation values above (or below) sea level for individual point locations. These data are most commonly provided in a square gridded network (a so-called digital elevation model, or DEM). Data for depicting the Earth's landscape can be acquired from mapping agencies such as the USGS.

*Section 15.2 considers approaches for symbolizing topography that have focused on a vertical (or plan) view, as though you are viewing the Earth from an airplane high above it. Methods covered include **hachures**, contour-based methods, the **physiographic method**, **shaded relief**, and various morphometric methods (these focus on the structure of the topography, e.g., **aspect** and **slope**). Assuming that the topographic surface is illuminated obliquely from the northwest, these methods utilize various approaches for distinguishing illuminated areas from areas in the shadows. Because the visual effect of shaded relief can be rather dramatic, and it has become a popular computer-based method, we cover some of the details of associated computational formulas.*

In section 15.3, we consider approaches for symbolizing topography that have focused on an oblique view of the

*Earth, including **block diagrams**, **panoramas**, and **draped images**. Although such approaches might not provide the geometric accuracy of a vertical view, their realism (and associated lack of abstraction) makes them an attractive option. Traditionally, only a single oblique view of a landscape was possible and so some information was of necessity hidden. Now, however, interactive software permits users to create their own oblique views. Another limitation was the time needed to create realistic oblique views, but digital design tools are now making it easier for cartographers to create realistic-looking views.*

*Although maps based on an oblique view have a distinctive 3-D appearance, they are still created on a two-dimensional surface (either paper or a graphics display device). It can be argued that a more effective approach is a **physical model** that is truly 3-D (we can hold and manipulate small models in our hands). Such physical models are the topic of section 15.4. Physical models have been around for hundreds of years, but only recently has technology enabled us to create such models from digital data.*

15.1 NATURE OF THE DATA

Fundamental data for depicting topography consists of elevation values above (or below) sea level for individual point locations. Although elevation data might be available at irregularly spaced control points (as with other smooth continuous phenomena), these data are most commonly provided in a square gridded network (a so-called *digital elevation model*, or *DEM*). This square gridded network is analogous to the grid resulting from either the inverse-distance or kriging approaches discussed in Chapter 14. Elevation data are sometimes also provided as a *triangulated irregular network (TIN)*, which is closely associated with the triangulation approach discussed in Chapter 14. The procedures for creating DEMs and TINs are described in GIS textbooks (see, e.g., Chapter 9 of Lo and Yeung 2002). DEM data can be acquired from various mapping agencies such as the USGS (*http://edc. usgs.gov/products/elevation/dem.html*).*

When symbolizing topography, you will often want to display not only the elevation of the landscape but also features found on the landscape (e.g., roads, streams, and land use). There are two basic approaches for portraying features found on the landscape. One is to utilize "raw" remotely sensed images, such as those obtained by the Landsat satellite; the other is to use maps that have been either derived from remotely sensed images (e.g., through visual interpretation) or acquired from ground surveys. An excellent source of such data for the United States is the EROS Data Center (*http://edc.usgs.gov/*).

15.2 VERTICAL VIEWS

In this section, we consider approaches for symbolizing topography that traditionally emphasized a vertical (or *plan*) view of the landscape. A vertical view was emphasized for two reasons. First, there was a desire to visualize the landscape of a particular region (say the Rocky Mountains in Colorado) in its entirety. Given the time and cost constraints of creating printed maps, this generally meant that a single map would be produced, and that to visualize the entire region, this map would have to be viewed vertically (an oblique view would of necessity hide some features). Second, there was a desire to show features in their correct geographic location. In this respect, cartographers distinguished between maps and paintings. For instance, Eduard Imhof (1982, 79–80) stated:

> The map is not only a picture; it is primarily, a means of *providing information* [italics in original]. It has other responsibilities than has painting.... Above all, it must include the depth of the valleys and the heights of mountains, not only in the visual sense but in the geometrical sense also.

To create the impression of a 3-D view, the bulk of the methods in this section assume that the surface is lit obliquely from the northwest. According to Imhof (1982, 178), a northwest light source is commonly used because of the preponderance of people who write with their right hand. When writing with the right hand, lighting is placed above and to the left to eliminate the possibility that a

shadow will be cast over the written material. Our experience in using a light in this situation causes us to expect a light in a similar orientation in other situations such as relief depiction. If an alternate light direction is used, then the relief might appear to be inverted (e.g., valleys might become mountains).†

15.2.1 Hachures

Hachures are constructed by drawing a series of parallel lines perpendicular to the direction of contours. Two forms of hachures were traditionally used in manual cartography (Yoeli 1985). In the first, known as *slope hachures*, the width of hachures was proportional to the steepness of the slope; thus, steeper slopes were darker. Because slope hachures generally failed to create the impression of a third dimension, *shadow hachures* were developed (Figure 15.1) in which an oblique light source was presumed. In this approach, the width of hachures was a function of whether the hachure was illuminated or in the shadow.

Although hachures normally are associated with the bygone era of manual cartography, Imhof (1982, 229) argued that they should not be forgotten.

> Hachures alone, without contours, are more capable of depicting the terrain than is [relief] shading alone.... Hachures also possess their own special, attractive graphic style. They have a more abstract effect than [relief] shading, and perhaps for this reason are more expressive.

Additionally, Pinhas Yoeli (1985, 112) argued that hachures could be used for a structural analysis of the landscape (e.g., to depict different degrees of slope, as in Figure 15.2).

The maps shown In Figures 15.1 and 15.2 were developed by Yoeli using the following rules recommended by Imhof:

1. Each hachure follows the direction of steepest gradient.

2. Hachures are arranged in horizontal rows.

3. The density of hachures is constant except for the horizontal plane, which is blank.

4. The length of hachures corresponds to the local horizontal distance between assumed contours.

5. Wider hachures are used on steeper slopes.

Imhof suggested that these rules were appropriate for large-scale maps, but they should be relaxed for small-scale maps (scales smaller than 1:500,000). Based on this notion,

---

* An extensive set of Web links to sources of digital spatial data for the United States can be found at *http://www.cast.uark.edu/local/hunt/ index.html*.

† It should be noted that by placing a method in this section, we do not mean that these approaches could *only* be used with a vertical view, but that they have been *most commonly* used to create vertical views.

FIGURE 15.1 Hachures created using Yoeli's computer-based procedure. (From Yoeli 1985, 123; courtesy of The British Cartographic Society.)

Patrick Kennelly and Jon Kimerling (2000) devised a method based on regularly spaced point symbols derived from a DEM. The resulting hachure maps (Figure 15.3) obey Rules 1, 3, and 5, but ignore Rules 2 and 4, which assume a contour framework for hachure construction. Kennelly and Kimerling's obliquely illuminated map avoids changing thickness or spacing of hachures by using white and black hachures on a gray background. Kennelly and

Kimerling also added arrowheads to the downslope end of hachure lines to remove ambiguity associated with which end of the hachure is upslope or downslope (Figure 15.4). Their method calculates a slope and aspect grid from the DEM, converts these grids to point themes, and spatially joins the point themes into one theme containing both slope and aspect information. Arrows are then drawn in the direction of aspect (Rule 1), with thickness being a function of slope (Rule 5). The hachure density is constant as all points derived from the DEM are evenly spaced (Rule 3).*

15.2.2 Contour-Based Methods

Several methods for portraying topography have been developed that involve manipulating contour lines. **Tanaka's method** (Tanaka 1950) involves varying the width of contour lines as a function of their angular relationship with a presumed light source; lines perpendicular to the light source have the greatest width, whereas those parallel to the light source are narrowest. Furthermore, contour lines facing the light source are drawn in white, whereas those in shadow are drawn in black (Figure 15.5). Although this technique was used as early as 1870 (Imhof 1982, 153), Kitiro Tanaka is generally credited with originating it because of the attractive maps he produced (Figure 15.6) and the mathematical basis he provided.

Like hachures, Tanaka's method originally was produced manually. Berthold Horn (1982) was one of the first to create a computer-based version of Tanaka's method. A visual analysis of Horn's maps suggests that Tanaka's method compares favorably with other computer-assisted methods for representing relief (see Horn, 113–123). More recently, Patrick Kennelly and Jon Kimerling (2001) suggested improvements to Tanaka's method. One of the attractive maps that resulted from their work appears in Color Plate 15.1. Here we see that they drew contours in a fashion similar to Tanaka, but added colors between the contours "to give the topography a somewhat realistic appearance" (p. 117).

J. Ronald Eyton's (1984b) **illuminated contours**, a raster (pixel-based) approach for computer-based contouring, was a simplification of Tanaka's method. In a fashion similar to Tanaka, Eyton brightened contours facing the light source and darkened contours in the shadow. Eyton's approach, however, differed in that the width of contour lines was not varied (to speed computer processing). The topography resulting from Eyton's approach is not as distinctive as Tanaka's (compare Figures 15.6 and 15.7) but still gives the impression of three dimensions, especially when viewed from a distance.

---

* Kennelly and Kimerling used a northeastern illumination rather than the usual northwestern illumination because they felt that "it highlighted a few features slightly better" (Patrick Kennelly, personal communication).

FIGURE 15.2 Hachures used for a structural analysis of the landscape, in this case depicting differing degrees of slope. (From Yoeli 1985, 124; courtesy of The British Cartographic Society.)

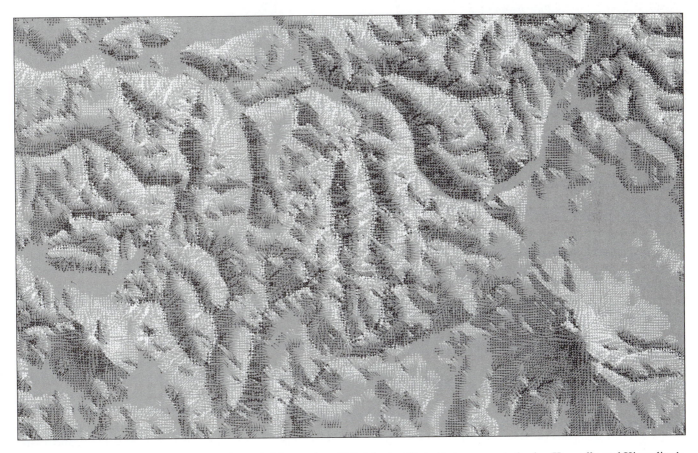

FIGURE 15.3 A small-scale map of the Cascade Mountains of Washington State that was created using Kennelly and Kimerling's hachure method. (Courtesy of Patrick Kennelly.)

FIGURE 15.4 An enlargement of a portion of Figure 15.3 showing the character of individual hachures. (Courtesy of Patrick Kennelly.)

Shadowed contours is another approach for representing topography by modifying contour lines. In this approach, contour lines facing the light source are drawn in a normal line weight, whereas contour lines in the shadow are thickened (Figure 15.8). This technique differs from Tanaka's principally in that the contours facing the light source are drawn in black rather than white (compare Figures 15.5 and 15.8). Figure 15.9 depicts a computer-assisted implementation of shadowed contours developed by Yoeli (1983).

15.2.3 Raisz's Physiographic Method

Based on the work of A. K. Lobeck (1924), which we consider later, Erwin Raisz (1931) developed the **physiographic method**, in which the Earth's geomorphologic features are represented by a set of standard, easily recognized symbols. A portion of Raisz's (1967) classic map *Landforms of the United States* is shown in Color Plate 15.2A and a sampling of his standard symbols is shown in Figure 15.10. Raisz argued that his physiographic map "appeals immediately to the average [person]. It suggests actual country and enables [the individual] to see the land instead of reading an abstract location diagram. It works on the imagination"

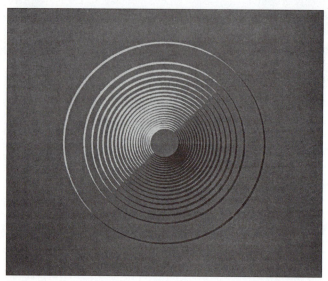

FIGURE 15.5 A simplified representation of Tanaka's method for depicting topography. The width of contour lines is a function of their angular relationship with a light source; contour lines facing the light source are drawn in white, whereas those in the shadow are drawn in black. (After Tanaka 1950. First published in *The Geographical Review* 40(3), p. 446. Reprinted with permission of the American Geographical Society.)

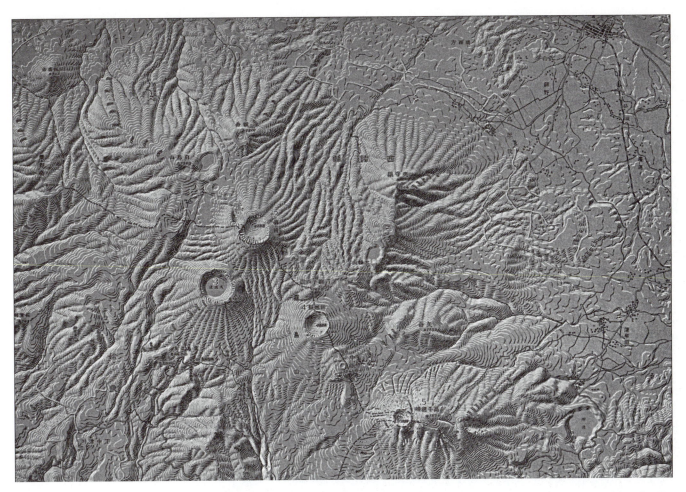

FIGURE 15.6 A map created using Tanaka's method for depicting topography. (After Tanaka 1950. First published in *The Geographical Review* 40(3), p. 451. Reprinted with permission of the American Geographical Society.)

(Raisz 1931, 303). To our knowledge, no one has ever automated Raisz's method, probably because it requires extensive knowledge of geomorphology and involves a subjective interpretation of the landscape.

15.2.4 Shaded Relief

Shaded relief (also known as **hill shading** or **chiaroscuro**) has long been considered one of the most effective methods for representing topography. Swiss cartographers (Imhof in particular) have been widely recognized for their superb work. Like the preceding methods, hill shading presumes an oblique light source: Areas facing away from the light source are shaded, and areas directly illuminated are not shaded. Slope can also be a factor, with steeper slopes shaded darker. When produced manually, shading was accomplished using pencil and paper or an airbrush. Color Plate 15.3 is an example of one of the maps that Imhof created, and Color Plate 15.2B is a portion of one that Richard Edes Harrison (1969) produced for the 1970 *U.S. National Atlas*.

Numerous people have contributed to automating shaded relief maps, with pioneering work done by Yoeli (1967; 1971), Brassel (1974), Hügli (1979), and Horn (1982). A discussion of sophisticated lighting algorithms that would apply to more than just shaded relief can be found in basic computer graphics texts such as Foley et al. (1996) and Rogers (1998). Today, shaded relief maps are a standard element of GIS software (e.g., ArcGIS) and software explicitly intended for mapping smooth continuous phenomena (e.g., Surfer).

The effectiveness of automated shaded relief maps is superbly illustrated by Gail Thelin and Richard Pike's (1991) relief map of the conterminous United States. (a small-scale version of the complete map is shown in Figure 1.9, and a portion of it is shown in Color Plate 15.2C).*

---

* To develop a full appreciation of the map shown in Figure 1.9, readers should examine the full-size printed version.

FIGURE 15.7   A map created using Eyton's illuminated contours in which contours facing the light source are drawn in white, whereas contours in the shadow are drawn in black; note that the width of contour lines is not varied. (Courtesy of J. Ronald Eyton.)

Thelin and Pike (3) noted that their map not only portrays well-known features such as the "structural control of topography in the folded Appalachian Mountains," but also less familiar features, such as the Coteau des Prairies, a flat-iron shaped plateau in South Dakota. Pierce Lewis (1992) described Thelin and Pike's map as follows:

> Let us not mince words. The United States Geological Survey has produced a cartographic masterpiece: a relief map of the coterminous United States which, in accuracy, elegance, and drama, is the most stunning thing since Erwin Raisz published his classic "Map of the Landforms of the United States." (p. 289)

As Raisz's map has been so widely used by geographers, it is natural to ask which is better, the USGS map or the Raisz map? Lewis (1992) argued that both maps have merit: The USGS map is "more like an aerial photograph" (298), allowing us to see the structure of the landscape as it actually exists (or at least within the constraints of the computer program). In contrast, the Raisz map represents one knowledgeable geographer's view of the landscape, with certain features emphasized, as Raisz deemed appropriate.

Another good illustration of automated shaded relief is Lloyd Treinish's (1994) rendition of the Earth's topography

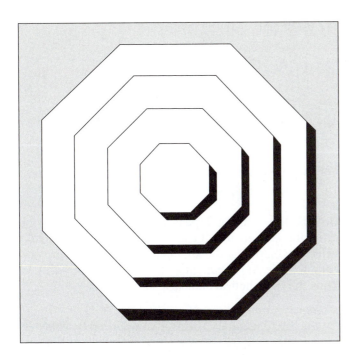

FIGURE 15.8 A simplified representation of shadowed contours in which contour lines facing a light source are drawn in black, whereas contour lines facing away from the light source produce a black shadow. (After Yoeli 1983. First published in *The American Cartographer* 10(2), p. 106. Reprinted with permission from the American Congress on Surveying and Mapping.)

FIGURE 15.9 A map created using shadowed contours. (After Yoeli 1983. First published in *The American Cartographer* 10(2), p. 109. Reprinted with permission from the American Congress on Surveying and Mapping.)

(Color Plate 15.4) that was developed using the IBM Visualization Data Explorer (DX; *http://www.research. ibm.com/dx/*). The attractiveness of this image is illustrated by the fact that a variant of the image was chosen for the cover of the December 1992 issue of *GIS World*, a popular magazine for those interested in GIS. The image was created by combining shaded-relief techniques with hypsometric tints representing elevation values above and below sea level. The shaded-relief technique was based on Gouraud shading, which interpolates shades between vertices on a polygon (or cell) boundary, and thus produces a smoother appearance than if the same shade is used throughout a polygon (Foley et al. 1996, 734–737). The color scheme for elevation values (see the top of Color Plate 15.4) was developed such that the portion above sea level had "the appearance of a topographic map" (Treinish 1994, 666), whereas the portion below sea level used darker shades of blue to represent greater depth.

Although fully automated shaded relief methods can produce highly effective renditions, they are sometimes deficient when compared to their manual counterparts. On this basis, Lorenz Hurni and his colleagues (2001) noted that interpretation of shaded relief maps can be enhanced through *user-controlled* local adjustments of the assumed position of the light source, vertical exaggeration,

brightness, elevation-dependent contrast, and the transition between diffuse reflection and aspect-based shading. Similarly, Wendy Price (2001) described the limitations of software tools intended to simulate the traditional airbrush, and suggested that the manual airbrush still has its uses.

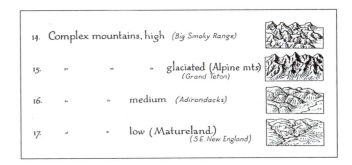

FIGURE 15.10 Examples of standard symbols that Raisz used to create his physiographic diagrams. (After Raisz 1931. First published in *The Geographical Review* 21(2), p. 301. Reprinted with permission of the American Geographical Society.)

Computation of Shaded Relief

A DEM is the basic input necessary for creating a computer-assisted shaded relief map. Fundamental calculations involve manipulating 3×3 "windows" of elevation values within the grid of the DEM. To illustrate this concept, we will use the window and hypothetical data shown in Figure 15.11. We will presume that our objective is to determine an appropriate gray tone for the central cell of the 3×3 window. The following steps are based on Eyton's (1991) formulas.*

Step 1. Compute the slope of the window. The slopes along the x and y axes, sl_x and sl_y, are

$$sl_x = \frac{Z(2,3) - Z(2,1)}{2D}$$

$$sl_y = \frac{Z(3,2) - Z(1,2)}{2D}$$

where the Z values represent digital elevation values as shown in Figure 15.11A, and D is the distance between grid points.

---

* For a discussion of alternative approaches for computing slope and aspect, see Chrisman (2002, 174–179). Studies of various approaches can be found in Hodgson (1998), Jones (1998), and Dunn and Hickey (1998).

For the sample data the results are

$$sl_x = \frac{115 - 105}{60} = 0.167$$

$$sl_y = \frac{114 - 104}{60} = 0.167$$

The overall slope, sl_o, is then defined as

$$\sqrt{(sl_x)^2 + (sl_y)^2}$$

For the hypothetical data, we have

$$\sqrt{(0.167)^2 + (0.167)^2} = 0.236$$

Because the resulting slope is expressed as a tangent (the ratio of rise over run), it can be converted to an angular value by computing the arc tangent ($\tan^{-1}$). The arc tangent for the hypothetical data is $\tan^{-1} 0.236 = 13.3°$. Thus, the hypothetical data have an overall slope of $13.3°$.

Step 2. Compute the down-slope direction or azimuth for the 3×3 window. This is accomplished by calculating a local angle (ϕ) between the overall slope and the slope in the x direction, and then converting this to an azimuth (an angle measured clockwise from a north base of $0°$). ϕ is defined as

$$\phi = \cos^{-1}\left[\frac{sl_x}{sl_o}\right]$$

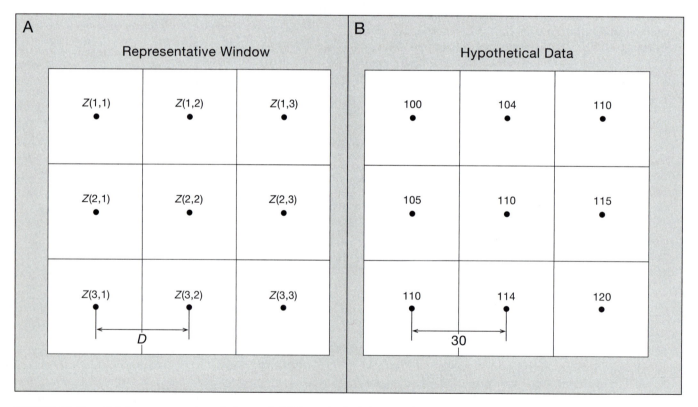

FIGURE 15.11   (A) A representative window and (B) hypothetical data used to compute digital shaded relief.

For the hypothetical data, we have

$$\phi = \cos^{-1}\left[\frac{0.167}{0.236}\right] = 45$$

To convert ϕ to an azimuth (θ), use the signs shown in Table 15.1. For the hypothetical data, we compute $\theta = 270° + 45° = 315°$. Note that this result confirms a visual examination of the 3×3 window in Figure 15.11: The window appears to face northwest (an azimuth of 315°).

Step 3. Estimate the reflectance for the central cell of the 3×3 window, the cell associated with $Z(2, 2)$. A simple but reasonably effective approach is to presume a *Lambertian reflector*, which reflects all incident light equally in all directions. The formula for Lambertian reflectance is

$$L_r = \cos(A_f - A_s) \cos E_f \cos E_s$$
$$+ \sin E_f \sin E_s$$

Where L_r = relative radiance (scaled 0.0–1.0)
A_f = azimuth of a slope facet (0° to 360°)
A_s = azimuth of the sun (0° to 360°)
E_f = elevation of a normal to the slope facet (90° minus the slope magnitude in degrees)
E_s = elevation of the sun (0° to 90°)

A visual representation of these terms is shown in Figure 15.12. If we presume solar azimuth and solar elevations of 315° and 45°, respectively, then the Lambertian reflectance for the hypothetical data is

$$L_r = \cos(315 - 315) \cos(90 - 13.3) \cos(45)$$
$$+ \sin(90 - 13.3) \sin(45)$$
$$= 0.850$$

Step 4. Convert the Lambertian reflectance value to either RGB or CMYK values appropriate for display or printing. Presuming a CRT display with intensities ranging from 0 to 255 for each color gun, RGB values would be calculated by multiplying the Lambertian reflectance by 255 (the same value would be used for each color gun

to produce a gray tone). For our hypothetical data the result would be 255 × 0.850 = 217 for each color gun. Thus, the central cell of the 3 × 3 window would have a relatively bright reflectance value.

15.2.5 Morphometric Techniques

The techniques that we have described thus far are intended to provide an overall view of topography. Often, however, it is useful to focus on particular morphometric (or structural) elements of topography. Aspect and slope are two of the most common structural elements that have been visualized. **Aspect** deals with the direction the land faces (e.g., north or south), whereas **slope** considers the steepness of the land surface. For example, a developer of a ski resort might desire north-facing aspects (to minimize snowmelt) and a diversity of slopes (to handle skiers' varied abilities). In this section we consider two approaches that have been developed for

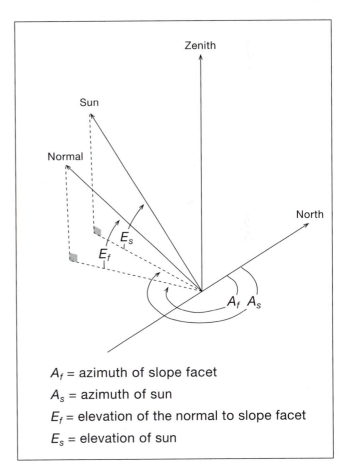

A_f = azimuth of slope facet

A_s = azimuth of sun

E_f = elevation of the normal to slope facet

E_s = elevation of sun

FIGURE 15.12 A visual representation of terms associated with computing Lambertian reflection. (From Eyton, 1990, *Cartographica*, p. 27. Reprinted by permission of University of Toronto Press Incorporated. © University of Toronto Press 1990.)

TABLE 15.1 Conversion of local angle (ϕ) to azimuth (θ) for calculations associated with computer-based shaded relief

Sign of Slope in $x(sl_x)$	Sign of Slope in $y(sl_y)$	Azimuth
Positive	Positive	$270° + \phi$
Positive	Negative	$270° - \phi$
Negative	Positive	$90° - \phi$
Negative	Negative	$90° + \phi$

handling aspect and slope—the first deals solely with aspect, whereas the second deals with both aspect and slope. We then consider a software package, LandSurf, that has been developed to visualize and analyze a broad range of morphometric elements.

Symbolizing Aspect: Moellering and Kimerling's MKS-ASPECT™ Approach

Harold Moellering and Jon Kimerling (1990) developed an approach for symbolizing aspect based on the opponent-process theory of color (see section 10.1.3). Their approach is known as MKS-ASPECT™ and is protected by two patents (No. 5,067,098 and No. 5,283,858) held by the Ohio State University Research Foundation in the name of Harold Moellering and A. Jon Kimerling. MKS-ASPECT treats aspect as a nominal phenomenon (a northwest aspect is *different* than a southeast aspect) and assigns colors to opposing aspects based on opposing color channels; thus, red and green, and blue and yellow are placed opposite one another. Mixtures of nonopposing hues can then be used for intermediate aspects (Figure 15.13).

Moellering and Kimerling argue that the **luminance** (or brightness) of a color should ideally be highest for a northwest aspect (assuming the usual light source from the northwest) and decrease to a minimum for a southeast aspect; because yellow is inherently a bright color, it is logical to use it for the northwest aspect. More specifically, they argued that colors should ideally fall on a curve represented by the following equation:

$$L = 50 \times [\cos(315 - \phi) + 1]$$

where ϕ is the azimuth of the aspect and L is the relative luminance as measured by a **colorimeter**, a device for measuring the brightness of a color (Figure 15.14).

Moellering and Kimerling selected their colors using the HLS (hue–lightness–saturation) color specification model. The HLS model is similar to the HSV model described in section 10.3.3, except that it is a double rather than a single hexcone (compare Figure 15.15 with Figure 10.17). Initially, Moellering and Kimerling created colors by holding the lightness and saturation of the HLS model constant at 50 (one-half the maximum) and 100 (the maximum), respectively, and simply varying hue. Unfortunately, because HLS is not a perceptually based model, they found this approach produced colors relatively far from the ideal curve shown in Figure 15.14. Only after considerable experimentation did they arrive at the set of eight colors plotted in Figure 15.14 and shown in Table 15.2. A map created using Moellering and Kimerling's color scheme is shown in Color Plate 15.5; in this case 128 colors are used (rather than 8) to provide greater color fidelity.

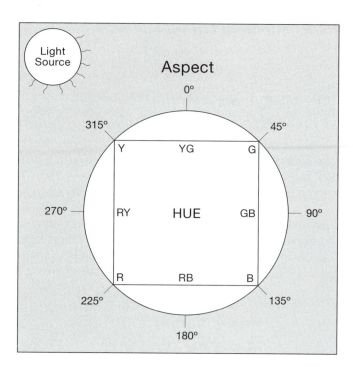

FIGURE 15.13 Using opponent process colors to represent aspect. (After Moellering and Kimerling 1990, 154.)

Symbolizing Aspect and Slope: Brewer and Marlow's Approach

Cynthia Brewer and Ken Marlow (1993) attempted to rectify Moellering and Kimerling's difficulty in working with HLS by employing the perceptually based HVC model (Figure 10.20), at the same time incorporating slope in addition to aspect. Like Moellering and Kimerling, Brewer and Marlow chose eight aspect hues. For each aspect, however, they also depicted three slope classes (5 to 20 percent, 20 to 40 percent, and >40 percent) as

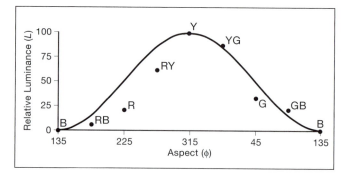

FIGURE 15.14 Relationship between aspect (as measured in degrees clockwise from a north value of 0) and relative luminance in Moellering and Kimerling's method for symbolizing aspect. The curve shows the ideal relationship, while dots represent the actual relationship. (After Moellering and Kimerling 1990, 158.)

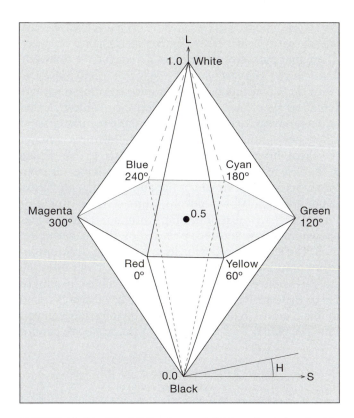

FIGURE 15.15 The HLS color model. Note that it is a double hexcone, as opposed to HSV, which is a single hexcone. (Compare with Figure 10.17.)

given the perceptual basis of HVC) because a wide range of lightness and saturation values are not possible for all hues; for example, it is not possible to create a variety of lightnesses and saturations between green and blue. The saturation (or chroma) of a slope class within an aspect is represented by the distance of a black dot from the center of the diagram; this distance can be determined using the scale labeled "Chroma" in the diagram. For example, for the blue hue, the third dot from the center of the diagram appears to have a saturation

differing levels of saturation; they represented the <5 percent slope category as gray for all aspects. In total, they showed 25 classes of information (8 aspects × 3 slope classes, plus the <5 percent slope category). The colors selected by Brewer and Marlow for aspect and slope are shown graphically in HVC coordinates in Figure 15.16. Each black dot in the graph represents one of the 25 colors used. The eight lines extending from the interior to the edge of the circle represent the hues selected for the eight aspects. The hues do not appear to be equally spaced (as one might suspect

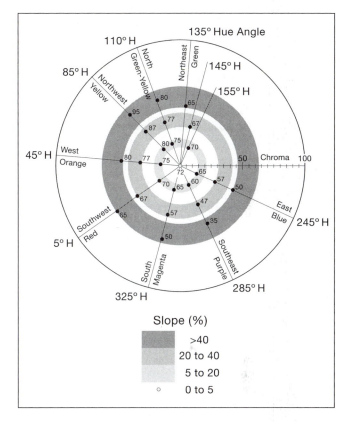

FIGURE 15.16 A plot of HSV colors used by Brewer and Marlow to depict aspect and slope; see text for explanation. (After Brewer and Marlow 1993, 331.)

TABLE 15.2 HLS, *Yxy*, and RGB specifications for Moellering and Kimerling's method for representing aspect*

Aspect	H	L	S	Y	x	y	R	G	B
0	70	50	100	37.5	.354	.544	213	255	0
45	120	40	100	15.1	.288	.584	0	204	0
90	190	40	100	10.4	.200	.257	0	170	204
135	240	40	100	2.3	.160	.087	0	0	204
180	290	40	100	4.7	.250	.139	170	0	204
225	0	50	100	10.6	.594	.357	255	0	0
270	50	50	100	27.1	.446	.476	255	213	0
315	60	50	100	42.4	.396	.514	255	255	0

*Aspect is measured in degrees clockwise from north, which is 0°. HLS values are based on the color model shown in Figure 15.15. *Yxy* values are taken from Moellering and Kimerling (1990, 158).

of approximately 45. The numbers printed alongside each dot represent the value (lightness) of each color. Note that higher lightness values tend to be found facing the direction of the light source (the northwest), and that lightness also varies with slope. Overall, Brewer and Marlow selected colors based on their knowledge of the HVC color space and the desire to have a suitable range of lightnesses and saturations available for each hue. HVC, *Yxy*, and RGB specifications for the 25 slope-aspect classes developed by Brewer and Marlow are shown in Table 15.3, and a map created using their color scheme is shown in Color Plate 15.6.

Wood's LandSerf

Although aspect and slope are the most common parameters utilized in examining the morphometry of topography, those interested in a detailed analysis (e.g., geomorphologists) utilize a richer set of parameters. For instance, two additional parameters are the *profile*

curvature, which describes the rate of change of slope, and *plan curvature*, which describes the rate of change of aspect. These four parameters can be used to define the following features: *pits*, *channels*, *passes*, *ridges*, *peaks*, and *planar regions*. Jo Wood has developed an intriguing package, LandSerf (*http://www.soi.city.ac.uk/~jwo/landserf/*), that allows you to visualize and analyze these parameters and associated features. Color Plate 15.7 illustrates some of the capability of LandSerf. In Color Plate 15.7A, we see a shaded relief map of the Columbia River area in Washington State in which the colors depict elevation using a green–brown–purple–white scheme. In Color Plate 15.7B, we see a slope map (with the steepest slopes shown in yellow and red) overlaid on the shaded relief map. The slope map obviously shows quite different information than the shaded relief map as it identifies the steepest slopes as being near major drainage basins. In Color Plate 15.7C, we see some of the key features found in this region: Channels are shown in blue, ridges in yellow, and planar regions in gray. At this scale, we cannot

TABLE 15.3 HVC, *Yxy*, and RGB specifications for Brewer and Marlow's method for representing aspect and slope * (After Brewer and Marlow 1993. First published in *Auto-Carto II* proceedings. Reprinted with permission from the American Congress on Surveying and Mapping.)

Aspect	H	V	C	Y	x	y	R	G	B
Maximum Slopes (>40 percent slope)									
0	110	80	56	55.3	.339	.546	132	214	0
45	135	65	48	35.2	.252	.486	0	171	68
90	245	50	46	19.8	.180	.191	0	104	192
135	285	35	51	10.3	.240	.135	108	0	163
180	325	50	60	19.6	.350	.198	202	0	156
225	5	65	63	33.7	.442	.319	255	85	104
270	45	80	48	53.8	.421	.421	255	171	71
315	85	95	58	81.1	.395	.512	244	250	0
Moderate Slopes (20 to 40 percent slope)									
0	110	77	37	50.6	.319	.451	141	196	88
45	145	67	33	37.0	.253	.400	61	171	113
90	245	57	31	25.7	.221	.238	80	120	182
135	285	47	34	17.4	.262	.209	119	71	157
180	325	57	40	25.1	.332	.240	192	77	156
225	5	67	42	34.3	.390	.316	231	111	122
270	45	77	32	47.3	.377	.384	226	166	108
315	85	87	39	63.7	.359	.445	214	219	94
Low Slopes (5 to 20 percent slope)									
0	110	75	19	45.4	.305	.381	152	181	129
45	155	70	16	39.4	.270	.348	114	168	144
90	245	65	15	32.7	.260	.284	124	142	173
135	285	60	17	27.5	.280	.272	140	117	160
180	325	65	20	32.6	.311	.282	180	123	161
225	5	70	21	37.2	.339	.317	203	139	143
270	45	75	16	41.8	.328	.348	197	165	138
315	85	80	19	49.7	.321	.375	189	191	137
Near-Flat Slopes (0 to 5 percent slope)									
None	0	72	0	39.0	.290	.317	161	161	161

*Aspect is measured in degrees clockwise from north, which is 0°.

see them, but a blow-up of the region would also show peaks (in red) and passes (in green). Finally, in Color Plate 15.7D, we see the effect of broadening the scale of analysis. In contrast to the images shown in Color Plate 15.7A–C, which were created by processing 3×3 windows of 30-meter DEM cells, Color Plate 15.7D was created using 65×65 windows of 30-meter cells. Obviously, the larger window size has generalized the landscape. (The border around the map occurs because of the large window size and lack of information surrounding the map.) We could continue the analysis by examining the character of the features (e.g., channels and ridges) associated with this larger scale landscape. Because space does not permit us to illustrate the full potential of Land-Serf here, we encourage you to explore the capability of this software on your own.

15.3 OBLIQUE VIEWS

In contrast to the preceding section, which focused on vertical views, several approaches for depicting topography have focused on oblique views. Here we consider three: block diagrams, panoramas, and "draped images."

15.3.1 Block Diagrams

In a fashion similar to the physiographic method described in the preceding section, **block diagrams** (Figure 15.17) are useful for examining the Earth's geomorphologic and geologic features, but with an oblique rather than a vertical view. Although block diagrams have their origins in the 18th century (Robinson et al. 1995), A. K. Lobeck (1958) is commonly associated with originating them because of his book *Block Diagrams*. Lobeck indicated that block diagrams have a twofold function: "First, to present pictorially the surface features of the ground; and second, to represent the underground structure" (p. 1). Obviously, a purely vertical view could not achieve this function. Lobeck recognized that cartographers might be concerned by the lack of accuracy in the block diagram (as certain features are purposely hidden, whereas others are exaggerated), but he felt that any inaccuracies were outweighed by the ease of understanding: "No conventions are needed to represent the topography. No explanation is needed to indicate the position of the geological cross-section. It carries its message directly to the eye and leaves a visual impression unencumbered by lengthy descriptions" (p. 2).

Like the physiographic method, a full automation of methods for creating block diagrams would be difficult because of the extensive knowledge base required and the subjective interpretation involved. Many parts of the process, however, have been automated. For instance, if sample points indicating the elevation of sedimentary rock layers are known, it is possible to interpolate between these points to estimate the bounds of individual layers. Color Plate 4.1 is an example of an image that could result from this process. A major advantage of automating block diagrams is that users are not restricted

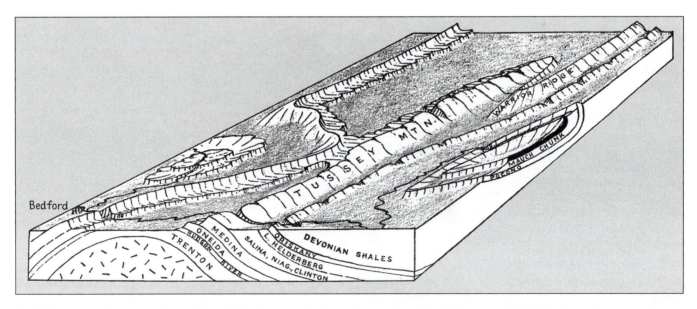

FIGURE 15.17 Example of a traditional block diagram. (From Figure 98, p. 69 of Lobeck, A. K. (1958) *Block Diagrams*. Second Edition. Amherst, MA: Emerson-Trussel Book Company.)

to a single view—programs such as EarthVision (developed by Dynamic Graphics; *http://www.dgi.com*) allow users to rotate the diagram to different positions and to explore the data by, for example, slicing the diagram at different locations. We consider the possibility of exploring 3-D surfaces in greater depth when we discuss Vis5D in Chapter 21.

15.3.2 Panoramas and Related Oblique Views

Panoramas provide a view of the landscape that we might expect from a painting (which is how traditional panoramas were created), but they also pay careful attention to geography so that we can clearly recognize known features. To illustrate, Color Plate 15.8 shows a classic panorama of Denali National Park in Alaska, created by one of the best known panoramists, Heinrich Berann, whose work Tom Patterson (2000) so ably described. Traditionally, developing panoramas required considerable artistic skill, but Patterson (1999) showed how an array of digital tools can allow those with limited artistic ability to create panoramas and other oblique views of the 3-D landscape. Figure 15.18 is an example of one of the intriguing oblique images that Patterson created. Patterson argued that

> Humans naturally tend to visualize physical landscapes in profile based on our grounded lives on the Earth's surface, rather than as flat maps ... [Thus] an obliquely viewed 3D landscape is ... probably more understandable to users, especially those with limited map reading skills. (p. 217)

Along these same lines, Patterson (2002) suggested that the *realism* inherent in oblique 3-D landscapes should

also be employed in maps that have a *vertical view*. If this is done, he argued that "Users can comprehend ... map information relatively effortlessly without explanation (such as text and legends)" and that more attractive maps will result, which is important in our "media-driven age." Patterson suggested four "rules" for creating realistic topographic maps having a vertical view:

1. *Remove lines* as they are rare in nature and can be visually distracting.
2. *Rasterize* by converting vector lines and fills to a pixel form. Again, this will make the map appear more natural.
3. *Modulate tones* rather than use flat area tones. How often do we see a lake that is perfectly uniform?
4. *Texturize* by adding "graphical noise" to areas, and even lines. He argued that this provides "a tactile appearance that mimics nature."

Color Plate 15.9 shows a map of Crater Lake National Park that Patterson created using such rules. Because space does not permit us to provide details of the approaches Patterson uses to implement these rules, we encourage you to examine his Web site (*http://www.nacis.org/cp/cp28/resources.html*), where he discusses these approaches.

15.3.3 Draped Images

In recent years, the most popular form of oblique view has been to drape remotely sensed images (or other information, such as land use and land cover) on a 3-D map of elevation (essentially this is a fishnet map, but we can't see the fishnet once the image is draped on the map of

FIGURE 15.18 An image of Yellowstone Lake, the largest mountain lake in North America. This shows the nature of attractive oblique 3-D images that can be created using modern digital tools. (Courtesy of National Park Service.)

elevation); the resulting maps are termed **draped images**. The popularity of this approach is indicated by the fact that several hundred packages have been developed for 3-D visualization of the terrain (see *http://www.tec.army. mil/TD/tvd/survey/survey_toc.html*).* As with modern forms of block diagrams, the problem of blockage is generally handled by either producing an animated fly-by (Figure 15.19) of the landscape or allowing users to rotate the draped image. Depending on the nature of the draped information, the resulting maps can appear quite realistic; for instance, with one-meter satellite imagery it is possible to identify individual cars. If you are using a software package to drape an image onto a 3-D elevation map, you should be aware that the character of the visualization can be enhanced by using Patterson's techniques described earlier, as many of his maps begin with elevation data and a draped image. Given the realism of draped images, a full discussion of them gets us into the area of virtual environments (VEs), which we cover in Chapter 24.

15.4 PHYSICAL MODELS

In the preceding two sections, we covered approaches for depicting topography that gave the illusion of a third dimension even though one did not actually exist, as the maps were either created on a piece of paper or on a graphics display screen, which are both two-dimensional. A natural alternative is to create a truly 3-D **physical model**. Although physical models have been around since the time of Alexander the Great, they have been used infrequently by cartographers because of the difficulty

---

* We have no way of knowing how many of these packages permit draping of images, but we presume that many of them do.

FIGURE 15.19 An example of a frame from a fly-by. The original illustration was in color, and was created using TruFlite software (*http://www.truflite.com/*). (Courtesy of Keith Clarke.)

of obtaining terrain information, the cost and time of creating them manually, and the difficulty of disposing of them. Today, such limitations no longer exist, as digital data are readily available, several technologies have been developed for converting the digital data into a solid model, and disposal is less of an issue given the range of available technologies (Caldwell 2001).

To illustrate the capability of physical models, we consider two examples. The first is the extensive set of colorful models that have been created for National Geographic's Explorer's Hall in Washington, DC, by Solid Terrain Modeling (STM; *http://www.stm-usa.com/*). STM has created a total of 17 models for the hall, with the largest (the Grand Canyon) measuring 52 × 6 feet. Based on DEM data, a machine cuts the basic structure of the model out of dense polyurethane foam; colorful satellite images (or any other image) are then printed on the foam model. Color Plate 15.10 shows two examples of models that have been created for Explorer's Hall. The Mt. Everest example (on the left) illustrates the dramatic 3-D character of the models, and the Hurricane Floyd example (on the right) illustrates the colorful images that can be depicted. Although the models in Explorer's Hall are too large for an individual to manipulate in his or her hands, the STM technology can be used to produce models of various sizes.

As a second example of physical models, consider those created by researchers at San Diego State University to understand faulting in California's Death Valley (Sides 1996). These models are being created using the San Diego Supercomputer Center's Tele-Manufacturing Facility (TMF), which is accessible via the Internet. With the TMF, each model is created by laminating individual pieces of paper, which the "looks, feels, and acts like it was made of wood" (Bailey 2002, 3). Mitra Fattahipour, one of the researchers, indicated that the resulting 17.8 × 17.8 cm (7 × 7 in.) model "provides new insight into fault geometries because you can see and feel the topographic details." Interestingly, the researchers created the model using an exaggeration factor much higher than cartographers would normally recommend (they used a factor of 14, whereas cartographers generally use no more than a factor of 5). The researchers, however, felt that this helped them visualize the topography and associated faulting. Although the researchers have also used animated fly-throughs and interactive 3-D visualizations to analyze faulting in the valley, they appear particularly enamored with the capability of the physical model.

There appears to be considerable excitement about the potential of physical models. Although one cannot interact with these models as one would on a graphics display (e.g., pointing to a location to determine the elevation), the ability to hold the model and manipulate it in one's hands appears to be a very desirable feature.

An interesting research project would be to compare the effectiveness of physical models with 3-D visualizations created on graphics displays. Such studies might help us build more effective 3-D visualizations for graphic displays.

SUMMARY

This chapter has examined methods for symbolizing the Earth's topography and associated landscape features. Basic data for depicting topography are generally provided in an equally spaced gridded format (as a *digital elevation model*, or *DEM*). DEMs and related data representing the Earth's landscape can be acquired from mapping agencies such as the USGS.

We divided methods for symbolizing topography into vertical views, oblique views, and physical models. Methods that have emphasized a vertical view include **hachures** (drawing a series of parallel lines perpendicular to the direction of contours), contour-based methods (e.g., **Tanaka's method**), and **shaded relief** (or **hill shading**). These methods assume that the topographic surface is illuminated, typically obliquely from the northwest. The various methods use different approaches to distinguish areas that appear illuminated from those that appear in the shadows. For example, in Tanaka's method, contour lines perpendicular to the light source have the greatest width, whereas those parallel to the light source are narrowest; furthermore, contour lines facing the light source are drawn in white, whereas those in shadow are drawn in black. Today, vertical views of topography are most commonly created using computer-based shaded relief, but recent software developed by Kennelly and Kimerling might revive interest in other methods for depicting a vertical view. We also considered morphometric methods as a special form of vertical view that emphasizes the structural elements of topography. For instance, Moellering and Kimerling developed an approach for symbolizing aspect based on the opponent-process theory of color, and Wood has developed the package LandSerf to visualize a variety of topographic features, including pits, channels, passes, ridges, peaks, and planar regions.

Approaches for creating oblique views include **block diagrams**, **panoramas**, and **draped images**. Software for creating block diagrams (e.g., EarthVision) has been available for quite some time, whereas utilizing computers to create panoramas is a more recent phenomenon. Particularly interesting is the work of Patterson, who has created attractive oblique views in the spirit of the famed panoramist Berann. Patterson also has illustrated how the realism normally depicted in oblique views can be utilized in vertical views. Draping remotely sensed images (or other information, such as land use and land cover) on a 3-D map of elevation is a common approach for creating oblique views, and hundreds of software packages have been created for the 3-D visualization of terrain.

Physical models are an attractive possibility for depicting topography because the models are truly 3-D. Thus, users can work with these models as they would other 3-D objects in the real world; for instance, you can hold a small physical model in your hands and easily view it from different angles. Physical models have been around since the time of Alexander the Great, but only in the last few years has technology become available to convert digital data into solid models. A good indication of the effectiveness of these models is the large number that National Geographic has portrayed in their Explorer's Hall in Washington, DC.

FURTHER READING

Ambroziak, B. M., Ambroziak, J. R., and Bradbury, R. (1999) *Infinite Perspectives: Two Thousand Years of Three-Dimensional Mapmaking.* New York: Princeton Architectural Press.

Provides maps illustrating much of the history of symbolizing topography and proposes a novel approach based on anaglyphs.

Barnes, D. (1999) "Creation of publication quality shaded-relief maps with ArcView GIS." *Cartographic Perspectives*, no. 32:65–69.

Explains methods for creating a shaded relief map in ArcView.

Caldwell, D. R. (2001) "Physical terrain modeling for geographic visualization." *Cartographic Perspectives*, no. 38:66–72.

Reviews modern approaches for creating physical models of the Earth's landscape and discusses several related cartographic issues.

Cartographic Perspectives no. 42, 2002.

Several papers in this special issue on "Practical Cartography" deal with symbolizing topography.

Ding, Y., and Densham, P. J. (1994) "A loosely synchronous, parallel algorithm for hill shading digital elevation models." *Cartography and Geographic Information Systems* 21, no. 1:5–14.

Describes a sophisticated method for creating shaded relief that incorporates both shadows cast by other terrain features and atmospheric scattering; the method was implemented on a parallel computer.

Eyton, J. R. (1991) "Rate-of-change maps." *Cartography and Geographic Information Systems* 18, no. 2:87–103.

Includes basic calculations for creating a shaded relief map, but also several more advanced concepts such as computing and mapping the convexity and concavity of a 3-D surface.

Foley, J. D., van Dam, A., Feiner, S. K., and Hughes, J. F. (1996) *Computer Graphics: Principles and Practice*. 2nd ed. Reading, MA: Addison-Wesley.

The notion of trying to simulate the appearance of topography by using a presumed light source is analogous to attempts to simulate the appearance of other phenomena, such as how a room looks when lit from a certain direction. Chapter 16 deals with simulating the appearance of these other phenomena.

Heller, M., and Neumann, A. (2001) "Inner-mountain cartography—from surveying towards information systems." *Proceedings, 20th International Cartographic Conference*, International Cartographic Association, Beijing, China, CD-ROM.

Describes software for visualizing caves; also see Gong et al. (2000).

Hobbs, F. (1995) "The rendering of relief images from digital contour data." *The Cartographic Journal* 32, no. 2:111–116.

Presents an algorithm for shaded relief that is appropriate for elevation data stored as contour lines. Also illustrates how multiple light sources can be used to enhance the visualization of terrain.

Hodgson, M. E. (1998) "Comparison of angles from surface slope/aspect algorithms." *Cartography and Geographic Information Systems* 25, no. 3:173–185.

Compares several algorithms for computing the slope and aspect of DEM cells; also see Jones (1998) and Dunn and Hickey (1998).

Horn, B. K. P. (1982) "Hill shading and the reflectance map." *Geo-Processing* 2, no. 1:65–144.

A dated, but thorough, review of automated approaches for shaded relief.

Hurni, L., Jenny, B., Dahinden, T., and Hutzler, E. (2001) "Interactive analytical shading and cliff drawing: Advances in digital relief presentation for topographic mountain maps." *Proceedings, 20th International Cartographic Conference*, International Cartographic Association, Beijing, China, CD-ROM.

Describes software that allows users to interact with a shaded relief map, and thus produce a result that simulates manual relief shading; also describes an approach for digital cliff drawing.

Imhof, E. (1982) *Cartographic Relief Presentation*. Berlin: Walter de Gruyter.

A classic text on methods for symbolizing terrain.

Patterson, T. (1999) "Designing 3D Landscapes." In *Multimedia Cartography*. ed. by W. Cartwright, M. P. Peterson, and G. Gartner, pp. 217–229. Berlin: Springer-Verlag.

Discusses digital approaches for creating obliquely viewed 3-D maps.

Patterson, T. (2002) "Getting real: Reflecting on the new look of National Park Service maps." Accessed June 21, 2003. *http://www.nps.gov/carto/silvretta/realism/index.html*.

Discusses digital approaches for creating *realistic* topographic maps when a vertical view is used; much of Patterson's work can be found at *http://www.nacis.org/cp/cp28/resources.html*.

Raper, J. (2000) *Multidimensional Geographic Information Science*. London: Taylor & Francis.

An advanced treatment of methods for mapping true 3-D data.

Thelin, G. P., and Pike, R. J. (1991) "Landforms of the conterminous United States—A digital shaded-relief portrayal." Map I-2206. Washington, DC: U.S. Geological Survey. Scale 1:3,500,000.

Describes the methods used to create Thelin and Pike's shaded relief map, along with an interpretation of physiographic features seen in the map; also see Pike and Thelin (1990–91; 1992).

Treinish, L. A. (1994) "Visualizations illuminate disparate data sets in the earth sciences." *Computers in Physics* 8, no. 6:664–671.

Discusses the creation of the dramatic rendition of the earth's topography shown in Color Plate 15.4, along with several other visualizations of spatial data.

16

Proportional Symbol Mapping

OVERVIEW

This chapter examines the **proportional symbol map** (or **graduated symbol map**), which is used to represent numerical data associated with point locations. Section 16.1 discusses two basic kinds of point data that can be displayed with proportional symbols: true and conceptual. **True point data** are actually measured at point locations; an example would be the production of oil wells. **Conceptual point data** are collected over areas (or volumes), but the data are conceived as being located at points for the purpose of symbolization; an example would be the number of oil wells in each state. Section 16.1 also considers how data standardization can be applied to proportional symbol maps (e.g., the murder rate for cities generally is a more useful measure than the raw number of murders).

Section 16.2 introduces two general kinds of proportional symbols: **geometric** and **pictographic** (respective examples would be circles and caricatures of people). Although geometric symbols have been more frequently used, pictographic symbols are becoming common because of the ease of creating them with map design software. Circles have been the most frequently used geometric symbol because they are visually stable, users prefer them, and they conserve map space. Three-dimensional symbols (e.g., spheres) traditionally have been frowned on because of the difficulty of visually estimating their size, but they can produce attractive, eye-catching graphics and can be useful for representing a large data range.

Section 16.3 deals with three methods of scaling (or sizing) proportional symbols: mathematical, perceptual, and range grading. In **mathematical scaling**, symbols are sized in direct proportion to the data; thus, a data value 20 times another is represented by an area (or volume) 20 times as large. In **perceptual scaling**, a correction is introduced to account for visual underestimation of larger symbols; thus, larger symbols are made bigger than

would normally be specified by mathematical scaling. Both mathematical and perceptual scaling yield an unclassed map. In **range grading**, a classed map results by grouping data into classes, and letting a single symbol size represent all data falling in a class.

Section 16.4 considers the problem of designing legends for proportional symbol maps. Basic decisions include deciding how symbols should be arranged in the legend and how many and what size of symbols should be used. In the **nested-legend arrangement**, smaller symbols are drawn within larger ones, whereas in the **linear-legend arrangement**, symbols are placed adjacent to each other in either a horizontal or vertical orientation.

Section 16.5 addresses the issue of symbol overlap. Small symbols cause little or no overlap, potentially resulting in a map devoid of spatial pattern; in contrast, large symbols create a cluttered map, making it difficult to interpret individual symbols. In this context, this section considers two issues: deciding how much overlap there should be, and how the overlap should be symbolized. Two basic solutions for symbolizing overlap are transparent and opaque symbols. **Transparent symbols** enable readers to see through overlapping symbols, easing the problem of estimating symbol size, whereas **opaque symbols** do not enable readers to see through overlapping symbols, but do enhance figure-ground contrast. Other solutions for handling overlap include **inset maps** portraying an enlarged scale of a congested area, the use of a **zoom function** in an interactive graphics environment, and the possibility of moving symbols away from the center of a congested area.

Finally, section 16.6 describes how redundant symbols can be used on proportional symbol maps. **Redundant symbols** represent a single attribute (e.g., magnitude of oil production) by two visual variables (e.g., size and value), *thus enabling users to perform specific tasks more easily and accurately. Because redundant symbols are arguably less effective for portraying general information, they are best used in a data exploration framework.*

16.1 SELECTING APPROPRIATE DATA

Proportional symbols can be used for two forms of point data: true and conceptual. **True point data** are actually measured at point locations; examples include the number of calls made from telephone booths and temperatures at weather stations. Although we generally use isarithmic maps to display continuous phenomena such as temperature (see Chapter 14), we can use proportional symbols to focus on the raw data, as they are collected at point locations. **Conceptual point data** are collected over areas (or volumes), but the data are conceived of as being located at points (e.g., the centroids of enumeration units) for purposes of symbolization. An example is the number of microbreweries and brewpubs in each state (Figure 16.1A).

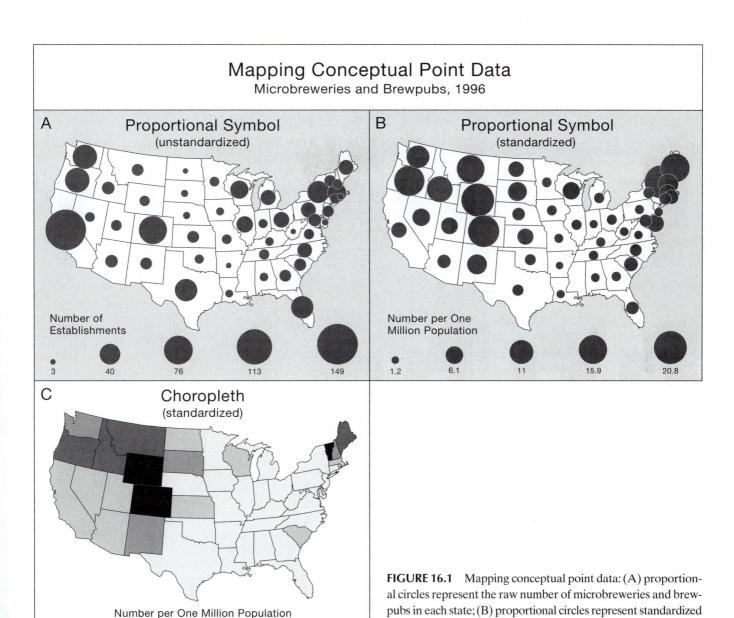

FIGURE 16.1 Mapping conceptual point data: (A) proportional circles represent the raw number of microbreweries and brewpubs in each state; (B) proportional circles represent standardized data—the number of microbreweries and brewpubs per 1 million people; (C) a choropleth map representing the standardized data shown in (B). (Data source: *http://www.beertown.org/.*)

Some data are not easily classified as either true or conceptual. For example, data associated with cities are collected over the areal extent of the city, but they normally are treated as occurring at point locations because at typical mapping scales cities are depicted as points. An example is the number of out-of-wedlock births to teenagers in major U.S. cities (Figure 16.2A).

The concept of data standardization discussed for choropleth maps (see Chapters 4 and 13) is also applicable to proportional symbol maps. As an example, consider the data for out-of-wedlock births. The unstandardized map (Figure 16.2A) is useful for showing the sheer magnitude of out-of-wedlock births, but care must be taken in interpreting any spatial patterns on this map because cities with a large population are apt to have a large number of out-of-wedlock births. (Note the strong visual correlation between Figure 16.2A and 16.2B, the population of major U.S. cities.)

One method of standardization is to compute the ratio of two raw-total attributes. In the case of the birth data, we can compute the ratio of out-of-wedlock births to the total number of births, obtaining the proportion of out-of-wedlock births in each city (this is shown in percentage form in Figure 16.2C). This map has a markedly different appearance than the unstandardized map, in large part because of a much narrower range of data: 3.3 to 22.7 percent, as opposed to 102 to 11,236 births. This map also illustrates the difficulty of using proportional circles when the data range is relatively narrow—here the largest value is more than six times the smaller, but

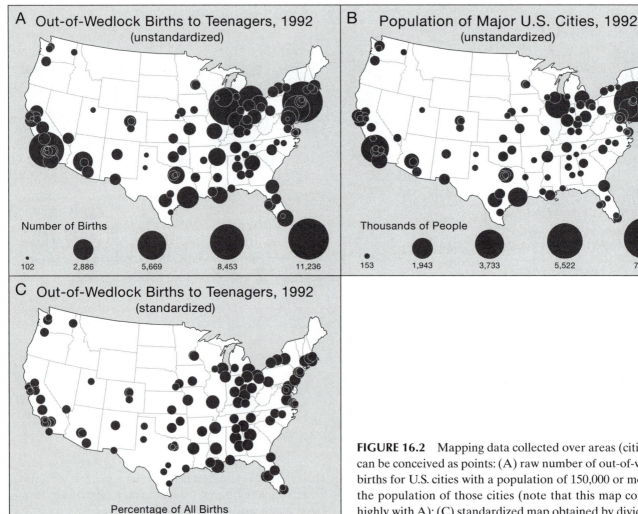

FIGURE 16.2   Mapping data collected over areas (cities) that can be conceived as points: (A) raw number of out-of-wedlock births for U.S. cities with a population of 150,000 or more; (B) the population of those cities (note that this map correlates highly with A); (C) standardized map obtained by dividing the number of out-of-wedlock births by the total number of births. (Data source: National Center for Health Statistics, 1995.)

the map does not immediately suggest this. In section 16.3.2, we discuss how formulas for circle sizes might be modified to handle this problem. Another solution might be to use bars rather than circles.

One can also argue that the microbrewery and brewpub data should be standardized, as the number of microbreweries and brewpubs is likely to be greater in more populous states. In this case, a useful standardization is the number of microbreweries and brewpubs per 1 million population (Figure 16.1B). Although standardized conceptual point data can be represented with proportional symbols, a choropleth map is more commonly used (Figure 16.1C) because conceptual point data are associated with areas.

16.2 KINDS OF PROPORTIONAL SYMBOLS

Proportional symbols can be divided into two basic groups: geometric and pictographic. **Geometric symbols** (circles, squares, spheres, and cubes) generally do not mirror the phenomenon being mapped, whereas **pictographic symbols** (heads of wheat, caricatures of people, diagrams of barns) do. Prior to the development of digital mapping, geometric symbols predominated because templates were readily available for manually constructing common geometric shapes. Today, digital mapping has eased the development of pictographic symbols: One can create the basic design for a symbol by hand, scan the symbol into a computer file, and then use design software to duplicate the symbol at various sizes. The map of beer mugs shown in Figure 16.3 was created using this approach. Alternatively,

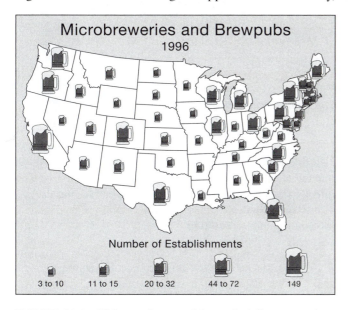

FIGURE 16.3 Using a pictographic symbol (beer mugs) to represent the number of microbreweries and brewpubs in each state. The data were range-graded and the sizes of symbols chosen by "eye."

one can find numerous pictographic symbols already in digital form in **clip art** files.

The ease with which readers can associate pictographic symbols with the phenomenon being mapped, their eye-catching appeal, and their increasingly greater ease of construction suggest that these symbols will appear more commonly on maps. Pictographic symbols, however, are not without problems. One is that when symbols overlap, they might be more difficult to interpret than geometric symbols (compare the northeastern portions of Figures 16.1A and 16.3). Another problem is that it might be more difficult to judge the relative sizes of irregular pictographic symbols (e.g., judging size relations among beer mugs is arguably more difficult than judging size relations among circles).

Circles have been the most frequently used geometric symbol. Arguments traditionally offered for using circles include the following:

1. Circles are visually stable.
2. Users prefer circles over other geometric symbols.
3. Circles (as opposed to, say, bars) conserve map space.
4. When constructing circles, it is easy to determine appropriate relations between data and circle size.
5. Circles are easy to construct.

The latter two advantages have largely disappeared with the rise of digital mapping: Formulas for calculating the sizes of geometric symbols can now be implemented in spreadsheets, and software can be used to automatically construct geometric symbols.

Traditionally, cartographers recommended against using 3-D geometric symbols (spheres, cubes, and prisms) because of the difficulty of both estimating their size and constructing them. Like pictographic symbols, however, these symbols can produce attractive eye-catching graphics (Figure 16.4). Additionally, 3-D geometric symbols can be useful for representing a large range in data; for example, in Figure 16.5 the small 3-D symbol is easily detected, whereas the corresponding two-dimensional symbol nearly disappears.

16.3 SCALING PROPORTIONAL SYMBOLS

16.3.1 Mathematical Scaling

Mathematical scaling sizes areas (or volumes) of point symbols in direct proportion to the data; thus, if a data value is 20 times another, the area (or volume) of a corresponding point symbol will be 20 times as large (Figure 16.6). We now consider some formulas for calculating symbol sizes; we deal with circles first because of their common use.

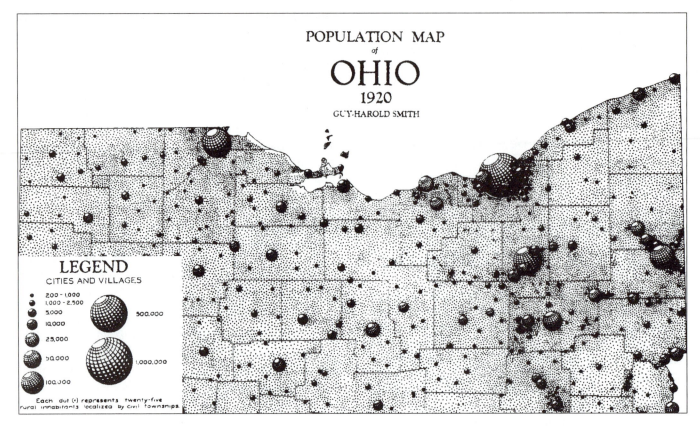

FIGURE 16.4   An eye-catching map created using 3-D geometric symbols. (After Smith, 1928. First published in *The Geographical Review,* 18(3), plate 4. Reprinted with permission of the American Geographical Society.)

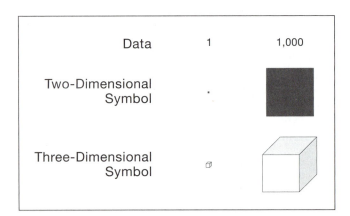

FIGURE 16.5   Attempts to portray a large range of data (note that the smallest 3-D symbol is more easily seen than the smallest two-dimensional symbol).

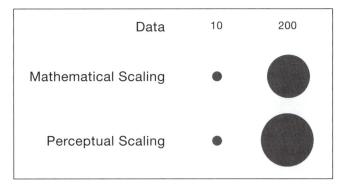

FIGURE 16.6   Mathematical versus perceptual scaling. In mathematical scaling, the areas of the circles are constructed directly proportional to the data; in perceptual scaling, the size of the larger circle is increased (using Flannery's exponent) to account for underestimation.

Remembering from basic math that the area of a circle is equal to πr^2, we can establish the relation

$$\frac{\pi r_i^2}{\pi r_L^2} = \frac{v_i}{v_L}$$

where r_i = radius of the circle to be drawn
r_L = radius of the largest circle on the map

v_i = data value for the circle to be drawn
v_L = data value associated with the largest circle

Note that this formula specifies circle areas in direct proportion to corresponding data values. The relation uses the largest radius (and largest data value) for one of the circles because proportional symbol maps are often constructed by beginning with a largest symbol size to minimize the effect of symbol overlap.

Because the values of π cancel, this equation reduces to

$$\frac{r_i^2}{r_L^2} = \frac{v_i}{v_L}$$

Taking the square root of both sides, we have

$$\frac{r_i}{r_L} = \left(\frac{v_i}{v_L}\right)^{0.5}$$

Finally, solving for r_i, we compute

$$r_i = \left(\frac{v_i}{v_i}\right)^{0.5} \times r_L$$

To apply this formula, consider how a radius for the circle representing Los Angeles was computed for Figure 16.2A. After some experimentation, we decided that the largest circle (New York) should have a radius of .2125 inches (.5398 cm). To determine the radius for Los Angeles, we inserted the number of out-of-wedlock births for Los Angeles and New York (8,507 and 11,236, respectively) into the formula, along with the largest radius as follows:

$$r_{\text{Los Angeles}} = \left(\frac{v_{\text{Los Angeles}}}{v_{\text{New York}}}\right)^{0.5} \times r_{\text{New York}}$$

$$r_{\text{Los Angeles}} = \left(\frac{8{,}507}{11{,}236}\right)^{0.5} \times .2125 = .1849$$

Formulas for other geometric symbols could be derived in a similar fashion. The results would be as follows:

SQUARES: $s_i = \left(\dfrac{v_i}{v_L}\right)^{0.5} \times s_L$

where s_i is the length of a side of a square to be drawn and s_L is the length of a side of the largest square;

BARS: $h_i = \left(\dfrac{v_i}{v_L}\right) \times h_L$

where h_i is the height of a bar to be drawn and h_L is the height of the tallest bar;

SPHERES: $r_i = \left(\dfrac{v_i}{v_L}\right)^{1/3} \times r_L$

where r_i is the radius of a sphere to be drawn and r_L is the radius of the largest sphere;

CUBES: $s_i = \left(\dfrac{v_i}{v_L}\right)^{1/3} \times s_L$

where s_i is the length of a side of a cube to be drawn and s_L is the length of a side of the largest cube.

In using these formulas, there are several issues that you need to consider. First, although these formulas are sometimes embedded in mapping software, you might find that you have to enter them into spreadsheets along with the raw data. Second, because these formulas typically involve taking the square or cube root of the ratio of data (raising the ratio to either the 1/2 or 1/3 power), mathematical scaling is sometimes referred to as *square root* or *cube root scaling*. We term the power to which the ratio of data is raised the *exponent for symbol scaling*.

Third, it is important to recognize that these formulas produce unclassed maps, as differing data values are depicted by differing symbol sizes. Classed, or **range-graded**, **maps** can be created by classing the data and letting a single symbol size represent a range of data values, but *unclassed* proportional symbol maps are more common. This might seem surprising given the frequency with which *classed* choropleth maps are used. The difference stems, in part, from the ease with which unclassed proportional symbol maps could be created in manual cartography (either an ink compass or a circle template could be used to draw circles of numerous sizes).

Fourth, you will find that some software (e.g., MapViewer) permits arbitrarily specifying the smallest and largest symbol sizes, with other symbols scaled proportionally between these symbols. For symbols based on area, the formulas are

$$z_i = \frac{v_i - v_S}{v_L - v_S}$$

$$A_i = Z_i(A_L - A_S) + A_S$$

where v_S and v_L = smallest and largest data values
Z_i = proportion of the data range associated with the data value v_i
A_S and A_L = smallest and largest areas desired
A_i = area of a symbol associated with the data value v_i

This approach produces an unclassed map, but the symbols are not scaled proportional to the data (i.e., a data value twice another does not have a symbol twice as large). An advantage of the approach, however, is that it can enhance the map pattern, just as do an arbitrary exponent and range grading, as described in the following sections.

16.3.2 Perceptual Scaling

Numerous studies have shown that the perceived size of proportional symbols does not correspond to their mathematical size; rather, people tend to underestimate the size of larger symbols. For example, in viewing the larger mathematically scaled circle in Figure 16.6, most people would estimate it to be less than 20 times as large as the smaller circle. If larger symbols are underestimated, it seems reasonable to suggest that formulas for mathematical scaling might be modified (or "corrected") to account for underestimation; this process is known as **perceptual** (or psychological) **scaling**.

Formulas for Perceptual Scaling

To develop formulas for perceptual scaling, it is useful to consider how researchers have summarized the results of experiments dealing with perceived size. The relation between actual and perceived size typically has been stated as a *power function* of the form

$$R = cS^n$$

where R = response (or perceived size)
S = stimulus (or actual size)
c = a constant
n = an exponent

To differentiate the exponent n from the one for symbol scaling introduced earlier, we term it the **power function exponent**.

The power function exponent is the key to describing the results of experiments involving perceived size. If size is estimated correctly, the exponent will be close to 1.0. Underestimation and overestimation are represented by exponents appreciably below and above 1.0. For example, an oft-cited study by James Flannery (1971) found that for circles $R = (0.98365)S^{0.8747}$; here the exponent of 0.8747 is indicative of underestimation.

The power function equation states what response arises from a certain stimulus. For constructing symbols, we need to know the reverse of that: what stimulus must be shown to get a certain response. Therefore, the power function must be transposed by dividing each side by c, and then raising each side to the $1/n$ power. The result is

$$S = c_1 R^{1/n}$$

where c_1 is a constant equal to $(1/c)^{1/n}$. In Flannery's study, the transpose was $S = (1.01902)R^{1.1432}$.* To simplify computations, the value of the constant (1.01902) in the transposed equation can be ignored because it is close to 1.0; thus we have $S = R^{1.1432}$. Because this equation expresses the relation between the areas of circles, and circles are constructed on the basis of radii, we need to take the square root of both sides, producing $S^{0.5} = R^{0.5716}$. Again for simplicity, the value 0.5716 has normally been rounded to 0.57. As a result, a perceptual scaling formula for circles is

$$r_i = \left(\frac{v_i}{v_L}\right)^{0.57} \times r_L$$

The result of using an exponent of 0.57 in the circle scaling formulas can be seen for the perceptually scaled circles in Figure 16.6; for most readers, the larger circle should now appear closer to 20 times larger than the smallest circle.

---

* Flannery did not raise $1/k$ to $1/n$, and so his constant differed slightly from the value reported here.

The magnitude of the power function exponent varies, depending on the symbol type. For squares, Crawford (1973) derived an exponent of 0.93 (which yields an exponent of 0.54 for symbol scaling), indicating that squares are estimated better than circles. For bars, Flannery found that underestimates were balanced by overestimates, and thus recommended no corrective formula. Although these results suggest that squares and bars should be used when precise estimates are desired, we recommend doing so with caution. A study by Slocum and his colleagues (in press) revealed that squares were not aesthetically pleasing (when compared with several geometric and pictographic symbols) and that bars were difficult to associate with a point location.

Power function exponents for "drawn" 3-D symbols have been appreciably lower than for two-dimensional symbols, indicating severe underestimation. For example, for spheres and cubes Ekman and Junge (1961) derived exponents of 0.75 and 0.74 (corresponding to exponents of 0.44 and 0.45 for symbol scaling). Interestingly, research by Ekman and Junge also indicated that when truly 3-D cubes were used (they were "made of steel with surfaces polished to a homogeneous, dull silvery appearance," p. 2), the exponent was 1.0. This result suggests that the manner in which 3-D symbols are portrayed in two-dimensional space might have an impact on the exponent. For example, if an interactive graphics program gives the impression of traveling through 3-D space, we might expect an exponent closer to 1.0.

Problems in Applying the Formulas

Unfortunately, there are numerous problems in applying formulas for perceptual scaling: (1) the value of a power function exponent can be affected by various experimental factors, (2) using a single exponent might be unrealistic because of the variation that exists between subjects and within an individual subject's responses, (3) the formulas fail to account for the spatial context within which symbols are estimated, and (4) experimental studies for deriving exponents have dealt only with acquiring specific map information (as opposed to considering memory and general information). We briefly consider each of these problems.

Kang-tsung Chang (1980) provided a good summary of the various experimental factors that can affect a power function exponent. One is whether subjects are asked to complete a ratio or magnitude estimation task. **Ratio estimation** involves comparing two symbols and indicating how much larger or smaller one symbol is than another (noting that "this symbol appears to be five times larger than this one"); on a map, this involves comparing symbols without consulting the legend. Flannery's exponent of 0.87 for circles was developed on this basis.

Magnitude estimation involves assigning a value to a symbol on the basis of a value assigned to another symbol; for example, if a single circle is included in a map legend, values for other circles on the map can be estimated by comparing them with the legend circle. Using this approach, Chang (1977) found that the use of a moderately sized circle in the legend led to an exponent of 0.73 for circles.

A second factor affecting the exponent is the size of the standard symbol against which other symbols are compared. For example, in magnitude estimation studies, Chang (1977) and Cox (1976) showed that the value of the exponent gradually increased as the size of the standard increased. Other experimental factors noted by Chang that influenced the exponent included the wording of the instructions to subjects, the range of stimuli presented, and the order in which estimates were made.

T. L. C. Griffin (1985) is one of several researchers who have noted that reporting a single exponent for an experiment hides the variation among and within subjects. Although overall Griffin found a power function exponent of 0.88 for circles (nearly identical to Flannery's 0.87), he stressed that the exponents for individuals varied from approximately 0.4 to 1.3. He also noted that "perceptual rescaling was shown to be inadequate to correct the estimates of poor judges, while seriously impairing the results of those who were more consistent" (p. 35).

One solution to the problems noted by Chang and Griffin is to apply no correction and stress the importance of a well-designed legend. For example, Chang recommended including "three standards [in the legend]—small, medium, [and] large"; he based this recommendation, in part, on a magnitude estimation study in which he found an exponent of 0.94 when three legend circles were used (Chang 1977). Chang (1980, 161) also recommended including the statement that "circle areas are made proportional to quantities," to encourage users to make estimates based on area.

Another solution to the problems noted by Chang and Griffin is to train users how to read proportional symbol maps. For example, Judy Olson (1975a) found that when readers were asked to make estimates of circle size and then were given feedback on the correct answers that the power function exponent was closer to 1.0. Olson also found, however, that only when practice was combined with perceptual scaling did the dispersion of errors also decrease, suggesting that training and perceptual scaling should be used in concert.

The importance of spatial context on developing a perceptual scaling formula can be seen by considering the **Ebbinghaus illusion** shown in Figure 16.7: The two circles in the middle of the surrounding circles are identical in size, but the one surrounded by larger circles appears smaller. Applying this principle to a map, a small circle surrounded by larger circles should appear smaller than it really is, whereas a large circle surrounded by smaller circles should appear larger than it really is. In an experimental study with maps, Patricia Gilmartin (1981) found that spatial context actually had these effects. On

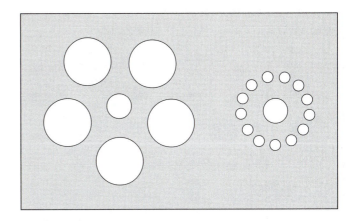

FIGURE 16.7 The Ebbinghaus illusion: The two circles in the middle of the surrounding circles are identical in size, but the one surrounded by larger circles appears smaller.

this basis, one could argue for a formula in which each circle is scaled to reflect its local context. Gilmartin argued against this, indicating it would "have the undesirable effect of weakening the overall pattern perception" (p. 162).

Another limitation of perceptual scaling is that experiments for deriving an appropriate exponent have focused solely on specific map information. In Chapter 1, we argued that the portrayal of general information is a more important function of maps. Thus, it seems that general tasks should be considered in developing an exponent, or at least that the effect of the exponent should be considered on the overall look of the map. To illustrate this point, consider Figure 16.8, which compares mathematically and perceptually scaled maps of the microbrewery and brewpub data. Here the largest circle size on each map has been held constant to minimize the effects of circle overlap on the two maps. Although a larger range of circle sizes appears on the perceptually scaled map, the patterns on the two maps are similar, suggesting that perceptual scaling has little effect on general information.

Rather than basing the exponent on specific circle estimates, Olson (1976a, 155–156; 1978) suggested that an arbitrary exponent might be used to enhance the recognition of spatial pattern. To illustrate, Figure 16.9 compares a mathematically scaled map (the exponent in the circle-scaling formula is set to 0.5) with one having an arbitrary exponent (the exponent is set to 1.0) for the standardized birth data. The narrow range of data on the mathematically scaled map (Figure 16.9A) makes it difficult to detect any pattern. In contrast, on the arbitrarily scaled map (Figure 16.9B) it is easier to detect a pattern; for example, note the lower percentage of out-of-wedlock births in the extreme south central part of the United States (in Texas).

A final limitation of perceptual scaling is that most of the relevant experiments have been based only on information *acquisition*, as opposed to *memory* for that information. Studies by psychologists (e.g., Kerst and Howard

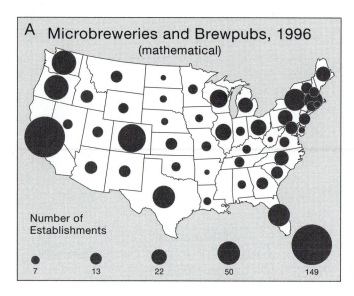
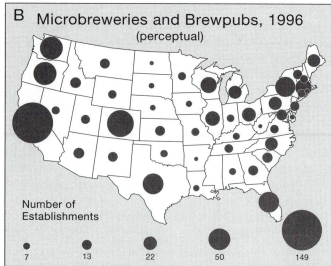

FIGURE 16.8   Effect of mathematical versus perceptual scaling on map pattern: (A) mathematically scaled map; (B) perceptually scaled map based on Flannery. Note that perceptual scaling appears to have little effect on the overall pattern when the largest circle on both maps is the same size.

1984) have revealed that underestimation of larger sizes occurs twice: once when acquiring information and once again when recalling that information. As a result, the exponent for memory is appreciably lower than that for acquisition (values obtained are approximately the square of that for acquisition). Although the exponent in the perceptual scaling formulas might be adjusted to account for this, we suspect that most cartographers would not make this adjustment because map readers are not expected to remember precise, specific information. It is, however, interesting to note that the lower exponent resulting from this approach would have an effect similar to that just suggested for enhancing spatial pattern.

Range-Graded Scaling

Range grading groups raw data into classes, and represents each class with a different sized symbol (as an example, if the data are grouped into five classes, then five symbol sizes are used). Three basic decisions must be made in range grading: the number of classes to be shown, the method of classification to be used, and the symbol sizes to be used for each class. The first two decisions are standard in any classification of numerical data and were discussed extensively in Chapter 5. (Figure 5.7 summarizes the advantages and disadvantages of each classification method.)

The sizes of range-graded symbols normally are selected to enhance the visual discrimination among classes. Figure 16.10 portrays two sets of symbols that cartographers have developed for this purpose. The first (Figure 16.10A) was developed by Hans Meihoefer

(1969) in a visual experiment. Meihoefer indicated that people "were able generally to differentiate" (p. 112) among all of these circles in a map environment. Arguing that some of Meihoefer's circles were difficult to discriminate, Borden Dent (1999, 181–183) used his personal experience in designing maps to develop the set shown in Figure 16.10B. For both sets, the mapmaker simply selects n adjacent circles, where n is the number of classes to be depicted on the map.

Range grading is considered advantageous because readers can easily discriminate symbol sizes and thus readily match map and legend symbols; another advantage is that the contrast in circle sizes might enhance the map pattern, in a fashion similar to the use of an arbitrary exponent described in the preceding section. To illustrate the latter, consider Figure 16.11, which compares range-graded and mathematically scaled maps for both the standardized birth data and the raw microbrewery and brewpub data.*In the case of the microbrewery and brewpub data, we see that range grading had relatively little effect on the spatial pattern because the range of circle sizes on the range-graded and mathematically scaled maps was similar. For the birth data, however, the results were dramatic, with range grading resulting in considerable overlap.

Because specific range-graded sizes have been recommended only for circles, individual mapmakers must

---

* In this case, the data were classed using Jenks's optimal approach, and Dent's five smallest circles were used.

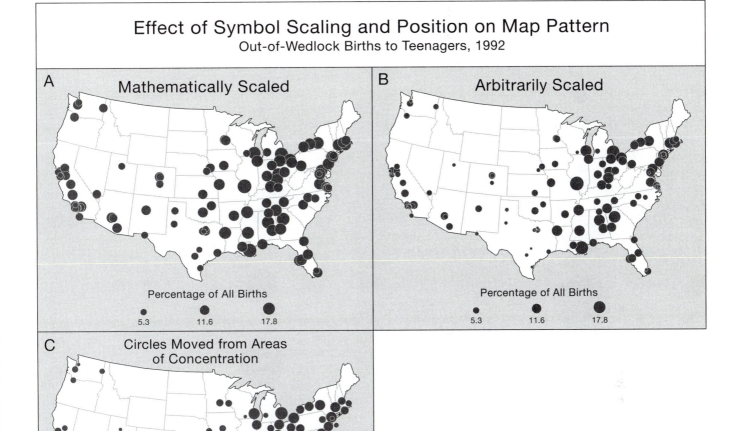

Effect of Symbol Scaling and Position on Map Pattern
Out-of-Wedlock Births to Teenagers, 1992

A Mathematically Scaled

Percentage of All Births

5.3   11.6   17.8

B Arbitrarily Scaled

Percentage of All Births

5.3   11.6   17.8

C Circles Moved from Areas of Concentration

Percentage of All Births

5.3   11.6   17.8

FIGURE 16.9 Manipulating symbol scaling and symbol position to enhance map pattern: (A) mathematically scaled map (the exponent in the circle scaling formula is set to 0.5); (B) an arbitrarily scaled map (the exponent in the circle scaling formula is set to 1.0); (C) circles are moved away from congested areas.

determine appropriate sizes for other symbol types. For example, to create the range-graded proportional-square map shown in Figure 16.12, distinctly different small and large squares were selected, and then three intermediate squares were specified so that their areas were evenly spaced between the smallest and largest squares.

Range grading is particularly desirable for pictographic symbols because their unusual shape often makes precise relationships between symbols awkward to compute (for the mapmaker) and to estimate (for the map reader). Thus, it makes sense to select intermediate-sized symbols by eye, as opposed to spacing them regularly on the basis of area or volume. This was the approach taken to construct the pictographs of beer mugs shown in Figure 16.3.

A disadvantage of range grading is that readers might misinterpret specific information if they do not pay careful attention to the legend. For example, a reader failing to examine the legend might say that the value for circle T in Figure 16.11A is considerably larger than the value for circle S (say, approximately 30 times as large), but the numbers specified in the legend indicate that the values differ only by a factor of about 4. Another disadvantage is that range grading creates clear differences in circle size, and thus potentially creates a pattern when there might not be a meaningful one. For example, for a data set with minimum and maximum values of 10 and 11 percent, respectively, range grading would create obvious differences among circles, even though a 1 percent difference might be of no interest or significance to the reader.

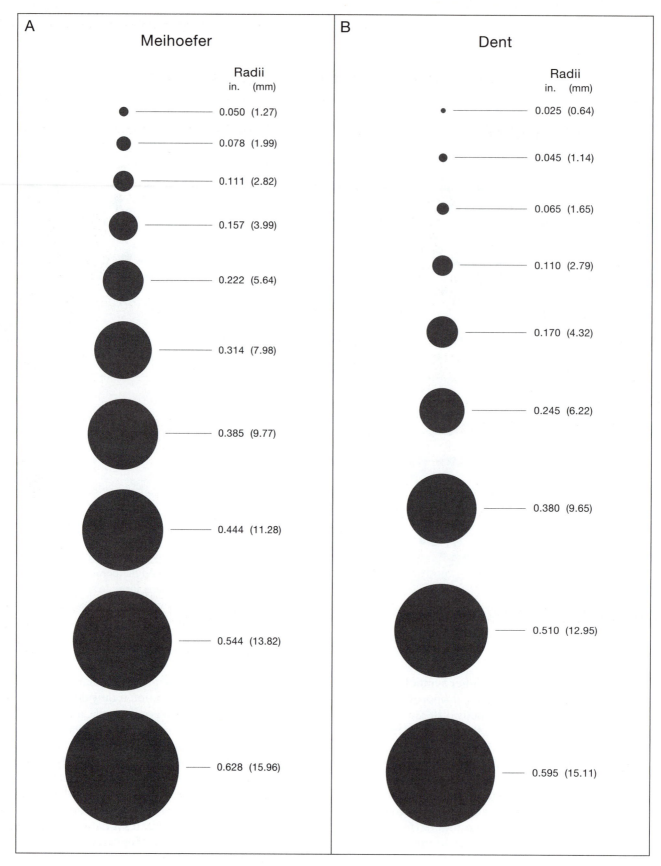

FIGURE 16.10 Potential circle sizes for range grading: (A) a set developed by Meihoefer in a visual experiment (after Meihoefer 1969, Figure 4); (B) a set developed by Dent based on practical experience (after Dent 1999, 183).

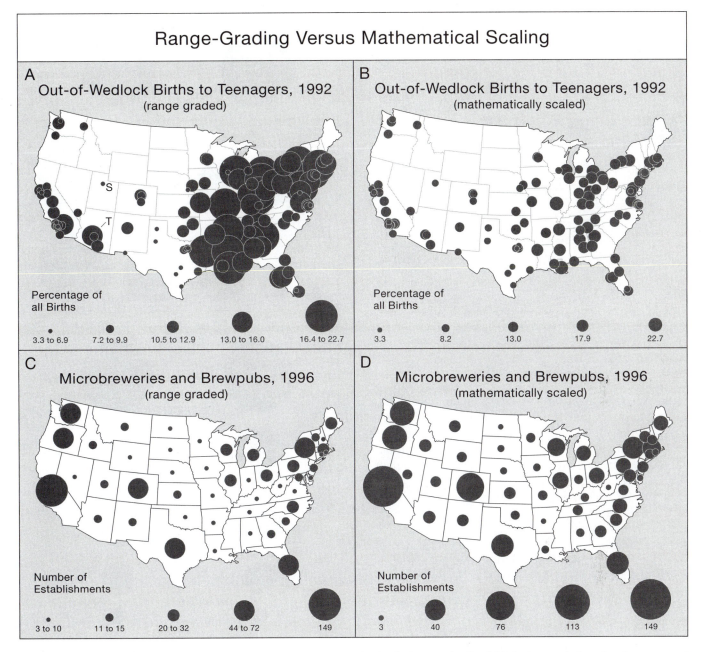

FIGURE 16.11 Range-graded and mathematically scaled maps for both the standardized birth data and the microbrewery and brewpub data. Maps A and C are range-graded based on Dent's five smallest circles. Maps B and D are mathematically scaled.

16.4 LEGEND DESIGN

There are two basic problems in designing legends for proportional symbol maps: deciding how the symbols should be arranged and determining which symbols should be included.

16.4.1 Arranging Symbols

Two basic legend arrangements are used on proportional symbol maps: nested and linear. In the **nested-legend arrangement**, smaller symbols are drawn within larger symbols, whereas in the **linear-legend arrangement**, symbols are placed adjacent to each other in either a horizontal or vertical orientation (Figure 16.13). An advantage of the nested arrangement is that it conserves map space. Note, however, that this might make it difficult to compare a symbol in the legend with a symbol on the map, as symbols in the legend (with the exception of the smallest) are covered by other symbols.

When a linear horizontal arrangement is used, you must decide whether the symbols should be ordered with

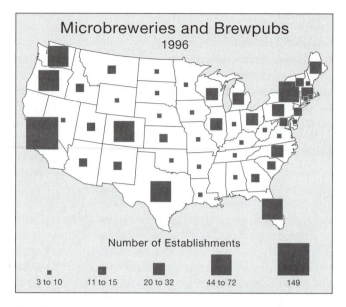

FIGURE 16.12   A range-graded map using squares as a point symbol.

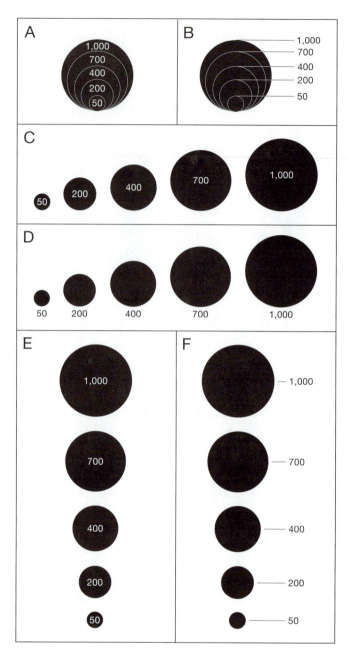

FIGURE 16.13   Various legend arrangements: (A) and (B) nested; (C) and (D) linear with a horizontal orientation; and (E) and (F) linear with a vertical orientation.

the smallest on the left and the largest on the right, or vice versa. Displaying larger symbols on the right is most desirable given that the traditional number line progresses from left to right. When a linear vertical arrangement is used, you similarly must decide whether symbols should be ordered with the largest at the top or the bottom. As with choropleth maps, we can use the argument that "people associate 'up' with 'higher' and 'higher' with larger data values" to justify placing larger symbols at the top. With range-graded maps, however, it makes sense to place smaller symbols at the top so that values increase from left to right and from top to bottom, matching the order in which we normally read text (see section 13.5 for a similar argument for choropleth maps).

16.4.2 Which Symbols to Include

With range grading, legend symbols are a function of the classes displayed on the map (e.g., a five-class map yields five legend symbols). For mathematical and perceptual scaling, there are two general methods for selecting legend symbols. One is to include the smallest and largest symbol actually shown on the map and then interpolate several intermediate-sized symbols (Figure 16.14A). A second method is to select a set of symbols that are most representative of those appearing on the map, which should minimize estimation error. The latter method can be implemented by applying Jenks's optimal classification to the raw data and then constructing legend symbols based on the median (or mean) of each class (Figure 16.14B). This method might also be refined to include circles representing the extremes in the data because of the difficulty

of extrapolating beyond symbols shown in the legend (Dobson 1974).

In addition to selecting one of these general methods for mathematical and perceptual scaling, the mapmaker must also decide how many symbols will be shown in the legend. One possibility would be to use three symbols, as recommended by Chang. Although three symbols might be sufficient for data with a small range, it makes sense to use more than three when the range is large. As a rule of thumb, we suggest using as many symbols as appear to be easily discriminated.

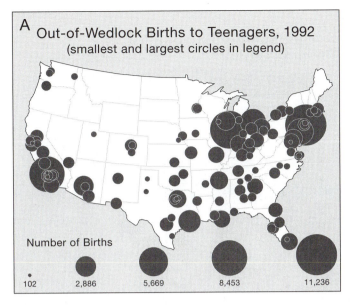

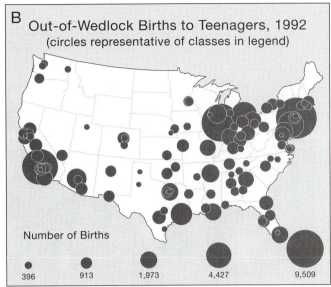

FIGURE 16.14   Approaches for selecting circles for the legend on an unclassed proportional symbol map: (A) the smallest and largest symbol actually shown on the map are used, along with several intermediate-sized symbols; (B) the raw data are classified, and a representative value of each class is used (in this case the median).

16.5 HANDLING SYMBOL OVERLAP

A major issue in proportional symbol mapping is deciding how large symbols should be, and consequently how much overlap there should be. Small symbols cause little or no overlap, and thus a map potentially devoid of spatial pattern; in contrast, large symbols create a cluttered map, making it difficult to interpret individual symbols. This section considers two issues: deciding how much overlap there should be and how the overlap should be symbolized.

16.5.1 How Much Overlap?

Unfortunately, there are no rules regarding the appropriate amount of overlap. Rather, cartographers have suggested subjective guidelines; for instance, Robinson et al. (1984) indicated that the map should appear "neither 'too full' nor 'too empty'" (p. 294). Examples of improper amounts of overlap are shown in Figures 16.15A and 16.15B. Although most cartographers would agree that such extreme cases should be avoided, there would be disagreement as to which map should be shown between the extremes. (Figures 16.15C and 16.15D are two possibilities.)

The role that overlap plays is determined to some extent by whether the map is to be used for communication or data exploration. In the case of communication, you probably will want to manipulate circle overlap to enhance the spatial pattern. For example, Figure 16.15C

might be a more appropriate choice than Figure 16.15D if you wish readers to note the concentration of microbreweries in the northeast part of the United States. In the case of data exploration, mapping software ideally should provide an option to easily change symbol sizes, such as the interactive option described by Slocum and Yoder (1996).

16.5.2 Symbolizing Overlap

Overlap can be handled using either transparent or opaque symbols. **Transparent symbols** enable readers to see through overlapping symbols, whereas **opaque symbols** are stacked on top of one another (see Figure 16.16). On black-and-white maps, the fill within both transparent and opaque symbols can be varied from white to solid black. Color can, of course, provide considerable flexibility in overlap, as different colors can be used for the boundary and interior of symbols.

Transparent symbols allow background information (e.g., a road network) to be seen beneath the symbols (Figures 16.16A, 16.16C, and 16.16E), and make it easier to determine the bounds of individual symbols, thus simplifying the estimation of symbol size. In contrast, opaque symbols enhance figure-ground contrast because the symbols tend to appear as a figure against the background of the rest of the map; this is most apparent for the gray and black symbols in Figure 16.16. The opaque approach also promotes a visual hierarchy, as the circles appear "above" other map information (e.g., a road network).

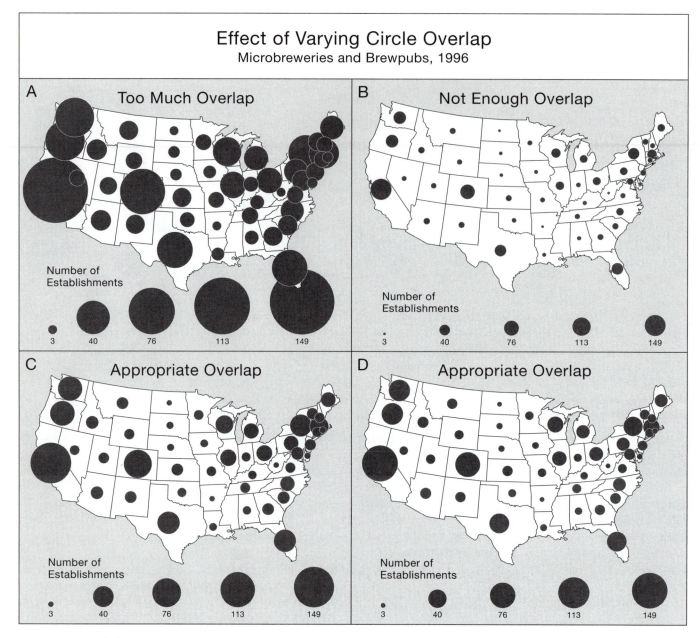

FIGURE 16.15   Effect of varying the amount of overlap: (A) too much overlap—the map appears crowded; (B) not enough overlap—the map appears empty; (C) and (D) are examples of maps having an appropriate amount of overlap.

These ideas are supported by the work of T. L. C. Griffin (1990) and Richard Groop and Daniel Cole (1978). Griffin found that people were about equally split in their preference for transparent and opaque symbols; transparent symbols were liked because they provided "maximum information," and those favoring opaque symbols "stressed the quality of clarity" (24). With respect to fill, Griffin found that for transparent circles, gray and black were about equally popular, whereas for opaque circles, black was most popular; almost no one liked a white fill. Groop and Cole found that transparent symbols were

more accurately estimated than opaque symbols, and that for opaque symbols there was a strong correlation between the amount of overlap and the error of estimation.

Other solutions for handling overlap include **inset maps**, which portray a congested area at an enlarged scale (see Chapter 11); the use of a **zoom function** in an interactive graphics environment; and the possibility of moving symbols slightly away from the center of congested areas. Zooming typically is implemented by enclosing the area to be enlarged by a rectangular box, clicking a mouse button, and having the outlined area fill the screen. MapTime

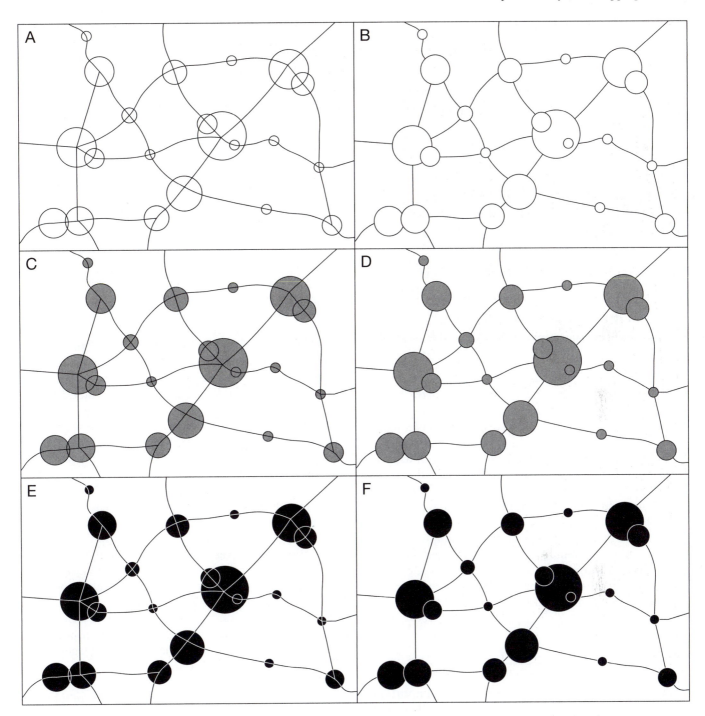

FIGURE 16.16 Transparent and opaque methods of handling overlap. Maps on the left are transparent, and maps on the right are opaque. From top to bottom different fills are shown, ranging from white to gray to black. (After Griffin 1990, 23.)

(discussed in Chapter 21) contains such a zoom function. Figure 16.9C illustrates the effect of moving symbols away from congested areas. In comparing Figure 16.9C with 16.9B, we can see that in Figure 16.9C it is easier to compare the sizes of individual circles and the spatial pattern is potentially more obvious.

16.6 REDUNDANT SYMBOLS

Redundant symbols portray a single attribute with two or more visual variables. For example, on proportional symbol maps the visual variables size and value could be used

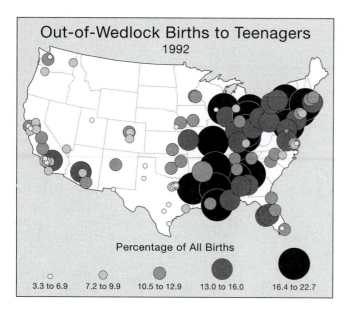

Out-of-Wedlock Births to Teenagers
1992

Percentage of All Births

| 3.3 to 6.9 | 7.2 to 9.9 | 10.5 to 12.9 | 13.0 to 16.0 | 16.4 to 22.7 |

FIGURE 16.17 Use of redundant symbols (the visual variables size and value) on proportional symbol maps.

(Figure 16.17). The logic of using redundant symbols is that it is easier to discriminate symbols on the basis of two visual variables than when a single visual variable is used. As a result, users are able to perform specific map tasks more easily. For example, in Figure 16.17 it is possible to differentiate the large circles from others, even though there is considerable overlap (compare with Figure 16.11A).*

Although redundant symbols promote the communication of specific information, they seldom are used on printed maps, presumably because mapmakers feel that they impede the communication of general information (it is difficult to combine symbols to form a region when the symbols differ in both size and value). Redundant symbols seem more appropriate for a data exploration environment in which multiple views of a data set are considered.

SUMMARY

In this chapter, we have covered a number of important principles involved in constructing **proportional symbol maps**. We have stressed that two kinds of data can be mapped using proportional symbols: **true point data**, which can actually be measured at point locations, and **conceptual point data**, which are collected over areas (or volumes) but are conceived as being located at points. Respective examples would be the water released by wells associated with center-pivot irrigation systems and

---

* For an experimental study dealing with redundancy on proportional symbol maps, see Dobson (1983).

the number of fatalities due to drunk driving in each state. It is often useful to standardize data intended for a proportional symbol map by considering raw totals for one attribute relative to values of another attribute. For example, we might divide the gallons of water pumped at a well by the area covered by the associated center-pivot irrigation system.

Two basic kinds of proportional symbols are possible: geometric and pictographic. **Geometric symbols** (e.g., circles and spheres) generally do not mirror the phenomenon being mapped, whereas **pictographic symbols** (e.g., drawings of oil derricks to represent oil production) do. Conventionally, geometric symbols have been used more frequently, but pictographic symbols are becoming common because of the ease of creating them in a digital environment. Three-dimensional symbols (e.g., spheres) traditionally have been frowned on because of the difficulty readers have in estimating their size, but they can produce attractive, eye-catching graphics and are useful for representing a large data range.

There are three methods for scaling (or sizing) proportional symbols: mathematical, perceptual, and range grading. In **mathematical scaling**, symbols are sized in direct proportion to the data; thus, a data value 10 times another is represented by an area (or volume) 10 times as large. In **perceptual scaling**, a correction is introduced to account for visual underestimation of larger symbols; thus, larger symbols are made bigger than would normally be specified by mathematical scaling. For example, when constructing proportional circles, a **Flannery correction** is normally applied so that larger circles appear even larger (a symbol scaling exponent of 0.57 is used). We noted a number of problems with applying such a correction (e.g., using a single exponent might be unrealistic because of the variation in perception that exists among subjects). In **range grading**, data are grouped into classes, and a single symbol size is used to represent all data falling in a class. Range grading is considered advantageous because readers can easily discriminate symbol sizes and thus readily match map and legend symbols, and because the contrast in circle sizes might enhance the map pattern.

Other important issues in proportional symbol mapping include legend design, symbol overlap, and symbol redundancy. Basic problems of legend design include the arrangement of symbols (nested or linear) and the number and size of symbols. A nested arrangement conserves map space, but it might make it more difficult to compare a symbol in the legend with a symbol on the map. In terms of circle overlap, small symbols cause little or no overlap, and thus potentially create a map devoid of spatial pattern; in contrast, large symbols create a cluttered map, making it difficult to interpret individual symbols. **Redundant symbols** represent a single attribute (say, the magnitude of production at water wells) by two visual variables (e.g., size and value), thus enabling users to perform specific tasks more easily and accurately.

Because redundant symbols are arguably less effective for portraying general information, they are best used in a data exploration framework.

FURTHER READING

Brewer, C. A., and Campbell, A. J. (1998) "Beyond graduated circles: Varied point symbols for representing quantitative data on maps." *Cartographic Perspectives*, no. 29:6–25.

Summarizes literature related to proportional symbols, and illustrates both univariate and bivariate point symbols.

Chang, K. (1980) "Circle size judgment and map design." *American Cartographer* 7, no. 2:155–162.

Summarizes various experimental factors that can affect a power function exponent.

Dobson, M. W. (1983) "Visual information processing and cartographic communication: The utility of redundant stimulus dimensions." In *Graphic Communication and Design in Contemporary Cartography*. ed. by D. R. F. Taylor, pp. 149–175. Chichester, England: Wiley.

An experimental study dealing with redundancy on proportional circle maps.

Flannery, J. J. (1971) "The relative effectiveness of some common graduated point symbols in the presentation of quantitative data." *Canadian Cartographer* 8, no. 2:96–109.

A frequently cited study for determining an appropriate exponent for scaling circles.

Gilmartin, P. P. (1981) "Influences of map context on circle perception." *Annals, Association of American Geographers* 71, no. 2:253–258.

Examines spatial context as a factor in estimating proportional symbol sizes.

Griffin, T. L. C. (1985) "Group and individual variations in judgment and their relevance to the scaling of graduated circles." *Cartographica* 22, no. 1:21–37.

Considers the notion that power function exponents may vary as a function of the individual.

Griffin, T. L. C. (1990) "The importance of visual contrast for graduated circles." *Cartography* 19, no. 1:21–30.

Covers various methods for handling overlap of proportional circles.

Kerst, S. M., and Howard, J. H. J. (1984) "Magnitude estimates of perceived and remembered length and area." *Bulletin of the Psychonomic Society* 22, no. 6:517–520.

Considers the notion that we should consider memory for information when developing symbol scaling exponents.

Lindenberg, R. E. (1986) *The Effect of Color on Quantitative Map Symbol Estimation.* Unpublished Ph.D. dissertation, University of Kansas, Lawrence, KS.

Examines the role of color in the adjustment of circle scaling exponents, and concludes that when all circles are the same color, Flannery's adjustment is appropriate.

Meihoefer, H. (1969) "The utility of the circle as an effective cartographic symbol." *Canadian Cartographer* 6, no. 2:105–117.

Develops an appropriate set of range-graded circles.

Olson, J. M. (1975a) "Experience and the improvement of cartographic communication." *Cartographic Journal* 12, no. 2:94–108.

Considers the notion that readers might be trained to improve their estimates of symbol size.

Olson, J. M. (1976a) "A coordinated approach to map communication improvement." *American Cartographer* 3, no. 2:151–159.

Considers how changes in both map design and reader training might be used to enhance map communication. Of particular interest is the discussion of adjustments to the symbol scaling exponent to enhance map pattern.

Patton, J. C., and Slocum, T. A. (1985) "Spatial pattern recall: An analysis of the aesthetic use of color." *Cartographica* 22, no. 3:70–87.

Examines the effect of circle color on the ability to recall the pattern of proportional circles.

Rice, K. W. (1989) *The Influence of Verbal Labels on the Perception of Graduated Circle Map Regions.* Unpublished Ph.D. dissertation, University of Kansas, Lawrence, KS.

An extensive study of the role that verbal labels play in the perception of regions on proportional circle maps.

Slocum, T. A. (1983) "Predicting visual clusters on graduated circle maps." *American Cartographer* 10, no. 1:59–72.

Describes a model for predicting the visual clusters that readers see on proportional circle maps.

17

Dot and Dasymetric Mapping

OVERVIEW

Although data frequently are collected for enumeration units and mapped at that level, the resulting maps can be misleading because the distribution of the underlying phenomenon often varies within enumeration units. Dot and dasymetric maps are two solutions to this problem. On a **dot map** one dot represents a certain amount of some phenomenon (e.g., one dot might represent a specified number of people), and dots are placed at locations within enumeration units where the phenomenon is apt to occur. Like the choropleth map, a **dasymetric map** utilizes areal symbols, but the bounds of the symbols do not necessarily match the bounds of enumeration units (e.g., a single enumeration unit might have a full range of gray tones representing differing levels of population density).

Section 17.1 considers the kinds of data appropriate for dot and dasymetric maps. Dot maps utilize raw totals such as the number of people living in each county of a state, whereas dasymetric maps are based on standardized data such as population density (obtained by dividing a raw population figure by the area over which the population is distributed). **Ancillary information** is also necessary for locating dots (for dot mapping) or defining zones (for dasymetric mapping). For instance, if making a map of population, we might avoid placing dots representing people in water bodies, presuming that people do not live on the water (although in some areas of the world this would not be true). Conventionally, cartographic texts have covered dot and dasymetric mapping in separate sections, reflecting the different kinds of data on which they are based and the different symbology used. We consider them together here because ancillary information is a critical element of both.

Section 17.2 briefly considers the notion of manual versus automated dot and dasymetric map production.

Traditionally, cartographers incorporated ancillary information for dot and dasymetric maps through a tedious manual process. Today, this process can be automated through the use of GIS and remote-sensing techniques.

The remaining sections of the chapter consider several applications of dot and dasymetric mapping. Section 17.3 considers procedures for creating a dot map of wheat harvested in Kansas (such a dot map was initially introduced in Chapter 4). Here we consider (1) how ancillary information can be utilized to select regions where dots will be placed, (2) the selection of **dot size** (how large each dot is) and **unit value** (the raw total represented by each dot), and (3) approaches for placing dots within regions. In section 17.4 we consider work that Cory Eicher and Cynthia Brewer (2001) undertook to compare the effectiveness of several automated dasymetric methods. The most effective method was a limiting variable approach that has its roots in early work by John K. Wright (1936). Section 17.5 considers a novel approach for mapping population density developed by Mitchel Langford and David Unwin (1994) that utilizes remote sensing, dasymetric symbolization, and a generalization (smoothing) operation.

It is important to note that this chapter focuses on the use of ancillary information to create detailed maps of data collected for enumeration units. A more detailed map can also be constructed through a mathematical manipulation of the data. Tobler's pycnophylactic method (section 14.6) is an example of this approach. Recall that Tobler's method began with a set of prisms (each raised to a height proportional to the data), which were sculpted to create a smooth surface. Although Tobler's method was introduced as a form of interpolation, it can also be thought of as a form of reallocation (personal communication, Tobler 1997) in the sense that population is reallocated to *different portions of an enumeration unit as a function of surrounding units. A similar mathematical approach has been developed by David Martin (1996) that considers not only the population of each enumeration unit, but also the centroid of the population within that unit.*

17.1 SELECTING APPROPRIATE DATA AND ANCILLARY INFORMATION

Dot and dasymetric maps are utilized when you have collected data for enumeration units, but wish to show that the underlying phenomenon is not uniform throughout the enumeration units. For instance, imagine that you have data on the number of elephants living in each country of Africa. You could make a choropleth map of such data (assuming that you standardized the data appropriately), but this map would be misleading because each country would have a uniform areal symbol, suggesting no variation in the density of elephants within a country. Dot and dasymetric maps solve this problem by using **ancillary information** to create a more detailed map. In the case of elephants, appropriate ancillary information might be the location of heavily vegetated areas, especially forests and wooded savannas, and protected areas, such as national parks.

Recall from section 4.4 that **dot maps** are created by letting one dot equal a certain amount of some phenomenon and then locating dots where that phenomenon is most likely to occur. Raw-total data (e.g., the number of elephants or the number of acres of wheat harvested) are normally the basis for dot mapping. In contrast, **dasymetric maps** use areal symbols to depict zones of uniformity (Figure 17.1). As with choropleth maps, dasymetric maps utilize standardized data; for in-

stance, we might compute a density figure for the elephant data by dividing the number of elephants by the area within which elephants are believed to be located.

Ancillary information used to locate dots (for dot mapping) or to define zones (for dasymetric mapping) is commonly split into limiting and related attributes. *Limiting attributes* place absolute limits on where dots or zonal boundaries can be placed. For example, generally, it would not make sense to place dots representing population within water bodies. *Related attributes* are those that are correlated with the phenomenon being mapped, but that do not place absolute limits on the location of dots or zonal boundaries. For example, certain types of soil in a region might favor a particular crop, but the crop might be found on all types of soil within the region.

17.2 MANUAL VERSUS AUTOMATED PRODUCTION

When maps were produced manually, creating dot and dasymetric maps was a tedious process. Initially, the cartographer had to collect a set of paper maps that provided an appropriate set of limiting and related attributes. Next, the limiting and related attributes had to be synthesized onto a single base map. This process was generally accomplished using an enlarging-reducing machine that allowed one to trace information from maps of varying scale onto the same base. For large, detailed maps, this could be a time-consuming process. The cartographer then had to decide how the combined information would be used to place dots or zonal boundaries (e.g., that more dots would be placed in one land slope category than another, or that no dots would be placed in wetlands). Finally, actual symbols would be placed on the map; in the case of the dot map, this was particularly time consuming. The techniques of GIS and remote sensing can alleviate much of the burden of the manual process. First, through its layering and associated analysis capability, GIS enables the automated processing of limiting and related attributes. Second, remote sensing, as an integral part of GIS, can be used both to define limiting and related attributes or, alternatively, to pinpoint the location of a particular phenomenon.

17.3 CREATING A DOT MAP

In Chapter 4, we introduced a dot map of wheat harvested in Kansas (Figure 4.9D is repeated in the lower right portion of Figure 17.2). This section considers three issues relevant to creating such a dot map: (1) determining regions within which dots should be placed, (2) selecting dot size and unit value, and (3) placing dots within selected regions. Although we consider these issues in the context of the wheat harvested example, they are generic to most dot-mapping problems.

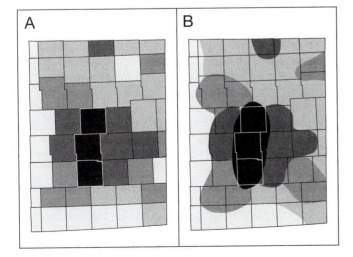

FIGURE 17.1   A hypothetical comparison of (A) choropleth and (B) dasymetric mapping.

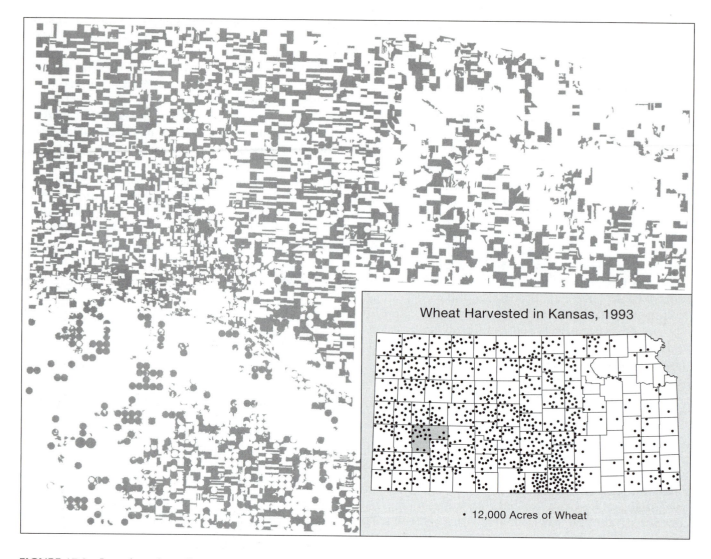

FIGURE 17.2 Locating wheat directly using remote sensing versus locating wheat based on cropland from a land use/land cover map. The upper left map depicts the distribution of wheat within Finney County based on an analysis of 30-meter resolution Landsat imagery, whereas the lower right map portrays the same distribution (the gray-shaded area) based on a 1-kilometer resolution land use/land cover map. (Courtesy of Kansas Applied Remote Sensing Program, University of Kansas.)

17.3.1 Determining Regions within Which Dots Should Be Placed

In constructing the dot map of wheat harvested, we first considered several attributes that might limit the distribution of wheat, and crops in general. Obvious *limiting* attributes included the location of water bodies (in the case of Kansas, there are a number of reservoirs), the location of urban areas (e.g., Wichita and Kansas City), and slope (tractors cannot be used on very steep slopes). Such limiting attributes might be found on paper maps and entered into a GIS as layers via *scanning* or *digitizing* (or they might already be available in digital form). The resulting layers could then be overlaid, and dots placed only in areas not constrained by the limiting attributes.

Rather than employ an overlay approach, we used a Kansas land use/land cover map developed by Jerry Whistler and his colleagues (1995) because it provided a convenient synthesis of limiting attributes. The land use/land cover map, which was created from Landsat Thematic Mapper remote-sensing data, split the landscape into 10 classes: 5 urban (residential, commercial-industrial, grassland, woodland, and water) and 5 rural (cropland, grassland, woodland, water, and other). The map defined directly two of the limiting attributes we had considered (urban areas and water bodies), and included two limiting attributes that we had not immediately thought about: woodland and grassland. Most important, however, was the cropland class, which indirectly handled the slope attribute and any other limiting and related attributes likely associated with cropland.

Although a mapmaker might be tempted to place all dots representing wheat within the cropland class, it is important to recognize that land use/land cover maps based on remote-sensing techniques are not without error. Whistler and his colleagues (1995, 773) estimated the overall accuracy of the Kansas land use/land cover map to be 85 to 90 percent or better, but individual categories within particular counties deviated considerably from this. For example, using data provided by Whistler et al., we found that only 53 percent of the cropland in Butler County was estimated to be correctly classified. To handle this error, we placed some dots representing cropland within grassland areas (the category typically confused with cropland on imagery). The number of dots so placed was a function of the estimated percentage of incorrectly classified cropland in a county: If 47 percent of the cropland was estimated to be misclassified as grassland, then 47 percent of the dots were placed in the grassland category within that county. Although these percentage figures suggest large numbers of dots were affected, they were not because there was generally little wheat harvested in counties with large error values; less than 5 percent of all dots were affected by this procedure.

The map shown in the lower right of Figure 17.2 resulted from using the dot placement procedure just discussed. We considered this map satisfactory for this book because our primary purpose was to show that a more detailed map could be developed using the dot-mapping method than the choropleth method (see Figure 4.9); there was no need to create the "perfect" map. In other situations, however, a mapmaker might wish to create a more detailed map. Let's now consider how this might be done.

One obvious limitation of the Kansas land use/land cover map is that it does not distinguish the location of individual crops. This limitation might be handled by considering ancillary information that could assist in locating wheat within the cropland areas on the land use/land cover map. One potential ancillary attribute would be precipitation, as it varies from a high of about 35 inches in eastern Kansas to only about 15 inches in western Kansas. Presuming an ideal amount of precipitation for wheat, a mathematical function could be developed that places a higher probability on locating dots near the ideal precipitation area. Ideally, such an approach also would consider other complicating attributes, such as the timing of the precipitation.

Another potential ancillary attribute would be irrigated cropland. The *1994 Kansas Farm Facts* specifies the acres of wheat harvested from irrigated land, so if the location of irrigated cropland could be determined, a percentage of dots could be placed in such areas equivalent to the percentage of wheat irrigated in that area. Remote sensing could be used for this purpose (Eckhardt et al. 1990), particularly in the case of center-pivot irrigation, which traces a distinctive circle on the landscape (Astroth et al. 1990). Still another possible ancillary attribute would be the distribution of soils within each county. One might suspect that certain soils would be more conducive to wheat production than other crops. In discussing this possibility with those knowledgeable about Kansas agriculture (a farmer and a county extension agent), we found that the selection of crops at the county level was more likely a function of farm policy and associated programs rather than soil type. For example, wheat would continue to be planted at a particular location because a farm program specified that the same crop must be planted for financial support to be retained.

Although ancillary information could assist in creating a more detailed dot map, remote sensing alone provides the capability to create the most precise map because of its ability to pinpoint the location of individual crops. For example, Stephen Egbert and his colleagues (1995) used remote sensing to differentiate wheat, corn, grain sorghum (milo), alfalfa, fallow land, shortgrass prairie, and sand-sage prairie in Finney County, Kansas, and found classification accuracies as high as 99 percent for wheat. They accomplished this by using imagery for three different time periods, as opposed to the single-date imagery employed for the Kansas land use/land cover map. The map shown in the upper left of Figure 17.2 is illustrative of the kind of detail they were able to attain. Although using remote sensing in this fashion can provide considerable detail, the greater amount of imagery incurs a greater expense, and the necessary verification (ground truth) data is also costly (Egbert, personal communication 1996). Thus, mapmakers wishing to use remote sensing must carefully consider its cost and benefits for a particular application.

17.3.2 Selecting Dot Size and Unit Value

Dot size (how large each dot is) and **unit value** (the count represented by each dot) are important parameters in determining the look of a dot map. Cartographers have argued that very small dots produce a "sparse and insignificant" distribution, and very large dots "give an impression of excessive density." Similarly, a small unit value produces a map that "gives an unwarranted impression of accuracy," whereas a large unit value results in a map that "lacks pattern or character" (Robinson et al. 1995, 498). Generally, it has been argued that dots in the densest area should just begin to coalesce (merge with one another).

J. Ross Mackay (1949) developed a graphical device known as the **nomograph** (Figure 17.3) to assist in selecting dot size and unit value. To use the nomograph, you must first calculate the number of dots per square

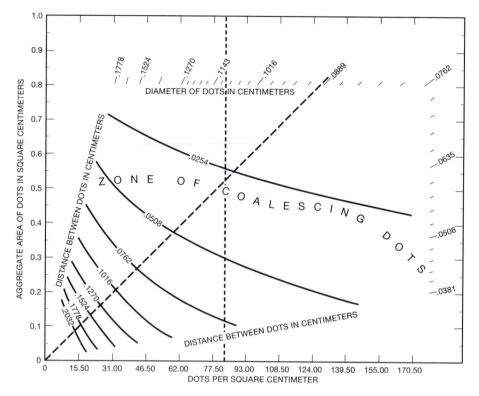

FIGURE 17.3 The nomograph—a device used for computing appropriate dot size and unit value. A vertical line is drawn representing a relatively dense area on the map, and a second line is drawn from the origin to a desired dot size. The distance between dots can be determined by where the two lines intersect. (After Mackay 1949. First published in *Surveying and Mapping* 9, p. 7. Reprinted with permission from the American Congress on Surveying and Mapping.)

centimeter for a sample area on your map. For example, imagine that we found our densest enumeration unit to be a county with 500,000 acres of wheat and an area of 6 square centimeters (at the mapped scale). If we presume a unit value of 1 dot equals 1000 acres, then the number of dots required would be 500 (500,000/1000), and the number of dots per square centimeter would be 83.3. (500/6). The dashed vertical line in Figure 17.3 represents the value of 83.3. If we now construct a line from the origin to a desired dot diameter, and note where this line intersects the vertical dashed line, we will have an indication of the approximate distance between the edges of dots. For instance, in Figure 17.3 we can see that a dot diameter of approximately 0.889 centimeters has been selected, and the distance between dots will be slightly greater than 0.0254 centimeters. Note that this distance falls in the "zone of coalescing dots." It should be noted that these calculations presume that dots are distributed in a relatively uniform hexagonal pattern.

In practice, some experimentation is generally required to select an appropriate dot size and unit value, regardless of whether the nomograph is used. For our wheat map, we selected several counties that had both a high percentage of cropland in wheat and a large value for acres of wheat harvested. Selecting counties on the basis of percentage of cropland in wheat ensured that dense areas of wheat harvested would be considered, and having a large value for acres harvested ensured that the dense area would consist of a relatively

large number of dots. We placed all necessary dots within the sampled counties for several dot sizes and unit values and then subjectively evaluated the resulting maps for coalescence. Once we found an acceptable dot size and unit value, we used a graphic design program (Freehand) to place dots.

One problem with the approach that we have described is that it fails to consider the difficulty that map readers have in correctly estimating the number and density of dots within subregions of a map. In a fashion analogous to the underestimation of proportional symbol size (see section 16.3.2), readers also underestimate the number and density of dots. Judy Olson (1977) indicated that it is possible to correct for this underestimation, but the effect on the overall pattern is small, and thus of questionable utility.

It is interesting to note that mapmakers can ignore the issue of dot size and unit value when the phenomenon is located explicitly via remote sensing (as in the Finney County example shown in Figure 17.2); in this case, the darkened pixels tell us precisely where the phenomenon is located. One disadvantage of this approach, however, is that it might be difficult to make a visual estimate of the magnitude of production, as individual pixels merge together. Of course, because the map is automatically generated, the computer can be used to make a precise estimate of the acres of wheat grown in any area (albeit with the estimate a function of the ability to correctly interpret the phenomenon via remotely sensed imagery).

17.3.3 Placing Dots within Regions

When dots were located manually, cartographers generally used one of three approaches for placing dots: uniform, geographically weighted, and geographically based. For each of these approaches, it was presumed that you had calculated the number of dots associated with particular enumeration units by dividing the raw total for the enumeration unit by the unit value. In the *uniform* approach (Figure 17.4A), an attempt was made to place dots associated with an enumeration unit in a relatively uniform fashion throughout the enumeration unit. Mackay (1949, 5) described this process as "placing the first dot near the center [of the enumeration unit] and then each successive dot in the largest remaining space." Perfect uniformity was not desirable because patterns in nature are not perfectly uniform. Pure randomness also was not desired because this could lead to unrealistic clustering and large gaps between some dots. Thus, the result might better be termed *uniform with a random component*. The end result simulated a choropleth map, as the boundaries of enumeration units were sometimes apparent (Figure 17.4A).

In the *geographically weighted* approach, dots were shifted (*weighted*) toward neighboring enumeration units of higher value, as shown in Figure 17.4B. This approach took advantage of the notion of spatial autocorrelation—that high values tend to be located near other high values. This was an obvious improvement over the uniform approach, as it tended to diminish the effect of county boundaries. It essentially simulated an isopleth map, based on the assumption of smooth continuous changes between enumeration units. In the *geographically based* approach (Figure 17.4C), dots were located based on the ancillary information that we have described previously, albeit using the uniform approach for areas within which dots were supposed to be placed.* This, of course, was the preferred approach, as it took advantage of geographic knowledge and produced the most detailed dot map.

Unfortunately, present-day software for dot mapping generally does not include satisfactory approaches for dot placement. Software vendors have developed dot-mapping techniques (commonly referred to as dot-density routines), but these techniques generally place dots in a purely random fashion, which, as we have said, can lead to unrealistic clusters and gaps in the dot pattern. These routines also tend to presume that the user wishes to place dots within enumeration units without considering ancillary information. Furthermore, there is the presumption that dots will be placed within polygons (a vector-based structure), as opposed to placing

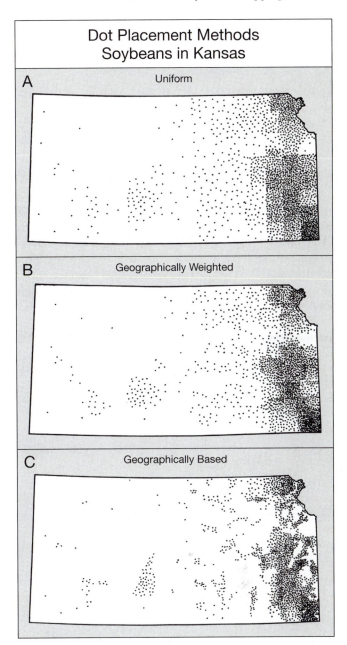

FIGURE 17.4 Approaches used by cartographers to place dots on a dot map: (A) uniform, (B) geographically weighted, and (C) geographically based.

them in a cell-based (raster) structure.† To illustrate the difficulty of working with a cell-based structure, consider the portion of the Kansas land use/land cover map shown in the left portion of Color Plate 17.1. Clearly, it is not easy to define polygons, as remote sensing techniques used to create the land use/land cover map have produced very detailed patterns. At the end of the following section,

---

* In its most sophisticated form, the geographically based approach could also include a weighting component.

† For a description of vector and raster techniques, consult a GIS textbook such as Lo and Yeung (2002).

we suggest an automated approach that might handle this problem.

An Approach for Automatic Dot Placement

In this section, we consider an approach for automated dot placement that would circumvent the problems of clusters and gaps resulting from purely random placement. The approach is based on a **dot-density shading** technique that Stephen Lavin (1986) developed for mapping continuous phenomena (e.g., rainfall). Lavin's approach requires four basic steps:

Step 1: Read an equally spaced gridded network of data values into the computer (Figure 17.5A). Such a grid is normally the end result of the inverse-distance (or *gridding*) approach for automated contouring (see section 14.3.2). Conceptually, each grid intersection is presumed to be surrounded by a square cell within which dots will be placed (Figure 17.5B).

Step 2: Compute the proportion of dot coverage, P_i, for each of the square cells. The formulas are

$$z_i = \frac{v_i - v_s}{v_L - v_s}$$

$$P_i = z_i(P_{max} - P_{min}) + P_{min}$$

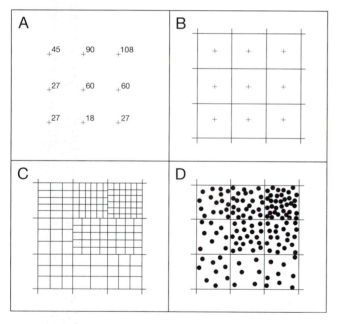

FIGURE 17.5 A depiction of Lavin's dot-density shading approach: (A) a matrix of grid points serves as input; (B) square cells are presumed to surround each grid point; (C) square cells are split into equally sized subcells, with the number of subcells a function of the number of dots to be placed; (D) a dot is placed within each subcell. (After Lavin 1986. First published in *The American Cartographer* 13(2), p. 145. Reprinted with permission of Stephen Lavin and the American Congress on Surveying and Mapping.)

where v_i = data value for a particular grid cell

v_s and v_L = smallest and largest data values in the grid

z_i = proportion of the data range associated with data value v_i

P_{min} and P_{max} = minimum and maximum desired proportions of dot coverage

P_i = proportion of dot coverage for data value v_i

Assuming that the minimum and maximum desired proportions of dot coverage are 0.08 and 0.50,* respectively, the following would be computed for the lower left grid point value 27 in Figure 17.5:

$$z_i = \frac{27 - 18}{108 - 18} = 0.10$$

$$P_i = 0.10(0.50 - 0.08) + 0.08 = 0.1220$$

Step 3: Compute the number of dots, N_d, to be placed within a square cell. The formula is

$$N_d = \frac{P_i \times A_c}{A_d}$$

where A_c is the area of the cell, and A_d is the area of a dot. For the lower left grid point, we have

$$N_d = \frac{0.1220 \times 0.1296}{0.0018} = 8.784 \approx 9$$

Step 4: Partition the square cells surrounding each grid point into n equally sized subcells, where n is as close as possible to N_d. In the case of the lower left grid point, the number of cells (9) matches N_d exactly (Figure 17.5C). The subcells serve as plotting boundaries for each dot: A dot is plotted randomly within each cell and cannot extend outside the cell (Figure 17.5D). A map resulting from this process is shown in Figure 17.6, along with a traditional isoline map.

Applying Lavin's dot-density shading to a dot map would involve the following steps. First, standardized values would have to be computed for each subregion within which dots are to be placed (subregions would be determined by limiting and related attributes). Standardization would be accomplished by dividing a raw-total value (say, acres of wheat) associated with a subregion by the area of that subregion. Second, a grid would be laid over the map. All grid points falling within a particular subregion would be assigned the standardized value for

---

* We chose the minimum and maximum values so that we could reproduce the dots in Lavin's illustrations. Normally, you would choose a maximum value that would cause dots in the densest region to coalesce. The minimum value would then be chosen so that the density of dots properly reflected the data relations on the map.

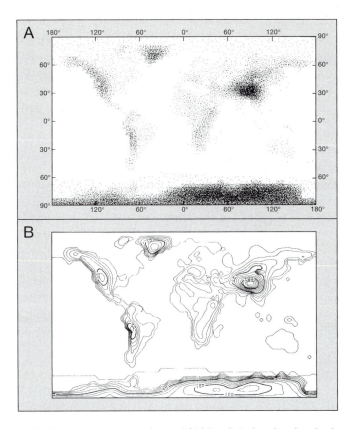

FIGURE 17.6   A comparison of (A) Lavin's dot-density shading approach and (B) the traditional isoline method. Maps are based on data for land elevations, and contour lines are in hundreds of feet. (After Lavin 1986. First published in *The American Cartographer* 13(2), p. 142. Reprinted with permission of Stephen Lavin and the American Congress on Surveying and Mapping.)

that subregion. Finally, steps 2 to 4 just described would be executed.

One problem with applying Lavin's approach to dot mapping is that subregions within which dots are to be placed are often in raster (pixel-based) form, as we have indicated for the Kansas land use/land cover map (Color Plate 17.1). In this case, note that individual subregions are frequently smaller than the dots that would be placed within them. In this case it might make more sense to develop a plotting algorithm based on pixels (rather than dots), such as the following:

Determine the proportion of cropland within an enumeration unit to be symbolized as wheat. (Divide acres of wheat harvested by acres of cropland.)
For each pixel within the enumeration unit:
If the pixel is cropland,
    generate a random number between 0 and 1.
    If the random number is ≤ the proportion of cropland in wheat, set the pixel to wheat.

The resulting map would appear more like the detailed map of Finney County shown in the upper left portion of Figure 17.2 than a traditional dot map. A limitation, however, is that it would not exhibit some of the regularities of the Finney County map, such as the circular features representing center-pivot irrigation.

17.4 EICHER AND BREWER'S COMPARISON OF DASYMETRIC METHODS

Cory Eicher and Cynthia Brewer (2001) were among the first to implement and compare several methods for dasymetric mapping in an automated context. In their study Eicher and Brewer created dasymetric maps of several population-related attributes (e.g., population density and density of homes of a specified value) for a 159-county region in parts of four states—Pennsylvania, West Virginia, Maryland, and Virginia—and the District of Columbia. For ancillary information, they used a land use/land cover approach analogous to the one we described for our dot map: In their case, they utilized a data set from the U.S. Geological Survey that split the 159-county region into 580 zones that fell into either urban, agricultural, woodland, forested, or water categories (the boundaries of the zones are shown in Figure 17.7). For the discussion to follow, we assume that population is the attribute to be mapped using the dasymetric method.

To allot population to dasymetric zones, Eicher and Brewer compared three methods: binary, three-class, and limiting variable. In the *binary method*, the land use/land cover categories were split into two groups: habitable and uninhabitable. The habitable group included urban, agricultural, and woodland categories, and the uninhabitable group consisted of the water and forested categories. One hundred percent of the population was assigned to the habitable group and 0 percent to the uninhabitable group. Thus, if a county consisted of 1,000 people, all of these people were assigned to the habitable group and a population density figure for the area associated with that group was calculated by dividing 1,000 by the area of the habitable categories in the county. Obviously, this approach produced some error in the resulting map, as the urban category was treated identically to the agricultural and woodland categories; furthermore, the binary approach was unrealistic because it presumed no population for the forested category. An advantage of the binary approach, however, was its simplicity.

In the *three-class method*, Eicher and Brewer first combined the agricultural and woodland categories into one category termed agricultural/woodland. They then presumed that population would be assigned to only three categories: urban, agricultural/woodland, and forested—hence the name *three-class* method. In addition to using three categories, Eicher and Brewer assigned different percentages

FIGURE 17.7 The 580 zones utilized by Eicher and Brewer in their study of dasymetric mapping. (After Eicher and Brewer 2001. First published in *Cartography and Geographic Information Science* 28(2), p. 127. Reprinted with permission from the American Congress on Surveying and Mapping. Courtesy of Cory Eicher and Cynthia Brewer.)

of population to each of the three categories. For counties with all three categories present, the percentages were as follows: 70 for urban, 20 for agricultural/woodland, and 10 for forested. Although an improvement over the binary method, the three-class method was limited in that no

consideration was made for the area falling in each category. Thus, a county with very little area in the urban category would still have 70 percent of its population allotted to that category, which could produce an unrealistically high population density for the associated mapped area.

In the *limiting variable method*, Eicher and Brewer used an approach first developed by John K. Wright (1936). In this approach, the population was first assigned so that the density of the three habitable categories just mentioned was identical. Thus, if a county consisted of 1,000 people and was 2 square kilometers in area, and only habitable categories were found in the county, a density of 1,000/2 or 500 was assigned for the entire county. Next, thresholds were set for the maximum density allowed in each habitable category: In the case of population, these were 15 and 50 people per square kilometer for the forested and agricultural/woodland categories, respectively. If the overall density figure exceeded these thresholds, then the densities would be set to the maximum threshold, and the remaining population would be assigned to other categories on the map. In our example, because the value 500 exceeds both thresholds, the excess people all would have to be assigned to the urban category.

To illustrate some of the computations involved in the limiting variable method, we'll consider a simple hypothetical example in which only two categories are present (urban and forested), the population to be assigned is 1,000, and the area of the county is 1 square kilometer (Figure 17.8A). In this case, the population density initially assigned is 1,000/1 or 1,000 people per square kilometer. Again, assuming a forested threshold of 15 people per square kilometer, the density for the urban category can be calculated using the following formula (Wright 1936):

$$D_n = \frac{D - D_m a_m}{1 - a_m}$$

where a region has been divided into two areas *n* and *m*, *D* is the overall density of the region (county in our case),

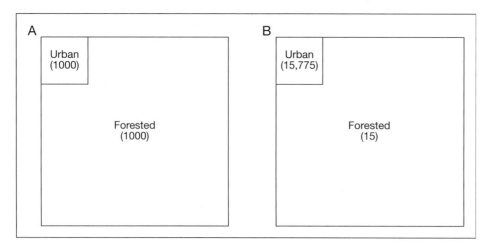

FIGURE 17.8 Computing population density figures for regions when a maximum threshold is assumed for a subregion: (A) the same density (1,000 people per square mile) is used for both regions; (B) a maximum value of 15 people per square mile is presumed for the forested region. See text for associated formula.

D_m is the estimated density of subregion m, a_m is the fractional area of region m (relative to the entire region), and D_n is the density of region n. In Figure 17.8B, the forested and urban regions would be equivalent to regions m and n, and we would compute a density of 15,775 people per square kilometer for the urban region as follows:

$$D_n = \frac{1{,}000 - (15)(.9375)}{1 - .9375} = 15{,}775$$

Eicher and Brewer evaluated the effectiveness of the three approaches for allotting population to dasymetric zones by comparing the population values computed for zones with those for U.S. Bureau of the Census block groups composing the zones. This constituted a reasonably precise check as the 580 zones were composed of more than 13,000 block groups. Using this approach, Eicher and Brewer found that the limiting variable method produced significantly better results (from a statistical standpoint) than the other two methods. The results were also visualized using maps such as those shown in Color Plate 17.2. In Color Plate 17.2A we see a dasymetric map of population density resulting from the binary method. Obviously, such a map portrays considerably more detail than a simple choropleth map of the data (Figure 17.9). In Color Plate 17.2B and 17.2C, we see error maps associated with the population density map. These error maps were computed by comparing the estimated values based on dasymetric mapping with the actual values for zones based on U.S. Bureau of the Census block groups composing each zone. The percent error map shows error relative to the total population of a zone, and the count error map simply shows the difference between the dasymetric estimates and the true values. Note the linear patterns running northeast to southwest (in the ridge-and-valley physiographic province) on the percent error map. Eicher and Brewer indicated that such patterns were a function of the binary method assigning zero population to forested areas (which is unrealistic). The limiting variable method did not exhibit such patterns.

Those of you with knowledge of GIS software might wish to further refine Eicher and Brewer's approach. For instance, they note that further work is needed to determine appropriate (1) threshold values for land use/land cover categories, (2) cell sizes if using a raster-based approach (as opposed to a vector-based one), and (3) approaches for analyzing the error on the resulting dasymetric maps (pp. 135–136).

17.5 LANGFORD AND UNWIN'S APPROACH FOR MAPPING POPULATION DENSITY

Traditionally, population density has been mapped using choropleth symbology because population data are readily available for enumeration units, and computer software frequently has the choropleth map as a major option (if not the only option). Mitchel Langford and David Unwin (1994, 22) noted three major problems with using choropleth maps for population data: (1) larger enumeration units tend to have lower population densities (and conversely, smaller enumeration units tend to have higher population densities); (2) enumeration units hide the data variation that exists within them; and (3) enumeration unit boundaries are arbitrary, and thus unlikely to be associated with actual discontinuities in population density. In response to these problems, Langford and Unwin developed a novel mapping method based on remote sensing, dasymetric mapping, and generalization techniques. First, they used Landsat Thematic Mapper imagery to split pixels within their study region (northern Leicestershire, United Kingdom) into two groups: "Residential Housing" and "All Other Categories." A comparison of the resulting map (Figure 17.10B) with the traditional choropleth map (Figure 17.10A) makes clear the variation in population density that can exist within enumeration units, and the fact that enumeration unit boundaries generally do not coincide with discontinuities in population density.

Next, Langford and Unwin created a dasymetric map by combining their remote sensing–based map with the population data associated with enumeration units. Standardized data for the dasymetric map were computed by dividing the population for each enumeration unit by the area of residential housing falling in that unit. They

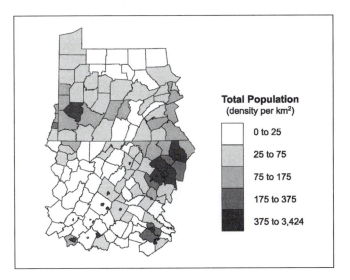

Total Population
(density per km²)

- 0 to 25
- 25 to 75
- 75 to 175
- 175 to 375
- 375 to 3,424

FIGURE 17.9 A choropleth map of population density for the 159-county region utilized by Eicher and Brewer. (After Eicher and Brewer 2001. First published in *Cartography and Geographic Information Science* 28(2), p. 128. Reprinted with permission from the American Congress on Surveying and Mapping. Courtesy of Cory Eicher and Cynthia Brewer.)

noted that the resulting density values varied considerably from those depicted on the choropleth map: The dasymetric map ranged from 2,820 to 14,900 people per square kilometer, and the choropleth map ranged from 50 to 10,000 people per square kilometer. The dasymetric map initially was symbolized by shading the residen-

tial housing pixels shown in Figure 17.10B with an intensity proportional to the standardized data associated with the enumeration unit within which the pixel fell. Unfortunately, Langford and Unwin found the resulting dasymetric map a "poor cartographic product ... because the reader [could] see individual pixels and

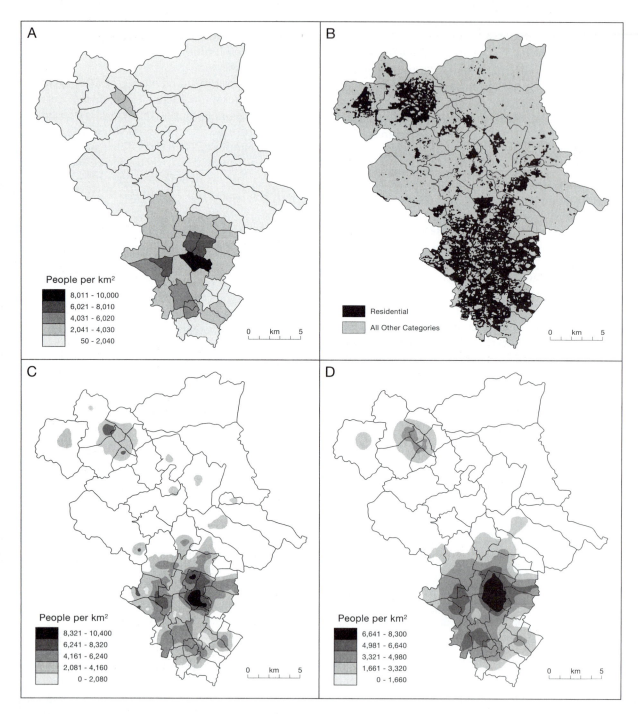

FIGURE 17.10   Langford and Unwin's approach for mapping population density: (A) a traditional choropleth map; (B) a binary map of "residential housing" and "all other categories" derived from remote sensing (a dasymetric map was created from this map by dividing the population for each enumeration unit by the area of residential housing falling in that unit); (C and D) generalized versions of the dasymetric map based on circular "windows" of 0.5 and 1 km search radius centered on each pixel. (From Langford and Unwin 1994; courtesy of The British Cartographic Society.)

too much fine spatial detail" (p. 24). To improve the visual quality of the dasymetric map, Langford and Unwin next applied a generalization operator. This was accomplished in three steps. First, they computed the population associated with a pixel by dividing the population of the enumeration unit by the number of residential housing pixels falling in that unit. Next they floated a circular "window" (having a search radius of either 0.5 or 1.0 kilometer) over each pixel and calculated the population falling within the window. Finally, a population density for each pixel was computed by dividing the population for the associated window by its area. When the resulting pixel values were symbolized as gray tones (Figure 17.10C), the map depicted relatively smooth changes in population density (as opposed to the fine irregularities of the dasymetric map).

Although Langford and Unwin's resulting map clearly provides a much more detailed depiction than a conventional choropleth map, their approach is not without problems. One is that remotely sensed categories have some error, as we indicated for the Kansas land use/land cover map. In more recent research, Fisher and Langford (1996, 308) suggested that this is not a serious problem, as "errors at the pixel level … can be large … without impacting the accuracy of estimates of regional amounts." Another problem is that remote sensing might not provide information concerning the type of housing unit (e.g., a single-family dwelling vs. high-rise apartment complex). Langford and Unwin argued that this is not a serious problem if "housing type is relatively uniform within the study region" (p. 24). A third problem is that the resulting map is a function of the search radius used (compare Figures 17.10C and 17.10D). This "problem" can also be viewed as an advantage, however, as it provides alternative views of the data (a form of data exploration).

SUMMARY

In this chapter, we have examined several principles involved in dot and dasymetric maps, which precisely map data commonly collected for enumeration units. **Dot maps** portray *raw-total data* (e.g., the number of people of Swedish descent living in counties in Wisconsin), and are constructed by letting each dot equal a specified value (e.g., one dot equals 100 people of Swedish descent). In contrast, **dasymetric maps** portray *standardized data* (e.g., the proportion of the population that is Swedish), and use area symbols to represent zones of uniformity. In both dot and dasymetric mapping, **ancillary information** is used to determine appropriate locations for the symbology. For example, you might have information available on where Swedes tended to settle in Wisconsin, and thus place dots on this basis. Of course, if you have data available for very fine enumeration units (e.g., census

tracts) and you are constructing a small-scale map, ancillary information might be unnecessary.

Remote sensing can be particularly useful in providing suitable ancillary information. For example, we saw that a land use/land cover map for Kansas created via remote sensing was useful for locating dots representing wheat. We also saw that remote sensing could be used to determine precisely where wheat was grown (in the case of Finney County, Kansas, an accuracy of 99 percent was obtained). Finally, in the case of Langford and Unwin's mapping of population density, we saw how remote sensing could be used to determine locations of "residential housing," which served as the basis for dasymetric mapping.

An important issue in dot mapping is selecting the **dot size** and **unit value**. Normally, dot size and unit value are selected so that dots in the densest area just begin to coalesce (merge with one another). The **nomograph** is a useful device for determining an appropriate dot size and unit value. The number of dots to place within a region is computed by dividing a raw total figure (e.g., acres of wheat harvested) associated with a region by the unit value. Although it is possible to adjust such numbers to account for the *perceived* number and density of dots, this is normally not done because the effect on the overall dot pattern is small. Another issue in dot mapping is the actual placement of dots on a map. In the manual realm, cartographers used three approaches for placing dots: *uniform, geographically weighted,* and *geographically based.* The latter is the preferred approach as it utilizes the notion of ancillary information. In computer-based mapping, so-called dot-density routines have been developed for placing dots, but these methods are undesirable because they (1) produce unrealistic clusters and gaps in the dot pattern and (2) presume that mapmakers have already formed polygons within which dots are to be placed. A potential solution to the first problem is to modify Lavin's **dot-density shading** developed for smooth continuous phenomena and apply it to the abrupt discrete phenomena characteristic of dot maps. The second problem might be handled by developing a raster-based plotting algorithm: The net result is that small square pixels would be plotted, as opposed to traditional round dots.

We considered several manual dasymetric methods that Cory Eicher and Cindy Brewer have implemented in an automated context. In the context of mapping population-related attributes, the most effective of these was a *limiting variable* method in which a uniform density was initially assumed, and then adjusted to account for maximum threshold values for various land use/land cover categories.

This chapter focused on how ancillary information can be used to create detailed maps of data collected for enumeration units. Bear in mind that a more detailed map can also be constructed through a mathematical manipulation of the data. Tobler's pycnophylactic method (described in section 14.6) is an example of one such approach.

FURTHER READING

Dent, B. D. (1999) *Cartography: Thematic Map Design*. 5th ed. Boston: McGraw-Hill.

> Chapter 8 covers basic principles of dot mapping. For more on basic principles, see pages 497–502 of Robinson et al. (1995).

Egbert, S. L., Price, K. P., Nellis, M. D., and Lee, R.-Y. (1995) "Developing a land cover modeling protocol for the high plains using multi-seasonal thematic mapper imagery." ACSM/ASPRS Annual Convention & Exposition, *Technical Papers, Vol. 3*, pp. 836–845.

> Describes how remote sensing can be used to differentiate a variety of crops.

Fisher, P. F., and Langford, M. (1996) "Modeling sensitivity to accuracy in classified imagery: A study of areal interpolation by dasymetric mapping." *Professional Geographer* 48, no. 3:299–309.

> Examines to what extent the accuracy of a dasymetric map is a function of the error in remotely sensed imagery.

Gerth, J. D. (1993) *Towards Improved Spatial Analysis with Areal Units: The Use of GIS to Facilitate the Creation of Dasymetric Maps*. Research paper for an M.A. degree, Ohio State University, Columbus, OH.

> Describes how ARC/INFO software can be used to create a dasymetric map. To acquire the paper, contact the Department of Geography at Ohio State University (e-mail: *geography@osu.edu*).

Langford, M., and Unwin, D. J. (1994) "Generating and mapping population density surfaces within a geographical information system." *Cartographic Journal* 31, no. 1:21–26.

> Provides detail on Langford and Unwin's dasymetric method.

Lavin, S. (1986) "Mapping continuous geographical distributions using dot-density shading." *American Cartographer* 13, no. 2:140–150.

> Describes a method for using dot symbols to represent smooth continuous phenomena.

Martin, D. (1996) "An assessment of surface and zonal models of population." *International Journal of Geographical Information Systems* 10, no. 8:973–989.

> Describes mathematical approaches for reallocating population within enumeration units. The method presumes that population density peaks at the centroid of population within an enumeration unit. A choropleth symbolization is used for the resulting cell-based data.

McCleary, G. F. J. (1969) *The Dasymetric Method in Thematic Cartography*. Unpublished Ph.D. dissertation, University of Wisconsin, Madison, WI.

> Covers the historical development of the dasymetric map and describes several forms of dasymetric mapping.

Monmonier, M. S., and Schnell, G. A. (1984) "Land use and land cover data and the mapping of population density." *International Yearbook of Cartography* 24:115–121.

> Indicates how the denominator in a ratio for choropleth mapping might be adjusted to produce a more meaningful map. For example, rather than dividing population by the total area of an enumeration unit, it could be divided by the area within which people were likely to live.

Olson, J. M. (1977) "Rescaling dot maps for pattern enhancement." *International Yearbook of Cartography* 17:125–136.

> Describes how the pattern on dot maps can be enhanced by increasing the contrast between areas of similar dot density.

Ryden, K. (1987) "Environmental Systems Research Institute mapping." *American Cartographer* 14, no. 3:261–263.

> Briefly describes how ARC/INFO was used to create a population map of various ethnic groups in California.

Wright, J. K. (1936) "A method of mapping densities of population: With Cape Cod as an example." *Geographical Review* 26, no. 1:103–110.

> A classic article on dasymetric mapping. Introduces a formula for "density of parts" that enables densities to be calculated for portions of an enumeration unit based on ancillary information.

18

Bivariate and Multivariate Mapping

OVERVIEW

In the preceding chapters, we have focused on univariate mapping—the display of individual attributes (or variables). Frequently, however, mapmakers need to display multiple attributes. For example, a climatologist might wish to simultaneously view temperature, precipitation, barometric pressure, and cloud cover for a geographic region. The cartographic display of multiple phenomena is known as **multivariate mapping**. If only two attributes are to be displayed, the process is termed **bivariate mapping**. Bivariate mapping is covered in section 18.1, and multivariate mapping is covered in section 18.2.

A fundamental issue in both bivariate and multivariate mapping is whether individual maps should be shown for each attribute (maps are compared) or whether all attributes should be displayed on the same map (maps are combined). Thus, the bivariate and multivariate sections of this chapter are divided into two subsections: one for map comparison and one for map combination. The map comparison section for bivariate mapping focuses on selecting an appropriate method of classification when comparing choropleth maps. Although choropleth maps have their limitations, they are commonly used in bivariate mapping, just as in univariate mapping.

The map combination section for bivariate mapping describes how choropleth maps can be overlain to create a **bivariate choropleth map**. Other symbols considered in this section include the **rectangular point symbol**, in which the width and height of a rectangle are made proportional to values of two attributes being mapped, and the **bivariate ray-glyph**, in which rays (straight-line segments) pointing to the left and right of a small central circle represent two attributes.

The multivariate mapping section begins by considering **small multiples**—the simultaneous comparison of three or more maps. The difficulty of synthesizing information depicted via small multiples has led to numerous approaches for combining attributes onto one map. Techniques discussed include the **trivariate choropleth map** (three choropleth maps are overlaid), the **multivariate dot map** (different-colored dots are used to represent multiple phenomena), and **multivariate point symbols** (a good example is the **Chernoff face**, in which various facial features are used to represent multiple attributes).

Although it is possible to combine a large number of attributes on a single map, it is often difficult to visually interpret the resulting symbols. A solution to this problem is to explore multivariate data in an interactive graphics environment. SLCViewer, a software package developed at the Deasy Geographics Laboratory at Penn State University, provides this sort of capability by allowing users to view data as small multiples or to combine up to three attributes using proportional point symbols, **weighted isolines** (isolines of varying width), and area shading. Software described elsewhere in this text that also provides exploratory capability for multivariate phenomena includes Project Argus, CommonGIS, HealthVisPCP, and VIS5D (see Chapter 21).

It should be noted that this chapter considers only methods for symbolizing multivariate data. An alternative is to manipulate the data statistically prior to symbolizing it. For example, in Chapter 5 we saw how **cluster analysis** could be used to group observations (say counties) based on their scores on a set of attributes. For a brief introduction to the use of cluster analysis and other statistical approaches for manipulating multivariate data in cartography, see Chang (1982).

18.1 BIVARIATE MAPPING

18.1.1 Comparing Maps

In this section, we consider approaches for bivariate mapping in which individual maps are shown for each attribute (maps are compared). We focus on choropleth maps because they have been a common choice for map comparison, just as in univariate mapping.

Choropleth Maps

As with a single choropleth map, an important consideration in comparing choropleth maps is deciding whether the data should be classed, and, if so, which method of classification should be used. To begin, we assume that we wish to class the data, and thus focus on the method of classification. Initially, we also assume that we wish to compare two attributes for a single point in time (say, median income and percent of the adult population with a college education, both collected for 2000). Later, we consider comparing maps for the same attribute collected over two time periods.

In selecting a method of classification, it is critical to consider the distribution of each attribute along the number line. If the attributes have differing distributions (e.g., if one is skewed and the other is normal), certain classification methods can lead to an inappropriate visual impression of correlation between the attributes. To illustrate, consider the hypothetical distributions shown in Figure 18.1A. Attribute 1 is clearly positively skewed, whereas attributes 2 and 3 appear to have normal distributions.[*] In Figure 18.1B, values of these attributes have been assigned spatially so that extremely high correlations result in each case (the correlation coefficients, r, appear in Figure 18.1C).[†]

Recalling from Chapter 5 that the optimal method of classification is often recommended because it minimizes classification error, it seems natural to ask whether the optimal method might also be used for map comparison. If the optimal method is applied to all three attributes shown in Figure 18.1B, we obtain the maps shown in Figure 18.1D. (Table 18.1 specifies the limits of each class.) Although these maps suggest positive associations between each pair of attributes, they do not support the high correlation coefficients found in Figure 18.1C (remember that 1.0 is the maximum possible value for r). The lack of a strong visual association between the skewed and normal distributions might not be surprising

---

[*] These visual assessments were confirmed by a Shapiro–Wilk test, an inferential test for normality (Stevens 1996, 245).

[†] Kendall's rank correlations (Burt and Barber 1996, 396–398), which are arguably more appropriate for skewed data, were 0.98 between attributes 1 and 2 and 1 and 3 and 1.00 between attributes 2 and 3.

TABLE 18.1 Class limits for each of the classification methods used in Figure 18.1

Optimal

	Attribute		
Class	1	2	3
1	0–5	26–37	126–134
2	13–16	40–53	137–142
3	23–25	56–68	144–149

Mean–Standard Deviation

	Attribute		
Class	1	2	3
1	None	26–32	126–129
2	0–16	37–56	131–144
3	23–25	63–68	147–149

Nested Means

	Attribute		
Class	1	2	3
1	0–2	26–37	126–132
2	3–5	40–45	133–137
3	13–16	47–53	138–142
4	23–25	56–68	147–149

Quantiles

	Attribute		
Class	1	2	3
1	0–2	26–41	126–133
2	3–5	42–47	134–139
3	13–25	48–68	140–149

(compare the optimal maps for attributes 1 and 2), but note that there is also a lack of visual association between maps of the normal distributions (attributes 2 and 3). The optimal method fails to reflect the high correlations between the attributes because it focuses on the precise distribution of the individual attributes along the number line.

Classification methods that are more appropriate for map comparison include mean–standard deviation, nested means, quantiles, and equal areas. As described in section 5.1.3, the mean–standard deviation method specifies class limits by repeatedly adding or subtracting the standard deviation from the mean. The results of applying the mean–standard deviation approach to the hypothetical distributions are shown in Figure 18.1D. The visual appearance of these maps supports the high correlation between the normal distributions, but not the high correlations between the skewed and normal

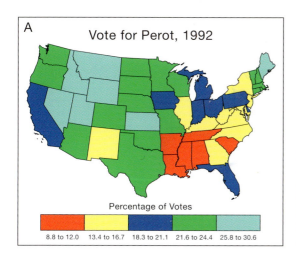

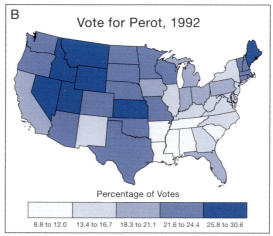

COLOR PLATE 1.1   The choropleth map: An example of a thematic map. Map A uses illogically ordered hues, whereas map B uses logically ordered shades of a single hue. Although map A might allow the reader to discriminate easily between individual states, it does not permit the reader to perceive the overall spatial pattern as readily as map B. (Data source: Famighetti 1993, p. 583.)

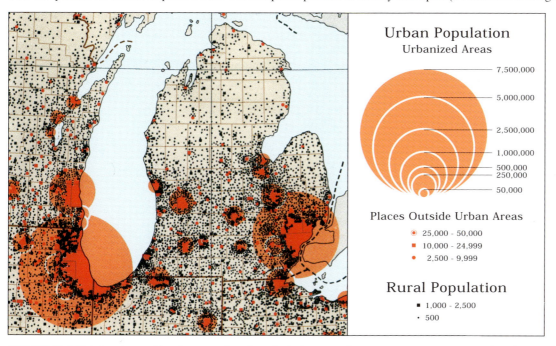

COLOR PLATE 1.2   A combined proportional symbol–dot map that attempts to represent what the population might look like in the "real world." (After U.S. Bureau of the Census 1970.)

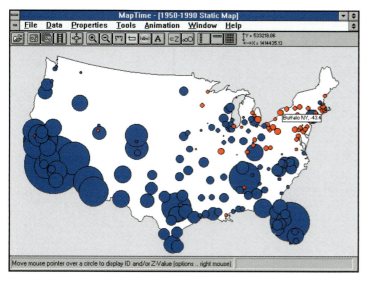

COLOR PLATE 1.3   A change map created using MapTime. Red and blue circles represent population decreases and increases, respectively, between 1950 and 1990. Note the distinctive region of population losses in the Northeast; this pattern was not revealed in an animation of the data over this time period. (Courtesy of Stephen C. Yoder.)

Plate 1

COLOR PLATE 1.4 A frame from an animated fly-through of downtown Lawrence, Kansas. (Courtesy of Jerome E. Dobson and his Geog 980 seminar students.)

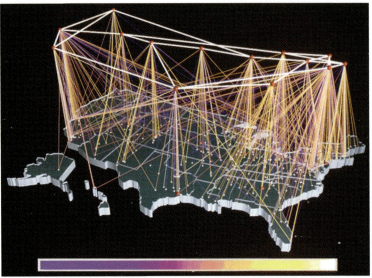

COLOR PLATE 1.5 An example of mapping Internet space—a visualization of the backbone of NSFNET, an early version of the Internet. Purple and white vertical lines represent low and high Internet traffic, respectively. (Courtesy of the National Center for Supercomputing Applications (NCSA) and the Board of Trustees of the University of Illinois.)

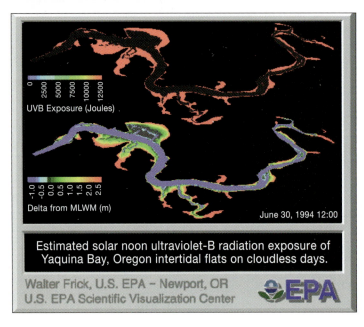

COLOR PLATE 1.6 An example of scientific visualization. Here the relationship between depth of water and ultraviolet B radiation is examined. The bottom image represents the deviation of a daily low tide (June 30, 1994) from the yearly average, and the upper image depicts the predicted ultraviolet B exposure at noon of the same day. Areas well above the average daily low tide (red, orange, and yellow) are associated with high ultraviolet B radiation (red). This work was done at the EPA's Scientific Visualization Center (*http://www.epa.gov/vislab/*).

Plate 2

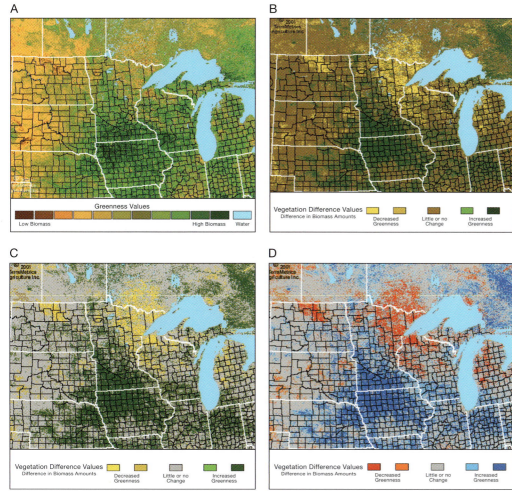

COLOR PLATE 1.7 Potential images to be used with the GreenReport: (A) a basic greenness map used to represent current vegetation conditions, (B) a change map in which the "Little or no Change" category might be confused with some of the categories on the basic greenness map, (C) a change map in which a gray is used for the "Little or no Change" category to avoid confusion, (D) a change map in which gray is used for the "Little or no Change" category, and colors not used on the greenness map are used to depict changes in greenness. (Courtesy of Kansas Applied Remote Sensing Program, University of Kansas.)

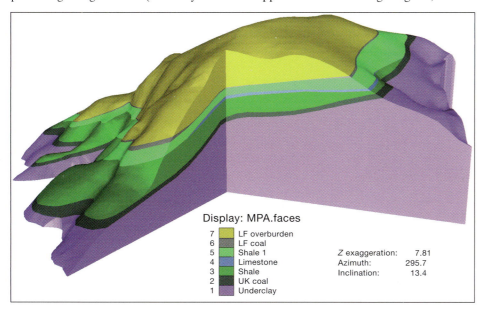

COLOR PLATE 4.1 3-D model of an open-pit coal mining site. (Data Source: Pennsylvania Department of Environment Protection; courtesy of Dynamic Graphics, Inc., Alameda, CA.)

Plate 3

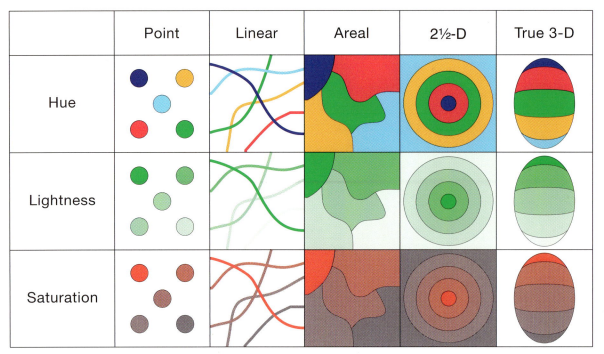

COLOR PLATE 4.2 Visual variables for colored maps. For visual variables for black-and-white maps, see Figure 4.3.

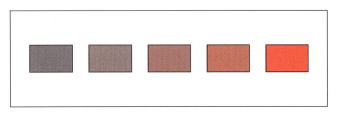

COLOR PLATE 4.3 An illustration of saturation, holding hue and value constant. The area symbols shown for saturation in Color Plate 4.2 are arranged from a desaturated red (gray) to a fully saturated red.

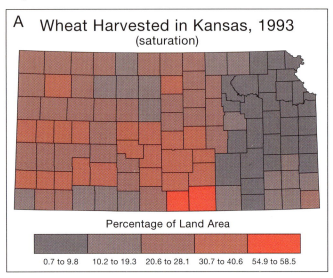

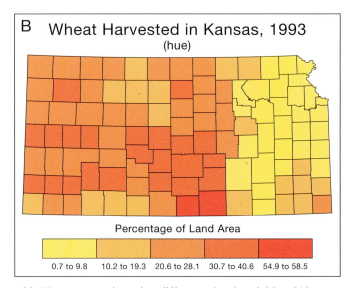

COLOR PLATE 4.4 Representing the percentage of wheat harvested in Kansas counties using different visual variables: (A) saturation, (B) hue. For black-and-white visual variables, see Figure 4.10.

Plate 4

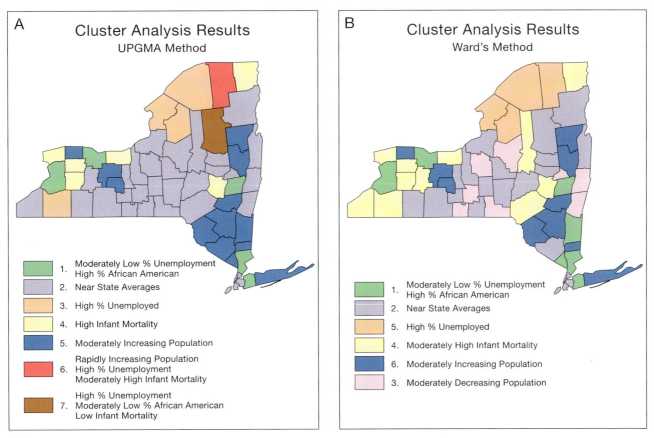

COLOR PLATE 5.1 Results of clustering the New York State data shown in Table 5.8 using (A) the UPGMA method, and (B) Ward's method. Numbers within the legend represent cluster numbers specified by the SPSS software used to generate the clusters.

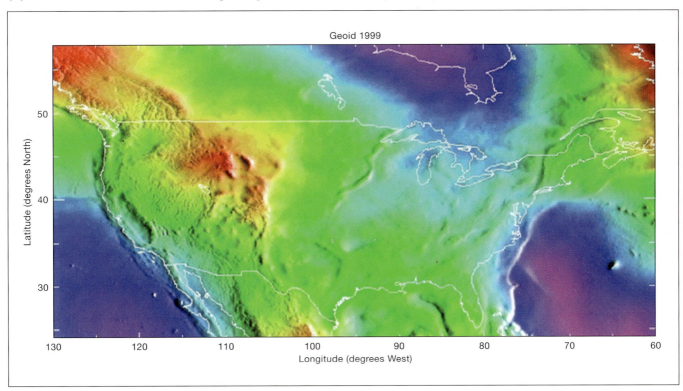

COLOR PLATE 7.1 GEOID99 is a model of the geoid in the United States for the conterminous United States. GEOID99 heights range from a low of –50.97 meters (167.2 feet) represented by magenta in the Atlantic Ocean to a high of 3.23 meters (10.6 feet) shown as red in portions of the Rocky Mountains and the Labrador Strait. (From the National Geodetic Survey; *http://www.ngs.noaa.gov/ GEOID/GEOID99/.*)

Plate 5

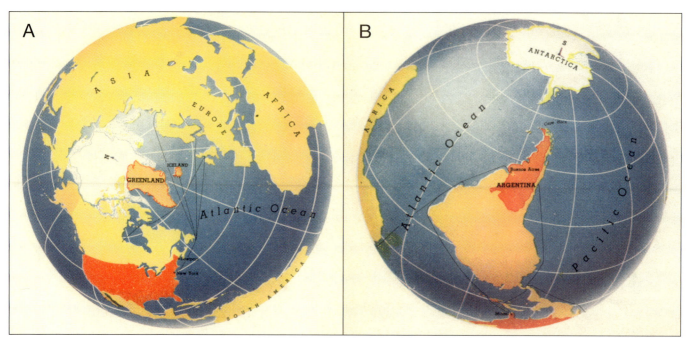

COLOR PLATE 9.1 The orthographic projection showing (A) North America and its proximity to Europe and (B) a view of South America and Antarctica. (From pp. 52–53 of Harrison, R. E. (1944) *Look at the World: The Fortune Atlas for World Strategy*. New York, NY: Alfred A. Knopf.)

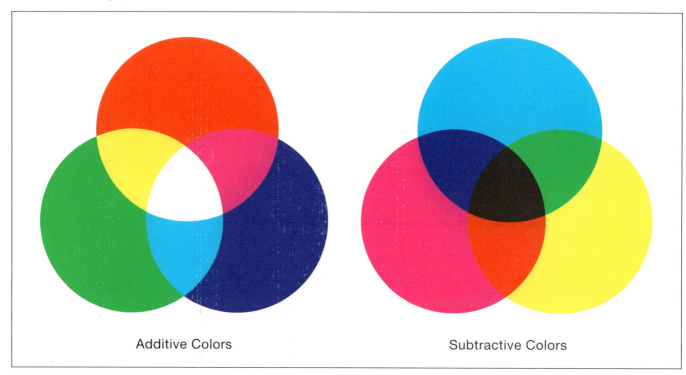

COLOR PLATE 10.1 Principles of additive and subtractive color. For additive color, overlapping red, green, and blue lights reveal how cyan, magenta, yellow, and white can be created. For subtractive color, the reverse is the case: cyan, magenta, and yellow combine to produce red, green, blue, and black. To obtain a true black with subtractive colors, it is often necessary to add a black layer.

Plate 6

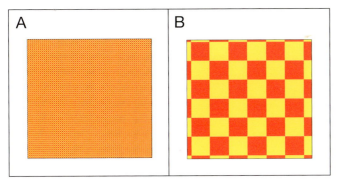

COLOR PLATE 10.2 The dithering process. (A) a shade created using dithering; (B) a blow-up of a portion of (A) showing the individual colors used to create the dithered color.

COLOR PLATE 10.3 The Munsell color model. (Courtesy of GretagMacbeth.)

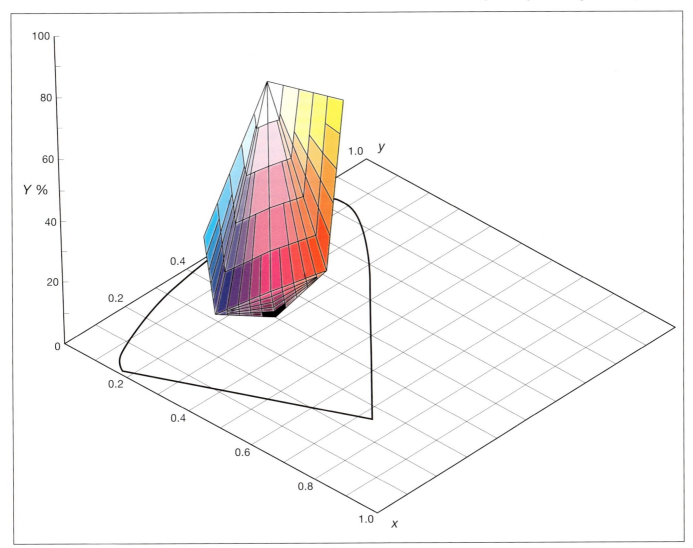

COLOR PLATE 10.4 A three-dimensional view of the Yxy CIE system. (Courtesy of A. Jon Kimerling.)

Plate 7

CURRENCY

Bolivar
Boliviano
Dollar
Franc
Guarani

Guilder
Nuevo Sol
Peso
Real
Sucre

Source: U.S. Department of Commerce.
National Trade Data Bank, May 6, 1999.

COLOR PLATE 11.1 Unique hues are preferred to line and dot patterns when symbolizing areal features with qualitative data. (Compare with Figure 11.12.)

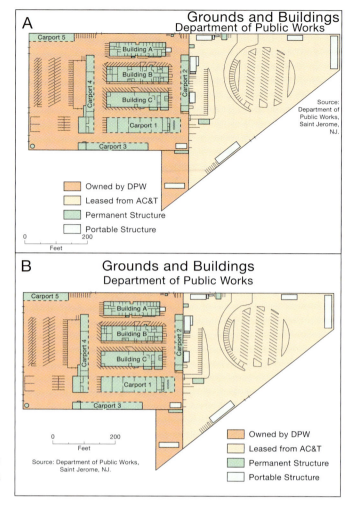

A

Grounds and Buildings
Department of Public Works

Carport 5

Building A

Building B

Building C

Carport 4

Carport 2

Carport 1

Carport 3

Source: Department of Public Works, Saint Jerome, NJ.

Owned by DPW
Leased from AC&T
Permanent Structure
Portable Structure

0   200
Feet

B

Grounds and Buildings
Department of Public Works

Carport 5

Building A

Building B

Building C

Carport 4

Carport 2

Carport 1

Carport 3

0   200
Feet

Source: Department of Public Works, Saint Jerome, NJ.

Owned by DPW
Leased from AC&T
Permanent Structure
Portable Structure

COLOR PLATE 11.3 (A) A poorly balanced design, in which map elements compete for space. (B) A well-balanced design, in which map elements exist in harmony and equilibrium.

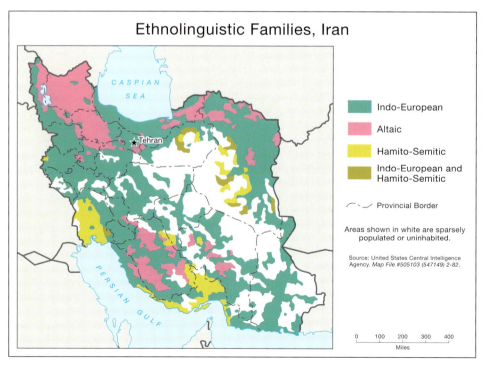

Ethnolinguistic Families, Iran

CASPIAN SEA

★ Tehran

PERSIAN GULF

Indo-European
Altaic
Hamito-Semitic
Indo-European and Hamito-Semitic

Provincial Border

Areas shown in white are sparsely populated or uninhabited.

Source: United States Central Intelligence Agency. *Map File #505103 (547149) 2-82.*

0   100   200   300   400
Miles

COLOR PLATE 11.2 Map resulting from completion of the design process, and based on the sketch map illustrated in Figure 11.29.

Plate 8

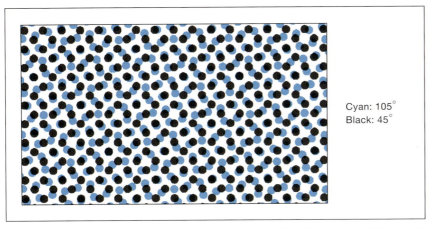

Cyan: 105°
Black: 45°

COLOR PLATE 12.1 Different halftone screen angles allow tints of black and cyan to mix on the page.

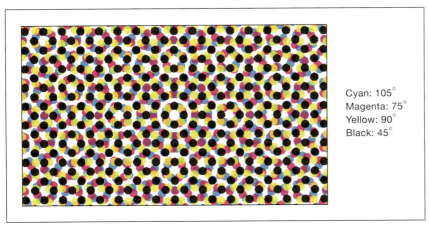

Cyan: 105°
Magenta: 75°
Yellow: 90°
Black: 45°

COLOR PLATE 12.2 Rosette pattern produced from halftone screens of the process colors, each at a different screen angle.

COLOR PLATE 12.3 An overlay proof of a map produced by the California State Automobile Association, San Francisco, California. © 2002 California State Automobile Association. Used by permission.

Plate 9

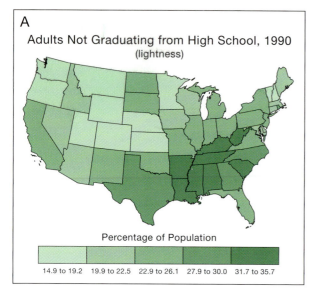

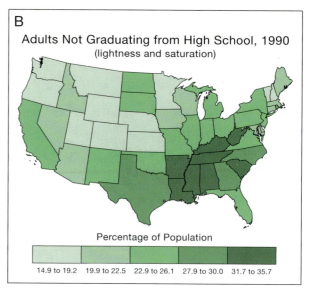

COLOR PLATE 13.1 A comparison of two sequential color schemes: (A) only lightness varies, (B) both lightness and saturation vary, with saturation increasing from the first to the third class and then decreasing for the latter two classes. (Colors based on CMYK values shown in Brewer 1989.)

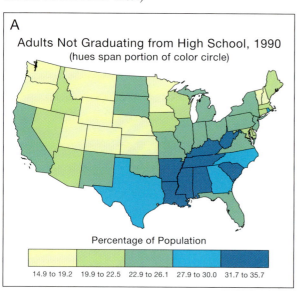

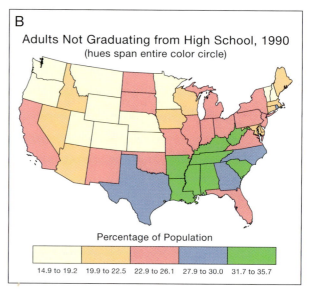

COLOR PLATE 13.2 A comparison of two sequential color schemes in which differing hues are used: (A) the hues span a portion of the color circle, (B) the hues span the entire color circle. (CMYK values for maps provided by Cynthia Brewer.)

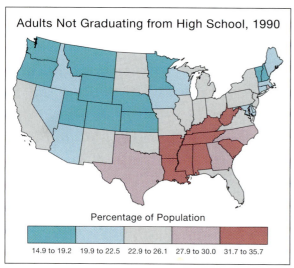

COLOR PLATE 13.3 An example of a diverging scheme in which two sequential schemes converge on a neutral gray.

Plate 10

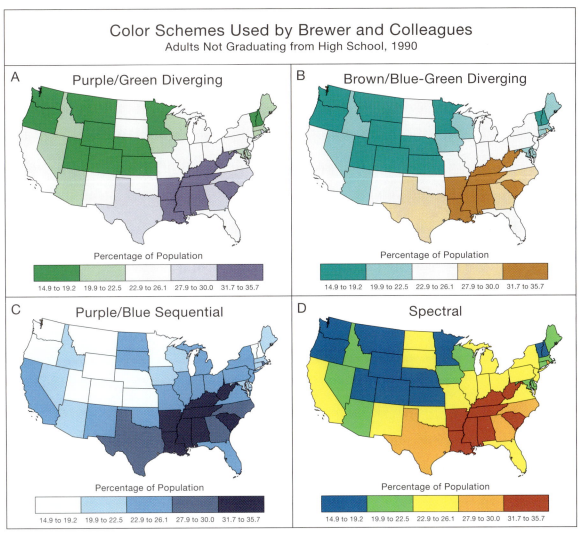

Color Schemes Used by Brewer and Colleagues
Adults Not Graduating from High School, 1990

A Purple/Green Diverging

Percentage of Population

14.9 to 19.2 | 19.9 to 22.5 | 22.9 to 26.1 | 27.9 to 30.0 | 31.7 to 35.7

B Brown/Blue-Green Diverging

Percentage of Population

14.9 to 19.2 | 19.9 to 22.5 | 22.9 to 26.1 | 27.9 to 30.0 | 31.7 to 35.7

C Purple/Blue Sequential

Percentage of Population

14.9 to 19.2 | 19.9 to 22.5 | 22.9 to 26.1 | 27.9 to 30.0 | 31.7 to 35.7

D Spectral

Percentage of Population

14.9 to 19.2 | 19.9 to 22.5 | 22.9 to 26.1 | 27.9 to 30.0 | 31.7 to 35.7

COLOR PLATE 13.4   Some of the effective color schemes tested in Brewer et al.'s (1997) work: (A) a purple/green diverging scheme, (B) a brown/blue–green diverging scheme, (C) a purple–blue sequential scheme, and (D) a spectral scheme. (CMYK values for maps provided by Cynthia Brewer.)

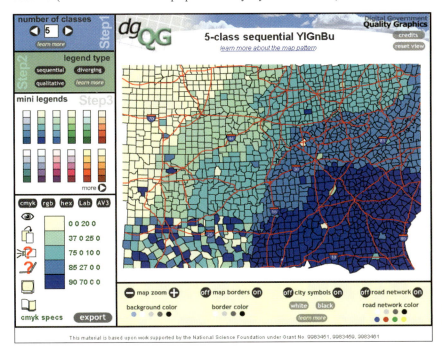

COLOR PLATE 13.5   ColorBrewer, a program that enables novices to easily select color schemes for choropleth maps. (Courtesy of Cynthia Brewer.)

Plate 11

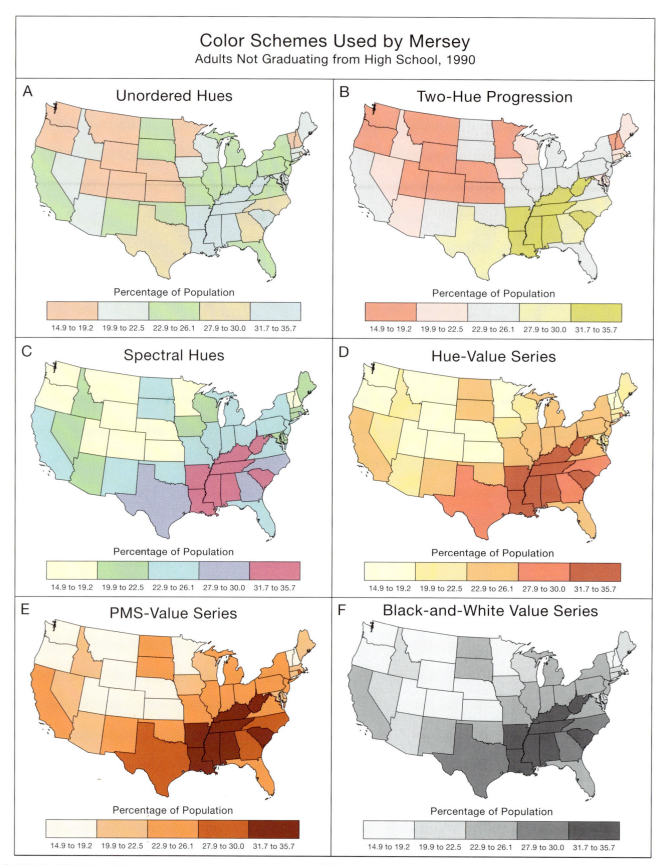

Color Schemes Used by Mersey
Adults Not Graduating from High School, 1990

A Unordered Hues

Percentage of Population

14.9 to 19.2   19.9 to 22.5   22.9 to 26.1   27.9 to 30.0   31.7 to 35.7

B Two-Hue Progression

Percentage of Population

14.9 to 19.2   19.9 to 22.5   22.9 to 26.1   27.9 to 30.0   31.7 to 35.7

C Spectral Hues

Percentage of Population

14.9 to 19.2   19.9 to 22.5   22.9 to 26.1   27.9 to 30.0   31.7 to 35.7

D Hue-Value Series

Percentage of Population

14.9 to 19.2   19.9 to 22.5   22.9 to 26.1   27.9 to 30.0   31.7 to 35.7

E PMS-Value Series

Percentage of Population

14.9 to 19.2   19.9 to 22.5   22.9 to 26.1   27.9 to 30.0   31.7 to 35.7

F Black-and-White Value Series

Percentage of Population

14.9 to 19.2   19.9 to 22.5   22.9 to 26.1   27.9 to 30.0   31.7 to 35.7

COLOR PLATE 13.6   Color schemes used in a study by Mersey (1990): (A) unordered hues, (B) a two-hue progression, (C) spectral hues, (D) a hue-value series, (E) a PMS-value series, and (F) a black-and-white value series. Colors shown are based on Mersey's specification of CMY and PMS colors. The PMS value series involved overprinting an orange-brown color with a black ink to increase the value range of the orange-brown color. Mersey found that the hue-value series worked best overall. (After Mersey 1990.)

Plate 12

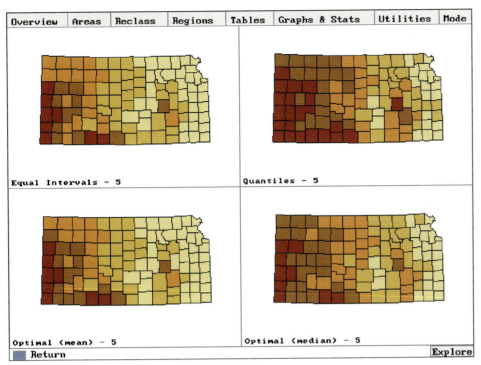

Overview | Areas | Reclass | Regions | Tables | Graphs & Stats | Utilities | Mode

Equal Intervals - 5

Quantiles - 5

Optimal (mean) - 5

Optimal (median) - 5

Return | Explore

COLOR PLATE 13.7 Using data exploration software to visualize the data in different ways, in this case, as four methods of classification (from the program ExploreMap).

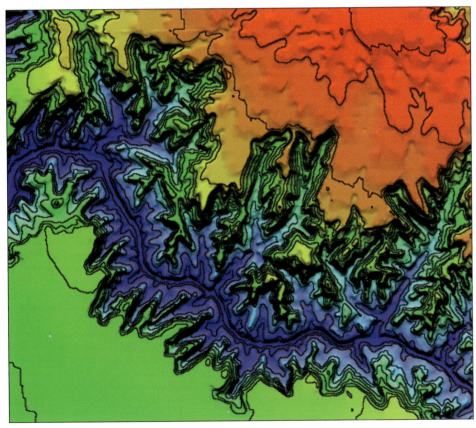

COLOR PLATE 14.1 A combination of contour lines and continuous spectral color scheme developed by Eyton (1990) are applied to the Grand Canyon. When viewed with special glasses (*http://www.chromatek.com/*), the color stereoscopic effect is exaggerated, producing a striking 3-D image. (Courtesy of J. Ronald Eyton.)

Plate 13

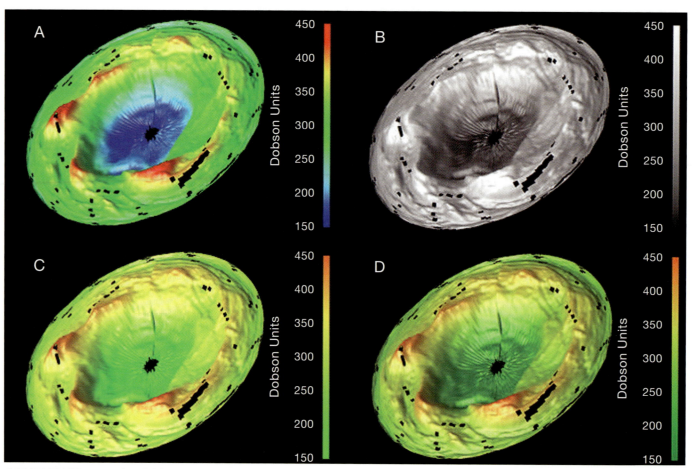

COLOR PLATE 14.2 Spatial frequency and the use of lightness and saturation. For all images, the height of the surface depicts the level of ozone in the upper troposphere and lower stratosphere. (A) a traditional spectral color scheme, (B) a lightness scheme captures artifacts of atmospheric circulation, which are of high spatial frequency, (C) a saturation scheme captures the depressed region (of ozone depletion) in the middle, (D) a combined lightness-saturation scheme captures both the high- and low-frequency features. (Courtesy of Bernice Rogowitz and Lloyd Treinish, IBM T. J. Watson Research Center.)

COLOR PLATE 15.1 Tanaka's approach is modified by adding colors between the contours to give the map a more realistic appearance (From Kennelly and Kimerling 2001. First published in Cartography and Geographic Information Science 28(2), p. 119. Reprinted with permission from the American Congress on Surveying and Mapping.)

Plate 14

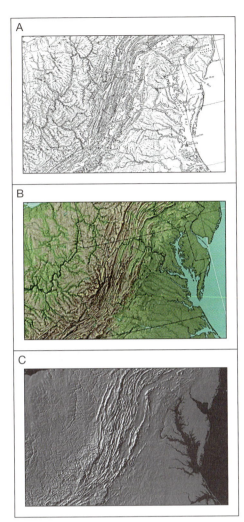

COLOR PLATE 15.3   An example of a shaded relief map that Imhof created by hand. (From Imhof, E., *Cartographic Relief Presentation*, Berlin, New York: Walter de Gruyter 1982.)

COLOR PLATE 15.2   A comparison of three methods of depicting topography: (A) Raisz's physiographic method; (B) Harrison's manual shaded-relief method; (C) a USGS digital shaded-relief portrayal developed by Thelin and Pike. (Map A copyright by Erwin Raisz, 1967. Reprinted with permission by RAISZ LANDFORM MAPS, 800-277-0047.)

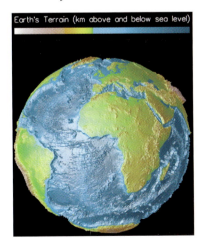

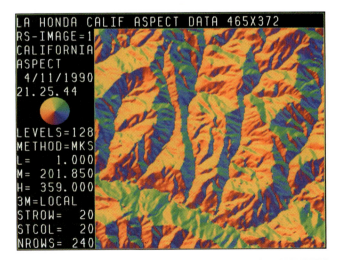

COLOR PLATE 15.4   A portrayal of the earth's topography developed by Treinish (1994) using the IBM Visualization Data Explorer (DX). The technique uses Gouraud shading to represent aspect and slope and hypsometric tints to represent elevation above or below sea level. (Courtesy of Lloyd Treinish, IBM T. J. Watson Research Center.)

COLOR PLATE 15.5   Moellering and Kimerling's MKS-ASPECT™ approach for symbolizing aspect based on the opponent-process theory of color. The image was created using 128 colors and was scanned from a slide provided by Moellering. (Courtesy of Harold Moellering and A. Jon Kimerling.)

Plate 15

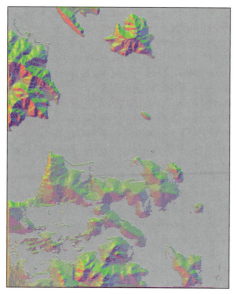

COLOR PLATE 15.6 Brewer and Marlow's method for portraying aspect and slope. The image is for the San Francisco North 7 and one-half minute USGS quad. Caution should be taken in interpreting these colors because the RGB values recommended by Brewer and Marlow have been converted to CMYK values for printing.

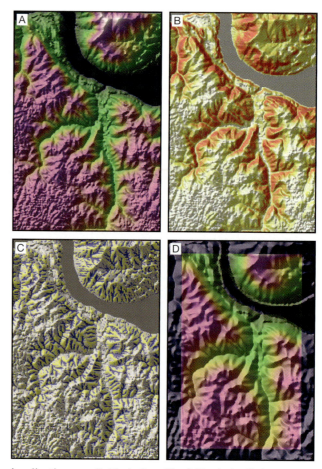

COLOR PLATE 15.7 Some of the visualizations available in LandSerf. Shaded relief maps are overlaid with: (A) a green–brown–purple–white color scheme depicting elevation ranges; (B) a white–yellow–red color scheme depicting slope; (C) features found in the region, with channels shown in blue, ridges in yellow, and planar regions in gray; and (D) a generalization of (A) using a 65 × 65 cell window. (Courtesy of Jo Wood.)

COLOR PLATE 15.8 Berann's classic hand-drawn panorama of Denali National Park, Alaska. (Courtesy of National Park Service, artist, Heinrich Berann.)

Plate 16

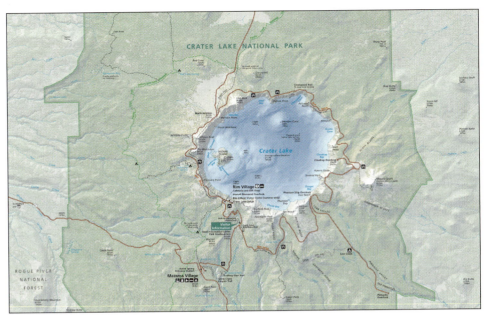

COLOR PLATE 15.9 A map of Crater Lake National Park that utilizes Patterson's rules for creating more realistic views of vertically viewed topographic maps. (Courtesy of National Park Service.)

COLOR PLATE 15.10 Examples of physical models created using technology developed by Solid Terrain Modeling: Mt. Everest is on the left, and Hurricane Floyd is on the right. (Courtesy of Solid Terrain Modeling.)

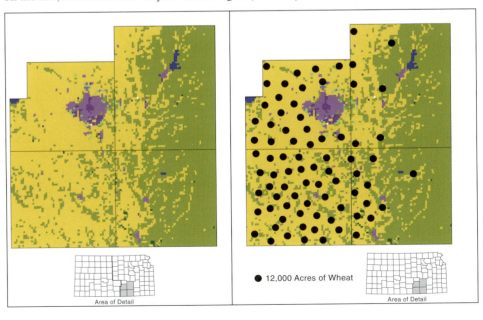

COLOR PLATE 17.1 A portion of the Kansas land use/land cover map illustrating the character of subregions within which dots were placed (cropland and grassland appear in yellow and green, respectively). The left map shows the character of the subregions, and the right map shows how dots were actually placed. (Courtesy of Kansas Applied Remote Sensing Program, University of Kansas.)

● 12,000 Acres of Wheat

Plate 17

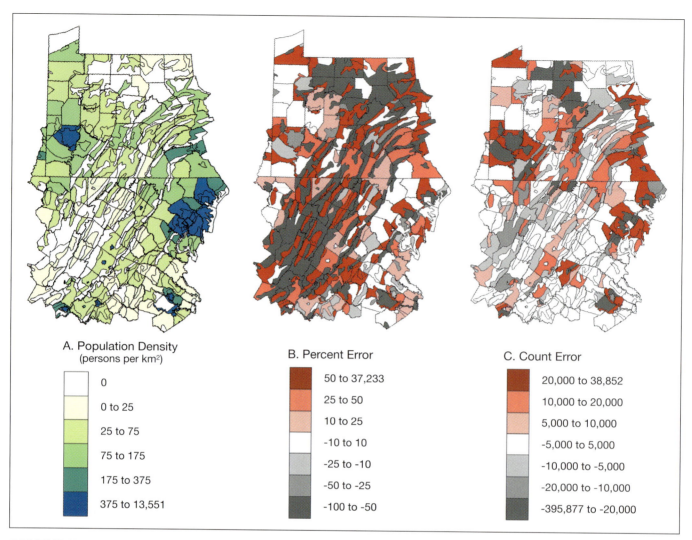

A. Population Density
(persons per km²)

	0
	0 to 25
	25 to 75
	75 to 175
	175 to 375
	375 to 13,551

B. Percent Error

	50 to 37,233
	25 to 50
	10 to 25
	-10 to 10
	-25 to -10
	-50 to -25
	-100 to -50

C. Count Error

	20,000 to 38,852
	10,000 to 20,000
	5,000 to 10,000
	-5,000 to 5,000
	-10,000 to -5,000
	-20,000 to -10,000
	-395,877 to -20,000

COLOR PLATE 17.2   Example of dasymetric and associated error maps resulting from Eicher and Brewer's work: (A) dasymetric map based on the binary method; (B) percent error map; (C) count error map. (After Eicher and Brewer 2001. First published in *Cartography and Geographic Information Science* 28(2), p.134. Reprinted with permission from the American Congress on Surveying and Mapping. Courtesy of Cory Eicher and Cynthia Brewer.)

Plate 18

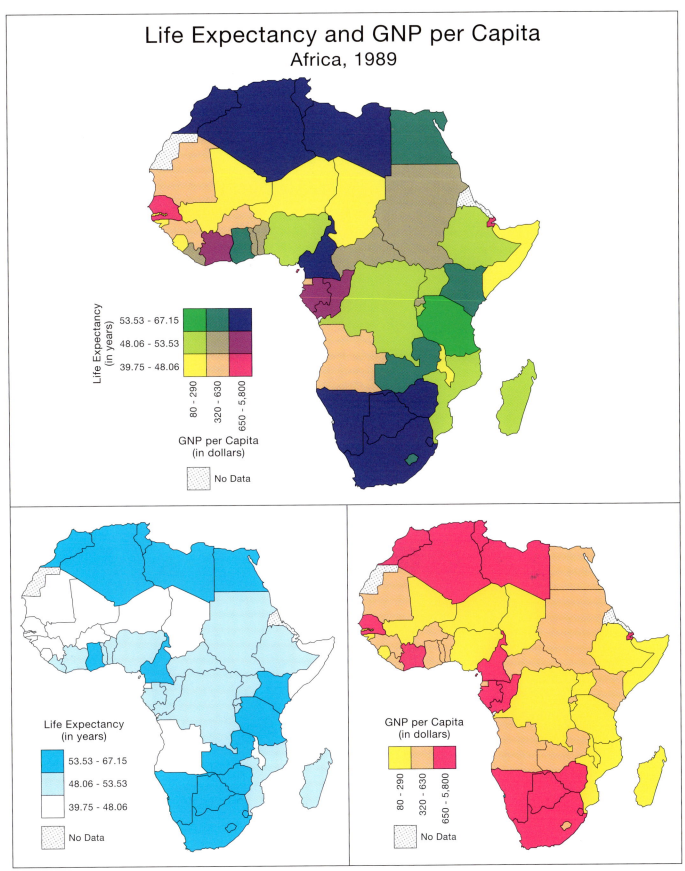

Life Expectancy and GNP per Capita
Africa, 1989

Life Expectancy (in years)
- 53.53 – 67.15
- 48.06 – 53.53
- 39.75 – 48.06

GNP per Capita (in dollars)
- 80 – 290
- 320 – 630
- 650 – 5,800

No Data

Life Expectancy (in years)
- 53.53 – 67.15
- 48.06 – 53.53
- 39.75 – 48.06
- No Data

GNP per Capita (in dollars)
- 80 – 290
- 320 – 630
- 650 – 5,800
- No Data

COLOR PLATE 18.1 A bivariate choropleth map and its component univariate maps. The color scheme was taken from Olson (1981, 269) and is similar to those popular on U.S. Bureau of the Census maps in the 1970s. A quantiles classification was used for each map because the GNP data were positively skewed. (Data Source: ArcView 3.)

Plate 19

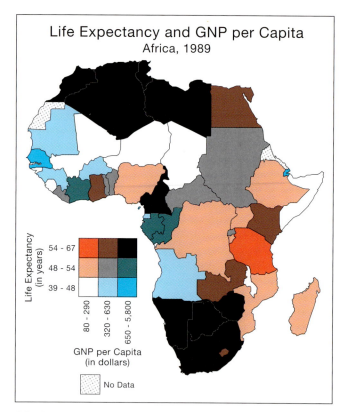

COLOR PLATE 18.2 A bivariate choropleth map based on the complementary colors red and cyan. (After Eyton 1984a.)

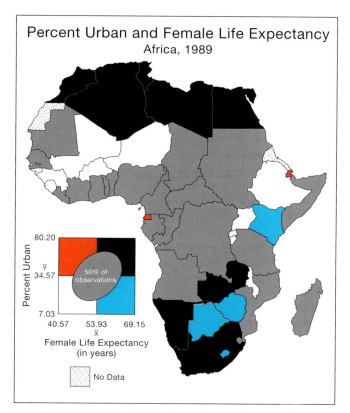

COLOR PLATE 18.3 A bivariate choropleth map using complementary colors in which legend classes are based on the means of the variables and an equiprobability ellipse enclosing 50 percent of the data. Because the ellipse is based on a bivariate normal distribution, only normally distributed data should be mapped using this method. (After Eyton 1984a; Data source: ArcView 3.)

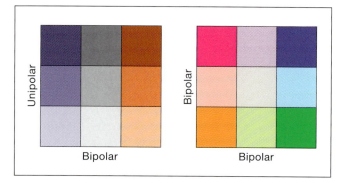

COLOR PLATE 18.4 Some color schemes suggested by Brewer for bivariate maps; the terms unipolar and bipolar refer to the type of data associated with each attribute.

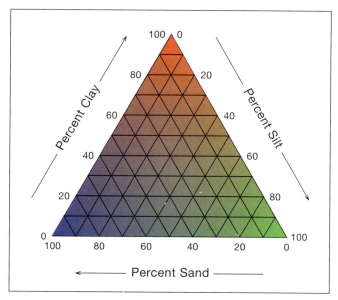

COLOR PLATE 18.5 An RGB color scheme for creating a trivariate choropleth map. (After Byron 1994, p. 126.)

Plate 20

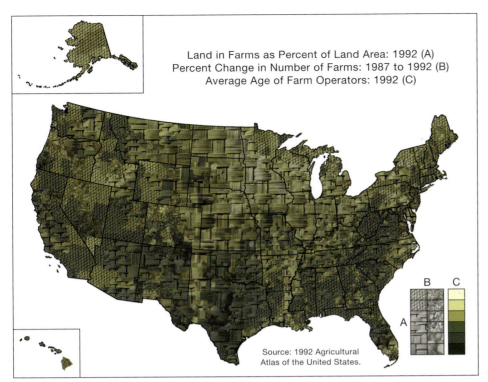

Land in Farms as Percent of Land Area: 1992 (A)
Percent Change in Number of Farms: 1987 to 1992 (B)
Average Age of Farm Operators: 1992 (C)

Source: 1992 Agricultural
Atlas of the United States.

COLOR PLATE 18.6 A trivariate choropleth map that uses pattern (or texture) for two attributes and a smooth colored tone for a third attribute. (After Interrante, V. (2000) "Harnessing natural textures for multivariate visualization." IEEE Computer Graphics and Applications 20, p. 9; © 2000 IEEE.)

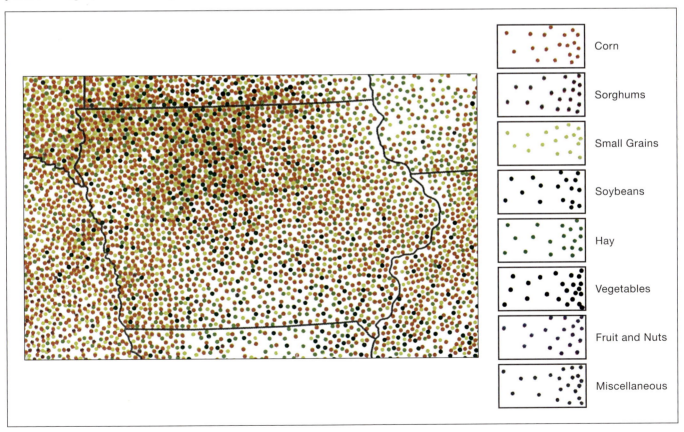

Corn

Sorghums

Small Grains

Soybeans

Hay

Vegetables

Fruit and Nuts

Miscellaneous

COLOR PLATE 18.7 A portion of a multivariate dot map constructed on the basis of pointillism. Because this map was scanned from the original map, colors shown must be considered approximations of the actual colors. (After Jenks, 1961.)

Plate 21

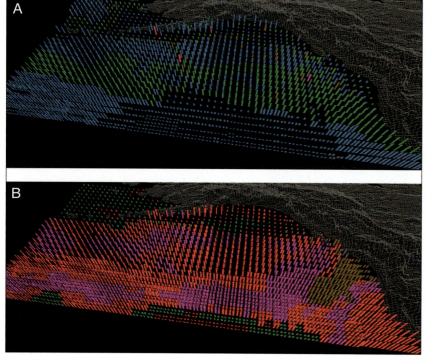

COLOR PLATE 18.8 Using pexels to depict multivariate data for the northern Pacific Ocean (A) in February and (B) June. Color of the pexels depicts the density of plankton (from low to high density, the scheme is blue, green, brown, red, and purple); height of the pexels depicts the ocean current (a higher pexel equals a stronger current); and spacing of the pexels depicts sea surface temperature (a tighter spacing indicates a warmer temperature). (Courtesy of Christopher Healey.)

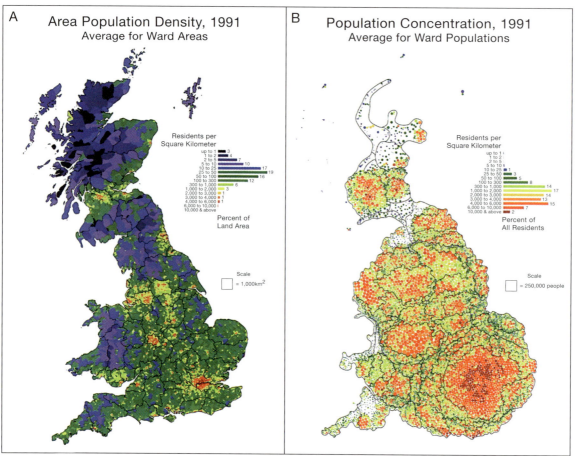

COLOR PLATE 19.1 Maps of population density in Britain using (A) a traditional equivalent projection and (B) a cartogram. The equivalent projection suggests that most of Britain is dominated by relatively lower population densities (the blues and greens), whereas the cartogram provides a detailed picture of the variation in population densities within urban areas. (From Dorling 1995a, p. xxxiii. Courtesy of Daniel Dorling.)

Plate 22

A

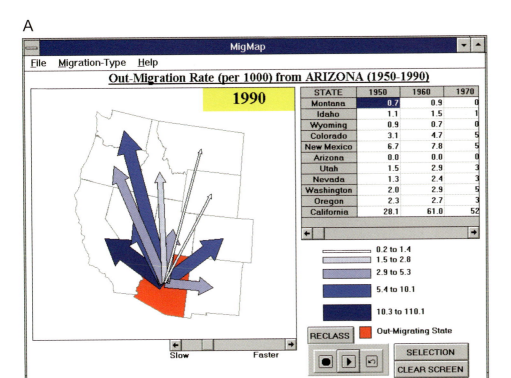

B

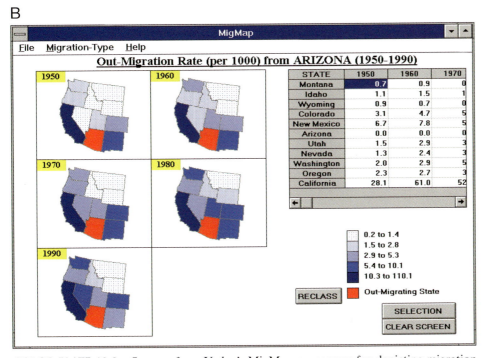

COLOR PLATE 19.2 Screens from Yadav's MigMap, a program for depicting migration flows: (A) a frame from an animation; (B) a small multiple based on choropleth symbols. Note the use of redundancy in (A) (migration is a function of both the width and color of the symbol). (Courtesy of Sunita Yadav-Pauletti.)

Plate 23

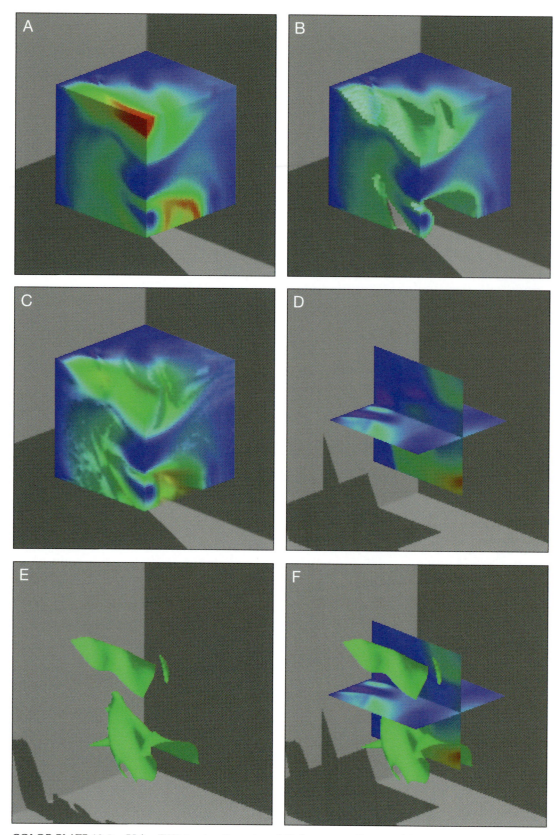

COLOR PLATE 19.3 Using T3D to visualize a true 3-D data set, upflows and downflows within a thunderstorm: (A) an opaque default rainbow color scheme is used; (B) a portion of the color scheme is made transparent; (C) the transparency of the color scheme is varied continuously; (D) examples of slices taken through the 3-D surface; (E) an example of an isosurface, the 3-D equivalent of a contour line; (F) combining isosurfaces and slices. (Images created with Noesys Visualization Pro, courtesy of Fortner Software LLC.)

Plate 24

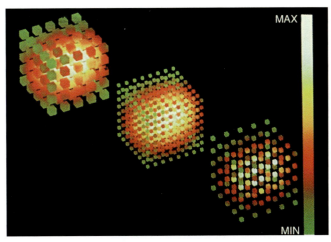

COLOR PLATE 19.4 Nielson and Hamann's tiny cubes method for visualizing true 3-D phenomena. (From Keller and Keller 1993, *Visual Cues: Practical Data Visualization*, p. 148; © 1993 IEEE.)

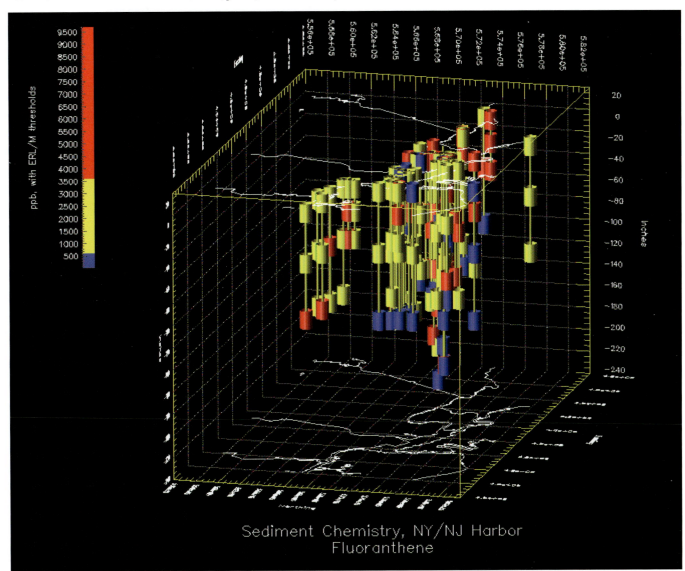

COLOR PLATE 19.5 Waisel's use of DX to visualize the concentration of fluoranthene in sediment in New York Harbor. Each cylinder represents a homogeneous sample within a core. (Courtesy of Laurie Waisel.)

Plate 25

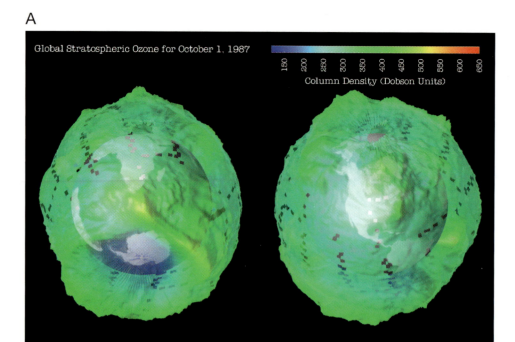

COLOR PLATE 20.1 Treinish's portrayal of the ozone hole over Antarctica: (A) the ozone symbolization is shown as an orthographic projection (this symbolization was used in an animation of the data); (B) small multiples depicting monthly averages of ozone during the spring warming for the Southern and Northern Hemispheres. (Courtesy of Lloyd Treinish, IBM T. J. Watson Research Center.)

Plate 26

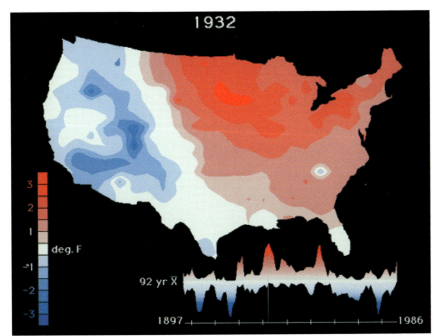

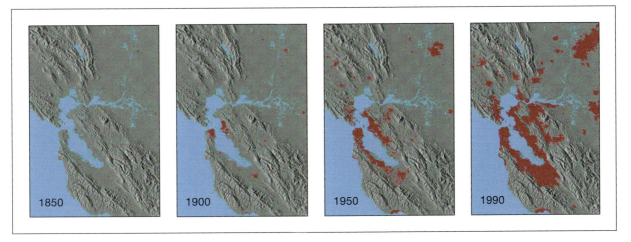

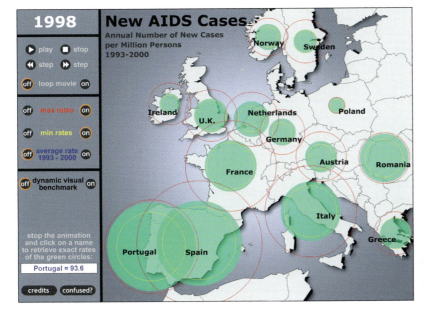

COLOR PLATE 20.2   A frame from Weber and Buttenfield's animation of U.S. surface temperatures. Note the use of a logical color progression, with reds and blues for positive and negative deviations, respectively. (Courtesy of Christopher R. Weber and Barbara P. Buttenfield.)

COLOR PLATE 20.3   Four frames from the USGS's animation of urban growth in the San Francisco–Sacramento region described in Buchanan and Acevedo (1996). Note the relatively smooth transitions between the frames. (Courtesy of William Acevedo.)

COLOR PLATE 20.4   An illustration of Harrower's (2002) use of visual benchmarks (the yellow and red circles). (Courtesy of Mark Harrower.)

Plate 27

A

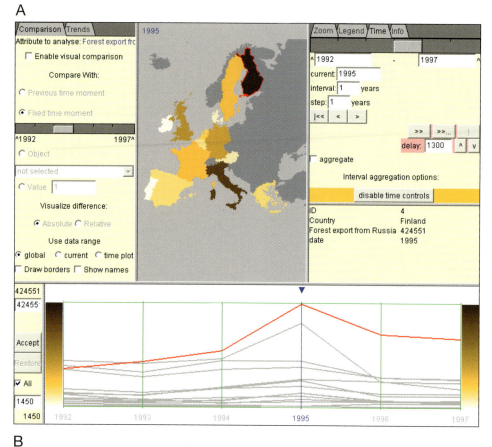

B

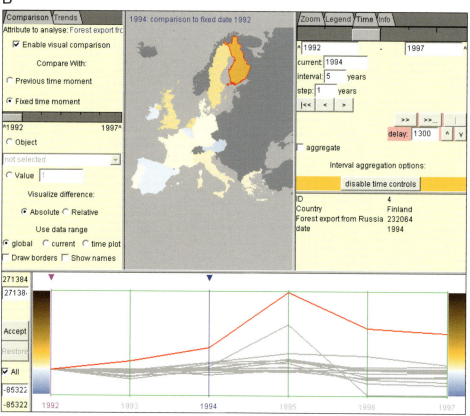

COLOR PLATE 20.5   Andrienko et al.'s (2001) interactive software that is intended to improve on "fixed" animations: (A) a frame from an animation of the raw data; (B) a frame from a "comparison" option, in which each year is compared against 1992. (Courtesy of Spatial Decision Support Team, Fraunhofer Institute AIS, *http://www.ais.fraunhofer.de/SPADE/.*)

Plate 28

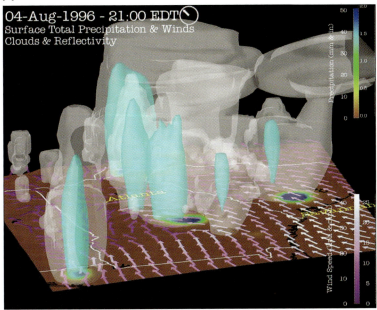

A

04-Aug-1996 - 21:00 EDT
Surface Total Precipitation & Winds
Clouds & Reflectivity

B

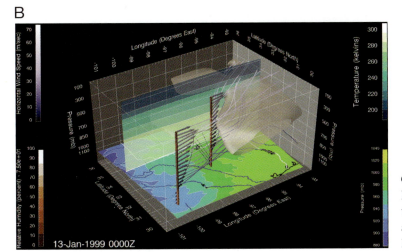

13-Jan-1999 0000Z

COLOR PLATE 20.6 Maps taken from animations in Deep Thunder, a research project for developing sophisticated weather prediction models and associated visualizations. (Courtesy of Lloyd Treinish, IBM T. J. Watson Research Center.)

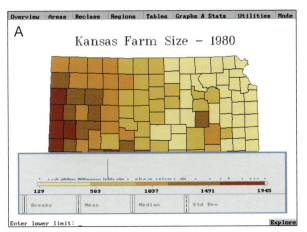

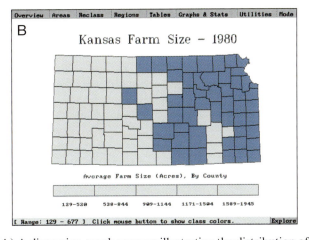

COLOR PLATE 21.1 The Subset option for ExploreMap: (A) A dispersion graph appears illustrating the distribution of the data. Note that the current class breaks, mean, median, and standard deviation can be toggled on; in this case, only the median is shown; (B) The user selects all values less than the median, and these are displayed in blue.

Plate 29

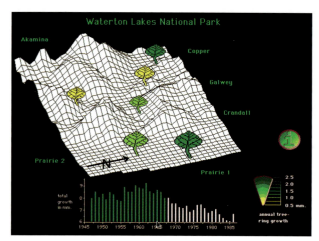

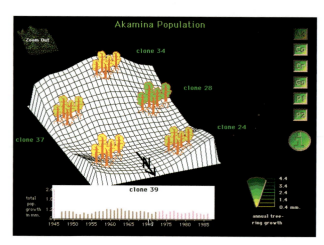

COLOR PLATE 21.2 A frame of an animation from Aspens. Size of leaf represents total cumulative tree growth for a study site, whereas leaf color represents incremental growth for each year. Clicking on a leaf reveals an animation for clones associated with that study site (Color Plate 21.3). (Courtesy of Barbara P. Buttenfield and Christopher R. Weber.)

COLOR PLATE 21.3 A frame of an animation obtained when a leaf in Color Plate 21.2 is clicked on. Size of trees represents total cumulative growth, whereas tree color represents incremental growth. Clicking on an icon in the upper right can lead to animations of clones for other study sites. (Courtesy of Barbara P. Buttenfield and Christopher R. Weber.)

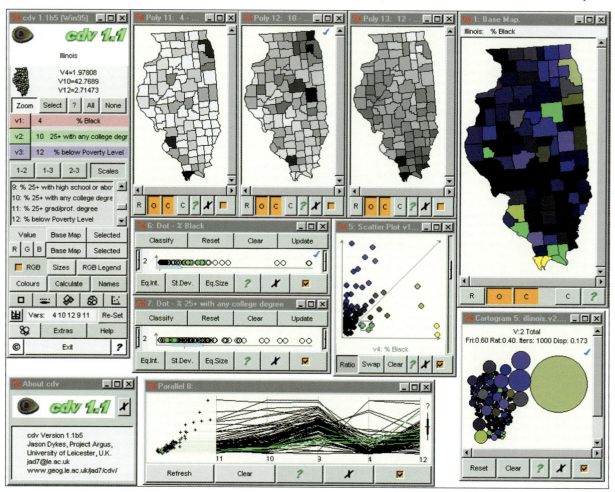

COLOR PLATE 21.4 A screen from the "enumerated" software within Project Argus. At the top of the screen three variables are shown: percent Black, percent of population 25 years and older with any college education, and percent below the poverty level. In the top right is a bivariate choropleth map of the first two variables; high values on these two variables are represented by increasing amounts of yellow and blue, respectively. In the central portion are point graphs and a scatterplot of the first two variables. The lower and lower right views show a parallel coordinate plot and Dorling's cartogram method, respectively. (Jason Dykes, Department of Geography, University of Leicester.)

Plate 30

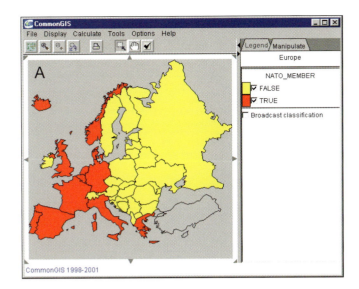

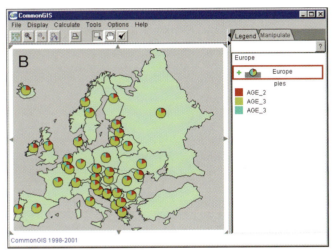

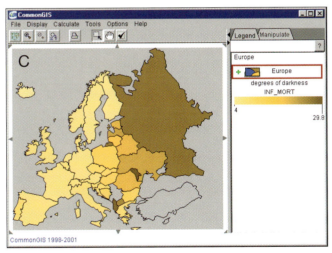

COLOR PLATE 21.5   Maps from CommonGIS corresponding to the mapping options shown in Figure 21.7: (A) NATO membership using the "colours" option, (B) percentage of people in three age categories using a "pie" option, and (C) infant mortality rate using a "degrees of darkness option," which produces an unclassed choropleth map. (Courtesy of Spatial Decision Support Team, Fraunhofer Institute AIS, *http://www.ais.fraunhofer.de/ SPADE/*.)

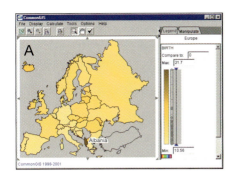

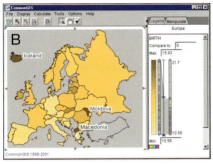

COLOR PLATE 21.6   Data exploration in CommonGIS involving the removal of an outlier: (A) an unclassed choropleth map of all data for birth rates in Europe, (B) the Albanian outlier is removed. (Courtesy of Spatial Decision Support Team, Fraunhofer Institute AIS, *http://www.ais.fraunhofer.de/SPADE/*.)

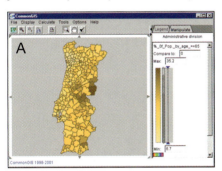

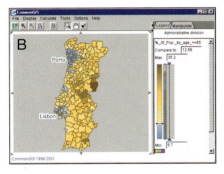

COLOR PLATE 21.7   Data exploration in CommonGIS using a dynamic diverging color scheme: (A) an unclassed choropleth map of the percent of the population greater than or equal to 65 in divisions of Portugal, (B) a diverging color scheme is applied to a portion of the data range, illustrating that the lowest values occur around the biggest cities of Portugal such as Lisbon and Porto. (Courtesy of Spatial Decision Support Team, Fraunhofer Institute AIS, *http://www.ais.fraunhofer.de/SPADE/*.)

Plate 31

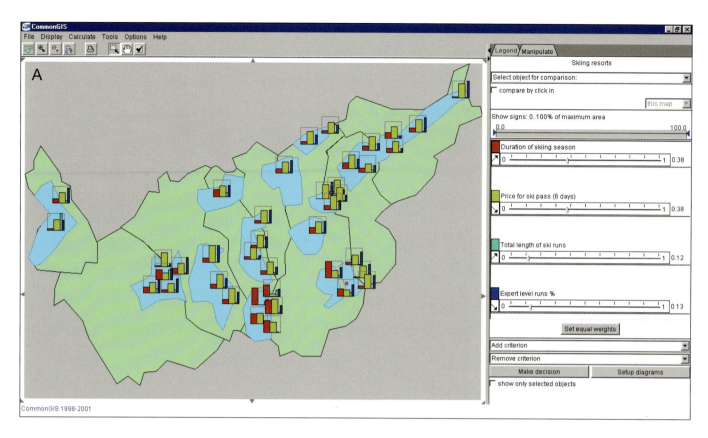

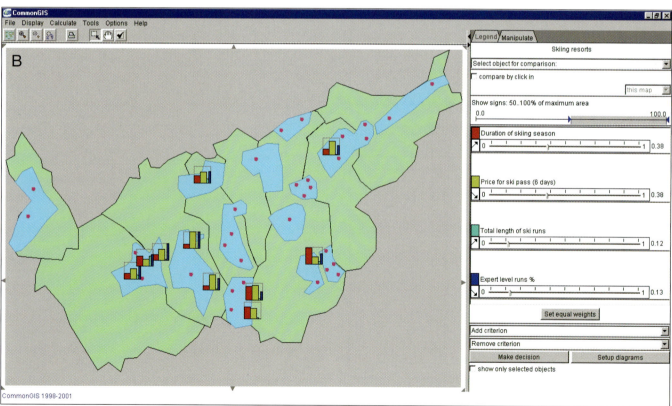

COLOR PLATE 21.8   Data exploration in decision making within CommonGIS: (A) using "utility bars" to display four attributes associated with choosing an ideal ski resort, (B) the display is simplified by hiding bars that constitute less than 50 percent of the theoretical best resort. (Courtesy of Spatial Decision Support Team, Fraunhofer Institute AIS, *http://www.ais.fraunhofer.de/SPADE/*.)

Plate 32

COLOR PLATE 21.9 The integration of analysis and display functions in ArcView 3.x, a popular GIS package: (A) hypothetical sales information by state is displayed as a choropleth map; (B) cities with a population over 80,000 are determined using a query builder and automatically displayed as blue circles; (C) cities within 300 miles of Atlanta are automatically displayed as yellow circles; (D) different spatial constraints are used (150,000 for population and 200 miles for distance from Atlanta). (Graphic image created using ArcView® GIS software. Source data: ESRI.)

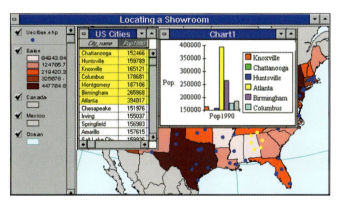

COLOR PLATE 21.10 Using ArcView to view data from a variety of perspectives: as a map, table, or chart. Note that cities highlighted on the map are highlighted in the table and also displayed in the chart. (Graphic image created using ArcView® GIS software. Source data: ESRI.)

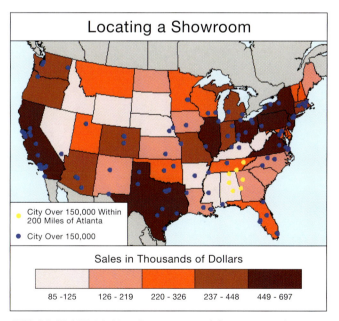

COLOR PLATE 21.11 A map created for presentation purposes. Cities in yellow that are located in states falling in the two lowest valued classes meet the three criteria specified for locating the showroom. (Graphic image created using ArcView® GIS software. Source data: ESRI.)

Plate 33

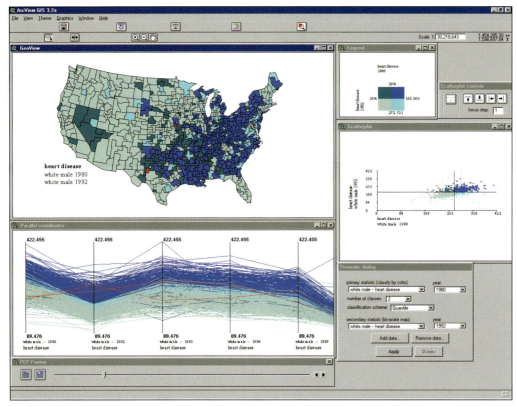

COLOR PLATE 21.12 The interface for Health VisPCP using heart disease for White men as an example. Major elements include a bivariate choropleth map (upper left), a scatterplot of the variables depicted on the map (middle right), and a parallel coordinate plot (lower left). (Courtesy of Robert Edsall.)

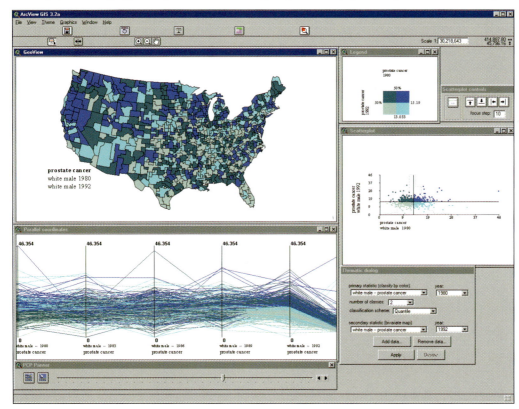

COLOR PLATE 21.13 The interface for Health VisPCP illustrating a more complex parallel coordinate plot (compare with Color Plate 21.12). (Courtesy of Robert Edsall.)

Plate 34

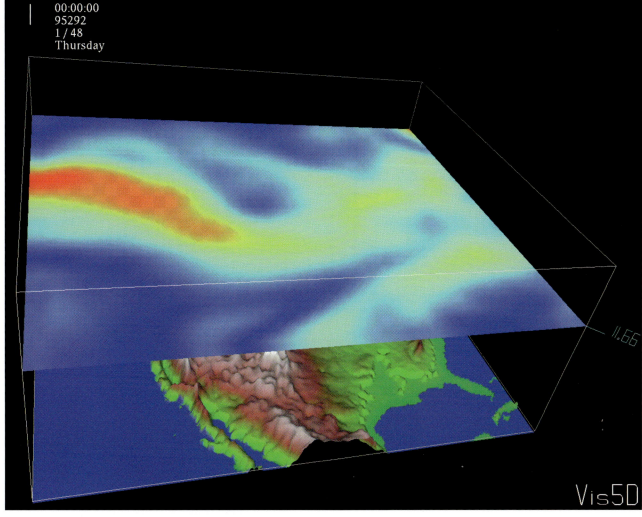

COLOR PLATE 21.14 In Vis5D, hypsometric tints are used to symbolize a horizontal slice through true 3-D wind speed data. Note the height of the slice is 11.66 kilometers, and thus the high-valued reddish-orange region represents a portion of the jet stream. (Courtesy of William L. Hibbard.)

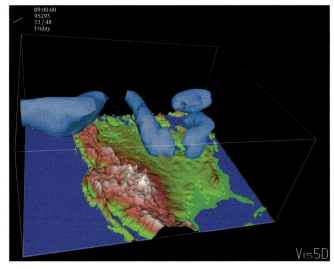

COLOR PLATE 21.15 An isosurface of wind speed created using Vis5D. An isosurface is a surface bounded by a particular value of an attribute; in this case, the 45 meter per second value for wind speed is used. (Courtesy of William L. Hibbard.)

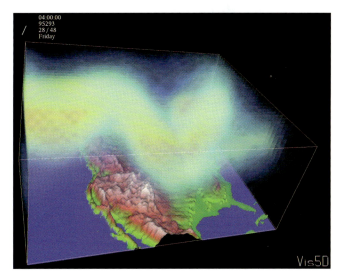

COLOR PLATE 21.16 Volume (or transparent fog) symbolization of wind speed created using Vis5D. Each point in the 3-D data is symbolized using a spectral color scheme. (Courtesy of William L. Hibbard.)

Plate 35

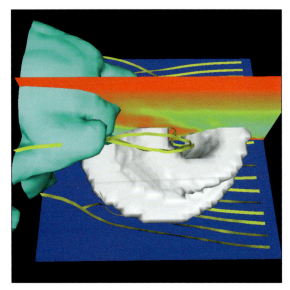

COLOR PLATE 21.17 Multiple variables depicted using Vis5D; see text for explanation. (From Hibbard, W. L., Paul, B. E., Santek, D. A., Dyer, C. R., Battaiola, A. L., and Voidrot-Martinez, Marie-F. 1994, *Interactive Visualization of Earth and Space Science Computations*, p. 67; © 1994 IEEE.)

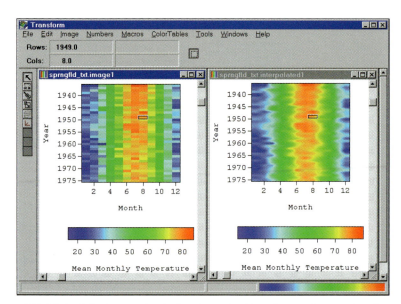

COLOR PLATE 21.18 A window from the Transform software showing unsmoothed and smoothed images (the left and right, respectively) of mean monthly temperatures for Springfield, Illinois. Years are shown along the *y* axis, and months of the year are shown along the *x* axis. Compare with the table of these data shown in Figure 21.9. (Image created with Noesys Visualization Pro, courtesy of Fortner Software LLC.)

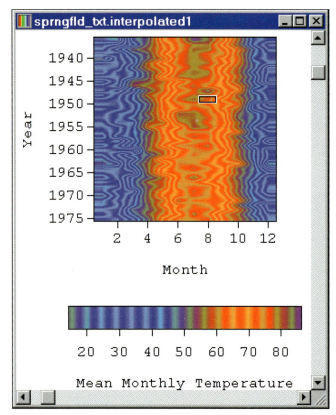

COLOR PLATE 21.19 A window from the Transform software showing a smoothed Lava Waves color scheme. Compare with the smoothed Rainbow scheme shown in Color Plate 21.18. (Image created with Noesys Visualization Pro, courtesy of Fortner Software LLC.)

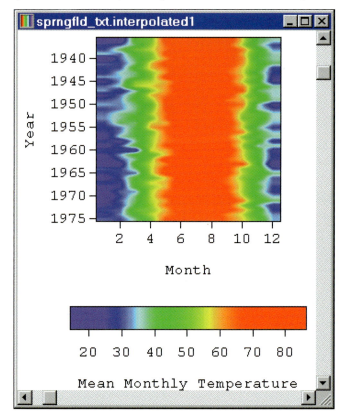

COLOR PLATE 21.20 A window from the Transform software showing a smoothed Rainbow color scheme shifted left and compressed (compare with Color Plate 21.18). (Image created with Noesys Visualization Pro, courtesy of Fortner Software LLC.)

Plate 36

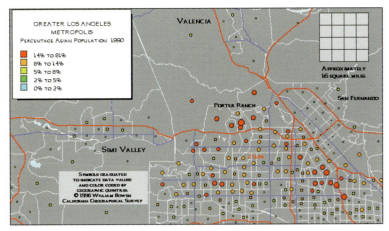

COLOR PLATE 22.1 A portion of a map from the Digital Atlas of California, which is part of William Bowen's Electronic Map Library. This California atlas emulates traditional paper atlases in the sense that no interaction with the maps is possible. (Courtesy of William Bowen.)

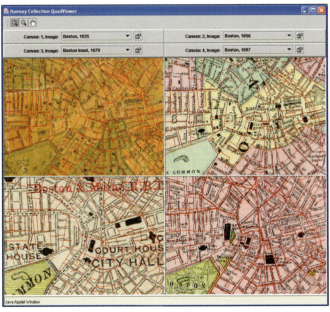

COLOR PLATE 22.3 Four historical maps of Boston created using the QuadViewer option within a GIS browser available in the David Rumsey Map Collection. Note that the modern street network is overlaid on each map. (Courtesy of the David Rumsey Map Collection, *www.davidrumsey.com.*)

COLOR PLATE 22.2 A portion of William Henry Holmes's 1882 rendition of the Grand Canyon—an example of the more than 8,800 maps available in the David Rumsey Map Collection. (Courtesy of the David Rumsey Map Collection, *www.davidrumsey.com.*)

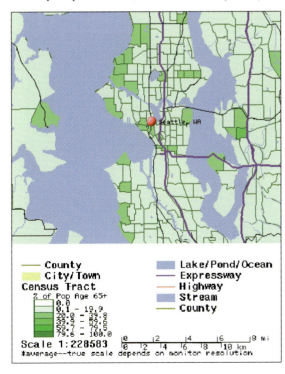

COLOR PLATE 22.4 A map of the Seattle, Washington, region created using the TIGER Map Server. Base data include water bodies and interstate highways, and the theme is the percent of the population 65 years and older within census tracts.

Plate 37

COLOR PLATE 22.5 A map of major hurricanes that passed over land in the 1960s—an example of a map that can be created using the Interactive Maps option within The National Atlas of the United States.

A

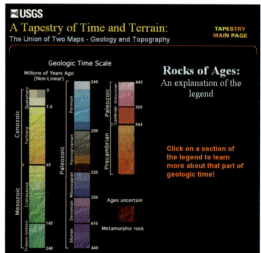

B

COLOR PLATE 22.6 Images from Tapestry of Time and Terrain, part of The National Atlas of the United States: (A) the "Color Legend" depicting the geologic time scale; (B) an image that appears when Precambrian is clicked in A.

Plate 38

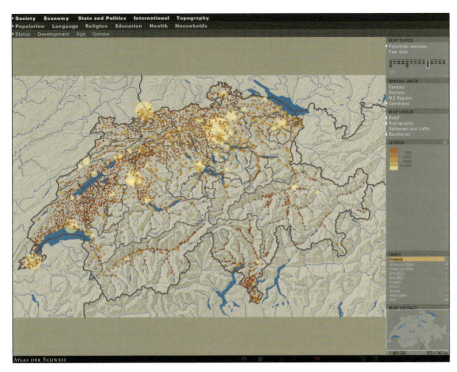

COLOR PLATE 22.7   A proportional symbol map from the Atlas of Switzerland at the commune level. This constitutes one frame (1950) from an animation that can be run from 1850 through 1990 (see the upper right corner). In this case, Relief, Hydrography, and Boundaries are shown, although each of these can be deselected. Also note the redundant symbology. (Courtesy of the Institute of Cartography, Zurich, Switzerland.)

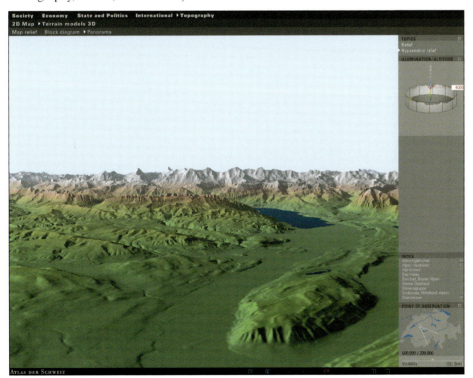

COLOR PLATE 22.8   A panoramic map from the Atlas of Switzerland. Note that the view is to the southeast, as indicated by the Point of Observation in the lower right of the image. (Courtesy of the Institute of Cartography, Zurich, Switzerland.)

Plate 39

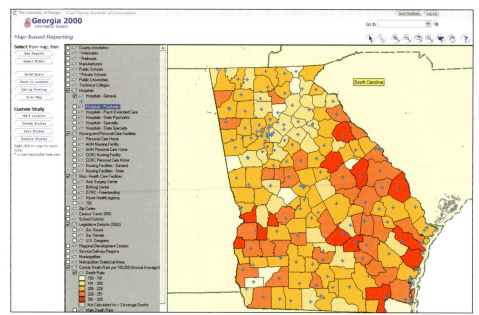

COLOR PLATE 22.9 A map created using the Georgia 2000 Information System. Note that multiple data sets can be mapped on top of one another, the map can be manipulated via zooming and panning, and individual locations can be queried. (Courtesy of the Carl Vinson Institute of Government, The University of Georgia.)

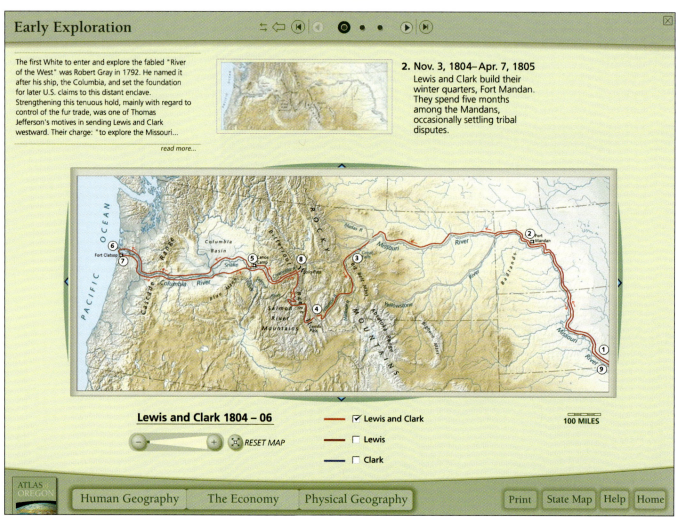

COLOR PLATE 22.10 A map from the electronic version of the Atlas of Oregon depicting Lewis and Clark's famous explorations. As suggested in the legend, any combination of routes can be shown. It is also possible to click on a number and receive historical information regarding that location. (Courtesy of the University of Oregon Press.)

Plate 40

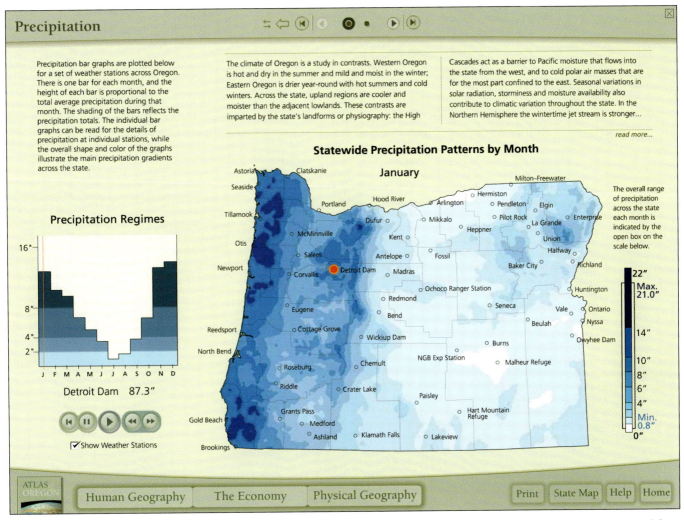

Precipitation bar graphs are plotted below for a set of weather stations across Oregon. There is one bar for each month, and the height of each bar is proportional to the total average precipitation during that month. The shading of the bars reflects the precipitation totals. The individual bar graphs can be read for the details of precipitation at individual stations, while the overall shape and color of the graphs illustrate the main precipitation gradients across the state.

The climate of Oregon is a study in contrasts. Western Oregon is hot and dry in the summer and mild and moist in the winter; Eastern Oregon is drier year-round with hot summers and cold winters. Across the state, upland regions are cooler and moister than the adjacent lowlands. These contrasts are imparted by the state's landforms or physiography: the High

Cascades act as a barrier to Pacific moisture that flows into the state from the west, and to cold polar air masses that are for the most part confined to the east. Seasonal variations in solar radiation, storminess and moisture availability also contribute to climatic variation throughout the state. In the Northern Hemisphere the wintertime jet stream is stronger...

read more...

Statewide Precipitation Patterns by Month

January

Precipitation Regimes

Detroit Dam   87.3"

☑ Show Weather Stations

The overall range of precipitation across the state each month is indicated by the open box on the scale below.

Human Geography   The Economy   Physical Geography

Print   State Map   Help   Home

COLOR PLATE 22.11   An illustration of the potential of animation and interaction within the electronic version of the Atlas of Oregon. Clicking on the large arrow in the lower left starts a month-by-month animation of isarithmic maps of precipitation for the state. Pointing the mouse cursor at a weather station displays a bar graph illustrating the precipitation regime for that station over the course of the year (on the left is one for Detroit Dam, which is highlighted in red). (Courtesy of the University of Oregon Press.)

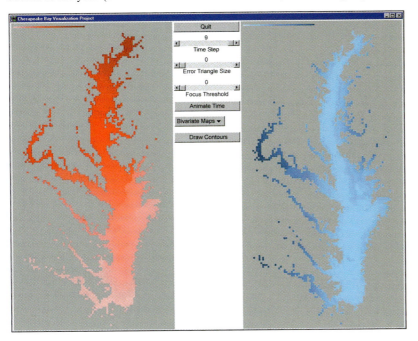

COLOR PLATE 23.1   The default display for RVIS. On the left is an isarithmic map resulting from interpolating between the 49 point locations for which dissolved inorganic nitrogen (DIN) values were collected. On the right is an uncertainty map based on a 95 percent confidence interval computed using kriging. Note that different hues are used for each map, and that lightness and saturation varies within each hue. (Courtesy of David Howard and Alan MacEachren.)

Plate 41

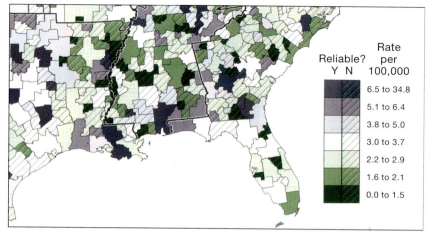

COLOR PLATE 23.2 MacEachren et al.'s (1998) approach for depicting uncertainty in which parallel lines are split down their length, with one side black and the other white. (From "Visualizing georeferenced data: Representing reliability of health statistics." *Environment and Planning A* 30: 1547–1561. Pion Limited, London. Courtesy of Alan MacEachren.)

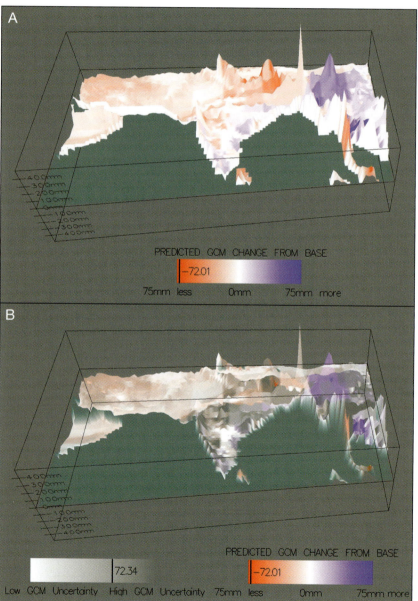

COLOR PLATE 23.3 Visualizations developed by Cliburn et al. (2002): (A) height of the surface shows water surpluses and deficits based on historical data, whereas the surface is colored to reflect changes anticipated based on average GCM model data; (B) transparency is used to depict areas that are uncertain—this is an intrinsic approach. (Courtesy of Daniel C. Cliburn.)

Plate 42

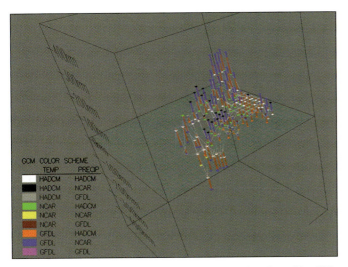

COLOR PLATE 23.4 Extrinsic visualization developed by Cliburn et al. (2002). Orange and purple bars represent the range of GCM predictions at a particular location, and small pyramid-like symbols at the end of bars denote which GCMs were associated with the extreme low or high point on a bar. (Courtesy of Daniel C. Cliburn.)

COLOR PLATE 24.1 An example of Ian Bishop's early work on visual realism—A simulation of a proposed lake created as a result of filling an existing open-cut coal mine with water. (Reprinted from *Visualization in Modern Cartography*, Ian Bishop, "Using Wavefront Technology's Advanced Visualizer Software to visualize environmental change and other data," pp. 99, 101–103, Copyright 1994, with permission from Elsevier.)

COLOR PLATE 24.2 An example of a wall-size display for creating a VE. The image portrays decision makers examining the uncertainty associated with a water balance model. (Courtesy of James R. Miller.)

COLOR PLATE 24.3 The user interface for Fuhrmann and MacEachren's software that utilized a flying saucer metaphor. On the left is the egocentric view; on the right is an overview map with a directional cone indicating the user's current position and orientation. (Courtesy of Sven Fuhrmann.)

Plate 43

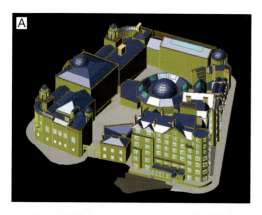

COLOR PLATE 24.5 Example of a virtual city for a portion of Washington, DC. (Courtesy of Urban Data Solutions, Inc.)

COLOR PLATE 24.4 Examples of the Bath virtual city: (A) one block of the more than 150 blocks included in the city, (B) an aerial view of the Abby in Bath. (Courtesy of CASA, University of Bath.)

COLOR PLATE 24.6 A scene from Virtual Los Angeles. (Courtesy of Bill Jepson, Director UCLA Urban Simulation Laboratory.)

Plate 44

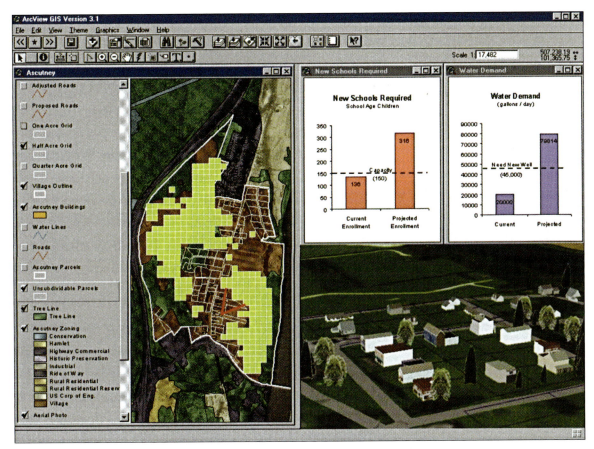

COLOR PLATE 24.7   A scene generated by Community Viz. (Source: ESRI.)

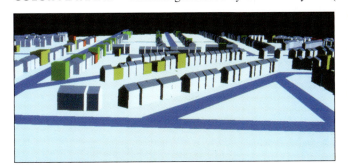

COLOR PLATE 24.8   An example of Martin and Higgs's use of realistic structures to depict abstract phenomena in which shape indicates type of property (terraced house, purpose-built flats, etc.), size indicates number of rooms, and color indicates use (rented rooms, private houses, business premises, etc.). (Courtesy of David Martin.)

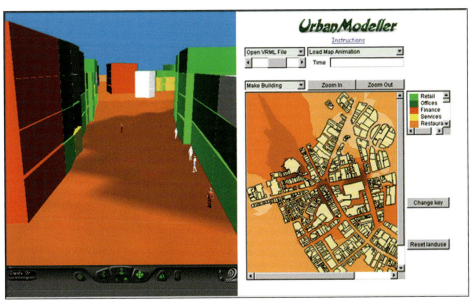

COLOR PLATE 24.9   A screen from Moore's Urban Modeller, which permits modeling the 3-D urban environment and animating pedestrian flows. (Courtesy of Kate Moore.)

Plate 45

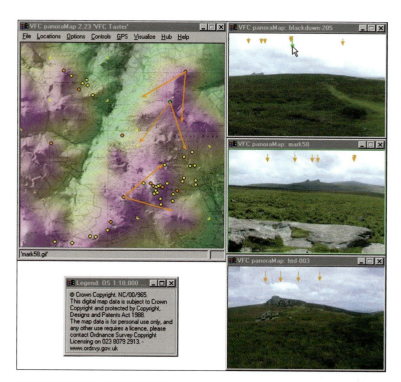

COLOR PLATE 24.10 How panoramic images created by panoraMap are utilized in the Virtual Field Course project. The orange arrows on the left correspond to the three views shown on the right. (Courtesy of Jason Dykes, City University; Reproduced from Ordnance Survey mapping on behalf of The Controller of Her Majesty's Stationary Office © Crown Copyright.)

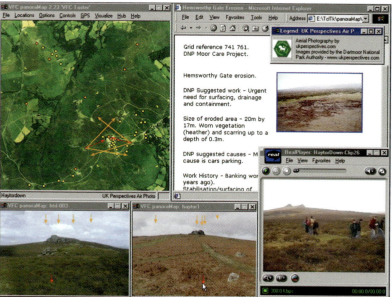

COLOR PLATE 24.11 An illustration of the power of panoraMap to link other kinds of information with panoramic images and to explore this information in a spatial context. The lower left views correspond to the orange arrows shown above. In the lower right is a georeferenced digital movie containing audio and video collected in the field, whereas the upper right is a hypertext document containing images, links, and other information. (Courtesy of Jason Dykes, City University.)

COLOR PLATE 24.12 A forest scene generated using Landscape Viewer, a component of Virtual Forest. (Courtesy of Buckley, D. and J. K. Berry, Virtual Forest Software for Advanced 3D Visualization of GIS Databases; see *www.innovativegis.com/basis*, select Virtual Forest.)

Plate 46

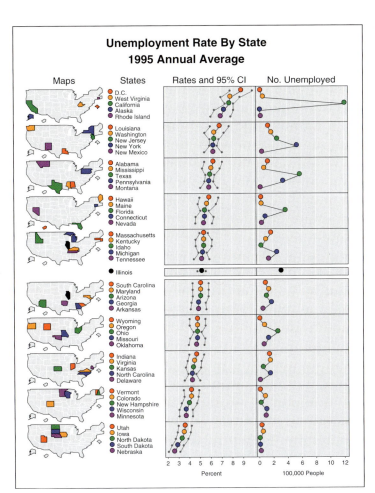

COLOR PLATE 25.1   Example of a linked micromap plot (LM plot). (Courtesy of Daniel Carr.)

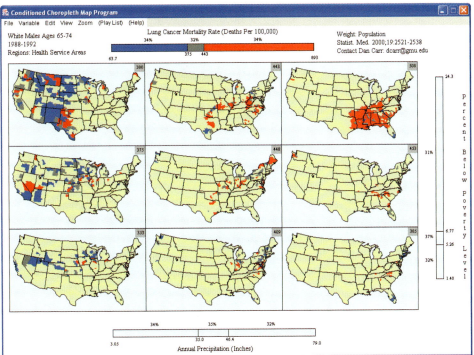

COLOR PLATE 25.2   Example of conditioned choropleth maps (CCmaps). The attribute being studied is the lung cancer mortality rate for White men aged 65 to 75. Potential explanatory attributes, annual precipitation, and percent below the poverty line are depicted from left to right, and from bottom to top, respectively. (Courtesy of Daniel Carr.)

Plate 47

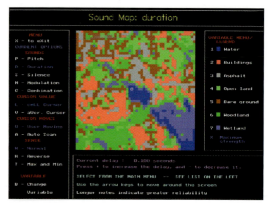

COLOR PLATE 25.3 A screen from Fisher's software for portraying the uncertainty in remotely sensed images via sound. As the cursor moves across the image, the user hears a sound representing the uncertainty associated with the current pixel location. In the case depicted here, a long duration would indicate a pixel with low uncertainty (high reliability). (Courtesy of Peter F. Fisher.)

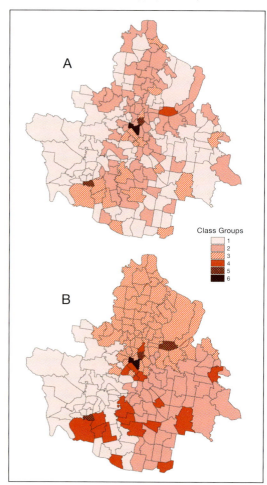

COLOR PLATE 25.5 Two maps resulting from applying Murray and Shyy's median clustering method: (A) weights of 1.0 and 0.25 are assigned to crime rates and distance, respectively; (B) weights of 1.0 and 0.95 are assigned to crime rates and distance. (An adaptation of Figures 4 and 5 from Murray and Shyy, 2000, "Integrating attribute and space characteristics in choropleth display and spatial data mining." *International Journal of Geographical Information Science* 14(7), pp. 649-667; courtesy of Taylor & Francis Ltd., *http://www.tandf.co.uk/journals*)

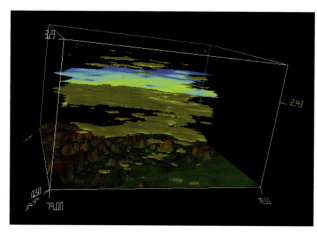

COLOR PLATE 25.4 An image viewed in MacEachren and his colleagues' collaborative work with ImmersaDesks. The bottom portion shows a 3-D portrayal of terrain; the middle portion represents temporal changes in precipitation (time increases with height), with olive green "clouds" depicting precipitation isosurfaces above a certain threshold value (Hurricane Agnes is depicted by a "blanket" of rain multiple days thick); and the top portion portrays temperature at the end of the hurricane event. (From Figure 11 (p. 24) of MacEachren and Brewer, 2004, "Developing a conceptual framework for visually-enabled geocollaboration," *International Journal of Geographical Information Science* 18(1), pp. 1-34; courtesy of Taylor and Francis Ltd., *http://www.tandf.co.uk/journals*)

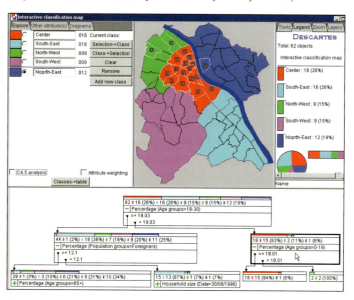

COLOR PLATE 25.6 An illustration of Natalia Andrienko and her colleagues' use of the C4.5 algorithm for spatial data mining. The portion of the decision tree highlighted with the thick black line is depicted on the map by small squares. (After Andrienko et al. 2001. First published in *Cartography and Geographic Information Science* 28(3), p. 158. Reprinted with permission from the American Congress on Surveying and Mapping.)

Plate 48

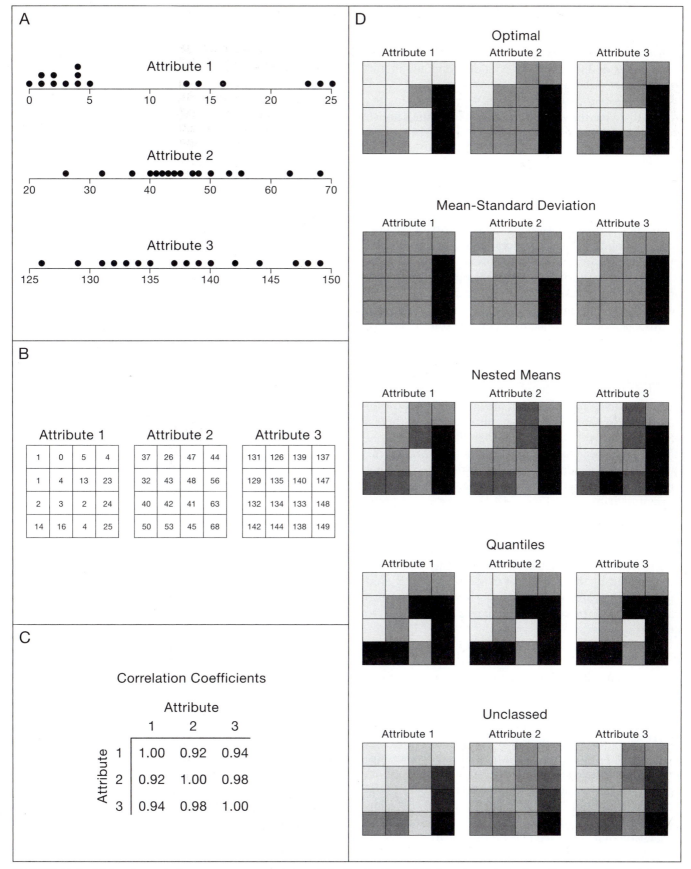

FIGURE 18.1 Using choropleth maps to compare geographic distributions: (A) three hypothetical attributes (attribute 1 is positively skewed, whereas attributes 2 and 3 are normal); (B) maps of the raw data for the three attributes; (C) correlation coefficients (*r* values) between each pair of attributes; (D) maps for differing methods of classification.

distributions. These results are not surprising given the fact that means and standard deviations generally should be used only with normal distributions.

In the **nested-means approach,** the mean of the data is used to divide the data into two classes—values above and below the mean. The resulting classes can be further subdivided by again computing the mean of each class, and this process can be repeated until no further subdivision is desired. The results of using nested means to create four classes are shown in Figure 18.1D. Like the mean–standard deviation approach, nested means seems to do a better job of portraying the similarity of the normal distributions. This is to be expected, as the mean used to define class breaks is an appropriate measure of central tendency for normal distributions, but not for skewed ones. Another weakness of nested means is that the number of classes can only be a power of 2 (2, 4, 8, 16, etc.); in this case, this complicates a comparison of this approach with the other methods of classification.

The quantiles method of classification places an approximately equal number of observations in each class based on the ranks of the data (see section 5.1.2). Because the classes resulting from the quantiles method are unaffected by the magnitudes of the data, the method is arguably appropriate for comparing differently shaped distributions. For example, for the hypothetical data, the quantiles method portrays high correlations between not only the normal distributions but also the skewed and normal distributions (Figure 18.1D). The **equal-areas method** of classification is similar in concept to quantiles, but rather than placing an equal number of observations in classes, an equal portion of the map area is assigned (the desired area in each class is simply the area of the map divided by the number of classes desired). If enumeration units are equal in size (as in the hypothetical data), the equal-area method produces a map identical to quantiles.

Up to this point, we have focused on comparing classed choropleth maps. As with univariate choropleth maps, it is natural to ask whether the issue of selecting an appropriate classification method might be obviated by simply not classing the data. To illustrate, consider the unclassed maps of the hypothetical data shown in Figure 18.1D. Here the visual impression seems similar to that for the mean–standard deviation and nested means approaches, as attributes 2 and 3 (the normal distributions) appear similar to one another, whereas attribute 1 (the skewed distribution) does not appear as highly correlated with the other attributes. This result leads one to the conclusion that unclassed maps should be used for comparative purposes only when the distributions have similar shapes.

At the beginning of this section, we assumed that we wished to compare two attributes for a single point in time (say, median income and percent of the adult population with a college education for 2000). For such data, we concluded that the optimal method is inappropriate. It should be noted, however, that if we wish to compare the same attribute for two points in time (say, median income for 1990 and 2000), the data could be combined into a single data set, and the optimal method could be applied to the combined set.

One limitation of this section is that our conclusions are based on a subjective interpretation of only those attributes shown in Figure 18.1. To alleviate this problem, let us now consider some of the more formal studies done by cartographers. Robert Lloyd and Theodore Steinke (1976; 1977) found that the visual correlation of maps is affected by the amount of blackness on each map (assuming that gray tones are used for symbolization); in other words, if maps A and B and C and D have the same statistical correlation, maps A and B will be judged more similar if their blackness levels are more similar than for maps C and D. As a result, Lloyd and Steinke argued for using equal areas (which we have indicated produces results similar to quantiles) when comparing choropleth maps.

Judy Olson undertook two studies to determine which classification methods appeared to most accurately preserve the correlation between two attributes. In Olson's (1972b) first study, she analyzed 300 pairs of attributes derived from theoretically normal distributions and found that quantiles was best at reflecting the correlation between the attributes. In her second study, Olson (1972a) analyzed 300 pairs of real-world attributes (which were primarily not normally distributed) and found that the mean–standard deviation and nested means methods were most effective. These results conflict with some of our own conclusions because of Olson's stress on broad relationships (she focused on a scatterplot of the correlation coefficient and a measure of rank correlation for all pairs of attributes) as opposed to looking at detailed graphs and maps of each distribution, as was done in Figure 18.1.

Studies by Michael Peterson (1979) and Jean-Claude Muller (1980b) dealt with people's visual comparison of both classed and unclassed choropleth maps. These studies concluded that people perceived similar correlations on pairs of classed and unclassed maps; as a result, the authors raised questions about the need to class data for choropleth mapping. Although the results of Peterson and Muller's studies contradict our recommendation to use unclassed maps only when comparing distributions having the same shape, we suspect that a detailed examination of data distributions would support our recommendation.

Strong support for our conclusion that quantiles is particularly appropriate for map comparison comes from the work of Cynthia Brewer and Linda Pickle (2002). In comparing a wide variety of classification methods in actual map comparison tasks, they found quantiles to be the most effective (a minimum boundary error method based on Cromley's (1996) work was a close second). They concluded that "quantiles seems to be one of the best methods for facilitating comparison as well as aiding general map-reading" (p. 679).

Miscellaneous Thematic Maps

Although it is common to hold the mapping method constant when comparing maps (e.g., show two choropleth maps or two isopleth maps), useful information often can be acquired by comparing two different kinds of thematic maps. This is especially true when one map is used to show raw totals and another standardized data. For example, consider Figure 18.2, which compares a proportional symbol map of the raw number of infant mortalities in New Jersey with a choropleth map of the infant mortality rate. The proportional symbol map suggests that the "problem" of infant mortality is in the northeastern part of the state (where the largest circles are). This map by itself, however, is not very meaningful because the pattern is likely a function of population (counties with more people are apt to have more infant deaths). In contrast, the choropleth map standardizes the raw mortality data by considering the number of deaths relative to the number of live births, and suggests that the "problem" is found in three areas of the state (represented by the shades in the highest class). Unfortunately, a high rate on the choropleth map might not be meaningful if there are also few deaths. Only when the two maps are viewed together can the complete picture emerge: The northernmost high-rate area is most problematic because it is located where the raw number of deaths is high.

18.1.2 Combining Two Attributes on the Same Map

In this section, we consider approaches for bivariate mapping in which two attributes are combined on the same map. Again, we focus on choropleth maps because of their common use.

Bivariate Choropleth Maps

In the 1970s, the U.S. Bureau of the Census developed the **bivariate choropleth map**—a method for combining two colored choropleth maps (see Meyer et al. 1975 for a summary of the technical methods used at that time). Color Plate 18.1 depicts a bivariate choropleth map using a color scheme similar to one used by the Bureau of the Census, along with each of the univariate distributions used to create the bivariate map.* A listing of the factors considered in constructing the Bureau of Census color schemes is given in Table 18.2.

Although bivariate choropleth maps were deemed a success from a technical standpoint, they received considerable criticism for their presumed failure to communicate either information about individual distributions or the correlation between them. In response to these criticisms, Judy Olson (1981) conducted an experimental

---

* Specifications for the colors used in Color Plate 18.1 were taken from Olson (1981, 269). For a discussion of the logic behind the selection of these colors, see Olson (1975b).

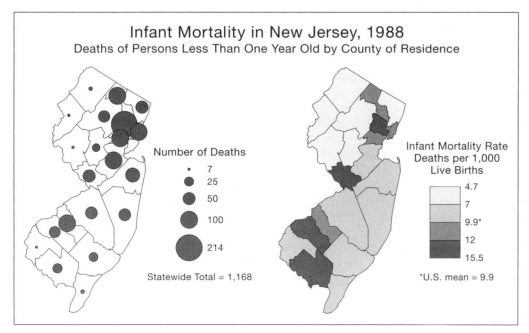

FIGURE 18.2 A comparison of proportional symbol and choropleth maps. The proportional symbol map provides information on the raw number of infant mortalities, whereas the choropleth map focuses on the rate of infant mortality. To understand the complete picture of infant mortality, both maps are necessary. (From Monmonier, *Mapping It Out: Expository Cartography for the Humanities and Social Sciences*, p. 166; © 1993 by The University of Chicago Press. All rights reserved. Published 1993.)

TABLE 18.2 Factors considered in developing color schemes for the U.S. Bureau of the Census bivariate choropleth maps

1. All colors must be distinguishable.
2. The transition of colors should progress smoothly in a visually coherent way.
3. The individual categories of each distribution should be visually distinguishable or coherent, and the two distributions as a whole should be separable from one another.
4. The arrangement of the colors presented in the legend should correspond to the arrangement of a scatter diagram.
5. Tones should progress from lighter to darker corresponding to a change in the numerical values from low to high.
6. Extreme values (legend corners) should be represented by pure colors.
7. There should be coherence in the triangle of cells above and below the main diagonals to show positive and negative residuals.
8. To convey relationship, positive diagonals (lower left to upper right) and negative diagonals (upper left to lower right) should have visual coherence.
9. The design of the color-coding scheme should take into account the difficulty in mentally sorting large numbers of colors in the legend.
10. The color scheme should relate to the data in such a way that the map relationship reflects as closely as possible the statistical relationship.
11. The crossed version of the map should be constructed as a direct combination of the specific sets of colors assigned to the two individual maps.
12. The combination of colors on the two individual maps should look like combinations of the specific colors involved.
13. The number of categories to be used should not exceed the number that can be dealt with by the reader. A 3 × 3 legend is both mechanically and visually simpler than a 4 × 4 arrangement and might actually convey more to the reader.
14. Alternatives to a rectangular arrangement to the legend should be considered. The rectangular form creates map interpretation problems and affects the message of the statistical relationship.

After Olson, 1975b, as specified by Eyton, J. R. (1984a) "Complementary-color two-variable maps." *Annals, Association of American Geographers,* 74, no. 3, p. 480. Courtesy of Blackwell Publishing.

study using color schemes similar to those used by the Bureau of the Census. In contrast to the earlier criticism, Olson found that bivariate maps provided "information about regions of homogeneous value combinations" and that users, rather than being confused by these maps, actually had a positive attitude about them (p. 275). Olson, however, stressed that a clear legend was critical to understanding bivariate maps, both bivariate and individual maps should be shown,* and an explanatory note should

be included describing the types of information that can be extracted (see p. 273 of Olson's work for an example).

As an alternative to the Bureau of the Census approach, J. Ronald Eyton (1984a) developed a bivariate method based on **complementary colors**, or colors that combine to produce a shade of gray.[†] Eyton chose red and cyan as the complementary colors for his maps, presumably because they produced an attractive map. Using the subtractive primary colors (CMY), Eyton created red by combining magenta (M) and yellow (Y), and cyan (C) was created directly from the cyan subtractive primary. Because overprinting colors in CMY does not produce a true gray, Eyton used black ink (K) for areas in which a true gray was desired. A bivariate map resulting from Eyton's process is shown in Color Plate 18.2.[◊]

Eyton argued that most of the factors listed in Table 18.2 were accounted for by his complementary method and that users appeared to understand the map more easily than one based on Bureau of the Census colors. A visual comparison of Color Plates 18.1 and 18.2 suggests that Eyton was correct. The Bureau of the Census color scheme implies nominal differences and requires careful examination of the legend, whereas Eyton's scheme appears more logically ordered and allows patterns on the map to be discerned more easily. Note, for example, the ease with which the reddish-brown values (values above the white–gray–black diagonal) can be found using the Eyton scheme, as compared with corresponding values using the Bureau of the Census color scheme.

In addition to suggesting that complementary colors be used, Eyton also recommended that the statistical parameters of the distributions should be considered in bivariate mapping. Specifically, Eyton made use of the reduced major axis and bivariate normality. Recall from section 3.3.2 that the reduced major axis method fits a regression line to a set of data such that the line bisects the regression lines of Y on X (Y is treated as the dependent attribute) and X on Y (X is treated as the dependent attribute; Figure 18.3). The reduced major axis is thus appropriate when it is not clear which attribute should be treated as the dependent attribute.

A distribution is considered **bivariate-normal** if the Y values associated with a given X value are normal and the X values associated with a given Y value are also normal. Generally, this condition is met if the individual distributions of X and Y are normal. Given a bivariate-normal distribution, it is possible to construct an **equiprobability ellipse** enclosing a specified percentage of the data. The equiprobability ellipse will be centered

---

* The univariate maps should be shown in black-and-white so that the two distributions can be readily compared. In Color Plate 18.1 they are shown in color so that the colors composing the bivariate map are clear.

[†]Colors opposite one another on the Munsell color circle (Figure 10.19) are complementary.

[◊] Eyton used cross-hatching (narrowly spaced horizontal and vertical lines) for intermediate shades, whereas we have chosen smooth tones.

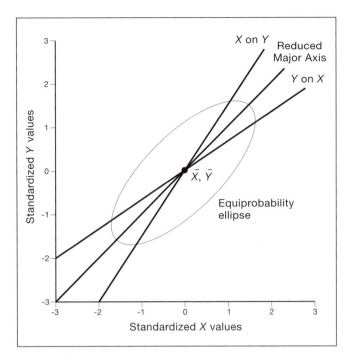

FIGURE 18.3 A reduced major axis and associated equiprobability ellipse. The reduced major axis is appropriate when one does not wish to specify a dependent attribute. The equiprobability ellipse can be used to enclose a specified percentage of the data associated with a bivariate normal distribution.

on the mean of the data and oriented in the direction of the reduced major axis (Figure 18.3). Using this ellipse and the means of the two attributes, Eyton created a bivariate map analogous to the one shown in Color Plate 18.3.* Eyton argued that the resulting map clearly contrasted observations near the means of the data (as defined by the 50 percent equiprobability ellipse) with extreme observations (those well above the mean on both attributes, those well below the mean on both attributes, and those high on one attribute and low on the other). For example, in Color Plate 18.3 we can clearly see that the countries of Botswana, Zimbabwe, Lesotho, and Kenya (those shown in cyan) have a low percent urban and high female life expectancy.

More recently, Cynthia Brewer (1994a) argued that color schemes for bivariate maps should be a function of whether the attributes are unipolar or bipolar in nature. For two unipolar data sets (e.g., percent urban and female life expectancy), she recommended using either a complementary scheme or two subtractive primaries; the latter produces a diagonal of shades of a constant hue,

as opposed to the grays resulting from complementary colors. (Brewer combined magenta and cyan to create a diagonal of purple-blues.) If one attribute is unipolar and the other bipolar, she recommended using a sequential scheme for the unipolar data and a diverging scheme for the bipolar data. If both are bipolar she recommended two diverging color schemes (Color Plate 18.4).

A potential concern in bivariate choropleth mapping is the number of classes used for each attribute. In the examples presented thus far we have used three classes for each attribute. Our thinking here parallels item 13 in Table 18.2, which states that "The number of categories to be used should not exceed the number that can be dealt with by the reader."

The approaches we have considered to this point for bivariate choropleth mapping have all been based on smooth (untextured) colors. As an alternative, both Laurence Carstensen (1982; 1986a; 1986b) and Stephen Lavin and J. Clark Archer (1984) created bivariate maps using cross-hatched shading consisting of horizontal and vertical lines (Figure 18.4). Interpretation of these maps focuses on the size and shape of the boxes formed by the cross-hatched lines (low values on both attributes are represented by large squares, whereas high values on both attributes are represented by small squares). A high positive correlation is represented on the map by a predominance of squares of varying sizes within enumeration units, and a high negative correlation is depicted by rectangles. In the case of Figure 18.4, we see a fair number of rectangles, which supports the correlation of $r = -.71$.

Both Carstensen and Lavin and Archer stressed that *unclassed* cross-hatched bivariate maps should be used (thus the legend in Figure 18.4 contains no class breaks). This suggestion contrasts with that of Eyton (1984a, 485–486), who found smooth-toned unclassed bivariate maps difficult to interpret. The reason that unclassed cross-hatched maps appear to be more effective is that the line-spacing attribute allows individual attributes to be seen. Carstensen (1982) found that cross-hatched bivariate maps were reasonably effective in communicating concepts about correlation, but noted two problems: the difficulty of shading small enumeration units and the unpleasant appearance of the symbology. These problems cause this method to be less frequently used than the color methods mentioned earlier.

Other Bivariate Maps

The preceding section focused on the use of choropleth symbology to depict two attributes on the same map. The **bivariate point symbol** is one alternative to choropleth symbology. Figure 18.5 depicts one form of bivariate point symbol, the **rectangular point symbol**, in which the width and height of a rectangle are made proportional

---

* Given the requirement of normality, we chose to map two normal distributions: percent urban and life expectancy for females.

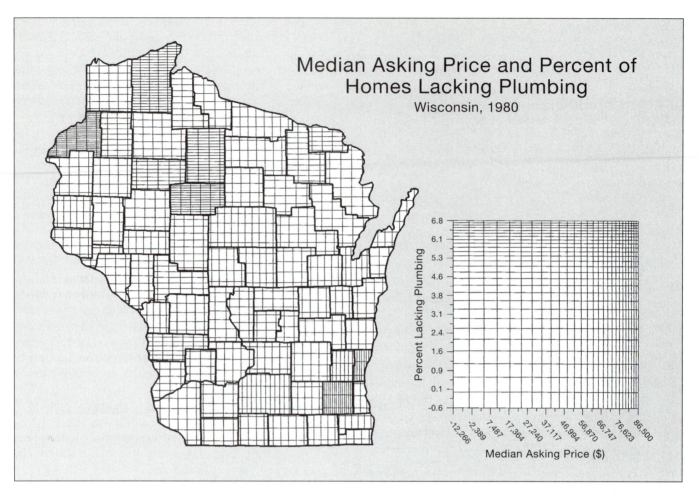

FIGURE 18.4 A bivariate choropleth map based on cross-hatching (the attributes are represented by horizontal and vertical lines of varying spacing). Note that each attribute is unclassed; negative values in the legend are a function of the major axis scaling used to fit the bivariate data. (From Carstensen 1986a, p. 36; courtesy of Laurence W. Carstensen.)

to each of the attributes being mapped. Stanton Wilhelm (1983) and Sean Hartnett (1987) each proposed this method as an alternative to cross-hatched symbology; Hartnett argued that examining rectangular point symbols is much easier than inspecting the small boxes formed by cross-hatched lines and that the resulting map is more aesthetically pleasing. Because point symbols are more readily associated with point locations than areas, bivariate point symbols might be particularly appropriate for mapping true point data for two attributes (say, the number of out-of-wedlock births to teenagers by city and number of hours of sex education for high school students by city).*

Another form of bivariate point symbol is the **bivariate ray-glyph**, which Daniel Carr and his colleagues (1992) used to examine the relationship between nitrate (NO_3) and sulfate (SO_4) concentrations in the eastern

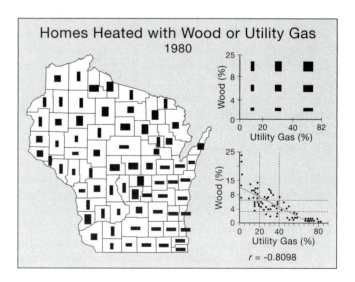

FIGURE 18.5 A bivariate map in which attributes are represented by the width and height of a rectangular point symbol. (Courtesy of Sean Hartnett.)

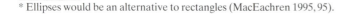

* Ellipses would be an alternative to rectangles (MacEachren 1995, 95).

United States and Canada. In Figure 18.6, rays (the straight-line segments) pointing to the left represent nitrate concentrations, and rays pointing to the right represent sulfate concentrations. High values on both attributes occur when both rays extend toward the top of the symbol, whereas low values occur when the rays extend toward the bottom of the symbol. An advantage of the ray-glyph is that the small symbols can be squeezed into a relatively restricted space. It seems likely, however, that the patterns represented by these symbols would be more difficult to interpret than those for the rectangular symbol.

A relatively common approach in bivariate mapping is to combine proportional and choropleth symbols, with the size of the proportional symbol used for raw-total data and a choropleth shade within the symbol used for standardized data. For example, this approach could be used to depict on a single map the infant mortality data presented as two maps in Figure 18.2. In section 16.6, we suggested that these same symbols could be used redundantly (both size and shading could be used to represent a single attribute). Obviously, a legend is critical in determining whether the symbols should be interpreted in a bivariate or redundant fashion.

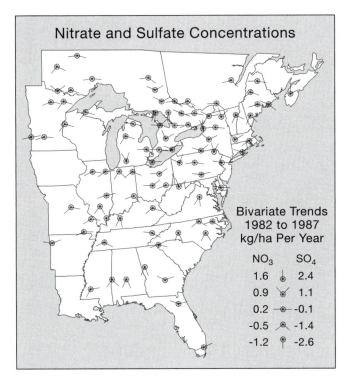

FIGURE 18.6 A bivariate map based on a ray-glyph symbol. Rays (straight-line segments) pointing to the right and left represent sulfate and nitrate concentrations, respectively. (After Carr et al. 1992. First published in *Cartography and Geographic Information Systems* 19(4), p. 234. Reprinted with permission from the American Congress on Surveying and Mapping.)

18.2 MULTIVARIATE MAPPING

18.2.1 Comparing Maps

If more than two attributes are shown, each as a separate map, the result is termed a **small multiple** (Figure 18.7). Edward Tufte (1990, 33) argued, "Small multiples, whether tabular or pictorial, move to the heart of visual reasoning—to see, distinguish, choose....Their multiplied smallness enforces local comparisons within our eyespan, relying on the active eye to select and make contrasts."

Although much can be gleaned from small multiples, they clearly have their limitations. A general problem is that comparing two particular points or areas across a set of attributes can be difficult: Try using Figure 18.7 to describe the nature of agriculture in the state of Michigan. In the case of choropleth maps, a problem is the difficulty of discerning small enumeration units (a number of countries would disappear for a small multiple of choropleth maps of Africa). Problems such as these have led researchers to develop methods for combining multiple attributes onto the same map.

18.2.2 Combining Three or More Attributes on the Same Map

Trivariate Choropleth Maps

The notion of overlaying two colored choropleth maps can be extended to three choropleth maps, thus producing a **trivariate choropleth map**. Ideally, this approach should be used only for three attributes that add to 100 percent; examples include soil texture (expressed as percent sand, silt, and clay) and voting data for three political parties (e.g., percent voting Republican, Democrat, and independent). Colors can be assigned to the three attributes using a variety of approaches: CMY (Brewer 1994a, 142), RGB (Byron 1994), and red, blue, and yellow primaries (Dorling 1993, 172). The result of using RGB is shown in Color Plate 18.5. The advantage of using attributes that add to 100 percent is that the resulting colors will be restricted to a triangular two-dimensional space. Brewer (1994a) argued that if the attributes do not sum to 100 percent, a 3-D cube-shaped legend will be required and the resulting map will be difficult to interpret.

A potential alternative for creating a trivariate choropleth map is to use *patterns* (or *textures*) as a substitute for smooth colored tones. For instance, Victoria Interrante (2000) created the multivariate map of agriculture for U.S. counties shown in Color Plate 18.6 by using texture for two attributes and color for a third attribute. One attribute (percent change in the number of farms) was represented by the type of texture (weave vs. rocks), and another attribute (land in farms) was represented by the

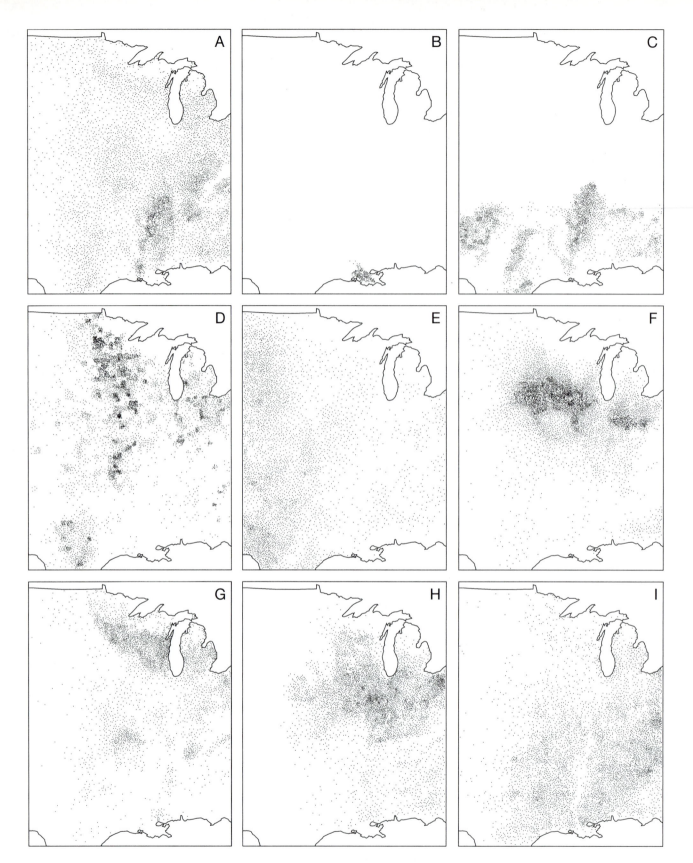

FIGURE 18.7 The small multiple: a method for multivariate mapping in which each attribute is depicted as a separate map. Maps depict the following agricultural data for 1954 for the central portion of the United States: (A) class V farms, (B) sugar cane, (C) cotton, (D) turkeys, (E) pasture, (F) hogs and pigs, (G) dairy, (H) expenditures for lime, and (I) residential farms. (After Lindberg 1987; courtesy of Mark B. Lindberg.)

size of the texture (e.g., small vs. large rocks). Color was used to depict the third attribute (average age of farm operators). Here we can see some of the difficulty of trying to interpret three attributes simultaneously (for instance, try describing the character of agriculture throughout Nevada on the three attributes), and trying to create a suitable legend (the legend shown here is a modification of Interrante's).

Multivariate Dot Maps

The notion of univariate dot mapping (described in Chapter 17) can be extended to create a **multivariate dot map** if a distinct shape or color of symbol is used for each attribute to be mapped.* In the case of color, George Jenks (1953b) introduced the notion of pointillism for multivariate dot mapping and developed two major maps based on it (Jenks 1961; 1962). Pointillism was a technique used by 19th-century painters to create various color mixtures by having the viewer visually combine very small dots of selected colors. Jenks applied this principle by letting different-colored dots represent various crops (or farm products; Color Plate 18.7); he argued that viewers could visually merge the separate colors to create mixtures, thus providing a more realistic view of the transitional nature of cropping practices often found in the landscape. Furthermore, Jenks noted that if the dots were large enough, the map could provide detail regarding the location of individual crops in selected areas. Jenks (1953b, 5) also provided some useful suggestions for dot colors:

1. Colors should remind the map reader of the crop that they represent.

2. High-value, low-acreage crops, such as tobacco or truck, should be of more intense hue than the more extensive and widely grown crops.

3. Selected minor crops, such as peanuts or soybeans, which tend to change the crop character of broader areas, should have colors of moderately high intensity.

Although considerable information can be obtained from multivariate dot maps, readers have difficulty determining the meaning of the areas of color mixture, because the legend contains only colors for individual crops. This problem might be solved if the map were viewed in an interactive data exploration environment in which individual map areas could be focused on and enlarged, and individual categories of dots could be turned on and off so that the relative contribution of each category could be determined. Richard Groop (1992) developed an automated method for creating colored dot maps, but unfortunately his method has not been published.

---

* For a discussion of the use of shape on multivariate dot maps, see Turner (1977).

A general question that can be asked of multivariate maps is how effective they are compared to a set of individual maps of each attribute (i.e., how a multivariate colored dot map would compare to a set of dot maps in small multiple format). Jill Rogers and Richard Groop (1981) evaluated this question by having readers identify regions on both a multivariate dot map (consisting of three categories) and its component univariate dot maps. Readers were asked to identify both "homogeneous regions" in which one category appeared to predominate and "mixed regions" in which there appeared to be a mixture of two or three categories. Rogers and Groop found that the multivariate dot map "was slightly more effective in communicating perceptions of both homogeneous and mixed regions" (p. 61). Their results, however, suggested that the perception of mixed regions on maps with more than three categories might not be effective.

Multivariate Point Symbol Maps

When multivariate data are depicted using a point symbol, the result is termed a **multivariate point symbol**. These symbols are obviously appropriate for point phenomena, but must also be used for areal phenomena because of the difficulty of creating multivariate areal symbols (note the limitations of the trivariate choropleth map described previously when only three attributes are shown). Two distinct forms of attributes are commonly mapped using multivariate point symbols: related (or additive) and nonrelated (or nonadditive). *Related attributes* are measured in the same units and are part of a larger whole. An example would be the percentages of various racial groups in a population: White, African-American, Asian/Pacific Islander, Native American, and so on. Such attributes can be depicted using the familiar **pie chart**, in which a circle is divided into sectors representing the proportion of each attribute.

Nonrelated attributes are measured in dissimilar units and thus are not part of a larger whole (e.g., percent urban and median income). Multivariate point symbols used to represent nonrelated attributes are commonly termed **glyphs**, several of which are illustrated in Figure 18.8. The **multivariate ray-glyph** or star (Figure 18.8A) is constructed by extending rays from an interior circle, with the lengths of the rays proportional to values of each attribute. When originally designed by Edgar Anderson (1960), rays were extended only from the top portion of the circle, but it has become common to extend rays in all directions.† If rays composing the multivariate ray-glyph

---

† Several terms have been used to describe this symbol. The term *star* comes from Borg and Staufenbiel (1992), whereas Anderson called them *metroglyphs*. We use the term *multivariate ray-glyph* because of their similarity to the bivariate ray-glyph introduced earlier.

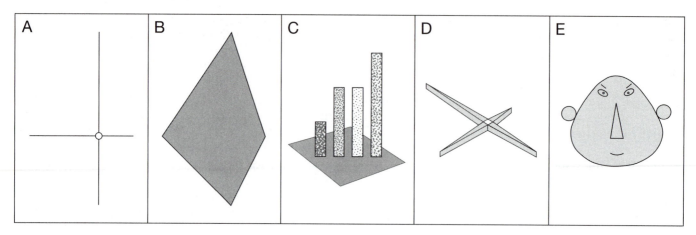

FIGURE 18.8   Examples of multivariate point symbols: (A) a multivariate ray-glyph or star: the length of rays are proportional to the values of attributes; (B) a polygonal glyph or snowflake: a polygon connects the endpoints of the rays shown in A; (C) three-dimensional bars: the height of bars is proportional to the magnitude of attributes; (D) data jacks: the spikes of the jack are proportional to the magnitude of each attribute; and (E) Chernoff faces: individual facial features (e.g., the size of the eyes) are associated with individual attributes.

are connected, a **polygonal glyph** or **snowflake** is created (Figure 18.8B).

Donna Cox (1990) and her colleagues at the National Center for Supercomputing Applications at the University of Illinois at Urbana-Champaign have developed several novel multivariate point symbols. One of these is **three-dimensional bars**, in which the height of bars is made proportional to the magnitude of various attributes (Figure 18.8C); in Cox's implementation, the individual bars were shown in different colors, as opposed to using the different textures seen in Figure 18.8C. Another technique is the **data jack** (Ellson 1990), in which triangular spikes are extended from a square central area and made proportional to the magnitude of each attribute (Figure 18.8D). As with bars, the spikes of jacks can be distinguished most easily if they are displayed in different colors. An advantage of the 3-D structure of jacks is that they can be viewed from arbitrary positions in 3-D space.

A particularly intriguing multivariate point symbol is the **Chernoff face** (Figure 18.8E), in which distinct facial features are associated with various attributes. For example, fatness of the cheeks might represent one attribute, whereas size of the eyes might represent another attribute. Figure 18.9, taken from Daniel Dorling's work with cartograms (see section 19.1.1) is illustrative of Chernoff faces. In this case, the width of the face represents mean house price (fat cheeks indicate more expensive housing); mouth style the percent adult employment (a smile for high employment); nose size the percent voting (a larger nose indicates a higher percent); eye size and position the percent employed in services (large eyes located near the nose indicate a high percent employed in services); and total area of the face the number of voters in a parliamentary constituency.

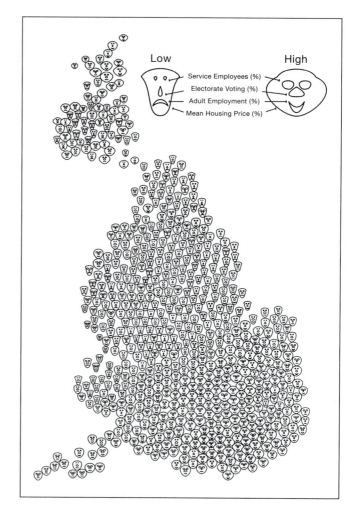

FIGURE 18.9   A cartogram in which Chernoff faces are used to display multivariate data. (*Environment and Planning B: Planning and Design* 1995, Vol. 22, pp. 269–290, Pion, London.)

For simplicity, only four attributes are depicted by the symbols shown in Figure 18.8. Several of these symbols, however, can be modified to represent many more attributes. For example, numerous rays can be added to the star and up to 20 different attributes can be represented by Chernoff faces (Wang and Lake 1978, 32–35).

An unusual form of multivariate point symbol is the "stick-figure" icons (Figure 18.10) used within Exvis, a system for exploratory visualization developed at the University of Lowell, Mass. Icons within Exvis were selected on the basis of *preattentive processing,* or "the ability to sense differences in shapes or patterns without having to focus attention on the specific characteristics that make them different" (Grinstein et al. 1992, 638). Up to 15 attributes could be mapped using one of these icons by varying the angles, lengths, and intensities of the limbs composing the icon.* Grinstein et al. noted that when the icons are dense enough, "they form a surface texture" and that "structures in the data are revealed as streaks, gradients, or islands of contrasting texture" (p. 638). A geographic example is shown in Figure 18.11, where five satellite images of the eastern portion of the Great Lakes have been combined into a single image. In this case, Stuart Smith and his colleagues (1991) argued that "We can see not only Lake Erie and Lake Ontario, but also Lake Huron, Georgian Bay, Lake Simcoe, and some of the smaller outlying lakes" (p. 197).

Another form of multivariate point symbol is the "perceptual texture elements" (or **pexels**) developed by Christopher Healey and his colleagues at North Carolina

FIGURE 18.11 A map created using Exvis. Five satellite images were combined to display water features in the eastern portion of the Great Lakes. (From Smith et al. 1991, p. 197; courtesy of SPIE—The International Society for Optical Engineering and Stuart Smith.)

State University (Healey and Enns 1999; Healey 2001). Pexels are small 3-D bars (similar to those utilized by Cox and her colleagues) that can be varied in height, spacing, and color.† To illustrate, consider Color Plate 18.8, which portrays three attributes related to plankton in the northern Pacific Ocean (plankton are an important food source for salmon). The density of plankton is indicated by color, with low to high plankton values indicated by blue, green, brown, red, and purple, respectively. The speed of ocean current is indicated by the height of pexels (taller pexels indicate a stronger current), and the magnitude of sea surface temperature is indicated by the spacing of pexels (a tighter spacing indicates a warmer temperature). In examining Color Plate 18.8A, we see that in February the density of plankton is relatively low (we see mostly blues and greens). At this time, ocean currents are relatively weak in the north central Pacific and along the south coast of Alaska (low bars are found at these spots), and most of the ocean is relatively cold (pexels are relatively widely spaced), although a warmer region can be seen in the south. In contrast, in June (Color Plate 18.8B) we see a higher density of plankton (reds and purples predominate), stronger currents

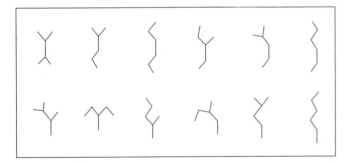

FIGURE 18.10 Some stick-figure icons (a form of multivariate point symbol) used in Exvis, a system for exploratory visualization. Multiple attributes can be mapped using one of these icons by varying the angles, lengths, and intensities of the limbs composing the icon. ("Visualization for Knowledge Discovery", G. Grinstein, J. C. J. Sieg, S. Smith, and M. G. Williams, from the *International Journal of Intelligent Systems*, Vol. 7. Copyright 1992 by Wiley Publishing, Inc. All rights reserved. Reproduced here by permission of the publisher.)

---

* Technically, they specified 17 attributes, but two of these are accounted for by the *x* and *y* coordinate location of the icon.

† Pexels can also vary in orientation, and the density can include a regularity parameter, but we do not illustrate these here.

seem to have shifted southward, and water is warmer further north, although the ocean off the Alaskan and British Columbia coasts is still relatively cold. These sorts of summaries come not only from looking at static images such as those shown here, but also from using interactive software that allows the user to view the 3-D image from different angles, as well as zoom in and pan around. It should be noted that Healey and his colleagues have undertaken numerous experiments to ensure that they are utilizing a reasonable set of parameters for pexels.

Interestingly, most of the multivariate point symbols just described were developed by those outside geography. Micha Pazner is one geographer who has begun to experiment with multivariate point symbols. With his students at the University of Western Ontario, Pazner has experimented with icons in which small fixed symbols within a matrix of pixels are varied in lightness or color (Lafreniere et al. 1996; Pazner and Lafreniere 1997). The resulting images appear rather complex in static printed form, but Pazner et al. argue that much can be learned from these images when they are explored in an interactive graphics environment. Mark Gahegan (1998) is another geographer who has worked with multivariate point symbols. One interesting notion that he suggested is the arrow symbol shown in Figure 18.12.

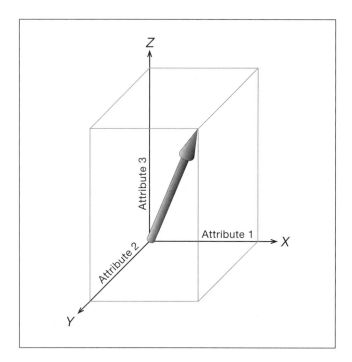

FIGURE 18.12   How an arrow symbol can be utilized to represent multivariate data. Three attributes can be represented, corresponding to the axes of 3-D space. Additionally, three other attributes could be represented by changing the color, width, and transparency of the arrow. (After Gahegan 1998, 46.)

When the origin of the arrow is placed at a location for which one wishes to represent multiple attributes, three attributes can then be represented by the position of the arrow in 3-D space. Although not illustrated in Figure 18.12, the color, width, and transparency of the arrow can be used to depict three additional attributes.

Specific versus General Information.   From Chapter 1, you will recall that two kinds of information can be acquired from univariate thematic maps: specific and general. These same kinds of information can also be acquired from multivariate thematic maps, but readers can examine the distribution of the attributes either individually or holistically (in combination). Thus, we can conceive of the kinds of information acquired from a multivariate map as a two-by-two matrix, with specific–general on one axis and nature of the attributes (individual or holistic) on the other axis. Although it is clear that this notion could be applied to a broad range of multivariate techniques, Elisabeth Nelson and Patricia Gilmartin (1996) illustrated it in the context of multivariate point symbols. To illustrate, consider Figure 18.13, which uses combined star and snowflake symbols to depict "quality of life" in South Carolina counties in 1992. An example of specific information for an individual attribute would be the desirability associated with a particular attribute within a county (that the southernmost county is highly desirable in terms of attribute 2, median income). A holistic form of specific information would involve comparing the size of one symbol with another (noting that the southernmost symbol is larger than the one to its immediate northwest). An example of general information for an individual attribute would involve examining the distribution of a single attribute across the state (e.g., examining the distribution of the attribute infant mortality)—note that for the combined star and snowflake symbol this is a difficult task. A form of holistic general information would involve examining the pattern of the size of symbols across the state (note that there appears to be a band of relatively less desirable counties running through the southern part of the state).

The apparent difficulty of visualizing the pattern of an individual attribute in Figure 18.13 raises the question of whether certain multivariate symbols might be more effective for one form of information than another. Nelson and Gilmartin (1996) evaluated this question by having map readers examine four types of multivariate point symbols: a modified Chernoff face, a circle divided into quadrants, a cross, and boxed letters representing attribute names (Figure 18.14). Each point symbol represented four attributes depicting quality of life on a map consisting of nine enumeration units. Nelson and Gilmartin found that all symbols were processed equally well if time to examine

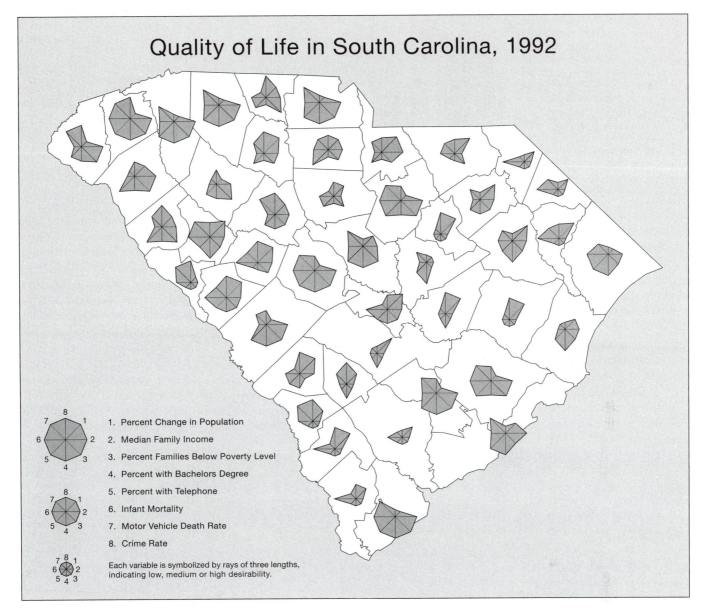

FIGURE 18.13   A multivariate map based on a combination of the star and snowflake symbols shown in Figure 18.8. (Data source: South Carolina State Budget and Control Board 1994.)

the map was not a factor. If time was a consideration, then boxed letters were most effective (had the fastest reaction time) followed by crosses, divided circles, and Chernoff faces. Nelson and Gilmartin stressed that certain symbols worked better for individual attribute questions, whereas other symbols worked better for holistic questions. In particular, Chernoff faces and letters worked best for individual attributes, whereas crosses and circles were more appropriate for holistic questions.

One problem with such experimental studies is that they often are unable to consider all of the factors that might affect map reader performance. In Nelson and Gilmartin's case, they mentioned aesthetics (e.g., that

Chernoff faces might be appropriate because of their attention-getting quality) and the number of attributes mapped (they noted that other researchers found Chernoff faces effective when displaying a large number of attributes) as potential factors to study. Other factors mapmakers should consider include the difficulty of discriminating attributes when small point symbols are used (imagine interpreting the boxed letter for 20 attributes), the number of enumeration units (Nelson and Gilmartin only considered a nine-county region), and the viewing environment (e.g., in a 3-D environment data jacks presumably would be more effective than, say, stars).

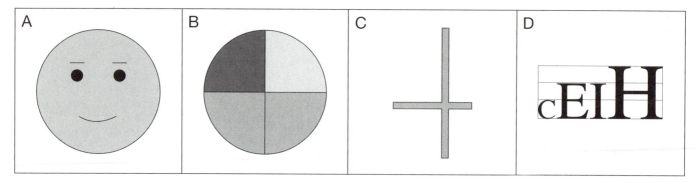

FIGURE 18.14 Multivariate symbols used in a study by Nelson and Gilmartin (1996): (A) a modified Chernoff face; (B) a circle divided into quadrants; (C) a cross; and (D) boxed letters representing attribute names.

Combining Different Types of Symbols

In combining three or more attributes on the same map, we have focused on using the same type of symbol for a particular application (we used an area symbol for the trivariate choropleth map and a point symbol for the multivariate point symbol map). It is also possible to combine various symbol types to display multivariate data. A good example of this is SLCViewer, a software package developed at the Deasy Geographics Laboratory at Pennsylvania State University by David DiBiase and his colleagues (1994a, 303–309). The purpose

of SLCViewer was to explore data produced by climate models. SLCViewer permitted the analyst to view up to four climatic attributes as small multiples or to overlay three attributes to create a multivariate map. For the multivariate map, point, line, and area symbolization could be overlaid. For example, Figure 18.15 shows the attributes mean annual evaporation, precipitation, and temperature displayed as proportional circles, weighted isolines, and choropleth shading, respectively. **Weighted isolines** were created by making the width of contour lines proportional to the data; note that this approach does not require labeling isolines because wider lines

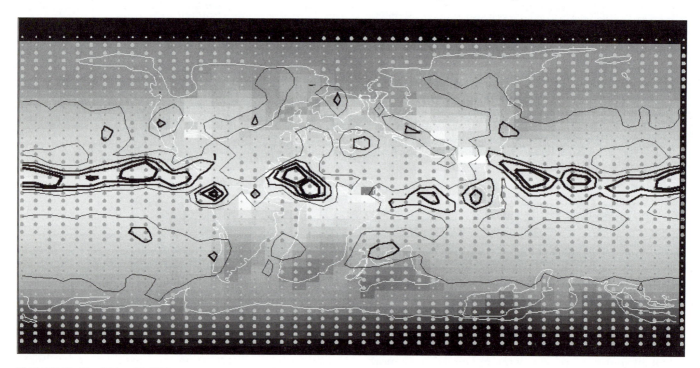

FIGURE 18.15 Using SLCViewer to create a multivariate map by combining point, line, and area symbolization. The attributes mean annual evaporation, precipitation, and temperature associated with a climate model are displayed as proportional circles, weighted isolines, and choropleth shading, respectively. (From DiBiase, D., Sloan, J. L., and Paradis, T. (1994b), "Weighted Isolines: An alternative method for depicting statistical surfaces." *The Professional Geographer* 46, no. 2, p. 219. Courtesy of Blackwell Publishing.)

are logically associated with more of the phenomena being mapped.* The obvious advantage of using different symbol types is that one symbol type will not conflict with another; thus, each attribute can be seen individually, and the attributes can be related to one another.

Separable versus Integral Symbols

When multivariate data are displayed on a single map, a question that sometimes arises is whether the symbols for each attribute are separable or integral. **Separable symbols** are those that can be attended to independently, thus allowing the map reader to focus on individual data sets. Cartogram size and lightness are considered separable

---

* In an experimental study, DiBiase et al. (1994b) found that weighted isolines were more effective than traditional labeled isolines and shadowed contours when the objective was to detect low and high areas in the data.

symbols, as illustrated in Figure 18.16A. Here we can ignore the shading of the cartograms and focus on their size, noting the higher number of tornadoes that appear to occur in Great Plains and Southern states. Alternatively, we can ignore the size of the states and focus on the shading, noting regions where mobile homes are common. **Integral symbols** are those that map readers tend to integrate, meaning that individual data sets cannot be attended to easily, but that it is possible to examine the correlation between the data sets. For instance, in Figure 18.16B we would have a difficult time visualizing the spatial pattern of one of the attributes, say percentage of murders committed with handguns, but we can see the correlation between the two types of murders—for instance, that the percentage of murders due to both handguns and knives tends to be moderate to high for many states along the U.S. east coast. The notion of separable and integral symbols was developed in psychology, but has recently been promoted and studied by a cartographer—Elisabeth

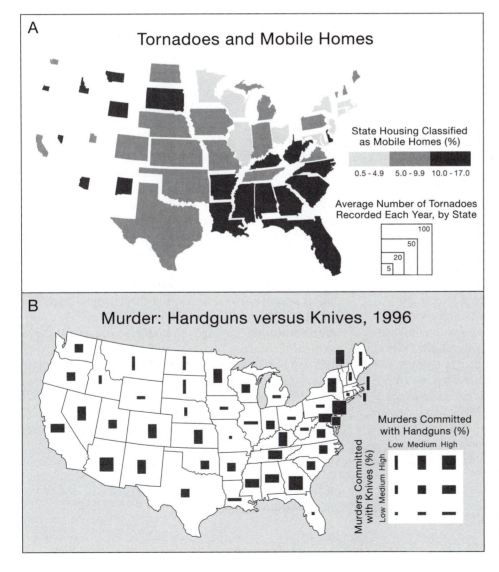

FIGURE 18.16 Examples of (A) separable and (B) integral symbols. (Courtesy of Elisabeth Nelson.)

Nelson (2000). Thus far, Nelson has focused on bivariate maps, as the examples here illustrate, but the concept could also be extended to multivariate maps.

SUMMARY

In this chapter we have covered a variety of methods for bivariate and multivariate mapping. Remember that **bivariate mapping** refers to the display of two attributes, whereas **multivariate mapping** refers to the display of three or more attributes. Both univariate and multivariate mapping can be accomplished through either map comparison (a separate map is created for each attribute) or by combining all attributes on the same map. For bivariate map comparison, we stressed the importance of using an appropriate method of data classification. Although optimal data classification is often recommended for univariate mapping, it is generally inappropriate for bivariate mapping because it focuses on the precise distribution of individual attributes along the number line. (An exception would be when the same attribute is compared for two different time periods.) The quantiles and equal-areas methods of classification are more appropriate choices for bivariate mapping, although the mean–standard deviation and nested means methods can be useful if the data are normally distributed. Unclassed maps can be used for map comparison, but only when the data distributions have a similar shape.

When two choropleth maps are overlain, the result is termed a **bivariate choropleth map**, a technique developed by the U.S. Bureau of the Census in the 1970s. Bivariate choropleth maps have been criticized because of their presumed failure to communicate information, although they can be effective if a clear legend is used, both bivariate and individual maps are shown, and an explanatory note is included. Eyton developed a logical color scheme for the bivariate choropleth map based on **complementary colors** (colors that combine to produce a shade of gray); Eyton also utilized a reduced major axis (an appropriate regression method when it is not clear which attribute should be treated as the dependent attribute) and an **equiprobability ellipse** (an ellipse that encloses a specified percentage of the data in a scatterplot so that values near the mean can be contrasted with more extreme observations). Brewer noted that color schemes for bivariate choropleth maps should be a function of whether the attributes are unipolar or bipolar. A cross-hatched symbology for bivariate choropleth mapping has been developed, but it is little used because of its coarse appearance.

Although choropleth maps have been the most frequently used symbology for combining two attributes on the same map, some interesting **bivariate point symbols** have been developed, including the **rectangular point symbol** (the width and height of a rectangle are varied), and the **bivariate ray-glyph** (straight-line segments point to either the right or left of a small central circle). It appears that the rectangular point symbol might be a suitable substitute for coarse cross-hatched symbology. An advantage of the ray-glyph is that it can be squeezed into a relatively restricted space. A bivariate point symbol also can be created by placing a choropleth shade within a proportional point symbol, but this technique should be used with caution as it could be confused with redundant symbology.

When more than two maps are displayed simultaneously, the result is termed a **small multiple**. Although small multiples can be useful for comparing the patterns on multiple maps, they are difficult to interpret when comparing subregions within each map (e.g., comparing the same small enumeration unit across a set of choropleth maps is not easy). A common alternative to the small multiple is the **multivariate point symbol** or **glyph**. Examples of glyphs include the **star** (multiple rays extend from a central circle), the **snowflake** (rays of the star are connected), **three-dimensional bars** (bars of varying height are placed alongside one another), **data jacks** (triangular spikes extend from a square central area), and **Chernoff faces** (distinct facial features are used). Although a considerable number of attributes can be represented with such methods, it is questionable whether map readers can understand the resulting symbols. Multiple attributes can also be combined on choropleth and dot maps to create **trivariate choropleth maps** and **multivariate dot maps**.

The difficulty of interpreting multivariate symbols has led to the development of software for exploring multivariate data. In the case of SLCViewer, an analyst can view up to four climatic attributes as a small multiple or overlay three attributes to create a multivariate map. The latter is a particularly intriguing option because proportional symbols, **weighted isolines** (the width of contour lines is proportional to the data), and choropleth shading are all included on the same map. In Chapter 21, we consider additional packages that can be used to explore multivariate data.

When displaying multiple attributes on the same map, the associated symbols can be termed either separable or integral. **Separable symbols** are those that a map reader can readily attend to independently, such as cartogram size and lightness, whereas **integral symbols** are those that can be integrated to examine the correlation between attributes, such as with the rectangular point symbol. We have only introduced the notion of separable and integral symbols here—for more details, you should examine Nelson's (2000) work.

FURTHER READING

Andrienko, G., and Andrienko, N. (2001) "Exploring spatial data with dominant attribute map and parallel coordinates." *Computers, Environment and Urban Systems* 25: 5–15.

> Discusses methods for mapping related variables using Descartes, which served as the foundation for the CommonGIS program discussed in Chapter 21.

Aspaas, H. R., and Lavin, S. J. (1989) "Legend designs for unclassed, bivariate, choropleth maps." *The American Cartographer* 16, no. 4:257–268.

> Examines the effect of legend designs on the interpretation of unclassed bivariate choropleth maps.

Brewer, C. A. (1994a) "Color use guidelines for mapping and visualization." In *Visualization in Modern Cartography*, ed. by A. M. MacEachren and D. R. F. Taylor, pp. 123–147. Oxford: Pergamon.

> Provides guidelines for using color in both univariate and multivariate mapping.

Brewer, C. A., and Campbell, A. J. (1998) "Beyond graduated circles: Varied point symbols for representing quantitative data on maps." *Cartographic Perspectives*, no. 29:6–25.

> Presents methods for creating bivariate point symbol maps.

Carstensen, L. W. J. (1982) "A continuous shading scheme for two-variable mapping." *Cartographica* 19, no. 3/4:53–70.

> Introduces cross-hatched shading for creating unclassed bivariate choropleth maps. Also see Lavin and Archer (1984).

Chang, K. (1982) "Multi-component quantitative mapping." *The Cartographic Journal* 19, no. 2:95–103.

> Considers statistical approaches (e.g., cluster analysis) for combining multivariate data.

Cox, D. J. (1990) "The art of scientific visualization." *Academic Computing* 4, no. 6:20–22, ff.

> Describes several multivariate symbolization methods that have been used in a range of disciplines.

DiBiase, D., Reeves, C., MacEachren, A. M., Von Wyss, M., Krygier, J. B., Sloan, J. L., and Detweiler, M. C. (1994a) "Multivariate display of geographic data: Applications in earth system science." In *Visualization in Modern Cartography,* ed. by A. M. MacEachren and D. R. F. Taylor, pp. 287–312. Oxford: Pergamon.

> An overview of various methods for displaying multivariate spatial data.

Eyton, J. R. (1984a) "Complementary-color two-variable maps." *Annals, Association of American Geographers* 74, no. 3:477–490.

> Discusses the implementation of a method for bivariate choropleth mapping based on complementary colors.

Hancock, J. R. (1993) "Multivariate regionalization: An approach using interactive statistical visualization." *AUTO-CARTO 11 Proceedings,* Minneapolis, MN, pp. 218–227.

> Introduces a method for multivariate regionalization based on interactive statistical visualization.

Healey, C. G., and Enns, J. T. (1999) "Large datasets at a glance: Combining textures and colors in scientific visualization." *IEEE Transactions on Visualization and Computer Graphics* 5, no. 2:145–167.

> Describes a method for creating pexels, a form of glyph for multivariate symbolization; includes a discussion of related perceptual experiments. See also Healey (2001).

Lindberg, M. B. (1987) *Dot Map Similarity: Visual and Quantitative.* Unpublished Ph.D. dissertation, University of Kansas, Lawrence, KS.

> Examines the issue of comparing dot maps, both in the visual and numerical sense.

Mersey, J. E. (1980) *An Analysis of Two-Variable Choropleth Maps.* Unpublished MS thesis, University of Wisconsin-Madison, Madison, WI.

> An experimental study of the effectiveness of bivariate choropleth maps.

Monmonier, M. S. (1975) "Class intervals to enhance the visual correlation of choroplethic maps." *The Canadian Cartographer* 12, no. 2:161–178.

> Introduces a method for enhancing the visual correlation of choropleth maps by modifying the boundaries of class intervals. Also see Monmonier (1976).

Monmonier, M. S. (1977) "Regression-based scaling to facilitate the cross-correlation of graduated circle maps." *The Cartographic Journal* 14, no. 2:89–98.

> Describes a technique for enhancing the visual correlation of proportional symbol maps.

Monmonier, M. (1993) *Mapping It Out: Expository Cartography for the Humanities and Social Sciences.* Chicago, IL: University of Chicago Press.

> Pages 227–241 describe methods for representing geographic correlation.

Nelson, E. S., and Gilmartin, P. (1996) "An evaluation of multivariate quantitative point symbols for maps." In *Cartographic Design: Theoretical and Practical Perspectives*, ed. by C. H. Wood and C. P. Keller, pp. 191–210. Chichester, England: John Wiley & Sons.

> Reviews various methods for creating multivariate point symbols and describes an experimental study designed to examine the effectiveness of some of these symbols.

Olson, J. M. (1981) "Spectrally encoded two-variable maps." *Annals, Association of American Geographers* 71, no. 2:259–276.

> An experimental study of the effectiveness of bivariate choropleth maps.

Pazner, M. I., and Lafreniere, M. J. (1997) "GIS Icon Maps." *1997 ACSM/ASPRS Annual Convention & Exposition, Volume 5* (Auto-Carto 13), Seattle, Washington, pp. 126–135.

> Describes the development of multivariate icons composed of fixed pixel patterns that vary in lightness or color.

Tufte, E. R. (1990) *Envisioning Information.* Cheshire, CT: Graphics Press.

> This book provides a wealth of methods for representing both nonspatial and spatial data. Chapter 4 deals entirely with small multiples.

Wang, P. C. C. (ed.) (1978) *Graphical Representation of Multivariate Data.* New York: Academic Press.

> A collection of chapters on the graphical representation of multivariate data. Many of the chapters focus on Chernoff faces.

19

Additional Techniques

OVERVIEW

Chapters 13 through 17 focus on particular thematic mapping techniques. The purpose of this chapter is to cover several additional thematic mapping techniques that are important, but for which we do not have space to dedicate separate chapters. The focus is on static maps, although we will see that some techniques are most effective in the exploration and animation realms.

*Section 19.1 introduces the **cartogram**, in which spatial geometry is distorted to reflect a theme; for example, sizes of countries might be made proportional to the population of each country. Conventionally, cartographers attempted to preserve the shape of enumeration units comprising a cartogram. More recently, Daniel Dorling has developed an algorithm in which uniformly shaped symbols (typically circles) represent each enumeration unit. A circle's size is a function of the magnitude of the phenomenon being mapped, which in Dorling's case is normally population. Within the circles, another theme (or themes) can be depicted, such as the percent of the adult population employed in manufacturing. Dorling contends that the resulting map more properly reflects the human geography of a region; this is in contrast to traditional choropleth maps, which he argues reflect the physical extent of the region.*

*Section 19.2 introduces **flow maps**, which utilize lines of varying width to depict the movement of phenomena between geographic locations; probably most familiar to you is the portrayal of flows between countries. Because general-purpose digital methods for depicting flows have yet to be fully developed, we focus on specialized software that has been devised for portraying migration and continuous vector-based flows (e.g., wind speed and direction).*

True 3-D phenomena *(e.g., level of carbon dioxide in the earth's atmosphere) are a challenge for mapping because they vary continuously in 3-D space; thus, a symbol*

*at one location can hide a symbol at another location. In section 19.3, we consider two approaches for symbolizing true 3-D phenomena: the software package T3D, which is intended for handling a broad range of true 3-D phenomena, and Laurie Waisel's use of the IBM Visualization Data Explorer (DX) to symbolize a particular phenomenon, the chemical composition of sediments deposited in New York Harbor. DX is an example of **visualization software**, which can be used to create visualizations for a wide variety of spatial data (true 3-D data being just one of them).*

*In the last portion of the chapter, we consider two less frequently used but relatively novel methods. In section 19.4, we introduce the **framed-rectangle symbol**, which involves changing the proportion of area filled within a rectangular frame of constant size. Framed-rectangle symbols have been proposed as an alternative to the choropleth map because readers can acquire accurate specific information from them. A problem with framed-rectangle symbols, however, is that as point symbols, they are difficult to associate with the areal extent of enumeration units. In section 19.5, we introduce the **chorodot map**, which combines the features of dot and choropleth maps by shading small squares ("dots") with varying intensities. It can be argued that a chorodot map is appropriate when a phenomenon falls near the middle of the discrete–continuous and abruptness–smoothness continua.*

19.1 CARTOGRAMS

When creating thematic maps, cartographers generally try to avoid distorting spatial relationships. For example, an **equivalent** (or equal area) **projection** is normally used for a dot map so that the density of dots is solely a function of the underlying phenomenon (and not the map projection). Sometimes, however, cartographers purposefully distort space based on values of an attribute; the resulting maps are known as **cartograms**. Probably the most common cartograms used in everyday life are **distance cartograms** in which real-world distances are distorted to reflect some attribute, such as the time between stops on subway routes: Here cartograms are appropriate because the time between (and order of) stops is more important than the actual distance between stops. In the geographic literature, **area cartograms** are the most common form of cartogram.* Area cartograms are created by scaling (sizing) enumeration units as a function of the values of an attribute associated with the enumeration units. For example, consider Figure 19.1, which illustrates the states and territories of Australia as they might appear on an equivalent map projection (A) and when scaled in direct proportion to the population of each state or territory (B). Although population is the most typical attribute portrayed on area cartograms, any ordinal or higher level attribute (unstandardized or standardized) can be used.

Two forms of area cartogram traditionally have been recognized: contiguous and noncontiguous. With *contiguous cartograms*, an attempt is made to retain the contiguity of enumeration units; as a result, the shape of units must be greatly distorted (as in Figure 19.1B). With *noncontiguous cartograms*, the shapes of enumeration units are retained, but gaps appear between units (as in Figure 19.2). Area cartograms are commonly used for their dramatic impact; in fact, it can be argued that an area cartogram should not be used if the relations between values of an attribute are similar to the relations between sizes of enumeration units, as the resulting map will not appear distorted. It should be recognized, however, that readers might have difficulty interpreting area cartograms because they look so different from other thematic maps, in which the size relations between enumeration units are similar to how they appear on a globe (in the case of equivalent projections the size relations will match the globe). To ease the interpretation of cartograms, mapmakers should retain characteristic features along boundaries of enumeration units (e.g., distinctive meanders of a river), include an inset map depicting actual areal relations of enumeration units (or, alternatively, the units should be labeled), and not use area cartograms if the anticipated readership is unfamiliar with the region depicted. Another potential way to enhance the understanding of cartograms is to smoothly change (as in an animation) between a standard equivalent projection and the cartogram. Alan Keahey (1999) developed such an algorithm, but to our knowledge the algorithm is not generally available.

---

* Dent (1999) termed these *value-by-area cartograms*; we have used the term *area cartogram* for simplicity.

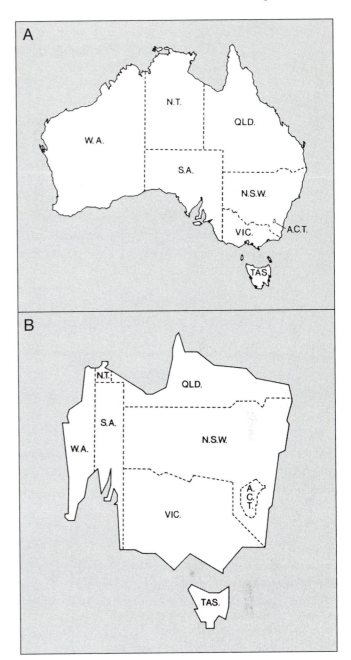

FIGURE 19.1 Creating a cartogram: (A) states and territories of Australia as they might appear on an equivalent projection; (B) a contiguous-area cartogram in which states and territories are scaled on the basis of 1976 population. (From Griffin 1983, p. 18; courtesy of T. L. C. Griffin.)

19.1.1 Dorling's Cartograms

Daniel Dorling (1993; 1994; 1995b; 1995c) has developed a novel algorithm for creating cartograms based on a uniformly shaped symbol (typically a circle). Dorling's algorithm begins by placing the uniformly shaped symbol in the center of each enumeration unit, with the

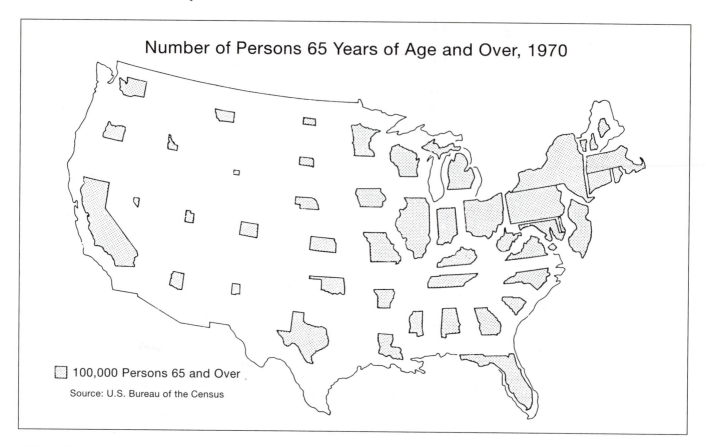

FIGURE 19.2 A noncontiguous area cartogram; shape of enumeration units is retained by introducing gaps between units. (After Olson, J. M. (1976b) "Noncontiguous area cartograms." *The Professional Geographer* 28, no. 4, p. 372. Courtesy of Blackwell Publishing.)

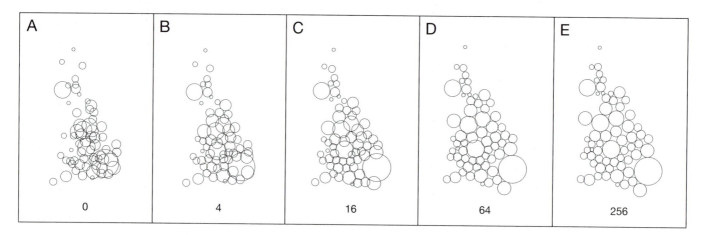

FIGURE 19.3 Dorling's algorithm for creating cartograms: (A) Uniformly shaped symbols (circles, in this case) are scaled on the basis of population, and are placed in the center of enumeration units on an equivalent projection; (B–E) the circles are gradually moved away from one another so that no two overlap. The numbers beneath each illustration represent the number of iterations in the algorithm. (After Dorling 1995b, p. 274.)

size of the symbol a function of population. Initially, symbols overlap one another because small enumeration units can have relatively large populations (Figure 19.3A). To eliminate overlap, an iterative procedure is executed in which symbols are gradually moved apart from one another (Figure 19.3B–19.3E). Wherever possible, points of contact between symbols reflect the points of contact between actual enumeration units, but sometimes it

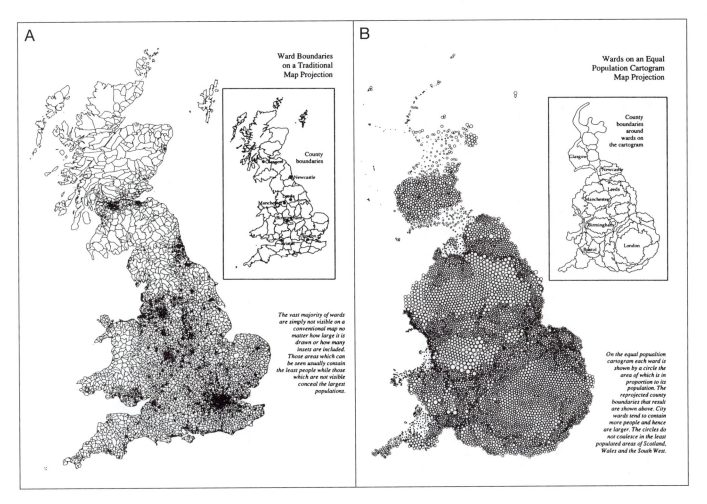

FIGURE 19.4 Applying Dorling's algorithm to the 9,289 wards of England and Wales: (A) boundaries as they would appear on a traditional equivalent projection; (B) a cartogram resulting from applying Dorling's algorithm. (From Dorling 1993, p. 171; courtesy of The British Cartographic Society.)

is not possible to meet this constraint. As a result, Dorling termed this a "noncontiguous form of cartogram."* The result of applying Dorling's algorithm to the 9,289 wards of England and Wales is shown in Figure 19.4. Figure 19.4A depicts the boundaries of wards as they would appear on a traditional equivalent projection, and Figure 19.4B displays the cartogram resulting from applying Dorling's algorithm. (The upper right portion of each figure portrays wards grouped into counties.) The basic difference between the maps is that small land areas with large populations are much more apparent on Dorling's cartogram than on the traditional map.[†] This is particularly apparent on the map of county boundaries: Note that London is a mere dot on the tra-

ditional map, but encompasses a substantial portion of Dorling's cartogram.

A major purpose of Dorling's algorithm is not to map population, but rather to serve as a base on which other attributes can be displayed. For example, Color Plate 19.1 portrays population density in Britain on both a traditional equivalent projection (the result is a choropleth map) and on Dorling's cartogram. Note the drastic difference in appearance between these two maps. The equivalent projection suggests that most of the country is dominated by relatively low population densities (the blues and greens), whereas the cartogram provides a detailed picture of the variation in population densities within urban areas. Dorling contended that his cartogram depicts the *human geography* of a region rather than focusing on its *physical extent*, as depicted on the choropleth map.

One obvious difference between Dorling's cartograms and traditional cartograms is that his provide no shape information for enumeration units. As a result, the addition of a conventional equivalent projection would seem

---

* A Pascal implementation of Dorling's algorithm can be found in Dorling 1995c, 373–378).

[†] No theme is displayed on the traditional map, as just the boundaries of wards appear; the dark areas represent where small wards bleed together.

essential when examining Dorling's cartograms, particularly if the map reader is unfamiliar with the region depicted. Another difference is that Dorling's cartograms typically show a large number of enumeration units, and thus considerable detail. As a result, readers should expect to study these maps meticulously. In fact, the complexity of Dorling's cartograms suggests that they might be examined most effectively in a data exploration environment. For example, imagine having both a traditional equivalent projection and the cartogram displayed on a computer screen at the same time. As the cursor is moved over an enumeration unit on the cartogram, this same enumeration unit could be highlighted and named on the equivalent projection (or vice versa). Alternatively, the user could zoom in to examine subregions in greater detail on both maps. We consider some related work that Dorling has done in this context in Chapter 20.

Although Dorling's algorithm for creating cartograms is original, it should be noted that he is not the only one to focus on the human geography of a region by mapping thematic attributes onto a cartogram base. The *Historical Atlas of Massachusetts* (Wilkie and Tager 1991) contains several examples of this approach; for example, page 91 of the atlas illustrates "Irish Ancestry in 1980" using a cartogram to show the population of each city and town in Massachusetts and green choropleth shading to represent the percentage of Irish ancestry in each city and town. In contrast to Dorling's cartograms, an attempt was made to maintain the shape of enumeration units in the Massachusetts atlas.

19.2 NOVEL METHODS FOR FLOW MAPPING

Flow maps utilize lines of differing width to depict the movement of phenomena between geographic locations. Although a wide variety of phenomena can be portrayed on flow maps, you are probably most familiar with those representing the flow of goods or movement of people between countries, such as the map of the slave trade shown in Figure 1.1. Unfortunately, basic flow maps have yet to be fully automated; thus, our focus in this chapter is on specialized software for portraying migration and continuous vector-based flows (e.g., wind speed).

19.2.1 Methods for Mapping Migration

Migration data are logical for digital portrayal because of the large number of movements that must be depicted. For example, for the 48 contiguous U.S. states there are 2,256 possible movements (assuming that all pairs of states are considered); if we consider U.S. counties, there are more than 9 million possible movements, and this does not even consider the attribute of time!

Waldo Tobler (1987) was one of the first to develop software for displaying migration flows.* One of the simpler options in Tobler's software was the depiction of one-way migration to or from a particular state by arrows of varying width (Figure 19.5). When one wishes to show the migration between all pairs of enumeration units simultaneously, a different approach is required because of the large number of arrows that result. The key is recognizing that many migration movements are small; deleting such movements will allow the reader to focus on more important movements. Tobler indicated that by deleting migration values below the mean, he generally was able to remove 75 percent or more of the flow lines, while deleting less than 25 percent of the migrants. For example, Figure 19.6 illustrates this approach for net migration for states in the United States from 1965 to 1970.

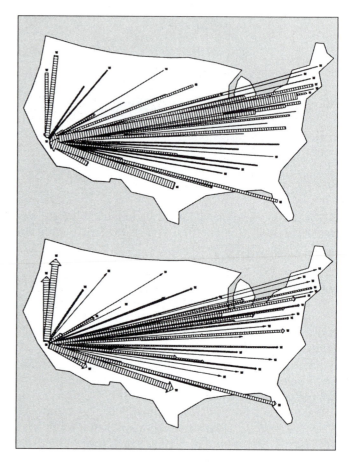

FIGURE 19.5 Migration to and from California, 1965–1970. (After Tobler 1987. First published in the *American Cartographer* 14(2), p. 160. Reprinted with permission from the American Congress on Surveying and Mapping.)

---

* A limited version of the software (known as FlowMap) is available from Professor Tobler.

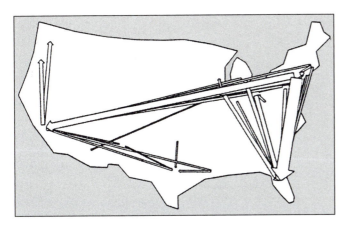

FIGURE 19.6 Net migration 1965–1970 for the 48 contiguous U.S. states, with flows below the mean net migration not shown. (Reprinted by permission from *Geographical Analysis*, Vol. 13, No. 1 (Jan. 1981). Copyright 1981 by Ohio State University Press. All rights reserved.)

Another interesting feature of Tobler's software was the ability to route data through enumeration units lying between the starting and ending points for migration, thus reflecting the route over which people were presumed to migrate. For example, migration data from New York to California would ideally have to be routed through Pennsylvania, Ohio, and numerous other states. Although the details of how Tobler achieved this are beyond the scope of this text (see Tobler 1981, 7–8 for a summary), it is interesting that he used some of the same concepts implemented in his pycnophylactic method (see section 14.6). Figure 19.7 illustrates the result for the migration data used for Figure 19.6.

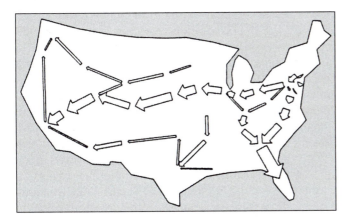

FIGURE 19.7 Migration data depicted in Figure 19.6 are rerouted to pass through states between the starting and ending points; again, flows below the mean are not shown. (Reprinted by permission from *Geographical Analysis*, Vol. 13, No. 1 (Jan. 1981). Copyright 1981 by Ohio State University Press. All rights reserved.)

One limitation of Tobler's software is that it was developed before graphical user interfaces became common and so did not exhibit the interaction characteristic of modern software. Examples of migration mapping software that have achieved some modicum of interaction include Tang's (1992) Visda and Yadav-Pauletti's (1996) MigMap. In Visda, Qin Tang symbolized migration data via choropleth maps. Users could point to an enumeration unit and receive a map of inmigration or outmigration with respect to that unit. Utilizing choropleth symbols avoids the overlap problem associated with Tobler's linear symbols, but choropleth symbols do not connote flows as readily as linear symbols. Sunita Yadav-Pauletti's MigMap symbolized migration data using both choropleth and arrow symbols, and utilized small multiples and animation to depict changes over time. Although it was possible to create a number of interesting maps with her software (Color Plate 19.2), an important limitation was the inability to input one's own migration data.*

19.2.2 Mapping Continuous Vector-Based Flows

Continuous vector-based flows are composed of two attributes, magnitude and direction, that can change at any geographic location. For example, wind is a continuous vector-based flow because at any point in the atmosphere we can consider the speed of the wind and the direction from which it blows. Traditionally, wind speed and direction were depicted using wind arrows (as in Figure 19.8). Stephen Lavin and Randall Cerveny (1987, 131) deemed wind arrows awkward because (1) readers must separate them visually to create the individual distributions of wind speed and direction, and (2) the symbols representing speed (the flaglike portion) are abstract and thus time-consuming to interpret. As an alternative, Lavin and Cerveny proposed **unit-vector density mapping**, which involves changing the density and orientation of short fixed-length line segments called unit vectors. The density and orientation of unit vectors are a function of the magnitude and direction of the vector-based flow being mapped. Typically, the line segments have an arrowhead to indicate movement, although Lavin and Cerveny argued that they can be used effectively without arrowheads.

Lavin and Cerveny's algorithm for unit-vector density mapping is analogous to the algorithm Lavin used in his dot-density shading technique (see section 17.3.3). The algorithm presumes that one has first created an equally spaced gridded network of vector-based flow values; thus, each of the crosses in Figure 19.9A would have two data values—one for vector magnitude and one for vector direction. Creating such data requires running one of the gridding approaches for interpolation discussed in

---

* Lee et al.'s (1994) XNV, intended for mapping airline flows, might also be useful for mapping migration data.

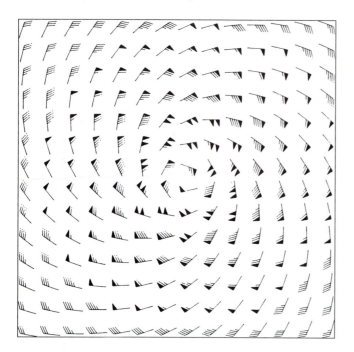

FIGURE 19.8 Wind arrows: a traditional method for depicting wind speed and direction. (From Lavin and Cerveny 1987, p. 132; courtesy of The British Cartographic Society.)

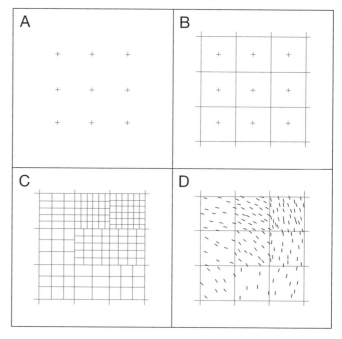

FIGURE 19.9 Lavin and Cerveny's unit-vector density method for mapping vector-based flows (data that have both magnitude and direction): (A) input is an equally spaced gridded network of magnitudes and directions; (B) a square cell is presumed around each grid point; (C) the cells in (B) are divided into subcells, the number of which is proportional to the magnitude of the data in each cell; (D) a unit vector is placed within each subcell, with the orientation corresponding to the direction of the data. (From Lavin and Cerveny 1987, p. 134; courtesy of The British Cartographic Society.)

Chapter 14 twice, once for vector magnitude and once for vector direction. Next, it is presumed that each grid intersection is surrounded by a square cell within which unit vectors will be placed (Figure 19.9B). Each of these square cells is then divided into subcells, with the number of subcells a function of the magnitude of the data being mapped (Figure 19.9C). Finally, individual unit vectors are placed within each subcell, with the orientation corresponding to the direction of the flow being mapped (Figure 19.9D).

As an example of the unit-vector density technique, consider the map of average January surface winds for the United States shown in Figure 19.10. Lavin and Cerveny noted that this map illustrates "many of the features sought on more traditional" maps, but avoids the difficulty of determining wind speed (p. 135). In particular, they noted (1) areas where arrows appear to converge or diverge, which are indicative of low- and high-pressure areas, respectively (in the extreme southwestern portion of the United States there is an apparent area of divergence); (2) the flow off the Great Lakes, which can lead to "lake effect" precipitation; and (3) the areas of high and low wind speed defined by the differing densities of unit vectors (e.g., the central high plains and portions of the Rocky Mountains, respectively).

A less obvious but intriguing use of unit-vector density mapping is the portrayal of topography. In this case, elevation is represented by the *density* of unit vectors, whereas aspect (the direction the slope faces) is depicted

by the *orientation* of the unit vector; no arrow symbol is necessary because topography does not move. The density of unit vectors readily identifies the regions of highest elevation, and the orientation of vectors indicates significant characteristics of aspect. To illustrate, Figure 19.11 contrasts present-day topography with the estimated situation 18,000 years ago. In Figure 19.11A, the density of unit vectors assists us in identifying major topographic features such as the Himalayas, the Greenland ice sheet, and the Andes. In Figure 19.11B, these areas are again apparent, but we notice other high-elevation areas, which reflect the fact that major ice sheets once covered much of North America and Europe. Also note the orientation of the unit vectors in Figure 19.11B; particularly distinctive are those radiating from source regions for glaciation in what is now Canada.

19.3 MAPPING TRUE 3-D PHENOMENA

In section 4.1.1, we indicated that two kinds of volumetric phenomena are possible: $2\frac{1}{2}$-D and true 3-D. $2\frac{1}{2}$-D

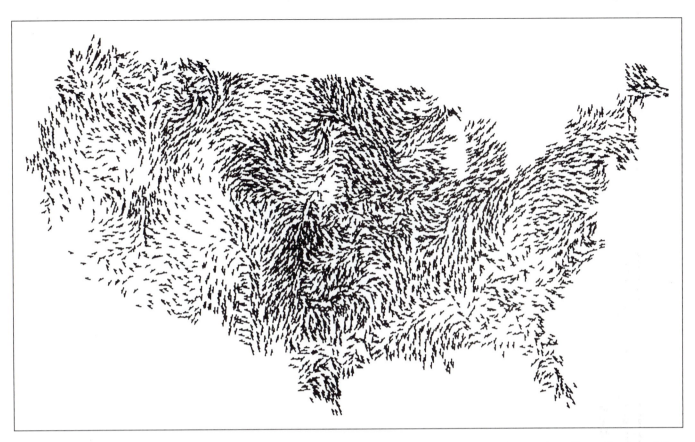

FIGURE 19.10   A unit-vector density map based on average January surface winds for 187 weather stations. (From Lavin and Cerveny 1987, p. 136; courtesy of The British Cartographic Society.)

phenomena have a single z value for each X and Y location (longitude and latitude), whereas true 3-D phenomena have multiple z values for each X and Y location. Here we consider two approaches for symbolizing true 3-D phenomena: (1) the software package T3D, which is intended for handling a broad range of true 3-D phenomena; and (2) Waisel's (1996) use of the IBM Visualization Data Explorer (DX) to symbolize a particular phenomenon, the chemical composition of sediment deposits in New York Harbor.

19.3.1 T3D: General-Purpose Software for Symbolizing True 3-D Phenomena

T3D, part of the software package Noesys, is an example of general-purpose software for symbolizing true 3-D phenomena. To illustrate some of T3D's capability, we use a sample data set of vertical wind velocity within a simulation of a thunderstorm that is distributed with T3D. We use a default "Rainbow" color scheme in which long wavelength colors (red, orange, and yellow) are assigned to upflows and short wavelength colors (blue and green) are assigned to downflows (Color Plate 19.3).

The basic problem in symbolizing true 3-D phenomena is that only the surface of the 3-D structure can be seen if opaque (nontransparent) shading is used, as in Color Plate 19.3A. Rather than seeing only the surface, we would like to "peer" into the data, which can be accomplished by varying the opacity of the shading. In the simplest case, this is achieved by making some portion of the data range transparent, and the remaining portion is left opaque. For example, in Color Plate 19.3B upflows are transparent and downflows are opaque. Alternatively, one can gradually change the opacity over a selected range of data. For example, in Color Plate 19.3C the opacity for the entire range of data gradually changes from completely transparent (for the upflows) to completely opaque (for the downflows). A second approach for peering into the data is to take a slice through the 3-D surface. By default, slices are completely opaque, although it is possible to vary their opacity, too. Color Plate 19.3D illustrates horizontal and vertical opaque slices through the thunderstorm data. A third approach for peering into the data is to create an **isosurface**, which is the 3-D equivalent of a contour line. Isosurfaces, like slices, are by default opaque (Color Plate 19.3E). It is also possible to combine these approaches; for example, Color Plate 19.3F illustrates a combination of the slice and isosurface methods.

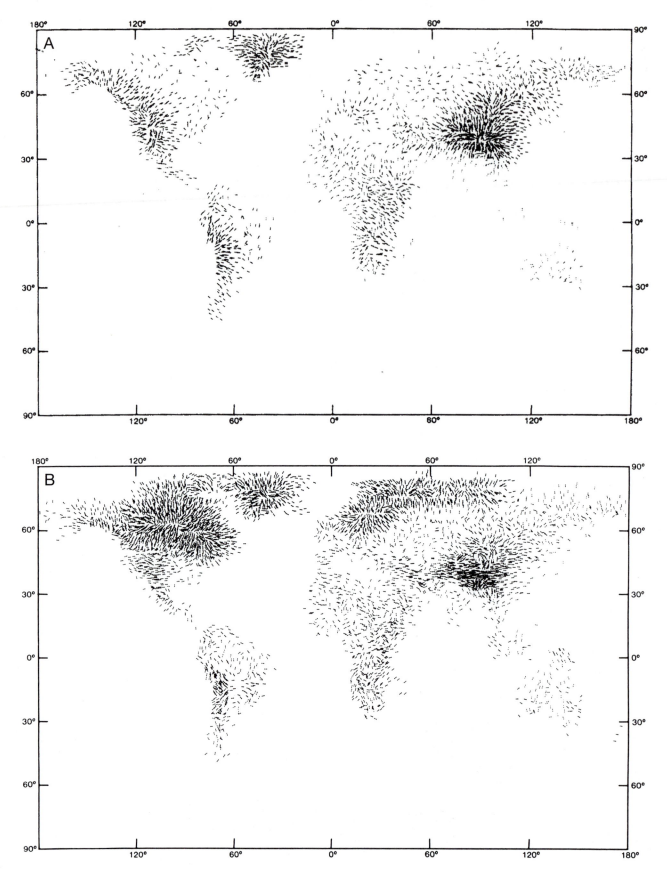

FIGURE 19.11 A comparison of unit-vector density maps for (A) present-day topography and (B) estimated topography 18,000 years before present. (From Lavin and Cerveny 1987, p. 139; courtesy of The British Cartographic Society.)

Other approaches that assist in visualizing true 3-D phenomena include rotating and animating the image. One limitation of rotation in T3D (at least for the PC version) is that users cannot rotate a fully rendered image in real time. Rather, one must select a desired rotation using a solid box enclosing the 3-D data set; only when the desired rotation has been selected is the complete image rendered. In Chapter 21 we consider another program, Vis5D, which does permit real-time rendering during rotation.* Animation can reveal hidden 3-D information by either showing a sequence of movements in a rotation of a 3-D surface or sequencing a set of slices (or isosurfaces). Users specify endpoints in the animation sequence, and intermediate frames are interpolated automatically. Still another option with T3D is to modify the color scheme. The use of color is described in detail when we consider the software package Transform (also a part of Noesys) in Chapter 21.

Although T3D provides a wide range of options for symbolizing true 3-D phenomena, it is not exhaustive. Another interesting possibility is the **tiny cubes method** of Nielson and Hamann (1990), in which small cubes are regularly sampled throughout the 3-D surface (Color Plate 19.4). Keller and Keller (1993, 148) provided suggestions on how these cubes might be used.

19.3.2 Waisel's Use of DX to Symbolize Sediment Deposits in New York Harbor

In this section we consider Laurie Waisel's (1996) use of the IBM Visualization Data Explorer (DX) to examine sediment chemistry in New York Harbor. DX is a form of **visualization software** that can be used to create visualizations for a wide variety of spatial data (true 3-D data being just one of them). Visualizing sediment chemistry in New York Harbor is an important consideration because the harbor must be dredged periodically to maintain a sufficient depth for ships. If scientists are able to visualize the level of contamination in the sediment, they might be able to decide whether costly decontamination efforts are necessary for a particular dredging operation. The basis for Waisel's visualizations was a set of core samples taken from the bottom of the harbor. Each core sample was partitioned into three sections, and each of these sections was mixed to create a homogeneous sample, which was then tested for levels of potential contaminants. Although a set of continuous 3-D data could have been interpolated from the homogeneous sample points (the result would be the sort of data used earlier for T3D), Waisel chose to map individual core samples and the homogeneous segments within those cores. An example of a typical map created by Waisel is shown in

Color Plate 19.5. Each cylinder in the map represents one of the three sections (homogeneous samples) within a soil core. Note that the cylinders have been color-coded based on National Oceanic and Atmospheric Administration (NOAA) reference values for sediment contamination: Effects Range Low (ERL) and Effects Range Medium (ERM). Blue cylinders are below ERL, greenish-yellow cylinders are above ERL and below ERM, and red cylinders are above ERM. Presumably, sediments associated with the red cylinders would be of greatest concern. Waisel used DX to examine sediment chemistry because of the considerable flexibility its visual programming environment provides. Although Waisel's work provides a good example of the interesting kinds of maps that can be developed with such software, based on her description of the technical challenges and our own experience working with DX it is apparent that creating such maps is more complex than using nonprogramming environments, such as the one T3D provides.

19.4 FRAMED-RECTANGLE SYMBOLS

The **framed-rectangle symbol** was developed by William Cleveland and Robert McGill (1984) as an alternative to the area shading conventionally used on choropleth maps (Figure 19.12). Each framed rectangle consists of a "frame" of constant size, within which a solid "rectangle" is placed (Figure 19.13). The tick marks on the side of the framed rectangle were added by Cleveland (1994, 209) to enhance the reader's ability to estimate correct data values. Note that framed-rectangle symbols allow similar values to be compared easily when those values would be difficult to distinguish with simple bars (Figure 19.14).

Cleveland and McGill argued that framed-rectangle symbols should be used in preference to choropleth shading because the estimation of data values is a simpler perceptual task (assigning values to the height of a rectangle is easier than assigning values to a shade), and because larger enumeration units will not have a larger visual impact. We can add to this the fact that estimation of data values on choropleth maps is adversely affected by simultaneous contrast (see Chapter 13). In an experimental study, Richard Dunn (1988) found that both the average error and variation in error was smaller for framed-rectangle symbols, and that on choropleth maps variation in error was a function of the size of enumeration units.

A limitation of Dunn's study was that it dealt only with specific (as opposed to general) information. This is critical because a primary reason for creating thematic maps is to portray general information. It is arguably difficult to extract general information from a framed-rectangle map because the framed rectangles are point symbols, which are not easy to associate with the areal extent of enumeration units. Imagine attempting to form regions

---

* Slicer Dicer, marketed by PIXOTEC (*http://www.slicerdicer.com/*), is another package for symbolizing true 3-D phenomena.

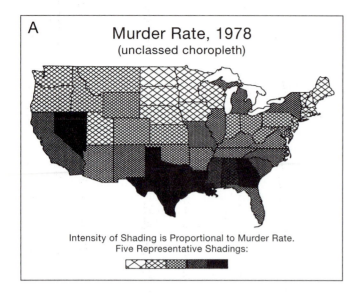

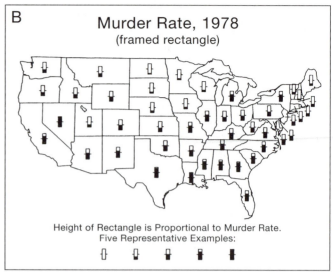

FIGURE 19.12   A comparison of (A) choropleth and (B) framed-rectangle maps. (Reprinted with permission from *The American Statistician.* Copyright 1988 by the American Statistical Association. All rights reserved.)

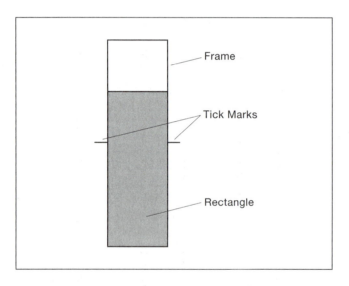

FIGURE 19.13   Components of a framed-rectangle symbol.

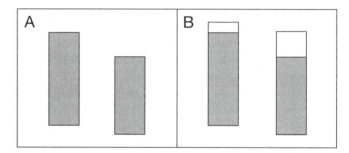

FIGURE 19.14   A comparison of (A) simple bars and (B) framed-rectangle symbols. Note the ease with which the framed-rectangle symbols can be differentiated. (After Cleveland and McGill 1984, p. 538.)

on the framed-rectangle map shown in Figure 19.12B and then having to recall those regions at a later time. Extracting general information would be especially problematic when framed rectangles must be drawn outside of small enumeration units to avoid overlap (as in the northeast portion of Figure 19.12B).

The framed-rectangle symbol, however, is particularly appropriate for mapping unstandardized data (raw totals) associated with enumeration units because point symbols normally are used for this purpose (see Chapter 16). As an example, Mark Monmonier (1992a) used framed-rectangle symbols in an animation of the number of daily newspapers subscribed to by states in the United States

over a 90-year period (a frame from his animation appears in Figure 19.15). Monmonier handled the problem of symbol overlap by adjusting the sizes of states so that smaller states would be more noticeable; note, however, that this approach distorts the shape of states, which might be disconcerting to readers.

19.5 THE CHORODOT MAP

In section 4.1.2, we introduced a range of models of geographic phenomena (Figure 19.16A) developed by Alan MacEachren and David DiBiase (1991). In creating these models, MacEachren and DiBiase also suggested symbolization methods that would be appropriate for each model (Figure 19.16B).* Most of

---

* Enumeration unit boundaries are shown on a number of these maps because MacEachren and DiBiase focused on data collected for enumeration units.

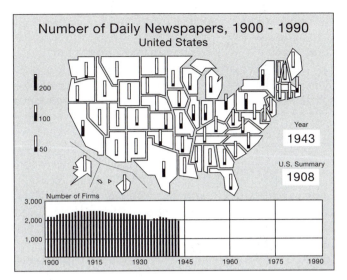

FIGURE 19.15 Use of framed-rectangle symbols to map unstandardized data (raw totals) for enumeration units. The map is a frame from an animation of the number of daily newspapers subscribed to from 1900 to 1990. (After Monmonier 1992. First published in *Cartography and Geographic Information Systems* 19(4), p. 249. Reprinted with permission from the American Congress on Surveying and Mapping and Mark Monmonier.)

these symbolization methods were conventional and have already been covered in this text. Others, notably the middle and middle right ones, were novel. MacEachren and DiBiase focused their discussion on the middle one,

which they termed the chorodot map. As the name implies, a **chorodot map** is a combination of choropleth and dot maps. The "dot" portion comes from the use of small squares within each enumeration unit, and the "choropleth" portion derives from the shading assigned to each square. MacEachren and DiBiase developed the chorodot map as a potential approach for illustrating a particular phenomenon—the spread of AIDS. Although they mapped the spread of AIDS using an animated series of isopleth maps (the animation can be found in DiBiase et al. 1991), they argued that chorodot maps might more appropriately illustrate the manner in which AIDS spreads—through sexual intercourse, sharing of needles, and blood transfusions. Figure 19.17 illustrates a chorodot map that MacEachren and DiBiase created of AIDS in Pennsylvania for 1988. In this case, a geometric classification was used to shade the squares: A light gray square represents 1 AIDS case; a medium gray square, 8 cases; and a dark gray square, 64 cases. Darker squares were placed only after lighter ones completely filled a county; thus no medium gray squares could be placed until the entire county was covered by light gray.

Because the chorodot map is a combination of the choropleth and dot maps, some of the problems associated with these maps arise in the construction of the chorodot map. For instance, the cartographer must determine how large the squares will be and where they will be placed, and what classification method will be

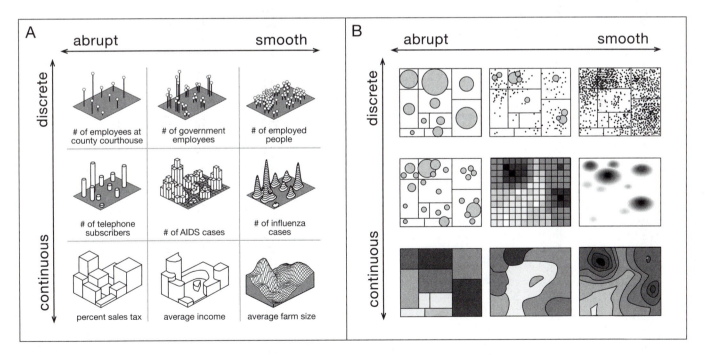

FIGURE 19.16 (A) MacEachren and DiBiase's models for representing geographic phenomena. (B) A set of symbolization methods that MacEachren and DiBiase argued would be appropriate for the models shown in A. (After MacEachren 1992, p. 16; courtesy of North American Cartographic Information Society and Alan MacEachren.)

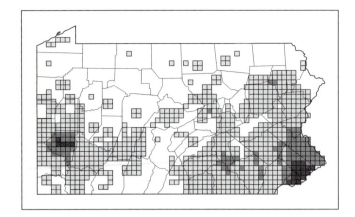

FIGURE 19.17   A chorodot map used to depict 1988 AIDS data for Pennsylvania counties. The number of cases represented by each cell is as follows: light gray (1 case), medium gray (8 cases), and dark gray (64 cases). (From MacEachren 1994a, p. 63; courtesy of Association of American Geographers and Alan MacEachren.)

used. MacEachren and DiBiase only introduced the notion of the chorodot map, leaving such questions to be explored by other cartographers.

In addition to introducing the chorodot map, MacEachren and DiBiase's work also provides an illustration of how a geographer's intended purpose might affect symbol selection. In MacEachren and DiBiase's case, they were developing maps of AIDS for Peter Gould, who had two basic goals: to portray the large number of people affected by AIDS (to alarm the map viewer) and to illustrate a "classic example of hierarchical diffusion (modeled as a smooth, continuous function)" (MacEachren and DiBiase, 225). As a result, they based their isopleth animation on the raw number of AIDS cases. MacEachren and DiBiase noted that isopleth maps technically were improper because they require standardized data (as we discussed in Chapter 4), but that the isopleth maps served Gould's goals well.

SUMMARY

In this chapter, we examined a variety of thematic mapping techniques not covered in earlier chapters. First, we introduced the **cartogram**, in which spatial geometry is distorted to reflect a theme. A popular form is the **area cartogram**, in which the sizes of enumeration units are scaled to reflect values of an attribute. For instance, we can portray the spatial distribution of population by scaling individual countries as a function of the population of each country. Area cartograms can be either *contiguous* or *noncontiguous*, depending on whether enumeration units are adjacent or not adjacent, respectively. Daniel

Dorling has developed a novel method for creating cartograms, in which circles of varying size represent the population of each enumeration unit. The resulting circles are shaded with an attribute of interest, for example, the percent of the population involved in the service sector of the economy. Dorling contends that the resulting map more properly reflects the human geography of a region because of its focus on people; this is in contrast to traditional equivalent projections (and the associated choropleth map), which reflect the physical extent of a region.

Second, we covered specialized software for **flow mapping**, specifically software for mapping migration flows and **continuous vector-based flows**. Waldo Tobler developed the first significant piece of software for displaying migration flows, showing that important migration movements could be highlighted by displaying only movements above the mean of the data. Tobler also developed an ingenious method for directing migration flows to reflect the route over which people were presumed to migrate. A limitation of Tobler's software was the lack of user interaction, but recently developed software, such as Tang's Visda and Yadav-Pauletti's MigMap, is beginning to handle this limitation. Stephen Lavin and Randall Cerveny's **unit-vector density mapping** provides a solution to mapping continuous vector-based flows (e.g., wind speed and direction). Unit-vector density mapping involves changing the density and orientation of short fixed-length line segments called unit vectors. The density of vectors is a function of the magnitude of the vector-based flow (e.g., a higher wind speed produces a higher density of vectors), whereas orientation is a function of the direction of flow (a north wind produces a north–south-oriented vector). Lavin and Cerveny suggest that unit-vector density symbology provides more information than traditional wind arrow symbology because users can readily interpret both wind speed and direction.

Third, we investigated software that can assist in interpreting true 3-D phenomena (e.g., water temperature throughout the 3-D extent of the ocean). T3D is illustrative of software for examining a broad range of true 3-D phenomena. T3D permits users to "peer" into true 3-D data by varying its transparency (or alternatively its opacity), slicing through the data, and viewing an **isosurface** (the 3-D equivalent of a contour line). T3D also permits users to rotate and animate a true 3-D data set, and to experiment with different color schemes. We will deal with its capability to manipulate color schemes in Chapter 21 when we consider a companion product, Transform. In the context of true 3-D phenomena, we also considered Laurie Waisel's use of the IBM Visualization Data Explorer (DX) to evaluate the chemical composition of sediment deposits in New York Harbor.

Finally, we considered two less frequently used, but relatively novel, methods: the framed-rectangle symbol and the chorodot map. The **framed-rectangle symbol**, a form of point symbol, is created by filling a rectangle in proportion to the data values being mapped (the approach is analogous to changing the height of a column of liquid in a thermometer). The framed-rectangle symbol has been promoted as an alternative to choropleth symbology because it eases the extraction of specific information. A disadvantage, however, is that general information is difficult to acquire because the point symbol does not match the areal extent of enumeration units.

The **chorodot map** combines features of dot and choropleth maps by shading small squares ("dots") with different intensities. Alan MacEachren and David DiBiase argued that the chorodot map falls midway on the abrupt–smooth and continuous–discrete continua shown in Figure 19.16B. The chorodot map would be appropriate for portraying the manner in which AIDS spreads—through sexual intercourse, sharing of needles, and blood transfusions.

FURTHER READING

Breding, P. (1998) *The Effect of Prior Knowledge on Eighth Graders' Abilities to Understand Cartograms.* Unpublished M.S. thesis, University of South Carolina, Columbia, SC.

> Describes an experiment on the effectiveness of cartograms; interestingly, one conclusion was that contiguous cartograms had a higher percentage of correct responses than proportional symbol maps, although the cartograms took longer to process.

Cleveland, W. S., and McGill, R. (1984) "Graphical perception: Theory, experimentation, and application to the development of graphical methods." *Journal of the American Statistical Association* 79, no. 387:531–554.

> Describes a variety of graphical and mapping techniques (one is the framed-rectangle symbol introduced in this chapter) and experiments to evaluate their effectiveness.

Cuff, D. J., Pawling, J. W., and Blair, E. T. (1984) "Nested value-by-area cartograms for symbolizing land use and other proportions." *Cartographica* 21, no. 4:1–8.

> Introduces a novel mapping technique in which nested cartograms are centered in enumeration units and scaled proportional to the percentage of various land uses.

Dent, B. D. (1999) *Cartography: Thematic Map Design.* 5th ed. Boston: McGraw-Hill.

> Chapters 11 and 12 provide an overview of cartograms and flow mapping, respectively.

Dorling, D. (1993) "Map design for census mapping." *The Cartographic Journal* 30, no. 2:167–183.

> Provides a thorough discussion of Dorling's novel cartogram method; for an interesting atlas comprised of cartograms, see Dorling (1995a).

Dougenik, J. A., Chrisman, N. R., and Niemeyer, D. R. (1985) "An algorithm to construct continuous area cartograms." *The Professional Geographer* 37, no. 1:75–81.

> Describes an algorithm for a contiguous-area cartogram. In contrast to Dorling's approach, Dougenik et al.'s approach tends to preserve the shape of enumeration units.

Gould, P. (1993) *The Slow Plague.* Cambridge, MA: Blackwell.

> Discusses the AIDS pandemic and methods for mapping it (see especially Chapter 9).

Gusein-Zade, S. M., and Tikunov, V. S. (1993) "A new technique for constructing continuous cartograms." *Cartography and Geographic Information Systems* 20, no. 3:167–173.

> Describes an algorithm for constructing a contiguous-area cartogram that tends to preserve shape. The authors argue that in

comparison to earlier algorithms (e.g., Dougenik et al.), this one "obtains better resultant images, especially in cases where a sharp difference in the initial distribution of the variable exists" (p. 172).

House, D. H., and Kocmoud, C. J. (1998) "Continuous cartogram construction." *Proceedings, Visualization '98*, Research Triangle Park, NC, pp. 197–204.

> Describes an algorithm for constructing a contiguous-area cartogram that tends to preserve shape. In contrast to the "pinching and ballooning" found in the Gusein-Zade and Tikunov algorithm (see preceding citation), the authors argue that their approach "preserves the distinctive shapes of [regions]" (p. 202).

Jackel, C. B. (1997) "Using ArcView to create contiguous and noncontiguous area cartograms." *Cartography and Geographic Information Systems* 24, no. 2:101–109.

> Presents Avenue code for contiguous- and noncontiguous-area cartograms. (Avenue is a programming language used in association with ArcView.)

Keim, D. A. (1998) "The Gridfit Algorithm: An efficient and effective approach to visualizing large amounts of spatial data." *Proceedings, Visualization '98*, Research Triangle Park, NC, pp. 181–188.

> Describes an approach in which individual plotted points (e.g., lightning strikes) are plotted near their actual location to avoid overplotting. The approach thus distorts point locations, as opposed to the area distortion common in cartograms.

Lee, J., Chen, L., and Shaw, Shih-L. (1994) "A method for the exploratory analysis of airline networks." *The Professional Geographer* 46, no. 4:468–477.

> Describes data exploration software for analyzing airline network flows.

Marble, D. F., Gou, Z., Liu, L., and Saunders, J. (1997) "Recent advances in the exploratory analysis of interregional flows in space and time." In *Innovations in GIS 4*, ed. by Z. Kemp, pp. 75–88. Bristol, PA: Taylor & Francis.

> Describes data exploration software developed at Ohio State University for analyzing interregional flows in both space and time. The software has the ability to map more than one attribute simultaneously, which is accomplished using "projection pursuit" concepts.

Nielson, G. M., and Hamann, B. (1990) "Techniques for the interactive visualization of volumetric data." *Proceedings Visualization '90*, San Francisco, CA, pp. 45–50.

> Describes various techniques for visualizing true 3-D phenomena.

Parks, M. J. (1987) *American Flow Mapping: A Survey of the Flow Maps Found in Twentieth-Century Geography Textbooks,*

Including a Classification of the Various Flow Map Designs. Unpublished M.A. thesis, Georgia State University, Atlanta, GA.

An extensive survey of various kinds of flow mapping.

Rittschof, K. A., Stock, W. A., Kulhavy, R. W., Verdi, M. P., and Johnson, J. T. (1996) "Learning from cartograms: The effects of region familiarity." *Journal of Geography* 95, no. 2:50–58.

An experimental study of the effectiveness of cartograms; also see Griffin (1983).

Thompson, W., and Lavin, S. (1996) "Automatic generation of animated migration maps." *Cartographica* 33, no. 2:17–28.

Summarizes methods for mapping migration and develops an animated method in which either small arrows or circles appear to move from one region to another.

Tikunov, V. S. (1988) "Anamorphated cartographic images: Historical outline and construction techniques." *Cartography* 17, no. 1:1–8.

Provides a history of cartograms.

20

Map Animation

OVERVIEW

*In this chapter, we focus on **animated maps**, or maps characterized by continuous change while the map is viewed. You are probably most familiar with animated maps from watching the weather segment of the nightly news or from watching The Weather Channel, but we will see that virtually any form of thematic symbology can be animated (albeit some are more effective than others). Although animation commonly is associated with recent technological advances, in section 20.1 we will see that the earliest map animations were produced in the 1930s, and that cartographers were discussing the potential of animation as early as the 1950s.*

*In section 20.2, we consider visual variables for animation and categories of animation. From Chapter 4, you will recall that visual variables refer to perceived differences in map symbols that are used to represent spatial data; examples include spacing, size, orientation, and shape. Animated maps can utilize these same visual variables, but they also use others, including **duration**, **rate of change**, **order**, **display date**, **frequency**, and **synchronization**. Animations can be split into categories according to whether the animation emphasizes (1) a change in a phenomenon's position or attribute, (2) the existence of a phenomenon at a particular location, or (3) an attribute of a phenomenon.*

Section 20.3 covers numerous examples of map animations, focusing on those that have been discussed in the refereed academic literature. For convenience, we have split examples of animations into two groups: those having "little or no interaction" and those "characterized by interaction." Animations shown on television or in a movie theater represent the extreme case of "no interaction," as the user has no control and thus must view the animation as a "movie." Animations shown in a computer environment often allow users to start and stop them, jump to an arbitrary frame, and play them forward

and backward, but we would still classify these as having "little" interaction. In contrast, some animations now permit considerable interaction, such as allowing the user to specify subsets of information to be displayed, linking the map with graphical representations, and allowing the user to interact with the display to acquire information. Software providing such potential is closely allied with the data exploration software we describe in Chapter 21.

As we discuss the animation examples, one issue that we raise is the difficulty of understanding many of them. We highlight this issue in section 20.4 by briefly considering some of the studies that have been done to evaluate the effectiveness of animation. Although those who have developed animations often seem enamored by them, studies to date do not clearly indicate the effectiveness of animation. We suspect that the effectiveness of animation ultimately will depend on numerous factors including the symbology used, the level of interaction that the user is permitted with the animation, whether the user has expertise in the domain being mapped, and the user's past experience with animation.

In recent years, numerous animations have been created that are not discussed in the refereed literature—we provide links to many of these on the home page for the book. Animations have also been developed in association with data exploration software, electronic atlases, and for depicting data uncertainty. We'll consider examples of these in Chapters 21, 22, and 23, respectively.

20.1 EARLY DEVELOPMENTS

Although animation is often viewed as a recent development, the notion has been around since the 1930s, and was promoted by cartographers as early as the 1950s, most notably by Norman Thrower (1959; 1961) who clearly was enamored by the potential of animation:

> By the use of animated cartography we are able to create the impression of continuous change and thereby approach the ideal in historical geography, where phenomena appear "as dynamic rather than static entities." Distributions which seem to be particularly well suited to animation include the spread of populations, the development of lines of transportation, the removal of forests, changing political boundaries, the expansion of urban areas, and seasonal climatic patterns. (1959, 10)

Because computers were only beginning to be developed in the 1950s, animations at that time were created using hand drawings ("cels") and a camera to record each frame of the animation.

The earliest computer-based animations included Waldo Tobler's (1970) 3-D portrayal of a population growth model for the city of Detroit, Michigan, and Hal Moellering's (1976) representation of the spatiotemporal pattern of traffic accidents. Mark Harrower and his colleagues (2000, 281) argued that work such as Moellering's illustrated the "power of map animation," as "the result was a clear representation of the daily cycle of accidents, with peaks during rush hour periods and troughs between those times, as well as the spatial pattern of weekday versus weekend accidents." Although Moellering continued to work with animation (we discuss this and his related work on data exploration in section 21.3.1), there were few other efforts in cartographic animation until the 1990s. Craig Campbell and Stephen Egbert (1990, 24–25) argued that several reasons accounted for this, including the lack of funds in geography departments, the lack of a "push toward the leading edges of technology" in these departments, a negative view of academics who develop and market software, and the difficulty of disseminating animations in standard print outlets. In the following sections, we see that many of these limitations appear to have been overcome.

20.2 VISUAL VARIABLES AND CATEGORIES OF ANIMATION

In 1991, David DiBiase and his colleagues created a videotape in which they formulated visual variables unique to animation and categorized various types of animation. This video was supplemented by an article (DiBiase et al. 1992), and the visual variables were later expanded on by Alan MacEachren (1995). In the following, we consider the character of these visual variables and the various categories of animation.

20.2.1 Visual Variables for Animation

In the videotape, DiBiase and his colleagues described how the visual variables for static maps (see section 4.3)

could be applied to animated maps. For example, they animated a choropleth map of New Mexico population from 1930 to 1980 to show how the visual variable lightness (represented by gray tones) could represent changes over time, and they transformed the state of Wisconsin from its real-world shape to that of a cow to illustrate how the visual variable shape could be used in animation. More important, however, they showed how animation requires additional visual variables beyond those used for static maps, namely duration, rate of change, and order.

In its simplest form, **duration** is defined as the length of time that a frame of an animation is displayed, with short duration resulting in a smooth animation and long duration a choppy one. More generally, a group of frames can be considered a scene (in which there is no change between frames), and we can consider the duration of that scene.* Because duration is measured in quantitative units (time), it can be logically used to represent quantitative data; for example, DiBiase and his colleagues animated electoral college votes for president, varying the duration of each scene in direct proportion to the magnitude of the victory in each election.

Rate of change is defined as m/d, where m is the magnitude of change (in position and attributes of entities) between frames or scenes, and d is the duration of each frame or scene. For example, Figure 20.1 illustrates different rates of change for the position of a point feature (A) and the attribute of circle size (B). The smoothness (or lack thereof) in an animation is a function of the rate of change. Either decreasing the magnitude of change between frames or decreasing the duration of each frame will decrease the rate of change and hence make the animation appear smoother.

Order is the sequence in which frames or scenes are presented. Normally, of course, animations are presented in chronological order, but DiBiase et al. argued that knowledge also can be gleaned by reordering frames or scenes. For example, they modified not only the duration of frames associated with the presidential election data, but also reordered them such that lower magnitudes of victory were presented first.

Alan MacEachren (1995) extended the visual variables for animation to include display date, frequency, and synchronization. He defined **display date** as the time some display change is initiated. For example, in an animated map of decennial population for U.S. cities, we might note that a circle first appears for San Francisco in 1850. **Frequency** is the number of identifiable states per unit time, or what MacEachren called *temporal texture*. To illustrate frequency, he noted its effect on *color cycling*, in which various colors are "cycled" through a symbol, as is often done on weather maps to show the flow of the

---

* DiBiase et al. borrowed the term *scene* from the work of Szego (1987).

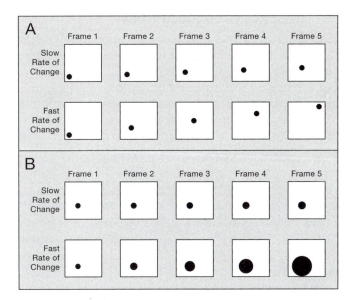

FIGURE 20.1 The visual variable rate of change is illustrated for (A) geographic position and (B) circle size. Rate of change is defined as *m/d,* where *m* is the magnitude of change between frames or scenes, and *d* is the duration of each frame or scene. For these cases, duration is presumed constant in each frame. (After DiBiase et al. 1992, 205.)

jet stream. **Synchronization** deals with the temporal correspondence of two or more time series. MacEachren (pp. 285–286) indicated that if the "peaks and troughs of … [two] time series correspond, the series are said to be 'in phase,'" or synchronized. For example, if we examine animations of precipitation and greenness (see Color Plate 1.7 for an example of a greenness map), we would expect these to be out of sync because greenness might not immediately respond to precipitation changes.

20.2.2 Categories of Animation

In addition to introducing visual variables for animation, DiBiase and his colleagues developed a useful categorization of animated maps as a function of whether the animation emphasizes (1) change in a phenomenon, (2) the location of a phenomenon, or (3) an attribute of a phenomenon.

Animations Emphasizing Change

For animations emphasizing change, DiBiase et al. focused on changes in the phenomenon's position or attribute. As an example of *change in position*, we can envision a moving point representing the mean center of the U.S. population for each decennial census. (See section 3.4.2 for formulas for the mean center.) A *change in an attribute* would be represented by the animated map

of New Mexico population mentioned previously. More recently, Mark Harrower (2001) expanded the notion of animations emphasizing change, suggesting that we also consider change in the shape, size, or extent of a phenomenon; for instance, we might animate changes in the shape and areal extent of a hurricane over time.

DiBiase et al. suggested that animations emphasizing change be further divided into three types: time series, re-expressions, and fly-bys. A **time series**, or an animation that emphasizes change through time, is most common. DiBiase et al. illustrated several time series, including a cartogram of world population growth and the spread of AIDS in Pennsylvania. Borrowing from the work of Tukey (1977), they defined **re-expression** as an "alternative graphic representation … whose structure has been changed through some transformation of the original data" (p. 209). Re-expression might involve choosing subsets of a time series (*brushing*), reordering a time series, or changing the duration of individual frames or scenes within a time series. To illustrate brushing, DiBiase et al. animated the longest sequence of victories by candidates of a single party in U.S. presidential elections (the Democratic party won all elections from 1932 to 1948). To illustrate reordering a time series and changing the duration of frames, they animated five global-climate-model predictions for Puebla, Mexico. In reordering the temperature predictions from those that varied the least to those that varied the most, and then pacing them such that those with the greatest variation had the greatest duration, they found "a trend of maximum uncertainty during the spring planting season of April, May, and June, a pattern that had previously gone unnoticed" (p. 211).

In a **fly-by**, the user is given the feeling of flying over a 3-D surface. A classic example was the video *L.A. The Movie,* which illustrated a high-speed flight over Los Angeles, California, and vicinity. Traditionally, creating fly-bys was beyond the scope of most mapmakers because of the considerable computing time required (*L.A. The Movie* required 130 hours of CPU time on a VAX 8600 mainframe). Today, this has changed as mapmakers now can create their own fly-bys on desktop computers. A wide range of software is available for this purpose (see sections 15.3.3 and 24.2). In addition to empowering mapmakers to create their own fly-bys, the most sophisticated hardware and software also now enable users to control the flight path while the surface is being viewed, as opposed to having the flight path governed by the mapmaker, which was the case with *L.A. The Movie.*

Animations Emphasizing Location

Although a static map can be used to *indicate* the location of phenomena, animation can assist in *emphasizing* location. For example, DiBiase and his colleagues used flashing-point symbols (circles) to emphasize the location of major earthquakes (a static rendition of this concept is shown in Figure 20.2). Flashing symbols do not necessarily help users interpret a distribution, but they do attract attention, and for earthquake data serve as a warning signal ("It happened once, it could happen here again").

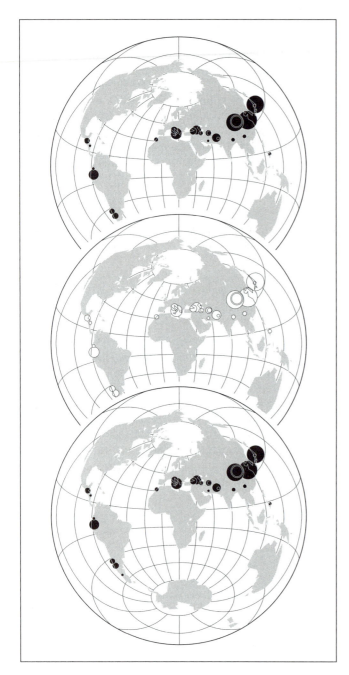

FIGURE 20.2 Using animation to emphasize location. Point symbols are flashed off and on to highlight the location of major earthquakes from 1900 to 1990. (After DiBiase et al. 1992. First published in *Cartography and Geographic Information Systems* 19(4), p. 207. Reprinted with permission from American Congress on Surveying and Mapping.)

Animations Emphasizing an Attribute

Animations can also emphasize an attribute by highlighting selected portions of it. An example would be **sequencing**, in which a map is displayed piece by piece. For instance, in section 21.3.2 we describe how the program ExploreMap can be used to present choropleth map classes in order from low to high. Later we consider a similar concept utilized by Gershon (1992) for isarithmic maps.

20.3 EXAMPLES OF ANIMATIONS

In this section we consider a range of animations that have been developed by cartographers and other graphic designers. Because these animations are impossible to depict here (we approximate a few with small multiples), we encourage you to examine the home page for this book, which provides information on where these animations can be acquired (or viewed), along with links to numerous other animations. We have divided the animations into two categories: those having little or no interaction and those characterized by interaction. The trend is toward interactive animations and related data exploration concepts.

20.3.1 Animations Having Little or No Interaction

Wilhelmson et al.'s Depiction of Thunderstorm Development

Robert Wilhelmson and his colleagues (1990) developed one of the more dramatic early animations, which depicted a numerical model of a developing thunderstorm via attractive ribbon-like streamers and buoyant balls (an online version can be found at *http://www.mediaport.net/ CP/CyberScience/BDD/fich_050.en.html*). Although Wilhelmson et al.'s animation was developed primarily for research purposes, it is apparent that similar animations could be used to teach students basic concepts about thunderstorm development and other meteorological concepts. Creating the thunderstorm animation required the effort of several researchers knowledgeable about storm development and the resources of the Visualization Services and Development Group at the National Center for Supercomputing Applications. Eleven months were required for development: Four scientific animators were involved, as were scriptwriters, artistic consultants, and postproduction personnel. Also required was a substantial hardware component: "about 200 [hours] of computer time on a Silicon Graphics Iris 4D/240 GTX using Wavefront software and special software written by the visualization group" (Wilhelmson et al. 1990, 33). Given today's improved software and hardware environment, such animations would take far less effort now, although sophisticated animations are still likely to require a group effort and a funding source.

Treinish's Portrayal of the Ozone Hole

Another striking early animation was Lloyd Treinish's (1992) *Climatology of Global Stratospheric Ozone,* which depicted the development of the ozone hole over Antarctica. Viewing an online version (*http://www.research.ibm. com/people/l/lloydt/ozone_video.mpg*),* you will see that levels of ozone (in columns of the earth's atmosphere) are symbolized using three redundant attributes: height above the earth's surface (as a $2\frac{1}{2}$-D surface), a traditional rainbow spectral scheme (red–orange–yellow–green–blue), and the degree of opacity. Thus, low ozone values appear in blue, close to the earth, and transparent, whereas high ozone values are red, far from the earth, and opaque. The ozone symbolization is shown wrapped around a global representation of the Earth, which is colored to represent the Earth's topography (Color Plate 20.1A). The complete animation consists of 4700 frames, one for each day over the period between 1979 and 1991.

Although Treinish's ozone animation is striking, it also illustrates some of the difficulty of interpreting animations over which the user has little control. There appear to be three reasons why this particular animation is difficult to follow. First, the entire 3-D structure (globe plus ozone symbolization) is frequently rotated during the animation to show interesting areas for certain times of year; as a result, it is not possible to focus on a specific region over the entire period of observation. Second, because the satellite that collected the data for the animation functioned on reflected sunlight, the Antarctic winter appears as a gap in the animation; unfortunately, it is difficult to differentiate this gap from the true ozone hole. Third, in its original videotape form (Treinish 1992), the user could not easily focus on a particular time segment of the animation. This latter problem has diminished somewhat in the online version, but the animation is still difficult to grasp. In spite of these problems, Treinish (1992) claimed that "the region of Antarctic ozone depletion each spring can be easily seen growing in size and severity over the last decade." We did not see this feature, but this might be due to our lack of familiarity with the ozone data, as DiBiase et al. (1992, 213) argued that those with expert knowledge about a problem are often more likely to benefit from animations. Those viewing the ozone animation might also have had access to small multiples of the data (Color Plate 20.1B), which make it possible to compare

one year directly with another and thus note that the ozone hole does appears larger in the last few years of the animation.

Weber and Buttenfield's Animation of Climate Change

In recent years, there has been considerable discussion concerning the increased level of carbon dioxide in the Earth's atmosphere and its potential impact on global warming. In considering this issue, it is natural to ask whether actual measured temperature data support the claim for global warming. To assist in answering this question, Christopher Weber and Barbara Buttenfield (1993) animated mean annual temperature data for the 48 contiguous United States over a 90-year period. In contrast to the two previous animations, which were initially available only as videotapes, Weber and Buttenfield's animation was distributed in diskette form, thus permitting the user greater control over the speed and direction of the animation.

Raw data for the animation consisted of monthly average temperatures for 344 weather stations from 1897 to 1986. Using these raw data, yearly average temperatures were computed from October to September "to preserve yearly highs (summers) and lows (winters) in seasonal groupings" (Weber and Buttenfield 1993, 144). Also, a three-year moving average was computed to smooth the data for purposes of animation. The actual animation was portrayed as positive or negative deviations from the mean of the 90-year period. Individual frames of the animation illustrate a logical use of color for bipolar data (Color Plate 20.2): Shades of red are used for positive deviations or temperatures warmer than normal, and shades of blue are used for negative deviations or temperatures cooler than normal. Each frame also contains a graph clearly summarizing deviations from the 90-year mean for each year, with the position of the current frame depicted by a line on the graph (see the lower right of Color Plate 20.2).

Weber and Buttenfield claimed that visual patterns in the animation matched those of other climatic research:

> The animation affirms a general cooling trend over the past 35 years, as well as a temporal pattern of cooling in the East and warming in the West in the last 20 years.... Isolines characterizing colder years are organized in a few clearly defined shapes; isoline patterns characterizing the warmest years are broken up, and these frames appear visually turbulent. (p. 147)

We, however, found the animation difficult to follow, largely because the pattern for one year did not merge smoothly into the pattern of a preceding year, but also possibly because of our lack of experience with spatiotemporal characteristics of climate change. This again is a case in which those familiar with the phenomenon might derive greater benefit from animation, and where

---

* For further discussion of the ozone data on which the animation is based, see Treinish and Goettsche (1991) and Treinish (1992).

small multiples might enable users to better understand the patterns in such data.

Animation of Urban Growth

Several researchers have developed animations of urban growth, including William Acevedo and Penny Masuoka (1997) and Michael Batty and David Howes (1996a; 1996b). In contrast to other animations that we have discussed, these animations seem relatively simple to follow because of the smooth transition between frames (Color Plate 20.3). The challenge does not seem to be in understanding the animation but in acquiring the data and creating the frames necessary for the animation. For instance, Acevedo and Masuoka discussed the problem of creating intermediate frames depicting urban growth—a simple algorithm based on linear distance produced "star-like outward growth along search directions" (pp. 427–428), and so a more complex algorithm was necessary.

Peterson's and Gershon's Nontemporal Methods

The preceding animations described all involved *temporal* data. In contrast, several researchers have experimented with animating *nontemporal* data. For instance, Michael Peterson's (1993) MacChoro II software permitted animating nontemporal univariate and multivariate data via choropleth symbolization (representative animations can be found at *http://maps.unomaha.edu/MP/Resume.html*). For univariate data, Peterson advocated animating both the classification method (say, mean–standard deviation, equal interval, quantile, and natural breaks) and the number of classes. For example, rather than displaying all four classification methods at once (as in Figure 20.3), MacChoro II displayed them in succession as an animation. For multivariate data, Peterson suggested using animation to examine the geographic trend of related attributes (those measured in the same units and part of a larger whole), and to compare unrelated attributes (those measured in dissimilar units and not part of a larger whole). With respect to related attributes, we might animate the percentage of population in various age groups (0–4 years old, 5–9 years old, etc.) within census tracts of a city. Peterson argued that such data "will usually show a clear regionalization in a city with older populations closer to the center and younger populations nearer the periphery. An animation . . . from low to high depicts high values 'moving' from the periphery to the center" (pp. 41–42). Two unrelated attributes would be median income and percent of the adult population with a high school education. Peterson argued that an animation of such attributes would detect similarities or differences in the two distributions (he recommended switching rapidly between the attributes).

More recently, Peterson (1999) promoted the idea of "active legends," in which the user controls the order of

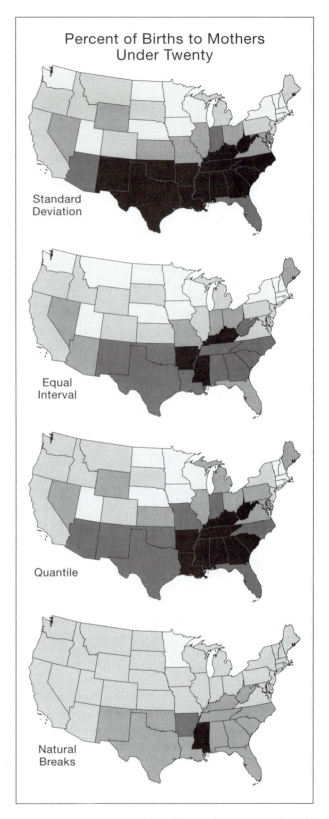

FIGURE 20.3 A small multiple illustrating four classification methods that Peterson animated. (After Peterson 1993. First published in *Cartography and Geographic Information Systems* 20(1), p. 41. Reprinted with permission from the American Congress on Surveying and Mapping.)

the animation by selecting individual legend boxes representing elements to be animated (Figure 20.4), thus moving away from a "movie" to a more interactive form of animation. Although Peterson's ideas for animating nontemporal data are intriguing, we found his animations difficult to follow. It seems that a small-multiples approach is more effective than animation for analyzing such nontemporal data, primarily because the transitions between frames of the animation are not smooth. In fact, Peterson (p. 41) demonstrated his ideas in static form using small multiples (as in Figure 20.3).

Nahum Gershon (1992) also experimented with nontemporal animations. One of Gershon's approaches involved an "animation of segmented components" in which users were shown an isarithmic map in a sequenced presentation: A high narrow range of temperature appeared first, followed by a progressively wider range until the entire range to be focused on appeared; then the process was reversed with progressively narrower ranges of temperature displayed until just the highest narrow range appeared again. Gershon indicated that users "perceived the existence of structures, their locations and shapes better than in the static display of the data" (p. 270). This result is a bit surprising given the lack of success some have found with sequencing on choropleth maps (see section 21.3.2). Gershon's superior results might be a function of his particular sequencing strategy; or possibly, sequencing is more appropriate for contour than for choropleth maps.

Gershon also had people examine the correlation between two attributes as he quickly switched between maps of the attributes. This approach is analogous to Peterson's suggestion for animating more than one attribute. In contrast to the difficulty we had understanding Peterson's animations, Gershon argued that his audience could "visually correlate structures appearing in both [maps]" (p. 272). We suspect that Gershon's superior results were due to the similar structure of his maps: They were both of January sea surface temperatures, but for different years. As a result, users could easily focus on the differences between the maps.

Lavin et al.'s Animation of Map Design Parameters

Stephen Lavin and his colleagues (1998) presented the notion of animating map design parameters, focusing on the effect that varying interpolation parameters has on an isarithmic map.* For instance, from section 14.3.2 you will recall that one of the parameters involved in inverse-distance interpolation is the power (or exponent) to which distances are raised. Those who have never worked with interpolation are often unsure what effect a particular exponent will have. Lavin and his colleagues argued that we can examine the effect of the exponent by creating a series of animated maps in which each map of the series has a slightly different exponent. Figure 20.5 illustrates this notion for six frames of a 120-frame animation; in the figure the exponent is varied from 1 to 6 in increments of 1, whereas in the actual animation the exponent was varied from 0.05 to 6, with increments of 0.05. When this was done, Lavin and his colleagues found that the vertical relief increased as the exponent increased, but that exponents above 4 produced little change, and that these findings held regardless of control point search and selection methods. Although Lavin and his colleagues focused on the inverse-distance parameter, they stressed that animation could be applied to other interpolation parameters, and more generally to other mapping techniques. With respect to the latter, Slocum and Yoder (1996) developed a proportional circle mapping program that would enable a map designer to dynamically change the largest circle and immediately see the impact on the overall map design.

Animations for Courseware and Lectureware

The bulk of the animation examples we discuss in this section were developed in association with refereed publications, which typically discuss the approaches used to create the animation and often comment on its presumed effectiveness. You should recognize, however,

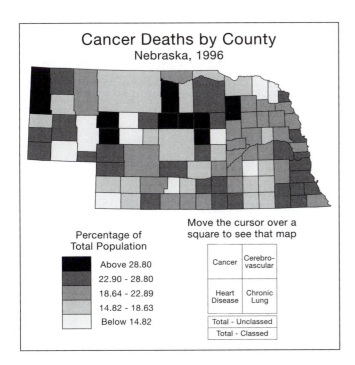

FIGURE 20.4 Peterson's (1999) notion of active legends: Users can control the order of frames within an animation by clicking on the desired legend boxes. (Courtesy of Michael P. Peterson.)

---

* Lodha et al. (1996a) and Mitas et al. (1997) also discussed using animation to illustrate the effect that varying interpolation parameters has on isarithmic mapping.

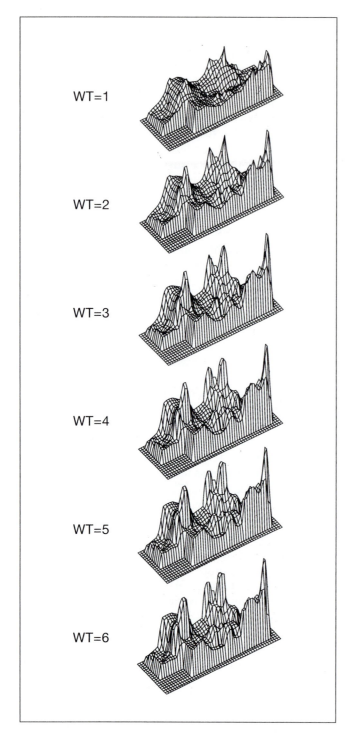

WT=1

WT=2

WT=3

WT=4

WT=5

WT=6

FIGURE 20.5 Lavin et al.'s (1998) notion of using an animated set of maps to depict the effect of the exponent in inverse-distance interpolation. The numbers indicate the weight of the exponent. (Courtesy of North American Cartographic Information Society.)

that a large number of animations have been developed outside of a refereed framework, particularly for educational purposes. In general, two forms of educational software are possible: courseware and lectureware. **Courseware**

would be software that assists students in learning concepts outside the classroom, whereas **lectureware** would be analogous software used in a more traditional lecture setting (Krygier et al. 1997). In practice, it appears that the bulk of educational software is developed as courseware, but then is also used in a lecture setting. Introductory physical geography textbooks, such as McNight and Hess (2004) and Strahler and Strahler (2003) commonly include courseware CDs, which depict numerous animations. Courseware materials, many of which contain animations, also have been released on the Web; one extensive source is the Digital Library for Earth System Education site (*http://www.dlese.org*). Because a wide range of courseware has been developed, it is not possible to easily classify this software in either of our two major categories based on the level of interaction. We have placed it in the little or no interaction category because many of the associated animations are characterized by limited interaction.

20.3.2 Animations Characterized by Interaction

Dorling's Work

Daniel Dorling (1992) provided some interesting thoughts on animation, experimenting with animation in three realms, which he designated "animating time," "animating space," and "three-dimensional animation." Although some of the animations he described appeared to fit our little-or-no-interaction category, we have included them in the interaction category because he supported the move toward greater interaction. Dorling's idea of animating time roughly equates to DiBiase et al.'s (1992) time-series animation. Dorling argued that animating time is not successful when animations involve changes in color at fixed locations due to "the brain's poor visual memory" (p. 223). His conclusion concurs with our difficulty of interpreting some of the animations we have considered (e.g., Weber and Buttenfield's animation of climate change, which involved changes in the color of contour maps at fixed locations, and Peterson's animation of nontemporal data associated with choropleth maps). However, Dorling argued that animating movement over time can be effective, noting that his most effective animation involved moving dots within a triangle representing the proportion of votes for three major political parties (pp. 218–219).

Animating space involves "panning and zooming . . . a large two-dimensional static image" (p. 216). To illustrate, Dorling described how a map of three occupational attributes in the more than 100,000 enumeration units making up British census districts could be examined:

What emerged was a picture of the way in which areas of affluence simultaneously surrounded, looped around, and appeared inside the places where the less prosperous lived.... One of the most interesting patterns occurs in inner London, where a thin, snakelike belt of "professionals" winds from north to south, between areas sharply differentiated as being dominated by the housing of those working in the lowest status occupations. (p. 217)

Although Dorling used the term "animation" to describe his interaction with the map, in the context of this book it is a form of data exploration: The instantaneous change in the data is a form of animation, but the control exerted by the user makes exploration a more appropriate term.

Dorling used the term "three-dimensional animation" to describe the animation of data in 3-D space; for instance, in section 21.3.9 we examine various 3-D approaches for depicting wind speed both horizontally and vertically using the software package Vis5D. As with animations of color at fixed locations, Dorling found 3-D animations confusing, arguing that although the computer can easily create 3-D graphics, it is more appropriate to represent data in two-dimensional form.

Dorling appeared to concur with our suggestion for using small multiples as a substitute for animation, but he indicated that this might require an animation of space (data exploration). He stated:

> Comparison requires at least one simultaneous view. The distribution of successive years' unemployment rates is often best shown on a single map, where symbols can become very small when a large number of areas are involved. Animating space can allow detailed spatial investigation of such a complex picture, which incorporates a temporal attribute statically. (p. 224)

In summary, although Dorling criticized certain aspects of animation, he felt it could be especially useful for enlivening presentations.

Monmonier's Graphic Scripts

Mark Monmonier (1992a) defined a **graphic script** as a series of dynamic maps, statistical graphics, and text blocks used to tell a story about a particular set of data. Monmonier intended graphic scripts as a form of data exploration, and so we have placed them in the "interaction" category. His initial implementation of these scripts, however, was as fixed animations ("movies"), as users could not control their speed or direction of play. To illustrate a graphic script, consider the portion of a script shown in Figure 20.6, which Monmonier used to examine the distribution and relationship of the attributes "Female Percentage of Elected Local Officials" and "Female Labor Force Participation Rate" for the 50 states of the United States. The complete script consisted of three "acts":

Act I: Introduce the Attributes

Act II: Variation and Covariation in Geographic Space and Attribute Space

Act III: Explore the Relationship with a Regression Model

Act I: Introduce the Attributes

I-1. Introduce dependent attribute "Female Percentage of Elected Local Officials, 1987," with brief title "Female Officials" and signature hue of red.

Text block: full title, brief title, and description; pause for reading . . .

Large map above rank-ordered bar graph, with each polygon linked to a bar.

Upward sweep by rank [Figure 20.2]; downward sweep by rank.

Upward sweep by value; downward sweep by value.

Highlight highest fifth; highlight lowest fifth.

Classify by quintiles (equal fifths); then blink-highlight each category (in sequence from highest category to lowest).

Classify by equal thirds; then blink-highlight each category.

Quintiles again; blink-highlight each of the nine census divisions . . .

I-2. Introduce independent attribute "Female Labor Force Participation Rate, 1987," with brief title "Females Working" and signature hue of blue.

Same layout and scenario as in I-1.

I-3. Compare geographic patterns by rapid alternation of the two attributes on a single map.

Large map above hue-coded titles for both attributes.

Classify by quintiles; 13 cycles: display dependent attribute on map, then display independent attribute on map.

FIGURE 20.6 A portion of a graphic script. (After Monmonier, M. 1992a. First published in *Cartography and Geographic Information Systems* 19(4), p. 254. Reprinted with permission from the American Congress on Surveying and Mapping.)

Figure 20.6 lists just the script for Act I, which we can see consists of three major steps: introduce the dependent attribute (I-1 in 20.1), introduce the independent attribute (I-2), and compare geographic patterns (I-3). Note that each step involves a series of animated maps. To illustrate, Figure 20.7 portrays a frame from "Upward sweep by rank; downward sweep by rank." The heights of individual bars beneath the map correspond to each of the values symbolized on the map. During an upward sweep, a bar would be lit in the graph and then on the map, beginning with the lowest ranked value and progressing to the highest; the opposite would be true for a downward sweep. In Chapter 21, we see that this approach is similar to the sequencing option used in the software ExploreMap.

Monmonier and Gluck (1994) found that users had difficulty understanding graphic scripts because they had no control over them (remember, they were

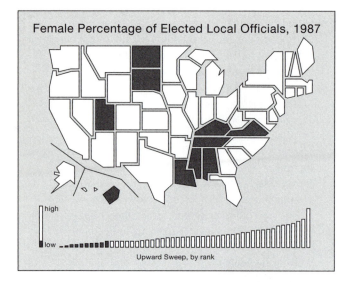

FIGURE 20.7 A frame from a graphic script illustrating "Upward sweep by rank; downward sweep by rank." Individual bars at the bottom correspond to values for each of the data shown on the map. During an upward sweep, a bar is lit in the graph and then on the map, beginning with the lowest ranked value and progressing to the highest; the opposite is true for a downward sweep. (After Monmonier, M. 1992a. First published in *Cartography and Geographic Information Systems* 19(4), p. 248. Reprinted with permission from the American Congress on Surveying and Mapping and Mark Monmonier.)

"movies"). As a result, Monmonier and Gluck suggested that graphic scripts should be stopped automatically at key points, continuing only when the viewer pushes a "resume" key. Alternatively, they suggested that a "rerun" key would provide greater flexibility by allowing a user to repeat the immediately preceding sequence. To justify our placement of scripts in the "interaction" category, capability would also have to be added to permit users to explore individual sections of the script. For example, when Figure 20.7 is presented, it should be possible for a user to easily select either an upward sweep by rank or a downward sweep by rank.

One problem that we encountered while viewing the scripts was the lack of a verbal commentary, which would explain what was being shown. Monmonier (1992a) indicated that the lack of verbal commentary was a function of the large storage space required by high-quality speech. Monmonier and Gluck avoided this problem by having a trained moderator provide a running commentary during their evaluation of the scripts. Apparently, this was essential, as one of the participants remarked "It was all logical, but without the narrative it would have been difficult. . . . If I had to read the description and watch what was going on, I would have had a hard time" (p. 43). Fortunately, those wishing to develop graphic scripts today will find that storage space is no longer an issue given the large disk drives that are available.

Harrower's Work

Mark Harrower has undertaken considerable work with animation; here we consider three studies that he has been involved in. In the first, Harrower and his colleagues (2000) analyzed EarthSystemsVisualizer (*http://www.geovista.psu.edu/grants/VisEarth/animations1.html*), a tool that they developed to facilitate undergraduate students' learning about global weather (Figure 20.8). Rather than simply showing "canned" animations, they wanted students to explore the underlying data using brushing and focusing tools. **Brushing** allows the user to highlight an arbitrary set of individual entities; for instance, the user might wish to see only animation frames associated with 6 A.M. In contrast, **focusing** allows the user to focus on a particular subrange of data; for instance, you might wish to see an animation of only a 48-hour period, as opposed to an entire week. An important characteristic of EarthSystemsVisualizer is the attempt to represent different kinds of time (linear and cyclic) with corresponding linear and cyclic legends (based on the work of Edsall and Peuquet (1997)). *Linear time* refers to the notion of general change over time, such as population growth for a city, whereas *cyclic time* refers to temporal changes that are repetitive in nature, such as diurnal changes in temperature. The bottom portion of Figure 20.8 shows examples of linear and cyclic legends, respectively. Important conclusions from Harrower et al.'s study were that students who understood the brushing and focusing tools performed the best, and that those without the tools performed better than those who did not understand the tools, suggesting that instructors must take great care in introducing such tools to students.

In a second study, Harrower (2001) promoted animation as a means for understanding remotely sensed images. He stated:

> Current change-detection techniques are insufficient for the task of representing the complex behaviors and motions of geographic processes because they emphasize the outcomes of change rather than depict the process of change itself. Cartographic animation of satellite data is proposed as a means of visually summarizing the complex behaviors of geographic entities. (p. 30)

To illustrate the potential of animation for remotely sensed imagery, Harrower developed a prototype system known as VoxelViewer (*http://www.geovista.psu.edu/members/harrower/voxelViewer.html*). Two key elements of VoxelViewer are tools that allow a temporal resolution ranging from 10 to 80 days and a spatial resolution (pixel size) ranging from 10 to 250 km. Such tools are useful for reducing spatial heterogeneity and short-term temporal fluctuations, ultimately producing smoother animations that capture general trends in the data. In an intriguing application, Harrower described how such tools can be used to examine the behavior of the Sahel,

FIGURE 20.8 The interface for EarthSystemsVisualizer (original appears in color). (Courtesy of Mark Harrower.)

a semi-arid region in northern Africa. Using a spatial resolution of 10 km and a temporal resolution of 40 days, Harrower stated:

> The northern and southern boundaries do not move in unison. Rather there is a temporal lag in which the southern boundary moves northward first (since the monsoon rains arrive from the south) compressing the region before the northern boundary expands into the desert. The southern boundary is also the first to retreat southward with the onset of the dry season. (p. 38)

Harrower noted that such a phenomenon could not be seen at other resolutions and would be missed if animation were not used.

In a third study, Harrower (2002) developed the notion of **visual benchmarks**, or reference points with which other frames of an animation can be compared. *Fixed* visual benchmarks remain constant from frame to frame; for example, Color Plate 20.4 illustrates fixed visual benchmarks for a proportional circle map—yellow and red circles represent the minimum and maximum number of new AIDS cases for each country over the time period of the animation (1993–2000). *Dynamic* visual benchmarks are more complicated because they change from frame to frame; for example, a "ghost image" of a previous frame might appear superimposed on the current frame. Harrower implemented visual benchmarks for both proportional symbol and isarithmic maps, finding that benchmarks were more effective on proportional symbol maps. Although Harrower found that benchmarks were not as effective as he had hoped, he noted that their effectiveness might be

enhanced through further use and training. Examining some of Harrower's animations (*http://www.geography. wisc.edu/~harrower/dissertation/index.html*) illustrates the character of interactivity that cartographers are now attempting to provide in map animations.

Andrienko et al.'s Work

Natalia Andrienko and her colleagues (2000; 2001) argued that we need to improve on traditional "fixed" animations. They stated:

> A simple "movie", e.g., showing the growth of city population by expanding circles that represent cities, is often enough for demonstration of a known temporal trend.... This is insufficient, however, for supporting exploration, i.e. revealing unknowns. An analyst needs special tools that can help to compare states at different moments, detect significant changes, assess magnitudes and speeds of the changes etc. (Andrienko et al. 2000, 217)

A software package that they have developed for this purpose is illustrated in Color Plate 20.5. Consider Color Plate 20.5A—on the right is a set of animation controls that allow considerable flexibility in animating choropleth maps, such as the step size between frames, the delay between frames, and whether the user steps frame by frame or the system automatically displays the frames. Below the map is a graph depicting the change in each individual enumeration unit over time. In this case, we see that Finland is highlighted and in the graph we can see the trend for forest exports from Russia to

Finland over the period from 1992 to 1997. Color Plate 20.5B further highlights the considerable flexibility of the software, as now "Enable visual comparison" has been checked in the upper left corner. Because "Fixed time moment" has also been selected, and the bar below that is positioned above "1992," each frame of the animation can be compared with the year 1992. Also notice that the graph below the map allows us to compare each year against 1992. Finally, note that a diverging color scheme has been selected to enhance the comparison. The complete flexibility of the software can be fully appreciated by experimenting with the online version at *http://borneo.ais.fraunhofer.de/descartes/time/AreaAnalysis/app/efis/*.

Another interesting piece of software developed by Andrienko and her colleagues (2000) is the depiction of the movement of storks across the continent of Africa using directed line segments of differing color. This animation is rather difficult to depict in static form but provides an interesting illustration of how linear movements can be depicted in an animation. Andrienko and her colleagues introduce the notion of varying both the interval of time depicted and the time step between individual frames; the result is that the directed line segments move across the map and change in length, depending on how long a stork stops at a particular location. An online version can be viewed at *http://ais.gmd.de/and/java/birds/*.

Deep Thunder

Deep Thunder is a research project being undertaken by Lloyd Treinish and his colleagues at IBM to develop sophisticated weather prediction models and associated visualizations (*http://www.research.ibm.com/weather/DT.html*). Two sample maps from Deep Thunder are shown in Color Plate 20.6. As you can see, the resulting images are typically 3-D visualizations of multiple atmospheric attributes. For instance, in the case of Color Plate 20.6A we see the following: a base map of the Atlanta and Savannah, Georgia, vicinity; arrows that have orientation and color indicating wind direction and speed at the surface; isarithmic information depicting precipitation; and a cloud structure depicted as a white translucent isosurface, with cyan isosurfaces in the interior indicating forecast rain shafts. Although Deep Thunder includes a substantial interactive component, an important element is animations, which are often viewed without extensive interaction. For example, Color Plate 20.6A was part of an animation that was used to predict the weather for the 1996 Centennial Olympic Games in Atlanta. Although Treinish and his colleagues obviously feel such animations are effective, their complexity again raises the issue of the usefulness of animation.

At several points in the preceding discussion, we have raised the question of whether animation works. Although those developing animations are often enamored of their capability, we (and others) often find them difficult to understand. Is it possible that the capability of animation has been overhyped? Such thoughts were shared by Barbara Tversky and her colleagues (2002), who undertook a meta-analysis of animation efforts that did not involve geospatial information. Although researchers had suggested that the animations Tversky and her colleagues evaluated were more effective than their static counterparts, Tversky et al. found that the animations often contained microsteps that were not available in the static displays, and thus a fair comparison was not possible. Tversky et al. also noted that although interactivity clearly has the potential to enhance an animation, studies have not yet been done that fully analyze the effects of such interactivity relative to static maps. Finally, they made the following observation:

> Animations must be slow and clear enough for observers to perceive movements, changes, and their timing, and to understand the changes in relations between the parts and the sequence of events. This means that animations should lean toward the schematic and away from the realistic, an inclination that does not come naturally to many programmers, who delight in graphic richness and realism.

These seem to be useful thoughts for those of you considering developing animations for others.

With maps, studies of animation have produced mixed results. We have already mentioned the success that Gershon (1992) appeared to have. A study often used to illustrate the benefits of animation is the work of Koussoulakou and Kraak (1992), who found that although the percentage of correct answers did not differ, animated maps were processed significantly faster than static maps when depicted as a small multiple. Koussoulakou and Kraak stressed that users might have done even better if users could interact with the animated maps. A study by Patton and Cammack (1996) also revealed support for animation on sequenced choropleth maps (see Chapter 21). In contrast, studies by Slocum and Egbert (1993), Johnson and Nelson (1998), and Cutler (1998) have found little difference between animated and static maps. For instance, in comparing an animated map with a small multiple, Cutler found that the animated version of a shaded isarithmic map had a significantly lower percentage of correct answers and a slightly slower processing time than the small multiple. Cutler noted "the strongest indicators of comprehension were not the type of map viewed, but the subjects' reading levels and their prior knowledge" (p. 63).

Clearly, more research is necessary if we are to determine whether animation is truly effective. We suspect that the effectiveness of animation will be a function of numerous factors, including the type of symbology (e.g., depicting temporal change on a proportional symbol map seems easier than on a choropleth map), how the animation is viewed (a "movie" in which the user has no control is obviously quite different than an animation in which the user can move forward, backward, and jump to different frames), whether the user has expertise in the domain being mapped (animated maps created using Deep Thunder seem complex to us, but they might not be to the trained meteorologist), and experience with animation (does one who uses animation on a regular basis become more adept at using it?).

SUMMARY

In this chapter, we examined several matters pertaining to animation. First, we noted that although animation commonly is associated with recent technological advances, the earliest map animations were produced in the 1930s, and cartographers began discussing the potential of animation in the 1950s. The first computer-based animations were produced in the 1970s, but it was not until the 1990s that computer-based animations became common.

The visual variables utilized for static maps can also be used on animated maps; for example, we might utilize proportional circles (the visual variable size) to depict changes in the population of census tracts over time. Animated maps, however, also utilize other visual variables, including **duration** (the length of time that a frame of an animation is displayed), **rate of change** (computed as m/d, where m is the magnitude of change between frames or scenes, and d is the duration of each frame or scene), **order** (the sequence in which frames or scenes are presented), **display date** (the time some display change is initiated), **frequency** (the number of identifiable states per unit time), and **synchronization** (the temporal correspondence of two or more time series). Animations can be categorized in terms of whether they emphasize change in a phenomenon, the location of a phenomenon, or an attribute of a phenomenon. Animations emphasizing change can be further divided into three types: **time series**, **re-expressions**, and **fly-bys**. A time series is the most common form of animation because animating time can serve as a scale model of real-world time. Some of the most interesting animations, however, do not involve time. For example, a fly-by, in which the viewer is given the impression of flying over a landscape, can be particularly dramatic.

We considered numerous examples of animations, extending from those that involved little or no interaction to those having considerable interaction. Examples of animations involving little or no interaction include Wilhelmson et al.'s depiction of thunderstorm development, Weber and Buttenfield's examination of climate change, and Peterson's nontemporal methods. Those who have developed animations involving considerable interaction include Harrower and Andrienko and her colleagues. In addition to creating interactive animations, Harrower is noteworthy for developing the notion of **visual benchmarks**, which enable reference points to be compared with other frames of an animation.

In examining animations for this chapter, we found many of them difficult to understand, raising the question of whether animation is truly effective. Apparently, we are not alone in this respect, as Tversky and her colleagues also raised this issue, noting that research supporting the use of animation over static renditions is often flawed (e.g., animations might contain microsteps that are not available in static displays). It appears that more research is needed to determine whether animation (and interactivity) actually enhances visualization. We suspect that the effectiveness of animation will be a function of numerous factors, including the type of symbology, the level of interaction that the user is permitted, whether the user has expertise in the domain being mapped, and the user's experience with animation.

Our focus in this chapter has been on stand-alone animations or on interactive software in which animation plays a major role. In subsequent chapters, we will see that animations can also play a part in data exploration software, electronic atlases, and the visualization of uncertainty.

FURTHER READING

Acevedo, W., and Masuoka, P. (1997) "Time-series animation techniques for visualizing urban growth." *Computers & Geosciences* 23, no. 4:423–435.

 Discusses the development of animations to depict the growth of urban areas. Also see Batty and Howes (1996a; 1996b).

Andrienko, N., Andrienko, G., and Gatalsky, P. (2001) "Exploring changes in census time series with interactive dynamic maps and graphics." *Computational Statistics* 16:417–433.

 Discusses highly interactive software for exploring spatiotemporal data.

Blok, C., Köbben, B., Cheng, T., and Kuterema, A. A. (1999) "Visualization of relationships between spatial patterns in time by cartographic animation." *Cartography and Geographic Information Science* 26, no. 2:139–151.

 Uses animation to examine the relationship between two attributes that change over time.

Campbell, C. S., and Egbert, S. L. (1990) "Animated cartography: Thirty years of scratching the surface." *Cartographica* 27, no. 2:24–46.

 A review of early animation efforts in cartography; explores the reasons for the lack of animation in cartography prior to 1990 and suggests future prospects for animation.

Cox, D. J. (1990) "The art of scientific visualization." *Academic Computing* 4, no. 6:20–22, 32–34, 36–38, 40.

Discusses a number of animations developed at the National Center for Supercomputing Applications, where so-called Renaissance Teams were formed.

Eddy, W. F., and Mockus, A. (1994) "An example of the estimation and display of a smoothly varying function of time and space—The incidence of the disease mumps." *Journal of the American Society for Information Science* 45, no. 9:686–693.

Discusses sophisticated mathematical approaches for creating a smooth animation of spatiotemporal data for enumeration units.

Gahegan, M. (1998) "Scatterplots and scenes: Visualization techniques for exploratory spatial analysis." *Computers, Environment and Urban Systems* 21, no. 1:43–56.

Describes sophisticated approaches for using animation to analyze multivariate relationships at one point in time.

Gersmehl, P. J. (1990) "Choosing tools: Nine metaphors of four-dimensional cartography." *Cartographic Perspectives* no. 5:3–17.

Discusses various ways in which software can be used to create animations; Gersmehl termed these "animation metaphors."

Karl, D. (1992) "Cartographic animation: Potential and research issues." *Cartographic Perspectives* no. 13:3–9.

Introduces a number of research issues associated with animation.

Krygier, J. B., Reeves, C., DiBiase, D., and Cupp, J. (1997) "Design, implementation and evaluation of multimedia resources for geography and earth science education." *Journal of Geography in Higher Education* 21:17–38.

Discusses the development and evaluation of multimedia resources for geography and earth science education.

Mitas, L., Brown, W. M., and Mitasova, H. (1997) "Role of dynamic cartography in simulations of landscape processes based on multivariate fields." *Computers & Geosciences* 23, no. 4:437–446.

Discusses the use of animation to illustrate (1) simulations of landscape processes and (2) the effect of changing parameters associated with interpolation methods.

Monmonier, M. (1990) "Strategies for the visualization of geographic time-series data." *Cartographica* 27, no. 1:30–45.

Discusses methods for visualizing spatiotemporal data; animation is considered one useful method.

Monmonier, M. (1992b) "Summary graphics for integrated visualization in dynamic cartography." *Cartography and Geographic Information Systems* 19, no. 1:23–36.

Presents static maps and graphics that can be used in association with animated maps.

Monmonier, M. (1996) "Temporal generalization for dynamic maps." *Cartography and Geographic Information Systems* 23, no. 2:96–98.

Describes methods for smoothing map animations.

Ogao, P. J., and Kraak, M.-J. (2001) "Geospatial data exploration using interactive and intelligent cartographic animations." *Proceedings, 20th International Cartographic Conference*, International Cartographic Association, Beijing, China, CD-ROM.

Suggests the possibility of *intelligent animation*, in which features within animations are created on the fly, as in a developing storm system.

Peterson, M. P. (1995) *Interactive and Animated Cartography*. Englewood Cliffs, NJ: Prentice Hall.

The present chapter in our book focuses on the appearance of animations, as opposed to how they are created. In contrast, Peterson's text provides information on creating animations; see particularly Chapters 7 through 9.

Treinish, L. A. (2003) "Web-based dissemination and visualization of mesoscale weather models for business operations." *Proceedings of the Nineteenth International Conference on Interactive Information and Processing Systems for Meteorology, Oceanography and Hydrology*, American Meteorological Society. Available at *http://www.research.ibm.com/weather/vis/web_apps.pdf*.

Discusses visualization aspects of Deep Thunder; for other related papers, see *http://www.research.ibm.com/people/l/lloydt/biblio.html*.

von Wyss, M. (1996) "The production of smooth scale changes in an animated map project." *Cartographic Perspectives*, no. 23:12–20.

Presents methods for selecting appropriate scales when creating an animation that involves scale change.

21

Data Exploration

OVERVIEW

Data exploration, like the larger notion of geographic visualization, is a private activity in which unknowns are revealed in a highly interactive environment. The first two sections of this chapter consider the goals and methods of data exploration. Broad goals include: (1) identifying the spatial pattern associated with a single attribute at one point in time; (2) comparing spatial patterns for two or more attributes at one point in time; (3) identifying how spatial patterns for a single attribute change over time; and (4) comparing spatial patterns for two or more attributes to see how they covary over time. In a sense, these same goals apply to static paper maps, but the interactive environment of data exploration software permits us to discover spatial patterns that might not be seen in a single static map.

*The methods of data exploration include manipulating the input data (e.g., standardizing data), varying the symbolization (e.g., changing the color scheme on a choropleth map), manipulating the user's viewpoint, highlighting portions of a dataset (**focusing** and **brushing** are two approaches), multiple views (when a large number of maps are shown, the result is a **small multiple**), animation, linking with other forms of display (e.g., tabular and graphical displays), access to miscellaneous resources (e.g., via the Web), how symbols are assigned to attributes on a multivariate map, and automatic map interpretation (e.g., **data mining**). No single data exploration software package will include all of these features, but we would expect that an effective package would include many of them.*

The third section of the chapter covers examples of data exploration software. We begin by considering some early data exploration software: Hal Moellering's 3-D mapping software, ExploreMap, and Aspens. This software is either no longer available or has definite limitations when compared to today's software, but it does

provide a useful historical perspective. Next, we cover more recent software that was developed primarily with geographers in mind: MapTime, Project Argus, CommonGIS, ArcView 3.x and ArcGIS. Finally, we'll consider software that, although not explicitly developed for geographers, clearly is useful for geographers: HealthVisPCP, Transform, and Vis5D. With the exception of the historical software, we have attempted to select software that is available (software developers in academia often create prototypes and do not distribute them), is inexpensive (all are available for free, except Transform, ArcView3.x, and ArcGIS), and can run in a relatively low-cost hardware environment (Vis5D and Project Argus run most effectively on high-end platforms, such as Silicon Graphics workstations).

21.1 GOALS OF DATA EXPLORATION

In Chapter 1, we indicated that, like the larger notion of geographic visualization, data exploration is a private activity in which unknowns are revealed in a highly interactive environment. What are these *unknowns*, or more specifically what are the goals of data exploration? Next we list four broad goals for data exploration. In each case, we provide an example for attributes associated with census tracts in Los Angeles, California:

1. Identify the spatial pattern for a single attribute at one point in time. (Example: What was the spatial pattern of housing costs in 2000?)

2. Compare spatial patterns for two or more attributes at one point in time. (Example: How did the spatial pattern for housing costs compare with that for the percentage of Hispanics in the population in 2000?)

3. Identify how the spatial pattern for a single attribute changes over time. (Example: How did the spatial pattern of housing costs change with each decennial census?)

4. Compare spatial patterns for two or more attributes to see how they covary over time. (Example: How has the relationship between housing costs and percentage of Hispanics changed over time?)

You might argue that these goals are no different than those for static paper maps, and this is true. One difference, however, is that the character of spatial patterns might not become clear until we manipulate the map in a highly interactive graphics environment. For instance, if for the first goal we make a single choropleth map of housing costs, our visual interpretation might be a function of the color scheme—we might visualize a quite different pattern using a red–blue diverging scheme than with a blue sequential scheme. In a data exploration environment, we can change the color scheme instantaneously, and thus learn something about the spatial pattern of housing costs that was "hidden" in the single-map solution.

Another difference between a paper map and a data exploration environment is that the latter provides linkages to other views of the data as tables, graphs, and numerical summaries. For instance, when viewing a bivariate map of housing costs and the proportion of Hispanics in the population, you might also utilize a scatterplot depicting the relationship between the attributes. Examining this scatterplot might help you better understand the spatial pattern appearing on the map. Alternatively, you might notice a group of points that cluster together on the scatterplot and wonder where these occur on the map. This might be accomplished by using a mouse to select the clustered points in the scatterplot, which would cause these same points to be highlighted on the map. Clearly, this function could not be accomplished using a static map.

21.2 METHODS OF DATA EXPLORATION

In this section we summarize a broad range of methods that we would hope to find in data exploration software. Obviously, individual software packages will not include all of the methods, as each piece of software normally has specific (and thus limited) purposes. As you read about (and hopefully use) some of the data exploration software that follows, you might find it useful to consider which of these methods have been implemented and how they are utilized. It should be noted that there is no widely agreed-on set of methods for exploring spatial data. The methods listed here were culled from the works of MacEachren et al. (1999c), Crampton (2002), Kraak (1998), and Buja et al. (1996).

21.2.1 Manipulating Data

Methods for manipulating data include standardizing data (e.g., adjusting raw totals for enumeration units to account for the area over which the totals are distributed), transforming attributes (e.g., taking logarithms of data), and various methods of data classification. The first two of these methods can be handled in spreadsheet programs, but it is easier for the user if they are included within data exploration software.

21.2.2 Varying the Symbolization

One approach for varying the symbolization is to change the general form of symbolization. For instance, in Figure 4.9 we showed how different views of a wheat-harvested data set could be obtained with four different methods of symbolization: choropleth, isopleth, proportional symbol, and dot. In a data exploration environment, such views can often be generated quickly, although certain symbolization methods (e.g., the dot map) are not trivial to create. It is also possible to vary the precise manner in which the symbolization is implemented. For instance, if a choropleth map is deemed appropriate, then numerous color schemes can be employed. As we saw in Chapter 13, alternative color schemes can lead to dramatically different-looking maps. The power of data exploration software is that not only can the overall color scheme be changed, but the detailed relation between color and data can be modified dynamically. For instance, with the CommonGIS software we can drag a breakpoint separating two colors in a diverging scheme and have the result displayed instantaneously.

21.2.3 Manipulating the User's Viewpoint

Manipulating the user's viewpoint refers to modifying the "view" that users have of the map, with the data and method of symbolization held constant. Perhaps the most common viewpoint manipulations are zoom and pan functions. Such functions are critical given the relatively small size of standard display screens and the large databases that we sometimes wish to display. Sophisticated software will automatically adjust the detail shown as a function of the degree of zoom (e.g., the names of census tracts might not be shown at a small scale, but would appear at a larger scale). The ability of users to change their viewpoint is of course critical with 3-D maps, as high points on the surface can obscure lower points. Also important with 3-D maps is the means by which a user navigates through the landscape (e.g., using a mouse vs. a joystick, and the associated on-screen controls that are available), and how the user is provided a sense of orientation. Because these notions are particularly relevant to virtual environments, we focus on them in Chapter 24.

21.2.4 Highlighting Portions of a Data Set

Fundamental to data exploration is the ability to highlight portions of a data set via either focusing or brushing. **Focusing** involves highlighting a subrange of numeric values. For example, you might want to focus on census tracts that are less than 25 percent Hispanic. In a sense, this is a form of data classification, as we divide the data into two classes: those tracts with less than 25 percent Hispanic, and those with 25 percent or more. **Brushing** involves highlighting an arbitrary set of spatial entities. For example, you might wish to display the percentage of Hispanics only for tracts that voted for a particular political party.

21.2.5 Multiple Views

Multiple views refers to displaying more than one map image on the computer screen at a time. One form of multiple view is the **small multiple**, in which maps are displayed for individual time steps or attributes. Thus, if we have data on housing costs for decennial censuses, we would include a map for each decennial census. Alternatively, for a particular decennial census, we might create separate maps for percent Hispanic, median income, housing costs, and percentage of students attending public schools. Another form of multiple view is to show the same data set using different symbolization methods, as was discussed in section 21.2.2. In general, multiple views take advantage of our ability to compare things visually.

21.2.6 Animation

Animation, which we discussed more fully in the preceding chapter, is often an integral part of data exploration software. The key with data exploration software is that the user has considerably more control over an animation than is possible with canned animation "movies." With data exploration software, users might be able to stop an animation, redefine the data–symbol relationship, and rerun the animation. We'll see an example of this with the MapTime software.

21.2.7 Linking Maps with Other Forms of Display

As we indicated earlier, the ability to link maps with tables, graphs, and numerical summaries is a critical aspect of data exploration software. Users should be able to highlight data in either tables, graphs, or maps and have the highlighted specifications carried through to the other forms of display. Numerical summaries provide a useful complement to various forms of graphical display. For a single attribute, numerical summaries often include measures of central tendency and dispersion, although in Chapter 3 we

argued that a measure of spatial autocorrelation also could be useful in determining whether a spatial pattern might have resulted by chance. For multiple attributes, more complex numerical summaries would be appropriate (e.g., for two numeric attributes the user might want to know the correlation coefficient).

21.2.8 Access to Miscellaneous Resources

Throughout this discussion we have assumed that you have already collected data that you wish to visualize. In practice, as you explore spatial data, particularly with either multiple attributes or spatiotemporal data, you might find that you wish to acquire other information. In some cases, this will be data in a format similar to what you have already acquired. For instance, in relating housing costs to percent of Hispanic population, you might want to examine public expenditures for education. In other cases, you might wish to collect a quite different form of data. For instance, in studying housing costs, you might want to see what the houses actually look like on the landscape; you would like to click on a census tract and see what typical houses look like in that tract. Ideally, it would be desirable if you could readily import these sorts of information directly into your data exploration software, possibly via the Web.

21.2.9 How Symbols Are Assigned to Attributes

With multiple attributes (a multivariate map), an important decision often involves which attributes are assigned to which symbols. For instance, for the combined star and snowflake glyph symbol (section 18.2.2), the appearance will depend on the assignment of attributes to rays of the glyph. In Figure 21.1, the eight attributes on the left are arbitrarily assigned, whereas on the right the attributes have been ordered from smallest to largest beginning at the top and progressing clockwise. As another example,

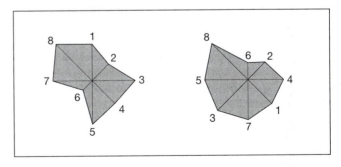

FIGURE 21.1 Different ways of assigning attributes to the combined star and snowflake glyph symbol. On the left the attributes are arbitrarily assigned; on the right they are assigned in order from the lowest to the highest attribute.

the order in which attributes are displayed on a parallel coordinate plot will influence the look of the plot. Referring back to Figure 3.9, remember that for highly positively correlated attributes, line segments on the parallel coordinate plot tend to be parallel, whereas for highly negatively correlated attributes, line segments tend to intersect one another in the middle of the graph.

21.2.10 Automatic Map Interpretation

In discussing data exploration thus far, we have focused on the ability of the user to interpret spatial patterns. In complex multiple attribute situations or with spatiotemporal data, it might be useful to utilize methods that automatically interpret the data, or at least assist the user in finding useful patterns in the data. This gets us into the notion of *data mining*, a relatively complex topic that we reserve for the last chapter.

21.3 EXAMPLES OF DATA EXPLORATION SOFTWARE

In this section, we cover numerous examples of data exploration software. First, we provide a historical perspective by considering some early attempts at data exploration: Moellering's 3-D mapping software, ExploreMap, and Aspens. Second, we consider more recent software that has been developed primarily with geographers in mind as potential users: MapTime, Project Argus, CommonGIS, ArcView 3.x, and ArcGIS. Third, we consider recent software that was not necessarily developed for geographers, but that certainly could be useful to them: HealthVisPCP, Vis5D, and Transform.

21.3.1 Moellering's 3-D Mapping Software

Hal Moellering's (1980) 3-D mapping software was an early attempt to demonstrate the potential of data exploration. With Moellering's software, a user could explore a 3-D digital elevation model (DEM) in real time, a process that Moellering termed "surface exploration."* Because Moellering's software was dependent on expensive hardware, he developed a video (Moellering 1978) to demonstrate software capabilities. In addition to illustrating surface exploration, the video portrayed two animations of spatiotemporal data: the growth of U.S. population from 1850 to 1970 and the diffusion of farm tractors in the United States from 1920 to 1970. These animations were created using attractive, colored 3-D prism maps.

Although Moellering's work suggested great potential for both data exploration and animation, for approximately 10 years following his effort there was little development in data exploration (at least by geogra-

phers), in part because many geographers lacked the necessary hardware and software. The lack of work in animation, in particular, was lamented by Craig Campbell and Stephen Egbert (1990). Fortunately, considerable advancements have taken place since 1990, as this and other chapters in this book illustrate. Today it is possible to generate maps similar to Moellering's on personal computers right on our desktop.

21.3.2 ExploreMap

Stephen Egbert and Terry Slocum (1992) developed ExploreMap to assist users in exploring data commonly mapped with choropleth symbolization. In developing ExploreMap, their intention was not to create the ultimate data exploration tool, but rather to provide a prototype that others might improve on. In contrast to Moellering's video distribution, ExploreMap and other software described in this chapter are distributed via diskette, CD-ROM, or over the Web.

ExploreMap operates in two basic modes: Design and Explore. Design is necessary for creating maps, whereas Explore includes methods for exploring data, and thus is our focus. Two of ExploreMap's functions, Areas and Overview, correspond to the two types of information that can be acquired from a map: specific and general (see section 1.2). To illustrate how Areas provides specific information, consider two of its options: Single Area and Ratio of Areas. The Single Area option is used to determine the exact value of an enumeration unit, either by pointing at it with the mouse or typing its name. In contrast, on a traditional paper map the value for an enumeration unit is determined by visually comparing a shade on a map with shades in the legend—assuming a classed map, the best estimate is a range of values shown in the legend.[†] Most current data exploration software includes an option similar to the Single Area option. The Ratio of Areas option is used to determine the ratio of values for two enumeration units: When a user selects two units, they are highlighted on the map, and the ratio of their values is shown at the bottom of the display (Figure 21.2). Such precise information is not available from paper maps, although it remains to be seen how worthwhile such an option is to users.

The Overview function within ExploreMap includes three options: Sequenced, Classes, and Subset. Rather than presenting the entire map at once (as on printed maps), the Sequenced option builds the map piece by piece: The title, geographic base, and legend title and boxes appear first (in sequence), and then each class is displayed (from low to high). Two arguments can be

---

* Technically, the data were $2\frac{1}{2}$-D, but exploration was done in 3-D space.

[†] Data values on a paper map can be displayed within enumeration units, but this detracts from the visualization of map pattern; moreover, values cannot be placed within small units.

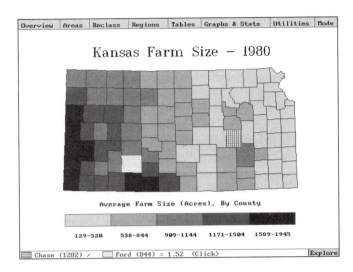

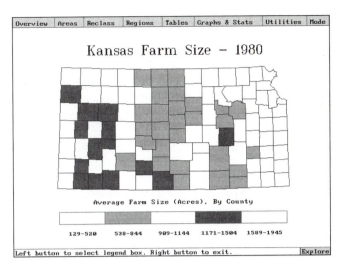

FIGURE 21.2   Using the Ratio of Areas option within ExploreMap to compare two enumeration units. The units (counties in this case) compared are highlighted on the map and the ratio of their values is shown at the bottom of the display.

FIGURE 21.3   The Classes option for ExploreMap. Any subset of classes can be selected by clicking on the legend boxes; in this case, classes 2 and 4 are shown.

made for **sequencing**: (1) it should enhance map understanding by providing users "chunks" of information (Taylor 1987), and (2) it should emphasize the quantitative nature of the data (that the data are ordered from low to high values). Experimental tests of sequencing, however, have produced mixed results. A study by Slocum et al. (1990) found that although more than 90 percent of readers favored sequencing over the traditional static (printed) approach, objective measures of map use tasks revealed no significant difference between sequenced and static approaches. As a result, Slocum et al. argued that sequenced maps might be preferred simply because of their novelty. A more recent study by David Patton and Rex Cammack (1996) revealed that sequencing is more effective when the time to study the map is limited.

The Classes option within Overview permits any combination of classes to be displayed; for example, Figure 21.3 portrays two classes of a five-class map. In an evaluation of ExploreMap, Egbert (1994, 87) found that the Classes option "was liked by all subjects without reservation"; subjects used Classes not only to identify the location of individual classes, but also to examine patterns and trends. Although useful, the Classes option is constrained by how the data are classed (in Figure 21.3, a five-class optimal map is used). One solution to this constraint is the Reclass function, which permits changing the method and number of classes. An option within Reclass, Compare Maps, permits showing up to four maps simultaneously (each map can have a different method of classification or number of classes). For example, Color Plate 13.8 compares four methods of classification for five-class maps: equal intervals, quantiles, optimal based on the mean, and optimal based on the median.

An alternative to reclassing the data is the Subset option, which permits focusing on an arbitrary range of data. To assist users in selecting data to be focused on, ExploreMap provides a dispersion graph (Color Plate 21.1A) illustrating the distribution of an attribute along the number line; the current class breaks, mean, median, and standard deviation can be shown on the dispersion graph (only the median is shown in Color Plate 21.1A). Users simply click on a desired range (or type a range of values), and the specified range is drawn in a highlighted color (blue is used in Color Plate 21.1B to display all values less than the median).

Although Subset enables users to highlight selected portions of a data set, it is limited in that all highlighted data are shown in a single color and real-time interaction with the dispersion graph is not possible. A more flexible approach would permit a range of shades for the highlighted color (say, a range of blues instead of a constant blue). This would enable users to focus on a subset of the data, while also seeing the variation within that subset. A more flexible approach would also allow the map to change dynamically as the mouse cursor is dragged back and forth along the number line. Interestingly, the latter capability was provided in the early Xmap system developed at the Massachusetts Institute of Technology by Joseph Ferreira and Lyna Wiggins (1990). Xmap used a "density dial" to highlight portions of a data set; Ferreira and Wiggins described the use of the density dial as follows:

> What makes the density dial a truly interactive visualization tool is its ability ... to change the group of cells that are highlighted in red ... as fast as we can move the mouse. We get an effect that is like watching a video. We see a moving sequence of red-shaded polygons going from the least dense tract to the

most dense tract, or vice versa. This speed—and the sense of motion that comes with it—enhances our ability to remember what we have just seen of the spatial pattern. (p. 71)

Unfortunately, Xmap was not made generally available, and so it is unlikely that you will have access to it. We will see, however, that the CommonGIS system provides similar functionality.

Other major functions in ExploreMap include Regions, Tables, Graphs & Stats, and Utilities. Whereas Subset allows users to focus on a subset of data along the *number* line, Regions permits focusing on a subregion of a *mapped* display of the data. For example, a politician interested in counties comprising an economic development region could use Regions to highlight enumeration units falling within that region. The Tables and Graphs & Stats functions focus on tabular, graphical, and statistical views of the data (as opposed to spatial views). For instance, Graphs & Stats can display a histogram and statistical parameters (Figure 21.4).

21.3.3 Aspens

Barbara Buttenfield and Christopher Weber (1994) developed Aspens to assist Dennis Jelinski (1987) in exploring the growth in trembling aspen trees at Waterton Lakes National Park in Alberta, Canada. More specifically, their purpose was to examine "the apparent contradiction that radial growth rates [in aspens] were higher where local environmental conditions (elevation, precipitation, and soils) were more harsh" (p. 11). Although Aspens was designed solely to handle Jelinski's data, it was intended as a prototype for illustrating the broader concept of **proactive graphics**, a term Buttenfield and Weber coined to describe software that enables users to initiate queries using icons (or symbolic representations) as opposed to words. Ideally, they argued proactive graphics should avoid steep learning curves and be "responsive to commands and queries...not... anticipated by system designers" (p. 10).

The philosophy of proactive graphics in Aspens is apparent in the opening screen of the study area (Color Plate 21.2), in which no pull-down menus are found (as, for example, appear in MapTime). Rather, it is presumed that users will become aware that links to information are most often available via the color green found on various icons. For example, clicking on the green "i" provides information about using the display, and clicking on the green symbol portion of the legend leads to further information about the legend. Interestingly, an animation begins as soon as the study area appears: Six leaves (corresponding to the six major study sites) change in size and color as each year of growth is highlighted in a bar graph; one frame from the animation is shown in Color Plate 21.2. Larger leaf sizes represent larger total cumulative growth, and a darker green leaf color represents a greater incremental growth for each year. Buttenfield and Weber argued that having the animation start automatically is appropriate because "The sooner graphical activity is recognized, the more quickly data will be explored" (p. 11).

A click on the name of a study site (e.g., Prairie 2 in Color Plate 21.2) leads to a screen of descriptive statistics and a histogram of the yearly incremental growth for that site. As we have suggested, linking statistical and graphical displays with maps is critical for exploratory analysis. Clicking on a leaf associated with a study site leads to a more detailed view of that site: the aspen clones making up the site (Color Plate 21.3). To distinguish the clones from the generalized study site, treelike symbols are used rather than leaves. The clones are animated in a manner analogous to the general study site, and descriptive statistics and a histogram of yearly growth rates can also be obtained by clicking on a clone identifier. Within any clone, it is possible to jump to another by clicking on one of the icons in the upper right of the screen.

Buttenfield and Weber recognized that Aspens is not truly proactive because it is not possible to generate queries and commands unanticipated by system designers. They also indicated that it would be desirable to incorporate photographs of study sites, add capabilities to reverse the animation and jump to an arbitrary year, and implement functions for saving data, maps, and text information to files. There is also a need for incorporating other data that might be related to tree growth as are elevation, precipitation, and soils. In spite of these limitations, Aspens is illustrative of an early attempt at data exploration.

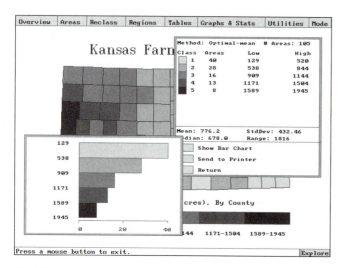

FIGURE 21.4   The Graphs & Stats option within ExploreMap. Graphical and statistical information provide an alternate view of the map of the data.

21.3.4 MapTime

MapTime permits users to explore temporal data associated with fixed point locations using three major approaches: animation, small multiples, and change maps. Stephen Yoder (1996), the developer of MapTime, argued that animation is a logical solution for showing changes over time because it "incorporates time itself in the presentation . . . the cartographic presentation [is not only] a scale model of space, but of time as well" (30). Animations of point data are easiest to interpret when data change relatively gradually over time. For example, an animation of the change in population of U.S. cities from 1790 to 1990 (distributed with MapTime) is easy to follow because city populations gradually increase and decrease. In contrast, an animation of the change in water quality at point locations along streams and rivers would exhibit much sharper increases and decreases, and thus be harder to interpret.

Animations in MapTime are constructed using a series of **key frames** (frames associated with collected data) and **intermediate frames** (frames associated with interpolated data). A distinct advantage of MapTime is that the number of intermediate frames can be varied. Thus, if one has data for every year from 1900 to 1940, but only for every other year between 1940 and 1980, intermediate frames might be used just for the 1940–1980 data.

In the context of temporal data, a **small multiple** consists of a set of maps, one for each time element for which data have been collected. As an example, Figure 21.5 portrays a small multiple of stream discharge for seven collection stations within a hypothetical drainage basin at 12-hour intervals over a four-day period (the data are distributed with MapTime). It is presumed that the

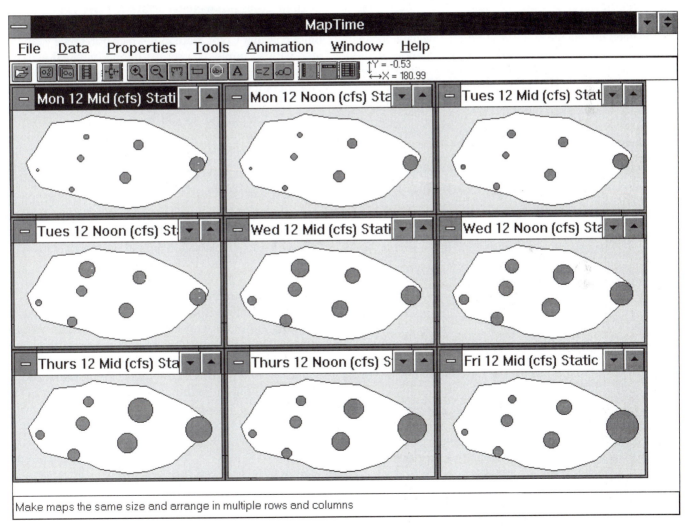

FIGURE 21.5 A small multiple created within MapTime. Each map represents one point in time for hypothetical stream discharge data at selected points. (Courtesy of Stephen C. Yoder.)

streams flow from west to east, that circle size is proportional to stream discharge (in cubic feet per second), and that heavy rainfall occurred between 6 P.M. Monday and 3 A.M. Tuesday throughout the region. The small multiple reveals an initial increase in discharge for upstream stations, followed by a later increase for downstream stations; essentially, a pulse of water moves through the system. The pulse of water seen in the small multiple is also detectable in an animation of the stream discharge data, but it is not easy to discern. Thus, small multiples are often a necessary complement to animations. Small multiples can also assist in contrasting two arbitrary points in time. For example, one might examine the beginning and ending multiples for stream discharge and note approximately how much change has taken place over the period at each location.

Change maps explicitly show the change that takes place between two points in time; in MapTime change can be computed in terms of the raw data (magnitude change), in percent form, or as a rate of change. The U.S. city population data from 1790 to 1990 provide a good illustration of the need for change maps. When these raw data are animated, one sees a major growth in city populations in the Northeast beginning in 1790, with an apparent drop in population for some of the largest Northeastern cities from about 1950 to the present. In contrast, a map showing the percent population change between 1950 and 1990 reveals a distinctive pattern of population decrease throughout most of the Northeast, as shown in Color Plate 1.3.

An interesting possibility is to create a small multiple of a series of change maps (an example of this for the U.S. population data from 1790–1990 is shown in Figure 21.6). In an evaluation of MapTime by users, Slocum and his colleagues (in press) found that a small multiple of change maps was favored over both animation and a simple raw small multiple for showing change over time. Animation, however, was deemed useful for examining general trends and providing a *sense* of change over time, whereas the raw small multiple was useful for comparing arbitrary time periods.

As with other exploration software, MapTime can determine precise values associated with a particular location or highlight a subset of the data. Additionally, a zoom feature enables users to enlarge a portion of the map. These features can be used to explore a single moment in time, or be implemented throughout an animation. Those who wish to experiment with different circle scaling methods (as described in Chapter 16) also will find that circle scaling exponents can easily be changed in MapTime.

21.3.5 Project Argus

Project Argus, a collaborative venture of research laboratories at the University of Leicester and Birkbeck College in England, was created to illustrate a broad range of possible data exploration functions. Jason Dykes was responsible for developing most of the software and has written papers (1995; 1996; 1997) summarizing its capability. Two major pieces of software are available at the Web site for Project Argus: one for data collected in the form of enumeration units (the "enumerated" software) and one for tourist data collected over time (the "time-space" software). Here we focus on the enumerated software.*

Like ExploreMap, the enumerated software is intended for data collected in the form of enumeration units, but it extends well beyond ExploreMap's capabilities. For example, the software can symbolize data in more than one form (choropleth maps, proportional circle maps, and Dorling's cartograms are possible) and display up to three attributes at a time. Interestingly, proportional circle maps are by default displayed using a redundant symbology of size and gray tones (see section 16.6); the gray tones can, however, also depict a second attribute, thus producing a bivariate map.

To illustrate some of the enumerated software's capability, consider Color Plate 21.4, which portrays a view created with a sample data set for Illinois included with the software. The three attributes mapped are percent Black, percent of the population 25 years and older with any college education, and percent below the poverty level. Across the top of the screen, the attributes are shown as individual choropleth maps. Choropleth maps are shown (as opposed to proportional circle maps) because the maps are based on standardized data. By default, unclassed choropleth maps are shown, although several classification methods are available within the software. From the discussion presented in section 18.1.1, we know that the visual correlation of choropleth maps can be adversely affected by using unclassed maps, but unclassed maps are a useful starting point for visual analysis.

In the middle portion of the screen, we see point graphs (one-dimensional scatterplots) of two of the attributes: percent Black and percent of the population 25 years and older with any college education. We can compare these plots with the previous maps and see that the overall light appearance of the percent Black map is a function of a positive skew (there are outliers toward the positive end of the number line). To the right of the point graphs is a scatterplot illustrating the relationship between the percent Black and education attributes (displayed along the x and y axes, respectively). We can see that when percent Black is low, a large range of education percentages is possible, but that the larger percent Black values are associated with lower percent education values.

---

* Technically, the enumerated software is termed the "cdv for enumerated data sets." We use "enumerated software" here for simplicity.

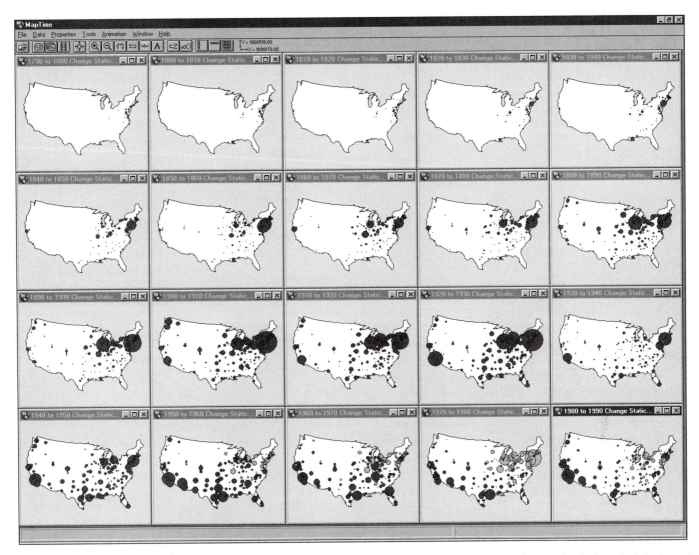

FIGURE 21.6 A small multiple of change maps created using MapTime. Population increases are shown by darker circles, whereas population decreases are shown by lighter circles. The population decreases in the northeastern United States from 1950 to the present were not easily recognized in either an animation or raw small multiple of the data. (Courtesy of Stephen C. Yoder.)

In the top right of the screen, we see a bivariate choropleth map (introduced in section 18.1.2) of the percent Black and education attributes. Higher values on these attributes are represented by increasing amounts of yellow and blue, respectively, whereas the lowest values are represented by the absence of these colors, or black in the extreme case. Note that the two highest percent Black values shown in the scatterplot (they have low scores on the education attribute) are depicted as a bright yellow in the extreme southern part of Illinois. Overall, the map is relatively dark because of the concentration of both attributes in the lower left portion of the scatterplot. Also note that the colors displayed on the map are shown within dots on the scatterplot.

We have focused on how two of the three attributes shown in Color Plate 21.4 might be analyzed. When

actually using the software, a much more complete analysis is, of course, possible. One interesting option is the use of a trivariate choropleth map (introduced in section 18.2.2), in which three choropleth maps are overlaid. It also must be recognized that the static nature of a book prevents a full understanding of the dynamic capability of the enumerated software. For example, the software permits designating an outlier on a point graph and highlighting that point on all other displays currently in view. Alternatively, in a fashion similar to ExploreMap, it is possible to highlight a subrange of values on an attribute. To illustrate these concepts, in Color Plate 21.4 a small cluster of dots has a box drawn around it, and these same observations are highlighted in green on other selected views. Also shown in Color Plate 21.4 is a parallel coordinate plot (see section 3.3.2 and section 21.3.8) and

Dorling's cartogram method (in this case illustrating the prominence of Cook County). Like much of the software in this chapter, these various display methods can best be understood by actually using the software.

21.3.6 CommonGIS

CommonGIS (*http://www.commongis.com*) is being developed to provide Web users easy access to visualizations of attribute data associated with enumeration units or point locations. Software development is funded by the CommonGIS Consortium, which is composed of seven public and private companies from five countries of the European Union. The roots of CommonGIS are largely the Descartes software (formerly IRIS), which was developed by Gennady and Natalia Andrienko (1999). Here we focus on the CommonGIS system, although you might find it interesting to contrast CommonGIS with Descartes, which is also freely available via the Web (*http://borneo. gmd.de/and/java/iris/*); a limitation of Descartes is that users cannot enter their own data.

An important characteristic of CommonGIS is the ability to automatically select the appropriate method of symbolization given a set of attribute data that a user wishes to map. To illustrate this notion, we utilize data for European countries and administrative divisions of Portugal that are distributed with CommonGIS. Selections of symbology in CommonGIS are based on the nature of data input by the user and, in the case of numerical data, the user's responses to a series of questions posed by CommonGIS. If the user attempts to map nonnumerical data, then CommonGIS assumes that the data are nominal and thus differing color hues are appropriate. For instance, if a user wishes to map membership in the North Atlantic Treaty Organization (NATO) using "TRUE" for Yes and "FALSE" for No, CommonGIS permits only a "colours" option (Figure 21.7A), which means that different hues indicate whether countries are members of NATO (Color Plate 21.5A).

To illustrate the types of questions that might be asked for numerical data, we consider a couple of examples. First, imagine that you have collected data for attributes depicting the percentage of the population of each country falling in each of three age categories. When you initially specify these attributes, CommonGIS permits you to select among five types of maps, as shown in Figure 21.7B. For sake of illustration, let's presume that you initially choose "pies" (corresponding to what cartographers normally term "pie charts"). CommonGIS will then ask you to "Select the attribute which includes all others" or allow you to select the option "No such attribute." The purpose of this question is to allow you to map raw totals, as opposed to the percentages we assumed in this case (to calculate a percentage for a wedge of the pie chart, CommonGIS would have to divide a raw total for a wedge by the grand total for the pie). Because you have percentages, you specify the option "No such attribute." CommonGIS then asks whether the attributes "represent

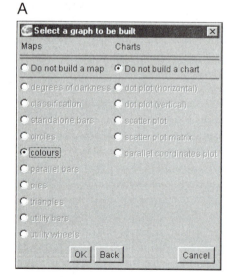

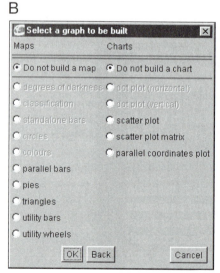

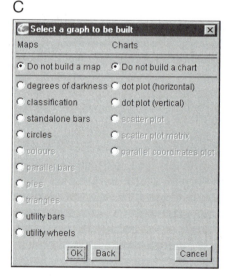

A   B   C

FIGURE 21.7   Menus from CommonGIS illustrating various mapping options available when the following kinds of data are input: (A) character data for membership in NATO, (B) three numeric attributes depicting the percentage of the population in various age categories, (C) a single numeric attribute—infant mortality rate. (Courtesy of Spatial Decision Support Team, Fraunhofer Institute AIS, *http://www.ais.fraunhofer.de/SPADE/*.)

non-overlapping parts of some whole." The purpose of this question is to ensure that the pieces of the pie chart do not overlap, and that they are related to one another. If you respond "Yes," CommonGIS produces a map of pie charts (Color Plate 21.5B). Alternatively, if you respond "No," CommonGIS responds with "No inclusion or additivity relationship between the attributes." The latter would be an appropriate response from CommonGIS if you had attempted to use a pie chart to map unrelated attributes (e.g., percent urban and median income).

As a second example for numerical data, consider mapping a single numeric attribute such as infant mortality rate. Here CommonGIS would provide a different set of initial symbolization options (Figure 21.7C). In this case, presume that you selected "degrees of darkness," which is CommonGIS's term for an unclassed choropleth map (Color Plate 21.5C). CommonGIS would ask no further questions as it is presumed an unclassed map could be used for any single numeric attribute. Unfortunately, the latter approach is problematic, as we have stressed the importance of standardizing data for choropleth maps; thus, we would argue that an additional question should be posed pertaining to whether the data are standardized. On the plus side, CommonGIS does include several methods for standardizing data.

In posing questions, CommonGIS provides the option of "memorizing" certain responses. Thus, if we use the three preceding age attributes and take the option that the responses be memorized, a pie chart immediately appears when the user initially clicks on "pies." This approach might be advantageous for novice users, as they will not have to deal with so many questions.

CommonGIS includes some interesting capabilities for exploring spatial data. Consider the unclassed choropleth map of birth rate for European countries shown in Color Plate 21.6A. Note that the "Manipulate" tab on the right side has been clicked, revealing a vertical point graph (CommonGIS terms this a "dot plot") of the data displayed alongside the range of possible colors representing birth rates. This point graph displays the distribution of the data along the number line—it just happens to be a vertical plot rather than the horizontal plot that we introduced in Chapter 3. Note that the point graph reveals a positively skewed distribution, largely due to a single outlier, Albania, which has a birth rate of 21.7. (Pointing to the outlier on the point graph causes that point and the country to be highlighted and the name of the country and value to appear next to the point.) Because of this outlier, most of the countries appear to be fairly similar in color, and it is difficult to visualize which countries have the highest birth rates after Albania. We can improve the ability to discriminate countries if we remove the outlier and extend the color scheme over the remaining countries. This is accomplished by dragging the small dark blue triangle

at the top of the point graph to just above the next highest observation. The result is shown in Color Plate 21.6B, where the outlier is depicted by a small brown triangle (in Albania) and a second point graph is drawn (to the left of the initial one) representing the reduced data set. Now we see that Moldova, Iceland, and Macedonia clearly have the highest birth rates after Albania.

In Chapter 13, we introduced the notion of diverging color schemes, in which two colors diverge from a central light point. In CommonGIS, diverging color schemes can be applied by selecting a "compare by click in" option, and then clicking on the legend or map to indicate the point from which colors will diverge. If a "dynamic map update" option is also selected, the diverging point can be dragged, allowing you to change the diverging point dynamically. Color Plate 21.7B suggests this notion for the percentage of the population greater than or equal to 65 in divisions of Portugal. As you move the horizontal marker along the legend from low to high data values, you can see "how blue shades (corresponding to low values) spread from the areas around the biggest cities of Portugal such as Lisbon and Porto, first along the coast and then into the inner parts of the country" (CommonGIS documentation, 2002). Note that this pattern is very difficult to detect in the nondiverging scheme shown in Color Plate 21.7A.

Data exploration also can be used in CommonGIS in the context of decision making, where several attributes can be utilized in arriving at a decision. For instance, imagine that you are planning a vacation and would like to select an appropriate ski resort in Wallis, Switzerland, and that you have data available on duration of the skiing season, price for a ski pass over six days, total length of ski runs, and the percentage of the runs that are classified as expert at each resort. In CommonGIS, these four attributes can be mapped via "utility bars" (Color Plate 21.8A). Here each attribute is represented by a different colored bar, with the height of the bar proportional to the value for a given ski resort divided by the maximum possible for all ski resorts for that attribute. Note that consideration is given to whether low or high values on an attribute are desirable, as indicted by the direction of the arrow in the "Manipulate" section of the display: An upward-trending arrow indicates high values are desirable (as for "Duration of skiing season"), whereas a downward-trending arrow indicates low values are desirable (as for "Price for ski pass (6 days)"). The width of each bar is proportional to the relative importance (or weight) assigned to the associated attribute. For instance, the bars for duration of the ski season and the price for a ski pass are wider in Color Plate 21.8A because the user has placed greater weights on those attributes (.38 for each). The frames surrounding each set of utility bars indicate a theoretical best ski resort, one that has the maximum possible score on each attribute (or minimum possible if low values are desirable). Data exploration is useful in this case, as the resulting map is difficult to interpret. One simple way to ease interpretation is to use the tool near the top of the legend. By dragging the small blue arrow near 0.0 to the right, bars that constitute less than 50 percent of the theoretical best resort are not displayed, thus greatly simplifying the distribution (Color Plate 21.8B). In ultimately selecting a desired ski resort, a user still must decide among several alternatives, but the choice is certainly simpler than if raw data for all ski resorts were considered.

21.3.7 ArcView 3.x and ArcGIS

GISs have commonly been defined on the basis of the order in which they process data. For example, Keith Clarke (1995, 13) defined GISs as "automated systems for the capture, storage, retrieval, analysis, and display of spatial data." Such a definition reflects, in part, the nature of early GIS software, which tended to separate the cartographic display component from the analysis of data. Typically, this separation was achieved using **command-line interfaces**, which required users to type commands specifying the processes to be performed; for example, a command to overlay two layers would be followed by a command to display the visual results of the overlay process. Such interfaces tended to be difficult to learn and the resulting maps were often of low quality. Today this situation has changed, as command-line interfaces have been replaced by **graphical user interfaces (GUIs)**, which integrate the analysis and display components of GISs and allow some of the interactivity characteristic of data exploration software. To illustrate the capability of GIS software, we describe a tutorial that has been distributed with the popular ArcView 3.x software.* In the tutorial, you are asked to imagine that you wish to build a new showroom for a company in the southeastern United States. The showroom should be in a state where sales were low the previous year, in a city with more than 80,000 people, and within 300 miles of Atlanta (a presumed regional distribution center).

Determining potential locations for the showroom can be implemented as a three-step process in ArcView. First, the sales information can be displayed via a choropleth map (Color Plate 21.9A); for our purposes, we assume that low sales are represented by the two lowest valued classes on the map (the two lightest tones of red).† This step involves no GIS analysis (unless data classification

is considered analysis), as we simply display a particular theme (sales) associated with each state. The second step is to determine which cities have a population greater than 80,000. This is accomplished using a "query builder" function in ArcView—all cities selected using the query builder are *automatically* shown on the map as blue circles of a fixed size (Color Plate 21.9B). (The default color produced by ArcView was actually green, but blue is used here because it contrasts better with the colors on the choropleth map.) The third step is to determine which cities with a population greater than 80,000 are within 300 miles of Atlanta. In calculating distances of cities from Atlanta, ArcView again *automatically* displays all of the desired cities in yellow (Color Plate 21.9C). Presumably, cities in yellow that are located in those states falling in the two lowest valued classes on the choropleth map meet the three criteria specified for locating the showroom.

Integrating analysis and display functions provides a number of advantages from a data exploration perspective. One advantage is that it enhances the efficiency of trying alternative what-if scenarios. For example, we might change the population and distance figures for locating a showroom to 150,000 and 200 miles, respectively, producing the map shown in Color Plate 21.9D. These sorts of changes and associated displays generally can be created in a matter of seconds. A second advantage is that the GIS analyst can view data from a variety of perspectives. In the case of the showroom problem, ArcView permits viewing the data as a map, table, or chart (Color Plate 21.10); note that values highlighted on the map in yellow are also highlighted in the table in yellow, and that this highlighted subset is depicted within the chart. A final advantage is that the GIS analyst can experiment with different symbol schemes to see if they reveal different patterns in the data. For example, the tutorial associated with the showroom example used a diverging color scheme (greens on one end and oranges on the other, with white in the middle). The diverging scheme might help the analyst locate the lower valued sales states more quickly.

In examining the illustrations in Color Plates 21.9 and 21.10, readers unfamiliar with ArcView might be bothered by the inability to see all of the numbers for the "Sales" portion of the "table of contents" (this is the left portion of the screen and is similar in appearance to what would normally be termed a legend). The logic of not showing all of the numbers is that the GIS analyst is presumably familiar with the data (is exploring the data). Also, the analyst can easily see all of the numbers by simply changing the size of the table of contents (by using the mouse to drag the edge of the gray shaded area). If, however, a map is to be made for presentation purposes, it is essential that GIS software provide a capability to create

---

* The tutorial comes from pp. 15–29 of *Using ArcView GIS*.

† Ideally, the sales information should be standardized to account for different populations of states, but here we use the raw data as provided in the tutorial.

a map with a more conventional appearance. In ArcView, presentation is implemented using a Layout option; the result of using the Layout option for the showroom problem is shown in Color Plate 21.11.*

Recently, ESRI released ArcGIS, a more sophisticated program that eventually will replace ArcView 3.x. For the immediate future, however, it appears that the two programs will coexist, especially because ArcView 3.x has a large pool of users and ArcGIS requires a more sophisticated platform to run on. Although we have not had a chance to fully evaluate ArcGIS, it clearly has greater flexibility than the traditional ArcView product. For instance, when creating choropleth maps, it is possible to display a histogram of the data along with class breaks (Figure 21.8); these class breaks can be dragged to new locations to examine the effect on the resulting display. A limitation of this approach, however, is that the results are not instantaneous, as in the CommonGIS program. Both ArcView 3.x and ArcGIS include a range of tools ("extensions" in ArcView) for visualizing spatial data; for instance, 3D Analyst can be used to examine surfaces in 3-D space. Although we won't discuss 3D Analyst here, you might wish to examine it in association with virtual environments, which we discuss in Chapter 24.

21.3.8 HealthVisPCP

HealthVisPCP, developed by Robert Edsall (2003a; 2003b) is an extension of HealthVis, which was created by Alan MacEachren and his colleagues (1998a) to explore spatiotemporal health statistics data. The basic interface for HealthVisPCP is shown in Color Plate 21.12. In the upper left, we see a bivariate choropleth map of heart disease mortality rates (for White men) for two time periods (1980 and 1992) for health service areas of the United States (an associated legend is shown to the right of the map in a separate window). Below the legend is a scatterplot of the heart disease attributes shown on the map; note that class breaks used on the bivariate map are also depicted on the scatterplot. Additionally, the scatterplot includes a faint histogram depicting the distribution of heart disease mortality for 1980, which is plotted along the x axis. In the lower left is the key feature of HealthVisPCP, a parallel coordinate plot (PCP), which provides a graphical portrayal of not only the attributes shown in the other windows, but additional ones (those for heart disease for White men in 1983, 1986, and 1989 are shown here, but others can be seen by moving the slider bar in the PCP Panner window). Note that the colors on the PCP correspond to those on the bivariate choropleth map, and thus enable the reader to see how certain groups of health service areas change over time.

Recall from Chapter 3 that when two attributes are highly positively correlated the lines crossing the attribute axes on the PCP will tend to be parallel. If we look at the left side of the PCP in Color Plate 21.12, we see that this generally is the case here, indicating that the ranking of health service areas in terms of heart disease mortality has remained relatively constant over time. The general downward trend of the lines also indicates that heart disease mortality was less of a problem in 1992 than in 1980 (note that the range of values on the vertical axes is identical, enabling the different temporal attributes to be readily compared). Although the overall trend in heart disease mortality is downward, we note in the scatterplot that there are some health service areas that do not fit this trend as well as others, and so we might want to focus on these. In the scatterplot we have selected two of these (in red), and they have been highlighted in the other windows. Note that in the PCP these are lines that indicate an increase in heart disease mortality from 1980 to 1992. If we had additional data on health service areas available, we would want to examine these two areas to see what might have created these potential "hot spots."

Color Plate 21.13 illustrates a much more complex PCP. Here we see that the different colored lines tend to cross one another, suggesting a lack of correlation between prostate cancer mortality for different temporal elements (we have ordered the attributes temporally, whereas for the heart disease example we placed 1992 adjacent to 1980).

21.3.9 Vis5D

Vis5D provides some intriguing possibilities for those wishing to explore true 3-D phenomena. This package was developed at the University of Wisconsin–Madison Space Science and Engineering Center, where it has been primarily used for visualizing data associated with numerical weather models. The "5-D" portion of the name comes from the notion that meteorological data have three spatial dimensions (latitude, longitude, and altitude), a time dimension, and a dimension for the attributes to be displayed (e.g., pressure, temperature, and wind speed). Although originally intended for meteorological data, Vis5D can display any data that can be expressed in the 5-D format; for example, oceanic data have three spatial dimensions (latitude, longitude, and depth), a time dimension, and multiple attributes (e.g., temperature, salinity, and current speed; Hibbard and Santek 1990).

To illustrate some of Vis5D's capability, we will use a 48-hour forecast data set for North America distributed with the software (V.GRID1.v5d). The data set consists of 29 attributes, with 49 timesteps for each attribute. Topographic base information also is provided with

---

* Some polishing of the final map was done in Freehand.

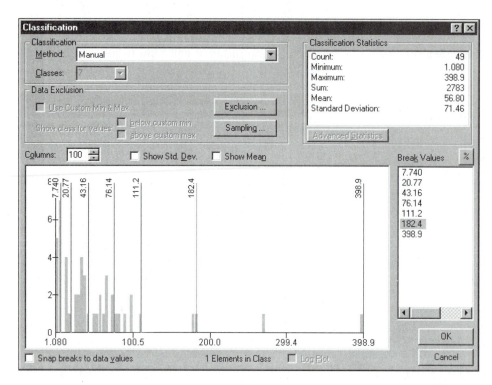

FIGURE 21.8 A combined histogram and class breaks display created by ArcGIS; the class breaks can be dragged to define new classes. (Source: ESRI.)

Vis5D (the base for North America is shown in Color Plate 21.14); the base obviously provides a necessary geographic perspective, with topography critical in meteorological applications.

One basic option in Vis5D is to create a slice through the 3-D space of an attribute in either a vertical or horizontal direction. The symbology for the slice can be either isolines or hypsometric tints (to represent the magnitude of a phenomenon) or vectors (to represent the flow of a phenomenon). Color Plate 21.14 illustrates a slice of hypsometric tints for wind speed for timestep 1 of the forecast data set. The default color scheme is a traditional spectral one (red–orange–yellow–green–blue), although users can create their own schemes. Note that the altitude of the slice is 11.66 kilometers, and thus the high-valued reddish-orange area represents a portion of the jet stream. Another basic option in Vis5D is to create an **isosurface**, a surface bounded by a particular value of an attribute. For example, Color Plate 21.15 illustrates the 45 meter per second isosurface for wind speed for timestep 33. Wind speeds inside the surface would be greater than 45 meters per second, whereas those outside would be less. We would expect high wind speeds to be associated with the jet stream, and the general shape of the isosurface confirms this notion.

A third option in Vis5D is "volume," in which the entire 3-D space for an attribute is symbolized at once. This might seem impossible due to symbol blockage, but it can be done if the symbology is treated as a transparent fog (if opacity is varied, as was done for T3D; see section 19.3.1). For example, Color Plate 21.16 illustrates the

volume option for wind speed for timestep 28 (a default spectral color scheme is again used). As before, the presumed location of the jet stream is indicated by its higher speed. Admittedly, the volume symbology is difficult to interpret in this static map, but Vis5D's interactive nature enables a user to "see through the fog" by varying his or her viewpoint in 3-D space. Potentially the most interesting feature of Vis5D is its real-time animation capability. Simply clicking on Animate causes the system to animate any currently visualized attribute. For example, after selecting the slice for wind speed shown in Color Plate 21.14, clicking on Animate reveals how the jet stream changes over the 48-hour period within that slice. One can, of course, stop the animation at any moment, and then explore a particular timestep further.

It is also important to realize that more than one attribute can be explored at a time. For example, William Hibbard and his colleagues (1994) illustrated William Gray's novel idea for generating energy by creating a permanent rainstorm over a hydroelectric generator (Color Plate 21.17). They described the process as follows:

The white object is a balloon 7 kilometers high in the shape of a squat chimney that floats in the air above a patch of tropical ocean. The purpose of the numerical experiment is to verify that once air starts rising in the chimney, the motion will be self-sustaining and create a perpetual rainstorm. The vertical color slice shows the distribution of heat (as well as the flow of heat when model dynamics are animated); the yellow streamers show corresponding flow of air up through the chimney; and the blue-green isosurface shows the precipitated cloud ice. (p. 66)

Such capability does not come without a price, as one limitation of Vis5D is that it is most effective on high-end workstations. This limitation is now changing, however, as PC-based systems (with the appropriate graphics cards) are now approaching the power of expensive workstations. Vis5D is particularly impressive given that it is available for free via the Internet; sophisticated visualization systems offering similar capability can cost thousands of dollars.

21.3.10 Transform

Transform is part of a larger software package known as Noesys (in Chapter 19 we introduced some of the capability of T3D, another component of Noesys). Although Transform contains routines for exploring data associated with both graphs and maps, we focus on graphs because they provide useful illustrations of Transform's unique capabilities, and are an important element of data exploration (along with maps, tables, and numerical summaries).

The data shown in Figure 21.9 represent a portion of mean monthly temperatures for Springfield, Illinois (the complete data set includes mean monthly temperatures for a 40-year period). When viewed in this tabular form, it is difficult to detect trends; we can see that one column has higher values than another, but it is difficult to visualize the trend over the course of a year, and especially difficult to detect any trends over the 40-year period. As an alternative to the tabular display, Transform permits users to view data as a graphical representation, or

"image." The left portion of Color Plate 21.18 shows one such representation. In this case, a traditional spectral or "Rainbow" color scheme has been used. Note that the trend over the course of a year is now more readily apparent, but detecting any trend over the 40-year period is still difficult due to variability in the data: An unusually cold or hot month appears to disrupt the overall pattern. One way to minimize such variation is to smooth the data. This is accomplished by treating cells in the left portion of Color Plate 21.18 as control points for contouring, much as we did for maps in Chapter 14. The result is shown in the right portion of Color Plate 21.18. Now we see the pattern over the year, but we also see a potential trend over the 40-year period, as more recent years appear to exhibit slightly cooler summers and Januarys.

In addition to representing data by images, Transform links data and images directly. For example, if one uses the mouse to point to a cell (or group of cells), that cell (or group of cells) is also highlighted in the image, as August 1949 is in Color Plate 21.18. Additionally, users can develop a graphical representation of trends along a row or column in the table, as shown in Figure 21.10.

An unusual feature of Transform is the capability to symbolize an image with what many cartographers would consider illogical color schemes. To see the effect of one of these schemes, compare Color Plates 21.18 and 21.19, which use the Rainbow and Lava Waves schemes, respectively. Based on the discussion of color schemes in Chapter 13, you should argue that the Lava Waves scheme is illogical: The raw data increase from low to high values, but the banded appearance of the color scheme does not logically relate to these values. Although not necessarily logical, Berton (1990) argued that "such banding effects . . . can provide excellent markers for transitional areas in the data" (p. 112).

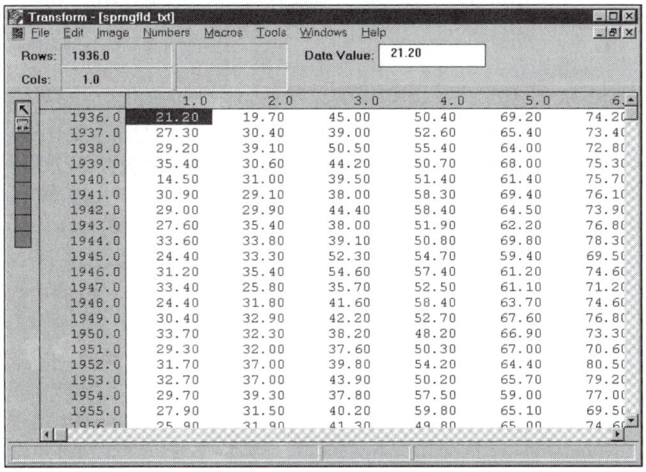

FIGURE 21.9 A window from Transform showing a portion of a table of mean monthly temperatures for Springfield, Illinois. Years are shown along the y axis, and months of the year are shown along the x axis. Compare with the image of these data shown in Color Plates 21.18–21.20. (Image created with Noesys Visualization Pro, courtesy of Fortner Software LLC.)

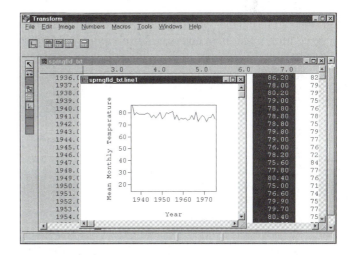

FIGURE 21.10 A graphical representation of the column for July temperatures shown in Figure 21.9. (Image created with Noesys Visualization Pro, courtesy of Fortner Software LLC.)

Another unusual feature of Transform is the Fiddle tool, which permits shifting or compressing colors within a scheme. When using the Fiddle tool, a movement of the cursor to the left or right shifts the color scheme in that direction, whereas a cursor movement up or down either stretches or compresses the scheme. For example, Color Plate 21.20 shows the Rainbow scheme shifted left and compressed. Although we know of no experiments involving the Fiddle tool in a map environment, it seems feasible that it, like the various color schemes, would assist in examining the "transitional areas in the data" noted by Berton.

21.3.11 Other Exploration Software

The limited space of a textbook does not permit us to discuss the full range of software that has been developed for exploring spatial data. In surveying the literature, some of the more interesting pieces that we have encountered include Miller's (1988) Great American History Machine (GAHM), MacDougall's (1992) Polygon Explorer, Shneiderman's (1999) work with "dynamic queries," and Roth et al.'s work (1997). Each of these is summarized briefly in the Further Reading section. You should also keep in mind that we have tried to focus on software that emphasizes the mapped display—although numerical (or statistical) summaries can be utilized, they generally are not focused on, nor are they fully integrated with the map display. In contrast, exploratory spatial data analysis (ESDA) methods do emphasize numerical summaries of the data and attempt to more fully integrate them with the map display. One limitation of these approaches is that they often require a deeper understanding of statistics than is typical for the introductory cartography student. For an overview of ESDA methods, see Anselin (1998).

SUMMARY

Data exploration, like the larger notion of geographic visualization, is a private activity in which unknown spatial patterns are revealed in a highly interactive environment.

Data exploration is achieved by using a variety of methods, including varying the symbolization (moving the breakpoint separating two colors in a diverging color scheme), highlighting subsets of data via either **focusing** (for a subrange of the data) or **brushing** (for an arbitrary set of spatial entities), creating multiple views (e.g., the **small multiple**), animation, and linking with other forms of display (e.g., tabular and graphical displays). The methods for data exploration continue to evolve—for instance, in Chapter 25 we consider the notion of **data mining**, which enables patterns to be detected automatically.

We have looked at a variety of software that has been developed for exploring spatial data. One program particularly illustrative of data exploration capabilities is CommonGIS. An interesting feature of CommonGIS is the ability to assist the user in selecting an appropriate method of symbolization. Other key features include eliminating the effect of outliers by extending a color scheme over only a portion of the data set, varying the breakpoint of a diverging color scheme (and seeing the result depicted dynamically), and visualizing the result of combining multiple attributes to arrive at a decision. Other exploration software, however, provides capabilities not found in CommonGIS. For instance, MapTime enables you to explore spatiotemporal data at fixed point locations via animation, small multiples, and change maps; ArcView 3.x and ArcGIS provide a wide variety of analytical functions that can be utilized to create visualizations of what-if scenarios; and Vis5D enables users to explore true 3-D data that have both temporal and multivariate components.

In this chapter, we have focused on data exploration software that emphasizes the mapped display—although numerical (or statistical) summaries can be utilized, they are not fully integrated with the map display. In contrast, exploratory spatial data analysis (ESDA) methods emphasize numerical summaries of the data and attempt to more fully integrate them with the map display.

FURTHER READING

Andrienko, G. L., and Andrienko, N. V. (1999) "Interactive maps for visual data exploration." *International Journal of Geographical Information Science* 13, no. 4:355–374.

Discusses development of the Descartes software; many ideas developed for Descartes have been implemented in CommonGIS.

Crampton, J. W. (2002) "Interactivity types in geographic visualization." *Cartography and Geographic Information Science* 29, no. 2:85–98.

Considers a range of tools for interacting with geospatial data.

Dykes, J. A. (1997) "Exploring spatial data representation with dynamic graphics." *Computers & Geosciences* 23, no. 4:345–370.

Discusses using the programming language Tcl/Tk to create Project Argus.

Haslett, J., and Power, G. M. (1995) "Interactive computer graphics for a more open exploration of stream sediment geochemical data." *Computers & Geosciences* 21, no. 1:77–87.

Describes using the data exploration software REGARD to examine the geochemistry of stream sediment.

Heywood, I., Oliver, J., and Tomlinson, S. (1995) "Building an exploratory multi-criteria modelling environment for spatial decision support." In *Innovations in GIS 2*, ed. by P. Fisher, pp. 127–136. Bristol, PA: Taylor & Francis.

> Describes the development of specialized software that can assist in exploring data commonly analyzed within a GIS.

Jankowski, P., Andrienko, N., and Andrienko, G. (2001) "Map-centered exploratory approach to multiple criteria spatial decision making." *International Journal of Geographical Information Science* 15, no. 2:101–127.

> Presents a sophisticated approach for exploring attributes involved in decision making.

Kraak, M.-J. (1998) "Exploratory cartography: Maps as tools for discovery." *ITC Journal* 1, no. 1:46–54.

> Considers numerous issues related to data exploration.

MacDougall, E. B. (1992) "Exploratory analysis, dynamic statistical visualization, and geographic information systems." *Cartography and Geographic Information Systems* 19, no. 4:237–246.

> Describes the data exploration software Polygon Explorer, which was developed for handling both univariate and multivariate data. Readers with knowledge of cluster analysis will find the capability to handle multivariate data intriguing.

MacEachren, A. M., and Ganter, J. H. (1990) "A pattern identification approach to cartographic visualization." *Cartographica* 27, no. 2:64–81.

> A classic paper on the need for visualization and data exploration in discovering patterns in spatial data.

Miller, D. W. (1988) "The great American history machine." *Academic Computing* 3, no. 3:28–29, 43, 46–47, 50.

> Summarizes the data exploration software GAHM, which was developed to encourage undergraduates to think like professional historians.

Moellering, H. (1980) "The real-time animation of three-dimensional maps." *The American Cartographer* 7, no. 1:67–75.

> Describes Moellering's early efforts to explore and animate spatial data. For a related video, see Moellering (1978).

Monmonier, M. (1991b) "Ethics and map design: Six strategies for confronting the traditional one-map solution." *Cartographic Perspectives* no. 10:3–8.

> Questions the ethics of using one map to display a set of spatial data and proposes data exploration as one of six solutions to the problem.

Roth, S. F., Chuah, M. C., Kerpedjiev, S., and Kolojejchick, J. A. (1997) "Toward an information visualization workspace: Combining multiple means of expression." *Human-Computer Interaction* 12, no. 1/2:131–185.

> Covers several sophisticated pieces of software (Visage, SAGE, SDM) for exploring spatial data.

Shneiderman, B. (1999) "Dynamic queries for visual information seeking." In *Readings in Information Visualization: Using Vision to Think,* ed. by S. K. Card, J. D. Mackinlay, and B. Shneiderman, pp. 236–243. San Francisco, CA: Morgan Kaufmann.

> Discusses some of Shneiderman and his colleagues' early work with "dynamic queries," which involved developing methods for obtaining rapid visual queries to databases. For more recent work by Shneiderman and his colleagues, see Fredrikson et al. (1999).

Slocum, T. A., Yoder, S. C., Kessler, F. C., and Sluter, R. S. (2000) "MapTime: Software for exploring spatiotemporal data associated with point locations." *Cartographica* 37, no. 1:15–31.

> Provides a detailed discussion of the development of MapTime and suggests potential improvements. For a user evaluation of MapTime, see Slocum et al. (in press).

Waniez, P. (2000) "Two software packages for mapping geographically aggregated data on the Apple Power Macintosh." *The Cartographic Journal* 37, no. 1:51–64.

> Describes Philexplo, a program for exploring spatial data. The program appears to be intended for someone who reads French.

22

Electronic Atlases and Multimedia

OVERVIEW

This chapter focuses on electronic atlases and the related area of multimedia. Section 22.1 considers the problem of defining an electronic atlas. It is helpful to begin by considering the nature of paper atlases. If asked to envision a **paper atlas**, many of us conceive of "a bound collection of maps focusing on a particular region or topic." Actually, modern paper atlases consist of not only maps, but also text, photographs, tables, and graphs, which, when combined, tell a story about a particular place or region. **Electronic atlases** can emulate such characteristics, or they can take advantage of the capabilities that computer-based systems provide, permitting animation, data exploration, and multimedia (including links to text, photographs, sound, and video). Furthermore, electronic atlases have the potential to permit users to create their own maps and analyze spatial data.

The bulk of the chapter (section 22.2) presents numerous examples of electronic atlases, which we have divided into four major categories: atlases developed by individuals, national atlases, state atlases, and commercial atlases and related software. Atlases developed by individuals include William Bowen's Electronic Map Library, which is actually a series of atlases of various places around the United States, along with a set of panoramic maps; Hugh Howard's atlas Death Valley, which features animations of physical processes in Death Valley; and David Rumsey's superb collection of historical maps, some of which have been linked to modern digital information. National atlases that we discuss include the Domesday Project, one of the earliest electronic national atlases; the U.S. Census Bureau's online mapping systems (TIGER and American Fact Finder); The National Atlas of the United States; and the Atlas of Switzerland Interactive.

Electronic state atlases can be broken down into two groups: formal and informal. Formal atlases normally are called "atlases," have cartographers involved in their development, and are characterized by a fair amount of planning in their content and design. Examples include The Interactive Atlas of Georgia and the Atlas of Oregon. Because the cost of creating a formal atlas can be high, many states are distributing informal collections of maps, some of which might not immediately bring to mind the notion of an atlas. To illustrate these informal collections, we discuss a series of online maps that are provided for the state of Georgia by the Georgia GIS Data Clearinghouse. As is the case with Georgia, GIS data support centers are often the source of such maps.

Atlases developed by individuals, nations, or states within the United States seldom are money-making ventures; in fact, national and state atlases often must be supported by the government. In contrast, commercial atlases and products that have atlas-like components generally have the goal of making a profit for the developers. As examples, we'll consider World Atlas, one of the earliest commercial ventures, the atlas-related features of Microsoft's® Encarta® Reference Library 2003, and the 2003 Grolier Multimedia Encyclopedia.

Although electronic atlases often contain multimedia information, the focus is on maps. In section 22.3, we briefly consider **multimedia systems**, where greater emphasis is placed on multimedia. Michael Shiffer and Shunfu Hu are two researchers who have promoted the use of multimedia.

There are several cautionary notes that should be kept in mind with reference to the atlases we will consider. First, it is important to recognize that the discussions of individual atlases are not intended as reviews, but rather to illustrate the *general character of both early and modern electronic atlases. Comprehensive reviews can be found in a variety of sources, including journals (e.g.,* Cartographic Perspectives*) and newsletters (e.g., the* Microcomputer Specialty Group of the Association of American Geographers Newsletter*). Second, bear in mind that many atlases that we discuss are updated frequently, and so the content might vary from what is presented here. Third, no attempt has been made to point out all the strengths and weaknesses of each atlas. A useful exercise would be to evaluate the correctness of symbology used in each of these atlases (e.g., you might consider whether color schemes for choropleth maps are appropriately used). Finally, it must be recognized that only a sampling of atlases is provided here, and that new electronic atlases are constantly being developed. Refer to the Web site for this book (*http://www.prenhall.com/slocum*) for information on other electronic atlases.*

22.1 DEFINING ELECTRONIC ATLASES

In attempting to define an electronic atlas, it is useful to consider the nature of paper atlases. One simple definition for a paper atlas is "a bound collection of maps focusing on a particular region or topic." Although this definition is appropriate for many traditional atlases, it fails to consider the nature of modern paper atlases, which contain not only maps, but also text, photographs, tables, and graphs and frequently tell a story about a particular place or region (Keller 1995). For example, we can contrast the classic *Goode's World Atlas* (Espenshade 1990) with the *Historical Atlas of Massachusetts* (Wilkie and Tager 1991). The former consists of a variety of thematic and general-reference maps, with virtually no text or other graphic material, whereas the latter introduces the history of Massachusetts through a multitude of maps, and includes ample text, graphs, and photographs. One important characteristic of both traditional and modern paper atlases is that they generally utilize a limited number of map scales and levels of generalization because of the physical constraints of the paper format; for example, a state atlas might focus on information at the county level, as opposed to showing information at both the county and township level.

In their simplest form, electronic atlases emulate the appearance of traditional paper atlases. An early example was the *Electronic Atlas of Arkansas* (Smith 1987b), which consisted of alternating screens of maps and text; no interaction was permitted with either the maps or the text. A more recent example is the *Digital Atlas of California* (available at *http://130.166.124.2/CApage1.html*), which consists solely of thematic maps of California census tracts and block groups (see section 22.2.1 for a more complete discussion). Like the Arkansas atlas, no interaction is possible with maps in

the California atlas. Although the digital emulation of paper atlases can be useful (e.g., the *Digital Atlas of California* is available for free to anyone who has access to the Internet), there are many other reasons for creating an electronic atlas. First, users can explore data in an interactive graphics environment. This could involve pointing to an enumeration unit to determine the associated data value, zooming in on a geographic region to get more detail, or comparing any two arbitrary maps in the database. Second, animated maps are possible; as discussed in Chapter 20, animated maps are a natural choice for depicting temporal change. Third, multimedia is readily accomplished: Not only can the text, photographs, tables, and graphs found in paper atlases be used, but sound, video, and even virtual environments (see Chapter 24) are possible. Fourth, users have the potential to create their own maps. Although in Chapter 1 we cautioned that allowing users to create their own maps might lead to improperly designed maps, such capability is essential for users who have collected their own data. Finally, an electronic atlas has the capability to analyze spatial data. As a simple example, an electronic atlas might be used to identify cities having both poor air quality and a high incidence of lung disease. A more sophisticated electronic atlas might include a broad range of analysis functions, such as those found in GIS software. Given this range of potential functionality, the following is one possible definition for an **electronic atlas**: a collection of maps (and associated database) that is available in a digital environment; sophisticated electronic atlases enable users to take advantage of the digital environment through a variety of means, such as Internet access, data exploration, map animation, and multimedia. Electronic atlases might also permit users to create their own maps and analyze spatial data. The key point is that electronic atlases permit users to manipulate maps and associated databases in ways not possible with traditional printed atlases.

22.2 EXAMPLES OF ELECTRONIC ATLASES

This section considers numerous examples of electronic atlases, which have been divided into four groups: atlases developed by individuals, national atlases, state atlases, and commercial atlases and related software. You will find that the bulk of these atlases are available at a modest cost (or for free via the Internet). Although some atlases we discuss are no longer available (the *Domesday Project, The Interactive Atlas of Georgia,* and *World Atlas*), we consider them to provide a historical perspective. Because the notion of what constitutes an electronic atlas will likely change as hardware and software evolve, we encourage you to examine these and other atlases and develop your own notion of what constitutes an electronic atlas.

22.2.1 Atlases Developed by Individuals

The Electronic Map Library

The *Electronic Map Library* (*http://geogdata.csun.edu/ library.html*), developed by William Bowen, consists of a series of atlases for various cities and states around the United States, along with a set of panoramic maps. To illustrate the nature of the atlases, we consider the *Digital Atlas of California* (*http://130.166.124.2/CApage1.html*), which consists of a large set of thematic maps depicting data at either the census tract or block group level for all of California. Maps in the California atlas are divided into the following major categories: Population, Citizenship, Income, Poverty, and Education. Within each category, numerous attributes can be mapped; for example, the Population category consists of 15 different attributes. Maps are included both for the entire state and for major metropolitan areas. A redundant symbology is used to depict themes: Proportional circles are scaled according to the magnitude of a theme in each census tract (or block group), and five colors are shown within each circle depicting a quintiles classification of the data (Color Plate 22.1).

A distinct advantage of having the *Electronic Map Library* available on the Internet is that it can be accessed for free by anyone with an Internet connection. Another advantage is that Professor Bowen can enlarge the library indefinitely; his only limitation is the storage space available on his server. A disadvantage of the library, however, is that no interaction with the maps is possible: The maps are largely available in GIF format, which is useful for browsing and downloading but not for providing information about the underlying data.

Death Valley: An Animated Atlas

Death Valley: An Animated Atlas, developed by Hugh Howard as part of a master's thesis (Howard 1995), provides an impressive introduction to Death Valley, a national park located in southern California. Death Valley is unique in that it has the lowest elevation (282 feet below sea level) and highest summer temperatures (commonly in excess of 115°F) in the United States. It also suffers from extreme aridity (less than 2 inches of rainfall per year, with potential evaporation rates in excess of 150 inches per year), and is home to some unusual plant and animal life (e.g., several species of pupfish that occur nowhere else in the world).

The Death Valley atlas provides a series of narratives that describe the attributes of Death Valley via maps (many of which are animated), photographs, graphs, and spoken text. The latter is particularly useful as it permits the viewer to focus on interpreting graphics without having to read text. The narratives include an Introduction and Quick Tour, and two major sections dealing with the physical and cultural environment.

Because static illustrations do not do the atlas justice, we describe how the atlas presents three factors (a rain-shadow effect, adiabatic heating, and a regional low-pressure system) involved in creating the hot, dry, and windy climate characteristic of Death Valley. The rain-shadow effect is caused by the Sierra Nevada and other mountain ranges, which prevent moist air from reaching the valley from the west. In the atlas, a shaded relief map depicts these mountain ranges, and moving arrows represent the moist air—the arrows dissolve as they pass over the mountains. Adiabatic heating occurs in the summer, as hot air rises from the valley and is replaced by cooler air from the surrounding mountains. This effect is illustrated in the atlas by moving colored arrows (red for rising hot air and blue for descending cool air). The regional low-pressure system is also a result of hot air rising from the valley; this rising air is replaced by hot, dry, tropical air from the Mexican Plateau. In the atlas, the rising hot air is illustrated by a set of red arrows moving upward. These animations (in the form of moving arrows) serve to imprint on the viewer a clearer understanding of these processes than static images could provide.

Admittedly, the Death Valley atlas is not without problems—one cannot reverse the direction of a narrative, and the text of the narrations cannot be printed. Overall, though, the atlas provides a thorough introduction to Death Valley via a number of effective animations, and serves as a good illustration of how spoken text in a computer-based environment can be used to assist in interpreting maps and graphs.

The *David Rumsey Map Collection*

The *David Rumsey Map Collection* (*http://www.davidrumsey. com/*) is an online set of more than 8,800 maps (and map-related items, such as globes) from David Rumsey's personal collection of more than 150,000 18th- and 19th-century maps of North and South America (Rumsey 2003). For those interested in historical maps, this is a tremendous resource, as most historical maps traditionally have been available only in paper form in libraries. Color Plate 22.2 illustrates an example from this impressive collection—in this case, William Henry Holmes's 1882 rendition of the Grand Canyon.

In addition to making the historical maps available online, Rumsey has begun to link some of the maps with modern geographic data, such as satellite images, and a variety of vector and raster data available from the U.S. Geological Survey. For instance, Color Plate 22.3 shows a QuadViewer option available within Rumsey's GIS browser. In this case, four historical maps of Boston are shown with the modern-day street network (in red) overlaid on them. Note that this image reveals some of the difficulty of linking historical and modern data, as the modern streets do not line up precisely with the historical data.

Rumsey (2003) recognized this problem, noting that "Many historical maps were created before the availability of modern mapping and surveying technologies, and some of the items were tourist maps that were visually surveyed or based on an artist's rendering." One nicety of the GIS browser is that one can zoom in on any of the four maps, and all four maps are shown at the same enlarged scale.

Another intriguing capability provided in Rumsey's *Map Collection* is the capability to create 3-D images by overlaying some of the historical images on digital elevation models (DEMs). For instance, Figure 22.1 shows a map of Yosemite Valley in 1879 overlaid on a DEM. Users who have the necessary graphics hardware can fly through these 3-D images, producing a virtual environment, a topic that we discuss more fully in Chapter 24.

22.2.2 National Atlases

The *Domesday Project*

In 1086, William the Conqueror enlisted a group of royal officers to conduct a survey of England, principally to determine land ownership and associated levels of taxation, but also to provide census-related and land-use information. The resulting survey was commonly referred to as the *Domesday Book* because one could not appeal a decision based on it. As a source of geographical and historical data on a national scale, Palmer (1986) indicates

that the *Domesday Book* was "unrivaled until the nineteenth century" (p. 279). In 1986, the *Domesday Project* was developed to commemorate the 900th anniversary of the *Domesday Book*. Rather than providing information for taxation, the *Domesday Project* was intended to give the general public a contemporary snapshot of the United Kingdom during the 1980s.

A key characteristic of the *Domesday Project* was the use of video disk technology, which permitted the inclusion of text, maps, and photographs in a user-friendly, highly interactive environment. The Community and National video disks included approximately 250 megabytes of digital data, 50,000 photographs, and 20 million words of text. The Community disk (or "people's database") contained information collected locally by more than 300,000 schoolchildren (Rhind and Mounsey 1986, 317). The National disk contained information normally collected in censuses, but also included a wealth of other data dealing with employment, disease, and the environment. In total, the National disk provided color choropleth mapping capability for "over 20,000 variables for 33 different sets of geographical areas at ten different levels of resolution" (Openshaw et al. 1986, 296).

Today, approximately 20 years after it was created, the *Domesday Project* is still recognized as one of the premier electronic atlases ever developed. Unfortunately, most of those outside of the United Kingdom were never

FIGURE 22.1   A map of Yosemite from 1879 (part of the Wheeler Survey of the western United States) overlaid on a modern DEM. The *David Rumsey Map Collection* permits users to fly through this environment. (Courtesy of the David Rumsey Map Collection, *www.davidrumsey.com.*)

able to appreciate its capability because the system was designed for specialized BBC microcomputers used in U.K. schools. Interestingly, the system was not marketed after the late 1980s because sales were disappointing (Siekierska and Taylor 1991, 12). These low sales appear to have been the result of the "read-only" character of the video disk technology and the out-of-date information collected in 1985 (Hocking and Keller 1993).

U.S. Census Bureau Online Mapping

For those wishing to map U.S. census data, the U.S. Census Bureau provides two major options: the *TIGER Map Server* and *American Fact Finder* (*http://www.census.gov/geo/www/maps/CP_OnLineMapping.htm*). *TIGER* is now a bit dated, as it was designed for mapping 1990 U.S. census data, but it is useful to contrast its capabilities with the newer *American Fact Finder*. *TIGER* permits users to create maps of base data (e.g., a grid of latitude and longitude, streets, and railroads) and/or thematic data (e.g., family income) for anywhere in the United States from the state level down to the block-group level. Completed maps are not part of the service but are a function of user desires; for example, users can easily change the scale, and select from 21 base data layers and 18 different themes to create their own map. Color Plate 22.4 illustrates a sample map for the Seattle, Washington, region. In this case, base data include water bodies and interstate highways, whereas the theme is the percent of the population 65 years and older within census tracts.

American Fact Finder was developed to distribute and map 2000 U.S. census data (although 1990 data can also be mapped). From the standpoint of thematic mapping, two major options are provided: Basic Facts and Maps. Basic Facts enables the user to create a thematic map of an attribute with minimal effort—the user simply specifies an attribute and level of geography (e.g., county or state level) and the map is created. No modifications to the resulting map are possible. In contrast, the Maps option allows considerable modification to the map once it initially is created (Figure 22.2). Users can zoom to 10 different levels (see the right portion of Figure 22.2) or reposition the map by specifying an address, zip code, or latitude/longitude. Clicking on the word Legend will reveal options for number of classes, the color scheme, and method of classification. As with the earlier *TIGER* system, a variety of base information also can be shown (available under Options as Boundaries and Features). A limitation of both *American Fact Finder* and *TIGER* is that only choropleth maps are possible; as we discussed in section 4.4.1, this is particularly problematic when raw totals are mapped, which is quite common with census data.

TIGER and *American Fact Finder* can be viewed as atlases in the sense that a large variety of maps of a particular area can be accessed reasonably quickly (creation times are generally on the order of seconds, although during busy Internet periods, several minutes might be required). The term *atlas* might be inappropriate, however, because maps in the system are not already created, but are developed interactively. Rather, the term **map server** potentially is more appropriate, as such systems "serve" maps to users. An extensive list of sites that serve maps can be found at *http://www.geoplace.com/gr/webmapping/sites.asp*.

The National Atlas of the United States

In 1970, the U.S. Geological Survey produced *The National Atlas of the United States of America*, a paper atlas depicting national conditions in the United States in the mid-1960s. Although this large and weighty atlas (all 12 pounds of it) proved to be useful and contained many intriguing maps (e.g., Color Plate 15.2B depicts a portion of Richard Edes Harrison's superb shaded relief map), it can be argued that a national paper atlas is limiting in today's rapidly changing digital age. Thus, in 1997 the United States began development of an electronic national atlas, known as *The National Atlas of the United States*. The Web site for the resulting atlas (*http://www.nationalatlas.gov/*) indicates that one of the key driving forces was to produce an atlas "intended as an essential reference for all computer users—not just the scientists and decision makers who are our traditional customers."

The National Atlas of the United States contains five major categories of mapping capability: Interactive Maps, Multimedia Maps, Printed Maps, Printable Maps, and Map Layers Data Warehouse. Interactive Maps allows you to map numerous types of broad thematic categories, including agriculture, biology, climate, environment, geology, history, people, transportation, and water. Each of these broad categories includes subcategories; for instance, climate includes average annual precipitation, major hurricanes that move over land, and tropical cyclones. Color Plate 22.5 illustrates a map of major hurricanes that passed over land ("landfalling" hurricanes) in the 1960s. Note how the track of each hurricane indicates that it moved quickly over water and then slowed over land, a fundamental principle of hurricane movement. Given our discussion of color in this book, you might think about whether the colors used to depict different speeds of the hurricane are appropriate.

One limitation of *The National Atlas of the United States* is that data for enumeration units are invariably mapped with the choropleth method. As we have seen with the U.S. Census Bureau products previously, this problem is not peculiar to this atlas. Like the U.S. Census Bureau products, *The National Atlas of the United States* contains a variety of base data, such as cities and towns, roads, and streams. A key characteristic of the atlas (and one shared with several other atlases reviewed here) is the ability to automatically control the amount

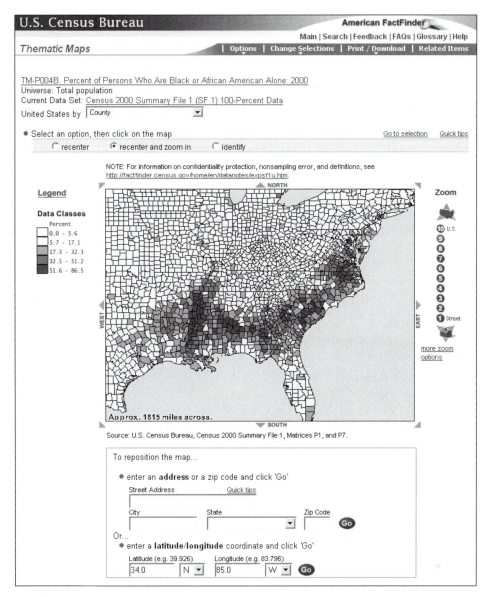

FIGURE 22.2   Example of a map produced using the U.S. Census Bureau's *American Fact Finder*. The attribute mapped is the percentage of people who identified themselves as "Black or African American alone" as opposed to "two or more races" (original in color).

of information displayed as a function of scale. For example, when "cities and towns" is selected and the entire country is shown, no cities appear. As you zoom in, you will find that more and more cities are displayed. This feature also illustrates the problem of labeling closely spaced cities: In the atlas all labels are assigned just above a city location; more elegant labeling routines are possible (see section 11.3.7).

The Multimedia Maps section of *The National Atlas of the United States* utilizes animation and interactivity to examine a variety of interesting topics. Particularly intriguing is a Tapestry of Time and Terrain, which combines Pike and Thelin's digital shaded relief map (which

we discussed in section 15.2.4) with a geologic map of the United States. One way of combining these maps is to gradually blend them in an animation: The view begins with just the geologic map shown and then relief information is gradually added; finally, the geologic information is removed until only the relief information appears. Although we found ourselves questioning the utility of this animation, we also found that the interactive portion of Tapestry of Time and Terrain provided many useful features, including the following options: Description of Features, Color Legend, Boundaries, and Play Puzzle of Regions. Color Legend illustrates quite well the power of interactive mapping, as one can click on a portion of the geologic time scale (Color Plate 22.6A) and see the location of associated rocks in the United States (Color Plate 22.6B). Other topics in the Multimedia Maps section illustrate the potential of linking text and photos with maps, and of course the potential of linking information via the Web (e.g., see the Animated Invasive Species Map, which depicts the invasion of zebra mussels in U.S. rivers and lakes).

Printed Maps allows users to order a variety of large-format printed maps such as a shaded relief map of North America and a map depicting electoral votes by political party and state for all presidential elections from 1789 to 2000. Printable Maps allows users to print maps of interest that are suitable for home printers (examples of topics include West Nile Virus and boundaries for congressional districts). Finally, the Map Layers Data Warehouse allows users to download actual data, which can then be imported into other programs, such as ArcView and ArcGIS. This is desirable for those wishing to do more sophisticated analyses and visualization. With this broad range of features, *The National Atlas of the United States* illustrates the considerable flexibility that electronic atlases can provide, especially when they are implemented in a Web format.

Atlas of Switzerland Interactive

The *Atlas of Switzerland Interactive*, a CD-ROM-based atlas, includes a wide variety of thematic maps for Switzerland and Europe, along with a series of topographic maps for Switzerland. The bulk of the thematic maps are either choropleth or proportional symbol (circles or squares), the choice being carefully based on whether the data are standardized or unstandardized, respectively. An interesting feature of proportional symbol maps is the use of redundant symbology, with both the size of the symbol and shading within the symbol used to represent the data. Although we suggested in Chapter 16 that redundant symbols would complicate the retrieval of general information, we found the redundant symbols in this atlas an attractive design option (Color Plate 22.7). Choropleth maps allow the user considerable flexibility in design options, including changing the number of classes (up to 20), the methods of classification (equal intervals and quantiles), and the color scheme. Although such options are useful for the experienced cartographer, they might overwhelm the novice. For instance, it is possible to drag the bounds of class limits in a histogram display of the data; novices might wonder why this would be useful.

An impressive feature of the *Atlas of Switzerland* is the capability to examine a fair amount of the data through time (e.g., in the case of Switzerland, population data are mapped at roughly 10-year intervals from 1850–1990) and at different spatial levels (cantons, districts, MS regions, and communes for Switzerland). Communes provide a

particularly detailed view, as there are approximately 3,000 in Switzerland (the precise number varying over time). Individual time elements can be selected by clicking on a number line, or an animation can be played. We found the animated proportional symbol maps relatively easy to follow (partly because the developers attempted to smooth transitions between time periods), but the animated choropleth maps were hard to understand (as we suggested in Chapter 20, choropleth maps are often awkward to animate). Given the limitations of animation, we suspect that users would also find a map comparison option useful. Another attractive feature of the atlas is that individual enumeration units are highlighted by a flashing boundary when the cursor is dragged over them—although cartographers often argue that flashing should be avoided, in this case it is quite effective. Individual enumeration units also can be highlighted by simply typing the name in an Index option; this is very useful when the user is not familiar with the geography of Switzerland.

Given the traditional strength of the Swiss in topographic mapping (see section 15.2.4), it is not surprising that the *Atlas of Switzerland* contains a wealth of topographic maps. Three basic options are possible for viewing shaded relief maps: Map Relief (vertical view), Block Diagram, and Panorama. When shaded relief is combined with hypsometric tints, some rather stunning images are possible (Color Plate 22.8). A distinct advantage of having such topographic maps available in a digital format is that users can vary lighting options (e.g., the source of light) to create alternative views of the landscape. Often, certain features in the landscape not apparent under one lighting option will be apparent under another.

Another useful characteristic of the *Atlas of Switzerland* is the Map Extract tool, a small reference map in the lower right corner (Color Plate 22.7), which enables users to easily select a desired portion of the map to be displayed. In contrast to the difficulty of manipulating some of the other atlases discussed here, we found this tool particularly useful for repositioning the focus of the map and zooming into a subregion.

22.2.3 State Atlases

Atlases for Georgia

The state of Georgia provides a useful example of changes that are taking place in the nature of electronic atlases. In the 1990s, the state of Georgia supported the development of *The Interactive Atlas of Georgia*, which was distributed by the Institute of Community and Area Development at the University of Georgia. To a certain extent, *The Interactive Atlas of Georgia* emulated traditional paper atlases, as it was possible to click on an arrow key in the menu and move through the atlas one screen (or page) at a time. In contrast to the static nature of printed atlases, however, it was possible to interact with each map in the atlas. The range of interactive tools was similar to those described for ExploreMap in Chapter 21. For example, a Where option permitted users to obtain specific information: Users selected a county name from a menu, which caused the associated county to be highlighted and the name and value to appear. This option worked not only for choropleth maps, which were common in the atlas, but also for a variety of other maps. For example, when clicking at an arbitrary location on a dot map of peanut production, the Where option displayed the county boundary within which the location fell, along with the total peanut production for that county. Another interactive tool was the Query option, which enabled users to specify ranges of data for multiple attributes, and thus map a set of counties that met a variety of criteria. For example, a user might locate counties having low crime rates and a large amount of money spent educating students.

Intended as an atlas rather than as a set of exploration tools for a particular kind of map, *The Interactive Atlas of Georgia* not surprisingly provided a broader range of options than ExploreMap; for example, a Cities option enabled users to locate more than 3,000 places, and a County Learn option assisted users in learning county names in Georgia. The Georgia atlas was limited, however, in that it did not contain text, photographs, or graphs; nor did it contain animation or multimedia capability. Such limitations were, in part, a function of the limited funding commonly available for state atlases, and because the atlas was developed in the mid-1990s.

Today, a variety of atlas-like products for the state of Georgia are distributed by the Georgia GIS Data Clearinghouse (*http://www.gis.state.ga.us/Clearinghouse/ clearinghouse.html*). As Table 22.1 illustrates, a wide variety of both static and interactive maps are available through the Clearinghouse. All of these maps are available online, with the exception of the Environmental Atlas of Georgia, which must be purchased as a set of CD-ROMs. The Thematic Map Catalog is indicative of the nature of the static maps available from the Clear-

TABLE 22.1 Map products available through the Georgia GIS Data Clearinghouse

Static maps
  Georgia Department of Transportation (GDOT) maps
  TIGER Map Service
  Environmental Atlas of Georgia
  Image Map Catalog
  Thematic Map Catalog
  3-Dimensional Map Catalog
Interactive maps
  Georgia 2000 Information System
  Georgia High-Speed Telecommunications Atlas
  Interactive Map Server

inghouse, as it consists of 13 previously created county-level choropleth maps of census data for the state (e.g., population). The Georgia 2000 Information System is indicative of interactive online maps; it provides maps of a range of point-based phenomena (e.g., hospitals) and area-based cancer rate data. As Color Plate 22.9 suggests, multiple data sets can be mapped on top of one another (although no specialized bivariate or multivariate maps are possible); the map can be manipulated via zooming and panning, and individual locations can be queried. Georgia is, of course, not the only state providing atlas-like products via the Web—the home page for this book provides links to numerous other states that are creating such products.

Atlas of Oregon

The *Atlas of Oregon* (Loy et al. 2001) is particularly interesting to consider because it recently has been published in both printed and electronic form. The printed version has received adulation from many; for instance, Ronald Abler, Executive Director of the Association of American Geographers, described it as follows:

> The first edition of the Atlas of Oregon was a masterpiece of its genre. The second edition is that and then some. Atlas of Oregon II constitutes a new benchmark for state atlases. For atlases of all kinds, it sets daunting standards for beauty and quality in cartographic and graphic presentation. The second edition's content is simultaneously penetrating and comprehensive; its structure is coherent and instructive; its maps and diagrams are simply stunning. (*Atlas of Oregon* CD-ROM)

The success of the Oregon atlas is due not only to the efforts of cartographers at the University of Oregon (William Loy, Aileen Buckley, and James Meacham) but also the cartographic editing capabilities of Stuart Allan, who is widely known for his expertise in map design. The 33.5 × 24.5 cm paper atlas consists of 301 pages divided into four major sections: Human Geography, The Economy, Physical Geography, and References. The first three sections include a vast array of thematic maps that can be used to illustrate effective symbolization and design. As such, the atlas could serve as a useful supplement to a cartography course focusing on design.

The electronic CD-ROM version of the *Atlas of Oregon* contains essentially the same set of maps as the printed version, but provides interactive and animated capabilities not possible in the static printed version. For instance, a series of maps entitled Early Exploration depicts the routes of explorers in the 1800s (Color Plate 22.10). In the printed atlas, all explorers' routes for a particular time period are shown on the same map, whereas in the electronic atlas it is possible to select any combination of routes, thus permitting the user to focus on certain explorations. On the map depicting Lewis and Clark's famous explorations, it is also possible to click on individual numbers and see a description of each place along their route. In the printed version of the atlas, these descriptions are all printed at once, whereas in the electronic version the description appears only for the location clicked on. It can be argued that clicking on a location (and having text displayed) immerses the user more fully in the learning process and reinforces the order in which events took place.

Another example illustrating the potential of interactivity and animation is the Precipitation topic, which enables users to view an animated series of shaded isarithmic maps of monthly precipitation. For any month in the animation, users can stop, point the mouse cursor at a weather station, and view a bar graph illustrating the precipitation regime for that station over the course of the year (Color Plate 22.11). Although such capabilities add a degree of dynamism to the atlas, they also raise the issue of how such information should best be presented. In the printed atlas, isarithmic maps for each month are printed as a small multiple and all precipitation regimes are shown on a single map at their actual locations (albeit in a smaller format). The argument can certainly be raised that the printed version permits a more effective visual analysis of the spatial pattern.

Another map series that provides dynamic capability, but also raises questions about the need for interactivity, is entitled Politics. This series allows users to examine choropleth and proportional symbol maps (depicting the percentage and number of votes for Republicans and Democrats) for each Presidential election. In the electronic version users select which election they want to see, whereas in the printed version the results for all elections are shown as a small multiple of choropleth maps. Although viewing individual elections seems useful, a map comparison option or a small multiple would have been a desirable addition to the electronic version.

An unusual feature of the electronic atlas that is available for some topics is a Magnify tool that can be used to enlarge a portion of a map. In most of the atlases that we discuss here, enlargement is accomplished using a zoom tool in which an area is selected and the selected area fills the entire map display window. In the case of the Magnify tool, only the square area within the tool is enlarged, while the rest of the map remains at the original scale. We found this approach an attractive option for examining detail while still keeping the overall map in mind.

22.2.4 Commercial Atlases and Related Software

World Atlas

The development of electronic atlases in the United States was heavily influenced by Richard Smith, who was a leader in creating the *Electronic Atlas of Arkansas*, the first state electronic atlas. Smith was also responsible for developing *World Atlas* one of the first widely distributed commercial electronic atlases. Two basic types of thematic maps were available in *World Atlas*: choropleth and proportional symbol (specifically, proportional squares were used). Choropleth maps were used for standardized data, whereas proportional symbol maps were used for raw count data (Figure 22.3).

Multimedia was incorporated into *World Atlas* as sound, video, and pictures. Audio features included national anthems, names of countries, and foreign phrases. Ideally, it would seem desirable to link these features directly with maps, but generally one had to point at the name of a country within a menu (as opposed to the map) to hear an associated audio feature. Because of video's large storage requirements, video clips were shown only for each major city in the world, and only a limited number were included for each city; furthermore, the clips were small (taking up approximately one-tenth of the screen) and lasted only a few seconds. A brief textual phrase announced the video and background music played during it. Examples from six clips for China included "Great Wall of China," "Streets of Beijing," and "Tiananmen Square Monument." When viewing the videos, we found ourselves wanting much more information, but realized that acquiring such information would be costly and that storage space was a major limitation at the time the atlas was developed. This limitation has become less critical as CD-ROM technology and associated compression software has improved, and as the Internet has become a common delivery mechanism for electronic atlases.

Microsoft's® Encarta® Reference Library 2003

Microsoft's Encarta Reference Library 2003 contains a host of map-related material, including a World Atlas, Historical Maps, Map Treks, and some specialized maps for Africa. The World Atlas provides a great variety of information for countries and cities around the world; for example, it boasts of nearly 1.8 million place names. The focal point of the World Atlas is an attractive perspective view of the world (the world appears as it would from space)—an orthographic projection is used, although a "Flat Map View" is also possible via a cylindrical projection. Users can easily rotate the perspective view by simply clicking and dragging the mouse, or they can click on a zoom tool to move quickly through a variety of scales.

More than 20 different map "styles" are possible in the World Atlas, including Political, Physical Features, Earth by Day or Night, Precipitation, Temperature, Languages, and Statistical. All of these are interesting from a geographic standpoint, but the Statistical style is of greatest relevance to the present text. With this style,

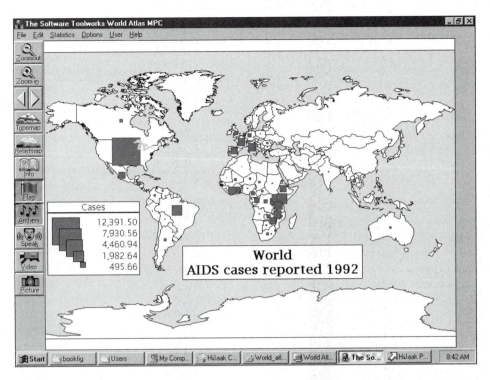

FIGURE 22.3 An example of a proportional symbol map from *World Atlas*. (©1998, The Learning Company, Inc.)

several hundred attributes can be mapped at the level of individual countries (no detail is possible within a country). Interestingly, all attributes are symbolized using unclassed choropleth maps (even when raw totals are mapped). For unipolar data, shades of orange are used, whereas for bipolar data a diverging scheme is used (dark orange to white to dark blue). The data can be symbolized in an unmodified form (as "linear"), or a "logarithmic" transform is possible; the latter will make positively skewed distributions more normal (see section 3.3.1) and thus enable lower data values to be more easily differentiated. Although many potential users of the atlas would be unfamiliar with this concept, we found the logarithmic maps much easier to read, as colors for lower data values could be differentiated. Those who find unclassed maps difficult to use might be comforted by the fact that precise values for individual countries can be obtained by simply pointing the mouse cursor at the desired country.

When the user zooms in on a particular country within World Atlas, various information can be obtained about the location pointed to (and the surrounding area). For instance, when viewing a map of Portugal, you can click on Lisbon and retrieve information about Lisbon, Western Europe, Southern Europe, the Iberian Peninsula, or Portugal. Presuming that you select Lisbon, you can view a Contents Page that provides basic information about Lisbon, such as a brief history, photos, news, weather, and Web links that provide additional information.

The Historical Maps section of the *Encarta Reference Library* includes both numerous static maps and approximately 20 animations focusing on either war-related topics or explorers' routes. We found ourselves contrasting these animations with those found in the *Grolier Multimedia Encyclopedia* discussed later. From this perspective, one clear limitation of the Historical Maps was the lack of sound (a verbal description), which made it difficult to follow many of the animations (interestingly, sound is used elsewhere in *Encarta*, such as for national anthems). On the plus side, we noted that animating explorers' routes did seem to make these more meaningful than a static image would provide. The Map Treks section of *Encarta* provides guided tours of various geographic features and concepts around the world. For example, for a Physical Features map type, clicking on Great Lakes allows you to get maps and information for the Great Lakes of Africa, Lake Baikal (the world's deepest inland body of water), the Great Lakes of North America, and Lake Titicaca (the largest freshwater lake in South America). The specialized maps for Africa in *Encarta* include animations of African slave trade routes; as is usual with animation, the question can be raised as to whether the animations enhance understanding. Finally, it should be noted that the *Encarta Reference Library* includes a set of Virtual Tours that include 360° panoramic photographs for a variety of places around the world. Such realism moves the library into the realm of virtual environments, as the intention is to give you the feeling of actually being in a place.

Overall, the *Encarta Reference Library* provides an impressive array of mapping capability. Because the bulk of the atlas can be installed on your hard drive, the program is of course more responsive than Web-based atlases, such as *The National Atlas of the United States*. The library is particularly illustrative of how maps can be linked with text, images, and sound. From a thematic mapping standpoint, the World Atlas is limited, however, in its range of classification and symbolization methods.

Grolier's Multimedia Encyclopedia

Multimedia encyclopedias are like paper encyclopedias in the sense that they provide considerable information about places, people, and events, but the multimedia variant is enhanced through the inclusion of sound, video, and the ease of linking various topics. As an example, the *2003 Grolier Multimedia Encyclopedia* includes an "Atlas" section that provides a variety of thematic maps of the world for individual countries and cities, and for the 50 U.S. states. These maps permit a limited degree of interactivity. For example, on a map showing numerous explorers' routes, it is possible to select one explorer's route or to show all routes simultaneously. As another example, it is possible to show a succession of colored contour maps of temperature by month for the United States by simply clicking on Next, creating a sort of animation under control of the user. In general, however, one cannot point to locations (e.g., countries) to get associated attribute values (e.g., population density).

The most interesting feature of the Grolier encyclopedia from a cartographic standpoint is the use of multimedia in association with animated maps, which David DiBiase (1994) assisted in designing. An example is "The American Revolution," which is particularly informative and enjoyable to watch. In this presentation, a map of the United States serves as a backdrop against which animated symbols portray the location of major battles and movement of troops. While the animation plays, an audio portion describes the battles and troop movements, along with major historical events. Like the Death Valley atlas, the combination of animated maps and audio is arguably more effective than a traditional printed atlas in which one must repeatedly move back and forth between text and maps. Also included in "The American Revolution" are photographs of key people and events. For example, when George Washington is mentioned, a picture of him appears. Although such pictures could detract from the animation, they are entertaining and provide a useful break for the viewer. The multimedia presentations include subtitles, which are essential for the hearing impaired and useful for users with older computers lacking sound cards. Overall, the multimedia presentations are particularly impressive given

that DiBiase indicated the maps were created before the audio portion was available.

22.3 MULTIMEDIA SYSTEMS

Although electronic atlases frequently contain links to multimedia information, the emphasis is still on the collection of maps stored in the atlas. In contrast, some mapping systems place much greater emphasis on multimedia aspects—we refer to these as **multimedia systems**. Two researchers who have been heavily involved in developing multimedia systems are Michael Shiffer and Shunfu Hu. Shiffer (1995; 2001; 2002) has promoted the use of multimedia in urban planning, terming the result *spatial multimedia*. In urban planning applications, a map (or maps) of the region under examination serves to orient planners and interested parties to the geography of the region. This map can then be queried at individual locations to get more information about those locations via a variety of multimedia (e.g., text, photographs, sound, and video). For example, if a city were deciding whether to allow a particular model of airplane to land at its airport, they could allow concerned citizens to click on various locations around the airport and hear what the noise of the plane would be like at those locations. Shiffer has stressed the role that multimedia systems can play in promoting community participation in urban planning, as individuals can provide input to plans as text, audio, and video annotation.

Shunfu Hu (1999; 2003) has promoted the use of multimedia in natural resource applications. For instance, Hu (1999) developed a multimedia system for examining the flora, fauna, and human activities in Everglades National Park. When completed, one could point to a location on the map and receive a variety of multimedia information; one might see a photograph of the forest found at that location along with text describing that particular forest community. Recently, Hu (2003) has promoted using the Web to provide greater access to multimedia systems.

SUMMARY

In this chapter, we have focused on electronic atlases and the related area of multimedia. In the simplest sense, an electronic atlas can be viewed as a collection of maps focusing on a particular region or topic that is available in digital form. More sophisticated **electronic atlases**, however, permit users to take advantage of the digital environment through a variety of means, such as Internet access, data exploration, map animation, and multimedia.

We examined four broad groups of electronic atlases: atlases developed by individuals, national atlases, state atlases, and commercial atlases and related software. Rather than summarize each category, we will make some general comments about the nature of electronic atlases. One is that the format of the atlas (e.g., CD-ROM, Web-based, or a combination of CD-ROM and the Web) obviously provides different capabilities. Purely CD-ROM atlases (e.g., the *Atlas of Oregon* and the *Atlas of Switzerland*) are advantageous in providing rapid access to maps and related information. In contrast, Web-based atlases (e.g., *The National Atlas of the United States*) have the potential for storing and accessing virtually unlimited amounts of information, but with today's technology, access speeds might seem painfully slow. Perhaps the ideal is the combination of CD-ROM (or an equivalent technology) with the Web (as in the *Encarta Reference Library*)—here maps can be retrieved nearly instantaneously, but the user also has access to the rich resources of the Web, which also can be retrieved quickly because the associated information generally does not involve downloading large spatial data sets.

We saw several ways in which thematic data sets are symbolized in electronic atlases. Some atlases (e.g., *American Fact Finder, The National Atlas of the United States*, and *Encarta*) stress the choropleth option, which leads to misinterpretation when unstandardized (raw total) data are mapped. On the plus side, however, some atlases carefully differentiate between standardized and unstandardized data; for instance, the *Atlas of Switzerland* utilized choropleth and proportional symbol maps, respectively. The *Atlas of Oregon* is noteworthy for illustrating a broad range of thematic mapping techniques.

Animation is a common feature of several of the atlases that we examined. Although developers of atlases invariably stress the potential benefits of animation, we found ourselves sometimes questioning its usefulness; choropleth animations, in particular, seem difficult to understand. If animation is to be included, it seems that a map comparison option or a small multiples approach also should be considered.

Allowing the user to interact with the map was a common feature of most of the atlases. A simple form of interaction is pointing to an enumeration unit and acquiring a value (as is done effectively in the *Atlas of Switzerland*). A slightly more complicated form of interaction is in the *Atlas of Oregon*, where particular explorers' routes can be selected, and additional information can be gleaned by selecting numbers along the routes. We feel that such interactivity might be more effective from a learning standpoint than viewing static printed information.

An obvious advantage of electronic atlases is the ability to examine a region at different scales. Several atlases not only permit viewing the data at different scales, but also automatically generalize the information displayed (e.g., *The National Atlas of the United States* and *Encarta*). In the context of scale, we saw various approaches for positioning and zooming in on a region. Particularly attractive options were the Magnify tool in the *Atlas of Oregon* and the Map Extract tool in the *Atlas of Switzerland*.

The *David Rumsey Map Collection* provides a good illustration of how historical maps can be integrated with modern digital information. Particularly intriguing is the ability to explore what were originally "flat" two-dimensional maps in a 3-D virtual environment. Because Rumsey has only begun to experiment with these approaches, one wonders what the future holds for the integration of historical and modern maps!

It is important to recognize that we have discussed only a sample of electronic atlases; the home page for this book provides information about a more complete set. Because electronic atlases are still evolving, readers will need to consider the capability of new atlases and decide how the definition for electronic atlases given here might ultimately be modified.

FURTHER READING

Buckley, A. (2003) "Atlas mapping in the 21st century." *Cartography and Geographic Information Science* 30, no. 2:149–158.

> Discusses the nature of both paper and electronic atlases in the 21st century.

Cartographic Perspectives no. 20, 1995 (entire issue).

> This special issue deals entirely with electronic atlases; several articles in the issue are summarized in this section.

Cartwright, W., Peterson, M. P., and Gartner, G. (eds.) (1999) *Multimedia Cartography*. Berlin: Springer-Verlag.

> Contains numerous chapters dealing with electronic atlases and multimedia.

DiBiase, D. (1994) "Designing animated maps for a multimedia encyclopedia." *Cartographic Perspectives* no. 19:3–7, 19.

> Describes the author's practical experience in developing animations for the *Grolier Multimedia Encyclopedia.*

Fung, K. (2001) "Atlas of Saskatchewan—a product of technological innovation." *Proceedings, 20th International Cartographic Conference*, International Cartographic Association, Beijing, China, CD-ROM.

> Discusses the *Atlas of Saskatchewan*, which will ultimately be available in paper, CD-ROM, and on the Web.

Hu, S. (1999) "Integrated multimedia approach to the utilization of an Everglades vegetation database." *Photogrammetric Engineering and Remote Sensing* 65, no. 2:193–198.

> Discusses creating a multimedia system for Everglades National Park.

Keller, C. P. (1995) "Visualizing digital atlas information products and the user perspective." *Cartographic Perspectives* no. 20:21–28.

> Discusses the potential for electronic atlases and promotes utilizing user surveys to determine the content and format of atlases.

Limp, W. F. (2002) "Web Mapping, 2002." *GeoWorld* 15, no. 3:30–32.

> Discusses a range of approaches for creating maps for the Web.

Ormeling, F. (1995) "New forms, concepts, and structures for European national atlases." *Cartographic Perspectives* no. 20:12–20.

> Discusses some of the capability that should be provided by electronic atlases.

Raper, J. (1997) "Progress towards spatial multimedia." In *Geographic Information Research: Bridging the Atlantic,* ed. by M. Craglia and H. Couclelis, pp. 525–543. London: Taylor & Francis.

> An early overview of issues associated with multimedia in geography.

Rystedt, B. (1995) "Current trends in electronic atlas production." *Cartographic Perspectives* no. 20:5–11.

> Briefly summarizes a variety of early electronic atlases.

Shiffer, M. J. (2001) "Spatial multimedia for planning support." In *Planning Support Systems,* ed. by R. K. Brail and R. E. Klosterman, pp. 361–385. Redlands, CA: ESRI Press.

> Discusses the use of multimedia in urban planning applications.

Siekierska, E. M., and Taylor, D. R. F. (1991) "Electronic mapping and electronic atlases: New cartographic products for the information era—the electronic atlas of Canada." *CISM Journal ACSGC* 45, no. 1:11–21.

> An early overview of electronic atlases, with emphasis on the *Electronic Atlas of Canada.*

Smith, R. M., and Parker, T. (1995) "An electronic atlas authoring system." *Cartographic Perspectives* no. 20:35–39.

> Presents the notion of developing software that is expressly designed to create electronic atlases. Smith (personal communication 1997) indicates that the software was never actually completed.

Transactions, Institute of British Geographers (new series) 11, no. 3, 1986.

> Contains numerous articles dealing with the *Domesday Project.*

Walker, T., Cartwright, W., and Miller, S. (2000) "An investigation into the methodologies of producing a web-based multimedia atlas of Victoria." *Cartography* 29, no. 2:51–64.

> Discusses a variety of issues associated with the production of an electronic atlas and reviews several electronic atlases.

23

Visualizing Uncertainty

OVERVIEW

We often think of maps as truthful representations of reality. For example, imagine viewing a state-level choropleth map of the United States entitled "Percent of the Population That Smokes." If a shade in the legend has class limits of 21 to 25 percent, then we are apt to presume that states with that shade do, in fact, fall in the 21- to 25-percent range. In reality, this probably would not be correct because the map likely would be based on a sample—thus, there would be a margin of error around the sampled value (see section 3.2). Such error in maps is a form of **uncertainty**; *in this chapter, we consider approaches for visualizing this uncertainty.*

In section 23.1, we consider elements of uncertainty. Uncertainty can arise from a variety of sources, including the raw data on which a map is based, the manner in which these data are processed, and the manner in which the visualization is created. Other terms are often used in place of uncertainty—popular choices include reliability and quality (if data are uncertain, they are also unreliable and of poor quality). The National Institute of Standards has proposed five categories for assessing data quality: **lineage**, **positional accuracy**, **attribute accuracy**, **completeness**, *and* **logical consistency**. *The categories of positional accuracy and attribute accuracy have received the most attention in the literature pertaining to visualization.*

In section 23.2, we briefly describe the general ways in which uncertainty can be depicted: Separate maps are created for an attribute and its associated uncertainty (maps are compared); the attribute and its uncertainty are displayed on the same map (maps are combined); and data exploration tools are utilized. Generally, the "maps combined" approach has been most commonly used, although several software packages have been developed for exploring uncertainty.

Section 23.3 considers various visual variables that can be utilized to depict uncertainty. Broadly speaking, we can group these into **intrinsic** *and* **extrinsic visual variables**, *depending on whether the variable is visually separable from the variable depicting the actual attribute (extrinsic variables are separable, whereas intrinsic variables are not). We will see that although some traditional visual variables (e.g., size and shape) can be utilized, specialized visual variables have been developed for depicting uncertainty, including* **crispness**, **resolution**, *and* **transparency**.

In section 23.4, we cover applications for visualizing uncertainty. By looking at numerous applications, we hope to give you a feel for the broad range of approaches that have been developed for visualizing uncertainty. Although cartographers and other graphic designers have created a multitude of approaches for visualizing uncertainty, you might find yourself asking whether these methods are effective: Can users understand the notion of uncertainty, its various visual depictions, and maybe more important, can they act on it? In this context, section 23.5 deals with some studies explicitly developed to examine the effectiveness of methods for depicting uncertainty.

23.1 ELEMENTS OF UNCERTAINTY

Uncertainty can arise from a variety of sources, including the raw data on which a map is based, the manner in which these data are processed, and the manner in which the visualization is created (Pang et al. 1997). As an example of *error in the raw data*, consider the attribute "percent foreign born," which is commonly provided by the U.S. Bureau of the Census. If we see a value of 25.6 percent foreign born in a census publication for a particular census tract, we are apt to think that this is the "correct" value for the tract, but an attribute such as this is based on sampling approximately one of every six housing units (U.S. Bureau of the Census 1994, A-2), and thus values are estimates of population values. Moreover, such values are recorded at a particular point in time, and are, of course, subject to the errors in estimating the size of the population itself, as we discussed in Chapter 1. As an example of *error in processing data*, consider the problem of interpolating values between known control points. In Chapter 14, we found that there were a variety of possible algorithms, each producing a potentially different set of interpolated values. Thus, for any particular location on the map, we can think of the set of interpolated values as constituting uncertainty in the data. Finally, an example of *error in the visualization* is the various approaches that are used to illuminate 3-D scenes (Pang et al. 1997).

Although the term **uncertainty** is commonly used in the literature to describe the potential variation in values of an attribute at a spatial location, the terms *quality* and *reliability* are also frequently used: *Uncertainty* is equated with *unreliability* and *poor quality*. The term *quality* should be used with care, as it has been used to refer to a variety of data characteristics. In particular, the U.S. Federal Information Processing Standard 173 (FIPS 173, National Institute of Standards and Technology 1994) lists five categories for assessing data quality: lineage, positional accuracy, attribute accuracy, completeness, and logical consistency. **Lineage** refers to the historical development of the digital data. For example, if the data were acquired from a paper map, you might want to know the scale and projection of the map, whether digitizing or scanning was used to generate the digital data, and when the data were converted from analog to digital form. **Positional accuracy** refers to the locational accuracy of geographic features, both horizontally and vertically. On a USGS topographic sheet, you might wish to examine the accuracy of a stream's position or the spot height of a mountain. **Attribute accuracy** refers to the accuracy of features found at particular locations. In remote sensing, you might be interested in the accuracy of a land use/land cover classification. **Logical consistency** describes "the fidelity of relationships encoded in the data structure of the digital spatial data" (National Institute of Standards and Technology 1994, 23). We might ask whether all polygons close correctly (in GIS terminology, this would be termed *topological correctness*). Finally, **completeness** includes "information about selection criteria, definitions used and other relevant mapping rules. For example, geometric thresholds such as minimum area or minimum width must be reported" (National Institute of Standards and Technology 1994, 24). These five categories and two additional ones (semantic accuracy and temporal information) were described in detail in *Elements of Data Quality*, which was published by the International Cartographic Association (ICA) Commission on Spatial Data Quality (Guptill and Morrison 1995). The categories of positional accuracy and attribute accuracy have received the most attention in the literature pertaining to visualization, and so we focus on them here.

23.2 GENERAL METHODS FOR DEPICTING UNCERTAINTY

Alan MacEachren (1992) suggested three general methods for depicting uncertainty. First, individual maps can be shown for both an attribute and its associated uncertainty. For example, we might create one map showing the "Percent of the Population That Smokes" and a second map showing a confidence interval for each enumeration unit depicted on the former map. Second, the attribute and its uncertainty can be displayed on the same map—given the appropriate visual variables (e.g., gray tones for one and parallel lines for the other), we can conceive of overlaying one map on the other. Note that these first two methods correspond to the "maps compared" and "maps combined" approaches presented for bivariate and multivariate mapping in Chapter 18. The third method is to use interactive data exploration tools that allow us to easily manipulate the display of both the attribute and its uncertainty. For example, the user might be permitted to toggle back and forth between the display of the attribute and its uncertainty. These methods will be further illustrated in section 23.4, where we consider examples of efforts to visualize uncertainty.

23.3 VISUAL VARIABLES FOR DEPICTING UNCERTAINTY

If we assume the "maps combined" approach, visual variables for depicting uncertainty can be broken into two broad groups: intrinsic and extrinsic (Gershon 1998). As the name implies, **intrinsic visual variables** are intrinsic to the display; for example, to depict uncertainty we might vary the saturation of colored tones on a choropleth map. **Extrinsic visual variables** involve adding objects to the display, "such as dials, thermometers, arrows, bars, [and] objects of different shapes" (Gershon, 44). Because many of the approaches developed by cartographers appear to have involved intrinsic visual variables, we consider these first.

One approach for depicting data uncertainty is to utilize the basic visual variables introduced in section 4.3. In some cases, these visual variables are suitable, whereas in others they can be confusing. A suitable usage would be Tissot's indicatrix (distortion ellipse), in which the visual variables size and shape represent the ability of various map projections to maintain correct size and angular relationships at point locations (Figure 23.1). Potentially confusing would be utilizing the visual variable size to depict the uncertainty of stream position (a wider line indicating greater uncertainty), as a wide line normally would be associated with greater discharge (Figure 23.2).

Of those visual variables introduced in section 4.3, MacEachren (1992) argued that saturation is particularly logical for depicting uncertainty, with "pure hues used for very certain information [and] unsaturated hues for uncertain information" (14). Referencing the work of Brown and van Elzakker (1993), MacEachren (1995, 440–441) noted that three levels of saturation can be used for up to 12 individual hues. Although portraying only three levels of data uncertainty might appear limiting, an analyst could utilize data exploration tools to apply the three levels to only a portion of the data.

In addition to the standard visual variables introduced in section 2.3, MacEachren (1995, 275–276)

FIGURE 23.2 The visual variable size (width of line) depicts data uncertainty associated with stream position. Here the level of data uncertainty might be misinterpreted as stream discharge. (After McGranaghan 1993, 17.)

proposed the visual variable **clarity**, which he argued could be subdivided into three additional visual variables: crispness, resolution, and transparency. **Crispness** refers to the sharpness of boundaries (or area fills): A crisp boundary would represent reliable data, whereas a fuzzy boundary would represent uncertain data. For example, Figure 23.3 illustrates the uncertain nature of the

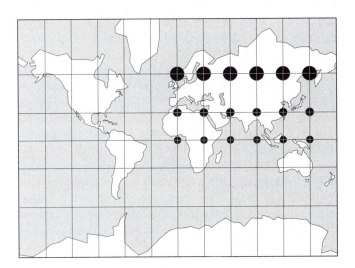

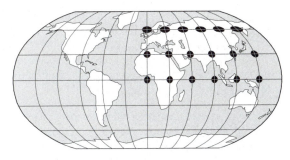

FIGURE 23.1 The visual variables size and shape represent the ability of various map projections to maintain correct size and angular relationships at point locations. Here the visual variables are easily understood. (Abler, Ronald F., Melvin G. Marcus, and Judy M. Olson, *Geography's Inner Worlds: Pervasive Themes in Contemporary American Geography.* Copyright © 1992 by Rutgers, The State University. Reprinted by permission of Rutgers University Press.)

FIGURE 23.3 The visual variable crispness depicts the uncertain nature of the boundary of grits in the United States.

boundary of where grits is consumed in the United States (grits is a southern food made of corn meal). **Resolution** is the level of detail in the spatial data underlying an attribute. For example, Figure 23.4 illustrates different resolutions for a raster database. **Transparency** is the ease with which a theme can be seen through a "fog" placed over that theme; reliable data can be easily seen through the fog, whereas uncertain data cannot (Figure 23.5).*

---

* For additional discussion on visual variables for depicting uncertainty, see McGranaghan (1993). An alternative visual variable that he proposed is *realism*, which refers to the "photorealism of the image" (p. 10).

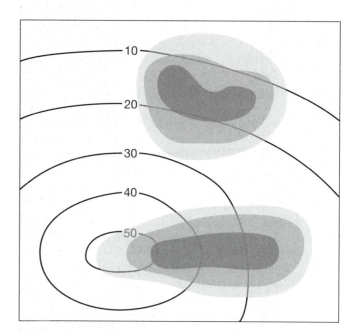

FIGURE 23.5   The visual variable transparency is used to depict the uncertainty of the data. A light "fog" indicates reliable data, whereas a dark fog indicates uncertain data. (After MacEachren 1992, 15.)

Alex Pang and his colleagues (1997) at the Santa Cruz Laboratory for Visualization & Graphics have been major proponents of using extrinsic visual variables to depict uncertainty. Wittenbrink et al. (1996) described this as "verity visualization," as it "suggests the quality of the data that is exactly what it purports to be or is in complete accord with the facts." Figure 23.6B is an example of a glyph that Wittenbrink et al. created for showing the uncertainty in wind speed direction and magnitude. Uncertainty in direction is indicated by width of the arrowhead, and extra arrowheads indicate uncertainty in magnitude. Note how this glyph compares with the traditional wind arrows shown in Figure 23.6A, which do not depict uncertainty. Although their glyph might appear complex, Wittenbrink and his colleagues showed that people could extract useful information from it.

Another example of glyphs is Helen Mitasova and her colleagues' (1993) efforts to map the uncertainty of interpolation methods. They stated:

> Smoothly interpolated surfaces, while visually pleasing and valuable for analysis of data, sometimes imply that the data has greater spatial resolution than is the case; by actually marking on the surface the location, value, and/or uncertainty of each measurement, the visualization contains much more valuable information. The use of glyphs helps to highlight which sampling sites were most important in calculating the interpolated concentrations.... Areas where the glyphs are small and round indicate areas of low uncertainty, while areas where the glyphs are elongated indicate areas where the data is sparse.

In the following section, we see additional ways in which visual variables can be used to depict uncertainty.

FIGURE 23.4   Example of differing resolutions for a raster database. (© Copyright 1995 by the Guilford Press. Reprinted with permission.)

23.4 APPLICATIONS OF VISUALIZING UNCERTAINTY

23.4.1 Howard and MacEachren's R-VIS Software

David Howard and Alan MacEachren (1996) developed the software R-VIS (*R*eliability *VIS*ualization) to enable environmental scientists and policy analysts to examine levels of dissolved inorganic nitrogen (DIN) in Chesapeake Bay. Basic data input to R-VIS is a set of 49 DIN values collected at point locations by a scientific vessel. Because DIN is presumed to be a smooth continuous phenomenon, the data for each point in time can be contoured using the approaches described in Chapter 14 (triangulation, inverse distance, and kriging). Howard and MacEachren chose kriging for R-VIS because kriging provides an uncertainty map expressed as confidence intervals (see section 14.3.3), in addition to a basic isarithmic map of the theme (DIN in the case of R-VIS).

The default display for R-VIS compares the interpolated DIN data for a particular point in time with an associated uncertainty map. Initially, Howard and MacEachren chose a lightness scheme (light to dark red) for the contoured data, and a saturation scheme (desaturated to saturated red) for the uncertainty map, but the schemes "did not provide enough contrast... with the saturation range being particularly ineffective" (p. 68). Thus, they ultimately used different hues for the two maps (red for the contour map and blue for the reliability map) and varied both the lightness and saturation within each map (Color Plate 23.1).

To combine the isarithmic and uncertainty maps, two basic approaches are used in R-VIS. One is termed the "overlay method," in which area shading is used for the isarithmic map and weighted isolines (isolines of varying width; see section 18.2.2) are used for the uncertainty map (Figure 23.7). Howard and MacEachren indicated that the overlay method "emphasize[s] the data while allowing analysts to check the reliability in map areas that seem to be particularly good or bad in terms of meeting the EPA dissolved inorganic nitrogen targets" (p. 70). Another approach for combining the isarithmic and uncertainty maps in R-VIS is termed a "merged display" in which each map is displayed with a unique color (or symbol). For example, using the bivariate choropleth method described in section 18.1.2, the isarithmic map can be shown with a lightness scheme and the uncertainty map with a saturation scheme. Alternatively, a merged display can be created by showing the isarithmic map as a lightness/saturation range (or as shades of gray) and the uncertainty map as varying amounts of fill within a raster grid cell. Howard and MacEachren noted that merged displays emphasize the *relationship* between the isarithmic and uncertainty maps.

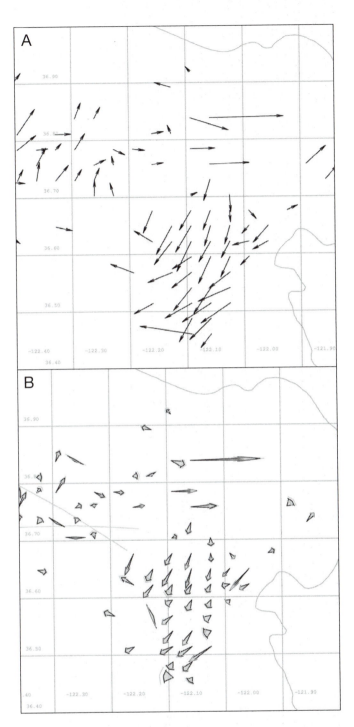

FIGURE 23.6 (A) A traditional approach for depicting wind speed and direction; (B) Wittenbrink et al.'s glyph approach for depicting uncertainty: the width of the arrowhead indicates uncertainty in direction, and extra arrowheads indicate uncertainty in speed. (After Wittenbrink, et al. (1996) "Glyphs for visualizing uncertainty in vector fields" *IEEE Transactions on Visualization and Computer Graphics* 2, p. 272; © 1996 IEEE.)

FIGURE 23.7 The "overlay method" within R-VIS. The isarithmic map is shown in shades of gray, and the uncertainty map is shown as weighted isolines. (Courtesy of David Howard and Alan MacEachren.)

Several exploration approaches also are available within R-VIS to examine uncertainty. One is to alternate isarithmic and uncertainty maps at the same screen location. This alternation can be under the control of the user or it can be animated. With respect to this approach, Howard and MacEachren indicated, "it is with multiple viewing that relationships between data and reliability distributions (if there are any) begin to become apparent" (p. 71). A second exploration approach is to high-

light some portion of the data set by dynamically specifying the range of either isarithmic or uncertainty values to be shown. For example, users might be interested in targeting areas that exceed a particular DIN concentration. A third exploration approach is to animate the isarithmic and uncertainty maps over time. Although Howard and MacEachren did not release R-VIS for general use (Howard, personal communication 1997), you will find it useful to study their description of the software, and thus consider the capability that might be included in software packages that depict uncertainty.

23.4.2 Fisher's Animation of Dot Map Uncertainty

Dot maps create an interesting problem for interpretation because the reader is inclined to think the phenomenon being mapped is actually found where the dots are located. To illustrate, imagine a dot map of bears in the United States, where each dot represents 25 bears. The naive viewer might assume that bears are found wherever dots are located on the map. Where dots are highly concentrated, there would be a high probability of bears being found, but where dots are dispersed, there would be a relatively low probability (high uncertainty) that a bear would be found at the location of a particular dot. Peter Fisher (1996) developed a method to depict this uncertainty via animation.

To understand Fisher's approach, we must first review how dot maps are constructed (see section 17.3 for a more complete discussion). The first step is to delineate regions (polygons) within which the phenomenon being mapped is located; ideally, this step considers ancillary information that can assist in determining appropriate locations for dots. Second, decisions are made on dot size and unit value (the count represented by each dot). Third, dots are placed within each region. Fisher followed these basic steps, with the exception that he did not consider ancillary information because of the slow computer hardware he used. First, an initial dot map was created by placing dots randomly within a minimum **bounding rectangle** surrounding each enumeration unit. A random dot falling within the bounding rectangle and within the enumeration unit was plotted on the map, whereas a dot falling within the bounding rectangle but outside the enumeration unit was not plotted. This initial map was then animated by selecting one dot at random, deleting that dot, and locating a replacement dot within the enumeration unit containing the deleted dot. The movement of dots on the resulting animated map was intended to illustrate the uncertainty associated with dot placement.

Although Fisher's notion of animating dot maps is intriguing, the animations themselves are disconcerting. One problem is that relatively dense enumeration units appear just as uncertain as less dense units, although intuitively it seems that the position of dots within a dense area should appear more certain. A second problem is that a purely random placement of dots can lead to strange concentrations of dots (note the circular arrangement of dots in the western portion of Connecticut in the top map in Figure 23.8). These problems might be ameliorated by a revised dot placement algorithm such as the modification of Lavin's (1986) dot density method described in section 17.3.3. It is also possible that a map comparison approach might be useful in stressing areas of greater uncertainty; for example, compare the two animation frames shown in Figure 23.8 and note the quite different appearance of dot distributions in the state of Vermont (an area of low density), and the relatively greater stability for New York State (an area of high density).

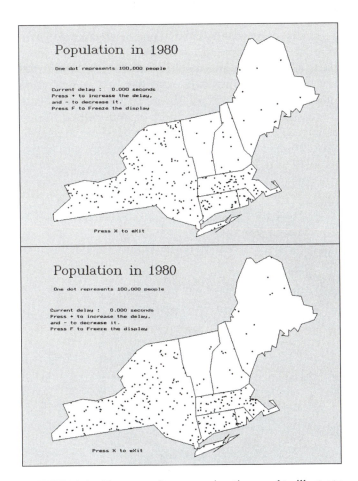

FIGURE 23.8   Two maps from an animation used to illustrate uncertainty. The number of dots within each data collection unit (state in this case) is held constant, but they are moved to different random locations. (After Fisher, P. F. 1996. First published in *Cartography and Geographic Information Systems* 23(4), p. 200. Reprinted with permission from the American Congress on Surveying and Mapping and Peter F. Fisher.)

23.4.3 Hunter and Goodchild's Mapping of DEM Uncertainty

As discussed in Chapter 15, topographic data are commonly available in the form of an equally spaced gridded network known as a digital elevation model (DEM). Such data can be readily contoured using standard mapping software such as Surfer. Gary Hunter and Michael Goodchild (1995) indicated that one important application of the resulting contours is the specification of a particular threshold elevation. For example, in floodplain mapping we might specify the contour above which building is permitted. Although contour maps are useful, users often are unaware of the uncertainty associated with DEMs, which can be measured using the root mean square error (RMSE) as follows:

$$\sqrt{\frac{\sum_{i=1}^{n}(Z_i - Z_i^*)^2}{n}}$$

where Z_i = elevation of the DEM at a sampled point

Z_i^* = true elevation at the sampled point

n = number of sampled points

Hunter and Goodchild (1995) described how the RMSE associated with the DEM can be used to specify a probability that the actual elevation value is above a particular threshold elevation. Their method is based on the assumption that the distribution of error around a particular contoured value is normal (see section 3.3.1) and that the standard deviation of this normal distribution is equal to the RMSE. This concept is illustrated in Figure 23.9. Here Z_t is the threshold elevation value of interest (e.g., the elevation above which building is permitted in a floodplain). Note that if the DEM value for a raster cell (Z_{cell}) is equal to the threshold (as in Figure 23.9A), then the probability of being above the threshold is .5. If, however, the DEM value for a cell is below or above the threshold (as in Figure 23.9B and C), then the area under the normal curve above the threshold must be computed. This can be accomplished using tables of probability published in basic statistics books (e.g., Burt and Barber 1996, 194–198), and can be implemented in GIS software (Hunter and Goodchild indicated the method appeared in the GIS software IDRISI).

Once the basic probabilities have been computed for each cell, contour maps of these probabilities can be constructed. Hunter and Goodchild illustrated several different ways of visualizing these probabilities. A simple black-and-white approach is shown in Figure 23.10. An obvious problem with this approach is that the neutral gray tone used to mask cells outside 2.5 percent to 97.5 percent probability values is identical to one of the gray tones falling within the 2.5 percent to 97.5 percent

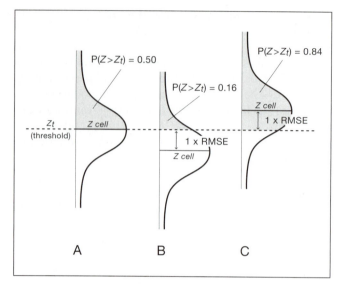

FIGURE 23.9   Determining the probability that the elevation for a raster cell in a DEM (Z_{cell}) is greater than a particular threshold (Z_t). The process involves computing the area under the normal curve above the threshold value. The standard deviation of the normal curve is a function of the root mean square error (RMSE) for the DEM. (After Hunter and Goodchild 1995, 532.)

probability range. As a solution to this problem, Hunter and Goodchild suggested several alternative color schemes (see Plate 1 of their paper). If the 50 percent probability value were treated as a logical breakpoint in the data, then the diverging color schemes proposed by Brewer (see section 13.3.1) also would be a logical solution to portraying such data.

23.4.4 MacEachren et al.'s Depiction of Uncertainty in Health Statistics Data

Epidemiologists often use maps to detect patterns in health-statistics-related data. For example, Deborah Winn and her colleagues (1981) described a situation in which a map of oral cancer death rates among women in the southern United States prompted a study that identified snuff dipping as a major risk factor for this type of cancer. Although maps such as Winn et al.'s can be useful in detecting patterns, they must be treated with caution when the populations for individual enumeration units are small, as the resulting death rates and ratios are unstable. To illustrate, consider two enumeration units with populations of 100 and 10,000: A single death due to cancer in each produces dramatically different rates (1 out of 100 vs. 1 out of 10,000). There are several solutions to this problem. One solution is to aggregate small populations, but this eliminates small-scale

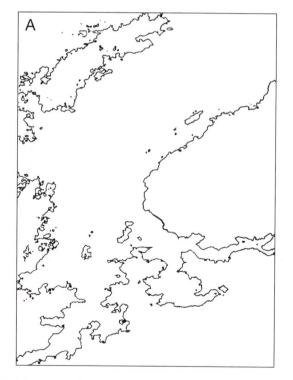

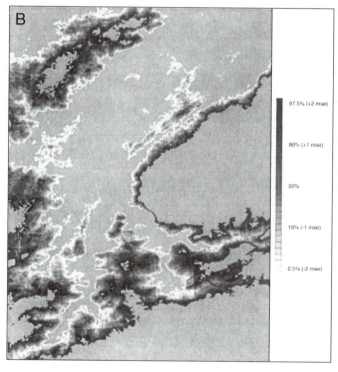

FIGURE 23.10   The contour line associated with a particular threshold (A) and the probabilities associated with exceeding that particular threshold (B). (Reproduced with permission, the American Society for Photogrammetry and Remote Sensing, *Photogrammetric Engineering and Remote Sensing* 61, no. 5, p. 529–537.)

variation that we might be interested in. A second solution is to not symbolize small population areas (leave them blank), but Lewandowsky et al. (1995) found that this interfered with pattern recognition on the resulting map. A third solution is to size each enumeration unit as a function of the population (i.e., use Dorling's cartogram; see section 19.1.1), but as we have noted, Dorling's cartograms can be difficult to interpret.

Given these problems, Alan MacEachren and his colleagues (1998b) promoted a fourth solution, embedding the uncertainty directly in the map. MacEachren and his colleagues utilized two approaches for embedding uncertainty. In one approach, the data and its uncertainty were made *visually separable* so that users could view one component and ignore the other. MacEachren and his colleagues indicated that they experimented with dozens of possibilities here, finally arriving at a set of "parallel lines split down their length, with one side black and the other white" (p. 1,551). This is illustrated in Color Plate 23.2, where we can see that dark lines appear if the underlying tone is light and light lines appear if the underlying tone is dark. In the second method, an attempt was made to make the data and its uncertainty *visually integral* so that users could view the components as a unit. This was accomplished by *shifting the hue* of diverging color schemes and *changing the saturation* of the spectral and sequential color schemes (see section 13.3.1 for a discussion of color schemes). In experiments with potential users, MacEachren and his colleagues found the overlaid parallel line approach most successful for depicting uncertainty, as the uncertainty information could "be disregarded when necessary in order to notice potentially important data clusters" and the method prompted "map readers to recognize those maps, or map sections, in which data should be used with caution" (p. 1,559). An advantage of visually integral colors, however, was that they prevented readers from seeing clusters of enumeration units that were uncertain.

23.4.5 The FLIERS Project

The FLIERS Project (Fuzzy Land Information from Environmental Remote Sensing) was an attempt by researchers at several European universities to develop software for depicting the uncertainty in remotely sensed images (the FLIERS home page can be found at *http://www.geog.le.ac.uk/fliers/index.html*). Although the FLIERS group was not the first to develop such software (van der Wel et al. 1998 provided an early flexible package), their software provides a wide range of visualization options, is available for free at the site just mentioned, and has been evaluated in user studies.

Uncertainty in remotely sensed imagery arises from several sources including subpixel mixing, spatial misregistration, and biases of particular sensors (Bastin et al. 2002). *Subpixel mixing* refers to the fact that individual square pixels in a remotely sensed image are often composed of a mixture of various land use/land cover categories; for example, a 30-meter pixel might include portions of a house, a grassy area, and a pond. Obviously, this produces uncertainty if we are trying to classify the pixel in one of three categories, say buildings, grassland, and water. From a data processing standpoint, one solution to this problem is to develop a probabilistic measure for each pixel belonging to particular land use categories; thus, the following probabilities might be computed for a pixel: building (40 percent), grassland (10 percent), and water (50 percent). One approach for determining such probabilities is known as *fuzzy classification* (Foody 1996).

Spatial misregistration refers to the notion that a remotely sensed image might not be perfectly registered with other images and maps of interest. For instance, we might wish to check our remotely sensed classification against a high-resolution photograph. If these two images are not registered precisely, then uncertainty will arise in our interpretation. *Biases of particular sensors* can be illustrated by the fact that the pixel-level information stored in a database does not actually match what was recorded by the sensor on the Earth's surface. In the case of Landsat TM, a traditionally popular sensor, a stronger signal is recorded in the center of the pixel, with surrounding pixels also providing input to the signal associated with the central pixel.

Although the FLIERS Project developed approaches for visualizing these three sources of uncertainty, for simplicity we discuss just the approaches used to examine subpixel mixing. In considering these approaches, we should first recognize how conventionally classified remotely sensed images are symbolized—normally, distinct colors are used for each land use/land cover category (as in Color Plate 25.3). With this in mind, three approaches were used to examine the uncertainty arising from sub pixel mixing: a static grayscale image, serial animation of a grayscale image and random animation. The *static grayscale image* was shown separately from the usual colored classified image and simply consisted of gray tones, with dark areas representing more uncertain information. *Serial animation of a grayscale image* involved animating a series of grayscale images, where each image portrayed a progressively larger portion of the probability values for a particular land use/land cover. For instance, we might begin with a 0–0.1 probability range for a particular land use/land cover and progress to a 0–1.0 range: In the beginning only a few pixels would appear, and in the end we would see uncertainty for all pixels. *Random animation* integrated the visualization of uncertainty with the colored raw land use/land cover map by randomly selecting pixels and then assigning a land cover as a random function of probability values (i.e., a pixel with high reliability would be more likely to remain constant, whereas a pixel with high uncertainty would continually change).

In Chapter 20, we raised the question of whether animation actually works. A study with users of FLIERS by Blenkinsop et al. (2000) supported some of our concern about the usefulness of animation. Overall, Blenkinsop et al. found that users tended to prefer the static grayscale image approach for depicting uncertainty. They noted, however, that animations were helpful "in providing a very effective first impression of uncertainty" and that "animations may, therefore, have their greatest usefulness at an early stage in the exploration of a data set" (p. 11). They also stressed the exploratory capability of FLIERS, noting that no single method could answer all desired questions and that the performance of users might improve with training and experience in issues of visualization and uncertainty.

23.4.6 Pang et al.'s Work with Uncertainty

Alex Pang and his colleagues (1997) have worked extensively with uncertainty, in both geographic and nongeographic contexts. We have already mentioned their work with glyphs for vector-based data (Wittenbrink et al. 1996). Another major area of research has been the depiction of uncertainty in interpolation on isarithmic maps (Lodha et al. 1996a). We touched on the notion of uncertainty in interpolation earlier in this chapter (see section 23.3) and when we considered Lavin et al.'s use of animation to depict the effect of the exponent in inverse-distance interpolation in Chapter 20. Rather than look at one parameter of a particular interpolation approach (as Lavin et al. did), Pang et al. focused on comparing one interpolation approach with another (e.g., we might compare inverse distance with kriging). Unfortunately, you would find Pang et al.'s work difficult to follow because we have not covered some of the interpolation methods they examined. Still, it is useful to consider some of their approaches for visualizing uncertainty, including displacement mapping, spot mapping, and glyphs. *Displacement mapping* involves randomly perturbing one of the interpolated surfaces in proportion to the uncertainty (Figure 23.11), whereas in *spot mapping*, regions of high relative difference between the surfaces appear spotted. Figure 23.12 illustrates one of several glyphs they developed—in this case a *volume-filling glyph* in which the volume between two surfaces is filled by spheres with radii that are proportional to the difference between the two surfaces. In Chapter 25, we consider efforts that they have undertaken to utilize sound to depict the uncertainty of interpolation.

Another area that Pang and his colleagues have worked on is the visualization of map projection distortion.

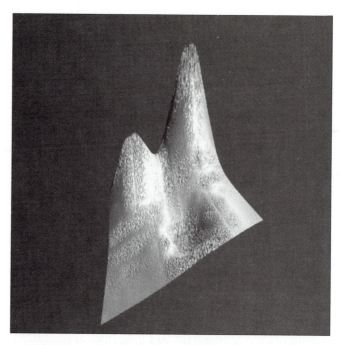

FIGURE 23.11   Displacement mapping: in comparing two interpolated surfaces, one surface is randomly perturbed in proportion to the uncertainty. (Courtesy of Suresh K. Lodha.)

FIGURE 23.12   Volume-filling glyph: The volume between two interpolated surfaces is filled by spheres with radii that are proportional to the difference between the two surfaces. (Courtesy of Suresh K. Lodha.)

We can think of different map projections as a form of uncertainty, as the nature of geographic representation varies as a function of the map projection. In this context, Brainerd and Pang (2001) developed the Interactive Map

Projections (IMP) software (*http://www.cse.ucsc.edu/research/slvg/map.html*). A key element of IMP is the *floating ring tool*, which is displayed on both the globe and a desired projection (a Mercator in the case of Figure 23.13). On the globe, the floating ring is displayed as a fixed-size circle, indicating no distortion, whereas on the Mercator projection, the ring reflects the nature of the distortion. For instance, in Figure 23.13B we see that the floating ring is relatively large near the pole, reflecting the greater areal distortion in the Mercator projection as one moves away from the Equator. Also note that the floating ring is not perfectly circular on the projection and that its center is offset, indicating that there is shape distortion for this region—this visualization is in contrast to Tissot's indicatrix, which reveals a perfect circle for the Mercator projection because there is no angular distortion at a point.

23.4.7 Cliburn et al.'s Visualization of Uncertainty in Water Balance Models

Daniel Cliburn and his colleagues (2002) have utilized various visualization techniques to examine the uncertainty of water balance models for terrestrial regions of the world. The basic output of a water balance model is an indication of whether each geographic location has a water surplus or deficit, along with an associated magnitude. Uncertainty arises because modelers must determine appropriate input parameters for the model (e.g., which historical temperature and precipitation data should be used) and which global circulation models (GCMs) should be used to predict the future water balance. Although GCMs do generally predict global warming (and thus less water available), there is spatial variation within models, with cooling occurring at some locations (and thus more water available).

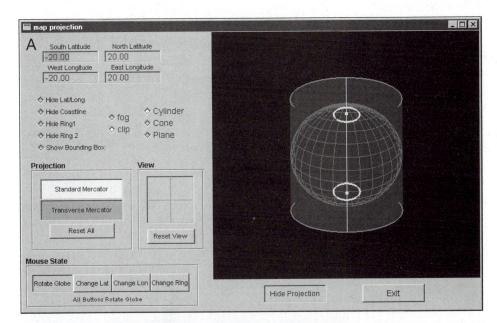

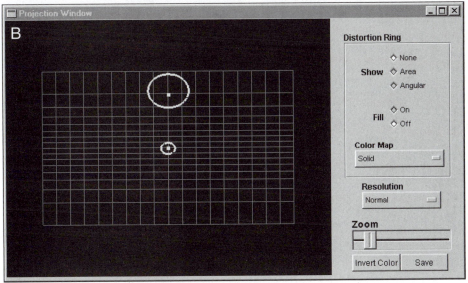

FIGURE 23.13 An image from the Interactive Map Projection (IMP) software; note how the Mercator projection distorts the size and shape of the floating ring tool in the Northern latitudes. (Courtesy of Jeffrey Brainerd and Alex Pang.)

Following Gershon's notion of intrinsic and extrinsic visual variables, Cliburn and his colleagues proposed a range of intrinsic and extrinsic methods for depicting uncertainty, with an emphasis on 3-D (technically $2\frac{1}{2}$-D) surfaces. Color Plate 23.3 is illustrative of some intrinsic approaches that Cliburn et al. utilized. In Color Plate 23.3A, the height of the 3-D surface provides an indication of water availability based on historical temperature and precipitation data. Areas of the surface above 0 mm (note the scale in the lower left) represent a water surplus, and areas below 0 mm represent a deficit. Obviously, it is not easy to see both surpluses and deficits in this static image; Cliburn et al. handled this problem by permitting users to manipulate the image (e.g., rotate and zoom in) via a wall-size display (Color Plate 24.2). Note that Color Plate 23.3A is colored using an orange–purple diverging scheme, indicating how the average of GCM models compares with the historical data; in this case, orange and purple indicate less and more water, respectively, in comparison to historical data. Recall that Brewer found the orange–purple scheme particularly effective from the standpoint of color naming, and that it avoided problems of color vision impairment and simultaneous contrast (Table 13.2). Color Plate 23.3B depicts the uncertainty in the GCM predictions using the method of transparency (here the more transparent the image, the greater the variation among the GCM models and hence the greater the uncertainty). Note how this affects our view of orange and purple peaks in the northeast part of Color Plate 23.3A—apparently, there is disagreement among the models on what will happen at these locations.

Color Plate 23.4 is an example of an extrinsic method utilized by Cliburn et al. In this case, orange and purple bars represent the range of GCM predictions at a particular location. For example, if only an orange bar appears, then at least one GCM predicted less water available (none predicted more water available), whereas if both orange and purple bars appear, then at least one GCM predicted less water and another predicted more water. Small pyramid-like symbols at the end of bars denote which GCMs were associated with the extreme low or high point on a bar (note the legend in the lower left). Obviously, the complexity of this image requires that users be able to manipulate it, an option that Cliburn et al. did provide.

23.5 STUDIES OF THE EFFECTIVENESS OF METHODS FOR VISUALIZING UNCERTAINTY

In this section, we briefly consider several studies that have examined the effectiveness of methods for visualizing uncertainty. Although several of the preceding applications also involved tests of effectiveness, these tests were generally not the major emphasis.

23.5.1 Evans's Study

Beverley Evans (1997) was one of the first to explore the effectiveness of displays of uncertainty from the standpoint of the map user. In an experiment designed to examine various methods for depicting uncertainty, Evans utilized four maps: a land use/land cover map, a map depicting areas of the land use/land cover map that had a reliability of 95 percent or more, a static bivariate map that combined the basic land use/land cover map with the 95 percent reliable map, and a dynamic map in which the land use/land cover map was "flickered" with the bivariate map. Evans asked three basic questions:

1. Will map users access and comprehend reliability information in graphic form when it is provided?

2. Which of four versions of graphically depicted reliability information will the map user prefer?: flickering . . . ,[the] static composite . . . , a map depicting only "95% reliable" information, or an interactive toggling between the mapped data and highly reliable information?

3. Is graphically depicted reliability information equally useful . . . for novices and experts? (p. 410)

Regarding the first question, users did access and comprehend the reliability information; moreover, they claimed to be willing to access it in similar situations in the future. Concerning the various methods for depicting reliability, the flickering and static methods were deemed about equally useful, although some found flickering annoying. The 95 percent reliable map was deemed difficult to use (as it was missing information), although some users found it useful to toggle between it and the land use/land cover map, thus creating their own sort of flickering. Finally, Evans found little difference between novices and experts, indicating that both groups were able to make use of reliability information.*

23.5.2 Leitner and Buttenfield's Study

Michael Leitner and Barbara Buttenfield (2000) examined how the decision making process is affected when uncertainty information is added to a map. Their decision making process involved locating first a park and then an airport, with the key information being how wetlands were represented. On one map all wetlands

---

* Our summary of Evans's study is based on a summary of her work by Cliburn (2001).

were grouped into one class and no uncertainty information was included. On a second map, the wetlands were split into three classes of information, but again no uncertainty information was included. On six additional maps, wetlands were again shown, with greater uncertainty depicted by the following visual variables: lighter value, darker value, coarser texture, finer texture, less saturated color, and more saturated color.

Leitner and Buttenfield analyzed the results both in terms of number of correct sitings and speed of response for the siting. In terms of the number of correct sitings, the addition of uncertainty information significantly increased the correct number, but only if a lighter value or finer texture was chosen to display greater uncertainty. Leitner and Buttenfield deemed this result surprising, as a "darker value is usually perceived as being more prominent" (p. 13). They argued, however, that their finding was likely a function of displaying the maps on a CRT rather than on paper.

From the standpoint of speed of response, Leitner and Buttenfield found that including uncertainty information resulted in either the same or a faster response time than when only basic wetland information was shown. This might seem surprising given the greater information content of the display, but Leitner and Buttenfield stated "It would seem that map certainty is understood as clarification rather than adding complexity to a map display" (p. 14). Regarding speed of response, they also found that a more saturated color (for greater uncertainty) produced the fastest response. This result was rather surprising given the findings for correct responses noted earlier.

23.5.3 Edwards and Nelson's Study

Laura Edwards and Elisabeth Nelson (2001) examined methods for depicting uncertainty on proportional symbol maps. In their test maps, the basic phenomenon being mapped ("Total Reports of Disease D in Costa Rica" in Figure 23.14) was depicted using circle size. Uncertainty was then illustrated using one of four approaches: a legend statement (Figure 23.14A), a reliability diagram (Figure 23.14B), focus (Figure 23.14C), and value (Figure 23.14D). Edwards and Nelson anticipated that the legend and reliability approaches would be most difficult to use because of their separation from the mapped phenomenon of interest. Moreover, they felt that the legend statement would be particularly difficult to process because of its nongraphic nature. Regarding focus and value, Edwards and Nelson felt that value would be more effective because of its familiarity and graphic "punch." They also noted the subtlety of the focus approach from a figure-ground perspective (p. 23).

Edwards and Nelson performed two basic experiments with maps similar to those shown in Figure 23.14. In the first experiment, users performed a *rapid pattern detection task* by identifying areas in which data values were perceived to be highest and areas in which data were perceived to be most certain. In the second experiment, users answered multiple-choice questions about the variation in the mapped phenomenon and its uncertainty, and indicated their confidence in the answers. The result from the experiments that stood out most clearly was the poor performance of the legend statement. A surprising result, however, was that the focus method outperformed the value method for depicting uncertainty. Edwards and Nelson explained this result as follows: "More certain data gets the graphic punch, at the expense of less certain data, so much so that perhaps it becomes a more effective means of displaying the two data sets in tandem" (p. 34).

SUMMARY

In this chapter, we have considered the notion of **uncertainty** and various approaches for visualizing this uncertainty. Uncertainty can arise from a variety of sources, including the raw data on which a map is based, the manner in which these data are processed, and the manner in which a visualization is created. The creation of an isarithmic map depicting precipitation recorded at weather stations provides a good illustration of various sources of uncertainty. Uncertainty in the raw data is illustrated by the possibility that incorrect values can be recorded at individual weather stations; uncertainty due to processing is reflected in the variety of interpolation methods that can be utilized to estimate values between weather stations (the control points); and uncertainty of visualization is a function of whether a classed or unclassed isarithmic map is created.

There are three general methods for depicting uncertainty: separate maps can be created for an attribute of interest and its associated uncertainty (maps can be compared); the attribute and its uncertainty can be displayed on the same map (maps can be combined); and data exploration tools can be utilized. Generally, the "maps combined" approach has been most commonly used, although a number of data exploration tools have been developed.

Visual variables for depicting uncertainty can be split into intrinsic and extrinsic ones. **Intrinsic visual variables** are intrinsic to the display; for example, uncertainty might be depicted by varying the transparency of colors composing an isarithmic map. **Extrinsic visual variables** involve adding objects to the display, such as the glyphs that Wittenbrink et al. (1996) and Pang et al. (1997) used for depicting the uncertainty of wind speed and direction.

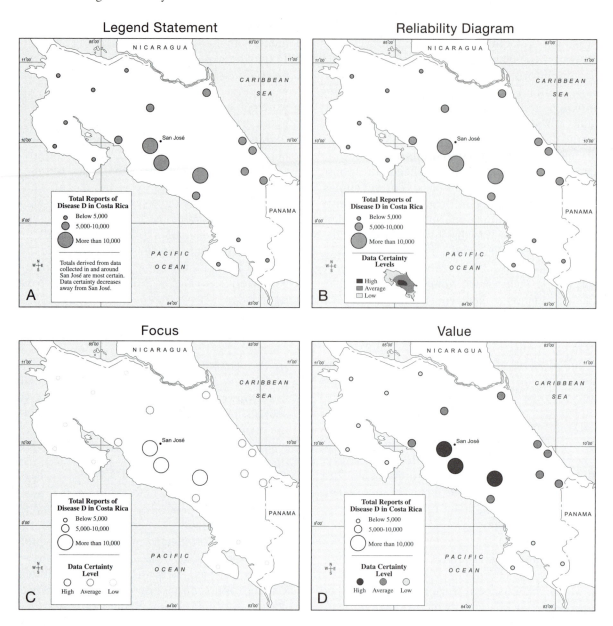

FIGURE 23.14   Approaches that Edwards and Nelson (2001) used to depict uncertainty on proportional symbol maps. (Courtesy of North American Cartographic Information Society.)

We covered a broad range of applications involving the visualization of uncertainty, including environmental data at point locations (DIN), dot maps, contours associated with DEMs, health statistics data, remotely sensed data, Pang and his colleagues' work (e.g., with map projections), and water balance models. Although the applications that we have discussed illustrate that many cartographers have attempted to visualize uncertainty, our impression is that the notion of uncertainty is often disregarded or forgotten. Software for the analysis and display of spatial data generally does not provide routines expressly intended for visualizing uncertainty; rather it is up to the user of the software to recognize uncertainty and to explicitly map this uncertainty.

It also must be recognized that relatively little is known about how users respond to the visualization of uncertainty. We have touched on some studies of visualizing uncertainty here, but these studies are all relatively recent and do not span the full range of uncertainty portrayal. Certainly, there is more work that needs to be done in this area.

FURTHER READING

Beard, M. K., Buttenfield, B. P., and Clapham, S. B. (1991) *NCGIA Research Initiative 7: Visualization of Spatial Data Quality. Technical Paper 91-26.* Santa Barbara, CA: National Center for Geographic Information and Analysis. Available at *http://www.ncgia.ucsb.edu/Publications/Tech_Reports/91/91-26.pdf.*

An early work that discusses many of the issues involved in visualizing uncertainty.

Cartographica 30, nos. 2/3, 1993.

Includes several articles dealing with the visualization of uncertainty.

Clarke, K. C., Teague, P. D., and Smith, H. G. (1999) "Virtual depth-based representation of cartographic uncertainty." *Proceedings of the International Symposium on Spatial Data Quality,* Hong Kong, pp. 253–259.

Discusses the possibility of using apparent visual depth to represent uncertainty via a head-mounted display (HMD).

Comenetz, J. (1999) *Cartographic Visualization of the Quality of Demographic Data.* Unpublished PhD dissertation, University of Minnesota, Minneapolis, MN.

Provides a thorough treatment of the errors that lead to uncertainty when visualizing demographic data.

Ehlschlaeger, C. R., Shortridge, A. M., and Goodchild, M. F. (1997) "Visualizing spatial data uncertainty using animation." *Computers & Geosciences* 23, no. 4:387–395.

Uses animation to visualize the uncertainty in digital elevation data.

Fisher, P. F. (1996) "Animation of reliability in computer-generated dots maps and elevation models." *Cartography and Geographic Information Systems* 23, no. 4:196–205.

An example of some of the work that Peter Fisher has done to visualize uncertainty. Also see Fisher (1993; 1994a; 1994b).

Goodchild, M., Chih-chang, L., and Leung, Y. (1994) "Visualizing fuzzy maps." In *Visualization in Geographical Information Systems,* ed. by H. M. Hearnshaw and D. J. Unwin, pp. 158–167. Chichester, England: Wiley.

Describes efforts to visualize the uncertainty in remotely sensed images.

Howard, D., and MacEachren, A. M. (1996) "Interface design for geographic visualization: Tools for representing reliability." *Cartography and Geographic Information Systems* 23, no. 2:59–77.

In addition to covering R-VIS (see section 23.4.1), this paper discusses general considerations in designing interactive software for visualizing spatial data.

Jervis, J. (2002) "Visualizing uncertainty in Earth Observing System satellite data." *Gridpoints* 3, no. 2:6–9.

Summarizes recent research at the NASA Advanced Supercomputing Division to visualize the uncertainty in remotely sensed data.

MacEachren, A. M. (1992) "Visualizing uncertain information." *Cartographic Perspectives* no. 13:10–19.

Covers basic principles for visualizing uncertainty; for more recent work, see MacEachren (1995).

Pang, A., Wittenbrink, C., and Lodha, S. (1997) "Approaches to uncertainty visualization." *The Visual Computer* 13, no. 8:370–380.

Covers a broad range of methods for visualizing uncertainty for both geographic and nongeographic data. For a more comprehensive overview and more recent work, see *http://www.cse.ucsc.edu/research/slvg/unvis.html.*

Wittenbrink, C. M., Pang, A. T., and Lodha, S. K. (1996) "Glyphs for visualizing uncertainty in vector fields." *IEEE Transactions on Visualization and Computer Graphics* 2, no. 3:266–279.

Introduces a range of glyphs for visualizing the uncertainty in vector-based data.

24

Virtual and Mixed Environments

OVERVIEW

From reading daily newspapers, watching television, or playing computer games, you are probably familiar with **virtual reality (VR)**—*it likely brings forth the image of donning a head-mounted display (HMD) and experiencing a simulation of reality. Although this notion is correct, in this chapter we will see that other terminology is often used and that a broader set of technologies are now possible. In section 24.1, we define terms that are closely associated with VR. A* **virtual environment (VE)** *is a 3-D computer-based simulation of a real or imagined environment that users are able to navigate through and interact with; for example, you might "fly through" the Grand Canyon while sitting at your computer. Such VEs of interest to geographers and other geoscientists can be termed geospatial virtual environments (GeoVEs). A* **mixed environment (ME)** *(or* **mixed reality [MR]**) *combines a real-world experience with a virtual representation, such as overlaying the boundaries of "old growth" forests on your view of a mountainous environment. Most commonly, MEs are referred to as* **augmented reality** *because the real world is dominant (virtual representations are used to augment reality).*

Section 24.2 describes some of the technologies for creating VEs, including **desktop**, **wall-size**, **head-mounted**, **drafting table–format**, *and* **room-format** *displays. Because these technologies are likely to change over time, you should check the home page for this book for links to the most recent developments. Section 24.3 discusses four "I" factors that are relevant in designing GeoVEs:* **immersion**, **interactivity**, **information intensity**, *and* **intelligence of objects**. *Immersion is closely tied with* **presence**, *which deals with your sense of being part of a VE.*

Various examples of GeoVE applications are covered in section 24.4, including virtual cities, the Virtual Field Course, the Digital Earth project, water-resource applications, and

virtual forests. Because GeoVEs have been developed recently, we also cover some related applications in Chapter 25. Given the novelty of GeoVEs, relatively little is known about how they should be designed and used. Thus, section 24.5 considers a number of research issues related to GeoVEs.

Section 24.6 covers ongoing developments in MEs of interest to geographers. We indicate "ongoing" because these are recent developments that have the potential to fundamentally transform the way geographers do fieldwork and interpret maps. Although geographers are excited about the potential offered by virtual and mixed environments, we should recognize that there are health, safety, and social issues associated with their use—these issues are considered in section 24.7.

24.1 DEFINING VIRTUAL AND MIXED ENVIRONMENTS

Early notions of **virtual reality (VR)** promoted the idea of simulating the tangible real world—the everyday world that we see, hear, touch, smell, taste, and move through. One of the first to recognize the potential of VR for geography was Ian Bishop (1994). Although Bishop did not use the term virtual reality and focused largely on vision, it was clear this was what he had in mind:

> The non-scientific audience . . . wants abstraction minimized, information content maximized . . . with the whole package digestible and non-threatening. This suggests the use of a visual realism approach [that] shows information consumers what will or might happen under a variety of conditions and permits them to explore the alternative environments using their natural sensory perceptions. (p. 61)

Color Plate 24.1 is an example of some of Bishop's early work—a simulation of a proposed lake created as a result of filling an existing open-cut coal mine with water.

Although it makes sense to simulate the tangible real world, there are instances in which we might need to display something intangible. For instance, imagine depicting the concentration of ozone in the atmosphere both vertically and horizontally at some particular point in time. Because we cannot see ozone, we must create an abstract representation of it. To achieve this, we might immerse individuals in a computer-simulated environment in which they travel through a virtual atmosphere, seeing the concentration of ozone as they move. The concentration of ozone could be depicted by the density of small balls having a uniform size and color, with a high density of balls indicating a high ozone concentration. Here the term virtual *reality* seems inappropriate, as humans cannot see ozone; as a result, researchers have chosen to use the term *virtual environment* to encompass both tangible and intangible simulations.

As a more abstract example, consider Figure 24.1, which was developed by Andrew Wood and his colleagues (1995) to portray the spatial structure of a portion of the Web. Here the structure is not based on geographic location, but rather on the relation between topics presented via the Web, with more similar topics shown together. Each sphere represents a Web page, with the links between spheres representing the links between Web pages, and the size of a sphere a function of the number of linking Web pages. The idea is that we could utilize the resulting 3-D space to avoid becoming "lost in hyperspace" (Dodge and Kitchin 2001b, 126). In this case, the ultimate reality is a set of zeros and ones in the computers used to depict this information, and so the term virtual *environment* is again appropriate.

Although we choose to use the term virtual environment (VE) rather than VR, it should be recognized that researchers, even within geography, do not agree on the definition and use of the terms VR and VE. For instance, the recent book *Virtual Reality in Geography*, edited by Peter Fisher and David Unwin, largely utilizes the term VR as opposed to VE. Others, such as Alan MacEachren and his colleagues (1999b), have utilized the term VE, defining the subset of interest to geographers as *geospatial* virtual environments (GeoVEs).

For our purposes, we define a **virtual environment** as a 3-D computer-based simulation of a real or imagined environment that users are able to navigate through and interact with. Normally, VEs are experienced visually, although ideally it should be possible to utilize the full range of senses, including sound, touch (**haptics**), smell, taste, and body movement. Using this definition, 3-D film animations in which the user does not control the point of view (e.g., the classic *L.A. The Movie* described in Chapter 20) would not be considered VEs.

Although early efforts focused on the potential of immersing oneself in a VE, more recent work has suggested that a combination of reality and VEs might also be useful. This notion is depicted in Figure 24.2, where we see a real world to VE continuum. The "real world" end of the continuum represents what we perceive in the real world, whereas the VE end represents a simulation of an environment. In between, we see a **mixed environment (ME)** (or **mixed reality [MR]**), which represents a combination of the real world and a virtual representation. The notion of an ME can be further subdivided into **augmented reality** and **augmented virtuality**, depending on whether the real world or virtual representation is emphasized.

In this chapter, we look at two examples of MEs of interest to geographers. One is Project Battuta (*http://www.statlab.iastate.edu/dg/*), which is developing approaches for collecting and using geospatial data in the field. Utilizing *wearable computers*, users are able to see

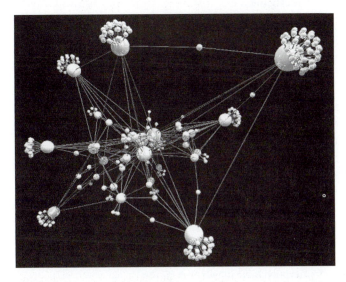

FIGURE 24.1 A depiction of the spatial structure of a portion of the Web. (Courtesy of Bob Hendley.)

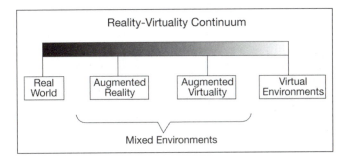

FIGURE 24.2 The real world to virtual environment continuum. (After Drascic and Milgram, 1996. In their illustration, Drascic and Milgram used the term "virtual reality" rather than "virtual environment" and "mixed reality" rather than "mixed environment.")

computer-simulated information overlaid on top of their real-world view. For instance, an urban geographer could go to the upper floor of a tall building, look out at the city, and see a map of crime incidence overlaid on the city. The second example of an ME is the work of Nick Hedley (2001; 2003), who displays a VE on a large card held in the user's hands. Users can rotate the card, just as they would rotate an object in the real world, thus providing a potentially more intuitive approach for examining 3-D landscapes than traditional mouse-based interfaces. Although Hedley and those involved with Project Battuta both describe their work as a form of augmented reality, it can be argued that Hedley's work places greater emphasis on interpreting the VE. We consider both of these examples in more detail in section 24.6.

24.2 TECHNOLOGIES FOR CREATING VIRTUAL ENVIRONMENTS

Technologies for creating VEs are important because they impact how we experience these environments. Nick Hedley (2001; 2003) argued that technology for creating VEs can be broken into three parts: the display, the hardware controls, and the GUI. For instance, with today's desktop computers, we might use a CRT display, a mouse and keyboard as controls, and Microsoft Windows® as the interface. Normally, in describing the different technologies for visualization purposes, emphasis is placed on the display, and so we do that here. We should bear in mind, however, that affordances offered by a particular display might be a function of the controls and graphical interface.

VEs can be created using five basic forms of display technologies: desktop, wall-size, head-mounted, room-format, and drafting table–format. The **desktop display** (e.g., CRT and laptop) is by far the most common approach for depicting VEs. Popular software packages for desktop displays, such as ArcView's 3D Analyst and ERDAS Imagine's Virtual GIS, include interactive 3-D mapping options, and hundreds of packages have been developed for creating 3-D displays, many of which provide an interactive capability (see *http://www.tec.army.mil/TD/tvd/survey/survey_toc.html*). Particularly important has been the development of the Virtual Reality Modeling Language (VRML), which enables VEs to be disseminated over the Web. A specialized version of VRML intended for the representation of geographical data is also now available (see *http://www.geovrml.org/*).

A **wall-size display** covers a large portion of a wall by tiling images created by multiple projectors. For instance, one at the University of Kansas measures 25 × 6 feet, covers 120° of the visual field, and provides 5,760 × 1,200 pixel resolution (Color Plate 24.2). As with desktop displays, wall-size displays often are viewed without any special

devices (e.g., the eyeglasses mentioned later for the CAVE).* Although a traditional keyboard and mouse can be used to interact with a wall-size display, other control devices are either being used or in the process of being developed. For instance, Guimbretière et al. (2001) and Davis and Chen (2002) described pen-based and laser pointer approaches, respectively. Such novel control devices are essential given the varied distances of users from the display and the need to handle simultaneous users. An overview of technical issues associated with wall-size displays (e.g., combining images from multiple projectors) can be found in the July/August 2000 issue of *IEEE Computer Graphics and Applications*.

The term **head-mounted display (HMD)** derives from the means used to create such a display—a helmet-like device is placed on the user's head that shields the real-world view and provides images of the VE to each eye (Figure 24.3A). Sophisticated HMDs provide separate images for the left and right eye, thus enabling a stereoscopic view, just as we have in the real world. The HMD includes a head-tracking device so that the view of the virtual representation is a function of the position of the user's head. Because the real world is not visible when using HMDs, nontraditional control devices such as a *dataglove* are utilized. HMDs have sometimes been criticized because they are a burden for users to wear (due to their weight), their resolution is low (a typical high resolution has been 640 × 480 addressable pixels), the field of view is narrow (the view is rectangular as in a traditional computer screen), and one cannot see other users (as in a collaborative environment).

Room-format and **drafting table–format displays** were developed in response to limitations of HMDs. Here we consider two well-known examples, the CAVE and the ImmersaDesk, which were developed by researchers at the Electronic Visualization Laboratory at the University of Illinois at Chicago. The CAVE provides a room-size view of a VE by projecting images onto three walls and the floor (Figure 24.3B).† As users stand or move in the CAVE, a head tracker on one user governs the view that all users see through stereo glasses; the tracker and glasses are much less obtrusive than traditional HMDs. Interaction with the VE is achieved through a 3-D mouse, which consists of a joystick for navigation and programmable buttons for interaction. One advantage of the CAVE is that users can see each other—in this sense the CAVE can be considered an ME, but the emphasis is clearly on viewing the VE and so it makes sense to specify it as VE display technology. The ImmersaDesk was developed as an inexpensive substitute for the CAVE. With the ImmersaDesk, the VE is projected

---

* An exception is the Infinity Wall (Czernuszenko et al. 1997).

† The software will permit projection onto six surfaces, but normally only four are used.

onto a single screen that is tilted at a 45° angle (the logic is that looking down in the CAVE is important—the tilt helps simulate this; Figure 24.3C). Again, a head tracker on one user governs the view that other users see, and a specialized 3-D mouse can be used to interact with the simulated environment.

With the exception of the desktop display, high-quality versions of these displays currently are beyond the reach of the typical geographer. Not only does the hardware for the display tend to be expensive, but high-end computers (e.g., Silicon Graphics) generally are required to power the display and specialized software is necessary. Although this is true today, keep in mind that technological advances are continually being made and the cost of computer-based technology is dropping, so we anticipate that many of these displays will be more generally available within a few years.

24.3 THE FOUR "I" FACTORS OF VIRTUAL ENVIRONMENTS

Based on the work of Michael Heim (1998), Alan MacEachren and his colleagues (1999a) proposed four "I" factors important in creating geospatial VEs: immersion, interactivity, information intensity, and intelligence of objects.

24.3.1 Immersion

Immersion is defined as "a psychological state characterized by perceiving oneself to be enveloped by, included in, and interacting with an environment that provides a continuous stream of stimuli and experiences" (Witmer and Singer 1998, 225). The notion of immersion is closely tied in with that of **presence**, which is defined "as the subjective experience of being in one place or environment when one is physically in another" (Witmer and Singer 1998, 225). Because a high level of immersion leads to a sense of presence, those discussing VEs often speak of the level of immersion, and we do the same. Keep in mind, though, that the notion of presence is important—one indication of its importance is that a journal dealing with VEs is named *Presence*.

Each of the displays we have discussed provides a different sense of immersion. Standard desktop displays provide the lowest sense of immersion, as both the real world outside the computer screen and the VE depicted on the screen are readily visible, and images do not look truly 3-D (because they are not presented stereoscopically).* For this reason, desktop displays are sometimes termed *nonimmersive*, but we prefer to think of them as

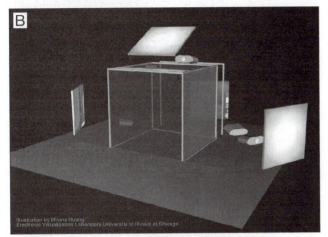
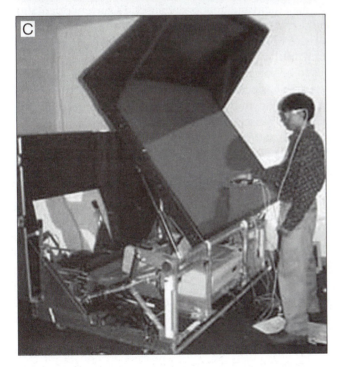

FIGURE 24.3 Examples of various displays for creating VEs: (A) head-mounted, (B) room-format (CAVE), (C) drafting table–format (ImmersaDesk). (Figure A courtesy of William Winn; Figure B Image of the CAVE® virtual reality theater courtesy of Milana Huang, Electronic Visualization Laboratory, University of Illinois at Chicago; Figure C courtesy of Electronic Visualization Laboratory, University of Illinois at Chicago.)

---
* Here we assume that no special technology has been used to enhance the appearance of 3-D images. Three-dimensional displays can be enhanced through color anaglyphic representations (red/cyan), polarizer displays, or shutter glass technology.

semi-immersive, as viewers often feel they are part of the VE. For instance, in using the Virtual Field Course software to be described subsequently, Jason Dykes (2002, 71–72) argued that the desktop display provides a "strong sense of immersion, [in which] users frequently move with the surface, duck whilst navigating around obstacles, and express their desire to 'go' somewhere, or the fact that they are 'lost'."

A wall-size display provides a greater sense of immersion than a desktop display because a greater portion of the visual field is covered, especially when one stands or sits close to the screen. This still must be termed a semi-immersive system, though, as images do not look truly 3-D (unless multiple projectors are used to create an image and specialized viewing glasses are used). In contrast, HMDs, the ImmersaDesk, and the CAVE all provide a fuller sense of immersion, as the real world can be hidden from view, images are presented stereoscopically, and senses other than vision (e.g., sound) are often used. The CAVE provides the greatest sense of immersion because a larger portion of the visual field is covered by the VE (the walls and floor are covered by the VE).

One interesting perspective on immersion and the various forms of display was offered by Edward Verbree and his colleagues (1999), who considered immersion in the context of the landscape planning process in the Netherlands. Verbree and his colleagues identified three stages in the landscape planning process, and an associated set of relevant maps: (1) orientation—2-D maps with a bird's-eye view; (2) design and modeling—3-D maps with abstract symbols; and (3) presentation and final decision making—3-D maps with photorealistic quality. Corresponding to these three stages and maps, they proposed three levels of immersion: a relatively nonimmersive traditional desktop display, an immersive display using a drafting table format, and the more fully immersive CAVE. They noted, however, that for convenience all mapping techniques should be capable of being viewed on any of the hardware. Clearly, research is needed to determine in what instances each type of display will be most useful for the broad range of possible geospatial tasks.

How important is a sense of immersion? One argument is that systems providing greater immersion will be more effective because users will utilize real-world cognitive processing strategies (Buziek and Döllner 1999). A counter argument is that cartography traditionally has been successful because it is abstract—we need a separation between the real world and its representation to make sense of the real world (Slocum et al. 2001). William Cartwright (1997) argued that the "reality" suggested by highly immersive VEs could be misleading, as it might be confused with "real reality." He stated, "using information derived in such a manner may lead the user to believe that Surreal Reality is in fact truly real. [As a result] knowledge-building and decision-making could

be made in an artificial world completely devoid of nature's checks and balances" (p. 451).

The display systems that we have described rely primarily on vision, and to a limited extent on sound. It is important to recognize that a sense of immersion is a function of more than vision and sound—it involves all of the senses, including touch (haptics), smell, taste, and body movement. Outside the field of geography, very sophisticated systems have been developed that incorporate multiple senses. For instance, the $50 million National Advanced Driving Simulator (NADS) housed at The University of Iowa can be used to simulate the "feel" of driving (Samuels 1999), and thus provide a safe means of evaluating the effect of alcohol, drugs, and fatigue on driving. Although geographers probably do not need such sophisticated systems, work needs to be done to explore the role of other senses in GeoVEs. Some of the senses currently being explored include sound, touch, hand gestures, and body movements. Because developments in these areas are relatively recent, we consider them in Chapter 25.

24.3.2 Interactivity

One basic issue in interacting with VEs is navigating through them and understanding one's current orientation and location; for instance, users often report being lost in VEs, as was reported by Dykes (2002, 71–72) in describing users of the Virtual Field Course. This is an issue that has yet to be solved by those developing VEs, although some research has been done. As an example, consider the work of Sven Fuhrmann and Alan MacEachren (2001), who observed geoscientists navigating a desktop GeoVE with a standard VRML browser. Because Fuhrmann and MacEachren found that users were often disoriented and thus not satisfied with the navigation and orientation capabilities provided, they suggested an alternative GUI based on a flying saucer metaphor. The notion of using metaphors is widely accepted in computer software development (for instance, a "folder" is used to store documents in the typical Windows environment). Fuhrmann and MacEachren provided several reasons for the flying saucer metaphor, including the following:

- Its movement characteristics are commonly understood and movement in all directions is expected to be possible.
- In science fiction, flying saucers operate in geographic environments.
- Expertise in controlling flying saucers is not an issue, because no user has training using the GUI.
- Because a flying saucer has no apparent moving parts, users think of it as sophisticated technology with simple-to-learn controls.
- Flying saucers are simple in shape, providing a clean user interface.

Fuhrmann and MacEachren tested the usability of their GUI and found that navigation was easier than with the standard VRML browser and that users liked the interface. A shortcoming, however, was that many users still claimed to be "lost." The problem was that users had a natural (egocentric) field of view of the VE but they were not able to orient themselves because of missing spatial information. To solve this problem, Fuhrmann and MacEachren added an overview map to the user interface showing the overall extent of the displayed landscape and a directional cone indicating the user's current position and orientation (Color Plate 24.3). In conducting a range of usability studies with this prototype, they found that users were less likely to be disoriented in a desktop GeoVE when provided with both the natural egocentric view and an enhanced overview map.

Another important issue concerning interactivity in VEs is determining appropriate *tools* for interacting with and modifying the VE. As a starting point, it seems reasonable to provide tools similar to those used outside VEs. For instance, based on section 21.2, we presume that tools should be available for brushing and focusing. However, the 3-D character of the VE implies that other tools should be available, such as those for picking up and rotating objects. Although a suitable set of tools for geography has yet to be developed, Joseph Gabbard and Deborah Hix (1997) summarized numerous interaction techniques for VEs that might serve as the basis for developing such tools.

A third issue concerned with interactivity in VEs is the nature of the control devices used to interact with the VE. As we indicated in the discussion of display technologies, various unconventional control devices are typically used to interact with VEs. Ideally, we need to evaluate the effectiveness of these devices and to consider the possibility of using speech to interact with the VE. This takes us into the realm of *multimodal interfaces*, a topic that we consider more fully in Chapter 25, as it also applies outside VEs.

24.3.3 Information Intensity

Information intensity deals with the level of detail (LOD) presented in the VE. One issue relevant to LOD is how much detail should be shown at a particular scale. If the purpose is to simulate reality, then the inclination is to show considerable detail, but how much detail is appropriate? For instance, in the virtual cities section we will see that developers strive for realistic images, but is such realism necessary? Martin Reddy (2001) developed an approach for specifying the amount of detail that should be shown, presuming that greater detail should appear where our vision is most sensitive to detail (in the fovea). His approach, however, is largely perceptual as opposed

to cognitive, meaning that it does not consider the information content of the image.

A second issue relevant to LOD is how it should change as scale changes. In Chapter 6, we presented a number of rules that are appropriate for generalizing abstract symbolization for two-dimensional maps, but these do not necessarily extend to realistic-looking VEs. Researchers have developed approaches for changing detail with scale (as part of the Digital Earth project described in a later section of the chapter), but these approaches deal largely with technical issues (Reddy et al. 1999), not with what users will necessarily need or want.

24.3.4 Intelligence of Objects

Intelligence of objects refers to the notion that objects within a VE should exhibit some degree of behavior or intelligence, as we would expect of real-world objects. For instance, if we attempt to move a small branch on a tree in a VE, we would expect the branch to move. Similarly, if we encounter a person (known as an **avatar** or *intelligent agent*) in a VE, we would like that person to exhibit intelligent behavior, such as guiding us to a desired location. Outside the field of geography, avatars have been utilized for a variety of purposes, such as teaching people how to work with machinery (Rickel and Johnson 1999) or for advertising and presentation (Noll et al. 1999). Within geography, this is not yet common, although researchers have noted the potential that avatars provide (e.g., see Cartwright 1999 and Crampton 1999).

24.4 APPLICATIONS OF GEOSPATIAL VIRTUAL ENVIRONMENTS

In this section, we consider a range of applications for GeoVEs. Numerous additional applications can be found in Peter Fisher and David Unwin's *Virtual Reality in Geography*, and the May 2001 issue of *Landscape and Urban Planning*.

24.4.1 Virtual Cities

Probably the most common form of geospatial VE is the *virtual city*, which typically takes the form of a highly detailed visual representation of buildings (and related structures) within the city. An early example was the virtual city developed for Bath in the United Kingdom under the direction of Alan Day (1994; see Color Plate 24.4 for some images). The purpose of the Bath virtual city was to assist in planning new developments in this historically sensitive city. For instance, Day (p. 375) described how a series of views and animations were set up to determine the most appropriate location for a new sports hall (see *http://www.bath.ac.uk/casa/completed/views.html* for an example of an animation). Strictly speaking, fixed views and animations would not meet our definition of a VE (because users cannot interact with the representation or select the particular route of navigation), but such work paved the way for later systems that do meet our definition of a VE.

Michael Batty and his colleagues indicate that more than 60 virtual cities have been developed (see *http://www.casa.ucl.ac.uk/3dcities/index.htm*). An example of one such city is shown in Color Plate 24.5. One criticism of virtual cities is that developers often stress the ability to create a realistic visual impression without considering the processes operating in the city (Batty et al. 2002). This problem, however, is changing, as several recently developed virtual cities provide more than a just a realistic visual impression.

A good example of a sophisticated virtual city is Virtual Los Angeles, which is being developed by William Jepson and his colleagues at UCLA (1998; 2001). The goal of Virtual Los Angeles is to develop a virtual view of the entire 4,000-square-mile Los Angeles Basin (Color Plate 24.6). Because this is such a large undertaking, Virtual Los Angeles is being developed in stages, with each stage encompassing from 1 to 15 square miles. Jepson and his colleagues are able to extend well beyond mere representational cities by utilizing a combination of a CAD package (AutoCAD), a 3-D modeling environment (MultiGen/Creator), a simulation tool (based on IRIS Performer and OpenGL), and a GIS (ArcView). For instance, the system can depict changes over time (e.g., grow trees in the urban landscape), track vehicles in real time and generate views from these vehicles, and perform basic query and display functions (e.g., highlight all buildings having a specified number of employees). Users can move through Virtual Los Angeles in either walk-through, drive-through, or fly-through modes. Presently, Virtual Los Angeles is available only to those at UCLA who have access to sophisticated Silicon Graphics equipment, but experiments are underway to port the system to a Windows platform.

A more recent example illustrating the sophistication of virtual cities is CommunityViz, which was developed under the direction of the Orton Family Foundation (*http://www.orton.org*) and the Environmental Simulation Center (*http://www.simcenter.org*; Kwartler and Bernard 2001). As opposed to the large-city examples we have been considering, CommunityViz was developed to assist rural communities in dealing with the rapid change often caused by suburbanization. CommunityViz consists of three modules built on the ArcView GIS system: Scenario Constructor, Policy Simulator, and TownBuilder 3D. Scenario Constructor "permits nontechnical users to cre-

ate land use scenarios, and to evaluate those scenarios against community objectives and constraints" (p. 289). Policy Simulator allows users to forecast land use change, along with associated demographic and economic changes, given a prospective policy change (e.g., a tax incentive to industry). TownBuilder 3D is most relevant to this discussion as it permits photorealistic 3-D models that can be examined in real time. To illustrate some of the capability of CommunityViz, consider Color Plate 24.7, which illustrates results of a proposed land development policy. Note that the image provides a conventional thematic map as a bird's-eye view (on the left), some simple graphs (upper right), and a realistic view of a portion of the landscape (lower right). CommunityViz is promoted as a tool that can assist members of the community in becoming involved in planning development—presumably, part of the reason this works is the ease of understanding the realistic views it provides.

Another example of ongoing work with virtual cities is that of researchers at the Centre for Advanced Spatial Analysis (CASA) at University College London (*http://www.casa.ucl.ac.uk/*). Their efforts include Wired Whitehall, a first step in creating a Virtual London, and the Hackney Building Exploratory Interactive, which is intended to allow the general public to see how cities can be designed via online public participation. Work at CASA is unusual in that it allows interactive manipulation of photorealistic images via the Web (normally, only coarse images can be handled via the Web). Like PanoraMap, to be described later, much of the interaction involves working with 360° panoramic images.

Some of those working with virtual cities have emphasized abstract phenomena and have thus diminished the importance of realism. For instance, David Martin and Gary Higgs (1997) used generalized building structures to represent socioeconomic data for the city of Cardiff in the United Kingdom. They accomplished this in two steps. First, they used a detailed database on individual properties to create eight different types of building representations (semidetached house, detached house, etc.). Then they colored the buildings to represent some nonphysical phenomenon of interest, such as the rateable value of the property or its use (Color Plate 24.8). This approach sometimes led to individual buildings overlapping one another and thus producing a single object, but Martin and Higgs deemed this acceptable because the goal was to illustrate the general appearance of the buildings.

Another example emphasizing abstract phenomena is Kate Moore's (2002) mapping of the vertical dimension of the city. Moore argues that the 3-D VE provides an opportune means for examining the 3-D character of a city. Traditionally, cartographers mapped cities in two dimensions even though they were aware that cities exhibit spatial variation vertically (e.g., rents charged on the ground floor differ from those on the top floor). Moore feels that the traditional horizontal outlook was partly a function of insufficient tools for visualizing the vertical component. Lack of data could also be considered a reason for not mapping the vertical dimension (e.g., the U.S. Bureau of the Census collects data for census blocks, not for layers within census blocks), but Moore suggests that the type of data collected has likely been a function of our traditional use of two-dimensional maps. In response, Moore has developed UrbanModeller to assist in visualizing the vertical structure of the city (Color Plate 24.9). In addition to modeling the 3-D character of cities, UrbanModeller also enables the animation of pedestrian flows. UrbanModeller is a component of the Virtual Field Course, which we cover in the next section.

24.4.2 The Virtual Field Course

The Virtual Field Course project, a collaborative effort among several researchers and universities in the United Kingdom, aims to introduce VEs and information technology to traditional field-based geography classes. Geographers from the University of Leicester, Birkbeck College, and City University London argue that a Virtual Field Course can assist in all facets of fieldwork, including preparation, collection of data, analysis of data, and debriefing (Dykes et al. 1999). Here we summarize two major software components (traVelleR and panoraMap) that can assist in examining these facets.

traVelleR fits closest to the notion of a VE that we have been working with, as it permits users to explore the 3-D character of the natural landscape where fieldwork is going to be undertaken or has already been done. It contains a number of features, however, that provide considerably more information and analytic capability than a simple visual analysis of a VE would offer. traVelleR has the ability to track users' movement through the VE on a two-dimensional map, vary the characteristics of the scene and query its properties, link multimedia representations, and display an animated route.

panoraMap uses linked georeferenced panoramic images to represent geographic locations. Color Plate 24.10 demonstrates how seamless 360° imagery can be used in this way. Discrete locations from which panoramic images have been taken are represented by orange symbols on the map (in the upper left of the plate). When particular panoramic images are selected from the map, a pair of wide arrows on the map depicts the direction and angular extent of the views, and a portion of each image is shown (in the right portion of the plate); thus, in this case the three images shown relate to the three pairs of orange arrows. The views are interactive, so moving the arrows on the map or dragging the images with the mouse will result in panning. Moving the mouse in any panorama

produces a longer, thinner arrow in the image and an associated symbol on the map showing the viewing angle of the identified feature. This means that features in the panoramas can be associated with data recorded on the map. The arrows at the top of each image in Color Plate 24.10 show the directions and distances (by length of the arrows) of other panoramas. These can be touched with the mouse to highlight the locations on the map (in green), and clicked to display the relevant view. The user is thus able to "navigate" between discrete locations across the landscape. The three panoramas shown in Color Plate 24.10 constitute a user-specified virtual route across part of Dartmoor National Park in the United Kingdom.

As with traVelleR, the real power of panoraMap is the ability to link other kinds of information with these panoramic images and to explore this information in a spatial context. Data can be georeferenced by clicking the map and using a selection interface to choose any file on the computer. The location of each file is then represented by a yellow symbol, which reveals the data when clicked. For instance, Color Plate 24.11 shows a session from an exercise on footpath erosion. Two panoramic views are centered on Haytor Rocks, an important and much visited local feature. A georeferenced digital movie (lower right) containing audio and video collected in the field is displayed along with a hypertext document (upper right) offering a series of images, links, and qualitative information on erosion around Haytor Rocks as recorded by the National Parks Authority. panoraMap explores the links between reality and VR by enabling users to plan routes for use with GPS receivers and plot them on returning from the field. The software also contains capabilities for graphical data analysis, meaning that numeric information can be assessed within the context provided by the qualitative data sources. So in the exercise shown in Color Plate 24.11, quantitative data recorded in the field can be visualized in combination with secondary qualitative data in a spatial context. The use of the GPS data before and after visiting the field emphasizes the links between the map and the physical environment and supports the "reality" metaphor of the interface.

Both traVelleR and panoraMap have been designed to take advantage of the VFC Hub mechanism (Dykes et al. 1999), which connects software to the Internet and means that data stored on remote servers can be incorporated into a session. For example, traVelleR can load maps and multimedia data found online (in one example application, audio files containing an expert's explanations were georeferenced at the locations of important geomorphological features, such as Haytor Rocks), and panoraMap can search online VFC databases for panoramas, add symbols to show their locations, and download and display the views when the symbols are clicked.

24.4.3 Digital Earth

Former Vice President Al Gore (1998) argued that the "desktop metaphor" employed by the Macintosh and Windows operating systems is insufficient for interacting with the vast amount of geographic data now available (e.g., one-meter satellite imagery). Instead, Gore called for a new metaphor—a multiresolution, 3-D Digital Earth:

> Imagine . . . a young child going to a Digital Earth exhibit at a local museum. After donning a head-mounted display, she sees Earth as it appears from space. Using a data glove, she zooms in, using higher and higher levels of resolution, to see continents, then regions, countries, cities, and finally individual houses, trees, and other natural and man-made objects. Having found an area of the planet she is interested in exploring, she takes the equivalent of a "magic carpet ride" through a 3-D visualization of the terrain. Of course, terrain is only one of the many kinds of data with which she can interact. Using the systems' voice recognition capabilities, she is able to request information on land cover, distribution of plant and animal species, real-time weather, roads, political boundaries, and population. She can also visualize the environmental information that she and other students all over the world have collected. . . . She is not limited to moving through space, but can also travel through time. After taking a virtual field-trip to Paris to visit the Louvre, she moves backward in time to learn about French history, perusing digitized maps overlaid on the surface of the Digital Earth, newsreel footage, oral history, newspapers and other primary sources.

Although it will likely be some time before Gore's full vision is achieved, numerous people and organizations are working on the concept of a Digital Earth (see *http://www.digitalearth.gov/*). Work thus far has focused on providing access to geospatial data via the Web. Much of this data is available in traditional static map form, or as an animation (as opposed to a VE). One approach that moves in the direction of VEs is The Globe Program (*http://www.globe.gov/*), which provides interactive 3-D visualizations of the Earth via the Web, but the focus here is on looking at patterns on the Earth's surface, not on providing the detail at different resolutions that Gore was envisioning. Another novel approach is the Digital Earth Workbench, which is housed at the National Aeronautics and Space Administration (NASA), where a drafting table–format display is being used to provide a sense of immersion (*http://webserv.gsfc.nasa.gov/DE-Workbench/*). A third novel approach is TerraVision (*http://www.ai.sri.com/TerraVision/*), which enables Web users to examine 3-D visualizations of terrain onto which remote sensed images and other information (e.g., cultural features) have been draped. Some important characteristics of TerraVision include the ability to: (1) handle very large data sets (on the order of terabytes); (2) access data of varying resolutions, thus permitting more detail to be seen as one zooms into a location; and (3) access data from multiple servers. TerraVision also contains some unusual navigation capabilities: a terrain-following option that can maintain a particular height above the Earth's surface, accounting for Earth curvature, and an altitude-based velocity, which varies speed as a function of the height above the surface (e.g., when traveling high above the surface, it makes sense to go faster). Normally, scenes created with TerraVision are viewed on a desktop display with a mouse as the control device; it is possible, however, to use HMDs and a *6-degrees-of-freedom* control device (a device that allows great flexibility in navigating 3-D space).

Several systems are becoming available that come closer to meeting Gore's vision. One of these is EarthViewer (*http://www.earthviewer.com*), a package that provides capabilities similar to TerraVision. Another system that should soon be available is ArcGlobe, a software component of ArcGIS, the flagship product of ESRI. ArcGlobe not only allows users to zoom into regions via remotely sensed images of varying resolution, but permits the creation of realistic landscapes along the lines discussed in this chapter.

24.4.4 Water-Resources-Related Applications

Several VEs have been developed for water-resources-related applications. Two of these, the Chesapeake Bay Virtual Environment (CBVE) and Virtual Puget Sound (VPS) have dealt with estuaries. We discuss CBVE here and consider VPS in a subsequent section of the chapter. Glen Wheless and his colleagues (Wheless et al. 1996a; Wheless et al. 1996b) developed CBVE for the ultimate purpose of visualizing potential changes in the ecosystem of Chesapeake Bay over both space and time. Chesapeake Bay is the largest estuary in the United States, serving as nursery grounds and spawning areas for many commercially important fish and shellfish. Wheless and his colleagues argued that understanding this ecosystem is critical given increasing urbanization, overfishing, and diseases that have affected some species of fish.

CBVE consisted of three modules: numerical simulation, visualization, and virtual reality. The purpose of the numerical simulation model was to examine the effect of tides, freshwater discharge, and winds on fish living in this fragile ecosystem. The visualization module was based on the Vis5D system, which we described in Chapter 21. The VR model utilized both the ImmersaDesk and CAVE systems. Although they did not conduct any usability tests of the software, Wheless and his colleagues clearly were enthralled with ImmersaDesk and CAVE:

> Aspects of the data not clearly recognizable with static two-dimensional or three-dimensional images became

immediately apparent simply because of our ability to intuitively navigate through the virtual environment so that features could be examined from different viewpoints. The complex relief of the bay bathymetry and associated topographical effects on the circulation in the bay became obvious . . . (Wheless et al. 1996a, 207)

Interestingly, Wheless and his colleagues also indicated that sound "heightened" their understanding of ecosystem processes:

We were able to navigate through the virtual environment to a location where we could clearly see an increase in biological distributions and hear the change in pitch or frequency corresponding to salinity value, thereby linking the biological behavior with a physical variable in a more enlightening way. (Wheless et al. 1996a, 205)

Another example of a water-resource-related VE is the Digital River Basin, which is being developed by the Mississippi RiverWeb Museum Consortium (*http:// archive.ncsa.uiuc.edu/Cyberia/RiverWeb/Projects/RWMu seum/*) to illustrate the dynamics of the Mississippi River and its associated watershed processes. For instance, the Science Museum of Minnesota has installed a River Pilot Simulator that enables individual visitors to experiment with the task of barge navigation. By utilizing a 52-inch plasma display, and providing "tiller" and "throttle" controls and engine "sound," users get a sense of actually piloting a barge (Johnston et al. 2000). Although this simulator does not provide the immersive capabilities of other systems (e.g., CBVE), it is suggestive of the direction that museums might eventually take in illustrating complex physical processes to the general public.

24.4.5 Virtual Forests

Virtual forests are another common application for VEs. As an example, we consider Virtual Forest (*http://www. innovativegis.com/basis/Vforest/default.htm*), which was developed for visualizing the effects of timber harvesting operations. Virtual Forest consists of two basic tools: Tree Designer and Landscape Viewer. Tree Designer allows users to design custom 3-D tree symbols (Figure 24.4), which can depict different species, seasons, and the overstory versus understory. Tree symbols can include a maturity definition that specifies how large a tree can become when mature, which is useful when simulating the growth of vegetation. Landscape Viewer allows users to populate a landscape with trees defined by Tree Designer—trees can be planted or harvested, depending on polygon parameters specified via ARC/INFO or ArcView. Realism is enhanced by simulating atmospheric effects (e.g., haze) and permitting texture maps to be overlaid. Animations are a common product of Virtual Forest, although the software

FIGURE 24.4 An example of a tree created using Tree Designer, a component of Virtual Forest. (Courtesy of Buckley, D. and J. K. Berry, Virtual Forest Software for Advanced 3D Visualization of GIS Databases; see *www.innovativegis.com/basis*; select Virtual Forest.)

permits interaction with the resulting landscape, thus meeting our definition of VEs. Color Plate 24.12 illustrates a sample image produced using Landscape Viewer.

24.4.6 Other Applications

Other applications of VEs include those in landscape and urban planning and virtual heritage. An excellent source of information on VEs and related visualization approaches relevant to landscape and urban planning can be found in volume 54 of *Landscape and Urban Planning* entitled "Our Visual Landscape: Analysis, Visualization and Protection." An overview paper by Stephen Ervin (2001) is useful for illustrating the current status of VEs and suggesting some avenues for research. *Virtual heritage* refers to VEs that enable users to experience the world's cultural and natural heritage without having to be at the actual site. The April/June 2000 issue of *IEEE Multimedia* is dedicated to this topic. Although virtual heritage is only marginally related to geography, developments in this area are illustrative of the many technological developments taking place in VEs.

24.5 RESEARCH ISSUES IN GEOSPATIAL VIRTUAL ENVIRONMENTS

From the preceding sections, it is clear that many people are excited about the capability that VEs are now providing or might provide in the future. Although we share this excitement, it must be recognized that VEs are novel, and so little is known about how they should be designed and effectively used. In response to these concerns, those affiliated with the International Cartographic Association's Commission on Visualization and Virtual Environments have developed the following set of research issues related to GeoVEs (Slocum et al. 2001):

- Determine the situations in which (and how) immersive technologies can assist users in understanding geospatial environments.
- Develop methods to assist users in navigating and maintaining orientation in GeoVEs.
- Develop suitable methods for interacting with objects in GeoVEs.
- Determine ways in which intelligent agents can assist users in understanding GeoVEs.
- Determine ways in which we can mix realism and abstraction in representations to influence cognitive processes involved in knowledge construction.
- Develop support for interpreting and understanding spatial trends and patterns in GeoVEs.

Researchers are now beginning to tackle some of these issues. We already mentioned Sven Fuhrmann and Alan MacEachren's work on designing user interfaces to support navigation and wayfinding in desktop GeoVEs. Considerable research has been undertaken by William Winn and his colleagues at the Human Interface Technology Laboratory (HIT Lab) at the University of Washington (Windschitl et al. 2002; Winn 2002; Winn et al. 2002). Much of this research focuses on Virtual Puget Sound (VPS), a VE that allows students to examine water movement, the behavior of tides, and salinity levels in Puget Sound. When using an HMD environment, Winn and his colleagues have found that *more successful* students use a more systematic approach, are more animated and active, use more virtual tools, comment more frequently on their progress, have less difficulty with the interface, and are more physically active. In comparing HMD and desktop approaches, Winn and his colleagues found that HMD users reported a greater sense of presence, which

in turn led them to learn more about the movement of water (the concept that water in the Sound circulates both horizontally and vertically). In the HMD, students could simply look around to see the water movement, whereas in the desktop display they had to manipulate a control device.

Based on their research and the work of others, Winn and his colleagues noted a number of advantages and disadvantages of VEs. Advantages of VEs include the following:

- They help students understand phenomena that are not directly accessible to the senses (e.g., varying salinity levels with water depth).
- Immersion helps students understand dynamic 3-D processes, but not static two-dimensional ones.
- They increase students' engagement in learning tasks.
- They enable rescaling objects and processes in time and space (e.g., a daily tidal cycle can be compressed to five seconds).
- They allow students to make errors that might be disastrous in the real world (e.g., experiments in fighting forest fires).

Disadvantages of VEs include the following:

- They require some training before learning can take place.
- They are not good for developing fine motor skills (because the skills used are often different from those used in the real world).
- Their side effects are unknown (e.g., if students believe something is real when in fact it is not, this could lead to confusion between fantasy and reality).

Although work by Winn and his colleagues forms a useful framework for studying VEs, they have only begun to deal with the research issues identified earlier, and so exciting research remains to be done.

24.6 DEVELOPMENTS IN MIXED ENVIRONMENTS

In this section, we consider some recent developments in MEs of interest to geographers. This section is short largely because the notion of MEs is novel. As the technology matures, such environments might become more important than GeoVEs.

24.6.1 Project Battuta

Project Battuta is a joint research project of Iowa State University and the University of California to develop modern approaches for collecting and using geospatial data in the field. The project is named after an early geographer, Ibn Battuta, who in the 1300s traveled to 44 countries, in the process covering more than 75,000 miles. Two key components of Project Battuta are a wearable computer and an associated augmented reality display package that enables users to overlay geospatial data and images on the landscape as one travels through the environment. Additionally, users will have real-time access to a vast array of distributed resources via a wireless Internet connection. When this system is fully developed, a geographer will not only be able to interpret the landscape as Battuta did, but be able to bring to bear a wealth of geospatial data to assist in that interpretation.

Figure 24.5 shows a version of the wearable computer currently being developed by Keith Clarke and his colleagues (2001) as part of Project Battuta. The user of this wearable computer can readily see the real world, as is obvious by the fact that her eyes are unimpeded by apparatus related to the computer. The exception is the small clip-on device attached to her glasses, which is a miniature computer display that provides the augmentation to the real world. In her hand, she holds a device that enables text input by pressing multiple keys to produce specific

FIGURE 24.5 The wearable computer currently being developed by Keith Clarke and his colleagues as part of Project Battuta. (Courtesy of Keith Clarke.)

letters and numbers. Although the input device looks cumbersome, Clarke indicates that with practice users can reach 35 words per minute. On her right shoulder, we see a miniature GPS receiver with embedded antenna that continuously records her position to within 2 to 7 meters, depending on sampling conditions. On her back, unobtrusively enclosed in her fishing-vest-like jacket, is a computer that drives the system. This computer weighs only four pounds and measures $6'' \times 5'' \times 2''$. Power is provided by two lithium-ion rechargeable batteries that support up to five hours of system run time each.

It is important to recognize that the technology for wearable computers is evolving rapidly. For instance, Clarke indicates that a promising new technology for creating augmentation is a retinal scanning technique developed by MicroVision (*http://www.mvis.com/*) that paints a computer image directly onto the retina. This approach is advantageous because it provides a high-resolution, high-contrast image that does not block the user's view.

Because the emphasis in Project Battuta is on augmenting the real world, the nature of the user interface and map display need to be different than traditional desktop systems or more immersive VEs. For example, Clarke noted that wire-frame equivalents of 3-D objects are appropriate (as opposed to the realism of VEs) because we can already see the real world. If using object-based augmentation, like arrows and wire frames, or text annotation, the major challenge for user interface design will be the display of information that best suits the user's need at a certain time in a certain location. Because the user's attention is not solely directed toward the computer, facilitation of a real-world task without information overload and distraction by the software is a key component in interface design.

Potential applications of augmented reality displays such as Project Battuta are numerous, as we have indicated earlier. Perhaps most important, however, are potential applications outside geography. For instance, Clarke described the following scenario for a rescue mission:

> The Santa Barbara police have received a report that a small plane that took off from the Santa Barbara airport . . . has disappeared . . . in the Santa Ynez Mountains. . . . A mobile unit is dispatched with the newly available GeoSearch system. Using the in-vehicle global positioning (GPS) receiver, the crew in the jeep on the way to the site download a real-time satellite image of the likely search area. . . . The GeoSearch operator downloads the imagery into her belt-worn mobile computer. . . . All looks OK as the operator stops the jeep at the point computed by a least cost path algorithm to offer the best stretcher-carry. . . . From the 1 cm resolution imagery, one injured person can be seen, and appears to have broken legs.

The GeoSearch operator sees the reassuring flashing orange dot of her own GPS location closing in on the image view of the location. . . . Twenty minutes later, at a spot recomputed on the fly as having the best combination of stretcher access and flat treeless terrain, a Medivac helicopter loads the injured and the team heads back to their jeep. The parting shot of the successful rescue team runs on the 11 o'clock news that evening, grainy through its having been shot with the infrared night camera necessary for a night-time rescue operation. (Clarke 1998, 120)

It appears that such practical usefulness of this technology is likely to move it forward, making it more accessible for geographic applications.

24.6.2 Work by Hedley and Colleagues

In addition to the work on VPS mentioned earlier, researchers at the HIT Lab at the University of Washington are doing intriguing research on advanced interface environments. Here we focus on work that Nick Hedley (2001; 2003) has been doing with individual users of "mixed reality" environments, and in the next chapter we consider work that Hedley and other researchers in the HIT Lab are doing in the context of collaborative geographic visualization and MEs.

One of the experiments that Hedley is conducting compares a standard desktop display with an augmented reality (AR) system. With the desktop display, users manipulate 3-D surfaces using a VRML browser (a shareware VRML viewer that runs in popular Internet browsers) and a mouse. The technology for the AR system is illustrated in Figure 24.6. In the lower left and upper right of Figure 24.6, we see a user wearing what appears to be a HMD, but this HMD is fundamentally different from those described earlier in the chapter. In this case, a miniature video camera is mounted on the HMD. The video stream captured by the camera is combined with a computer-generated 3-D image and displayed to each eye via the HMD. The image that appears when looking through the HMD (lower right of Figure 24.6) results from real-time tracking of the patterns shown on the card in the upper left of Figure 24.6 by AR software (ARToolkit). The AR software recognizes unique marker patterns in the view from the camera, and places virtual 3-D objects on them in the output view. The result is that users can attach virtual 3-D objects to real, manipulable objects (cards, paper), and hold the card in their hands and manipulate it just as they would manipulate objects in the real world. Contrast this notion with how we interact with 3-D objects in a desktop display—to rotate an object, we click a button and move the mouse. Novice users often find this task awkward, presumably because it does not correspond to how they would examine a 3-D object in the real world.

In his experiment, Hedley spent about 250 hours studying a total of 100 people who examined 3-D surfaces using these two different approaches. His initial findings suggest that AR interfaces allow people to complete basic spatial problem-solving activities more

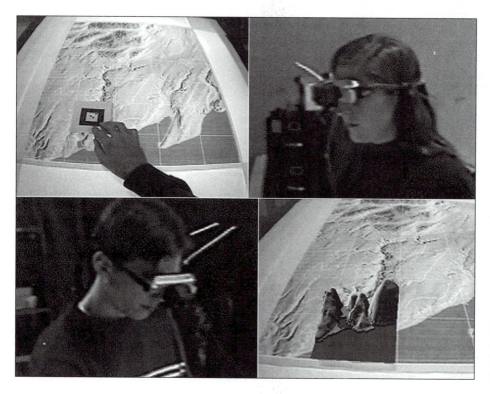

FIGURE 24.6 The augmented reality system utilized by Nick Hedley in the HIT Lab at the University of Washington. (Courtesy of Nick Hedley.)

quickly and more accurately than by using desktop interfaces for the same 3-D geographic visualizations. He has also revealed that different combinations of interface, user, and 3-D visualization can affect spatial problem solving. From an interface perspective, AR is a 3-D display (it is in the room with you) of a 3-D model, whereas a desktop interface is a two-dimensional display of a 3-D model. This means that in AR we can inspect the virtual 3-D model like a real object, gaining a better understanding of it by rotating it in our hands, or leaning closer to inspect it. This raises some interesting possibilities for visualizing spatial phenomena and gaining deeper insight into them through more natural interfaces. Hedley is also working with other colleagues to develop and study AR interfaces for geographic education (Shelton and Hedley 2002) and public visualization tools (Woolard et al. 2003).

24.7 HEALTH, SAFETY, AND SOCIAL ISSUES

In the research issues section for GeoVEs, we noted one disadvantage of VEs is that their side effects are unknown. This idea is related to the broader notion of health, safety, and social issues associated with VEs and MEs. Health and safety issues can range from something as simple as tripping over a power cord while immersed in a VE to discomfort associated with wearing an HMD to **cybersickness**, a form of motion sickness that can result from exposure to VEs. Although cybersickness and related problems are commonly associated with particular hardware (notably HMDs), such problems are not necessarily restricted to HMDs (Wann and Mon-Williams 1997, 54).

The use of VEs raises numerous social issues, such as (Stanney et al. 1998):

- How will interaction in the virtual world modify behavior?
- Will people turn their backs on the real world and become discontented zombies wandering around synthetic worlds that fulfill their whims?
- Will people avoid reality and real social encounters with peers and become addicted to escapism?

Because MEs are relatively new, little has been said about health, safety, and social issues related to them. Presumably, they would share some of the same concerns of VEs. For instance, imagine the potential danger of trying to walk through a landscape in which a computer image is draped on the actual landscape. It is certainly conceivable that the computer image might occlude a feature in the landscape that must be seen for safe navigation. How might one's interaction with other people be affected by the fact that others know you are seeing information that they cannot see?

SUMMARY

This chapter has examined how interactive computer graphics technologies can assist us in better understanding 3-D geospatial environments. The approaches range from those that are a simulation of the real world (a **virtual environment [VE]** results) to those in which our view of the real world is combined with computer-based information (a **mixed environment [ME]** results). An example of a geospatial VE would be if you were to take a tour through a simulation of how the 3-D landscape changed during major glaciations of North America. (Technically, we restrict the term VE to the situation in which you have some control over the tour; if you have no control, then the tour would be termed an *animation*.) An example of an ME would be if you were to look at the present-day landscape and see past glacial features overlaid on the landscape. Thus far, VEs are more common than MEs, but in the future MEs might prove to be more useful, particularly for geographic fieldwork.

We considered several technologies for creating VEs, including **desktop**, **wall-size**, **head-mounted**, **drafting table–format**, and **room-format** displays. A distinct advantage of the latter three is that they allow you to become more immersed in the VE—you might actually feel a sense of **presence**, of being in the VE, when in fact it is artificial. Today, most VEs are experienced visually, although it is possible to utilize the full range of senses, including sound, touch (**haptics**), smell, taste, and body movement.

We discussed four "I" factors that are relevant in designing geospatial VEs: **immersion**, **interactivity**, **information intensity**, and **intelligence of objects**. Although there is a tendency to think that a higher degree of immersion is desirable (as in donning an HMD that shields you from the real world), some researchers have found that users seem to work effectively with standard desktop systems, which have a low degree of immersion. Key issues in interacting with the geospatial VE are navigating through it and understanding our current orientation and location. Presumably this is difficult because when we enter a VE we lose the normal benchmarks found in the real world. Information intensity refers to the level of detail (LOD) present in the GeoVE. Our lack of experience with VEs makes it difficult to judge what an appropriate LOD should be or how it should change with scale. Intelligence of objects refers to the notion that objects in a VE can exhibit intelligence, just as they would in the real world. Potentially very useful are **avatars** (images of people) that could help guide us through a VE and answer questions for us about the VE.

We looked at several applications of GeoVEs. One common application is the *virtual city*, which often stresses providing realistic visual representation of buildings (and related structures). This approach has been criticized because it fails to consider processes operating in the city, although some applications, such as Virtual Los Angeles and CommunityViz, clearly provide more than just realistic representations.

Given the novelty of VEs, there are numerous associated research issues. The International Cartographic Association's Commission on Visualization and Virtual Environments has identified many research issues, and William Winn and his colleagues at the Human Interface Technology Laboratory at the University of Washington have conducted a substantial amount of research with GeoVEs. Although Winn and his colleagues have been able to identify a number of advantages and disadvantages of geospatial VEs, it should be recognized that their research has focused largely on HMDs, and thus more research is needed.

In the realm of MEs, we looked at two applications: Project Battuta and the work of Nick Hedley. Project Battuta is a form of **augmented reality (AR)**, in which the real world is dominant and computer-based information is overlaid on it (e.g., the preceding example in which glacial features were overlaid on the real-world landscape). Hedley's work is also a form of AR, but in this case greater emphasis is placed on interpreting the VE (the VE is displayed on a large card held in the users' hands, and users can rotate the card, just as they would rotate an object in the real world).

Finally, we touched on health, safety, and social issues associated with VEs and MEs. Health and safety issues can range from something as simple as tripping over a power cord to **cybersickness**, a form of motion sickness. Social issues deal with how use of VEs and MEs will affect our view of the real world and our interaction with others. For instance, will there be a tendency for some to escape into VEs, and will it be difficult to interact with someone who utilizes an ME when you do not have access to that same environment?

FURTHER READING

Batty, M., Chapman, D., Evans, S., Haklay, M., Kueppers, S., Shiode, N., Smith, A., and Torrens, P.M., (2001) "Visualizing the city: Communicating urban design to planners and decision makers." In *Planning Support Systems*, ed. by R. K. Brail and R. E. Klosterman, pp. 405–443. Redlands, CA: ESRI Press.

Summarizes technologies for creating virtual cities and describes numerous virtual cities; also covers recent developments in virtual cities at the Centre for Advanced Spatial Analysis (CASA).

Buckley, D. J., Ulbricht, C., and Berry, D. J. (1998) "The Virtual Forest: Advanced 3-D visualization techniques for forest management and research." *ESRI User Conference*, San Diego, CA. Available at *http://www.innovativegis.com/products/vforest/index.html*.

Discusses the need to visualize the effects of timber harvesting, various approaches that have been taken, and a recent approach known as the Virtual Forest.

Câmara, A. S., and Raper, J. (eds.) (1999) *Spatial Multimedia and Virtual Reality*. London: Taylor & Francis.

Chapters 11 and 13 deal with GeoVEs; the former involves a VE for the Delft University of Technology Campus, whereas the latter considers a "virtual GIS room."

Computer Graphics 31, no. 2, 1997.

Special issue covering a range of technologies for creating VEs.

Computers & Graphics 20, no. 4, 2000.

Special issue focusing on technical issues related to "virtual reality and 3D GIS."

Costello, P. (1997) "Health and safety issues associated with virtual reality—A review of current literature."

Reviews health and safety issues associated with VEs.

Feiner, S. K. (2002) "Augmented reality: A new way of seeing." *Scientific American* 286, no. 4:48–55.

Provides an overview of technical issues involved in creating augmented reality, along with a range of applications.

Fisher, P., and Unwin, D. (eds.) (2002) *Virtual Reality in Geography*. London: Taylor & Francis.

Covers a range of applications for VEs, including "virtual landscapes," "virtual cities," and "'other' worlds." The emphasis is on desktop displays. As an example of other worlds, Martin Dodge describes AlphaWorld, an artificial Web-based VE.

Goodchild, M. F. (2000) "Cartographic futures on a Digital Earth." *Cartographic Perspectives*, no. 36:3–11.

Considers what the future of digital cartography might hold and some issues associated with a Digital Earth. Also see papers by Pickles and Rhind that appear in issue no. 37 of this journal.

IEEE Computer Graphics and Applications 20, no. 4, 2000 (bulk of issue).

Focuses on wall-size displays, dealing largely with technical issues involved in generating them.

IEEE Multimedia 7, no. 2, 2000.

Special issue on virtual heritage.

Landscape and Urban Planning 54, 2001.

Special issue on VEs in landscape and urban planning.

MacEachren, A. M., Edsall, R., Haug, D., Baxter, R., Otto, G., Masters, R., Fuhrmann, S., and Qian, L. (2000) "Virtual environments for geographic visualization: Potential and challenges." In *Workshop on New Paradigms in Information Visualization and Manipulation (NPIVM '99)*, ed. by D. S. Ebert and C. D. Shaw, pp. 35–40. New York: The Association for Computing Machinery. Available at *http://www.geovista.psu.edu/publications/NPIVM99/ammNPIVM.pdf*.

Introduces the four "I" factors of geospatial VEs and describes a demonstration project for the ImmersaDesk involving the relationship of temperature to topography and precipitation.

Neves, J. N., and Câmara, A. (1999) "Virtual environments and GIS." In *Geographical Information Systems*, ed. by P. A. Longley, M. F. Goodchild, D. J. MaGuire, and D. W. Rhind, pp. 557–565. New York: Wiley.

Provides an overview of VEs and GIS.

Raper, J. (2000) *Multidimensional Geographic Information Science*. London: Taylor & Francis.

Pages 219–224 consider the use of VEs to depict geomorphological change on Scolt Head Island, England; much of this book deals with theoretical issues related to 3-D mapping, including the temporal aspect.

Slocum, T. A., Blok, C., Jiang, B., Koussoulakou, A., Montello, D. R., Fuhrmann, S., and Hedley, N. (2001) "Cognitive and usability issues in geovisualization." *Cartography and Geographic Information Science* 28, no. 1:61–75.

Introduces some research issues related to GeoVEs.

Stanney, K. M., Mourant, R. R., and Kennedy, R. S. (1998) "Human factors issues in virtual environments: A review of the literature." *Presence* 7, no. 4:327–351.

Reviews human factors issues related to VEs.

25

Ongoing Developments

OVERVIEW

As discussed in Chapter 1, the discipline of cartography has changed considerably since the 1960s, evolving from a discipline based on pen and ink to one based on computer technology. As the field continues to evolve, it is important to keep pace with ongoing developments. The purpose of this chapter is to examine some of these developments, including Daniel Carr and his colleagues' work with linked micromap plots and conditioned choropleth maps, using sound to interpret spatial data, collaborative geovisualization, multimodal interfaces, information visualization, and spatial data mining.

Linked micromap plots (LM plots) focus on a series of small maps termed *micromaps*, which divide a single spatial distribution into pieces. By using micromaps, LM plots focus on local pattern perception as opposed to the overall spatial pattern. LM plots also permit the display of statistical information such as confidence intervals for each observation. In a fashion similar to LM plots, *conditioned choropleth maps* (CCmaps) split a spatial pattern for a single attribute into a series of choropleth maps. In the case of CCmaps, however, rows and columns of maps are used, with the rows and columns corresponding to two attributes that might explain the attribute that is displayed on the choropleth map series.

From a cartographic standpoint, two basic forms of sound are possible: realistic and abstract. *Realistic sounds* are those with meaning based on our past experience, such as hearing the sound of the ocean when watching a map depicting ocean currents. *Abstract sounds* are those with a meaning that is not clear without using a legend; for example, clicking on counties on a choropleth map might produce sounds of differing loudness as a function of the percentage of people voting for a Republican candidate. John Krygier has proposed numerous *abstract sound*

variables analogous to the visual variables discussed in earlier chapters (examples of abstract sound variables would be loudness and pitch). We look at several applications of the use of sound in cartography; particularly interesting is the Haptic Soundscapes project in which Daniel Jacobson, Reginald Golledge, and their colleagues have begun experimenting with touch (*haptics*) in addition to sound.

Collaborative geovisualization (*or geocollaboration*) refers to geovisualization activities in which more than one individual is involved in the visualization process. For instance, you might be located at one university and wish to discuss the visualization aspects of a climate-change model with a researcher at another university. Ideally, you should both be able to manipulate the model from your respective locations and see the results of what the other person is doing. We consider three applications of collaborative geovisualization: Nick Hedley and his colleagues' work with augmented reality (AR) and a specialized projection table known as HI-SPACE; Alan MacEachren and his colleagues' work with ImmersaDesks; and visualization efforts at the National Center for Atmospheric Research (NCAR) Visualization Lab.

Although most of us today are still working with Windows and associated mouse-based interfaces, researchers are experimenting with novel interfaces that utilize speech, lip movements, pen-based gestures, free-hand gestures, and head and body movements. *Multimodal interfaces* are novel interfaces that involve two or more of these techniques; for instance, you might point to a location without touching the screen (a free-hand gesture) and say, "Show me all homes within 100 meters of this location." Some of the most exciting work in multimodal interfaces is being done at the GeoVISTA Center at Penn State University, *where researchers are developing collaborative geovisualization systems for emergency management situations.*

Information visualization involves the visualization and analysis of nonnumerical abstract information such as the nature of topics that are discussed on the front page of a newspaper over a month-long period. Spatialization is the process of converting such abstract information to a spatial framework in which visualization is possible. We consider some work that two geographers, André Skupin and Sara Fabrikant, have done in this realm.

Spatial data mining is a simplified term used for spatial applications involving knowledge discovery in databases (KDD), which refers to exploratory approaches for finding interesting patterns in databases. Spatial data mining is particularly useful for large databases because the computational approaches that are used can uncover patterns that visual processing might miss. We'll look at two applications of spatial data mining, one that utilizes a sophisticated cluster analysis method to process data that would normally be displayed as a choropleth map, and one that helps us determine which thematic attributes are associated with previously defined regions within a city.

One of the difficulties in writing a book in a rapidly changing discipline is that it is virtually impossible to provide up-to-date information. To assist in acquiring information on ongoing developments, the last section of the chapter provides lists of journals, conferences, and useful Internet sites. The home page for this book (http://www.prenhall.com/slocum) should also be consulted for more recent information.

25.1 CARR AND HIS COLLEAGUES' WORK

Daniel Carr and his colleagues have developed a number of visualization techniques that have a statistical emphasis. In this section, we consider two of their contributions: linked micromap plots and conditioned choropleth maps.

25.1.1 Linked Micromap Plots

Linked micromap plots (LM plots; Carr and Pierson 1996; Carr et al. 1998) consist of three basic elements, as illustrated in Color Plate 25.1: a series of small maps termed *micromaps* (the column labeled "Maps"), a legend (the column labeled "States"), and one or more related graphics (the third and fourth columns). The primary attribute of interest (in this case "unemployment rate by state") is displayed in the micromaps and plotted as colored dots in the column labeled "Rates and 95% CI." For instance, if we look at the bottom micromap, we can see that North Dakota has an unemployment rate between 3 and 4 percent. The term *linked* comes from the notion that information in one column can be related to another column by linking

the corresponding color; thus, the micromap gives us a location for North Dakota, the first graph its unemployment rate, and the last graph the number unemployed.

Rather than providing a depiction of the overall spatial pattern (as on a traditional choropleth map), the purpose of LM plots is to focus on local pattern perception. This is achieved by ordering the data from high to low (from top to bottom in Color Plate 25.1) and depicting subsets of these ordered data in each micromap. Again, looking at the bottom map, we can see that four of the states with the lowest unemployment rates are adjacent to one another and located in the north central United States.

Note that the 50 states (and Washington, DC) have been split into groups of five, with the exception of Illinois, which is placed by itself in the middle. Carr and Pierson (1996, 19) argued "that groups of five facilitate counting and still allow quick label and value matching by relative position." Placing Illinois by itself is logical if we remember that the middle value in an ordered set of data is the median (see Chapter 3). We can thus easily contrast states above the median from those below the median.

In looking at the micromaps, you might have noticed that individual states appear distorted, and that some appear larger or smaller than we might expect on an equivalent (equal-area) map projection. Carr and his colleagues did this intentionally so that smaller states could be more easily identified, using a notion developed by Mark Monmonier (see section 19.4). Even with such modifications, however, small enumeration units will be difficult to detect; thus, Carr and his colleagues suggest that 50 enumeration units is about the largest number that can be handled in an LM plot in static (printed) form.

Note that the "Rates and 95% CI" column includes a set of gray lines on either side of the line connecting the colored dots. These are the 95 percent confidence intervals, a measure of uncertainty in the data. Confidence intervals are necessary because the data are based on a sample; thus, rather than saying that the unemployment rate for Washington, DC, is 8 percent, we should say that we are 95 percent certain that the unemployment rate is between approximately 8 and 10 percent (see the top row in the color plate). Including such information illustrates the statistical bent of Carr and his colleagues.

The logic of including the graph "No. Unemployed" is that it places the data in context, as it can be argued that a high unemployment rate is more important if it affects a large number of people. Thus, we can see that California (the extreme right green dot) not only has a high unemployment rate, but that 1.2 million people are affected. In contrast, Washington, DC, has the highest unemployment rate, but less than 100,000 people are affected.

Xusheng Wang and his colleagues (2002) developed a Web-based program for exploring LM plots associated with cancer data for the United States (see *http://graphics.gmu.edu/~xwang/cancer4/).* This interactive program has a number of advantages over static printed LM plots (as suggested by Figure 25.1), including the ability to display a large number of enumeration units (via scrolling and zooming), display different cancer attributes (lung and bronchus cancer vs. cervical cancer), view the data at either the state or county level, sort on different columns (e.g., sort the states as opposed to the attribute of interest), change the way a confidence interval is displayed, change the way enumeration units are shown when not emphasized in a particular micromap (in the top portion of Figure 25.1 areas above the median are shown in white, whereas the opposite is the case in the bottom portion of the figure), and change the color linking individual units (spectral, sequential, and diverging schemes are possible, with sequential desirable when printing in black and white). These features are best appreciated by actually using the program, and so we encourage you to experiment with the given Web site.

25.1.2 Conditioned Choropleth Maps

Carr and his colleagues (2000b; 2002) developed **conditioned choropleth maps** (CCmaps) to permit a visual examination of attributes that might explain the spatial pattern of various forms of cancer. To illustrate, consider Color Plate 25.2, which provides a set of CCmaps intended to assist in explaining lung cancer mortality rates for White men aged 65 to 74. In the legend along the top of the color plate, you can see that lung cancer mortality rates are depicted in three colors: blue for low values, gray for moderate values, and red for high values. Values at the bottom of the legend are lung cancer mortality rates, whereas values above the legend are the percent of White men aged 65 to 74 in each class. In this example, the attributes

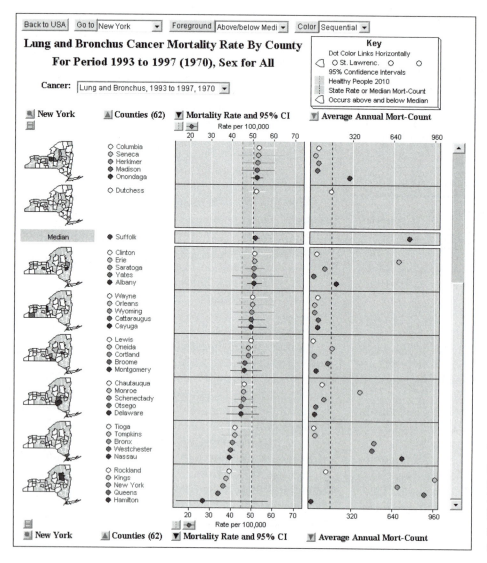

FIGURE 25.1 Example taken from Wang and his colleagues' Web-based program for creating linked micromap plots (LM plots). (Courtesy of Xusheng Wang; for more recent work involving micromap plots, see *http://statecancerprofiles.cancer.gov/micromaps/*.)

considered for their relationship with lung cancer are percent below the poverty level and annual precipitation. Although precipitation might seem like a strange attribute to relate to lung cancer, Carr and his colleagues (2002) argued that high precipitation was historically advantageous for growing tobacco, could lead to greater time spent indoors, and the associated high humidity could increase chemical activity on the lung surface. The annual precipitation data are depicted in the three columns of maps (health service areas having low, medium, and high values are shown from left to right, with the values for each column indicated in the legend at the bottom of Color Plate 25.2). In an analogous fashion, the percent below poverty level data are shown in the three rows of maps (health service areas having low, medium, and high values are displayed from bottom to top, as indicated in the legend to the right). We can see that a large number of health service areas (these are counties or aggregates of counties) with high lung cancer mortality rates are associated with high precipitation and a high percent below the poverty level (the red region in the southeast of the United States shown in the upper right map). This association (or correlation) does not necessarily mean that there is a cause-and-effect relationship, but it does suggest that these and other related attributes should be examined for explaining the high mortality rates in the Southeast.*

As with LM plots, CCmaps are best appreciated by experimenting with associated software, which can be found at *http://www.galaxy.gmu.edu/~dcarr/ccmaps/*. Here you will find that it is possible to dynamically change the class limits on any of the three attributes and thus develop different visualizations of the relationships among the attributes. The ability to change class limits fits well with the notion of data exploration software, as discussed in Chapter 21.

25.2 USING SOUND TO INTERPRET SPATIAL DATA

In this section, we consider how researchers have used sound to interpret spatial data. From a cartographic standpoint, John Krygier (1994) noted two basic forms of sound: realistic and abstract. **Realistic sounds**, those sounds with meaning based on our past experience, can be divided into speech (or narration) and mimetic sound icons (or "earcons"). A simple example of speech is having the computer speak the value for an enumeration unit rather than having it displayed as text. Another example of speech is the audio portion of the animations included in multimedia encyclopedias, as discussed in

Chapter 22 (e.g., you might hear, "You are now looking at major Allied troop movements during the Battle of the Bulge"). An example of a mimetic sound icon would be the sound of fire and wind in an animation illustrating changes in a forested landscape.

Abstract sounds have no obvious meaning, and thus require a legend to explain their use. For example, imagine a map of census tracts with the title "Median Income, 1997," in which different magnitudes of loudness represent different incomes (a mouse click on a high-income tract would produce a louder sound than a low-income tract). To understand the magnitudes of loudness, a legend would have to be provided indicating that a higher magnitude represents a higher income (if the reader were blind, the reader would have to be told this). Our emphasis in this section is on abstract sounds. The process of creating abstract sounds is sometimes referred to as **sonification**.

At this point, you might wonder why it makes sense to use sound to represent spatial data, when maps can be interpreted with vision. Most obvious is the fact that blind (or visually impaired) individuals cannot make use of vision. Although tactile maps have been created for the blind, they have numerous limitations, including the cost of creating them, the difficulty of construction, their limited portability, the need to identify features using the fingertip or palm, and the difficulty of labeling features (Jacobson et al. 2002). Sound can also be useful for the normally sighted for numerous reasons. First, sound can be useful when the work environment requires that vision be used for other things besides map interpretation (as in an aircraft cockpit). Second, sound can assist in interpreting a map when visual interpretation is not clear. For instance, Suresh Lodha and his colleagues (2000) found that the comparison of regions on choropleth maps was enhanced when both vision and sound were used. Third, sound has the potential for use in multivariate mapping, as different characteristics of sound (e.g., loudness and pitch) can be assigned to different attributes. Finally, sound might be useful as a means of orienting the user (a sound from the northern portion of the map indicates that you should look there for something interesting), and in 3-D displays where visual information might be occluded.†

Analogous to the visual variables that we discussed in section 4.3, Krygier (1994, 154–146) identified the following **abstract sound variables**:

- *Location:* The location of a sound in two- or three-dimensional space
- *Loudness:* The magnitude of a sound
- *Pitch:* The highness or lowness (frequency) of a sound
- *Register:* The relative location of a pitch in a given range of pitches

---

* The numbers in the upper right of each map are the weighted average of lung cancer mortality rate for health service areas appearing on that map.

---

† For a more extensive discussion of reasons for using sound, see Kramer (1994b) and Lodha et al. (1996b).

- *Timbre:* The general prevailing quality or characteristic of a sound
- *Duration:* The length of time a sound is (or is not) heard
- *Rate of change:* The relation between the durations of sound and silence over time
- *Order:* The sequence of sounds over time
- *Attack/decay:* The time it takes a sound to reach its maximum or minimum

As Figure 25.2 shows, Krygier argued that the bulk of these variables would be effective for ordinal-level data.

Krygier (1993) created a video that illustrated several applications of how abstract sound variables might be used. In one application, he combined a choropleth animation of AIDS in the United States from 1980 to 1995 with a loudness variable representing the total number of AIDS cases in each year. When viewing the resulting animation, one gets the feeling of an impending disaster, which is exactly what Krygier intended. In this same animation, Krygier also used pitch to represent the per-

centage increase of new AIDS cases each year (a lower pitch represented a lower percentage increase). Krygier (1994) noted, "The pitch can be heard 'settling down' as the percentage increase drops and steadies in the late 1980s. An anomaly can be heard in 1991 where the animation switches from actual AIDS cases to model predicted AIDS cases" (p. 157). We felt that viewers would readily hear the "settling down," but that it would be more difficult to detect the anomaly; this might be a case where experience would be advantageous (a musician would have an advantage over a nonmusician).

In another application, Krygier created a multivariate display by combining traditional visual variables (lightness and size) with sound variables. For the traditional visual variables, Krygier superimposed proportional circles on a choropleth map. The proportional circles depicted "median income," whereas the choropleth map displayed the "percentage of population not in the labor force." For the sound variables, Krygier used a pitch within three different octaves (a register) to display a "drive to work index," and three levels of pitch within each of the three octaves to represent the "percentage poor." Values for the sound variables were obtained by pointing at enumeration units. For example, for an enumeration unit with a long drive to work and low rate of poverty, a high-octave pitch followed by a low pitch within that octave could be heard. Krygier (1994) indicated that "After a short period of [use] ... it becomes relatively easy to extract the four data variables" (p. 158).

Several other researchers also have experimented with sound. Peter Fisher (1994a) developed software that utilized sound to portray the uncertainty in remotely sensed images. With Fisher's software, a cursor could be moved across a remotely sensed image under either user or automatic control (Color Plate 25.3). As the cursor moved, the user heard a sound representing the uncertainty associated with the current pixel location. For example, if duration were used as the sound variable, a long duration indicated a pixel with a low uncertainty (high reliability). In using Fisher's software, we found it relatively easy to determine the uncertainty associated with specific pixels, but difficult to determine the uncertainty of groups of pixels. This problem might be obviated by also using visual methods to depict uncertainty. For example, we might utilize both sound and a set of gray tones to depict uncertainty (say, with dark areas representing more uncertain information, as in the FLIERS Project described in section 23.4.5).

In addition to the study mentioned earlier, Lodha and his colleagues (1996b) also have used sound to visualize the uncertainty in interpolation on isarithmic maps. As discussed in section 23.4.6, Lodha and his colleagues compared one interpolation approach with another (say, inverse distance with kriging). In discussing their results,

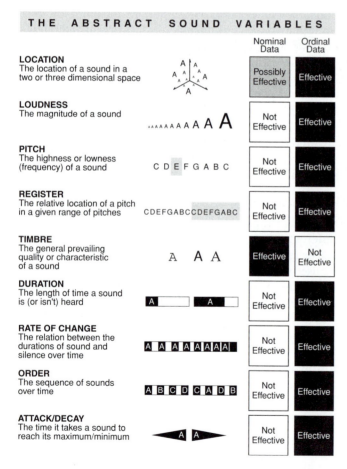

FIGURE 25.2 Abstract sound variables. (Reprinted from Krygier, J. B., *Sound and Geographic Visualization,* Copyright 1994, p. 153, with permission from Elsevier Science.)

they clearly felt that sound provided information not attainable from a visual display. For example, they created a 3-D display in which both the height and color of glyphs were mapped to uncertainty (the higher and whiter the glyph, the greater the uncertainty), and noted that it was "hard to glean information from a glyph" due to "the projection of the graphic." However, when they mapped the uncertainty to pitch and presented the glyphs in sequence from low to high uncertainty, they found the results enhanced "the understanding of quantification of uncertainty considerably" (p. 192).

Recently, Daniel Jacobson, Reginald Golledge, and their colleagues have begun experimenting with sound and touch (**haptics**) as substitutes for (or enhancements to) visual displays in a project known as Haptic Soundscapes (Jacobson et al. 2002; Jacobson 2004). Their Web site (*http://soundscapes.geog.ucsb.edu/*) contains numerous examples of some of their efforts. For instance, when we examined the site in December 2003, we noted two variants of a county-level map for California. In one, when the mouse cursor was moved over a county, the name of the county was spoken, and when the cursor was pushed, the population for the county was spoken. In the other, when the cursor was moved over the county, the pitch and volume were varied to reflect the population of the county (a high pitch and high volume denoted a high population). Jacobson and his colleagues indicated that they plan to add haptic effects to the first variation. As an illustration of the potential use of haptics, Jeong and Jacobson (2002) analyzed the ability of users to interpret crime rate data for states in the United States using only sound (musical tones were used), only haptics (different vibrations were created using a force-feedback mouse), or a combination of the two. Although the experiment showed that the haptic display was more effective, Jeong and Jacobson cautioned that, the musical tone … may not represent the auditory display well" (p. 203).

25.3 COLLABORATIVE GEOVISUALIZATION

Although we often think of maps and mapping software as being used by individuals, there clearly are instances in which collaboration among individuals is important. For example, a grade school teacher might ask students to interpret the pattern of earthquake activity on a printed map. If the map is of sufficient size, several students might gather around the map and discuss the pattern they see. When such collaborative activities involve visualization, we refer to them as **collaborative geovisualization** (or **geocollaboration**). If we consider the full realm of printed and computer-based maps, then we can conceive of collaboration that takes place at the *same* or a *different* time, and at the *same* or a *different* place. This produces a two-by-two matrix of possible instances for collaboration

	Same Time	Different Time
Same Place	Urban planning meeting	Strategic military planning
Different Place	Scientists collaborate with decision-makers	School project involving classwork and fieldwork

FIGURE 25.3 Collaborative geovisualization can take place at the *same* or *different* time (the columns) or the *same* or *different* place (the rows). (After Slocum et al. 2001. First published in *Cartography and Geographic Information Science* 28(1), p. 67. Reprinted with permission from the American Congress on Surveying and Mapping.)

(Figure 25.3). For instance, an urban planning meeting generally would involve people collaborating in the same time and place, whereas a school project involving both classwork and fieldwork could require collaboration at different times and places. In this section, we consider several examples of collaborative geovisualization being undertaken by cartographers and other geoscientists.

25.3.1 Hedley and His Colleagues' Work

In the previous chapter, we considered the work of Nick Hedley and his colleagues in the realm of augmented reality (AR). Here we discuss their related effort in collaborative geovisualization (Hedley et al. 2002), which utilizes their previous work with AR, but also permits computer interpretation of natural gestures, and allows a range of views—from the real-world to AR to an immersive virtual environment (VE). As described in the preceding chapter, Hedley and his colleagues created an AR through the use of a head-mounted display (HMD) with an attached video camera and associated software—when users look at a card containing a unique pattern of marks, a 3-D image appears at that location. Natural gestures are interpreted via a specialized projection table known as the *human information workspace* (or HI-SPACE) developed by Richard May.* An infrared light source is placed beneath the table and a camera above the table records the infrared light. When users gesture (e.g., by pointing with their finger) or place an object on the table, the infrared light is blocked, producing dark images that can be interpreted by computer software. Initially, users can view a paper map placed on the table

---

* For more information, see *http://www.hitl.washington.edu/people/rmay/hispace.html.*

without any specialized apparatus. If users then don HMDs and examine a marked card placed on the map, they all see a 3-D model of terrain at that location (Figure 25.4). This model can then be picked up by one individual (by picking up the card) and passed around to others, all of whom will see the 3-D AR view. Other types of geographic information also can be seen by placing the appropriately marked cards on the map next to a particular terrain model (Figure 25.5). If a handheld Sony Glasstron viewer is used rather than the HMD, it is possible to achieve both AR and immersive VE views (Figure 25.6). The Sony Glasstron contains an inertial tracker that enables head orientation to be interpreted and a pressure pad that can be used to move through the VE.

Hedley and his colleagues found that users preferred the Sony Glasstron to the more traditional hands-free HMD "because it was easy to share views with others [and] because it reduced ... incidences of nausea" (Hedley et al. 2002, 127). Remember that health issues was one of the concerns we noted for VEs in Chapter 24. Hedley et al. noted, however, that users of the Glasstron often wanted to move backward in the VE (as in computer games), which was something that they had not allowed. Hedley also noted several limitations of their initial collaborative effort (known as AR PRISM): Users could not annotate the data or place their own virtual models in the AR scene, there was no support for viewing two-dimensional imagery (e.g., remotely sensed images), virtual models might disappear if a marker card was obscured, and using a paper map as a base prevented zooming and panning (p. 128–129). Presently, they are responding to these problems by developing a new system known as GI2VIZ.

FIGURE 25.4 Collaborative geovisualization is achieved by Hedley and his colleagues by using HMDs containing a video camera and a specialized projection table known as HI-SPACE. (Courtesy of Nicholas R. Hedley, Mark Billinghurst, Lori Postner, Richard May, and Hirokazu Kato, "Explorations in the Use of Augmented Reality for Geographic Visualization," *Presence* 11:2 (April 2002), pp. 119–133. © 2002 by the Massachusetts Institute of Technology.)

25.3.2 MacEachren and His Colleagues' Work with ImmersaDesks

Alan MacEachren and his colleagues (2000; 2004) experimented with collaborative geovisualization by using ImmersaDesks to examine precipitation and temperature relationships in association with Hurricane Agnes, which was a major weather event in 1972. They attempted both

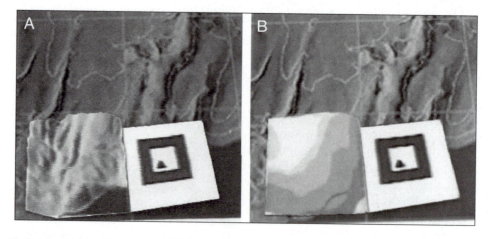

FIGURE 25.5 In Hedley and his colleagues' collaborative geovisualization approach, other types of geographic information can be seen by placing the appropriately marked cards on the map next to a terrain model that is being examined. Here we see a soil card placed next to a terrain model (A); when the soil card is interpreted, we get the result in (B). (Courtesy of Nicholas R. Hedley, Mark Billinghurst, Lori Postner, Richard May, and Hirokazu Kato, "Explorations in the Use of Augmented Reality for Geographic Visualization," *Presence* 11:2 (April 2002), pp. 119–133. © 2002 by the Massachusetts Institute of Technology.)

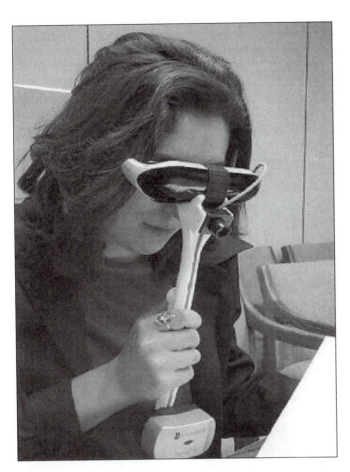

FIGURE 25.6 A handheld Sony Glasstron viewer that enables both AR and immersive VE views. (Courtesy of Nicholas R. Hedley, Mark Billinghurst, Lori Postner, Richard May, and Hirokazu Kato, "Explorations in the Use of Augmented Reality for Geographic Visualization," *Presence* 11:2 (April 2002), pp. 119–133. © 2002 by the Massachusetts Institute of Technology.)

same-place and different-place collaborations (for the same time) by having one group of researchers work with an ImmersaDesk at Pennsylvania State University, while another group worked with an ImmersaDesk at Old Dominion University (see Chapter 24 for a discussion of the ImmersaDesk). Color Plate 25.4 provides one example of the images utilized in the Hurricane Agnes study. The bottom portion of the color plate shows a 3-D portrayal of terrain; the middle portion represents temporal changes in precipitation rather than elevation, with olive green "clouds" depicting precipitation isosurfaces above a certain threshold value (Hurricane Agnes is depicted by a "blanket" of rain multiple days thick); and the top portion portrays temperature at the end of the precipitation event. By examining images such as these, researchers were able to see the "substantially reduced

temperature across the basin following Hurricane Agnes, as the huge quantities of water dumped on the region slowly evaporated" (MacEachren et al. 2000, 39). Although MacEachren and his colleagues (2004) felt that useful information was gleaned with the ImmersaDesks, they noted a number of problems, including the use of stereo glasses, limited the ability to utilize gestures and facial expressions; only one individual could have control of the display, and that individual had to hand a wand and control glasses to another desiring control; optimal viewing was best for the individual in control; and communication between the groups was only possible via a speaker telephone (Internet tools proved unreliable). Given limitations such as these, MacEachren and his colleagues (2004) have begun experimenting with other technologies for collaborative geovisualization. One of these is the HI-SPACE technology mentioned previously; another is a wall-size display that can generate stereo images viewable with polarized glasses.

25.3.3 NCAR's Visualization Lab and the Grid

The National Center for Atmospheric Research (NCAR) Visualization Lab provides a good illustration of how researchers are beginning to utilize collaborative geovisualization technologies. The centerpiece of the lab is a 24×9-foot wall-size display driven by a supercomputer with dedicated, parallel graphics engines. Users can view the screen without specialized apparatus, or they can don glasses that enable them to see stereo images (as MacEachren and his colleagues promoted). Lynda Lester (2002) stressed that such technology allows users to interact with other groups in similar labs around the world. These interactions "can take the form of large-scale distributed meetings, collaborative research, seminars, symposia, lectures, tutorials, and training" (p. 4). An important element of any such collaboration is the ability to share massive amounts of data and associated applications that are distributed throughout the world—this is achieved using the *Grid,* "a body of technologies that enables distributed supercomputers, storage systems, data sources, and applications to be used as a unified resource regardless of location" (p. 3). The key is that understanding the 3-D spatiotemporal character of the atmosphere will require interdisciplinary efforts of groups located around the world and that the necessary data and applications will not necessarily be found at one place.

25.4 MULTIMODAL INTERFACES

At present, the bulk of our work with personal computers is accomplished via a windows-icons-menus-pointing device (WIMP) interface. Although WIMP interfaces are likely to remain common for the next 5 to 10 years, researchers are experimenting with a variety of new interfaces that eventually might replace the WIMP paradigm. Novel ways of interacting with computers include speech, lip movements, pen-based gestures, free-hand gestures, and head and body movements. Speech refers to using spoken commands (e.g., "Draw a line from New York City to Albany"), which requires *speech recognition* hardware and software; because speech recognition is imperfect, it is often supplemented by computer-vision methods for interpreting lip movements. A *pen-based gesture* refers to using a pen-like device to specify locations or provide written input by *touching* a computer display or input device. In contrast, a *free-hand gesture* uses just the arm and hand to specify a location (or region) *without touching* the display. Obviously, free-hand gestures require considerably greater computer processing, as computer-vision methods must be used to determine the location intended by the user, as opposed to the *x* and *y* coordinates that result from touching a display in a pen-based gesture. Sharon Oviatt (2002; 2003) of the Oregon Graduate Institute of Science & Technology is one of those who has championed novel interaction methods. Oviatt has promoted the notion of **multimodal interfaces**, or an interface composed of two or more novel modes of interaction. Thus far, the most common multimodal interfaces have utilized a combination of either speech and pen-based gestures or speech and lip movements. For example, when using speech and a pen-based gesture, the user might point to the state of New York and say, "Shade this region in blue."

Multimodal interfaces have numerous advantages. One is that the associated novel forms of interaction are arguably more desirable than WIMP interfaces because they are more natural—imagine being able to simply say "Show me the location of houses costing between $75,000 and $125,000" as opposed to having to learn and traverse the menus of a WIMP interface. Second, multimodal interfaces allow a wider spectrum of users to utilize computers; for example, a visually impaired user might prefer speech, whereas a child might prefer a pen-based gesture. A third advantage is the ability to use certain interaction modes in specialized situations; for example, in the field you might want your hands free, and so speech is an obvious choice, whereas interaction during a meeting might require the quiet of a pen-based gesture. A fourth advantage is the need to reduce the task difficulty and errors resulting from using only one mode. For example, in speech-based input, users often have difficulty expressing themselves, particularly for spatial tasks. Oviatt (1997, 120) illustrated this point in a classic early study involving interactive maps. To specify the location of open space with just speech, a user stated, "Add an open space on the north lake to b- include the north lake part of the road

and north." When both pen-based gestures and speech were permitted, the user used a pen to circle the area and stated, "Open space." Overall, Oviatt found that when using multimodal interfaces, tasks were simplified, they were completed more quickly, and there were fewer errors. Moreover, people preferred multimodal interaction.

Oviatt and her colleagues (2000) described numerous applications involving multimodal interfaces. QuickSet is one that has involved the use of maps, both for military and health care applications. Regarding the military application, Cohen and his colleagues (1997, 35) noted, "Whereas previous manual methods for initializing scenarios resulted in a large number of people spending more than a year in order to create a division-sized scenario, a 60,000+ entity scenario recently took a single … user 63 hours, most of which was computation." Although the improvements for QuickSet were not quite as dramatic when compared to an automated WIMP approach, Oviatt et al. (2000, 287) indicated "a nine fold reduction in entity creation time."

Rajeev Sharma and his colleagues (2003) at Pennsylvania State University and Advanced Interface Technologies have done considerable work with multimodal interfaces involving speech and free-hand gestures. One of their early systems, known as XISM (Kettebekov et al. 2000), was used to simulate an urban emergency response system. In this case, a single user played the role of an emergency center operator who dispatched vehicles to particular locations (Figure 25.7). The user's gestures were recorded by a camera located at the top of the

FIGURE 25.7   The interface for the multimodal XISM used as an emergency response system. The user provided input by speech and free-hand gesture (note the microphone dome hanging from the ceiling and the camera placed above the display). (Courtesy of Rajeev Sharma.)

display, and the user's speech was recorded by a microphone dome hanging from the ceiling. Sophisticated computer hardware and software was then used to interpret the resulting speech and gestures. Emergencies were indicated in the display by the animation of symbols and associated audible alarm signals. The user would respond with an appropriate verbal command (e.g., "Acknowledge this") and then use speech and Gesture to dispatch the vehicles (e.g., the user might say "Dispatch an ambulance from this station to that location" while gesturing to the station and location). Because free-hand gestures could be too coarse at some scales, the system permitted zooming via speech and gesture (e.g., the user would say "Zoom here" and make the appropriate gesture).

More recently, Sharma and his colleagues, in concert with Alan MacEachren and his colleagues in the Geo-VISTA Center at Penn State, have begun developing a collaborative geovisualization system for emergency management situations commonly dealt with in a GIS framework (e.g., disaster relief associated with a hurricane; Rauschert et al. 2002; Sharma et al. 2003). Sharma and his colleagues argued that decision makers need to interact with disaster-related software in a natural manner, as opposed to utilizing GIS specialists who are familiar with the menus and software of traditional GIS interfaces. Their system, known as the Dialogue-Assisted Visual Environment for Geoinformation (DAVE_G), builds on XISM by using both speech and free-hand gesture, but in a collaborative framework. Interpreting the input of multiple users is handled, in part, by assigning separate microphones and cameras to each collaborator. The greater complexity of the system also is dealt with by carefully analyzing the actions of decision makers in real-world emergency operations centers before developing the software.

One key characteristic of ongoing work in multimodal interfaces is that it is multidisciplinary. For example, work at Penn State involves those in the departments of computer science, geography, and information sciences and technology. Clearly, the complexity of multimodal interfaces, particularly in the collaborative realm, necessitates a group research effort.

25.5 INFORMATION VISUALIZATION

As introduced in section 1.5, **information visualization** involves the visualization and analysis of nonnumerical abstract information. To create a visualization of nonnumeric information, graphic designers frequently utilize a spatial or geographic metaphor—a process known as **spatialization**. For example, Figure 1.7 illustrates a topographic surface of online news reports associated with the Kosovo crisis. The notion is that people should feel comfortable in interpreting this surface because of their familiarity with topographic surfaces.

André Skupin and Sara Fabrikant (2003) are two geographers who have done extensive work in information visualization.* An example of Skupin's work is his attempt to visualize the relationships among more than 2000 abstracts submitted to the 1999 annual meeting of the Association of American Geographers (Skupin 2002). Given that the discipline of geography normally is divided into three broad areas (human geography, physical geography, and techniques), Skupin wondered how these three areas, and their relations, might be expressed in the relation among abstracts, and also whether other divisions of the discipline might be appropriate. One visualization that Skupin created appears in Figure 25.8, where we see two-dimensional "maps" of the relationships among topics found in the abstracts. The left map portrays

---

* For examples of work outside geography, see Card et al. (1999), Keim (2001), and Eick (2001).

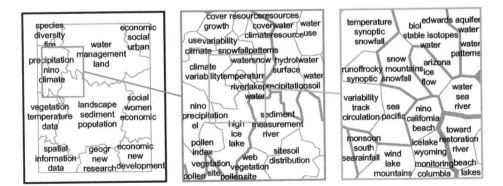

FIGURE 25.8   A spatialization of topics found in abstracts submitted to the 1999 meeting of the Association of American Geographers: Map A represents the most generalized view, whereas maps B and C show the detail that is possible when zooming to various levels. (After Skupin 2002, "A cartographic approach to visualizing conference abstracts," *IEEE Computer Graphics and Applications* 22, no. 1, p. 56; © 2002 IEEE.)

the most generalized version, in which 10 regions arise, whereas the right maps show the detail that is possible when maps of 100 and 800 regions are used, respectively. In discussing Figure 25.8A, Skupin stated:

> ...human geography occupies the right half...aspects of physical geography dominate the upper left quadrant. The processing and modeling of geographic data dominate the lower left quadrant. The cluster labeled geogr/new/research ...contains many abstracts that deal with the teaching of geography, such as research into the development of new teaching tools and techniques. Issues surrounding resource management dominate the top of the map. This is a heterogeneous area at the intersection of human and natural environments. (p. 56)

It would be interesting to see how visualizations such as these would evolve over time: Imagine comparing the visualization shown in Figure 25.8 with one developed 5 or 10 years later.

An example of Fabrikant's work is the research that she undertook with Barbara Buttenfield to develop spatializations of documents found in GEOREF, an online set of documents pertaining to literature in the geological sciences (Fabrikant and Buttenfield 2001). Figure 25.9 illustrates two different spatializations of 100 documents found in GEOREF developed by Fabrikant and Buttenfield. In Figure 25.9A we see a two-dimensional map of point symbols for the 100 documents, where each symbol is labeled according to a first-level keyword found in GEOREF, and each symbol is shaded according to the number of other documents in GEOREF that contain the same keyword. Documents that are near one another in the two-dimensional map are presumably more similar to one another in content than those that are further away. In Figure 25.9B, we see a 3-D view of the same data portrayed in Figure 25.9A: Here the height of the surface represents the frequency with which various keywords are used in other documents (in this case, the labels are topical index terms taken from the GEOREF thesaurus). Fabrikant and Buttenfield argued that spatializations such as these serve as "tools to provide an overview of available information, to discover relationships between items in a data archive, and to filter non-relevant pieces of information" (p. 263). Although these spatializations might seem abstract, Fabrikant and Buttenfield have found that users glean considerable information from them because they are able to take advantage of the geographic metaphor (e.g., that nearby things are more similar to one another).

A question that you might be asking is how do we convert nonnumerical abstract information into the spatializations depicted here? One key step is the selection of an appropriate *spatial layout* technique, such as multidimensional scaling (Skupin and Fabrikant 2003).

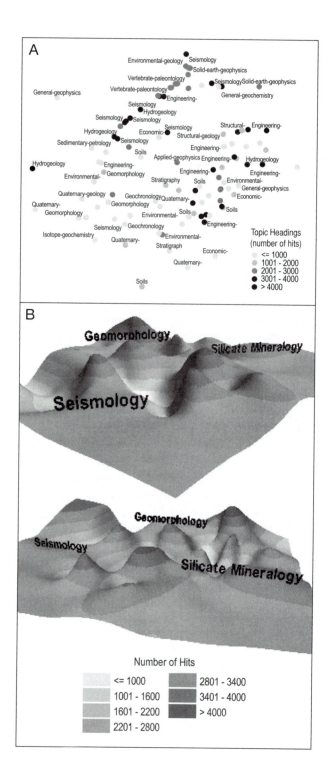

FIGURE 25.9 Spatializations of 100 documents found in GEOREF, an online bibliographic database: (A) a 2-D map depicting the number of other documents that share keywords with those documents plotted in *x* and *y*; (B) 3-D maps of the same information. (From Fabrikant, S. I., and Buttenfield, B. P. (2001). "Formalizing semantic spaces for information access." *Annals of the Association of American Geographers* 91, no. 2, pp. 272, 273. Courtesy of Blackwell Publishing.)

Unfortunately, space does not permit us to delve into such techniques here. These techniques are commonly discussed in intermediate or advanced courses in quantitative methods, so we encourage you to develop a solid background in quantitative methods.

25.6 SPATIAL DATA MINING

In recent years, computer scientists and other IT professionals have struggled with extracting knowledge from very large data sets. For example, a health-related database might have millions of patient records, with tens or even hundreds of attributes for each patient. Traditional statistical approaches such as correlation and regression can be utilized to examine such data, but these approaches have a high computational burden and normally are utilized only for theory-driven hypothesis testing. As a result, researchers have developed the technique known as **knowledge discovery in databases** (KDD), which refers to exploratory approaches for finding interesting patterns in databases and ultimately uncovering knowledge about those databases. Given that the intention is to *mine* useful information (and ultimately knowledge) from these databases, *data mining* is a term that is often used instead of KDD; moreover, when the database is inherently spatial, the term **spatial data mining** is often used. We should keep in mind, however, that technically data mining is considered a subset of KDD. In general, KDD involves *data selection* (determining the appropriate records and attributes to consider), *preprocessing* (eliminating noise and duplicate records, and selecting methods for handling missing data), *data reduction and projection* (reducing the number of records or attributes under consideration), *data mining* (using specialized algorithms to search for patterns in the data), and *interpretation and reporting* (interpreting findings, hopefully uncovering knowledge about the database).*

We now consider two examples that illustrate the notion of spatial data mining. Due to space limitations and the fact that we have not delved deeply into KDD, we have purposely selected examples that involve few observations and do not require completing all of the steps involved in KDD—thus, our focus is on the data mining step.† For our first example, we consider the work of Alan Murray and Tung-Kai Shyy (2000), who wished to examine property crime rates in the 178 suburbs of Brisbane, Australia. Murray and Shyy initially mapped crime rates for the suburbs using a traditional choropleth map

(they used the Fisher–Jenks optimal classification method). They argued, however, that this approach was inappropriate because traditional classification methods are aspatial, considering only the values of crime rates and not the spatial relationships of the associated suburbs, a notion that we introduced in section 5.2. Murray and Shyy then argued that both crime rates and the spatial relationships between suburbs could be considered by using *bicriterion median clustering*, a variant of a *median* cluster analysis approach commonly used in spatial data mining (see section 5.3 for a discussion of cluster analysis methods). They accounted for spatial relationships between suburbs by including the distance between neighboring suburbs as an attribute in the clustering method (in addition to the attribute property crime rates).

The key point about bicriterion median clustering is that Murray and Shyy were able to experiment with assigning different weights to each of the attributes. To illustrate, Color Plate 25.5 portrays two different runs of their cluster method: In Color Plate 25.5A the crime rate and distance attributes are weighted 1.0 and 0.25, respectively, whereas in Color Plate 25.5B crime rate and distance are weighted 1.0 and 0.95. Obviously, distance plays a greater role in Color Plate 25.5B, with the result being more homogeneous regions. Murray and Shyy suggested that their method ultimately must be used in a data exploration framework in which the weights for crime rate and distance can be experimented with, along with the number of classes (note that six classes were used in Color Plate 25.5). In examining the maps in Color Plate 25.5, also note that particular crime rates are not assigned to each class (the classes are simply numbered 1 to 6), and that the classes do not appear to be logically ordered visually, as we have promoted in earlier chapters (in addition to lightness, texture is also used to discriminate the classes). The reason for these characteristics is that class limits under the bicriterion median clustering can overlap one another, which is normally not the case on choropleth maps. Although overlapping classes are likely to complicate interpretation, they do permit a more thorough integration of spatial relationships into the clustering process.◊

As a second example of spatial data mining, we consider the efforts of Natalia Andrienko and her colleagues (2001b) to study regional groupings of administrative districts in Bonn, Germany. The five colored regions in Color Plate 25.6 represent one possible way in which a geographer might group the districts of Bonn. To examine the character of these regions, Andrienko and her colleagues

---

* For further discussion of the notion of KDD and data mining and their relation to traditional statistical analysis, see Miller and Han (2001b) and Shneiderman (2002).

† For a more sophisticated example, see Guo et al. (2003).

◊ Although the same symbology is used for both maps in Color Plate 25.5, note that the range of values falling in a particular class would not necessarily be the same on the two maps. For another cluster analysis approach that resulted in overlapping classes, see Rowles (1991).

submitted the district identifiers (which region the district is a member of) and numerous thematic attributes associated with the districts to the spatial data mining algorithm C4.5 (Quinlan 1993). The objective of C4.5 is to determine which attributes (and subranges thereof) are most closely associated with previously identified classes (in our case the five regions).

To illustrate the nature of the C4.5 algorithm, consider the resulting *decision tree* shown below the map of Bonn in Color Plate 25.6. At the top of the decision tree, we see that "Percentage (Age groups = 18–30)," or the percentage of the population 18 to 30 years old, is the attribute that initially best discriminates among the five regions, with the breakpoint being 19.03. Looking at the left portion of the second level of the decision tree, we see that a total of 44 districts have less than or equal to 19.03 percent of the population 18 to 30 years old, whereas the right portion of the second level reveals that 18 districts have greater than 19.03 percent of the population 18 to 30 years old. Most important, the right portion consists largely of the "Center" region (83 percent of the 18 districts are in this region). Again, looking at the left portion of the second level, we can see that "Percentage (Population groups = Foreigners)," or the percentage of foreigners in the population, breaks the 44-district group into two subgroups. In the third level we can see that one of the subgroups has a percentage of foreigners greater than 12.1, and is largely associated with the "South-East" region. Finally, again looking at the right portion of the second level, we see that "Percentage (Age groups = 0–18)," or percentage of the population 0 to 18 years old, splits the 18-district group into two groups: Here the two North-West districts have a higher percentage 0 to 18 years old. To summarize, the C4.5 method has revealed that the Center and South-East portions of Bonn appear to have different population structures than other areas: The Center has a high percentage of people in the 18- to 30-year-old range, whereas the South-East has a high percentage of foreigners.

Andrienko and her colleagues argue that spatial data mining is best used in conjunction with interactive visualization techniques. Thus, each portion of the decision tree can be depicted on the map by clicking on the desired portion of the tree. For instance, in Color Plate 25.6 note that the right portion of the second level of the tree has a thick black line surrounding it, indicating that it has been clicked on. Looking at the map, we can see that the 18 associated districts have been identified on the map by small square boxes. Admittedly, novel spatial data mining methods are more complicated than traditional thematic maps, but they have the potential to uncover patterns in the data that we might not normally detect (imagine performing a similar task with the 3,000+ counties of the United States for 100 attributes).

25.7 KEEPING PACE WITH RECENT DEVELOPMENTS

One problem that we had in writing this book was keeping pace with ongoing developments: The field of cartography is changing so fast that any text becomes dated as soon as it is written. We have handled this problem to some degree by providing updated information for each chapter on the home page for the book. Those wishing detailed information on current research topics should also consider reviewing recent issues of the journals and proceedings listed in Table 25.1. Note that journals in the table are split into *primary* and *secondary* ones. Most of the articles within a publication in the primary list have a cartographic component, whereas only some of the articles in the secondary list focus on cartography. In addition to reviewing recent publications, you might also find it useful to attend conferences that include topics of interest to cartographers (Table 25.2). Such conferences frequently include expositions illustrating the latest technological advances in hardware and software. Being aware of such advances is important, as the field of cartography is arguably technology-driven.

It is also important to be aware of research groups and organizations that can provide useful information on ongoing developments. The National Center for Geographic Information and Analysis (NCGIA; *http://www.ncgia.ucsb.edu/*) has produced numerous publications of interest to cartographers. More recently, the University Consortium for Geographic Information Science (UCGIS; *http://www.ucgis.org/*), a group of more than 70 universities, professional organizations, and affiliates, has developed a research agenda (*http://www.ucgis.org/priorities/research/2002researchagenda.htm*), much of which is relevant to cartographers. The International Cartographic Association (ICA; *http://www.icaci.org/*) has 22 ongoing commissions of relevance to cartography; just some of these that are relevant to this book include Generalization and Multiple Representation, Map Projections, Maps and the Internet, Mountain Cartography, National and Regional Atlases, and Visualization and Virtual Environments. Finally, you might wish to examine the Carto Project, an undertaking by ACM SIGGRAPH (Association for Computing Machinery, Special Interest Group on Computer Graphics) to explore how "viewpoints and techniques from the computer graphics community can be effectively applied to cartographic and spatial data sets" (*http://www.siggraph.org/~rhyne/carto/*).

TABLE 25.1 Journals and conference proceedings containing articles dealing with thematic cartography and geographic visualization

Primary journals

Cartographic Perspectives
Cartographica
Cartography
Cartography and Geographic Information Science
The Cartographic Journal

Secondary journals

Annals, Association of American Geographers
Byte
Computer Graphics
Computers & Geosciences
Environment and Planning A
Environment and Planning B: Planning and Design
Geographical Analysis
Geospatial Solutions
GEOWorld
IEEE Computer Graphics & Applications
IEEE Transactions on Visualization and Computer Graphics
International Journal of Geographical Information Science
International Journal of Human-Computer Studies
Journal of Geography
Journal of the American Statistical Association
Landscape and Urban Planning
Photogrammetric Engineering and Remote Sensing
Presence
Progress in Human Geography
Statistical Computing & Statistical Graphics Newsletter
The American Statistician
The Professional Geographer
Transactions in GIS
Transactions, Institute of British Geographers

Proceedings

ACM SIGGRAPH (Association for Computing Machinists, Special Interest Group on Graphics)
ICA (International Cartographic Association)
Innovations in GIS (Proceedings of the U.K. National Conference on GIS Research: GISRUK)
International Symposium on Spatial Data Handling (International Geographical Union)
Visualization (IEEE)

Note: For proceedings, the name of the sponsoring organization is provided in parentheses.

TABLE 25.2 Conferences having topics of interest to cartographers

AAG (Association of American Geographers)
ACM SIGGRAPH (Association for Computing Machinists, Special Interest Group on Graphics)
ACSM/ASPRS Annual Convention (American Congress on Surveying and Mapping)
ESRI International User Conference (ESRI)
GIScience (University Consortium for Geographic Information Science)
ICA (International Cartographic Association)
International Symposium on Spatial Data Handling (International Geographical Union)
NACIS (North American Cartographic Information Society)
Visualization (IEEE)

SUMMARY

In this chapter we have examined a number of ongoing developments in the evolving discipline of cartography. One development is Daniel Carr and his colleagues' work with **linked micromap plots** (LM plots) and **conditioned choropleth maps** (CCmaps). LM plots split a spatial distribution into a series of micromaps, which enable map readers to focus on local patterns as opposed to the overall spatial pattern. In addition, LM plots enable graphical and statistical information (e.g., confidence intervals) to be linked with spatial information. Whereas LM plots involve placing an equal number of observations on each micromap, CCmaps utilize other attributes to split (or condition) a thematic attribute of interest into individual maps. For instance, we saw how the attributes percent below the poverty level and precipitation could be used to explain the distribution of lung cancer. Like many novel techniques today, LM plots and CCmaps are most effectively utilized in a data exploration framework.

A second ongoing development is the use of sound to interpret spatial data. Although we normally think of maps as *visual* entities, sound is obviously useful for the visually impaired, and can be useful for the normally sighted (for instance, when work conditions preclude the use of vision, as in an aircraft cockpit). We saw that **abstract sound variables** such as loudness, pitch, and duration could be used to depict thematic information; for instance, we might create a bivariate map by letting one attribute be represented by loudness and another by pitch. Daniel Jacobson, Reginald Golledge, and their colleagues currently are experimenting with sound and touch (**haptics**) in the Haptic Soundscapes project.

Collaborative geovisualization is another development that stretches our traditional view of maps. Normally, we think of maps as being utilized by individuals, but clearly there are many instances in which collaboration is essential (think of schoolchildren who wish to visualize the migration of the Monarch butterfly or climatologists who wish to visualize the impacts of different models of global warming). We looked at several applications of collaborative geovisualization. In one, Nick Hedley and his colleagues experimented with a variety of high-tech capabilities, including augmented reality (AR), an immersive virtual environment (VE), and a specialized projection table known as HI-SPACE. In another, Alan MacEachren and his colleagues experimented with ImmersaDesks; they noted a number of disadvantages with the ImmersaDesks, suggesting that other technologies such as HI-SPACE and a wall-size display could be more useful.

In addition to developing geographic visualization software that will allow users to collaborate, researchers are experimenting with novel user interfaces that utilize speech, lip movements, pen-based gestures, free-hand gestures, and head and body movements. **Multimodal interfaces** are novel interfaces that involve two or more of these techniques. Multimodal interfaces are desirable because they are more natural (imagine simply saying "Find all sites within 30 meters of the river" as opposed to learning and traversing a set of menus), they allow a wider spectrum of computer users, they permit users to interact with a computer in whatever mode is convenient (if you are using your hands in the field, then speech is an obvious choice), and they reduce the errors resulting from using only one mode of interaction. Particularly exciting work in multimodal interfaces is being done at the Geo-VISTA Center at Penn State University, where researchers are developing collaborative geovisualization systems for emergency management situations.

Normally, we think of thematic maps as being used to depict the distribution of some *geographic* phenomenon. In the realm of **information visualization** researchers are stretching our notion of what constitutes a map by creating maps of *nonnumerical abstract* information. **Spatialization** is the process of converting such abstract information to a spatial framework. Sara Fabrikant and Barbara Buttenfield, two geographers who have developed information visualization applications, argue that users can glean useful information from such visualizations because they are able to take advantage of the geographic metaphor (e.g., that nearby things are more similar to one another).

Large databases consisting of millions of records and tens or hundreds of attributes often challenge our abilities to visualize information. As a result, researchers are utilizing **knowledge discovery in databases** (KDD) or, more simply, **spatial data mining** techniques to discover interesting patterns in these databases. As simple examples of spatial data mining, we considered Alan Murray and Tung-Kai Shyy's clustering approach for integrating the distance between enumeration units in a choropleth mapping problem and Natalia Andrienko and her colleagues' use of the C4.5 algorithm to determine which attributes are associated with subregions of a city.

Because the discipline of cartography continues to evolve, any book such as this will be out of date as soon as it is written. Thus, to keep up-to-date, you should read major journals and proceedings in which cartographers publish, attend conferences of interest to cartographers, and be aware of research groups and organizations that provide useful information on ongoing developments.

FURTHER READING

Barrass, S. (1994) "A perceptual framework for the auditory display of scientific data." *Proceedings of the Second International Conference on Auditory Display, ICAD '94*, Santa Fe, NM, pp. 131–145.

Presents a model for sound that is analogous to the 3-D models for color that are constructed based on hue, saturation, and lightness.

Carr, D. B., Olsen, A. R., Pierson, S. M., and Courbois, J.-Y. P. (2000a) "Using linked micromap plots to characterize Omernik ecoregions." *Data Mining and Knowledge Discovery* 4, no. 1:43–67.

Applies the technique of linked micromap plots to characterizing ecoregions in the United States.

Cartography and Geographic Information Science 28, no. 1, 2001 (entire issue).

This special issue focuses on research challenges in geographic visualization.

Communications of the ACM 45, no. 8, 2002 (entire issue).

This special issue focuses on data mining; pages 54–58 deal with scientific applications including geospatial data.

Hedley, N. R., Billinghurst, M., Postner, L., May, R., and Kato, H. (2002) "Explorations in the use of augmented reality for geographic visualization." *Presence* 11, no. 2:119–133.

Discusses techniques for collaborative geovisualization that involve augmented reality and the HI-SPACE projection table.

Koua, E. L. (2003) "Using self-organizing maps for information visualization and knowledge discovery in complex geographic datasets." *Proceedings of the 21st International Cartographic Conference*, Durban, South Africa, pp. 1694–1702.

Uses a form of artificial neural network, the Self-Organizing Map (SOM), for spatial data mining.

Kramer, G. (ed.) (1994a) *Auditory Display: Sonification, Audification, and Auditory Interfaces*. Reading, MA: Addison-Wesley.

A compendium of papers dealing with using sound to display data.

Krueger, M. W., and Gilden, D. (1999) "KnowWare™: Virtual reality maps for blind people." In *Medicine Meets Virtual Reality*, ed. by J. D. Westwood, H. M. Hoffman, R. A. Robb, and D. Stredney, pp. 191–197. Amsterdam: IOS Press.

Describes a specialized computer-based technology for permitting the blind to identify the names of countries of the world using touch and sound (the user touches the country and the computer speaks the name).

Krygier, J. B. (1994) "Sound and geographic visualization." In *Visualization in Modern Cartography*, ed. by A. M. MacEachren and D. R. F. Taylor, pp. 149–166. Oxford: Pergamon.

A good introduction to the use of sound in cartography.

MacEachren, A. M., and Brewer, I. (2004) "Developing a conceptual framework for visually-enabled geocollaboration." *International Journal of Geographical Information Science* 18, no. 1:1–34.

Presents a theoretical framework for studying collaborative geovisualization. For a review of early work in collaborative geovisualization, see MacEachren (2000; 2001).

MacEachren, A., Dai, X., Hardisty, F., Guo, D., and Lengerich, G. (2003) "Exploring high-d spaces with multiform matrices and small multiples." *Proceedings of the International Symposium on Information Visualization*, Seattle, WA, pp. 31–38.

Presents advanced approaches for visualizing multivariate data building on the notions of small multiples and scatterplot matrices.

MacEachren, A. M., Wachowicz, M., Edsall, R., and Haug, D. (1999c) "Constructing knowledge from multivariate spatiotemporal data: Integrating geographical visualization with knowledge discovery in database methods." *International Journal of Geographical Information Science* 13, no. 4:311–334.

A classic paper on spatial data mining that links geographic visualization and knowledge discovery in databases.

Miller, H. J., and Han, J. (eds.) (2001a) *Geographic Data Mining and Knowledge Discovery*. London: Taylor & Francis.

A collection of readings on spatial data mining and knowledge discovery in databases.

Olson, G. M., and Olson, J. S. (2000) "Distance matters." *Human-Computer Interaction* 15:139–178.

Although the focus is outside geography, considers several key factors that affect successful collaboration, such as differing time zones and cultures.

Oviatt, S. (2003) "Multimodal interfaces." In *The Human-Computer Interaction Handbook: Fundamentals, Evolving Technologies and Emerging Applications*, ed. by J. Jacko and A. Sears, pp. 286–304. Mahwah, NJ: Lawrence Erlbaum.

Provides an overview of multimodal interfaces.

Peterson, M. P. (ed.) (2003) *Maps and the Internet*. Amsterdam: Elsevier Science.

A series of papers dealing with the role of the Internet in modern cartography.

Sharma, R., Yeasin, M., Krahnstoever, N., Rauschert, I., Cai, G., Brewer, I., MacEachren, A., and Sengupta, K. (2003) "Speech-gesture driven multimodal interfaces for crisis management." *Proceedings of the IEEE* 91, no. 9:1327–1354.

Provides an overview of Sharma and his colleagues' work on multimodal interfaces, including those using collaborative geovisualization.

Shepherd, I. D. H. (1994) "Multi-sensory GIS: Mapping out the research frontier." *Proceedings of the 6th International Symposium on Spatial Data Handling (Advances in GIS Research)*, Edinburgh, Scotland, pp. 356–390.

Provides an overview of how our various senses might be used in computer interfaces for interpreting spatial data.

Steinbach, M., Tan, P.-N., Kumar, V., Klooster, S., and Potter, C. (2002) "Temporal data mining for the discovery and analysis of ocean climate indices." *KDD-2002, The Eighth ACM SIGKDD International Conference on Knowledge Discovery and Data Mining*, Edmonton, Canada. Available at *http://www-users.cs.umn.edu/~kumar/papers/kdd_tele_9.pdf*.

Uses data mining techniques to predict the effect of oceans on the climate over land surfaces.

Appendix A

LENGTHS OF ONE DEGREE LATITUDE AND LONGITUDE

This appendix provides two tables listing the lengths of one degree of latitude and longitude. Table A-1 lists lengths of one degree of latitude as measured along any meridian at one-degree intervals. Table A-2 lists the lengths of one degree of longitude at one-degree intervals as measured along a given parallel. In both tables, the lengths in statute miles and nautical miles were derived directly from the meter value using conversions where 1 meter = 0.00062137119 statute mile and 1 meter = 0.0005399568 nautical mile. The lengths of one degree of latitude and longitude were derived through a Visual Basic program using double precision, but have been rounded here to two decimal places. The equations used in the program were provided courtesy of daan Strebe.

TABLE A-1 Lengths of one degree of latitude along a meridian at one-degree intervals

Latitude	Statute Miles	Nautical Miles	Meters	Latitude	Statute Miles	Nautical Miles	Meters
0	68.7077	59.7053	110574.28	27	68.8501	59.8291	110803.52
1	68.7079	59.7055	110574.61	28	68.8600	59.8377	110819.45
2	68.7085	59.7061	110575.63	29	68.8702	59.8465	110835.76
3	68.7096	59.7070	110577.32	30	68.8805	59.8555	110852.44
4	68.7110	59.7082	110579.68	31	68.8911	59.8647	110869.46
5	68.7129	59.7099	110582.71	32	68.9019	59.8741	110886.81
6	68.7152	59.7119	110586.41	33	68.9128	59.8836	110904.46
7	68.7179	59.7142	110590.77	34	68.9240	59.8933	110922.39
8	68.7210	59.7169	110595.79	35	68.9353	59.9031	110940.57
9	68.7246	59.7200	110601.45	36	68.9467	59.9131	110959.00
10	68.7285	59.7234	110607.76	37	68.9583	59.9231	110977.64
11	68.7328	59.7272	110614.71	38	68.9700	59.9333	110996.48
12	68.7375	59.7313	110622.29	39	68.9818	59.9436	111015.48
13	68.7426	59.7357	110630.49	40	68.9937	59.9539	111034.63
14	68.7481	59.7404	110639.29	41	69.0057	59.9643	111053.91
15	68.7539	59.7455	110648.70	42	69.0177	59.9748	111073.28
16	68.7601	59.7509	110658.69	43	69.0298	59.9853	111092.74
17	68.7667	59.7566	110669.26	44	69.0419	59.9958	111112.24
18	68.7736	59.7626	110680.39	45	69.0541	60.0064	111131.78
19	68.7809	59.7689	110692.07	46	69.0662	60.0169	111151.32
20	68.7885	59.7755	110704.29	47	69.0784	60.0275	111170.84
21	68.7964	59.7824	110717.03	48	69.0905	60.0380	111190.32
22	68.8046	59.7896	110730.27	49	69.1025	60.0485	111209.74
23	68.8131	59.7970	110744.01	50	69.1145	60.0589	111229.06
24	68.8220	59.8047	110758.22	51	69.1265	60.0693	111248.28
25	68.8311	59.8126	110772.89	52	69.1383	60.0796	111267.35
26	68.8405	59.8207	110787.99	53	69.1501	60.0898	111286.27

Continued

TABLE A-1 *Continued*

Latitude	Statute Miles	Nautical Miles	Meters	Latitude	Statute Miles	Nautical Miles	Meters
54	69.1617	60.0999	111305.00	73	69.3435	60.2578	111597.53
55	69.1732	60.1099	111323.53	74	69.3501	60.2636	111608.25
56	69.1846	60.1198	111341.83	75	69.3564	60.2691	111618.38
57	69.1958	60.1295	111359.88	76	69.3624	60.2743	111627.93
58	69.2069	60.1391	111377.65	77	69.3679	60.2791	111636.87
59	69.2177	60.1486	111395.13	78	69.3731	60.2836	111645.19
60	69.2284	60.1578	111412.29	79	69.3779	60.2877	111652.88
61	69.2388	60.1669	111429.11	80	69.3823	60.2915	111659.94
62	69.2491	60.1758	111445.57	81	69.3863	60.2950	111666.35
63	69.2591	60.1845	111461.66	82	69.3898	60.2981	111672.11
64	69.2688	60.1929	111477.34	83	69.3930	60.3009	111677.21
65	69.2783	60.2012	111492.61	84	69.3958	60.3033	111681.64
66	69.2875	60.2092	111507.44	85	69.3981	60.3053	111685.40
67	69.2964	60.2170	111521.81	86	69.4000	60.3070	111688.49
68	69.3051	60.2245	111535.71	87	69.4015	60.3083	111690.89
69	69.3134	60.2317	111549.12	88	69.4026	60.3092	111692.60
70	69.3214	60.2387	111562.03	89	69.4032	60.3097	111693.64
71	69.3291	60.2454	111574.40	90	69.4034	60.3099	111693.98
72	69.3365	60.2518	111586.24				

TABLE A-2 Lengths of one degree of longitude along a parallel at one-degree intervals

Parallel	Statute Miles	Nautical Miles	Meters	Parallel	Statute Miles	Nautical Miles	Meters
0	69.1707	60.1077	111319.49	35	56.7238	49.2917	91288.17
1	69.1603	60.0986	111302.65	36	56.0251	48.6845	90163.69
2	69.1289	60.0713	111252.13	37	55.3093	48.0625	89011.67
3	69.0766	60.0259	111167.95	38	54.5766	47.4257	87832.46
4	69.0034	59.9623	111050.13	39	53.8272	46.7745	86626.40
5	68.9093	59.8805	110898.71	40	53.0613	46.1090	85393.86
6	68.7943	59.7806	110713.72	41	52.2792	45.4294	84135.19
7	68.6585	59.6626	110495.23	42	51.4811	44.7358	82850.76
8	68.5020	59.5266	110243.28	43	50.6672	44.0286	81540.97
9	68.3247	59.3726	109957.97	44	49.8378	43.3079	80206.19
10	68.1267	59.2005	109639.36	45	48.9932	42.5739	78846.84
11	67.9081	59.0106	109287.56	46	48.1335	41.8268	77463.30
12	67.6690	58.8027	108902.65	47	47.2590	41.0670	76056.00
13	67.4093	58.5771	108484.76	48	46.3700	40.2945	74625.35
14	67.1292	58.3337	108033.99	49	45.4668	39.5096	73171.79
15	66.8288	58.0726	107550.49	50	44.5497	38.7126	71695.75
16	66.5081	57.7939	107034.39	51	43.6188	37.9037	70197.68
17	66.1672	57.4977	106485.83	52	42.6745	37.0832	68678.02
18	65.8063	57.1841	105904.98	53	41.7171	36.2512	67137.23
19	65.4254	56.8531	105292.01	54	40.7469	35.4081	65575.77
20	65.0247	56.5049	104647.09	55	39.7641	34.5541	63994.13
21	64.6042	56.1395	103970.40	56	38.7691	33.6894	62392.77
22	64.1641	55.7571	103262.15	57	37.7621	32.8144	60772.19
23	63.7046	55.3577	102522.54	58	36.7435	31.9292	59132.86
24	63.2256	54.9416	101751.77	59	35.7135	31.0342	57475.30
25	62.7275	54.5087	100950.09	60	34.6725	30.1296	55800.00
26	62.2103	54.0592	100117.71	61	33.6208	29.2157	54107.48
27	61.6741	53.5934	99254.89	62	32.5588	28.2928	52398.25
28	61.1192	53.1112	98361.87	63	31.4866	27.3611	50672.82
29	60.5457	52.6128	97438.91	64	30.4048	26.4210	48931.74
30	59.9538	52.0984	96486.28	65	29.3135	25.4727	47175.53
31	59.3436	51.5682	95504.26	66	28.2132	24.5166	45404.73
32	58.7153	51.0222	94493.14	67	27.1041	23.5529	43619.88
33	58.0691	50.4607	93453.21	68	25.9867	22.5818	41821.53
34	57.4052	49.8838	92384.79	69	24.8612	21.6038	40010.23

Continued

TABLE A-2   *Continued*

Parallel	Statute Miles	Nautical Miles	Meters	Parallel	Statute Miles	Nautical Miles	Meters
70	23.7280	20.6191	38186.54	81	10.8562	9.4338	17471.35
71	22.5875	19.6280	36351.02	82	9.6585	8.3930	15543.78
72	21.4399	18.6308	34504.24	83	8.4577	7.3496	13611.39
73	20.2858	17.6278	32646.76	84	7.2544	6.3039	11674.77
74	19.1253	16.6194	30779.15	85	6.0488	5.2562	9734.52
75	17.9589	15.6058	28902.01	86	4.8413	4.2069	7791.25
76	16.7869	14.5874	27015.89	87	3.6323	3.1563	5845.56
77	15.6097	13.5645	25121.40	88	2.4221	2.1048	3898.05
78	14.4277	12.5373	23219.10	89	1.2113	1.0526	1949.33
79	13.2412	11.5063	21309.60	90	0.0000	0.0000	0.00
80	12.0506	10.4716	19393.49				

Appendix B

USING THE CIE L*u*v* UNIFORM COLOR SPACE TO CREATE EQUALLY SPACED COLORS

Section 13.4.1 describes several approaches for specifying color schemes for choropleth maps. One approach is to specify endpoints for a color scheme in RGB form, convert these to the 1976 CIE L*u*v* uniform color space, interpolate colors in the uniform color space, and then convert the interpolated colors back to RGB. This can be accomplished using the six steps listed here. (When RGB colors are referred to, it is assumed that a gamma correction has already been performed.)

Step 1. *Specify the desired endpoints of a color scheme in RGB form.* This might be accomplished directly using RGB values, or HSV might be used (because HSV values are readily converted into RGB values, as shown in Travis 1991, 82).

Step 2. *Using equations provided by Travis (pp. 93–96), convert the RGB values for the endpoints to their CIE tristimulus equivalents (X, Y, and Z).*

Step 3. *Convert the tristimulus values to the L*u*v* uniform color space using the following equations (Wyszecki and Stiles 1982, 165):*

$$L^* = 116 \left(Y/Y_n\right)^{1/3} - 16 \text{ for } Y/Y_n > 0.008856$$

$$L^* = 903.3 \left(Y/Y_n\right) \text{ for } Y/Y_n \le 0.008856$$

$$u^* = 13L^*(u' - u'_n)$$

$$v^* = 13L^*(v' - v'_n)$$

with $u' = \dfrac{4X}{X + 15Y + 3Z}$ $v' = \dfrac{9Y}{X + 15Y + 3Z}$

$u'_n = \dfrac{4X_n}{X_n + 15Y_n + 3Z_n}$ $v'_n = \dfrac{9Y_n}{X_n + 15Y_n + 3Z_n}$

where *X, Y, Z* are the tristimulus values of a color to be converted, and X_n, Y_n, Z_n are the tristimulus values of the **white point** of the display (when all color guns fire at the maximum intensity).

Step 4. *Linearly interpolate between the L*u*v* values.* For example, if you computed L* values of 50 and 60 for the endpoints, the L* values for the intermediate classes for a five-class map would be 52.5, 55, and 57.5.

Step 5. *Convert the interpolated L*u*v* values back to their tristimulus equivalents using the following equations (Tajima 1983, 313):*

$$Y = \left(\frac{L^* + 16}{116}\right)^3 \cdot Y_n \text{ for } Y/Y_n \le 0.008856,$$

$$Y = \frac{L^* \cdot Y_n}{903.3} \text{ for } Y/Y_n \le 0.008856$$

$$X = \frac{9}{4} \frac{(u^* + 13v'_n L^*)}{(v^* + 13v'_n L^*)} Y$$

$$Z = \left(\frac{39L^*}{v^* + 13v'_n L^*} - 5\right) Y - \frac{X}{3}$$

Step 6. *Using equations provided by Travis (p. 96), convert the tristimulus values to their RGB equivalents.*

Glossary

abrupt phenomena: phenomena that change abruptly over geographic space, such as sales tax rates for each state in the United States.

abstract sound variables: analogous to **visual variables**, these are variables used to depict abstract sounds (e.g., the location and loudness of sound).

abstract sounds: sounds that have no obvious meaning and thus require a legend for interpretation (e.g., a mouse click on a census tract produces a loudness value as a function of income).

accommodation: the process in which the eye automatically changes the shape of the lens to focus on an image.

additive colors: colors that are visually added (or combined) to produce other colors; for example, red, green, and blue are additive colors used in CRT displays.

aggregation: a generalization operation that involves the joining together of multiple point features, such as a cluster of buildings.

aliasing: a staircase appearance to straight lines caused by displaying lines on a coarse raster graphics display.

amalgamation: a generalization operation that combines nearby polygons (e.g., a series of small islands in close proximity with size and detail that cannot be depicted at a small scale).

anaglyphs: two images are created, one in green or blue, and one in red; when viewed through special anaglyphic glasses, a three-dimensional view is produced.

analytical cartography: a theoretical branch of cartography developed by Waldo Tobler that deals with analytical topics such as cartographic data models, coordinate transformations and map projections, interpolation, and generalization.

ancillary information: information used to more accurately map data associated with enumeration units (e.g., when making a dot map of wheat based on county-level data, we avoid placing dots in bodies of water).

animated maps: maps characterized by continuous change, such as the daily weather maps shown on television depicting changes in cloud cover.

anomalous trichromats: a less serious form of color vision impairment in which people use three colors to match any given color; contrast with **dichromats**.

application software: software programs designed for specific purposes, such as graphic design, GIS, and remote sensing.

area cartogram: a map in which the areal relationships of enumeration units are distorted on the basis of an attribute (e.g., the sizes of states are made proportional to the number of deaths due to AIDS).

areal phenomena: geographic phenomena that are two-dimensional in spatial extent, having both length and width, such as a forested region; data associated with enumeration units can also be considered areal phenomena, because each unit is an enclosed area.

arrangement: a visual variable in which the marks making up symbols are arranged in various ways, such as breaking up lines into dots and dashes to create various forms of political boundaries.

aspect: (1) placement of a projection's center with respect to the Earth's surface; common aspects include equatorial, polar, and oblique; (2) the direction that a topographic slope faces (e.g., a northwest aspect would face the northwest).

attribute: a theme or variable you might wish to display on a map (e.g., the predicted high temperature for each major U.S. city).

attribute accuracy: one means of assessing data quality; refers to the accuracy of features found at particular locations (e.g., the accuracy of a land use/land cover classification).

augmented reality: a combination of a real-world experience with a virtual representation in which the real world is dominant, such as overlaying the boundaries of "old growth" forests on your view of a mountainous environment; a subset of a **mixed environment**.

available space: areas on a map that are empty in which map elements can be placed.

avatar: a representation of a person in a virtual environment.

axis of rotation: an imaginary line running through the North and South Poles about which the Earth rotates.

azimuth: refers to specifying direction on a map and is customarily measured in a clockwise fashion starting with geographic north as the origin and passing through 360°.

azimuthal projection: a projection in which all directions or azimuths along straight lines radiating from the map projection's center represent great circles on the Earth; directions are preserved from the center of the map to any other point on the map.

balance: the organization of map elements and empty space resulting in visual harmony and equilibrium.

balanced data: two phenomena coexist in a complementary fashion, such as the percentage of English and French spoken in Canadian provinces.

bar scale: the map element that is a graphical expression of scale, resembling a small ruler.

base information: graphical symbols and type that provide a geographic frame of reference for thematic symbols.

bipolar and ganglion cells: cells that merge the input arriving from rods and cones in the eye.

bipolar data: data characterized by either a natural or meaningful dividing point, such as the percentage of population change.

bivariate choropleth map: the overlay of two univariate choropleth maps (e.g., one map could be shades of cyan and the other shades of red).

bivariate correlation: a numerical method for summarizing the relationship between two interval or ratio-level attributes.

bivariate mapping: the cartographic display of two attributes such as median income and murder rate for census tracts within a city.

bivariate-normal: when y values associated with a given x value are normal, and the x values associated with a given y value are also normal.

bivariate point symbol: a point symbol used to portray two attributes simultaneously (e.g., representing two attributes by the width and height of ellipses).

bivariate ray-glyph: used to map two attributes by extending straight line segments to either the right or left of a small central circle.

bivariate regression: the process of fitting a line to two interval or ratio-level attributes; the *dependent attribute* is plotted on the y axis, and the *independent attribute* is plotted on the x axis.

block diagram: a technique developed by A.K. Lobeck for depicting the three-dimensional structure of geomorphologic and geologic features.

boundary error: if classed data are conceived as a prism map, boundary error describes how close the resulting cliffs come to matching cliffs on an unclassed prism map of the data.

bounding rectangle: a rectangle that just touches and completely encloses an arbitrary shape.

box plot: a method of exploratory data analysis that illustrates the position of various numerical summaries along the number line (e.g., minimum, maximum, median, and lower and upper quartiles).

brushing: when exploring data, a user is able to highlight an arbitrary set of spatial entities; for instance, a user might display the percentage of Hispanics only for tracts that voted for a particular political party.

callout: a method for labeling a spatial feature; the effect of overprinting is minimized by using a combination of a mask and leader line.

Cartesian coordinates: a coordinate system in which the locations of points in a plane are referenced with respect to the x and y distance from two perpendicular intersecting axes forming an origin.

cartogram: a map that purposely distorts geographic space based on values of a theme (e.g., making the size of countries proportional to population).

cartographic design: a partly mental, partly physical process in which maps are conceived and created.

cartographic scale: based on the **representative fraction**; a large cartographic scale portrays a small portion of the Earth (contrast with **geographic scale**).

case: describes how a developable surface touches a reference globe's surface; the two common cases are tangent and secant.

cathode ray tube (CRT):   a graphics display in which images are created by firing electrons from an electron gun at phosphors, which emit light when they are struck.

central meridian: the line of longitude at the center of the map about which the projection is symmetrical.

centrifugal force: the tendency for an object that is positioned on and in motion along a curved path to move away from the center of that rotating body.

centroid: the "balancing point" for a geographic region.

change map: a map representing the difference between two points in time.

Chernoff face: distinct facial features that are associated with individual attributes (e.g., a broader smile depicts cities with a higher per-capita expenditure on public schools).

chiaroscuro: see **shaded relief**.

chorodot map: a combination of choropleth and dot maps; the "dot" portion derives from using small squares within enumeration units, whereas the "choropleth" portion derives from the shading assigned to each square.

choropleth map: a map in which enumeration units (or data collection units) are shaded with an intensity proportional to the data values associated with those units (e.g., census tracts shaded with gray tones with an intensity that is proportional to population density).

chroma: see **saturation**.

chronometer: a highly accurate mechanical timekeeping device that was developed to determine longitude.

CIE: an international standard for color; allows cartographers to precisely specify a color so that others can duplicate that color.

clarity: a term used to summarize the following visual variables for depicting data uncertainty: **crispness, resolution**, and **transparency**.

class: the types of developable surfaces used to create a projection, including cylindrical, conic, and planar.

class interval: in creating an equal interval map, the width that each class occupies along the number line.

classed map: a map in which data are grouped into classes of similar value, and the same symbol is assigned to all members of a class (e.g., a data set with 100 different data values might be depicted using only five shades of gray).

clip art: pictures that are available in a digital format; such pictures can serve as the basis for creating **pictographic symbols**.

cluster analysis: a mathematical method for grouping observations (say, counties) based on their scores on a set of attributes.

CMYK: a system for specifying colors in which cyan (C), magenta (M), yellow (Y), and black (K) are used, as is commonly done in offset printing.

cognition: the mental activity associated with map reading and interpretation, including the initial perception of the map, our thought processes, prior experience, and memory.

collaborative geovisualization: geovisualization activities in which more than one individual are involved in the visualization process; for instance, individuals located at different universities might manipulate and visualize a climate-change model (ideally, the results of one individual's manipulations can be seen by another individual immediately).

collapse: a generalization operation involving a conversion of geometry; for instance, a complex urban area might be collapsed to a point due to scale change, and symbolized with a circle.

color composite: a digital color proof created on a color laser, inkjet, thermal-wax transfer, dye sublimation, or other printer.

color copy machine: a device intended for image duplication, but often adapted for use as a printing device.

color laser printer: a type of laser printer that incorporates several colors, normally the process colors, which are mixed on the page.

color lookup tables: used in association with a frame buffer to make rapid changes in color in a graphics display.

color management system: a software application that identifies differences in gamuts among devices and corrects the variations in color introduced by each device.

color models: refers to various approaches for specifying color (e.g., **RGB** vs. **CMYK**).

color profile: an electronic file that describes the manner in which a particular device introduces color variations.

color ramping: a method for specifying colors on a graphics display in which the user selects two endpoints, and the computer automatically interpolates intermediate colors.

color separation: a film negative that represents just one of the base colors that a map is composed of.

color stereoscopic effect: occurs when colors from the long-wavelength portion of the electromagnetic spectrum appear nearer to a map reader than colors from the short-wavelength portion.

color vision impairment: the notion that some individuals do not see the range of colors that most people do; males are most commonly affected, with red and green colors being most frequently confused.

colorimeter: a physical device for measuring color; typically, color specifications are given in the CIE system.

command-line interface: a computer interface in which the user types commands specifying processes to be performed.

communication: the transfer of known spatial information to the public; typically, this is done via paper maps.

compaction index (CI): a measure of shape involving the ratio of the area of the shape to the area of a circumscribing circle.

complementary colors: colors that combine to produce a shade of gray, such as cyan and red.

completeness: one means of assessing data quality; includes information about selection criteria and definitions used (e.g., the minimum area needed to define a water body as a lake).

compromise projection: a map projection that does not possess any specific property, but rather strikes a balance among various projection properties.

computer-to-plate: a variation of traditional offset lithography, in which the film negative component is bypassed entirely; the printing plate is digitally imaged by a platesetter directly from a digital map file.

conceptual point data: data that are collected over an area (or volume) but conceived as being located at a point for the purpose of symbolization (e.g., the number of microbreweries in each state).

conditioned choropleth maps (CCmaps): a spatial pattern for a single attribute is split into a series of choropleth maps in which the rows and columns of maps correspond to two attributes that might explain the attribute that is displayed in the choropleth map series.

cone: one of three developable surfaces—produces the conic class of projections.

cones: a type of nerve cell within the retina that functions in relatively bright light and is responsible for color vision.

conformal projection: a map projection for which angular relationships around any point correspond to the same angular relationship on the Earth's surface.

conic projection: a map projection in which the graticule is conceptually projected onto a cone that is either tangent or secant to the developable surface; in most instances, conic projections represent meridians as straight lines and parallels as smooth curves about the North Pole.

constrained extended local processing routines: simplification algorithms that search beyond the immediate neighboring points and evaluate larger sections of lines.

constrained ratio: level of measurement for data sets that are constrained to a fixed set of numbers, such as probability (the range is 0 to 1) or percentages (the range is 0 to 100).

continuous phenomena: geographic phenomena that occur everywhere, such as the distribution of snowfall for the year in Wisconsin or sales tax rates for each state in the United States.

continuous tone: full-color printed output achieved without the use of screening techniques.

continuous-tone map: a map that uses a large number of hypsometric tints to create an unclassed isarithmic map.

continuous vector-based flow: a flow composed of two attributes, magnitude and direction, that can change at any point (e.g., at any point in the atmosphere, we can compute the speed and direction from which the wind blows).

contour lines: these represent the intersection of horizontal planes with the three-dimensional surface of a smooth continuous phenomenon.

contour map: see **isarithmic map**.

contrast: visual differences between map features that allow for differentiation of features, and to imply relative importance.

control points: points that are used as a basis for interpolation on an isarithmic map, such as weather station locations.

cophenetic correlation coefficient: in a cluster analysis, measures the correlation between raw resemblance coefficients (Euclidean distance values in the case of UPGMA) and resemblance coefficients derived from the dendrogram (normally referred to as the *cophenetic* coefficients).

cornea: the protective outer covering of the eye.

correlation coefficient: a numerical expression of the relationship between two interval or ratio-level attributes.

counts: a level of measurement in which individual objects, such as people, are counted.

courseware: software used to assist students in learning concepts outside the classroom.

crispness: a visual variable for depicting data quality; a sharp (or crisp) boundary would represent reliable data, whereas a fuzzy one would represent uncertain data.

cross-validation: a method for evaluating the accuracy of interpolation for isarithmic mapping; involves removing a control point from the data to be interpolated and using other control points to estimate a value at the location of the removed point.

cybersickness: a form of motion sickness that can result from exposure to virtual environments.

cyclical: a level of measurement appropriate for phenomena that have a cyclical character, such as angular measurements (the cycle is 360°).

cylinder: one of three developable surfaces—produces the cylindrical class of projections.

cylindrical projection: a map projection in which the graticule is conceptually projected onto a cylinder that is either tangent or secant to the developable surface; in most instances, cylindrical projections represent meridians and parallels as straight lines intersecting each other at 90°.

dasymetric map: like choropleth maps, area symbols are used to represent zones of uniformity, but the bounds of zones need not match enumeration unit boundaries.

data exploration: examining data in a variety of ways to develop different perspectives of the data; for example, we might view a choropleth map in both its unclassed and classed form. Data exploration can also be equated with **visualization**.

data jack: triangular spikes are drawn from a square central area and made proportional to the magnitude of each attribute being mapped.

data quality: the notion that data for thematic maps are often subject to some form of error; see **uncertainty**.

data resolution: indicates the granularity of the data that is used in mapping (e.g., 1-meter remote sensing data is considered high resolution).

data source: the map element that describes the origin of the data represented on a map.

data splitting: a method for evaluating the accuracy of interpolation for isarithmic mapping; control points are split into two groups, one to create the contour map, and one to evaluate its accuracy.

deconstruction: the process of analyzing a map (or text) to uncover its hidden agendas or meanings.

degrees: a basic unit of measure utilized in specifying longitude and latitude; the Earth encompasses 360° of longitude and 180° of latitude.

Delaunay triangles: the type of triangles commonly used in triangulation, a method of interpolation for isarithmic maps; the longest side of any triangle is minimized and thus the distance over which interpolation takes place is minimized.

dendrogram: a tree-like structure illustrating the resemblance coefficient values at which clusters in a cluster analysis combine.

descriptive statistics: used to describe the character of a sample or population, such as computing the mean square footage of a sample of 50 stone houses.

desktop display: a technology in which a display device is placed on desktop or table (e.g., a CRT or laptop).

developable surface: a surface that can be flattened to form a plane without compressing or tearing any part of it; three commonly used developable surfaces for map projections are the cylinder, cone, and plane.

dichromats: a more serious form of color vision impairment in which people use two colors to match any given color; contrast with **anomalous trichromats**.

digital elevation model (DEM): topographic data commonly available as an equally spaced gridded network.

digital proof: a proof created without film negatives.

digital television: a relatively new standard for high-definition television (HDTV) that provides higher-resolution images and better sound quality.

digital-to-analog converter (DAC): a device within a CRT that converts a digital value to an analog voltage value that is applied to an electron gun.

digital versatile disc (DVD): a high-capacity digital medium for storing and playing moving images and sound.

direct-to-press: a variation of traditional offset lithography in which the film negative and traditional printing plate components are bypassed entirely.

discrete phenomena: geographic phenomena that occur at isolated locations, such as water towers in a city.

dispersion graph: data are grouped into classes, and the number of values falling in each class are represented by stacked dots along the number line.

displacement: a generalization operation that involves moving features apart to prevent coalescence (e.g., when a highway and railroad follow a coastline in close proximity).

display date: one of the visual variables for animated maps; refers to the time that some display change in an animation is initiated (e.g., in an animation of population for U.S. cities, a circle for San Francisco appears in 1850).

distance cartogram: a map in which real-world distances are distorted to reflect some attribute (e.g., distances on a subway map might be distorted to reflect travel time).

distortion: involves altering the size of the Earth's landmasses and arrangement of the Earth's graticule when they are projected to the two-dimensional flat map.

dithering: colors are created by presuming that the reader will perceptually merge different colors displayed in adjacent pixels (on a graphics display) or dots (on a printer).

diverging scheme: a sequence of colors for a choropleth or isarithmic map in which two hues diverge from a common light hue or neutral gray, as in a dark red-light red-gray-light blue-dark blue scheme.

dot-density shading: a technique for mapping smooth continuous phenomena (e.g., precipitation) in which closely spaced dots depict high values of the phenomenon, whereas widely separated dots depict low values.

dot gain: phenomenon in which ink is forced to spread out due to pressures exerted between cylinders on an offset lithographic printing press.

dot map: a map in which small symbols of uniform size (typically solid circles) are used to emphasize the spatial pattern of a phenomenon (e.g., one dot might represent 1000 head of cattle).

dot size: how large dots are on a dot map.

drafting table-format display: a virtual environment is achieved by projecting images onto a large table-like screen that is tilted at a 45° angle; a head tracker on one user governs the view that others see and a specialized 3D mouse can be used to interact with the simulated environment.

draped image: a remotely sensed image (or other information, e.g., land use and land cover) is draped on a 3D map of elevation.

duration: one of the visual variables for animated maps; refers to the length of time that a frame of an animation is displayed.

dye-sublimation printer: a printing device (normally full-color) that incorporates heat sources to convert solid dyes into gases, and applies them to the print medium.

Ebbinghaus illusion: a circle surrounded by large circles will appear smaller than the same-size circle surrounded by small circles.

electromagnetic energy: a waveform having both electrical and magnetic properties; refers to the way that light travels through space.

electromagnetic spectrum: the complete range of wavelengths possible for electromagnetic energy.

electron gun: the device used within a CRT to fire electrons at phosphors, which emit light when they are struck.

electronic atlas: a collection of maps (and databases) available in a digital environment; sophisticated electronic atlases enable users to take advantage of the digital environment through Internet access, data exploration, map animation, and multimedia.

ellipsoid: see **oblate spheroid**.

Encapsulated PostScript (EPS): a subset of the PostScript page description language that allows digital maps and other documents to be transported between software applications, and between different types of computers.

enhancement: a generalization operation that involves a symbolization change to emphasize the importance of a particular object (e.g., the delineation of a bridge over an existing road is often portrayed as a series of cased lines).

enumeration unit: a data collection unit such as a county or state.

equal areas: a method of data classification in which an equal portion of the map area is assigned to each class.

equal intervals: a method of data classification in which each class occupies an equal portion of the number line.

Equator: the origin for the system of latitude (0° latitude); serves as the dividing line between the Northern and Southern Hemispheres.

equatorial aspect: positioning the developable surface over the reference globe such that the Equator becomes the central latitude.

equidistant projection: a map projection for which distances from one point (usually the projection's center) to all other points are preserved compared to those distances on the Earth's surface.

equiprobability ellipse: if data are bivariate normal, an ellipse can be drawn that encloses a specified percentage of the data.

equivalent (equal area) projection: a map projection for which areas are preserved compared to the same areas on the Earth's surface.

exaggeration: a generalization operation in which the character of an object is amplified (e.g., exaggerating the mouth of a bay that would close under scale reduction).

expert system: a software application that incorporates rules derived from experts to make decisions and solve problems.

exploratory data analysis: a method for analyzing statistical data in which the data are examined graphically in a variety of ways, much as a detective investigates a crime; this is in contrast to fitting data to standard forms, such as the normal distribution.

exploratory spatial data analysis (ESDA): data exploration techniques that accompany a statistical analysis of spatial data.

eXtensible Markup Language (XML): an approach to programming Web content that is better structured, more flexible, and more robust than HTML.

extrinsic visual variables: when mapping uncertainty, these are visual variables that involve adding objects to the display, such as dials and thermometers.

figure-ground: methods of accentuating one object over another, based on the perception that one object stands in front of another, and appears to be closer to the map user.

File Transfer Protocol (FTP): a set of methods defining the manner in which electronic files and documents are sent and retrieved via the Internet.

film negative: a representation of a map, or one color of a map, on clear plastic film; composed of black, gray tones, and clear areas, and represents the opposite of what ultimately will be printed.

Fisher–Jenks algorithm: a method for classifying data in which an optimal classification is guaranteed by *essentially* considering all possible classifications of the data.

fishnet map: a map in which a fishnet-like structure provides a three-dimensional symbolization of a smooth continuous phenomenon.

Flannery correction: a method of adjusting the sizes of proportional circles to account for the perceived underestimation of larger circle sizes; a symbol scaling exponent of .57 is used.

flattening constant: expresses the degree to which an ellipse deviates from a circle; computed as $(a - b) / a$, where a and b are the semimajor and semiminor axes of the ellipse.

flow map: a map used to depict the movement of phenomena between geographic locations; generally, this is done using "flow lines" of varying thickness.

fluorescent inks: specialized printing inks that produce brilliant, intense color.

fly-by: a form of animation in which the viewer is given the feeling of flying over a landscape.

focusing: when exploring data, a user is able to highlight a subrange of numeric values; for example, a user might focus on census tracts that are less than 25 percent Hispanic.

font: a set of all alphanumeric and special characters of a particular type family, style, and size.

four-color process printing: the combination of the process colors (CMYK) with screening techniques, allowing for a very wide range of colors in print reproduction.

fovea: the portion of the retina where visual acuity is the greatest.

frame buffer: an area of memory that stores a digital representation of colors appearing on the screen.

frame line: the map element—normally a rectangle—that encloses all other map elements.

framed-rectangle symbol: a point symbol that consists of a "frame" of constant size, within which a solid "rectangle" is placed; the greater the data value, the greater the proportion of the frame that is filled by the rectangle.

frequency: one of the visual variables for animated maps; refers to the number of identifiable states per unit time, as in color cycling used to portray the jet stream.

fuzzy categories: a level of measurement in which category memberships are fuzzy, as in some remote sensing classification procedures.

gamma function: the relation between the voltage of a color gun and the associated luminance of the CRT display.

gamut: a range of colors produced by a computer, printing, or display device.

general information: the type of information that one acquires when examining spatial patterns on thematic maps.

general-reference map: a type of map used to emphasize the *location* of spatial phenomena (e.g., a United States Geological Survey [USGS] topographic map).

generalization: process of reducing the information content of maps due to scale change, map purpose, intended audience, or technical constraints.

geocentric latitude: latitude as computed on a reference ellipsoid is measured by an angle that results when a line at the Earth's surface is drawn intersecting the plane of the Equator at the Earth's center.

geocollaboration: see **collaborative geovisualization**.

geodesy: the field of study investigating the Earth's size and shape.

geodetic datum: a combination of a specific reference ellipsoid and geoid (also commonly referred to as a datum).

geodetic latitude: latitude as computed on a reference ellipsoid is measured by an angle that results when a perpendicular line at the reference ellipsoid's surface is drawn toward the Earth's center.

geographic brush: as areas of a map are selected (e.g., using a mouse), corresponding dots are highlighted within a scatterplot matrix.

geographic information systems (GIS): automated systems for the capture, storage, retrieval, analysis, and display of spatial data.

geographic scale: the extent of a study area, such as a neighborhood, city, region, or state; here large scale indicates a large area (contrast with **cartographic scale**).

geographic visualization (or geovisualization): a private activity in which previously unknown spatial information is revealed in a highly interactive computer graphics environment.

geoid: describes Earth's two-dimensional curved surface that would result if the oceans were allowed to flow freely over the continents, creating a single undisturbed water body; the surface would have undulations reflecting the influence of gravity.

geometric symbols: symbols that do not necessarily look like the phenomenon being mapped, such as using squares to depict barns of historical interest; contrast with **pictographic symbols**.

geoslavery: the use of GPS and GIS technologies to monitor and control the movement of people.

geostatistical methods: numerical summaries in which spatial location is an integral part (e.g., **centroid** and **spatial autocorrelation** measures).

gerrymandering: the purposeful distortion of legislative or congressional districts for partisan benefit.

Gestalt principles: descriptions of the manner in which humans see the individual components of a graphical image, and organize them into an integrated whole.

global algorithms: simplification algorithms that process the entire line at once (e.g., the Douglas–Peucker method).

glyphs: a term applied to multivariate point symbols when the attributes being mapped are in dissimilar units (not part of a larger whole).

goodness of absolute deviation fit (GADF):   a measure of the accuracy of a classed choropleth map when the median is used as a measure of central tendency.

goodness of variance fit (GVF):   a measure of the accuracy of a classed choropleth map when the mean is used as a measure of central tendency.

Gouraud shading: a sophisticated method for creating shaded relief; shades are *interpolated* between vertices of a polygon (or cell) boundary, and thus a smoother appearance is produced than when the same shade is used throughout a polygon.

graduated symbol map: see **proportional symbol map**.

graphic display: used to describe the computer screen (and associated color board) on which a map is displayed in softcopy form.

graphic script: a series of dynamic maps, statistical graphics, and text blocks used to tell a story about a set of data.

graphical user interface (GUI):   a type of computer interface in which the user specifies tasks to be performed by pointing to the desired task, typically by using a mouse.

graticule: latitude and longitude taken in combination.

gray curves (or grayscales):   a graph (or equation) expressing the relation between printed area inked and perceived blackness.

great circle: represents the line along which a plane intersects the Earth's surface when the plane passes through the Earth's center.

gridding: a term applied to the inverse distance method of interpolation in which values are estimated at equally spaced grid locations; it should be borne in mind, however, that kriging also uses a grid.

grouped-frequency table: constructed by dividing the data range into equal intervals and tallying the number of observations falling in each interval.

hachures: a method for depicting topography in which a series of parallel lines is drawn perpendicular to the direction of contours.

halftone screening: the application of ink or toner in a pattern of equally spaced cells of variable size to create tints from a base color.

halo: an extended outline of letters in a type label.

haptics: the use of touch in virtual environments (e.g., a user feels different vibrations when a force-feedback mouse is placed over map locations having differing crime rates).

head-mounted display (HMD):   helmet-like device that shields a person from the real-world view and provides images of a virtual environment to each eye.

hexagon bin plot: a method of data display in which a set of hexagons is placed over a conventional scatterplot, and the hexagons are filled as a function of the number of dots falling within them.

high-fidelity process colors: colors based on the traditional process colors (CMYK), but including two or three additional colors that are mixed on the page on a printing device.

hill shading: see **shaded relief**.

histogram: a type of graph in which data are grouped into classes, and bars are used to depict the number of values falling in each class.

honoring the control point data: a term applied if, after an interpolation is performed, there is no difference between the original value of a control point and the value of that same point on the interpolated map.

horizontal datum: a specific reference ellipsoid to which accurate latitude and longitude values are referenced.

HSV: a system for specifying color in which hue (H), saturation (S), and value (V) are used; although the system utilizes common color terminology, colors are not equally spaced in the visual sense.

hue: along with lightness and saturation, one of three components of color; it is the dominant wavelength of light, such as red versus green.

HVC: a system for specifying color developed by Tektronix; the terms hue (H), value (V), and chroma (C) are used, and colors are equally spaced from one another in the visual sense.

hypermedia: a sophisticated form of multimedia in which various forms of media can be linked in ways not anticipated by system designers.

Hypertext Markup Language (HTML):   a relatively simple, text-based programming language that is used to define the content and appearance of Web pages.

Hypertext Transfer Protocol (HTTP):   a set of methods defining the manner in which electronic files and documents are sent and retrieved via the World Wide Web.

hypsometric tints: the shaded areas sometimes used between contour lines on an isarithmic map (e.g., using shades of blue between contour lines to depict increasing rainfall).

iconic memory: refers to the initial perception of an object (in our case, a map or portion thereof) by the retina of the eye; contrast with **short-term visual store** and **long-term visual memory**.

illuminated contours: a raster approach for depicting topography in which contours facing a light source are brightened, whereas those in the shadow are darkened.

imagesetter: a very high-resolution, large-format laser printer that produces film negatives.

immersion: a psychological state in which a user is enveloped by, included in, and is able to interact with an environment that provides a continuous stream of stimuli and experiences.

independent point routines: simplification algorithms that select coordinates based on their position along the line, nothing more (e.g., an *n*th point routine might select every third point).

induction: see **simultaneous contrast**.

inferential statistics: used to make an inference (or guess) about a population based on a sample.

information acquisition: the process of acquiring spatial information *while* a map is being used; contrast with **memory for mapped information**.

information visualization: involves the visualization and analysis of nonnumerical abstract information such as the topics that are discussed on the front page of a newspaper over a month-long period.

ink-jet printer: a printing device (normally full-color) that incorporates nozzles, or "jets," that squirt ink onto the print medium.

inset: the map element consisting of a smaller map used in the context of a larger map (e.g., a map that is used to show a congested area in greater detail).

integral symbols: the individual symbols representing separate attributes cannot be attended to easily, but it is possible to examine the correlation between the data sets (e.g., a rectangular point symbol, where the height represents one attribute and the width another attribute).

intellectual hierarchy: a ranking of symbols and map elements according to their relative importance.

intermediate frames: in an animation, the frames (or maps) associated with interpolated data; contrast with **key frames**.

International Color Consortium (ICC): an entity that has developed a standard for a vendor-neutral, cross-platform color profile, and architectures for color management.

International Dateline: the line coinciding approximately with the 180th meridian; when crossing it a day is lost or gained depending on the direction of travel.

International Meridian Conference: a conference held at Washington, DC, in 1884 that established Greenwich, England as the origin (0°) for the system of longitude.

Internet: a global network of computers that permits sharing of data and resources.

interpolation: in isarithmic mapping, the estimation of unknown values between known values of irregularly spaced control points.

interquartile range: the absolute difference between the 75th and 25th percentiles of the data.

interrupted projection: a projection having several central meridians where each central meridian (usually located over a landmass) creates a lobe.

interval: a level of measurement in which numerical values can be assigned to data, but there is an arbitrary zero point, such as for SAT scores.

intrinsic visual variables: when mapping uncertainty, these are visual variables that are intrinsic to the display; for example, we might vary the saturation of colored tones on a choropleth map.

inverse-distance: a method of interpolating data for isarithmic maps; a grid is overlaid on control points, and estimates of values at grid points are an inverse function of the distance to control points.

isarithmic map: a map in which a set of isolines (lines of equal value) are interpolated between points of known value, as on a map depicting the amount of snowfall.

isometric map: a form of isarithmic map in which control points are true point locations (e.g., snowfall measured at weather stations).

isopleth map: a form of isarithmic map in which control points are associated with enumeration units (e.g., assigning population density values for census tracts to the centers of those tracts and then interpolating between the centers).

isosurface: a surface of equal value within a true 3-D phenomenon (e.g., a surface representing a wind speed of 30 miles per hour in the jet stream).

Java: a platform-independent, object-oriented programming language that allows users to interact with maps and mapping applications via the World Wide Web.

JavaScript: a platform-independent, object-oriented programming language similar to Java, but simpler and less robust.

Jenks–Caspall algorithm: an optimal data classification that is achieved by moving observations between classes through trial-and-error processes known as *reiterative* and *forced* cycling.

kerning: the variation of space between two adjacent letters.

key frames: in an animation, the frames (or maps) associated with collected data; contrast with **intermediate frames**.

knowledge discovery in databases (KDD): refers to exploratory approaches for finding interesting patterns in databases, and ultimately uncovering knowledge about those databases.

kriging: a method of interpolating data for isarithmic maps that considers the spatial autocorrelation in the data.

labeling software: software developed for the automated positioning of type.

laminate proof: a separation-based proof consisting of overlying transparent film sheets that are melded into a single sheet.

large-scale map: a map depicting a relatively small portion of the Earth's surface.

latitude: an imaginary line that crosses the Earth's surface in an east to west fashion; latitude is measured in degrees, minutes, and seconds north or south of the Equator.

latitude of origin: the latitude that is set as the center of a map projection.

leading: the vertical space between lines of type; equivalent to line spacing.

lectureware: software used to assist students in learning material in a traditional lecture setting.

legend: the map element that defines graphical symbols.

legend heading: the map element that is used to explain a map's legend and theme.

lens: the focusing mechanism within the eye.

letter spacing: the space between each letter in a word.

level of measurement: refers to the different ways that we measure attributes; we commonly consider nominal, ordinal, interval, and ratio levels.

lightness: along with hue and saturation, one of three components of color; refers to how dark or light a color is, such as light blue versus dark blue.

lineage: one means of assessing data quality; refers to the historical development of the data.

linear-legend arrangement: a form of legend design for proportional symbol maps; symbols are placed adjacent to each other in either a horizontal or vertical orientation.

linear phenomena: geographic phenomena that are one-dimensional in spatial extent, having length but essentially no width, such as a road on a small-scale map.

linked micromap plots (LM plots): a series of small maps termed micromaps, which divide a single spatial distribution into pieces; graphical and statistical information is also included, and the emphasis is on local pattern perception.

liquid crystal displays (LCDs): a form of graphic display device in which liquid crystals are sandwiched between two glass plates; commonly used in laptops and overhead projection devices.

lobe: A portion of an interrupted projection described by a central meridian usually displaying a specific landmass, such as South America.

local area network (LAN): a series of computers within a limited geographic area that are connected, and allow for the sharing of data and resources.

local processing routines: simplification algorithms that utilize immediate neighboring points in assessing whether a point should be deleted.

location: a visual variable that refers to the possibility of varying the position of symbols, such as dots on a dot map.

logical consistency: one means of assessing data quality; describes the fidelity of relationships encoded in the structure of spatial data (e.g., are polygons topologically correct?).

long-term visual memory: an area of memory that does not require constant attention for retainment; contrast with **iconic memory** and **short-term visual store**.

longitude: an imaginary line originating at the poles that crosses the Earth's surface in a north to south fashion; longitude is measured in degrees, minutes, and seconds east or west of the Prime Meridian.

loxodromes: lines that intersect all meridians at the same angle.

luminance: brightness of a color as measured by a physical device.

magnitude estimation: (1) a method for constructing grayscales in which a user estimates the lightness or darkness of one shade relative to another; (2) a method for determining the perceived size of proportional symbols in which a value is assigned to one symbol on the basis of a value assigned to another symbol.

manual interpolation: in isarithmic mapping, the estimation of unknown values by "eye."

map communication model: a diagram or set of steps summarizing how a cartographer imparts spatial information to a group of map users.

map complexity: a term used to indicate whether the set of elements composing a map pattern appear simple or intricate (complex); for example, does the pattern of gray tones on a choropleth map appear simple or intricate?

map design research: research focusing on which mapping techniques are most effective and why.

map dissemination: the distribution of reproduced maps in physical or electronic form.

map editing: the critical evaluation and correction of every aspect of a map.

map elements: graphical units that maps are composed of, including the title, legend, and scale.

map projection: a systematic transformation of the Earth's graticule and landmass from the curved two-dimensional surface to a planimetric surface.

map reproduction: the printing of a map on paper or similar medium (print reproduction), or the electronic duplication of a map in digital form (nonprint reproduction).

map server: a computer connected to the Internet, which allows the user to access and interact with digital maps using Web browsers.

map user: the person for whom a particular map has been designed and produced.

mapped area: the map element representing the geographic region of interest.

mask: a method for labeling a spatial feature; the effect of overprinting is reduced by placing a polygon (e.g., a white rectangle) underneath type, but above underlying graphics.

mathematical scaling: sizing proportional symbols in direct proportion to the data; thus, if a data value is 40 times another, the area (or volume) of a symbol will be 40 times as large.

maximum breaks: a method of data classification in which the largest differences between ordered observations are used to define classes.

maximum contrast shades: on a choropleth map, areal shades are selected so that they have the maximum possible contrast with one another.

mean: the average of a data set, computed by adding all values and dividing by the number of values.

mean center: a measure of central tendency for point data; computed by independently averaging the x and y coordinate values of all points.

mean–standard deviation: a method of data classification in which the mean and standard deviation of the data are used to define classes.

measures of central tendency: used to specify a value around which data are concentrated (e.g., the **mean** and **median**).

median: the middle value in an ordered set of data.

memory for mapped information: remembering spatial information that was previously seen in mapped form; contrast with **information acquisition**.

merging: a generalization operation involving the fusing of groups of line features, such as parallel railway lines, or edges of a river or stream.

meridian: see **longitude**.

minimum error projection: a projection having the least amount of total distortion of any projection in that same class according to a specified mathematical criterion.

minutes: a unit of measurement of the sexagesimal degree system; each degree contains 60 minutes.

mixed environment (ME) (or mixed reality [MR]): combines a real-world experience with a virtual representation; most commonly, the real world is dominant—this is termed **augmented reality**.

mode: the most frequently occurring value in a data set.

modifiable areal unit problem: the notion that the magnitude of the correlation coefficient (r) can be affected by the level at which data have been aggregated.

monochromatic composite: a digital proof created by printing a map in black, white, and gray tones, typically on a laser printer.

monochromatic laser printer: a "black-and-white" printing device that utilizes a focused light source, a metal drum, toner, and a heat source.

multimedia: the combined use of maps, graphics, text, pictures, video, and sound.

multimedia encyclopedias: like paper encyclopedias, these provide information about places, people, and events, but they are enhanced through the use of sound, video, and ease of linking various topics.

multimedia systems: interactive mapping systems that place an emphasis on multimedia aspects.

multimodal interfaces: interfaces that involve two or more novel approaches for interacting with the computer; for instance, you might point to a location without touching the screen (a free-hand gesture) and say, "show me all homes within 100 meters of this location."

multiple regression: a statistical method for summarizing the relationship between a dependent attribute and a series of independent attributes.

multivariate dot map: a dot map in which distinct symbols or colors are used for each attribute to be mapped (e.g., wheat, corn, and soybeans could be represented by green, blue, and red dots).

multivariate mapping: the cartographic display of three or more attributes; for example, we might simultaneously map ocean temperature, salinity, and current speed.

multivariate point symbol: the depiction of three or more attributes using a point symbol.

multivariate ray-glyph: a term used when rays are extended from a small interior circle, with the lengths of the rays made proportional to values associated with each attribute being mapped.

Munsell: a system for specifying color in which colors are identified using hue, value, and chroma; colors are equally spaced from one another in the visual sense.

Munsell curve: a grayscale commonly used for smooth (untextured) gray tones.

natural breaks: a method of data classification in which a graphical plot of the data (e.g., a histogram) is examined to determine natural groupings of data.

neat line: the map element—normally a rectangle—that is used to crop the mapped area.

nested-legend arrangement: a form of legend design for proportional symbol maps in which smaller symbols are drawn within larger symbols.

nested means: a method of data classification in which the mean is used to repeatedly divide the data set; only an even number of classes is possible.

nominal: a level of measurement in which data are placed into unordered categories, such as classes on a land use/land cover map.

nomograph: a graphical device that can assist in selecting dot size and unit value on a dot map.

normal distribution: a term used to describe a bell-shaped curve formed when data are distributed along the number line.

North American Datum of 1927 (NAD27): a common datum for accurate maps of North America defined by the Clarke 1866 reference ellipsoid and a geoid specification having Meades Ranch, Kansas, as the point where the geoid and the reference ellipsoid contact each other.

North American Datum of 1983 (NAD83): a datum for accurate maps of North America defined by the Geodetic Reference System (GRS80) reference ellipsoid and a geoid located at the Earth's center of mass.

North Pole: the geographic location in the Northern Hemisphere with a latitude of 90° North that serves as a point around which the Earth's axis rotates.

North Star: the star (called Polaris) that happens to be located directly above the North Pole.

Northern Hemisphere: the portion of the Earth extending from the Equator to the North Pole.

numerical data: a term used to describe data associated with the interval and ratio levels of measurement, as numerical values are assigned to the data.

oblate spheroid: refers to the notion that the Earth bulges at the Equator and compresses at the poles due to the centrifugal force caused by rotation.

oblique aspect: a map projection with a center that is aligned somewhere between the Equator and a pole.

offset lithographic printing press: a mechanical printing device that provides excellent print quality, high printing speed, and a significant decrease in the cost per unit as the number of copies increases.

offset lithography: a form of lithography in which ink is transferred to an intermediate printing surface before being transferred to the print medium.

on-screen display: the least expensive proofing method, involving the viewing of a digital map on a graphic display; also known as soft proofing.

one-dimensional scatterplot: see **point graph**.

opaque symbols: proportional symbols that do not permit the reader to see base information beneath the symbols; readers also must infer the maximum extent of each symbol.

opponent-process theory: the theory that color perception is based on a lightness–darkness channel and two opponent color channels: red–green and blue–yellow.

optic nerve: the nerve that carries information from the retina to the brain.

optimal: a method of classification in which like values are placed in the same class by minimizing an objective measure of classification error (e.g., by minimizing the sum of absolute deviations about class medians).

optimal interpolation: a term sometimes applied to kriging because, in theory, it minimizes the difference between the estimated value at a grid point and the true (or actual) value at that grid point.

order: one of the visual variables for animated maps; refers to the sequence in which frames or scenes of an animation are presented.

ordinal: a level of measurement in which data are ordered or ranked but no numerical values are assigned, such as ranking states in terms of where you would like to live.

ordinary kriging: a form of kriging in which the mean of the data is assumed constant throughout geographic space (there is no trend or drift in the data).

orientation: (1) the indication of direction on a map, normally taking the form of an arrow or a graticule; (2) a visual variable in which the directions of symbols (or marks making up symbols) are varied, such as changing the orientation of short line segments to indicate the direction from which the wind blows.

outliers: unusual or atypical observations.

overlay proof: a separation-based proof consisting of a stack of transparent film sheets that are bound on one edge like a book.

overprinting: a phenomenon that occurs when a block of type is placed on top of another graphical object.

overview error: if classed data are conceived of as a prism map, overview error is the difference in volume between this map and an unclassed prism map of the data.

page description data: digital data consisting of a set of printing instructions describing every graphical and textual component of a map.

page description language: a particular data structure used to describe page description data.

panorama: provides a view of the landscape that we might expect from a painting (which is how traditional panoramas were created), but careful attention is given to geography so that we can clearly recognize known features.

paper atlas: a bound collection of maps; modern paper atlases also include text, photographs, tables, and graphs, which when combined tell a story about a place or region.

parallel: see **latitude**.

parallel coordinate plot: a method of graphical display in which attributes are depicted by a set of parallel axes, and observations are depicted as a series of connected line segments passing through the axes.

partitioning: a method for constructing grayscales in which a user places a set of areal shades between white and black such that the resulting shades will be visually equally spaced.

perceptual scaling: a term used when proportional symbols are sized to account for the perceived underestimation of large symbols; thus large symbols are drawn larger than suggested by the actual data.

perceptually uniform color models: see **uniform color models**.

perspective height: a visual variable involving a perspective three-dimensional view, such as using raised sticks (or lollipops) to represent oil production at well locations.

pexels: a multivariate point symbol consisting of small 3-D bars that can be varied in height, spacing, and color.

physical model: a truly 3-D model of the Earth's landscape (we can hold and manipulate small models in our hands).

physiographic method: a method for depicting topography developed by Raisz in which the Earth's geomorphological features are represented by standard, easily recognized symbols.

pictographic symbols: symbols that look like the phenomenon being mapped, such as using diagrams of barns to depict the location of barns of historical interest; contrast with **geometric symbols**.

picture elements: see **pixels**.

pie chart: a modification of the proportional circle in which a portion of the circle is filled in (e.g., to represent percentage of land area from which wheat is harvested).

pixels: the individual picture elements composing a raster image on a graphics display device.

planar projection: a map projection in which the graticule is conceptually projected onto a plane that is tangent or secant to the Earth's surface.

plane: one of the three developable surfaces—produces the planar class of projections.

platesetter: a device that produces a latent image on a printing plate through photographic means.

plotting coordinates: a coordinate system used to draw latitude and longitude on a planimetric surface.

point graph: a type of graph in which each data value is represented by a small point symbol plotted along the number line.

point-in-polygon test: a test that is made to determine whether a point falls inside or outside a polygon.

point of projection: where an imaginary light source is located when creating a projection.

point phenomena: geographic phenomena that have no spatial extent and can thus be termed "zero-dimensional," such as the location of oil well heads.

polar aspect: a map projection with a center that coincides with one of the poles.

polyconic projection: a class of projections that conceptually involve the placement of many cones over the reference globe, each with its own standard line producing curved meridians (except for a straight central meridian) and parallels represented as circular arcs.

polygonal glyph: a symbol formed by connecting the rays of a multivariate ray-glyph.

population: the total set of elements or things one could study, such as all stone houses within a city.

Portable Document Format (PDF): a file format that allows digital maps and other documents to be transported between software applications, between different types of computers, and is capable of embedding features such as hyperlinks, movies, and keywords.

positional accuracy: one means of assessing data quality; refers to the correctness of location of geographic features.

PostScript: the *de jure* standard page description language, produced by Adobe Systems, Inc.

power function exponent: the exponent used for the stimulus in the power function equation relating perceived size and actual size; this exponent is used to summarize the results of experiments involving perceived size of proportional symbols.

prepress: the phase of high-volume map reproduction consisting of various technologies and procedures that make offset lithographic printing possible.

presence: the subjective experience of being in one place or environment when one is physically in another.

press check: an inspection of the quality of print produced by an offset lithographic printing press, shortly before a press run begins.

primary visual cortex: the first place in the brain where all of the information from both eyes is handled.

Prime Meridian: the internationally agreed-upon meridian with a value of 0° that runs through the Royal Observatory located in Greenwich, England.

principal components analysis: a method for succinctly summarizing the interrelationships occurring within a set of attributes.

print controller: a stand-alone raster image processor (RIP) that is often used to allow a color copy machine to be used as a printing device.

Printer Control Language (PCL): a standard page description language produced by Hewlett Packard, Inc.

printer driver: software that converts digital map data from the native format of application software into page description data.

printing plate: a sheet of aluminum (or polyester) containing a latent image, which is mounted on a roller in an offset lithographic printing press.

printing unit: an organization of cylinders on an offset lithographic printing press that transfers one base color of a map to the print medium.

prism map: a map in which enumeration units are raised to a height proportional to the data associated with those units.

proactive graphics: software that enables users to initiate queries using icons (or symbolic representations) as opposed to words.

process colors: the subtractive primary colors (CMY), plus black (K); these colors are mixed on the page on a printing device.

programming languages: users write computer code to develop software applications.

projection selection guideline: a set of rules and suggestions developed by experts that are intended to help novices select an appropriate map projection.

prolate spheroid: refers to the historical notion that the Earth bulges at the poles and is compressed at the Equator; today it is known that an oblate spheroid (bulging at the Equator) is reality.

proof: a preliminary representation of what a final reproduced map will look like.

proportional symbol map: point symbols are scaled in proportion to the magnitude of data occurring at point locations, such as using circles of varying sizes to represent city populations.

pseudocylindrical projection: a projection that represents lines of latitude as straight parallel lines (similar to cylindrical projections), but the meridians are curved lines converging to the poles, which are represented as either lines or points.

pupil: the dark area in the center of the eye.

pycnophylactic interpolation: a method of interpolating data for isopleth maps; the data associated with enumeration units are conceived as prisms, and the volume within those prisms is retained in the interpolation process.

quantiles: a method of data classification in which an equal number of observations is placed in each class.

quasi-continuous tone method: a method for unclassed mapping in which only a portion of the data is shown with a smooth gradation of tones; very low and very high data values are represented by the lightest and darkest tones, respectively.

R: the stated radius of the reference globe.

random error: the notion that if we repeatedly measure a value for an enumeration unit, we will likely get a different value each time.

range: the maximum minus the minimum of the data.

range-graded map: a term used to describe a classed proportional symbol map.

raster: an image composed of pixels created by scanning from left to right and from top to bottom.

raster data model: a general approach for representing geographic features in digital form that uses rows and columns of square grid cells, or pixels.

raster image processor (RIP): software, or a combination of software and hardware, that interprets page description data and converts them into a raster image that can be processed directly by a printing device.

rate of change: one of the visual variables for animated maps; defined as m/d, where m is the magnitude of change between frames or scenes, and d is the duration of each frame or scene.

ratio: a level of measurement in which numerical values are assigned to data and there is a nonarbitrary zero point, such as with the percentage of forested land.

ratio estimation: a method for determining the perceived size of proportional symbols; two symbols are compared, and the viewer indicates how much larger or smaller one is than the other (e.g., one symbol appears five times larger).

raw table: a form of tabular display in which the actual data are listed from lowest to highest value.

raw totals: raw numbers that have not been standardized to account for the area over which the data were collected (e.g., we might map the raw number of acres of tobacco harvested in each county, disregarding the size of the counties).

realistic sounds: sounds that have meaning based on our past experience; contrast with **abstract sounds**.

receptive field: a single ganglion cell corresponding to a group of rods or cones; receptive fields are circular and overlap one another.

rectangular point symbol: the width and height of a rectangle are made proportional to each of two attributes being mapped.

redistricting: the process of assigning voting precincts to legislative or congressional districts to equalize voter representation.

reduced major-axis approach: the process of fitting a line to two attributes when we do not wish to specify a dependent attribute.

redundant symbols: used to portray a single attribute with two or more visual variables, such as representing population for cities by both size of circle and gray tone within the circle.

re-expression: an animation created by modifying the original data in some manner, such as choosing subsets of a time series or reordering a time series.

reference ellipsoid: a smooth mathematically defined figure with a semimajor axis that is longer than the semiminor axis and is designed to approximate the Earth's true shape to a better degree than would a spherical model.

reference globe: a conceptual spherical model of the Earth reduced to the same scale as the final map where the scale of the reference globe is defined by its radius R.

refinement: a generalization operation that involves reducing a multiple set of features such as roads, buildings, and other types of urban structures to a simplified representation or a "typification" of the objects.

refresh: a term used to describe how images on a CRT must be constantly redisplayed (refreshed) because phosphors making up the screen have a low persistence.

registration: the alignment of base colors in multicolor printing.

remote sensing: the acquisition of information about the Earth (or other planetary body) from high above its surface via reflected or emitted radiation; most commonly done via satellites.

representative fraction: expresses the relationship between map and Earth distances and is expressed as a ratio of map units/earth units (e.g., an RF of 1:25,000 indicates that one unit on the map is equivalent to 25,000 units on the surface of the Earth).

resemblance coefficients: in cluster analysis, the values that express the similarity of each pair of observations (e.g., the Euclidean distance between two observations).

resolution: (1) for CRT displays, the number of addressable pixels; for printers, the number of dots per inch; (2) a visual variable for depicting data quality; refers to the level of precision in the spatial data underlying the attribute (e.g., varying the size of raster cells to represent uncertainty in the data).

retina: the portion of the eye on which images appear after passing through the lens.

RGB: a system for specifying color in which red (R), green (G), and blue (B) components are used, as for red, green, and blue color guns on a CRT.

rhumb lines: see loxodromes.

rods: a type of nerve cell within the retina that functions in dim light and plays no role in color vision.

room-format display: provides a room-size view of a virtual environment by projecting images onto walls and the floor (e.g., the CAVE); a head tracker on one user governs the view that all users see through stereo glasses.

rosette: a pattern of halftone cells (consisting of the process colors) that results when moiré patterns are avoided through the appropriate specification of halftone screen angles.

rough drafts: preliminary printouts of a map that are used to evaluate an evolving design.

sample: the portion of the population that is actually examined, such as sampling only 50 stone houses out of 5,000 existing within a city.

satellite geodesy: the use of Earth-orbiting satellites to derive very accurate information regarding the Earth's size and shape.

saturation: along with hue and lightness, one of three components of color; can be thought of as a mixture of gray and a pure hue: very saturated colors have little gray, whereas desaturated colors have a lot of gray.

Scalable Vector Graphics (SVG): a file format based on XML that is an open, object-oriented standard for Web-based vector graphics.

scale: provides an indication of the amount of reduction that has taken place on a map.

scale factor (SF): the ratio of the scale found at a particular location on a map compared to the stated scale of the map; departure from a scale factor of 1.0 indicates distortion.

scatterplot: a diagram in which dots are used to plot the scores of two attributes on a set of x and y axes.

scatterplot brushing: subsets of data within a scatterplot matrix are focused on by moving a rectangular "brush" around the matrix; dots falling within the brush are highlighted in all scatterplots.

scatterplot matrix: a matrix of scatterplots is used to examine the relationships among multiple attributes.

scientific visualization: the use of visual methods to explore either spatial or nonspatial data; the process is particularly useful for large multivariate data sets.

screen angle: the angle at which lines of halftone cells are oriented.

screen frequency: the spacing of halftone cells within a given area, expressed in lines per inch (lpi).

screening: the lightening of graphics to reduce visual weight; this is accomplished by reducing the amount of ink or toner applied to the print medium.

secant case: describes a developable surface that intersects the reference globe along two separate lines, usually two parallels of latitude.

seconds: a unit of measurement in the sexagesimal degree system; each minute contains 60 seconds.

semivariance: a measure of the variability of spatial data in various geographic directions (e.g., we might want to determine whether temperature values are more similar to one another along a north–south axis than along an east–west one).

semivariogram: a graphical expression of semivariance; the distance between points is shown on the x axis and the semivariance is shown on the y axis.

sentence case: type composed of lowercase letters with the first letter of each sentence set in uppercase.

separable symbols: symbols representing attributes can be attended to independently, thus allowing the map reader to focus on individual data sets (e.g., cartogram size and lightness are separable).

separation-based proof: a proof created from color separations (film negatives), one for each base color.

sequencing: displaying a map piece by piece (e.g., displaying the title, geographic base, legend title, and then each class in the legend).

sequential scheme: a sequence of colors for a choropleth or isarithmic map in which colors are characterized by a gradual change in lightness, as in varying lightnesses of a blue hue.

serifs: short extensions at the ends of major letter strokes.

service bureau: a business that specializes in the creation of products such as film negatives, printing plates, and proofs, and provides print-related services.

sexagesimal system: the system for specifying locations on the Earth's surface in degrees, minutes, and seconds (a base 60 system).

sextant: a mechanical device used to determine latitude on the Earth's surface.

shaded relief: a method for depicting topography; areas facing away from a light source are shaded, whereas areas directly illuminated are not; slope might also be a factor, with steeper slopes shaded darker.

shadowed contours: a method for depicting topography; contour lines facing a light source are drawn in a normal line weight, whereas those in the shadow are thickened.

shape: a visual variable in which the form of symbols (or marks making up symbols) is varied, such as using squares and triangles to represent religious and public schools.

short-term visual store: an area of memory that requires constant attention (or activation) to be retained; contrast with **iconic memory** and **long-term visual memory**.

simplification: involves the elimination or "weeding" of unnecessary coordinate data when generalizing a line.

simultaneous contrast: when the perceived color of an area is affected by the color of the surrounding area.

size: (1) a visual variable in which the magnitudes of symbols (or marks making up symbols) are varied (e.g., using circles of different magnitudes to depict city populations); (2) how large a typeface is; normally expressed in points, where a point is 1/72 of an inch.

sketch map: a rough, generalized hand drawing that represents a developing idea of what a final map will look like.

skewed distribution: when data are placed along the number line, the distribution appears asymmetrical; most common is a *positive skew,* in which the bulk of data are concentrated toward the left and there are a few outliers on the right.

slope: the steepness of a topographic surface.

small circle: represents the line along which a plane intersects the Earth's surface when the plane does not pass through the Earth's center.

small multiple: many small maps displayed to show the change in an attribute over time or to compare many attributes for the same time period.

small-scale map: a map depicting a relatively large portion of the Earth's surface.

smooth phenomena: phenomena that change gradually over geographic space, such as the distribution of snowfall for the year in Wisconsin.

smoothing: shifts the position of points making up a line to improve the appearance of the line.

snowflake: see **polygonal glyph**.

sonification: using sound to represent spatial data.

South Pole: the geographic location in the Southern Hemisphere with a latitude of 90° South that serves as the point around which the Earth's axis rotates.

Southern Hemisphere: the portion of the Earth extending from the Equator to the South Pole.

spacing: a visual variable in which the distance between marks making up a symbol is varied, such as shading counties on a choropleth map with horizontal lines of varied spacing.

spatial autocorrelation: the tendency for like things to occur near one another in geographic space (technically, this is termed *positive* spatial autocorrelation).

spatial data mining: a simplified term used for *spatial applications* involving **knowledge discovery in databases (KDD)**.

spatial dimension: a term that describes whether a phenomenon can be conceived of as points, lines, areas, or volumes.

spatialization: the process of converting abstract information to a spatial framework in which visualization is possible (e.g., converting the nature of topics that are discussed on the front page of a newspaper over a month-long period to a spatial framework).

specific information: the information associated with particular locations on a thematic map.

spectral scheme: a sequence of colors for a choropleth or isarithmic map in which colors span the visual portion of the electromagnetic spectrum (sometimes referred to as a ROYGBIV scheme).

splining: using a mathematical function to smooth a contour line.

spot colors: colors that are mixed before they reach the printing device.

standard deviation: the square root of the average of the squared deviations of each data value about the mean.

standard distance (SD): a measure of dispersion for point data; essentially, a measure of spread about the mean center of the distribution.

standard line: a line on a map projection with a scale factor of 1.0, thus distortion is zero.

standard point: a point on a map projection with a scale factor of 1.0, thus distortion is zero.

standardized data: a term used when raw totals are adjusted to account for the area over which the data are collected; for example, we might divide the acres of tobacco harvested in each county by the areas of those counties.

star: see **multivariate ray-glyph**.

statistical map: see **thematic map**.

stem-and-leaf plot: a method for exploratory data analysis in which individual data values are broken into "stem" and "leaf" portions; the result resembles a histogram but provides greater detail for individual data values.

stereo pairs: a term describing when two maps of an area are viewed with a stereoscope, which enables the reader to see a three-dimensional view.

Stevens curve: a grayscale that is sometimes used for smooth (untextured) gray tones; some cartographers have recommended that it be used for unclassed maps.

stochastic screening: the application of ink or toner in a pattern of very small, pseudo-randomly spaced dots of uniform size, to create tints from a base color.

subtitle: the map element that is used to further explain the title.

subtractive primary colors: colors (cyan, magenta, and yellow) associated with printing inks that can be mixed to create a wide range of colors.

synchronization: one of the visual variables for animated maps; refers to the temporal correspondence of two or more time series.

tabular error: if classed data are conceived as a prism map, tabular error is the difference in height between prisms on this map and those on an unclassed prism map of the data.

Tanaka's method: a method for depicting topography in which the width of contour lines is varied as a function of their angular relationship with a light source; also, contour lines facing the light source are white, whereas those in shadow are black.

tangent case: describes a developable surface that intersects the reference globe along one line, usually a parallel of latitude.

thematic map: a map used to emphasize the *spatial distribution* (or *pattern*) of one or more geographic attributes (e.g., a map of predicted snowfall amounts for the coming winter in the United States).

thematic mapping software: software that is expressly intended for the creation of thematic maps; MapViewer is a traditional example.

thematic symbols: graphical symbols that directly represent a map's theme or topic.

thermal-wax transfer printer: a printing device (normally full-color) that incorporates heat sources, which melt wax-based inks and apply them to the print medium.

Thiessen polygons: polygons enclosing a set of control points such that all arbitrary points within a polygon are closer to the control point associated with that polygon than to any other polygon.

three-dimensional bars: the height of three-dimensional-looking bars are made proportional to each of the attributes being mapped.

three-dimensional scatterplot: a diagram that uses small point symbols in which the scores for observations on three attributes are plotted on a set of x, y, and z axes; viewing all points in the plot requires interactive graphics.

time series: an animation that emphasizes change through time.

tint: a lighter version of a base color.

tint percentage: the degree to which the appearance of an ink or toner is lightened when employing screening techniques.

tiny cubes method: a method in which small cubes are regularly spaced throughout a 3-D phenomenon to provide the viewer with a feel for how the phenomenon changes throughout three-dimensional space.

Tissot's indicatrix: a graphical device plotted on a map projection that illustrates the kind and extent of distortion at infinitesimally small points.

title: the map element that provides a succinct description of a map's theme.

title case: type composed of lowercase letters with the first letter of each word set in uppercase.

transparency: a visual variable for depicting data quality; refers to the ease with which a theme can be seen through a "fog" placed over that theme.

transparent symbols: proportional symbols that enable readers to see base information beneath a symbol and to see the maximum extent of each proportional symbol.

trapping: a series of techniques used to minimize the effects of misregistration in multicolor printing.

triangulation: (1) a method of surveying where two points having known locations are established and the distance between them is measured; angles from this baseline are then measured to a distant point, creating a triangle, which can then serve as a basis for other triangles; (2) a method of interpolating data for isarithmic maps in which original control points are connected by a set of triangles.

trichromatic theory: the theory that color perception is a function of the relative stimulation of blue, green, and red cones.

trivariate choropleth map: the overlay of three univariate choropleth maps; this approach should be used only when attributes add up to 100 percent, such as percent voting Republican, Democrat, and Independent.

true point data: data that can actually be measured at a point location, such as the number of calls made at a telephone booth over the course of a day.

true 3-D phenomena: a form of volumetric phenomena in which each longitude and latitude position has multiple attribute values depending on the height above or below a zero point (e.g., the level of ozone in the Earth's atmosphere varies as a function of elevation above sea level).

$2\frac{1}{2}$-D phenomena: a form of volumetric phenomena in which each point on the surface is defined by longitude, latitude, and a value above (or below) a zero point (e.g., the earth's topography).

type: words that appear on a map.

type family: a group of type designs that reflect common characteristics and a common base name.

type size: the size of type, expressed in points.

type style: variations of type design within a given type family.

typeface: type of a particular type family and style.

typography: the art or process of specifying, arranging, and designing type.

uncertainty: the notion that data for thematic maps are often subject to some form of error; uncertain data are also considered to be unreliable and of poor quality.

unclassed map: a map in which data are not grouped into classes of similar value and thus each data value can theoretically be represented by a different symbol (e.g., a data set with 100 different data values might be depicted using 100 different shades of gray on a choropleth map).

unconstrained extended local processing routines: simplification algorithms that search large sections of a line, with the search terminated by the geomorphological complexity of the line, not by an algorithmic criterion.

uniform color model: refers to a variation of the CIE color model in which colors are equally spaced in the visual sense.

unipolar data: data that have no dividing point and do not involve two complementary phenomena, such as per-capita income in African countries; contrast with **bipolar** and **balanced data**.

unit value: the count represented by each dot on a dot map.

unit-vector density map: a map in which the density and orientation of short fixed-length segments (unit vectors) are used to display the magnitude and direction, respectively, of a continuous vector-based flow, such as wind speed and direction.

universal kriging: a form of kriging that accounts for a trend (or drift) in the data over geographic space.

unweighted pair-group method using arithmetic averages (UPGMA): a method for computing resemblance coefficients when clusters in a cluster analysis are combined; involves averaging the coefficients between observations in the newly formed cluster with coefficients for observations in existing clusters.

value: see **lightness**.

variable scale projection: a map projection with a scale that is manipulated in specific ways to emphasize certain land areas at different scales.

vector: the manner in which images are created as we would draw a map by hand: the hardware moves to one location and draws to the next.

vector data model: a general approach for representing geographic features in digital form that uses discrete points, lines, and polygons.

verbal scale: the map element that is a textual expression of scale, such as "one inch to the mile."

vertical datum: a reference surface (often mean sea level) for elevations on the Earth's surface.

videocassette: an analog tape format for storing and playing moving images and sound.

virtual environment: a three-dimensional computer-based simulation of a real or imagined environment that users are able to navigate through and interact with.

virtual reality: the use of computer–based systems to create lifelike representations of the real world.

visible light: the portion of the electromagnetic spectrum to which the human eye is sensitive.

visual angle: the angle formed by lines projected from the top and bottom of an image through the center of the lens of the eye.

visual benchmarks: reference points with which other frames of an animation can be compared (e.g., for a proportional circle map animation of AIDS by country, yellow and red circles could represent the minimum and maximum number of new AIDS cases for each country over the time period of the animation).

visual hierarchy: the graphical representation of an intellectual hierarchy.

visual pigments: the light-sensitive chemicals found within rods and cones.

visual variables: the perceived differences in map symbols that are used to represent spatial phenomena.

visual weight: the relative amount of attention that a map feature attracts.

visualization software: software that is expressly intended for developing visualizations of spatial data.

wall-size display: a technology in which the graphic display covers a large portion of a wall; this is typically accomplished by combining multiple projected images.

Ward's method: commonly used hierarchical method of cluster analysis in which clusters are combined that lead to the smallest increase in sums of squares (SS) about the means of clusters.

wavelength of light: the distance between two wave crests associated with electromagnetic energy.

weighted isolines: the width of contour lines is made proportional to the associated data; thus labeling of contour lines is not required.

Williams curve: a gray scale that is recommended for coarse (textured) areal shades.

word spacing: the space between words.

World Wide Web: a system of Internet file servers that support the Hypertext Transfer Protocol.

xerography: a general method of image duplication involving a focused light source, a metal drum, toner, and a heat source.

Zip-a-Tone: in traditional manual cartography, areal shades could be cut from preprinted sheets of Zip-a-Tone and stuck to a base map.

zoom function: in an interactive graphics environment, an area to be focused on is enlarged, commonly by enclosing a desired area with a rectangular box and clicking the mouse.

References

Acevedo, W., and Masuoka, P. (1997) "Time-series animation techniques for visualizing urban growth." *Computers & Geosciences* 23, no. 4:423–435.

Adams, J. M., and Dolin, P. A. (2002) *Printing Technology* (5th ed.). Albany, NY: Delmar.

Adobe Systems Incorporated. (1997) *The Adobe PostScript Printing Primer*. San Jose, CA: Author.

Agfa Corporation. (1997) *The Secrets of Color Management*. Agfa-Gevaert N.V., Septestraat 27, B-2640 Mortsel, Belgium.

Agfa Corporation. (1999) *From Design to Distribution in the Digital Age*. Agfa-Gevaert N.V., Septestraat 27, B-2640 Mortsel, Belgium.

Agfa Corporation. (2000) *An Introduction to Digital Color Printing*. Agfa-Gevaert N.V., Septestraat 27, B-2640 Mortsel, Belgium.

Aldenderfer, M. S., and Blashfield, R. K. (1984) *Cluster Analysis*. Beverly Hills, CA: Sage.

Ambroziak, B. M., Ambroziak, J. R., and Bradbury, R. (1999) *Infinite Perspectives: Two Thousand Years of Three-Dimensional Mapmaking*. New York: Princeton Architectural Press.

American Cartographic Association. (1986) *Which Map Is Best: Projections for World Maps*. Bethesda, MD: American Congress on Surveying and Mapping.

American Cartographic Association. (1988) *Choosing a World Map: Attributes, Distortions, Classes, Aspects*. Bethesda, MD: American Congress on Surveying and Mapping.

American Cartographic Association. (1991) *Matching the Map Projection to the Need*. Bethesda, MD: American Congress on Surveying and Mapping.

Anderson, E. (1960) "A semigraphical method for the analysis of complex problems." *Technometrics* 2, no. 3:387–391.

Andrienko, G. L., and Andrienko, N. V. (1999) "Interactive maps for visual data exploration." *International Journal of Geographical Information Science* 13, no. 4:355–374.

Andrienko, G., and Andrienko, N. (2001) "Exploring spatial data with dominant attribute map and parallel coordinates." *Computers, Environment and Urban Systems* 25:5–15.

Andrienko, N., Andrienko, G., and Gatalsky, P. (2000) "Supporting visual exploration of object movement." *Proceedings of the Working Conference on Advanced Visual Interfaces AVI*, ACM Press, Palermo, Italy, pp. 217–220, 315.

Andrienko, N., Andrienko, G., and Gatalsky, P. (2001a) "Exploring changes in census time series with interactive dynamic maps and graphics." *Computational Statistics* 16:417–433.

Andrienko, N., Andrienko, G., Savinov, A., Voss, H., and Wettschereck, D. (2001b) "Exploratory analysis of spatial data using interactive maps and data mining." *Cartography and Geographic Information Science* 28, no. 3:151–165.

Anselin, L. (1998) "Exploratory spatial data analysis in a geocomputational environment." In *Geocomputation: A Primer,* ed. by P. A. Longley, S. M. Brooks, R. McDonnell, and B. MacMillan, pp. 77–94. Chichester, England: Wiley.

Arnheim, R. (1974) *Art and Visual Perception: A Psychology of the Creative Eye*. Los Angeles: University of California Press.

Arntson, A. E. (2003) *Graphic Design Basics* (4th ed.). Fort Worth, TX: Harcourt Brace College Publishers.

Aspaas, H. R., and Lavin, S. J. (1989) "Legend designs for unclassed, bivariate, choropleth maps." *The American Cartographer* 16, no. 4:257–268.

Astroth, J. H. J., Trujillo, J., and Johnson, G. E. (1990) "A retrospective analysis of GIS performance: The Umatilla Basin revisited." *Photogrammetric Engineering and Remote Sensing* 56, no. 3:359–363.

Bachi, R. (1973) "Geostatistical analysis of territories." *Bulletin of the International Statistical Institute, Proceedings of the 39th Session*, Vienna, Austria, pp. 121–132.

Bachi, R. (1999) *New Methods of Geostatistical Analysis and Graphical Representation: Distributions of Populations over Territories.* New York: Kluwer Academic/Plenum.

Bailey, M. J. (2002) "Tele-Manufacturing: Rapid prototyping on the Internet with automatic consistency checking." *http://www.sdsc.edu/tmf/Whitepaper/whitepaper.html.*

Barnes, D. (1999) "Creation of publication quality shaded-relief maps with ArcView GIS." *Cartographic Perspectives*, no. 32:65–69.

Barnett, V., and Lewis, T. (1994) *Outliers in Statistical Data* (3rd ed.). Chichester, England: Wiley.

Barrass, S. (1994) "A perceptual framework for the auditory display of scientific data." *Proceedings of the Second International Conference on Auditory Display, ICAD '94*, Santa Fe, NM, pp. 131–145.

Barrault, M. (2001) "A methodology for the placement and evaluation of area map labels." *Computers, Environment and Urban Systems* 25:33–52.

Barrett, R. E. (1994) *Using the 1990 U.S. Census for Research.* Thousand Oaks, CA: Sage.

Bastin, L., Fisher, P. F., and Wood, J. (2002) "Visualizing uncertainty in multi-spectral remotely sensed imagery." *Computers & Geosciences* 28, no. 3:337–350.

Batty, M., Chapman, D., Evans, S., Haklay, M., Kueppers, S., Shiode, N., Smith, A., Torrens, P. M. (2001) "Visualizing the city: Communicating urban design to planners and decision makers." In *Planning Support Systems*, ed. by R. K. Brail and R. E. Klosterman, pp. 405–443. Redlands, CA: ESRI Press.

Batty, M., Fairbairn, D., Ogleby, C., Moore, K. E., and Taylor, G. (2002) "Introduction." In *Virtual Reality in Geography*, ed. by P. Fisher and D. Unwin, pp. 211–219. London: Taylor & Francis.

Batty, M., and Howes, D. (1996a) "Exploring urban development dynamics through visualization and animation." In *Innovations in GIS 3*, ed. by D. Parker, pp. 149–161. Bristol, PA: Taylor & Francis.

Batty, M. and Howes, D. (1996b) "Visualizing urban development." *Geo Info Systems* 6, no. 9:28–29, 32.

Beard, M. K., Buttenfield, B. P., and Clapham, S. B. (1991) *NCGIA Research Initiative 7: Visualization of Spatial Data Quality. Technical Paper 91-26.* Santa Barbara, CA: National Center for Geographic Information and Analysis. Available at *http://www.ncgia.ucsb.edu/Publications/Tech_Reports/91/91-26.pdf.*

Bemis, D., and Bates, K. (1989) "Color on temperature maps." Unpublished manuscript, Department of Geography, Pennsylvania State University, University Park, PA.

Bertin, J. (1981) *Graphics and Graphic Information-Processing.* Berlin: Walter de Gruyter.

Bertin, J. (1983) *Semiology of Graphics: Diagrams, Networks, Maps.* Madison: University of Wisconsin Press.

Berton, J. A. J. (1990) "Strategies for scientific visualization: Analysis and comparison of current techniques." *Extracting Meaning from Complex Data: Processing, Display, Interaction. Proceedings, SPIE*, Volume 1259, Santa Clara, CA, pp. 110–121.

Birdsall, S. S., and Florin, J. W. (1992) *Regional Landscapes of the United States and Canada.* New York: Wiley.

Birren, F. (1983) *Colour.* London: Marshall Editions Limited.

Bishop, I. (1994) "The role of visual realism in communicating and understanding spatial change and process." In *Visualization in Geographical Information Systems*, ed. by H. M. Hearnshaw and D. J. Unwin, pp. 60–64. Chichester, England: Wiley.

Blenkinsop, S., Fisher, P., Bastin, L., and Wood, J. (2000) "Evaluating the perception of uncertainty in alternative visualization strategies." *Cartographica* 37, no. 1:1–13.

Blok, C., Köbben, B., Cheng, T., and Kuterema, A. A. (1999) "Visualization of relationships between spatial patterns in time by cartographic animation." *Cartography and Geographic Information Science* 26, no. 2:139–151.

Board, C. (1984) "Higher order map-using tasks: Geographical lessons in danger of being forgotten." *Cartographica* 21, no. 1:85–97.

Bockenhauer, M. H. (1994) "Culture of the Wisconsin Official State Highway Map." *Cartographic Perspectives*, no. 18:17–27.

Bojorquez, T. (1995) "The color-laser promise." *MacUser* 11, no. 3:91–97.

Borg, I., and Staufenbiel, T. (1992) "Performance of snowflakes, suns, and factorial suns in the graphical representation of multivariate data." *Multivariate Behavioral Research* 27, no. 1:43–55.

Bowditch, N. (1995) *The American Practical Navigator.* Bethesda, MD: National Imagery and Mapping Agency.

Bowmaker, J. K., and Dartnall, H. J. A. (1980) "Visual pigments of rods and cones in a human retina." *Journal of Physiology* 298:501–511.

Boyce, R. R., and Clark, W. A. V. (1964) "The concept of shape in geography." *The Geographical Review* 54, no. 4:561–572.

Brainerd, J., and Pang, A. (2001) "Interactive map projections and distortion." *Computers & Geosciences* 27:299–314.

Brassel, K. (1974) "A model for automatic hill shading." *The American Cartographer* 1, no. 1:15–27.

Brassel, K., and Weibel, R. (1988) "A review and conceptual framework of automated generalization." *International Journal of Geographical Information Systems* 2, no. 3:229–244.

Brassel, K. E., and Utano, J. J. (1979) "Design strategies for continuous-tone area mapping." *The American Cartographer* 6, no. 1:39–50.

Breding, P. (1998) "The effect of prior knowledge on eighth graders' abilities to understand cartograms." Unpublished MS thesis, University of South Carolina, Columbia, SC.

Brewer, C. A. (1989) "The development of process-printed Munsell charts for selecting map colors." *The American Cartographer* 16, no. 4:269–278.

Brewer, C. A. (1992) "Review of colour terms and simultaneous contrast research for cartography." *Cartographica* 29, no. 3/4:20–30.

Brewer, C. A. (1994a) "Color use guidelines for mapping and visualization." In *Visualization in Modern Cartography*, ed. by A. M. MacEachren and D. R. F. Taylor, pp. 123–147. Oxford, England: Pergamon.

Brewer, C. A. (1994b) "Guidelines for use of the perceptual dimensions of color for mapping and visualization." In *Color Hard Copy and Graphic Arts III*, ed. by J. Bares, pp. 54–63. Bellingham, WA: The International Society for Optical Engineering.

Brewer, C. A. (1996) "Guidelines for selecting colors for diverging schemes on maps." *The Cartographic Journal* 33, no. 2:79–86.

Brewer, C. A. (1997) "Spectral schemes: Controversial color use on maps." *Cartography and Geographic Information Science* 24, no. 4:203–220.

Brewer, C. A. (2001) "Reflections on mapping Census 2000." *Cartography and Geographic Information Science* 28, no. 4:213–235.

Brewer, C. A., and Campbell, A. J. (1998) "Beyond graduated circles: Varied point symbols for representing quantitative data on maps." *Cartographic Perspectives*, no. 29:6–25.

Brewer, C. A., Hatchard, G. W., and Harrower, M. A. (2003) "ColorBrewer in print: A catalog of color schemes for maps." *Cartography and Geographic Information Science* 30, no. 1:5–32.

Brewer, C. A., MacEachren, A. M., Pickle, L. W., and Herrmann, D. (1997) "Mapping mortality: Evaluating color schemes for choropleth maps." *Annals, Association of American Geographers* 87, no. 3:411–438.

Brewer, C. A., and Marlow, K. A. (1993) "Color representation of aspect and slope simultaneously." *Auto-Carto 11 Proceedings*, Minneapolis, MN, pp. 328–337.

Brewer, C. A., and Pickle, L. (2002) "Evaluation of methods for classifying epidemiological data on choropleth maps in series." *Annals of the Association of American Geographers* 92, no. 4:662–681.

Brown, A., and van Elzakker, C. P. J. M. (1993) "The use of colour in the cartographic representation of information quality generated by a GIS." *Proceedings, 16th Conference of the International Cartographic Association*, Cologne, Germany, pp. 707–720.

Brown, L. A. (1990) *The Story of Maps*. Dover, DE: Dover Publications.

Bruno, M. H. (ed.). (2000) *Pocket Pal: A Graphic Arts Production Handbook* (8th ed.). Memphis, TN: International Paper Company.

Bucher, F. (1999) "Using extended exploratory data analysis for the selection of an appropriate interpolation model." In *Geographic Information Research: Trans-Atlantic Perspectives*, ed. by M. Craglia and H. Onsrud, pp. 391–403. London: Taylor & Francis.

Buckley, A. (2003) "Atlas mapping in the 21st century." *Cartography and Geographic Information Science* 30, no. 2:149–158.

Buckley, D. J., Ulbricht, C., and Berry, D. J. (1998) "The Virtual Forest: Advanced 3-D visualization techniques for forest management and research." *ESRI User Conference*, San Diego, CA. Available at *http://www.innovativegis.com/products/vforest/index.html*.

Bugayevskiy, L. M., and Snyder, J. P. (1995) *Map Projections: A Reference Manual*. London: Taylor & Francis.

Buja, A., Cook, D., and Swayne, D. F. (1996) "Interactive high-dimensional data visualization." *Journal of Computational and Graphical Statistics* 5, no. 1:78–99.

Burrough, P. A., and McDonnell, R. A. (1998) *Principles of Geographical Information Systems*. Oxford, England: Oxford University Press.

Burt, J. E., and Barber, G. M. (1996) *Elementary Statistics for Geographers* (2nd ed.). New York: Guilford.

Buttenfield, B. P. I. (1990) "NCGIA Research Initiative 3: Multiple representations." *http://www.ncgia.ucsb.edu/research/initiatives.html#i3*.

Buttenfield, B. P. (1991) "A rule for describing line feature geometry." In *Map Generalization: Making Rules for Knowledge Representation*, ed. by B. P. Buttenfield and R. B. McMaster, pp. 150–171. London: Longman.

Buttenfield, B. P., and Mark, D. M. (1991) "Expert systems in cartographic design." In *Geographic Information Systems: The Microcomputer and Modern Cartography*, ed. by D. R. F. Taylor, pp. 129–150. Oxford, England: Pergamon.

Buttenfield, B. P., and Weber, C. R. (1994) "Proactive graphics for exploratory visualization of biogeographical data." *Cartographic Perspectives*, no. 19:8–18.

Buziek, G., and Döllner, J. (1999) "Concept and implementation of an interactive, cartographic virtual reality system." *Proceedings of the 19th International Cartographic Conference*, Ottawa, Canada, pp. Section 5:88–99 (CD-ROM).

Byron, J. R. (1994) "Spectral encoding of soil texture: A new visualization method." *GIS/LIS Proceedings*, Phoenix, AZ, pp. 125–132.

Caldwell, D. R. (2001) "Physical terrain modeling for geographic visualization." *Cartographic Perspectives*, no. 38:66–72.

Caldwell, P. S. (1979) "Television news maps: An examination of their utilization, content, and design." Unpublished PhD dissertation, University of California at Los Angeles, Los Angeles, CA.

Caldwell, P. S. (1981) "Television news maps: The effects of the medium on the map." *Technical Papers, American Congress on Surveying and Mapping, 41st annual meeting*, Washington, DC, pp. 382–392.

Câmara, A. S., and Raper, J. (eds.). (1999) *Spatial Multimedia and Virtual Reality*. London: Taylor & Francis.

Campbell, C. S., and Egbert, S. L. (1990) "Animated cartography: Thirty years of scratching the surface." *Cartographica* 27, no. 2:24–46.

Canters, F. (2002) *Small-Scale Map Projection Design*. New York: Taylor & Francis.

Canters, F., and Decleir, H. (1989) *The World in Perspective: A Directory of World Map Projections*. New York: Wiley.

Card, S. K., Mackinlay, J. D., and Shneiderman, B. (eds.). (1999) *Readings in Information Visualization: Using Vision to Think*. San Francisco, CA: Morgan Kaufmann.

Cardin, J., Castellanos, A., and Romano, F. J. (2001) *PDF Printing and Publishing: The Next Revolution After Gutenberg*: Agfa Corporation.

Carr, D. B. (1991) "Looking at large data sets using binned data plots." In *Computing and Graphics in Statistics*. ed. by A. Buja and P. A. Tukey, pp. 7–39. New York: Springer-Verlag.

Carr, D. B. (1994) "A colorful variation on box plots." *Statistical Computing & Statistical Graphics Newsletter* 5, no. 3:19–23.

Carr, D. B., Olsen, A. R., Courbois, J.-Y. P., Pierson, S. M., and Carr, D. A. (1998) "Linked micromap plots: names and described." *Statistical Computing & Statistical Graphics Newsletter* 9, no. 1:24–32.

Carr, D. B., Olsen, A. R., Pierson, S. M., and Courbois, J.-Y. P. (2000a) "Using linked micromap plots to characterize Omernik ecoregions." *Data Mining and Knowledge Discovery* 4, no. 1:43–67.

Carr, D. B., Olsen, A. R., and White, D. (1992) "Hexagon mosaic maps for display of univariate and bivariate geographical data." *Cartography and Geographic Information Systems* 19, no. 4:228–236, 271.

Carr, D. B., and Pierson, S. M. (1996) "Emphasizing statistical summaries and showing spatial context with micromaps." *Statistical Computing & Statistical Graphics Newsletter* 7, no. 3:16–23.

Carr, D. B., Wallin, J. F., and Carr, D. A. (2000b) "Two new templates for epidemiology applications: Linked micromap plots and conditioned choropleth maps." *Statistics in Medicine* 19: 2521–2538.

Carr, D. B., Zhang, Y., and Li, Y. (2002) "Dynamically conditioned choropleth maps: Shareware for hypothesis generation and education." *Statistical Computing & Statistical Graphics Newsletter* 13, no. 2:2–7.

Carroll, L. (1893) *Sylvie and Bruno Concluded*. New York: Dover Publications (1988).

Carstensen, L. W. (1986a) "Bivariate choropleth mapping: The effects of axis scaling." *The American Cartographer* 13, no. 1:27–42.

Carstensen, L. W. (1986b) "Hypothesis testing using univariate and bivariate choropleth maps." *The American Cartographer* 13, no. 3:231–251.

Carstensen, L. W. J. (1982) "A continuous shading scheme for two-variable mapping." *Cartographica* 19, no. 3/4:53–70.

Carstensen, L. W. J. (1987) "A comparison of simple mathematical approaches to the placement of spot symbols." *Cartographica* 24, no. 3:46–63.

Cartographic Panel. (1950) "Discussion of Cartographic Panel at Worcester meetings of the Association of American Geographers." *The Professional Geographer* 2, no. 6.

Cartwright, W. (1997) "New media and their application to the production of map products." *Computers & Geosciences* 23, no. 4:447–456.

Cartwright, W. (1999) "Extending the map metaphor using web delivered multimedia." *International Journal of Geographical Information Science* 13, no. 4:335–353.

Cartwright, W., Peterson, M. P., and Gartner, G. (eds.). (1999) *Multimedia Cartography*. Berlin: Springer-Verlag.

Cartwright, W., and Stevenson, J. (2000) "A toolbox for publishing maps on the World Wide Web." *Cartography* 29, no. 2:83–95.

Castner, H. W., and Robinson, A. H. (1969) *Dot Area Symbols in Cartography: The Influence of Pattern on Their Perception*. Washington, DC: American Congress on Surveying and Mapping.

Chainey, S., and Stuart, N. (1998) "Stochastic simulation: An alternative interpolation technique for digital geographic information." In *Innovations in GIS 5*, ed. by Steve Carver, pp. 3–24. London: Taylor & Francis.

Chamberlin, W. (1947) *The Round Earth on Flat Paper: Map Projections Used by Cartographers*. Washington, DC: National Geographic Society.

Chang, K. (1977) "Visual estimation of graduated circles." *The Canadian Cartographer* 14, no. 2:130–138.

Chang, K. (1980) "Circle size judgment and map design." *The American Cartographer* 7, no. 2:155–162.

Chang, K. (1982) "Multi-component quantitative mapping." *The Cartographic Journal* 19, no. 2:95–103.

Chrisman, N. (2002) *Exploring Geographic Information Systems* (2nd ed.). New York: Wiley.

Chrisman, N. R. (1998) "Rethinking levels of measurement for cartography." *Cartography and Geographic Information Systems* 25, no. 4:231–242.

Christ, F. (1976) "Fully automated and semi-automated interactive generalization, symbolization and light drawing of a small scale topographic map." *Nachrichten aus dem Karten-und Vermessungswesen*, Uhersetzunge, Heft nr. 33:19–36.

Clark, S. M., Larsgaard, M. L., and Teague, C. M. (1992) *Cartographic Citations, A Style Guide*. Chicago: American Library Association, Map Geography Table, MAGERT Circular No. 1.

Clark, W. A. V., and Hosking, P. L. (1986) *Statistical Methods for Geographers*. New York: Wiley.

Clarke, K. C. (1995) *Analytical and Computer Cartography* (2nd ed.). Englewood Cliffs, NJ: Prentice Hall.

Clarke, K. C. (1998) "Visualizing different geofutures." In *Geocomputation: A Primer*, ed. by P. A. Longley, S. M.

Brooks, R. McDonnell, and B. Macmillan, pp. 119–137. Chichester, England: Wiley.

Clarke, K. C. (2001) "Cartography in a mobile Internet age." *International Cartographic Conference*, Beijing, China, CD-ROM.

Clarke, K. C., Teague, P. D., and Smith, H. G. (1999) "Virtual depth-based representation of cartographic uncertainty." *Proceedings of the International Symposium on Spatial Data Quality*, Hong Kong, pp. 253–259.

Cleveland, W. S. (1993) *Visualizing Data*. Summit, NJ: Hobart Press.

Cleveland, W. S. (1994) *The Elements of Graphing Data* (rev. ed.). Summit, NJ: Hobart Press.

Cleveland, W. S., and McGill, M. E. (1988) *Dynamic Graphics for Statistics*. Belmont, CA: Wadsworth.

Cleveland, W. S., and McGill, R. (1984) "Graphical perception: Theory, experimentation, and application to the development of graphical methods." *Journal of the American Statistical Association* 79, no. 387:531–554.

Cliburn, D. C. (2001) "Representing multiple uncertainties and evaluating usability in a spatial decision support system." Unpublished PhD dissertation, University of Kansas, Lawrence, KS.

Cliburn, D. C., Feddema, J. J., Miller, J. R., and Slocum, T. A. (2002) "Design and evaluation of a decision support system in a water balance application." *Computers & Graphics* 26:931–949.

Cloke, P., Philo, C., and Sadler, D. (1991) *Approaching Human Geography: An Introduction to Contemporary Theoretical Debates*. London: Paul Chapman.

Cohen, P. R., Johnston, M., McGee, D., Oviatt, S., Pittman, J., et al. (1997) "QuickSet: Multimodal interaction for distributed applications." *Proceedings of the Fifth ACM International Multimedia Conference*, Seattle, WA, pp. 31–40.

Comenetz, J. (1999) "Cartographic visualization of the quality of demographic data." Unpublished PhD dissertation, University of Minnesota, Minneapolis, MN.

Costello, P. (1997) "Health and safety issues associated with virtual reality—A review of current literature." Available at *http://www.agocg.ac.uk/reports/virtual/37/report37.htm*.

Coulson, M. R. C. (1978) "Potential for variation: A concept for measuring the significance of variation in size and shape of areal units." *Geografiska Annaler* 60B:48–64.

Coulson, M. R. C. (1987) "In the matter of class intervals for choropleth maps: With particular reference to the work of George F. Jenks." *Cartographica* 24, no. 2:16–39.

Cox, C. W. (1976) "Anchor effects and estimation of graduated circles and squares." *The American Cartographer* 3, no. 1:65–74.

Cox, D. J. (1990) "The art of scientific visualization." *Academic Computing* 4, no. 6:20–22, 32–34, 36–38, 40.

Crampton, J. W. (1999) "Online mapping: Theoretical context and practical applications." In *Multimedia Cartography*, ed. by W. Cartwright, M. P. Peterson, and G. Gartner, pp. 291–304. Berlin: Springer-Verlag.

Crampton, J. W. (2002) "Interactivity types in geographic visualization." *Cartography and Geographic Information Science* 29, no. 2:85–98.

Crawford, P. V. (1973) "The perception of graduated squares as cartographic symbols." *The Cartographic Journal* 10, no. 2:85–88.

Cressie, N. (1990) "The origins of kriging." *Mathematical Geology* 22, no. 3:239–252.

Cressie, N. A. C. (1993) *Statistics for Spatial Data* (rev. ed.). New York: Wiley.

Cromley, R. G. (1995) "Classed versus unclassed choropleth maps: A question of how many classes." *Cartographica* 32, no. 4:15–27.

Cromley, R. G. (1996) "A comparison of optimal classification strategies for choroplethic displays of spatially aggregated data." *International Journal of Geographical Information Systems* 10, no. 4:405–424.

Cromley, R. G., and Mrozinski, R. D. (1997) "An evaluation of classification schemes based on the statistical versus the spatial structure properties of geographic distributions in choropleth mapping." *1997 ACSM/ASPRS Annual Convention & Exposition, Technical Papers*, Volume 5 (Auto-Carto 13), Seattle, WA, pp. 76–85.

Cromley, R. G., and Mrozinski, R. D. (1999) "The classification of ordinal data for choropleth mapping." *The Cartographic Journal* 36, no. 2:101–109.

Cuff, D. J. (1973) "Colour on temperature maps." *The Cartographic Journal* 10, no. 1:17–21.

Cuff, D. J., and Bieri, K. R. (1979) "Ratios and absolute amounts conveyed by a stepped statistical surface." *The American Cartographer* 6, no. 2:157–168.

Cuff, D. J., Pawling, J. W., and Blair, E. T. (1984) "Nested value-by-area cartograms for symbolizing land use and other proportions." *Cartographica* 21, no. 4:1–8.

Cutler, M. E. (1998) "The effects of prior knowledge on children's abilities to read static and animated maps." Unpublished MS thesis, University of South Carolina, Columbia, SC.

Czernuszenko, M., Pape, D., Sandin, D., DeFanti, T., Dawe, G. L., et al. (1997) "The ImmersaDesk and Infinity Wall Projection-based virtual reality displays." *Computer Graphics* 31, no. 2:46–49.

Davis, J., and Chen, X. (2002) "LumiPoint: Multi-user laser-based interaction on large tiled displays." *Displays* 23, no. 5: 205–211. Available at *http://graphics.stanford.edu/papers/multiuser/*.

Davis, J. C. (1975) "Contouring algorithms." *AUTO-CARTO II, Proceedings of the International Symposium on Computer-Assisted Cartography*, pp. 352–359.

Davis, J. C. (1986) *Statistics and Data Analysis in Geology* (2nd ed.). New York: Wiley.

Davis, J. C. (2002) *Statistics and Data Analysis in Geology* (3rd ed.). New York: Wiley.

Day, A. (1994) "From map to model: The development of an urban information system." *Design Studies* 15, no. 3:366–384.

DeBraal, J. P. (1992) "Foreign ownership of U.S. agricultural land through December 31, 1992." *United States Department of Agriculture, Economic Research Service, Statistical Bulletin 853.*

Declercq, F. A. N. (1995) "Choropleth map accuracy and the number of class intervals." *Proceedings of the 17th International Cartographic Conference*, Volume 1, Barcelona, Spain, pp. 918–922.

Declercq, F. A. N. (1996) "Interpolation methods for scattered sample data: Accuracy, spatial patterns, processing time." *Cartography and Geographic Information Systems* 23, no. 3:128–144.

Deetz, C. H., and Adams, O. S. (1945) *Element of Map Projections with Applications to Map and Chart Construction* (5th ed.). Washington, DC: U.S. Department of Commerce Coast and Geodetic Survey.

Defense Mapping Agency. (1981) *Glossary of Mapping, Charting, and Geodetic Terms* (4th ed.). Washington, DC: Department of Defense.

De Genst, W., and Canters, F. (1996) "Development and implementation of a procedure for automated map projection selection." *Cartography and Geographic Information Systems* 23, no. 3:145–171.

DeLucia, A. A., and Hiller, D. W. (1982) "Natural legend design for thematic maps." *The Cartographic Journal* 19, no. 1:46–52.

Dent, B. D. (1996) *Cartography: Thematic Map Design* (4th ed.). Dubuque, IA: William C. Brown.

Dent, B. D. (1999) *Cartography: Thematic Map Design* (5th ed.). Boston: McGraw-Hill.

Derrington, A., Lennie, P., and Krauskopf, J. (1983) "Chromatic response properties of parvocellular neurons in the macaque LGN." In *Colour Vision: Physiology and Psychophysics*, ed. by J. D. Mollon and L. T. Sharpe, pp. 245–251. London: Academic Press.

De Valois, R. L., and Jacobs, G. H. (1984) "Neural mechanisms of color vision." In *Handbook of Physiology (Section 1: The Nervous System)*, ed. by J. M. Brookhart and V. B. Mountcastle, pp. 425–456. Bethesda, MD: American Physiological Society.

DiBiase, D. (1990) "Visualization in the earth sciences." *Earth and Mineral Sciences* 59, no. 2:13–18.

DiBiase, D. (1994) "Designing animated maps for a multimedia encyclopedia." *Cartographic Perspectives*, no. 19:3–7, 19.

DiBiase, D., Krygier, J., Reeves, C., MacEachren, A., and Brenner, A. (1991) "Elementary approaches to cartographic animation." Videotape. University Park, PA: Deasy GeoGraphics Laboratory, Department of Geography, Pennsylvania State University.

DiBiase, D., MacEachren, A. M., Krygier, J. B., and Reeves, C. (1992) "Animation and the role of map design in scientific visualization." *Cartography and Geographic Information Systems* 19, no. 4:201–214, 265–266.

DiBiase, D., Reeves, C., MacEachren, A. M., Von Wyss, M., Krygier, J. B., Sloan, J. L., Detweiler, M. C. (1994a) "Multivariate display of geographic data: Applications in Earth system science." In *Visualization in Modern Cartography*, ed. by A. M. MacEachren and D. R. F. Taylor, pp. 287–312. Oxford, England: Pergamon.

DiBiase, D., Sloan, J. L. I. I., and Paradis, T. (1994b) "Weighted isolines: An alternative method for depicting statistical surfaces." *The Professional Geographer* 46, no. 2:218–228.

Ding, Y., and Densham, P. J. (1994) "A loosely synchronous, parallel algorithm for hill shading digital elevation models." *Cartography and Geographic Information Systems* 21, no. 1:5–14.

Dobson, J. E., and Fisher, P. F. (2003) "Geoslavery." *IEEE Technology and Society Magazine* 22, no. 1:47–52.

Dobson, M. W. (1973) "Choropleth maps without class intervals?: A comment." *Geographical Analysis* 5, no. 4:358–360.

Dobson, M. W. (1974) "Refining legend values for proportional circle maps." *The Canadian Cartographer* 11, no. 1:45–53.

Dobson, M. W. (1980a) "Perception of continuously shaded maps." *Annals, Association of American Geographers* 70, no. 1:106–107.

Dobson, M. W. (1980b) "Unclassed choropleth maps: A comment." *The American Cartographer* 7, no. 1:78–80.

Dobson, M. W. (1983) "Visual information processing and cartographic communication: The utility of redundant stimulus dimensions." In *Graphic Communication and Design in Contemporary Cartography: Progress in Contemporary Cartography,* ed. by D. R. F. Taylor, pp. 149–175. Chichester, England: Wiley.

Dodge, M., and Kitchin, R. (2001a) *Atlas of Cyberspace.* Harlow, England: Addison-Wesley.

Dodge, M., and Kitchin, R. (2001b) *Mapping Cyberspace.* London: Routledge.

Dorling, D. (1992) "Stretching space and splicing time: From cartographic animation to interactive visualization." *Cartography and Geographic Information Systems* 19, no. 4:215–227, 267–270.

Dorling, D. (1993) "Map design for census mapping." *The Cartographic Journal* 30, no. 2:167–183.

Dorling, D. (1994) "Cartograms for visualizing human geography." In *Visualization in Geographical Information Systems*, ed. by H. M. Hearnshaw and D. J. Unwin, pp. 85–102, plates 7–10. Chichester, England: Wiley.

Dorling, D. (1995a) *A New Social Atlas of Britain.* Chichester, England: Wiley.

Dorling, D. (1995b) "The visualization of local urban change across Britain." *Environment and Planning B: Planning and Design* 22, no. 3:269–290.

Dorling, D. (1995c) "Visualizing changing social structure from a census." *Environment and Planning A* 27:353–378.

Dougenik, J. A., Chrisman, N. R., and Niemeyer, D. R. (1985) "An algorithm to construct continuous area cartograms." *The Professional Geographer* 37, no. 1:75–81.

Drascic, D., and Milgram, P. (1996) "Perceptual issues in augmented reality." *Proceedings, SPIE*, Volume 2653, San Jose, CA, pp. 123–134. Available at *http://gypsy.rose.utoronto.ca/people/david_dir/SPIE96/SPIE96.html*.

Dunn, M., and Hickey, R. (1998) "The effect of slope algorithms on slope estimates within a GIS." *Cartography* 27, no. 1:9–15.

Dunn, R. (1988) "Framed rectangle charts or statistical maps with shading: An experiment in graphical perception." *The American Statistician* 42, no. 2:123–129.

Dykes, J. (1995) "Cartographic visualization for spatial analysis." *Proceedings of the 17th International Cartographic Conference*, Volume 1, Barcelona, Spain, pp. 1365–1370.

Dykes, J. (1996) "Dynamic maps for spatial science: A unified approach to cartographic visualization." In *Innovations in GIS 3*, ed. by D. Parker, pp. 177–187, Color Plates 4–5. Bristol, PA: Taylor & Francis.

Dykes, J. (2002) "Creating information-rich virtual environments with geo-referenced digital panoramic imagery." In *Virtual Reality in Geography*, ed. by P. Fisher and D. Unwin, pp. 68–92. London: Taylor & Francis.

Dykes, J., Moore, K., and Wood, J. (1999) "Virtual environments for student fieldwork using networked components." *International Journal of Geographical Information Science* 13, no. 4:397–416.

Dykes, J. A. (1994) "Visualizing spatial association in area-value data." In *Innovations in GIS*, ed. by M. F. Worboys, pp. 149–159. Bristol, PA: Taylor & Francis.

Dykes, J. A. (1997) "Exploring spatial data representation with dynamic graphics." *Computers & Geosciences* 23, no. 4:345–370.

Eastman, J. R. (1986) "Opponent process theory and syntax for qualitative relationships in quantitative series." *The American Cartographer* 13, no. 4:324–333.

Eckert, M. (1908) "On the nature of maps and map logic." *Bulletin of the American Geographical Society* 40, no. 6:344–351. Translated by W. Joerg.

Eckhardt, D. W., Verdin, J. P., and Lyford, G. R. (1990) "Automated update of an irrigated lands GIS using SPOT HRV imagery." *Photogrammetric Engineering and Remote Sensing* 56, no. 11:1515–1522.

Eddy, W. F., and Mockus, A. (1994) "An example of the estimation and display of a smoothly varying function of time and space—The incidence of the disease mumps." *Journal of the American Society for Information Science* 45, no. 9:686–693.

Edmondson, S., Christensen, J., Marks, J., and Shieber, S. M. (1996) "A general cartographic labelling algorithm." *Cartographica* 33, no. 4:13–23.

Edsall, R., and Peuquet, D. (1997) "A graphical user interface for the integration of time into GIS." Available at *http://www.geog.psu.edu/~edsall/research/ACarticle.html*.

Edsall, R. M. (2003a) "Design and usability of an enhanced geographic information system for exploration of multivariate health statistics." *The Professional Geographer* 55, no. 2:146–160.

Edsall, R. M. (2003b) "The parallel coordinate plot in action: Design and use for geographic visualization." *Computational Statistics and Data Analysis*, 43, no. 4:605–619.

Edwards, L. D., and Nelson, E. S. (2001) "Visualizing data certainty: A case study using graduated circle maps." *Cartographic Perspectives*, no. 38:19–36.

Egbert, S. L. (1994) "The design and evaluation of an interactive choropleth map exploration system." Unpublished PhD dissertation, University of Kansas, Lawrence, KS.

Egbert, S. L., Price, K. P., Nellis, M. D., and Lee, R.-Y. (1995) "Developing a land cover modelling protocol for the high plains using multi-seasonal thematic mapper imagery." *1995 ACSM/ASPRS Annual Convention & Exposition, Technical Papers*, Volume 3, Charlotte, NC, pp. 836–845.

Egbert, S. L., and Slocum, T. A. (1992) "EXPLOREMAP: An exploration system for choropleth maps." *Annals, Association of American Geographers* 82, no. 2:275–288.

Ehlschlaeger, C. R., Shortridge, A. M., and Goodchild, M. F. (1997) "Visualizing spatial data uncertainty using animation." *Computers & Geosciences* 23, no. 4:387–395.

Eicher, C. L., and Brewer, C. A. (2001) "Dasymetric mapping and areal interpolation: Implementation and evaluation." *Cartography and Geographic Information Science* 28, no. 2:125–138.

Eick, S. G. (2001) "Visualizing online activity." *Communications of the ACM* 44, no. 8:45–50.

Ekman, G., and Junge, K. (1961) "Psychophysical relations in visual perception of length, area and volume." *The Scandinavian Journal of Psychology* 2, no. 1:1–10.

Ellson, R. (1990) "Visualization at work." *Academic Computing* 4, no. 6:26–28, 54–56.

Environmental Systems Research Institute Incorporated. (2002) *Geographic Information Systems for Java, an ESRI White Paper*. Redlands, CA: Author.

Ervin, S. M. (2001) "Digital landscape modeling and visualization: A research agenda." *Landscape and Urban Planning* 54: 49–62.

Espenshade, E. B. J. (1990) *Goode's World Atlas* (19th ed.). Chicago: Rand McNally.

Evans, B. J. (1997) "Dynamic display of spatial data-reliability: Does it benefit the map user?" *Computers & Geosciences* 23, no. 4:409–422.

Evans, I. S. (1977) "The selection of class intervals." *Transactions, Institute of British Geographers (New Series)* 2, no. 1:98–124.

Everitt, B. S., Landau, S., and Leese, M. (2001) *Cluster Analysis*. (4th ed.). London: Arnold.

Eyton, J. R. (1984a) "Complementary-color two-variable maps." *Annals, Association of American Geographers* 74, no. 3:477–490.

Eyton, J. R. (1984b) "Raster contouring." *Geo-Processing* 2:221–242.

Eyton, J. R. (1990) "Color stereoscopic effect cartography." *Cartographica* 27, no. 1:20–29.

Eyton, J. R. (1991) "Rate-of-change maps." *Cartography and Geographic Information Systems* 18, no. 2:87–103.

Eyton, J. R. (1994) "Chromostereoscopic maps." *Cartouche* Autumn/Winter Special Issue:15.

Fabrikant, S. I. (2003) "Commentary on 'A history of twentieth-century American academic cartography'." *Cartography and Geographic Information Science* 30, no. 1:81–84.

Fabrikant, S. I., and Buttenfield, B. P. (2001) "Formalizing semantic spaces for information access." *Annals of the Association of American Geographers* 91, no. 2:263–280.

Fairchild, M. D. (1998) *Color Appearance Models*. Reading, MA: Addison-Wesley.

Famighetti, R. (1993) *The World Almanac and Book of Facts 1994*. Mahwah, NJ: Funk & Wagnalls.

Feiner, S. K. (2002) "Augmented reality: A new way of seeing." *Scientific American* 286, no. 4:48–55.

Ferreira, J. J., and Wiggins, L. L. (1990) "The density dial: A visualization tool for thematic mapping." *Geo Info Systems* 1, no. 0:69–71.

Fisher, P., and Unwin, D. (eds.). (2002) *Virtual Reality in Geography*. London: Taylor & Francis.

Fisher, P. F. (1993) "Visualizing uncertainty in soil maps by animation." *Cartographica* 30, no. 2&3:20–27.

Fisher, P. F. (1994a) "Hearing the reliability in classified remotely sensed images." *Cartography and Geographic Information Systems* 21, no. 1:31–36.

Fisher, P. F. (1994b) "Visualization of the reliability in classified remotely sensed images." *Photogrammetric Engineering and Remote Sensing* 60, no. 7:905–910.

Fisher, P. F. (1996) "Animation of reliability in computer-generated dots maps and elevation models." *Cartography and Geographic Information Systems* 23, no. 4:196–205.

Fisher, P. F., and Langford, M. (1996) "Modeling sensitivity to accuracy in classified imagery: A study of areal interpolation by dasymetric mapping." *The Professional Geographer* 48, no. 3:299–309.

Fisher, W. D. (1958) "On grouping for maximum homogeneity." *Journal of the American Statistical Association* 53, December:789–798.

Flanagan, T. J., and Maguire, K. (1992) *Sourcebook of Criminal Justice Statistics 1991*. Washington, DC: U.S. Department of Justice, Bureau of Justice Statistics.

Flannery, J. J. (1971) "The relative effectiveness of some common graduated point symbols in the presentation of quantitative data." *The Canadian Cartographer* 8, no. 2:96–109.

Fleming, P. D. (2002) "Offset lithography." Available at *www.wmich.edu/~ppse/Offset/*.

Foley, J. D., van Dam, A., Feiner, S. K., and Hughes, J. F. (1996) *Computer Graphics: Principles and Practice* (2nd ed., C ed.). Reading, MA: Addison-Wesley.

Foody, G. M. (1996) "Approaches to the production and evaluation of fuzzy land cover classification from remotely-sensed data." *International Journal of Remote Sensing* 17, no. 7:1317–1340.

Forrest, D. (1999a) "Developing rules for map design: A functional specification for a cartographic-design expert system." *Cartographica* 36, no. 3:31–52.

Forrest, D. (1999b) "Geographic information: Its nature, classification, and cartographic representation." *Cartographica* 36, no. 2:31–53.

Fotheringham, A. S. (1997) "Trends in quantitative methods I: Stressing the local." *Progress in Human Geography* 21, no. 1:88–96.

Fotheringham, A. S. (1998) "Trends in quantitative methods II: Stressing the computational." *Progress in Human Geography* 22, no. 2:283–292.

Fotheringham, A. S. (1999) "Trends in quantitative methods III: Stressing the visual." *Progress in Human Geography* 23, no. 4:597–606.

Fredrikson, A., North, C., Plaisant, C., and Shneiderman, B. (1999) "Temporal, geographical and categorical aggregations viewed through coordinated displays: A case study with highway incident data." *Proceedings of the ACM Workshop on New Paradigms in Information Visualization and Manipulation*, Kansas City, KS, pp. 26–34.

Freeman, H. (1995) "On the automated labeling of maps." In *Shape, Structure and Pattern Recognition*, ed. by D. Dori and A. Bruckstein, pp. 432–442. Singapore: World Scientific.

Fuhrmann, S., and MacEachren, A. M. (2001) "Navigation in desktop geovirtual environments: Usability assessment." *Proceedings, 20th International Cartographic Conference*, International Cartographic Association, Beijing, China, CD-ROM.

Fung, K. (2001) "Atlas of Saskatchewan—A product of technological innovation." *Proceedings, 20th International Cartographic Conference*, International Cartographic Association, Beijing, China, CD-ROM.

Gabbard, J. L., and Hix, D. (1997) *A Taxonomy of Usability Characteristics in Virtual Environments*. Blacksburg: Virginia Polytechnic Institute and State University. Available at *http://csgrad.cs.vt.edu/~jgabbard/ve/documents/taxonomy.pdf*.

Gahegan, M. (1998) "Scatterplots and scenes: Visualization techniques for exploratory spatial analysis." *Computers, Environment and Urban Systems* 21, no. 1:43–56.

Garo, L. A. B. (1998) "Color theory." Available at *http://www.uncc.edu/lagaro/cwg/color/index.html*.

Gartner, G. (ed.). (2002) *Maps and the Internet 2002*. Vienna, Austria: Geowissenschaftliche Mitteilungen, Band 60, TU Wien, Institute of Cartography and Geomedia Technique.

Gershon, N. (1992) "Visualization of fuzzy data using generalized animation." *Proceedings, Visualization '92*, Boston, MA, pp. 268–273.

Gershon, N. (1998) "Visualization of an imperfect world." *IEEE Computer Graphics and Applications* 18, no. 4:43–45.

Gersmehl, P. J. (1990) "Choosing tools: Nine metaphors of four-dimensional cartography." *Cartographic Perspectives*, no. 5:3–17.

Gerth, J. D. (1993) "Towards improved spatial analysis with areal units: The use of GIS to facilitate the creation of dasymetric maps." Unpublished MA paper, The Ohio State University, Columbus, OH.

Gilmartin, P., and Shelton, E. (1989) "Choropleth maps on high resolution CRTs/The effect of number of classes and hue on communication." *Cartographica* 26, no. 2:40–52.

Gilmartin, P. P. (1981) "Influences of map context on circle perception." *Annals, Association of American Geographers* 71, no. 2:253–258.

Golden Software, Inc. (2002) *Surfer 8 User's Guide*. Golden, CO: Author.

Goldstein, E. B. (2002) *Sensation and Perception* (6th ed.). Pacific Grove, CA: Wadsworth.

Gong, J., Lin, H., and Yin, X. (2000) "Three-dimensional reconstruction of the Yaolin Cave." *Cartography and Geographic Information Science* 27, no. 1:31–39.

Goodchild, M., Chih-chang, L., and Leung, Y. (1994) "Visualizing fuzzy maps." In *Visualization in Geographical Information Systems*, ed. by H. M. Hearnshaw and D. J. Unwin, pp. 158–167. Chichester, England: Wiley.

Goodchild, M. F. (2000) "Cartographic futures on a Digital Earth." *Cartographic Perspectives*, no. 36:3–11.

Goodchild, M. F., and Quattrochi, D. A. (1997) "Scale, multiscaling, remote sensing, and GIS." In *Scale in Remote Sensing and GIS*, ed. by D. A. Quattrochi and M. F. Goodchild, pp. 1–11. New York: Lewis.

Gore, A. (1998) "The Digital Earth: Understanding our planet in the 21st century." Available at *http://www.earthscape.org/p1/goa01/*.

Gould, P. (1993) *The Slow Plague*. Cambridge, MA: Blackwell.

Griffin, T. L. C. (1983) "Recognition of areal units on topological cartograms." *The American Cartographer* 10, no. 1:17–29.

Griffin, T. L. C. (1985) "Group and individual variations in judgment and their relevance to the scaling of graduated circles." *Cartographica* 22, no. 1:21–37.

Griffin, T. L. C. (1990) "The importance of visual contrast for graduated circles." *Cartography* 19, no. 1:21–30.

Griffith, D. A. (1993) *Spatial Regression Analysis on the PC: Spatial Statistics Using SAS*. Washington, DC: Association of American Geographers.

Griffith, D. A., and Amrhein, C. G. (1991) *Statistical Analysis for Geographers*. Englewood Cliffs, NJ: Prentice Hall.

Grinstein, G., Sieg, J. C. J., Smith, S., and Williams, M. G. (1992) "Visualization for knowledge discovery." *International Journal of Intelligent Systems* 7:637–648.

Groop, R. E. (1992) "Dot-density crop maps of the Midwest." Abstract, *Association of American Geographers Annual Meeting*, San Diego, CA, p. 89.

Groop, R. E., and Cole, D. (1978) "Overlapping graduated circles: Magnitude estimation and method of portrayal." *The Canadian Cartographer* 15, no. 2:114–122.

Guilford, J. P., and Smith, P. C. (1959) "A system of color-preferences." *The American Journal of Psychology* 72, no. 4:487–502.

Guimbretière, F., Stone, M., and Winograd, T. (2001) "Fluid interaction with high-resolution wall-size displays." *UIST'2001*, Volume 2002, Orlando, FL. Available at *http://interactivity.stanford.edu/publications.html*.

Guo, D., Peuquet, D. J., and Gahegan, M. (2003) "ICEAGE: Interactive clustering and exploration of large and high-dimensional geodata." *GeoInformatica* 7, no. 3:229–253.

Guptill, S. C., and Morrison, J. L. (1995) *Elements of spatial data quality*. Kidlington, Oxford, England: Elsevier Science.

Gusein-Zade, S. M., and Tikunov, V. S. (1993) "A new technique for constructing continuous cartograms." *Cartography and Geographic Information Systems* 20, no. 3:167–173.

Hammond, R., and McCullagh, P. (1978) *Quantitative Techniques in Geography: An Introduction* (2nd ed.). Oxford, England: Clarendon.

Hancock, J. R. (1993) "Multivariate regionalization: An approach using interactive statistical visualization." *AUTO-CARTO 11 Proceedings*, Minneapolis, MN, pp. 218–227.

Harding, G. H. (1951) "A possible solution to the problems of surveying and mapping education." *Surveying and Mapping* 11, no. 2:104–106.

Harley, J., and Woodward, D. (1992) *The History of Cartography: Cartography in the Traditional Islamic and South Asian Societies*. Chicago: University of Chicago Press.

Harley, J. B. (1989) "Deconstructing the map." *Cartographica* 26, no.2: 1–20.

Harrison, R. E. (1944) *Look at the World: The Fortune Atlas for World Strategy*. New York: Knopf.

Harrison, R. E. (1950) "Cartography in art and advertising." *The Professional Geographer* 2, no. 6:12–15.

Harrison, R. E. (1969) Shaded Relief, Map. United States Department of the Interior, Geological Survey. In *The National Atlas of the United States of America, 1970*, pp. 56–57. (Scale 1:7,500,000.)

Harrower, M. (2001) "Visualizing change: Using cartographic animation to explore remotely-sensed data." *Cartographic Perspectives*, no. 39:30–42.

Harrower, M., and Brewer, C. A. (2003) "ColorBrewer.org: An online tool for selecting colour schemes for maps." *The Cartographic Journal* 40, no. 1:27–37.

Harrower, M., MacEachren, A., and Griffin, A. L. (2000) "Developing a geographic visualization tool to support earth science learning." *Cartography and Geographic Information Science* 27, no. 4:279–293.

Harrower, M. A. (2002) "Visual benchmarks: Representing geographic change with map animation." Unpublished PhD dissertation, The Pennsylvania State University, University Park, PA.

Hartigan, J. A. (1975) *Clustering Algorithms*. New York: Wiley.

Hartnett, S. (1987) "Employing rectangular point symbols in two-variable maps." Abstract, *Association of American Geographers Annual Meeting*, Portland, OR, p. 39.

Haslett, J., and Power, G. M. (1995) "Interactive computer graphics for a more open exploration of stream sediment geochemical data." *Computers & Geosciences* 21, no. 1: 77–87.

Healey, C. G. (2001) "Combining perception and impressionist techniques for nonphotorealistic visualization of multidimensional data." *SIGGRAPH 2001 Course 32: Nonphotorealistic Rendering in Scientific Visualization* (Los Angeles, California): 20–52. Available at *http://www.csc.ncsu.edu/faculty/healey/abstract/pubs.html#avi.02.*

Healey, C. G., and Enns, J. T. (1999) "Large datasets at a glance: Combining textures and colors in scientific visualization." *IEEE Transactions on Visualization and Computer Graphics* 5, no. 2:145–167.

Hedley, N. R. (2001) "Virtual and augmented reality interfaces: Empirical findings and implications for spatial visualization." *Proceedings, 20th International Cartographic Conference*, International Cartographic Association, Beijing, China, CD-ROM.

Hedley, N. R. (2003) "3D geographic visualization and spatial mental models." Unpublished PhD dissertation, University of Washington, Seattle, WA.

Hedley, N. R., Billinghurst, M., Postner, L., May, R., and Kato, H. (2002) "Explorations in the use of augmented reality for geographic visualization." *Presence* 11, no. 2:119–133.

Heim, M. (1998) *Virtual Realism.* New York: Oxford University Press.

Heller, M., and Neumann, A. (2001) "Inner-mountain cartography—From surveying towards information systems." *Proceedings, 20th International Cartographic Conference*, International Cartographic Association, Beijing, China, CD-ROM.

Helmholtz, H. von. (1852) "On the theory of compound colors." *Philosophical Magazine* 4:519–534.

Hering, E. (1878) *Zur Lehre vom Lichtsinne.* Vienna, Austria: Gerold.

Herzog, A. (1989) "Modeling reliability on statistical surfaces by polygon filtering." In *Accuracy of Spatial Databases*, ed. by M. Goodchild and S. Gopal, pp. 209–218. London: Taylor & Francis.

Heywood, I., Oliver, J., and Tomlinson, S. (1995) "Building an exploratory multi-criteria modelling environment for spatial decision support." In *Innovations in GIS 2*, ed. by P. Fisher, pp. 127–136. Bristol, PA: Taylor & Francis.

Hibbard, B., and Santek, D. (1990) "The VIS-5D system for easy interactive visualization." *Proceedings, Visualization '90*, San Francisco, CA, pp. 28–35.

Hibbard, W. L., Paul, B. E., Santek, D. A., Dyer, C. R., Battaiola, A. L., Voidrot-Martinez, M. (1994) "Interactive visualization of earth and space science computations." *Computer* 27, no. 7:65–72.

Hobbs, F. (1995) "The rendering of relief images from digital contour data." *The Cartographic Journal* 32, no. 2:111–116.

Hocking, D., and Keller, C. P. (1993) "Alternative atlas distribution formats: A user perspective." *Cartography and Geographic Information Systems* 20, no. 3:157–166.

Hodgson, M. E. (1998) "Comparison of angles from surface slope/aspect algorithms." *Cartography and Geographic Information Systems* 25, no. 3:173–185.

Horn, B. K. P. (1982) "Hill shading and the reflectance map." *Geo-Processing* 2, no. 1:65–144.

House, D. H., and Kocmoud, C. J. (1998) "Continuous cartogram construction." *Proceedings, Visualization '98*, Research Triangle Park, NC, pp. 197–204.

Howard, D., and MacEachren, A. M. (1996) "Interface design for geographic visualization: Tools for representing reliability." *Cartography and Geographic Information Systems* 23, no. 2:59–77.

Howard, H. (1995) "Death Valley: An animated atlas." Unpublished MA thesis, San Francisco State, San Francisco, CA.

Hsu, M. (1981) "The role of projections in modern map design." *Cartographica* 18, no. 2:151–186.

Hu, S. (1999) "Integrated multimedia approach to the utilization of an Everglades vegetation database." *Photogrammetric Engineering and Remote Sensing* 65, no. 2:193–198.

Hu, S. (2003) "Web-based multimedia GIS: Exploring interactive maps and associated multimedia information on the Internet." In *Maps and the Internet*, ed. by M. P. Peterson, pp. 335–344. Amsterdam: Elsevier.

Hubel, D. H. (1988) *Eye, Brain, and Vision.* New York: Scientific American Library.

Hudson, J. C. (1992) "Scale in space and time." In *Geography's Inner Worlds: Pervasive Themes in Contemporary American Geography*, ed. by R. F. Abler, M. G. Marcus, and J. M. Olson, pp. 280–300. New Brunswick, NJ: Rutgers University Press.

Hügli, H. (1979) "Vom Geländemodell zum Geländebild. Die Synthese von Schattenbildern, Vermessung, Photogrammetrie." *Kulturtechnik* 77:245–249.

Hunt, R. W. G. (1987a) *Measuring Colour.* Chichester, England: Ellis Horwood.

Hunt, R. W. G. (1987b) *The Reproduction of Colour in Photography, Printing & Television.* Tolworth, England: Fountain Press.

Hunter, G. J., and Goodchild, M. F. (1995) "Dealing with error in spatial databases: A simple case study." *Photogrammetric Engineering and Remote Sensing* 61, no. 5:529–537.

Hurni, L., Jenny, B., Dahinden, T., and Hutzler, E. (2001) "Interactive analytical shading and cliff drawing: Advances in digital relief presentation for topographic mountain maps." *Proceedings, 20th International Cartographic Conference*, International Cartographic Association, Beijing, China, CD-ROM.

Hurvich, L. M. (1981) *Color Vision.* Sunderland, MA: Sinauer Associates.

Hurvich, L. M., and Jameson, D. (1957) "An opponent-process theory of color vision." *Psychological Review* 64, no. 6:384–404.

Hyatt, J. (2000) "Stochastic screening in cartographic applications." *Cartouche* 40:14–19.

Iliffe, J. C. (2000) *Datums and Map Projections.* Boca Raton, FL: CRC Press.

Imhof, E. (1975) "Positioning names on maps." *The American Cartographer* 2, no. 2:128–144.

Imhof, E. (1982) *Cartographic Relief Presentation*. Berlin: Walter de Gruyter.

Inselberg, A. (1985) "The plane with parallel coordinates." *The Visual Computer* 1:69–91.

Interrante, V. (2000) "Harnessing natural textures for multivariate visualization." *IEEE Computer Graphics and Applications* 20, no. 6:6–11.

Isaaks, E., and Srivastava, R. M. (1989) *Applied Geostatistics*. New York: Oxford University Press.

Itten, J. (1973) *The Art of Color: The Subjective Experience and Objective Rationale of Color*. New York: Van Nostrand Reinhold.

Jackel, C. B. (1997) "Using ArcView to create contiguous and noncontiguous area cartograms." *Cartography and Geographic Information Systems* 24, no. 2:101–109.

Jacobson, D. (2004) "Haptic soundscapes: Developing novel multi-sensory tools to promote access to geographic information." In *WorldMinds: Geographical Perspectives on 100 Problems*, ed. by D. G. Janelle, B. Warf, and K. Hansen, pp. 99–103. Dordrecht, the Netherlands: Kluwer.

Jacobson, R. D., Kitchin, R., and Golledge, R. (2002) "Multi-modal virtual reality for presenting geographic information." In *Virtual Reality in Geography*, ed. by P. Fisher and D. Unwin, pp. 382–400. London: Taylor & Francis.

Jacoby, W. G. (1997) *Statistical Graphics for Univariate and Bivariate Data*. Thousand Oaks, CA: Sage.

Jacoby, W. G. (1998) *Statistical Graphics for Visualizing Multivariate Data*. Thousand Oaks, CA: Sage.

Jankowski, P., Andrienko, N., and Andrienko, G. (2001) "Map-centered exploratory approach to multiple criteria spatial decision making." *International Journal of Geographical Information Science* 15, no. 2:101–127.

Jankowski, P., and Nyerges, T. (1989) "Design considerations for MaPKBS—Map Projection Knowledge-Based System." *The American Cartographer* 16, no. 2:85–95.

Jelinski, D. E. (1987) "Intraspecific diversity in trembling aspen in Waterton Lakes National Park, Alberta: A biogeographical perspective." Unpublished PhD dissertation, Simon Fraser University, Burnaby, British Columbia, Canada.

Jenks, G. F. (1953a) "An improved curriculum for cartographic training at the college and university level." *Annals of the Association of American Geographers* 43, no. 4:317–331.

Jenks, G. F. (1953b) "Pointillism as a cartographic technique." *The Professional Geographer* 5, no. 5:4–6.

Jenks, G. F. (1961) Crop patterns in the United States: 1959. Map, U.S. Bureau of the Census. (Scale 1:5,100,000.)

Jenks, G. F. (1962) Livestock and livestock products sold in the United States: 1959. Map, U.S. Bureau of the Census. (Scale 1:5,100,000.)

Jenks, G. F. (1963) *Discussion from 1963 Summer Institutes for College Teachers*. Seattle, WA.

Jenks, G. F. (1977) *Optimal data classification for choropleth maps. Occasional Paper No. 2.* Lawrence: Department of Geography, University of Kansas.

Jenks, G. F. (1991) "The history and development of academic cartography at Kansas: 1920–80." *Cartography and Geographic Information Systems* 18, no. 3:161–166.

Jenks, G. F., and Caspall, F. C. (1971) "Error on choroplethic maps: Definition, measurement, reduction." *Annals, Association of American Geographers* 61, no. 2:217–244.

Jensen, J. R. (1996) *Introductory Digital Image Processing: A Remote Sensing Perspective* (2nd ed.). Upper Saddle River, NJ: Prentice Hall.

Jeong, W., and Jacobson, D. (2002) "Haptic and auditory display in multimodal information systems." In *Touch in Virtual Environments: Haptics and the Design of Interactive Systems*, ed. by M. L. McLaughlin, J. P. Hespanha, and G. S. Sukhatme, pp. 194–204. Upper Saddle River, NJ: Prentice Hall.

Jepson, W., and Friedman, S. (1998) "SimCity of Angels." *Civil Engineering* 68, no. 6:44–47.

Jepson, W. H., Liggett, R. S., and Friedman, S. (2001) "An integrated environment for urban simulation." In *Planning Support Systems*, ed. by R. K. Brail and R. E. Klosterman, pp. 387–404. Redlands, CA: ESRI Press.

Jervis, J. (2002) "Visualizing uncertainty in Earth Observing System satellite data." *Gridpoints* 3, no. 2:6–9.

João, E. M. (1998) *Causes and Consequences of Map Generalization*. Bristol, PA: Taylor & Francis.

Johnson, H., and Nelson, E. S. (1998) "Using flow maps to visualize time-series data: Comparing the effectiveness of a paper map series, a computer map series, and animation." *Cartographic Perspectives*, no. 30:47–64.

Johnston, D. M., Curtis, D., and Sonin, J. (2000) "The Digital River Basin: A science and technology program for informal education." *Proceedings, Hydroinformatics 2000*, Iowa City, IA. Available at *http://www.gis.uiuc.edu/riverweb/DRB/HydroInfo.pdf*.

Jones, C. B., Bundy, G. L., and Ware, J. M. (1995) "Map generalization with a triangulated data structure." *Cartography and Geographic Information Systems* 22, no. 4:317–331.

Jones, K. H. (1998) "A comparison of algorithms used to compute hill slope as a property of the DEM." *Computers & Geosciences* 24, no. 4:315–323.

Kähkönon, J., Lehto, L., Kilpeläinen, T., and Sarjakoski, T. (1999) "Interactive visualization of geographical objects on the Internet." *International Journal of Geographical Information Science* 13, no. 4:429–428.

Kansas Agricultural Statistics. (1994) *Kansas Farm Facts*. Topeka: Kansas State Board of Agriculture.

Karl, D. (1992) "Cartographic animation: Potential and research issues." *Cartographic Perspectives*, no. 13:3–9.

Keahey, T. A. (1999) "Area-normalized thematic views." *Proceedings of the 19th International Cartographic Conference*, Ottawa, Canada, pp. Section 6:12–21 (CD-ROM).

Keim, D. A. (1998) "The Gridfit Algorithm: An efficient and effective approach to visualizing large amounts of spatial data." *Proceedings, Visualization '98*, Research Triangle Park, NC, pp. 181–188.

Keim, D. A. (2001) "Visual exploration of large data sets." *Communications of the ACM* 44, no. 8:39–44.

Keller, C. P. (1995) "Visualizing digital atlas information products and the user perspective." *Cartographic Perspectives*, no. 20:21–28.

Keller, P. R., and Keller, M. M. (1993) *Visual Cues: Practical Data Visualization*. Los Alamitos, CA: IEEE Computer Society Press.

Kennedy, S. (1994) "Unclassed choropleth maps revisited/ Some guidelines for the construction of unclassed and classed choropleth maps." *Cartographica* 31, no. 1:16–25.

Kennelly, P., and Kimerling, A. J. (2000) "Desktop hachure maps from digital elevation models." *Cartographic Perspectives*, no. 37:78–81.

Kennelly, P., and Kimerling, A. J. (2001) "Modifications of Tanaka's illuminated contour method." *Cartography and Geographic Information Science* 28, no. 2:111–123.

Kenny, J. (2000) "Six color process." *Label and Narrow Web*, no. July/August. Available at *http://www.labelandnarrow web.com/july001.htm.*

Kerst, S. M., and Howard, J. H. J. (1984) "Magnitude estimates of perceived and remembered length and area." *Bulletin of the Psychonomic Society* 22, no. 6:517–520.

Kettebekov, S., Krahnstoever, N., Leas, M., Polat, E., Raju, H., Schapira, E., Sharma, R. (2000) "i2Map: Crisis management using a multimodal interface." *Proceedings of the ARL Federated Laboratory 4th Annual Symposium*, College Park, MD, CD-ROM.

Kilpeläinen, T. (1997) *Multiple Representation and Generalization of Geo-DataBases for Topographic Maps*. Helsinki, Finland: Finnish Geodetic Institute.

Kimerling, A. J. (1975) "A cartographic study of equal value gray scales for use with screened gray areas." *The American Cartographer* 2, no. 2:119–127.

Kimerling, A. J. (1985) "The comparison of equal-value gray scales." *The American Cartographer* 12, no. 2:132–142.

Kimerling, A. J. (1989) "Cartography." In *Geography in America*, ed. by G. S. Gaile and C. J. Willmott, pp. 686–717. Columbus, OH: Merrill.

Kish, G. (1950) "Teaching of cartography in the United States and Canada: Results of a preliminary survey." *The Professional Geographer* 2, no. 6:20–22.

Kolácný, A. (1969) "Cartographic information—A fundamental concept and term in modern cartography." *The Cartographic Journal* 6:47–49.

Kosslyn, S. M. (1994) *Elements of Graph Design*. New York: Freeman.

Koua, E. L. (2003) "Using self-organizing maps for information visualization and knowledge discovery in complex geographic datasets." *Proceedings of the 21st International Cartographic Conference*, Durban, South Africa, pp. 1694–1702.

Koussoulakou, A., and Kraak, M. J. (1992) "Spatio-temporal maps and cartographic communication." *The Cartographic Journal* 29, no. 2:101–108.

Kraak, M.-J. (1998) "Exploratory cartography: Maps as tools for discovery." *ITC Journal* 1, no. 1:46–54.

Kraak, M.-J., and Brown, A. (eds.). (2001) *Web Cartography: Developments and Prospects*. London: Taylor & Francis.

Kraak, M.-J., and Ormeling, F. J. (1996) *Cartography: Visualization of Spatial Data*. Essex, England: Addison Wesley Longman.

Kramer, G. (ed.). (1994a) *Auditory Display: Sonification, Audification, and Auditory Interfaces*. Reading, MA: Addison-Wesley.

Kramer, G. (1994b) "An introduction to auditory display." In *Auditory Display: Sonification, Audification, and Auditory Interfaces*, ed. by G. Kramer, pp. 1–77. Reading, MA: Addison-Wesley.

Krueger, M. W., and Gilden, D. (1999) "KnowWare™: Virtual reality maps for blind people." In *Medicine Meets Virtual Reality*, ed. by J. D. Westwood, H. M. Hoffman, R. A. Robb, and D. Stredney, pp. 191–197. Amsterdam: IOS Press.

Krygier, J. (1993) "Sound and cartographic design." Videotape. University Park: Deasy GeoGraphics Laboratory, Department of Geography, Pennsylvania State University.

Krygier, J. B. (1994) "Sound and geographic visualization." In *Visualization in Modern Cartography*, ed. by A. M. MacEachren and D. R. F. Taylor, pp. 149–166. Oxford, England: Pergamon.

Krygier, J. B., Reeves, C., DiBiase, D., and Cupp, J. (1997) "Design, implementation and evaluation of multimedia resources for geography and earth science education." *Journal of Geography in Higher Education* 21:17–38.

Kumler, M. P. (1994) "An intensive comparison of triangulated irregular networks (TINs) and digital elevation models (DEMs)." *Cartographica* 31, no. 2:1–99.

Kumler, M. P., and Groop, R. E. (1990) "Continuous-tone mapping of smooth surfaces." *Cartography and Geographic Information Systems* 17, no. 4:279–289.

Kwartler, M., and Bernard, R. N. (2001) "CommunityViz: An integrated planning support system." In *Planning Support Systems*, ed. by R. K. Brail and R. E. Klosterman, pp. 285–308. Redlands, CA: ESRI Press.

Lafreniere, M., Pazner, M., and Mateo, J. (1996) "Iconizing the GIS image." *Proceedings GIS/LIS '96*, Denver, CO, pp. 591–606.

Lam, N. S.-N. (1983) "Spatial interpolation methods: A review." *Cartography and Geographic Information Systems* 10, no. 2:129–149.

Langford, M., and Unwin, D. J. (1994) "Generating and mapping population density surfaces within a geographical information system." *The Cartographic Journal* 31, no. 1:21–26.

Lavin, S. (1986) "Mapping continuous geographical distributions using dot-density shading." *The American Cartographer* 13, no. 2:140–150.

Lavin, S., and Archer, J. C. (1984) "Computer-produced unclassed bivariate choropleth maps." *The American Cartographer* 11, no. 1:49–57.

Lavin, S., Rossum, S., and Slade, S. R. (1998) "Animation-based map design: The visual effects of interpolation on the appearance of three-dimensional surfaces." *Cartographic Perspectives*, no. 29:26–34.

Lavin, S. J., and Cerveny, R. S. (1987) "Unit-vector density mapping." *The Cartographic Journal* 24, no. 2:131–141.

Lee, J., Chen, L., and Shaw, S.-L. (1994) "A method for the exploratory analysis of airline networks." *The Professional Geographer* 46, no. 4:468–477.

Leitner, M., and Buttenfield, B. P. (2000) "Guidelines for the display of attribute certainty." *Cartography and Geographic Information Science* 27, no. 1:3–14.

Leonard, J. J., and Buttenfield, B. P. (1989) "An equal value gray scale for laser printer mapping." *The American Cartographer* 16, no. 2:97–107.

Lester, L. (2002) "Metaportal for a megatrend: New Visualization Lab is up and running." *UCAR Quarterly*, Summer. Available at *http://www.ucar.edu/communications/quarterly/summer02/metaportal.html.*

Lewandowsky, S., Behrens, J. T., Pickle, L. W., Herrmann, D. J., and White, A. (1995) "Perception of clusters in mortality maps: Representing magnitude and statistical reliability." In *Cognitive Aspects of Statistical Mapping*, ed. by L. W. Pickle and D. J. Herrmann, pp. 107–132. Washington, DC: Centers for Disease Control, National Center for Health Statistics.

Lewis, P. (1992) "Introducing a cartographic masterpiece: A review of the U.S. Geological Survey's digital terrain map of the United States." *Annals, Association of American Geographers* 82, no. 2:289–304.

li, M. Y. (1998) "Research in statistical methods for mobile population (SMMP)." *The Cartographic Journal* 35, no. 2:155–164.

Lim, J. S. (1998) "Digital television: Here at last." *Scientific American* 278, no. 5:78–83.

Limp, W. F. (2002) "Web Mapping, 2002." *GeoWorld* 15, no. 3:30–32.

Lindberg, M. B. (1987) "Dot map similarity: Visual and quantitative." Unpublished PhD dissertation, University of Kansas, Lawrence, KS.

Lindberg, M. B. (1990) "Fisher: A Turbo Pascal unit for optimal partitions." *Computers & Geosciences* 16, no. 5:717–732.

Lindenberg, R. E. (1986) "The effect of color on quantitative map symbol estimation." Unpublished PhD dissertation, University of Kansas, Lawrence, KS.

Lloyd, R., and Steinke, T. (1977) "Visual and statistical comparison of choropleth maps." *Annals, Association of American Geographers* 67, no. 3:429–436.

Lloyd, R. E., and Steinke, T. R. (1976) "The decisionmaking process for judging the similarity of choropleth maps." *The American Cartographer* 3, no. 2:177–184.

Lo, C. P., and Yeung, A. K. W. (2002) *Concepts and Techniques of Geographic Information Systems*. Upper Saddle River, NJ: Prentice Hall.

Lobeck, A. K. (1924) *Block Diagrams*. New York: Wiley.

Lobeck, A. K. (1958) *Block Diagrams* (2nd ed.). Amherst, MA: Emerson-Trussell.

Lodha, S. K., Joseph, A. J., and Renteria, J. C. (2000) "Audio-visual data mapping for GIS-based data: An experimental evaluation." In *Workshop on New Paradigms in Information Visualization and Manipulation (NPIVM '99)*, ed. by D. S. Ebert and C. D. Shaw, pp. 41–48. New York: The Association for Computing Machinery.

Lodha, S. K., Sheehan, B., Pang, A. T., and Wittenbrink, C. M. (1996a) "Visualizing geometric uncertainty of surface interpolants." *Graphics Interface*, pp. 238–245. Available at *http://www.cse.ucsc.edu/research/slvg/surf.html.*

Lodha, S. K., Wilson, C. M., and Sheehan, R. E. (1996b) "LISTEN: Sounding uncertainty visualization." *IEEE Visualization '96*, San Francisco, CA, pp. 189–195.

Loy, W. G. (ed.), Allan, S., Buckley, A. R., and Meacham, J. E. (authors). (2001) *Atlas of Oregon*. Eugene: University of Oregon Press.

MacDougall, E. B. (1992) "Exploratory analysis, dynamic statistical visualization, and geographic information systems." *Cartography and Geographic Information Systems* 19, no. 4:237–246.

MacEachren, A., Dai, X., Hardisty, F., Guo, D., and Lengerich, G. (2003) "Exploring high-d spaces with multiform matrices and small multiples." *Proceedings of the International Symposium on Information Visualization*, Seattle, WA.

MacEachren, A. M. (1982a) "Map complexity: Comparison and measurement." *The American Cartographer* 9, no. 1:31–46.

MacEachren, A. M. (1982b) "The role of complexity and symbolization method in thematic map effectiveness." *Annals, Association of American Geographers* 72, no. 4:495–513.

MacEachren, A. M. (1985) "Accuracy of thematic maps: Implications of choropleth symbolization." *Cartographica* 22, no. 1:38–58.

MacEachren, A. M. (1992) "Visualizing uncertain information." *Cartographic Perspectives*, no. 13:10–19.

MacEachren, A. M. (1994a) *Some Truth with Maps: A Primer on Symbolization & Design*. Washington, DC: Association of American Geographers.

MacEachren, A. M. (1994b) "Visualization in modern cartography: Setting the agenda." In *Visualization in Modern Cartography*. ed. by A. M. MacEachren and D. R. F. Taylor, pp. 1–12. Oxford, England: Pergamon.

MacEachren, A. M. (1995) *How Maps Work: Representation, Visualization, and Design*. New York: Guilford.

MacEachren, A. M. (1999) "Cartography, GIS, and the World Wide Web." *Progress in Human Geography* 22, no. 4:575–585.

MacEachren, A. M. (2000) "Cartography and GIS: Facilitating collaboration." *Progress in Human Geography* 24, no. 3:445–456.

MacEachren, A. M. (2001) "Cartography and GIS: Extending collaborative tools to support virtual teams." *Progress in Human Geography* 25, no. 3: 431–444.

MacEachren, A. M., Boscoe, F. P., Haug, D., and Pickle, L. W. (1998a) "Geographic visualization: Designing manipulable maps for exploring temporally varying georeferenced statistics." *Proceedings, Information Visualization '98*, IEEE Computer Society, Research Triangle Park, NC, pp. 87–94.

MacEachren, A. M., Brewer, C. A., and Pickle, L. W. (1998b) "Visualizing georeferenced data: Representing reliability of health statistics." *Environment and Planning A* 30:1547–1561.

MacEachren, A. M., and Brewer, I. (2004) "Developing a conceptual framework for visually-enabled geocollaboration." *International Journal of Geographical Information Science* 18, no. 1:1–34.

MacEachren, A. M., Buttenfield, B. P., Campbell, J. B., DiBiase, D. W., and Monmonier, M. (1992) "Visualization." In *Geography's Inner Worlds: Pervasive Themes in Contemporary American Geography*, ed. by R. F. Abler, M. G. Marcus, and J. M. Olson, pp. 99–137. New Brunswick, NJ: Rutgers University Press.

MacEachren, A. M., and DiBiase, D. (1991) "Animated maps of aggregate data: Conceptual and practical problems." *Cartography and Geographic Information Systems* 18, no. 4:221–229.

MacEachren, A. M., Edsall, R., Haug, D., Baxter, R., Otto, G., Masters, R., Fuhrmann, S., Qian, L. (1999a) "Exploring the potential of virtual environments for geographic visualization." Available at *http://www.geovista.psu.edu/publications/aag99vr/fullpaper.htm*.

MacEachren, A. M., Edsall, R., Haug, D., Baxter, R., Otto, G., Masters, R., Fuhrmann, S., Qian, L. (2000) "Virtual environments for geographic visualization: Potential and challenges." In *Workshop on New Paradigms in Information Visualization and Manipulation (NPIVM '99)*, ed. by D. S. Ebert and C. D. Shaw, pp. 35–40. New York: The Association for Computing Machinery.

MacEachren, A. M., and Ganter, J. H. (1990) "A pattern identification approach to cartographic visualization." *Cartographica* 27, no. 2:64–81.

MacEachren, A. M., and Kraak, M.-J. (1997) "Exploratory cartographic visualization: Advancing the agenda." *Computers & Geosciences* 23, no. 4:335–343.

MacEachren, A. M., Kraak, M.-J., and Verbree, E. (1999b) "Cartographic issues in the design and application of geospatial virtual environments." *Proceedings of the 19th International Cartographic Conference*, Ottawa, Canada, pp. Section 5: 108–116 (CD-ROM). Available at *http://www.geovista.psu.edu/publications/ica99/*.

MacEachren, A. M., and Mistrick, T. A. (1992) "The role of brightness differences in figure-ground: Is darker figure?" *The Cartographic Journal* 29, no. 2:91–100.

MacEachren, A. M., and Taylor, D. R. F. (1994) *Visualization in Modern Cartography*. Oxford, England: Pergamon.

MacEachren, A. M., Wachowicz, M., Edsall, R., and Haug, D. (1999c) "Constructing knowledge from multivariate spatiotemporal data: Integrating geographical visualization with knowledge discovery in database methods." *International Journal of Geographical Information Science* 13, no. 4:311–334.

Mackay, J. R. (1949) "Dotting the dot map." *Surveying and Mapping* 9, no.1:3–10.

Mak, K., and Coulson, M. R. C. (1991) "Map-user response to computer-generated choropleth maps: Comparative experiments in classification and symbolization." *Cartography and Geographic Information Systems* 18, no. 2:109–124.

Maling, D. H. (1989) *Measurements from Maps: Principles and Methods of Cartometry*. Oxford, England: Pergamon.

Maling, D. H. (1992) *Coordinate Systems and Map Projections* (2nd ed.). Oxford, England: Pergamon.

Marble, D. F., Gou, Z., Liu, L., and Saunders, J. (1997) "Recent advances in the exploratory analysis of interregional flows in space and time." In *Innovations in GIS 4*, ed. by Z. Kemp, pp. 75–88. Bristol, PA: Taylor & Francis.

Martin, D. (1996) "An assessment of surface and zonal models of population." *International Journal of Geographical Information Systems* 10, no. 8:973–989.

Martin, D., and Higgs, G. (1997) "The visualization of socio-economic GIS data using virtual reality tools." *Transactions in GIS* 1, no. 4:255–266.

Matlin, M. W. (2002) *Cognition* (5th ed.). Fort Worth, TX: Harcourt College Publishers.

Maxwell, B. A. (2000) "Visualizing geographic classifications using color." *The Cartographic Journal* 37, no. 2:93–99.

McCleary, G. F. (1983) "An effective graphic 'vocabulary'." *IEEE Computer Graphics & Applications* 3, no. 2:46–53.

McCleary, G. F. J. (1969) "The dasymetric method in thematic cartography." Unpublished PhD dissertation, University of Wisconsin-Madison, Madison, WI.

McCormick, B. H., DeFanti, T. A., and Brown, M. D. (1987) "Visualization in scientific computing." *Computer Graphics* 21, no. 6.

McCullagh, M. J. (1988) "Terrain and surface modelling systems: Theory and practice." *Photogrammetric Record* 12, no. 72:747–779.

McDonnell, P. W. (1991) *Introduction to Map Projections* (2nd ed.). Rancho Cordova, CA: Landmark Enterprises.

McGranaghan, M. (1989) "Ordering choropleth maps symbols: The effect of background." *The American Cartographer* 16, no. 4:279–285.

McGranaghan, M. (1993) "A cartographic view of spatial data quality." *Cartographica* 30, no. 2&3:8–19.

McGranaghan, M. (1996) "An experiment with choropleth maps on a monochrome LCD panel." In *Cartographic Design: Theoretical and Practical Perspectives*, ed. by C. H. Wood and C. P. Keller, pp. 177–190. Chichester, England: Wiley.

McKnight, T. L., and Hess, D. (2004) *Physical Geography: A Landscape Appreciation*. Upper Saddle River, NJ: Pearson Education.

McManus, I. C., Jones, A. L., and Cottrell, J. (1981) "The aesthetics of color." *Perception* 10, no. 6:651–666.

McMaster, R., and McMaster, S. (2002) "A history of twentieth-century American academic cartography." *Cartography and Geographic Information Science* 29, no. 3:305–321.

McMaster, R. B. (1986) "A statistical analysis of mathematical measures for linear simplification." *The American Cartographer* 13, no. 2:330–346.

McMaster, R. B. (1987) "The geometric properties of numerical generalization." *Geographical Analysis* 19, no. 4:103–116.

McMaster, R. B. (1989a) "The integration of simplification and smoothing routines in line generalization." *Cartographica* 26, no. 1:101–121.

McMaster, R. B. (1989b) "Introduction to 'Numerical Generalization in Cartography'." *Cartographica* 26, no. 1:1–6.

McMaster, R. B., and McMaster, S. (2003) "Response." *Cartography and Geographic Information Science* 30, no. 1:85–87.

McMaster, R. B., and Monmonier, M. (1989) "A conceptual framework for quantitative and qualitative raster-mode generalization." *GIS/LIS'89 Proceedings*, Volume 2, Orlando, FL, pp. 390–403.

McMaster, R. B., and Shea, K. S. (1992) *Generalization in Digital Cartography*. Resource Publications in Geography. Washington, DC: Association of American Geographers.

McMaster, R. B., and Thrower, N. J. W. (1987) "The training of academic cartographers in the United States: Tracing the routes." *Proceedings of the International Cartographic Association*, Morelia, Mexico, pp. 345–359.

Meihoefer, H.-J. (1969) "The utility of the circle as an effective cartographic symbol." *The Canadian Cartographer* 6, no. 2:105–117.

Mersey, J. E. (1980) "An analysis of two-variable choropleth maps." Unpublished MS thesis, University of Wisconsin-Madison, Madison, WI.

Mersey, J. E. (1990) "Colour and thematic map design: The role of colour scheme and map complexity in choropleth map communication." *Cartographica* 27, no. 3:1–157.

Meyer, M. A., Broome, F. R., and Schweitzer, R. H. J. (1975) "Color statistical mapping by the U.S. Bureau of the Census." *The American Cartographer* 2, no. 5:100–117.

Miller, D. W. (1988) "The great American history machine." *Academic Computing* 3, no. 3: 28–29, 43, 46–47, 50.

Miller, E. (1999) "Spatio-temporal geostatistical kriging." In *Geographic Information Research: Trans-Atlantic Perspectives*, ed. by M. Craglia and H. Onsrud, pp. 375–389. London: Taylor & Francis.

Miller, E. J. (1997) "Towards a 4D GIS: Four-dimensional interpolation utilizing kriging." In *Innovations in GIS 4*, ed. by Z. Kemp, pp. 181–197. Bristol, PA: Taylor & Francis.

Miller, H. J., and Han, J. (eds.). (2001a) *Geographic Data Mining and Knowledge Discovery*, London: Taylor & Francis.

Miller, H. J., and Han, J. (2001b) "Geographic data mining and knowledge discovery: An overview." In *Geographic Data Mining and Knowledge Discovery*, ed. by H. J. Miller and J. Han, pp. 3–32. London: Taylor & Francis.

Mitas, L., Brown, W. M., and Mitasova, H. (1997) "Role of dynamic cartography in simulations of landscape processes based on multivariate fields." *Computers & Geosciences* 23, no. 4:437–446.

Mitas, L., and Mitasova, H. (1999) "Spatial interpolation." In *Geographical Information Systems*, ed. by P. A. Longley, M. F. Goodchild, D. J. MaGuire, and D. W. Rhind, pp. 481–492. New York: Wiley.

Mitasova, H., Brown, W., Gerdes, D. P., Kosinovsky, I., and Baker, T. (1993) "Multidimensional interpolation, analysis and visualization for environmental modeling." *GIS/LIS Proceedings*, Volume 2, Minneapolis, MN, pp. 550–556.

Moellering, H. (1976) "The potential uses of a computer animated film in the analysis of geographical patterns of traffic crashes." *Accident Analysis & Prevention* 8:215–227.

Moellering, H. (1978) "A demonstration of the real time display of three dimensional cartographic objects." Videotape. Columbus: Department of Geography, Ohio State University.

Moellering, H. (1980) "The real-time animation of three-dimensional maps." *The American Cartographer* 7, no. 1:67–75.

Moellering, H., and Kimerling, A. J. (1990) "A new digital slope-aspect display process." *Cartography and Geographic Information Systems* 17, no. 2:151–159.

Moellering, H., and Rayner, J. N. (1982) "The dual axis fourier shape analysis of closed cartographic forms." *The Cartographic Journal* 19, no. 1:53–59.

Monmonier, M. (1989) "Geographic brushing: Enhancing exploratory analysis of the scatterplot matrix." *Geographical Analysis* 21, no. 1:81–84.

Monmonier, M. (1990) "Strategies for the visualization of geographic time-series data." *Cartographica* 27, no. 1:30–45.

Monmonier, M. (1991a) "Cartography at Syracuse University." *Cartography and Geographic Information Systems* 18, no. 3:205–207.

Monmonier, M. (1991b) "Ethics and map design: Six strategies for confronting the traditional one-map solution." *Cartographic Perspectives* no. 10:3–8.

Monmonier, M. (1992a) "Authoring graphic scripts: Experiences and principles." *Cartography and Geographic Information Systems* 19, no. 4:247–260, 272.

Monmonier, M. (1992b) "Summary graphics for integrated visualization in dynamic cartography." *Cartography and Geographic Information Systems* 19, no. 1:23–36.

Monmonier, M. (1993) *Mapping It Out: Expository Cartography for the Humanities and Social Sciences*. Chicago: University of Chicago Press.

Monmonier, M. (1996) "Temporal generalization for dynamic maps." *Cartography and Geographic Information Systems* 23, no. 2:96–98.

Monmonier, M. (1999a) *Air Apparent: How Meteorologists Learned to Map, Predict, and Dramatize Weather*. Chicago: The University of Chicago Press.

Monmonier, M. (1999b) "Coping with qualitative-quantitative data in meteorological cartography: Standardization, ergonomics, and facilitated viewing." *Proceedings of the 19th International Cartographic Conference*, Ottawa, Canada, pp. Section 6: 97–104 (CD-ROM).

Monmonier, M. (2001) *Bushmanders & Bullwinkles*. Chicago: The University of Chicago Press.

Monmonier, M. (2002) *Spying with Maps: Surveillance Technologies and the Future of Privacy*. Chicago: University of Chicago Press.

Monmonier, M., and Gluck, M. (1994) "Focus groups for design improvement in dynamic cartography." *Cartography and Geographic Information Systems* 21, no. 1:37–47.

Monmonier, M. S. (1975) "Class intervals to enhance the visual correlation of choroplethic maps." *The Canadian Cartographer* 12, no. 2:161–178.

Monmonier, M. S. (1976) "Modifying objective functions and constraints for maximizing visual correspondence of choroplethic maps." *The Canadian Cartographer* 13, no. 1:21–34.

Monmonier, M. S. (1977) "Regression-based scaling to facilitate the cross-correlation of graduated circle maps." *The Cartographic Journal* 14, no. 2:89–98.

Monmonier, M. S. (1980) "The hopeless pursuit of purification in cartographic communication: A comparison of graphic-arts and perceptual distortions of graytone." *Cartographica* 17, no. 1:24–39.

Monmonier, M. S. (1982) "Flat laxity, optimization, and rounding in the selection of class intervals." *Cartographica* 19, no. 1:16–27.

Monmonier, M. S. (1985) *Technological Transition in Cartography*. Madison: The University of Wisconsin Press.

Monmonier, M. S., and McMaster, R. B. (1990) "The sequential effects of geometric operators in cartographic line generalization." *International Yearbook of Cartography* 30:93–108.

Monmonier, M. S., and Schnell, G. A. (1984) "Land use and land cover data and the mapping of population density." *International Yearbook of Cartography* 24:115–121.

Montello, D. R. (2002) "Cognitive map-design research in the twentieth century: Theoretical and empirical approaches." *Cartography and Geographic Information Science* 29, no. 3:283–304.

Moore, K. E. (2002) "Visualizing data components of the urban scene." In *Virtual Reality in Geography*, ed. by P. Fisher and D. Unwin, pp. 257–269. London: Taylor & Francis.

Morgenstern, D., and Seff, J. (2002) "Macworld's ultimate buyer's guide: Monitors." *Macworld*, February: 56–73.

Morrill, R. L. (1981) *Political Redistricting and Geographic Theory*. Washington, DC: Association of American Geographers.

Morrison, J. L. (1974) "A theoretical framework for cartographic generalization with emphasis on the process of symbolization." *International Yearbook of Cartography* 14:115–127.

Morrison, J. L. (1984) "Applied cartographic communication: Map symbolization for atlases." *Cartographica* 21, no. 1:44–84.

Muehrcke, P. C., Muehrcke, J. O., and Kimerling, A. J. (2001) *Map Use: Reading, Analysis, and Interpretation* (revised 4th ed.). Madison, WI: JP Publications.

Mulcahy, K. A., and Clarke, K. C. (2001) "Symbolization of map projection distortion: A review." *Cartography and Geographic Information Science* 28, no. 3:167–181.

Muller, J. C. (1974) "Mathematical and statistical comparisons in choropleth mapping." Unpublished PhD dissertation, University of Kansas, Lawrence, KS.

Muller, J.-C. (1979) "Perception of continuously shaded maps." *Annals, Association of American Geographers* 69, no. 2:240–249.

Muller, J.-C. (1980a) "Perception of continuously shaded maps: Comment in reply." *Annals, Association of American Geographers* 70, no. 1:107–108.

Muller, J.-C. (1980b) "Visual comparison of continuously shaded maps." *Cartographica* 17, no. 1:40–51.

Mulugeta, G. (1996) "Manual and automated interpolation of climatic and geomorphic statistical surfaces: An evaluation." *Annals, Association of American Geographers* 86, no. 2:324–342.

Murray, A. T., and Shyy, T.-K. (2000) "Integrating attribute and space characteristics in choropleth display and spatial data mining." *International Journal of Geographical Information Science* 14, no. 7:649–667.

National Institute of Standards and Technology. (1994) *Federal Information Processing Standard 173 (Spatial Data Transfer Standard Part 1, Version 1.1)*. Washington, DC: U.S. Department of Commerce.

National Ocean Service. (1983) *Geodesy for the Layman* (5th ed.). Washington, DC: National Oceanic and Atmospheric Administration. Available at *http://164.214.2.59/GandG/geolay/toc.htm*.

National Oceanic and Atmospheric Administration. (1989) *North American Datum of 1983*. ed. by C. Schwarz. Washington, DC: U.S. Department of Commerce.

Nelson, E. S. (1996) "A cognitive map experiment: Mental representations and the encoding process." *Cartography and Geographic Information Systems* 23, no. 4:229–248.

Nelson, E. S. (2000) "The impact of bivariate symbol design on task performance in a map setting." *Cartographica* 37, no. 4:61–78.

Nelson, E. S., and Gilmartin, P. (1996) "An evaluation of multivariate quantitative point symbols for maps." In *Cartographic Design: Theoretical and Practical Perspectives*, ed. by C. H. Wood and C. P. Keller, pp. 191–210. Chichester, England: Wiley.

Neumann, A., and Winter, A. M. (2001) "Time for SVG—Towards high quality interactive web-maps." *Proceedings, 20th International Cartographic Conference*, International Cartographic Association, Beijing, China, CD-ROM.

Neves, J. N., and Câmara, A. (1999) "Virtual environments and GIS." In *Geographical Information Systems*, ed. by P. A. Longley, M. F. Goodchild, D. J. MaGuire, and D. W. Rhind, Volume 1, pp. 557–565. New York: Wiley.

Nickerson, B. G., and Freeman, H. R. (1986) "Development of a rule-based system for automated map generalization." *Proceedings, Second International Symposium on Spatial*

Data Handling, Williamsville, NY: International Geographical Union Commission on Geographical Data Sensing and Processing, Seattle, WA, pp. 537–556.

Nielson, G. M., and Hamann, B. (1990) "Techniques for the interactive visualization of volumetric data." *Proceedings Visualization '90*, San Francisco, CA, pp. 45–50.

Noll, S., Paul, C., Peters, R., and Schiffner, N. (1999) "Autonomous agents in collaborative virtual environments." *IEEE 8th International Workshops on Enabling Technologies: Infrastructure for Collaborative Enterprises (WET ICE '99)*, IEEE Computer Society, Stanford, CA, pp. 208–215.

Nyerges, T., and Jankowski, P. (1989) "A knowledge base for map projection selection." *The American Cartographer* 16, no. 1:29–38.

Odland, J. (1988) *Spatial Autocorrelation*. Newbury Park, CA: Sage.

Ogao, P. J., and Kraak, M.-J. (2001) "Geospatial data exploration using interactive and intelligent cartographic animations." *Proceedings, 20th International Cartographic Conference*, International Cartographic Association, Beijing, China, CD-ROM.

Olea, R. A. (1994) "Fundamentals of semivariogram estimation, modeling, and usage." In *Stochastic Modeling and Geostatistics*, ed. by J. M. Yarus and R. L. Chambers, pp. 27–35. Tulsa, OK: American Association of Petroleum Geologists.

Olea, R. A. (1999) *Geostatistics for Engineers and Earth Scientists*. Boston: Kluwer Academic.

Olson, G. M., and Olson, J. S. (2000) "Distance matters." *Human–Computer Interaction* 15:139–178.

Olson, J. (1972a) "Class interval systems on maps of observed correlated distributions." *The Canadian Cartographer* 9, no. 2:122–131.

Olson, J. M. (1972b) "The effects of class interval systems on choropleth map correlation." *The Canadian Cartographer* 9, no. 1:44–49.

Olson, J. M. (1975a) "Experience and the improvement of cartographic communication." *The Cartographic Journal* 12, no. 2:94–108.

Olson, J. M. (1975b) "The organization of color on two-variable maps." *AUTO-CARTO II, Proceedings of the International Symposium on Computer-Assisted Cartography*, pp. 289–294, 251, 264–266.

Olson, J. M. (1976a) "A coordinated approach to map communication improvement." *The American Cartographer* 3, no. 2:151–159.

Olson, J. M. (1976b) "Noncontiguous area cartograms." *The Professional Geographer* 28, no. 4:371–380.

Olson, J. M. (1977) "Rescaling dot maps for pattern enhancement." *International Yearbook of Cartography* 17:125–136.

Olson, J. M. (1978) "Graduated circles." *The Cartographic Journal* 15, no. 2:105.

Olson, J. M. (1981) "Spectrally encoded two-variable maps." *Annals, Association of American Geographers* 71, no. 2:259–276.

Olson, J. M., and Brewer, C. A. (1997) "An evaluation of color selections to accommodate map users with color-vision impairments." *Annals, Association of American Geographers* 87, no. 1:103–134.

Openshaw, S., Wymer, C., and Charlton, M. (1986) "A geographical information and mapping system for the BBC Domesday optical discs." *Transactions, Institute of British Geographers (New Series)* 11, no. 3:296–304.

Ormeling, F. (1995) "New forms, concepts, and structures for European national atlases." *Cartographic Perspectives*, no. 20:12–20.

Orwell, G. (1984) *1984*. London: Secker & Warburg.

Oviatt, S. (1997) "Multimodal interactive maps: Designing for human performance." *Human–Computer Interaction* 12:93–129.

Oviatt, S. (2002) "Breaking the robustness barrier: Recent progress on the design of robust multimodal systems." *Advances in Computers* 56:305–341.

Oviatt, S. (2003) "Multimodal interfaces." In *The Human–Computer Interaction Handbook: Fundamentals, Evolving Technologies and Emerging Applications*, ed. by J. Jacko and A. Sears, pp. 286–304. Mahwah, NJ: Lawrence Erlbaum Associates.

Oviatt, S., Cohen, P., Wu, L., Vergo, J., Duncan, L., et al. (2000) "Designing the user interface for multimodal speech and pen-based gesture applications: State-of-the-art systems and future research directions." *Human–Computer Interaction* 15:263–322.

Palmer, J. J. N. (1986) "Computerizing Domesday Book." *Transactions, Institute of British Geographers (New Series)* 11, no. 3:279–289.

Pang, A., Wittenbrink, C., and Lodha, S. (1997) "Approaches to uncertainty visualization." *The Visual Computer* 13, no. 8:370–380.

Pantone Incorporated. (2002) *Designer Hexachrome Primer*. Pantone Incorporated, N.J.

Parks, M. J. (1987) "American flow mapping: A survey of the flow maps found in twentieth century geography textbooks, including a classification of the various flow map designs." Unpublished MA thesis, Georgia State University, Atlanta, GA.

Paslawski, J. (1983) "Natural legend design for thematic maps." *The Cartographic Journal* 20, no. 1:36–37.

Patterson, T. (1999) "Designing 3D landscapes." In *Multimedia Cartography*, ed. by W. Cartwright, M. P. Peterson, and G. Gartner, pp. 217–229. Berlin: Springer-Verlag.

Patterson, T. (2000) "A view from on high: Heinrich Berann's panoramas and landscape visualization techniques for the U.S. National Park Service." *Cartographic Perspectives*, no. 36:38–65.

Patterson, T. (2002) "Getting real: Reflecting on the new look of National Park Service maps." Available at *http://www.nps.gov/carto/silvretta/realism/index.html*.

Patton, D. K., and Cammack, R. G. (1996) "An examination of the effects of task type and map complexity on sequenced and static choropleth maps." In *Cartographic Design: Theoretical and Practical Perspectives*, ed. by C. H. Wood and C. P. Keller, pp. 237–252. Chichester, England: Wiley.

Patton, J. C., and Slocum, T. A. (1985) "Spatial pattern recall/An analysis of the aesthetic use of color." *Cartographica* 22, no. 3:70–87.

Pazner, M. I., and Lafreniere, M. J. (1997) "GIS icon maps." *1997 ACSM/ASPRS Annual Convention & Exposition, Volume 5* (AUTO-CARTO 13), Seattle, WA, pp. 126–135.

Pearson, F. (1984) *Map Projection Methods*. Blacksburg, VA: Sigma Scientific.

Peddie, J. (1994) *High-Resolution Graphics Display Systems*. New York: Windcrest/McGraw-Hill.

Peters, A. (1983) *The New Cartography*. New York: Friendship Press.

Peterson, M. P. (1979) "An evaluation of unclassed crossed-line choropleth mapping." *The American Cartographer* 6, no. 1:21–37.

Peterson, M. P. (1980) "Unclassed choropleth maps: A reply." *The American Cartographer* 7, no. 1:80–81.

Peterson, M. P. (1992) "Creating unclassed choropleth maps with PostScript." *Cartographic Perspectives*, no. 12:4–6.

Peterson, M. P. (1993) "Interactive cartographic animation." *Cartography and Geographic Information Systems* 20, no. 1:40–44.

Peterson, M. P. (1995) *Interactive and Animated Cartography*. Englewood Cliffs, NJ: Prentice Hall.

Peterson, M. P. (1997) "Cartography and the Internet: Introduction and research agenda." *Cartographic Perspectives*, no. 26:3–12.

Peterson, M. P. (1999) "Active legends for interactive cartographic animation." *International Journal of Geographical Information Science* 13, no. 4:375–383.

Peterson, M. P. (ed.). (2003) *Maps and the Internet*. Amsterdam: Elsevier.

Pike, R. J., and Thelin, G. P. (1990–91) "Mapping the nation's physiography by computer." *Cartographic Perspectives*, no. 8:15–24.

Pike, R. J., and Thelin, G. P. (1992) "Visualizing the United States in computer chiaroscuro." *Annals, Association of American Geographers* 82, no. 2:300–302.

Pinto, I., and Freeman, H. (1996) "The feedback approach to cartographic areal text placement." In *Advances in Structural and Syntactical Pattern Recognition*, ed. by P. Perner, P. Wang, and A. Rosenfeld, pp. 341–350. Berlin: Springer-Verlag.

Plazanet, C. (1995) "Measurement, characterization and classification for automated line feature generalization." *AUTO-CARTO 12 (Volume 4 of ACSM/ASPRS '95 Annual Convention & Exposition Technical Papers)*, Bethesda, MD, pp. 59–68.

Powell, D. S., Faulkner, J. L., Darr, D. R., Zhu, Z., and Mac-Cleery, D. W. (1992) "Forest resources of the United States, 1992." United States Department of Agriculture, Forest Service, Rocky Mountain Forest and Range Experiment Station, *General Technical Report RM-234 (Revised)*.

Price, W. (2001) "Relief presentation: Manual airbrushing combined with computer technology." *The Cartographic Journal* 38, no. 1:107–112.

Quinlan, J. R. (1993) *C4.5: Programs for Machine Learning*. San Mateo, CA: Morgan Kaufmann.

Raisz, E. (1931) "The physiographic method of representing scenery on maps." *The Geographical Review* 21, no. 2:297–304.

Raisz, E. (1938) *General Cartography*. New York: McGraw-Hill.

Raisz, E. (1950) "Introduction" [to a special Cartography issue]. *The Professional Geographer* 2, no. 6:9–11.

Raisz, E. (1962) *Principles of Cartography*. New York: McGraw-Hill.

Raisz, E. (1967) Landforms of the United States. Map (Scale approximately 1:4,500,000.)

Raper, J. (1997) "Progress towards spatial multimedia." In *Geographic Information Research: Bridging the Atlantic*, ed. by M. Craglia and H. Couclelis, pp. 525–543. London: Taylor & Francis.

Raper, J. (2000) *Multidimensional Geographic Information Science*. London: Taylor & Francis.

Raskin, R., Funk, C., and Willmott, C. (1997) "Interpolation over large distances using Spherekit." *ACSM/ASPRS Annual Convention & Exposition, Technical Papers, Volume 5* (AUTO-CARTO 13), Seattle, WA, pp. 419–428.

Rauschert, I., Agrawal, P., Fuhrmann, S., Brewer, I., Wang, H., Sharma, R., Cai, G., MacEachren, A. M. (2002) "Designing a human-centered, multimodal GIS interface to support emergency management." *ACM GIS'02, 10th ACM Symposium on Advances in Geographic Information Systems*, McLean, VA. Available at *http://www.geovista.psu.edu/grants/nsf-itr/pubs.html*.

Reddy, M. (2001) "Perceptually optimized 3D graphics." *IEEE Computer Graphics and Applications* 21, no. 5:2–9.

Reddy, M., Leclerc, Y., Iverson, L., and Bletter, N. (1999) "TerraVision II: Visualizing massive terrain databases in VRML." *IEEE Computer Graphics and Applications* 19, no. 2:30–38.

Rhind, D., and Mounsey, H. (1986) "The land and people of Britain: A Domesday record, 1986." *Transactions, Institute of British Geographers (New Series)* 11, no. 3:315–325.

Rice, K. W. (1989) "The influence of verbal labels on the perception of graduated circle map regions." Unpublished PhD dissertation, University of Kansas, Lawrence, KS.

Rickel, J., and Johnson, W. L. (1999) "Animated agents for procedural training in virtual reality: Perception, cognition, and motor control." *Applied Artificial Intelligence* 13, no. 4/5:343–382.

Rittschof, K. A., Stock, W. A., Kulhavy, R. W., Verdi, M. P., and Johnson, J. T. (1996) "Learning from cartograms: The effects of region familiarity." *Journal of Geography* 95, no. 2:50–58.

Robertson, P. K. (1991) "A methodology for choosing data representations." *IEEE Computer Graphics & Applications* 11, no. 3:56–67.

Robeson, S. M. (1997) "Spherical methods for spatial interpolation: Review and evaluation." *Cartography and Geographic Information Systems* 24, no. 1:3–20.

Robinson, A. (1974) "A new projection: Its development and characteristics." *International Yearbook of Cartography* 14:145–155.

Robinson, A. H. (1952) *The Look of Maps*. Madison: University of Wisconsin Press.

Robinson, A. H. (1956) "The necessity of weighting values in correlation analysis of areal data." *Annals, Association of American Geographers* 46, no. 2:233–236.

Robinson, A. H. (1979) "Geography and cartography then and now." *Annals of the Association of American Geographers* 69, no. 1:97–102.

Robinson, A. H. (1982) *Early Thematic Mapping in the History of Cartography*. Chicago: The University of Chicago Press.

Robinson, A. H. (1991) "The development of cartography at the University of Wisconsin-Madison." *Cartography and Geographic Information Systems* 18, no. 3:156–157.

Robinson, A. H., Morrison, J. L., Muehrcke, P. C., Kimerling, A. J., and Guptill, S. C. (1995) *Elements of Cartography* (6th ed.). New York: Wiley.

Robinson, A. H., and Sale, R. D. (1969) *Elements of Cartography* (3rd ed.). New York: Wiley.

Robinson, A. H., Sale, R. D., and Morrison, J. L. (1978) *Elements of Cartography* (4th ed.). New York: Wiley.

Robinson, A. H., Sale, R. D., Morrison, J. L., and Muehrcke, P. C. (1984) *Elements of Cartography* (5th ed.). New York: Wiley.

Rogers, D. F. (1998) *Procedural Elements for Computer Graphics*. Boston: McGraw-Hill.

Rogers, J. E., and Groop, R. E. (1981) "Regional portrayal with multi-pattern color dot maps." *Cartographica* 18, no. 4:51–64.

Rogerson, P. A. (2001) *Statistical Methods for Geography*. London: Sage.

Rogowitz, B. E., and Treinish, L. A. (1996) "Why should engineers and scientists be worried about color?" Available at *http://www.research.ibm.com/people/l/lloydt/color/color.HTM*.

Romano, F. J. (ed.). (1997) *Delmar's Dictionary of Digital Printing and Publishing*. Albany, NY: Delmar.

Romesburg, H. C. (1984) *Cluster Analysis for Researchers*. Belmont, CA: Lifetime Learning.

Roth, S. F., Chuah, M. C., Kerpedjiev, S., and Kolojejchick, J. A. (1997) "Toward an information visualization workspace: Combining multiple means of expression." *Human–Computer Interaction* 12, no. 1/2:131–185.

Rowles, R. A. (1991) "Regions and regional patterns on choropleth maps." Unpublished PhD dissertation, University of Kentucky, Lexington, KY.

Rumsey, D. (2003) "Tales from the vault: Historical maps online." *Common-place* 3, no. 4. Available at *http://www.common-place.org/vol-03/no-04/tales/*.

Rundstrom, R. A. (ed.). (1993) "Introducing cultural and social cartography." *Cartographica* 30, no. 1.

Ryden, K. (1987) "Environmental Systems Research Institute mapping." *The American Cartographer* 14, no. 3:261–263.

Rystedt, B. (1995) "Current trends in electronic atlas production." *Cartographic Perspectives*, no. 20:5–11.

Sampson, R. J. (1978) *SURFACE II Graphics System*. Lawrence: Kansas Geological Survey.

Samuels, S. (1999) "Driving forward." *Spectator* 33, no. 1:1.

Sharma, R., Yeasin, M., Krahnstoever, N., Rauschert, I., Cai, G., Brewer, I., MacEachren, A., Sengupta, K. (2003) "Speech-gesture driven multimodal interfaces for crisis management." *Proceedings of the IEEE* 91, no. 9:1327–1354.

Shelton, B. E., and Hedley, N. R. (2002) "Using augmented reality for teaching earth–sun relationships to undergraduate geography students." *Proceedings of the First IEEE International Augmented Reality Toolkit Workshop*, Darmstadt, Germany. Available at *http://depts.washington.edu/pettt/papers/shelton-hedley-art02.pdf*.

Shepherd, I. D. H. (1994) "Multi-sensory GIS: Mapping out the research frontier." *Proceedings of the 6th International Symposium on Spatial Data Handling (Advances in GIS Research)*, Edinburgh, Scotland, pp. 356–390.

Sheppard, E., and McMaster, R. B. (eds.). (2004) *Scale and Geographic Inquiry: Nature, Society, and Method*. Oxford, England: Blackwell.

Sherman, J. C. (1987) "Interview." *The American Cartographer* 14, no. 1:75–87.

Sherman, J. C. (1991) "The development of cartography at the University of Washington." *Cartography and Geographic Information Systems* 18, no. 3:169–170.

Shiffer, M. J. (1995) "Environmental review with hypermedia systems." *Environment and Planning B: Planning and Design* 22, no. 3:359–372.

Shiffer, M. J. (2001) "Spatial multimedia for planning support." In *Planning Support Systems*, ed. by R. K. Brail and R. E. Klosterman, pp. 361–385. Redlands, CA: ESRI Press.

Shiffer, M. J. (2002) "Spatial multimedia representations to support community participation." In *Community Participation and Geographic Information Systems*, ed. by W. J. Craig, T. M. Harris, and D. Weiner, pp. 309–319. London: Taylor & Francis.

Shneiderman, B. (1999) "Dynamic queries for visual information seeking." In *Readings in Information Visualization: Using Vision to Think*, ed. by S. K. Card, J. D. Mackinlay, and B. Shneiderman, pp. 236–243. San Francisco, CA: Morgan Kaufmann.

Shneiderman, B. (2002) "Inventing discovery tools: Combining information visualization with data mining." *Information Visualization* 1, no. 1:5–12.

Shortridge, B. G. (1979) "Map reader discrimination of lettering size." *The American Cartographer* 6, no. 1:13–20.

Sides, S. (1996) "Puzzling out plate tectonics in 3-D." *Gather/Scatter*, Spring. Available at *http://www.sdsc.edu/GatherScatter/GSspring96/polit.html*.

Siekierska, E. M., and Taylor, D. R. F. (1991) "Electronic mapping and electronic atlases: New cartographic products for the information era—The electronic atlas of Canada." *CISM Journal ACSGC* 45, no. 1:11–21.

Skupin, A. (2002) "A cartographic approach to visualizing conference abstracts." *IEEE Computer Graphics and Applications* 22, no. 1:50–58.

Skupin, A., and Fabrikant, S. I. (2003) "Spatialization methods: A cartographic research agenda for non-geographic information visualization." *Cartography and Geographic Information Science* 30, no. 2:99–119.

Slocum, T. A. (1983) "Predicting visual clusters on graduated circle maps." *The American Cartographer* 10, no. 1:59–72.

Slocum, T. A., Blok, C., Jiang, B., Koussoulakou, A., Montello, D. R., Fuhrmann, S., Hedley, N. R. (2001) "Cognitive and usability issues in geovisualization." *Cartography and Geographic Information Science* 28, no. 1:61–75.

Slocum, T. A., and Egbert, S. L. (1993) "Knowledge acquisition from choropleth maps." *Cartography and Geographic Information Systems* 20, no. 2:83–95.

Slocum, T. A., and McMaster, R. B. (1986) "Gray tone versus line plotter area symbols: A matching experiment." *The American Cartographer* 13, no. 2:151–164.

Slocum, T. A., Robeson, S. H., and Egbert, S. L. (1990) "Traditional versus sequenced choropleth maps: An experimental investigation." *Cartographica* 27, no. 1:67–88.

Slocum, T. A., Sluter, R. S., Kessler, F. C., and Yoder, S. C. (in press) "A qualitative evaluation of MapTime, A program for exploring spatiotemporal point data." *Cartographica*.

Slocum, T. A., and Yoder, S. C. (1996) "Using Visual Basic to teach programming for geographers." *Journal of Geography* 95, no. 5:194–199.

Slocum, T. A., Yoder, S. C., Kessler, F. C., and Sluter, R. S. (2000) "MapTime: Software for exploring spatiotemporal data associated with point locations." *Cartographica* 37, no. 1:15–31.

Smith, J. R. (1997) *Introduction to Geodesy: The History and Concepts of Modern Geodesy*. New York: Wiley.

Smith, N. (1987a) "Academic war over the field of geography: The elimination of geography at Harvard, 1947–1951." *Annals of the Association of American Geographers* 77, no. 2:155–172.

Smith, R. D. (1987b) "Electronic Atlas of Arkansas: Design and operational considerations." *Proceedings of the 13th International Cartographic Conference*, Volume IV, Morelia, Mexico, pp. 161–167.

Smith, R. M., and Parker, T. (1995) "An electronic atlas authoring system." *Cartographic Perspectives*, no. 20:35–39.

Smith, S., Grinstein, G., and Pickett, R. (1991) "Global geometric, sound, and color controls for iconographic displays of scientific data." *Extracting Meaning from Complex Data: Processing, Display, Interaction II*, SPIE - The International Society for Optical Engineering, Volume 1459, Santa Clara, CA, pp. 192–206.

Snyder, J. P. (1987) *Map Projections: A Working Manual*. Washington, DC: U.S. Geological Survey.

Snyder, J. P. (1993) *Flattening the Earth: Two Thousand Years of Map Projections*. Chicago: University of Chicago Press.

Snyder, J. P. (1994) *Map Projections: A Working Manual*. Washington, DC: U.S. Government Printing Office.

Snyder, J. P., and Steward, H. (1988) *Bibliography of Map Projections*. Washington, DC: U.S. Geological Survey.

Snyder, J. P., and Voxland, P. M. (1989) *An Album of Map Projections*. Washington, DC: U.S. Geological Survey.

Sorbel, D. (1995) *Longitude: The True Story of a Lone Genius Who Solved the Greatest Scientific Problem of His Time*. New York: Viking Penguin.

South Carolina State Budget and Control Board. (1994) *South Carolina Statistical Abstract*. Columbia, SC: Office of Research and Statistical Services.

Spradlin, K. L. (2000) "An evaluation of user attitudes toward classed and unclassed choropleth maps." Unpublished MA thesis, University of Kansas, Lawrence, KS.

Stanney, K. M., Mourant, R. R., and Kennedy, R. S. (1998) "Human factors issues in virtual environments: A review of the literature." *Presence* 7, no. 4:327–351.

Steenblik, R. A. (1987) "The chromostereoscopic process: A novel single image stereoscopic process." *True Three-Dimensional Imaging Techniques and Display Technologies. Proceedings, SPIE*, Volume 761, pp. 27–34.

Steinbach, M., Tan, P.-N., Kumar, V., Klooster, S., and Potter, C. (2002) "Temporal data mining for the discovery and analysis of ocean climate indices." *KDD-2002, The Eighth ACM SIGKDD International Conference on Knowledge Discovery and Data Mining*, Edmonton, Canada. Available at *http://www-users.cs.umn.edu/~kumar/papers/kdd_tele_9.pdf*.

Stevens, J. (1996) *Applied Multivariate Statistics for the Social Sciences* (3rd ed.). Mahwah, NJ: Lawrence Erlbaum Associates.

Stevens, S. S., and Galanter, E. H. (1957) "Ratio scales and category scales for a dozen perceptual continua." *Journal of Experimental Psychology* 54, no. 6:377–411.

Strahler, A., and Strahler, A. (2003) *Introducing Physical Geography*. (3rd ed.). New York: Wiley.

Szego, J. (1987) *Human Cartography: Mapping the World of Man*. Stockholm, Sweden: Swedish Council for Building Research.

Tajima, J. (1983) "Uniform color scale applications to computer graphics." *Computer Vision, Graphics, and Image Processing* 21, no. 3:305–325.

Tanaka, K. (1950) "The relief contour method of representing topography on maps." *The Geographical Review* 40, no. 3:444–456.

Tang, Q. (1992) "From description to analysis: An electronic atlas for spatial data exploration." *Proceedings of ASPRS/ACSM/RT 92 Convention*, Vol. 3, pp. 455–463.

Taylor, D. R. F. (1987) "Cartographic communication on computer screens: The effect of sequential presentation of map information." *Proceedings of the 13th International Cartographic Conference*, Volume 1, Morelia, Mexico, pp. 591–611.

Taylor, J., Tabayoyon, A., and Rowell, J. (1991) "Device-independent color matching you can buy now." *Information Display* 7, no. 4/5:20–22, 49.

Thelin, G. P., and Pike, R. J. (1991). Landforms of the conterminous United States—A digital shaded-relief portrayal, Map I-2206. Washington, DC: U.S. Geological Survey. (Scale 1:3,500,000.)

Thibault, P. (2002) "Cartographic generalization of fluvial features." Unpublished PhD dissertation, University of Minnesota, Minneapolis, MN.

Thomas, E. N., and Anderson, D. L. (1965) "Additional comments on weighting values in correlation analysis of areal data." *Annals, Association of American Geographers* 55, no. 3:492–505.

Thompson, W., and Lavin, S. (1996) "Automatic generation of animated migration maps." *Cartographica* 33, no. 2:17–28.

Thrower, N. J. W. (1959) "Animated cartography." *The Professional Geographer* 11, no. 6:9–12.

Thrower, N. J. W. (1961) "Animated cartography in the United States." *International Yearbook of Cartography*, pp. 20–29.

Thrower, N. J. W. (2003) "To the editor." *Cartography and Geographic Information Science* 30, no. 3:295.

Tikunov, V. S. (1988) "Anamorphated cartographic images: Historical outline and construction techniques." *Cartography* 17, no. 1:1–8.

Tissot, N. A. (1881) *Memoire sur la Representation des Surfaces et les Projections des Cartes Geographiques*. Paris: Gautier-Villars.

Tobler, W., Deichmann, U., Gottsegen, J., and Maloy, K. (1997) "World population in a grid of spherical quadrilaterals." *International Journal of Population Geography* 3:203–225.

Tobler, W. R. (1970) "A computer movie simulating urban growth in the Detroit region." *Economic Geography* 46, no. 2(Supplement):234–240.

Tobler, W. R. (1973) "Choropleth maps without class intervals?" *Geographical Analysis* 5, no. 3:262–265.

Tobler, W. R. (1976) "Analytical cartography." *The American Cartographer* 3, no. 1:21–31.

Tobler, W. R. (1979) "Smooth pycnophylactic interpolation for geographic regions." *Journal of the American Statistical Association* 74, no. 367:519–536.

Tobler, W. R. (1981) "A model of geographical movement." *Geographical Analysis* 13, no. 1:1–20.

Tobler, W. R. (1987) "Experiments in migration mapping by computer." *The American Cartographer* 14, no. 2:155–163.

Travis, D. (1991) *Effective Color Displays: Theory and Practice*. London: Academic.

Treinish, L. A. (1992) "Climatology of global stratospheric ozone." Videotape, IBM Corporation.

Treinish, L. A. (1994) "Visualizations illuminate disparate data sets in the earth sciences." *Computers in Physics* 8, no. 6:664–671.

Treinish, L. A. (2003) "Web-based dissemination and visualization of mesoscale weather models for business operations." *Proceedings of the Nineteenth International Conference on Interactive Information and Processing Systems for Meteorology, Oceanography and Hydrology*, American Meteorological Society. Available at *http://www.research.ibm.com/weather/vis/web_apps.pdf*.

Treinish, L. A., and Goettsche, C. (1991) "Correlative visualization techniques for multidimensional data." *IBM Journal of Research and Development* 35, no. 1/2:184–204.

Trifonoff, K. M. (1994) "Using thematic maps in the early elementary grades." Unpublished PhD dissertation, University of Kansas, Lawrence, KS.

Trifonoff, K. M. (1995) "Going beyond location: Thematic maps in the early elementary grades." *Journal of Geography* 94, no. 2:368–374.

Tufte, E. R. (1983) *The Visual Display of Quantitative Information*. Cheshire, CT: Graphics Press.

Tufte, E. R. (1990) *Envisioning Information*. Cheshire, CT: Graphics Press.

Tukey, J. W. (1977) *Exploratory Data Analysis*. Reading, MA: Addison-Wesley.

Turner, E. J. (1977) "The use of shape as a nominal variable on multipattern dot maps." Unpublished PhD dissertation, University of Washington, Seattle, WA.

Tversky, B., Morrison, J. B., and Betrancourt, M. (2002) "Animation: Does it facilitate?" *International Journal of Human–Computer Studies* 57, 247–262.

Unwin, D. (1981) *Introductory Spatial Analysis*. London: Methuen.

U.S. Bureau of the Census. (1970) "Population distribution, urban and rural, in the United States: 1970." *United States Maps, GE-50, No. 45*.

U.S. Bureau of the Census. (1994) *County and City Data Book: 1994*. Washington, DC: U.S. Government Printing Office.

U.S. Bureau of the Census. (1995) *Agricultural Atlas of the United States*. Washington, DC: U.S. Government Printing Office.

U.S. Department of Commerce. (2001) "Centers of population computation for 1950, 1960, 1970, 1980, 1990 and 2000." Available at *http://www.census.gov/geo/www/cenpop/calculate2k.pdf*.

U.S. Geological Survey. (1970) *The National Atlas of the United States of America*. Washington, DC: U.S. Geological Survey.

U.S. Geological Survey. (1989) *North American Datum of 1983, Map Data Conversion Tables*. Denver, CO: Author.

van der Wel, F. J. M., van der Gaag, L. C., and Gorte, B. G. H. (1998) "Visual exploration of uncertainty in remote-sensing classification." *Computers & Geosciences* 24, no. 4:335–343.

Velikonja, J. (1997) "John C. Sherman, May 3, 1916 – October 21, 1996." Available at *http://faculty.washington.edu/~krumme/faculty/sherman.html*.

Verbree, E., Maren, G. V., Germs, R., Jansen, F., and Kraak, M.-J. (1999) "Interaction in virtual world views—Linking

3D GIS with VR." *International Journal of Geographical Information Science* 13, no. 4:385–396.

Veve, T. D. (1994) "An assessment of interpolation techniques for the estimation of precipitation in a tropical mountainous region." Unpublished MA thesis, The Pennsylvania State University, University Park, PA.

Viggiano, J. A., and Hoagland, W. J. (1998) "Colorant selection for six-color lithographic printing." *IST/SID 1998 Color Imaging Conference*, pp. 112–115.

Visvalingam, M., and Williamson, P. J. (1995) "Simplification and generalization of large scale data for roads: A comparison of two filtering algorithms." *Cartography and Geographic Information Systems* 22, no. 4:264–275.

von Wyss, M. (1996) "The production of smooth scale changes in an animated map project." *Cartographic Perspectives*, no. 23:12–20.

Waisel, L. (1996) "Three-dimensional visualization of sediment chemistry in the New York harbor." *Communique (Data Explorer Newsletter)* 4, no. 1:1–3.

Walker, T., Cartwright, W., and Miller, S. (2000) "An investigation into the methodologies of producing a web-based multimedia atlas of Victoria." *Cartography* 29, no. 2:51–64.

Wang, P. C. C. (1978) *Graphical Representation of Multivariate Data*. New York: Academic.

Wang, P. C. C., and Lake, L. T. G. E. (1978) "Application of graphical multivariate techniques in policy sciences." In *Graphical Representation of Multivariate Data*, ed. by P. C. C. Wang, pp. 13–58. New York: Academic.

Wang, X., Chen, J. X., Carr, D. B., Bell, B. S., and Pickle, L. W. (2002) "Geographic statistics visualization: Web-based linked micromap plots." *Computing in Science & Engineering* 4, no. 3:90–94.

Wang, Z., and Ormeling, F. (1996) "The representation of quantitative and ordinal information." *The Cartographic Journal* 33, no. 2:87–91.

Waniez, P. (2000) "Two software packages for mapping geographically aggregated data on the Apple Power Macintosh." *The Cartographic Journal* 37, no. 1:51–64.

Wann, J. P., and Mon-Williams, M. (1997) "Health issues with virtual reality displays: What we do know and what we don't." *Computer Graphics* 31, no. 2:53–57.

Ware, J. M., and Jones, C. B. (1997) "A multiresolution data storage scheme for 3D GIS." In *Innovations in GIS 4*, ed. by Z. Kemp, pp. 9–24. Bristol, PA: Taylor & Francis.

Weber, C. R., and Buttenfield, B. P. (1993) "A cartographic animation of average yearly surface temperatures for the 48 contiguous United States: 1897–1986." *Cartography and Geographic Information Systems* 20, no. 3:141–150.

Webster, R., and Oliver, M. A. (2001) *Geostatistics for Environmental Scientists*. Chichester, England: Wiley.

Wegman, E. J. (1990) "Hyperdimensional data analysis using parallel coordinates." *Journal of the American Statistical Association* 85, no. 411:664–675.

Weibel, R. (1992) "Models and experiments for adaptive computer-assisted terrain generalization." *Cartography and Geographic Information Systems* 19, no. 3:133–153.

Wertheimer, M. (1958) "Principles of perceptual organization." In *Readings in Perception*, ed. by D. C. Beardslee and M. Wertheimer, pp. 115–135. Princeton, NJ: Van Nostrand.

WhatTheyThink.com. (2002) "Proofing options: Survey examines current stats and future trends." Available at *members.whattheythink.com/allsearch/article.cfm?id=6396*.

Wheless, G. H., Lascara, C. M., Valle-Levinson, A., Brutzman, D. P., Sherman, W., Hibbard, W. L., Paul, B. E. (1996a) "The Chesapeake Bay Virtual Environment (CBVE): Initial results from the prototypical system." *The International Journal of Supercomputer Applications and High Performance Computing* 10, no. 2/3:199–210.

Wheless, G. H., Lascara, C. M., Valle-Levinson, A., Brutzman, D. P., Sherman, W., Hibbard, W. L., Paul, B. E. (1996b) "Virtual Chesapeake Bay: Interacting with a coupled physical/biological model." *IEEE Computer Graphics and Applications* 16, no. 4:52–57.

Whistler, J. L., Egbert, S. L., Jakubauskas, M. E., Martinko, E. A., Baumgartner, D. W., and Lee, R. (1995) "The Kansas State Land Cover Mapping Project: Regional scale land use/land cover mapping using Landsat Thematic Mapper data." 1995 ACSM/ASPRS Annual Convention & Exposition, Volume 3 (Technical Papers), Charlotte, North Carolina, pp. 773–785.

Wilford, J. N. (1982) *The Mapmakers: The Story of the Great Pioneers in Cartography from Antiquity to the Space Age*. New York: Vintage Books.

Wilhelm, S. D. (1983) "Two symbols for unclassed two-variable mapping: The bivariate box and the ratio bar." Unpublished MA thesis, University of North Carolina, Chapel Hill, NC.

Wilhelmson, R. B., Jewett, B. F., Shaw, C., Wicker, L. J., Arrott, M., Bushell, C. B., Bajuk, M., Thingvold, J., Yost, J. B. (1990) "A study of a numerically modeled storm." *The International Journal of Supercomputer Applications* 4, no. 2:20–36.

Wilkie, R. W., and Tager, J. (1991) *Historical Atlas of Massachusetts*. Amherst: The University of Massachusetts Press.

Windschitl, M., Winn, W., Hedley, N., Fruland, R., and Postner, L. (2002) "Using immersive visualizations to promote the understanding of complex natural systems: Learning inside Virtual Puget Sound." Available at *http://depts.washington.edu/edtech/jrst.html*.

Winn, D. M., Blot, W. J., Shy, C. M., Pickle, L. W., Roledo, A., Fraumeni, J. F. J. (1981) "Snuff dipping and oral cancer among women in the southern United States." *New England Journal of Medicine* 304, no. 13:745–749.

Winn, W. (2002) "What can students learn in artificial environments that they cannot learn in class?" *First International Symposium, Open Education Faculty*, Anadolu University, Turkey.

Winn, W., Windschitl, M., Fruland, R., and Lee, Y.-L. (2002) "Features of virtual environments that contribute to learners' understanding of earth science." *Annual Meeting of the National Association for Research in Science Teaching*, New Orleans, LA.

Witmer, B. G., and Singer, M. J. (1998) "Measuring presence in virtual environments: A presence questionnaire." *Presence* 7, no. 3:225–240.

Wittenbrink, C. M., Pang, A. T., and Lodha, S. K. (1996) "Glyphs for visualizing uncertainty in vector fields." *IEEE Transactions on Visualization and Computer Graphics* 2, no. 3:266–279.

Wood, A., Drew, N., Beale, R., and Hendley, B. (1995) "HyperSpace: Web browsing with visualization." *Proceedings of the Third International World Wide Web Conference*, Darmstadt, Germany, pp. 21–25. Available at *http://www.igd.fhg. de/archive/1995_www95/proceedings/posters/35/index.html*.

Wood, C. H. (2000) "A descriptive and illustrated guide for type placement on small scale maps." *The Cartographic Journal* 37, no. 1:5–18.

Wood, C. H., and Keller, C. P. (1996) *Cartographic Design: Theoretical and Practical Perspectives*. Chichester, England: Wiley.

Wood, D., and Fels, J. (1986) "Designs on signs/Myth and meaning in maps." *Cartographica* 23, no. 3:54–103.

Wood, D., and Fels, J. (1992) *The Power of Maps*. New York: Guilford.

Woodward, D., and Harley, J. (1987) *The History of Cartography: Cartography in Prehistoric, Ancient and Medieval Europe and the Mediterranean*. Chicago: University of Chicago Press.

Woodward, D., and Harley, J. (1994) *History of Cartography: Cartography in the Traditional East and Southeast Asian Societies*. Chicago: University of Chicago Press.

Woodward, D., and Lewis, M. (1998) *The History of Cartography: Cartography in the Traditional African, American, Arctic, Australian, and Pacific Societies*. Chicago: University of Chicago Press.

Woolard, A., Lalioti, V., Hedley, N., Carrigan, N., Hammond, M., Julien, J. (2003) "Case studies in application of augmented reality in future media production." *Proceedings of the Second IEEE Symposium on Mixed and Augmented Reality (ISMAR '03)*, Tokyo, Japan, pp. 294–295. Available at *http://csdl.computer.org/comp/proceedings/ismar/2003/2006/00/20060294.pdf*.

Wright, J. K. (1936) "A method of mapping densities of population: With Cape Cod as an example." *The Geographical Review* 26, no. 1:103–110.

Wright, J. K. (1942) "Map makers are human." *The Geographical Review* 32, no.4:527–544.

Wyszecki, G., and Stiles, W. S. (1982) *Color Science: Concepts and Methods, Quantitative Data and Formulae* (2nd ed.). New York: Wiley.

Yadav-Pauletti, S. (1996) "MIGMAP, a data exploration application for visualizing U.S. Census migration data." Unpublished MA thesis, University of Kansas, Lawrence, KS.

Yang, Q. H., Snyder, J. P., and Tobler, W. R. (2000) *Map Projection Transformation: Principles and Applications*. London: Taylor & Francis.

Yoder, S. C. (1996) "The development of software for exploring temporal data associated with point locations." Unpublished MA thesis, University of Kansas, Lawrence, KS.

Yoeli, P. (1967) "The mechanization of analytical hill shading." *The Cartographic Journal* 4, no. 2:82–88.

Yoeli, P. (1971) "An experimental electronic system for converting contours into hill shaded relief." *International Yearbook of Cartography* 11:111–114.

Yoeli, P. (1972) "The logic of automated map lettering." *The Cartographic Journal* 9, no. 2:99–108.

Yoeli, P. (1983) "Shadowed contours with computer and plotter." *The American Cartographer* 10, no. 2:101–110.

Yoeli, P. (1985) "Topographic relief depiction by hachures with computer and plotter." *The Cartographic Journal* 22, no. 2:111–124.

Young, T. (1801) "On the theory of light and colours." *Philosophical Transactions of the Royal Society of London* 92:12–48.

Zhan, F. B., and Buttenfield, B. P. (1995) "Object-oriented knowledge-based symbol selection for visualizing statistical information." *International Journal of Geographical Information Systems* 9, no. 3:293–315.

Zoraster, S. (1991) "Expert systems and the map label placement problem." *Cartographica* 28, no. 1:1–9.

Index